Microbiology

DIVERSITY, DISEASE, AND THE ENVIRONMENT

Microbiology
DIVERSITY, DISEASE, AND THE ENVIRONMENT

Abigail A. Salyers / Dixie D. Whitt

University of Illinois, Urbana-Champaign

FITZGERALD SCIENCE PRESS

Bethesda, Maryland

Fitzgerald Science Press, Inc.
Editorial Offices
4814 Auburn Ave.
Bethesda, Maryland 20814
(301) 654-6884
www.fitzscipress.com

Fitzgerald Science Press, Inc.
Orders and Fulfillment:
P.O. Box 605
Herndon, VA 20172
(800) 869-4409
Fax (703) 661-1501

Cover image: Colorized Transmission Electron Micrograph of *Escherichia coli* K 12 by Kwang Kim. Used with permission of Photo Researchers, Inc.

Library of Congress Cataloging-in-Publication Data

Salyers, Abigail A.
Microbiology: diversity, disease, and the environment / Abigail A. Salyers and Dixie D. Whitt.
p. cm.
Includes bibliographical references and index.
ISBN 1-891786-01-6 (alk. paper)
1. Microbiology. I. Whitt, Dixie D. II. Title.

QR41.2.S235 2000
579—dc21 00-055129

Production Management: Susan Graham
Editor: Nancy Knight
Art: Precision Graphics
Composition: Precision Graphics
Interior Design: Susan Brown Schmidler
Cover Design: Rock Creek Publishing Group
Printer: Courier, Kendallville

Printed and bound in the United States of America
00 01 02 — 10 9 8 7 6 5 4 3 2 1

This book is dedicated
to the memory of
the late Holger Jannasch
and also to
Ed Leadbetter,
two scientists who embody the
spirit of microbiology
past and future

Contents

Preface

THIS BOOK WAS WRITTEN FOR AN INTRODUCTORY COURSE in general microbiology taught to undergraduates majoring in the life sciences and applied sciences. Many of these students were preparing for careers in the health sciences. Portions of this book were also adapted from an extensive set of notes we had developed for an introductory microbiology course we have taught to first-year medical students. The goal was not only to provide a text that was readable and concise but also to convey the excitement of modern microbiology. The viewpoint of the text is that of scientists currently working in the field. This means that, in addition to well-established facts about microorganisms and classic methods of studying them, new understandings and new methods that are still under development are also major foci of attention.

The field of microbiology is tremendously dynamic. Accordingly, we think the best way to understand it is to look into the future as well as the past. The risk of this approach, of course, is that some of the concepts and approaches presented here may be altered in the future. Such still-developing areas of understanding are identified clearly in the text. Some students object to being taught things that may be subject to change. This view, although understandable, misses the very nature of scientific progress. All concepts and approaches are subject to scrutiny and modification. Scientists learn as much from their mistakes as from their successes. Even someone who does not plan to make a career in the biological sciences can benefit from understanding that progress in the sciences occurs by fits and starts and is not the seamless, always-forward progression that scientists and public policy makers sometimes suggest.

An unusual feature of this book is the use of an approach that could be called layering. Rather than cover a concept once and then go on to something else, this book often introduces a concept in one chapter and then returns to it at a deeper level or in a different context in later chapters. In our experience, this is the most effective way to provide students with a solid understanding of complex topics that may appear at first to be more straightforward than they are.

Another unusual feature of this text is that it is unashamedly anthropomorphic, in the sense that it emphasizes the effects of microbes on humans.

Because most newcomers to microbiology have never seen a microbe, except for a few photographs that convey very little, microbes seem like abstract constructs unless they are placed in the context of everyday human experience. Thus, the effects of microbes on human health and well being are a central focus of the book. We think this approach provides a more concrete understanding of what microbes are and do, an understanding that helps to create a better basis for a broader understanding of the importance of microbes in the larger environment, a subject that is covered in the latter part of the book. We are also guilty of anthropomorphizing the microbes themselves. That is, microbes sometimes are spoken of as wanting to achieve a particular goal. Purists will object to treating microbes as if they are tiny, one-celled humans, but the truth is that scientists are limited by their own human experience and tend to take an anthropomorphized view of microbes, even when they know it can sometimes lead them astray. The defense of this approach is that it is merely a useful way of generating hypotheses about microbes, hypotheses that can then be tested rigorously. The results of the test of a hypothesis can quickly indicate when a scientist is on the wrong trail. At times our writing style might strike some readers as irreverent and almost light-hearted. We plead guilty again. Our intent was to prepare a text that students could study from efficiently and even read with interest and pleasure. We are unabashedly passionate about contemporary microbiology and trust that our excitement comes across to the readers. This is a wonderful time to be a microbiologist, and we are determined to share this perception with anyone who will listen.

At the end of each chapter is a list of questions for review and reflection. These are meant not only to test a student's mastery of concepts covered in the chapter but also to encourage the student to compare and contrast these concepts in ways not covered in the text. The Chapter at a Glance section at the end of each chapter frequently makes such comparisons to help bring concepts together in a meaningful way. In some cases, the end-of-chapter questions lead the student to material not covered in the text. Such questions are intended to show how material presented in the text can be extrapolated to other areas. The ability to apply material presented in this text to other areas is valuable for keeping up with the rapidly changing field of microbiology and for confronting new microbial challenges as they appear.

Acknowledgements

No matter how enthusiastic two people are about the field of microbiology and about conveying the excitement of such an ever-changing area to students, they can't bring it all together into a basic textbook such as this without assistance from other scientists and teachers of microbiology. We wish to thank the following experts for taking the time to provide helpful suggestions and encouragement as we transformed class handouts into this final product: Albert Balows, formerly of the Centers for Disease Control and Prevention; Clifford Bond, Montana State University; Marcia L. Cordts, University of Iowa; Jerald Ensign, University of Wisconsin, Madison; Lynn W. Enquist, Princeton University; Leanne H. Field, University of Texas, Austin; Peter Gilligan, University of North Carolina Hospitals and Medical School; Elizabeth Godrick, Boston University; Peter E. Jablonski, Northern Illinois University; Edward R. Leadbetter, University of Connecticut; Stanley R. Maloy, University of Illinois, Urbana-Champaign; Karin L.

McGowan, Children's Hospital of Pennsylvania and the University of Pennsylvania; Anne Morris Hooke, Miami University, Ohio; Jayne Robinson, University of Dayton, Ohio; Thomas M. Schmidt, Michigan State University; Fred C. Tenover, Centers for Disease Control and Prevention; Gregory S. Whitt, University of Illinois, Urbana-Champaign. The Web site listings were prepared by Ralph Sorensen, Gettysburg College. We also thank James Slauch, University of Illinois, Urbana-Champaign and Grant Hamilton, a medical student at the University of Illinois, for providing early versions for some of the figures that appear in the text.

We want to thank the students in a 1996 Discovery Course at the University of Illinois who provided the inspiration for this book. The students in this course were Katie Chivington, Alina Chu, George Harris, Kerry Hinck, Andrea Isner, Nicole Karras, Otto Lee, Paul Mann, Rustin Robbins, Jennifer Rovel, Katherine Stachniw, and Angela Yee.

We also want to thank the students at the University of Illinois enrolled in Microbiology 200 (Fall, 1998 and Fall, 1999) who suffered patiently through earlier versions of this text and offered many helpful suggestions. Peter Gilligan deserves special thanks for his many helpful comments at different stages in the writing process and for providing a number of photographs. AS thanks the members of her research group for their patience during times when work on this book interfered with timely reading of their manuscripts. DW thanks Greg for being so supportive and for making helpful suggestions throughout the entire process.

Publishing a textbook requires more than authors and reviewers—it requires the dedication of editors, managers, designers, artists, typesetters, and many others. We would like to thank these professionals for their tremendous efforts in producing this book. Susan Graham, Production Manager at Fitzgerald Science Press, skillfully oversaw the whole process and managed to coordinate the activities of a diverse group of individuals at many different levels. Nancy Knight, our editor, appreciated our bad jokes, but, nevertheless, deleted them from the manuscript. Susan Schmidler designed the interior of the book. The cover was designed by Rock Creek Publishing Group. Karen Hawk, project manager at Precision Graphics, was responsible for coordinating the typesetting and art. The artist, Kitty Auble, at Precision Graphics, transformed (often magically) our sketches into clear figures. We also thank Hope Page the editorial assistant at Fitzgerald Science Press. Finally, we thank our publisher, Patrick Fitzgerald of Fitzgerald Science Press. He was always enthusiastic and encouraging. His role varied from that of head cheerleader to slave driver depending on the circumstances. At times he became so involved in the project that we had to show him a rolled-up newspaper to keep him in line! However, it was in large part due to his efforts that this textbook saw the light of day.

A Web site for this book can be found at www.fitzscipress.com/mdde/.

This Web site contains links to many interesting microbiology sites (listed in Appendix 2) on the World Wide Web as well as coverage of updates and new scientific findings relevant to this book.

Introduction

FOR MOST PEOPLE, MICROBES ARE OUT OF SIGHT AND THUS OUT OF MIND. However, increasing human control of microbial activities, especially during the last century and a half, has had a major and continuing impact on everyone alive today. Humans first began to domesticate microbes many thousands of years ago, with the discovery of fermented foods such as cheese and sausage and fermented beverages. Some folk remedies used microbes. For example, Scandinavian peasants would treat sick cows by giving them part of the cud from a healthy cow, in effect transplanting from the healthy cow to the sick cow a new infusion of the microbes on which the cow depends for digestion of its food. Such discoveries were haphazard and often not very successful, because people did not understand the basis for the processes they observed. The discovery by early scientists of microorganisms and their enormous impact on all aspects of human life opened a new era of biology in which prevention and cure of devastating infectious diseases became possible. New insights into the beneficial aspects of microorganisms have created the biotechnology industry and introduced many useful new products and processes into general use. Studies of microorganisms have also opened up the secrets of the genetic code, the blueprint for all living things. Revealing these secrets has forever changed the direction of biology and has led to new concepts and products that will improve human health and welfare for many years to come. This book seeks to reveal the exciting world of microbes as viewed by scientists currently working in the field and to convey the challenges and future promises awaiting scientists of the future.

The First Microbial Revolution: The Germ Theory of Disease, Vaccination, and Antimicrobial Compounds

For most of their history, humans have been completely at the mercy of infectious diseases. Without an understanding of how infectious diseases were spread, rational measures for control of disease could not be formulated and implemented. The discovery that microbes caused most of the diseases that had devastated human populations allowed public health officials for the first time to trace the source of specific infectious diseases and determine how they were spread. This knowledge made it possible to enact effective measures to prevent the spread of disease. Cleaning up the water supply, promoting better nutrition, reducing crowding, improving sanitation, and pasteurization of milk are examples of public health measures that saved millions of lives and dramatically bolstered human health.

The development of vaccination by 19th-century scientists, such as Jenner and Pasteur, opened up a new level of control over some important diseases, including smallpox, anthrax, polio, diphtheria, whooping cough, measles, mumps, and other diseases. With the elimination of smallpox in the 20th century, an infectious disease was eradicated completely for the first time in human history. Polio and measles are now close to eradication. Yet another step forward in the control of infectious diseases was the discovery of antimicrobial compounds, such as antibiotics, that allowed physicians to cure such dangerous diseases as pneumonia, tuberculosis, and malaria. Antibiotics and disinfectants also made possible the routine use of surgery to treat diseases such as cancer and heart disease. Before the discovery of aseptic surgical techniques and antibiotics, postsurgical infections made most surgeries too risky for routine use.

The Second Revolution: Molecular Biology and Genetic Engineering

Starting in the 1960s, microbiologists introduced yet another revolution with the discovery of the mechanisms of such central life processes as DNA replication, transcription, and translation. This discovery was the beginning of what is now called molecular biology or biology explained in terms of molecules. These first insights into molecular biology came from studies of bacteria, but the basic paradigms have proved to hold true, with few changes, for plant and animal cells.

Understanding the molecular basis of such processes as DNA replication made possible the advent of genetic engineering, a technology that has transformed biology by allowing scientists to dissect biological phenomena on the molecular level and to manipulate organisms genetically. Most of the research currently underway in such fields as neurobiology, human and plant genetics, and pharmacology was made possible by the tools and concepts discovered by the genetic engineering techniques that came out of studies by microbiologists. Genetic engineering has not only revolutionized all fields of traditional biology but has also penetrated such seemingly distant areas as forensic medicine and archaeology. DNA tests are now a routine part of criminal investigations, and archaeologists are learning more about humans and other animals in earlier times by extracting and analyzing their DNA. The era of genome sequencing promises to provide major new insights into the functioning of all living things, from bacteria to humans. This era began with the understanding of the genetic makeup that determines the composition and function of living things. Information obtained from genome sequences should provide new drugs to combat infectious microbes and treat human genetic disorders. Genetic engineering is helping to create new breeds of plants and animals that will dramatically improve animal health and agricultural productivity.

The Third Revolution: Discovery of the Extent of Microbial Diversity

The discovery of basic molecular paradigms and genetic engineering was facilitated by the decision of microbiologists to focus their attention on a few model organisms, such as the bacterium *Escherichia coli* and, later, the yeast *Saccharomyces cerevisciae*. This focus, however, had the effect of obscuring a feature of the microbial world that had been suspected by earlier microbiologists but was almost forgotten during the early genetic engineering era: that the vast majority of biodiversity of life on earth resided in the microbial world. The rediscovery of the enormous and almost entirely untapped wealth of microbial diversity and an emerging understanding of the importance of microbial activities for the maintenance of plant and animal life have begun to open up new areas of investigation. New solutions to environmental problems, such as pollution, new industrial products and processes, and new strategies for using beneficial microbes to improve the health of animals and plants are only a few of the benefits expected from future mining of microbial diversity. Even geologists have gotten into the act and have begun to discover ways in which the microbial world influences many geological processes previously thought to be abiotic.

Microbial diversity has manifested itself in ways that affect human health, both for good and ill. New infectious diseases continue to emerge as changing human practices, from the installation of air-conditioning to the use of cancer chemotherapy, create new opportunities for microbes to cause disease, including microbes that were not previously known to be capable of infecting humans. On the positive side, scientists are beginning to discover that some diseases previously classified as chronic and incurable, such as gastric ulcers and periodontal disease, are caused by bacteria and thus can be treated successfully with antibiotics. These discoveries have encouraged scientists to re-examine other chronic diseases, including heart disease and Alzheimer's disease, for possible microbial causes that would suggest new cures for these conditions.

The Next Revolution Is Up to You

The impressive track record of microbiologists in the past is only the beginning. Major challenges remain. One is decoding the genome sequences of animals and plants. The basic dictionary on which such a decoding is based will be written from information obtained from the genomes of microbes. Microbes were the ancestors of all living things on Earth today, and many of their genes have been carried forward into the larger and more recently evolved life forms. As a result of their short generation times, microbes are easier to manipulate and study than plants and animals. Much of the information that will teach us how to turn genome sequences into information about the function of genes in a living organism will come from studies of microbes and their genome sequences. Making the transition from DNA sequence to function will be a difficult and complicated undertaking and will dominate future research for years to come. Learning how to turn this information into new cures and new products also will be challenging.

A second major challenge is to find ways of understanding and exploiting the genetic and metabolic diversity of the microbial world. Whereas over 50% of mammalian species and 20% of arthropod species have been identified, it has been estimated that fewer than 0.2% of microbial species are known. Moreover, from what is already known, it is clear that the range of metabolic diversity in the microbial world is far greater than that seen in all animals and plants. New, previously unsuspected, microbial activities are discovered every day, and the task of fully enumerating the total spectrum of microbial activities and their impacts will be challenging and arduous, but full of promise.

Microorganisms are our ancestors. They are the life-support system of this planet and the key to life on other planets. They are sometimes our enemies, but more often our friends. They are the ultimate renewable resource and a potential gold mine of new processes and products. Surely, we need to know as much about them as we can.

microbial worlds

I

Most people go through life unaware of the existence of microbes—unless a microbial infection lays them low. Yet every minute of every day, without realizing it, people experience directly and indirectly the effects of microbial activities. In fact, microbes make the difference between a living planet and a dead one. They recycle dead biomass. They form the base of food chains. They make major contributions to the composition of the atmosphere. Although scientists often speak of "the" microbial world, there are actually many microbial worlds. The microbial population that normally occupies the human large intestine is different from that found in the intestine of a cow or pig or that found in the soil around the roots of plants. This text will survey many of these microbial worlds and the ways they influence the well-being of larger life forms—like us.

The first four chapters describe the types of microbes, their role in the origin of life on earth, and the ways in which their activities affect and are affected by human activities. In Chapters 5–10, the genetic and physiological features of various types of microbes are described in more detail. This material establishes a basis for later chapters that explore further the microbial worlds introduced in this section.

1 Earth: Planet of the Microbes

The moon is earth without microbes. A live planet is a microbial planet.

IS THERE LIFE ON OTHER PLANETS? This question never ceases to capture the imagination of people on Earth. Although no evidence of life beyond Earth has been found, such a discovery would certainly be revolutionary. Imagine then, how scientists felt when the first microscopes revealed the existence of a previously undetected living world—the world of microorganisms. In many ways, this revelation proved to be every bit as revolutionary as discovering life on other planets would be today. Early microbiologists, however, did not fully appreciate the importance of microbes. It was not until the 1800s that scientists began to realize that microorganisms, despite their small size, play a large role in human well being.

The Microbial World Around Us

Microorganisms as a cause of disease

A major breakthrough occurred when microbiologists discovered that long-feared diseases, such as plague, tuberculosis, cholera, and typhoid fever, were caused by microorganisms. Today, it is difficult to comprehend the sweeping changes this discovery made possible. To get some idea of its impact, consider the public response to epidemics of the deadly diarrheal disease cholera, both before and after the discovery that cholera was caused by microorganisms. We know now that the microorganism that causes cholera is spread by contaminated water. But to the people of the early 19th century, it might as easily have been spread by personal contact, through the air, or even by rats. Identifying a microorganism as the cause of cholera made it possible to establish that water was the culprit. Moreover, it became possible to differentiate "good" water from "bad" and to determine what treatments (adding chemicals or boiling) would make contaminated water safe to drink. For the first time, public officials could take effective preventive measures to stop the spread of an infectious disease. These preventive measures remain our best protection against infectious diseases today.

Later, in the 1940s and 1950s, scientists were able to use a growing body of knowledge about disease-causing microorganisms to develop compounds that could inhibit the growth of or kill microorganisms without adversely affecting the infected person. The age of **antimicrobials** had arrived. Antimicrobials made possible many of the medical miracles we now take for granted. Surgery and cancer chemotherapy, for example, would be too risky for any but the most desperate patients were it not for the antimicrobial compounds that control the infections such patients often develop.

Industrial use of microorganisms

Even before their existence was suspected, microorganisms had been used for centuries to make milk into cheese, grape juice into wine, and for many other applications. With the advent of the microscope, scientists were able to show that microorganisms were responsible for what had become familiar phenomena. As with infectious diseases, knowing more about the microorganisms responsible made it possible to improve and control these processes. Scientists also realized that if microorganisms could carry out such useful functions as the conversion of sugar to ethanol, then they might well be capable of performing many other commercially useful processes. Microorganisms are now used to produce a variety of compounds, ranging from amino acids to solvents. The development of new technologies for controlling microorganisms and their genes has created the **biotechnology industry,** which has already produced an impressive array of exciting new products. Human efforts to domesticate microorganisms have entered a new era.

Microbes: Earth's Life Support System

We are still learning new things about the microbial world. Only recently, for example, did scientists begin to appreciate how important microorganisms are in the establishment and continuation of life on Earth. The earth is a closed system (give or take an occasional meteorite). For life to continue, dead things must be broken down into their component parts to provide the ingredients needed to grow new living things. Microorganisms are the principal recyclers of biomass on Earth. They degrade dead biomass to its component parts, which can then be used by living organisms. Without the recycling activities of microorganisms, life on Earth would quickly come to a halt. And microorganisms do not always limit themselves to recycling dead biological material. Through their ability to utilize metals, some microorganisms can also recycle nonbiological items (Box 1–1).

Microorganisms also provide biomass that other creatures use for food. Microorganisms are at the base of all the earth's food chains. In the oceans, a collection of photosynthetic microorganisms called **phytoplankton** directly or indirectly feeds all marine animals. On land, microorganisms in the soil provide nitrogen sources and other services for plants. Trees, for example, benefit from the close association between their roots and microorganisms. The microorganisms get nutrients from the tree roots and secrete chemicals that cause the roots to develop more small offshoots, thus increasing the absorptive capacity of the tree's root system. Other plants, such as alfalfa and soybeans, go a step further and actually incorporate microorganisms into their root cells. Inside these cells, bacteria convert nitrogen gas from the atmosphere into nitrogen compounds that can be used by the plants. Microorganisms could exist without animals and plants, but animals and plants depend absolutely on the microbial world for continued survival.

Even the composition of the atmosphere is influenced by microorganisms, which produce and consume atmospheric gases such as methane, carbon dioxide, oxygen, and nitrogen. Only recently in the evolutionary

box

1–1 *The Last Passengers on the Titanic*

In 1998, *The Titanic* was a blockbuster film that spurred renewed public interest in the story of the doomed ocean liner. The luxury and romance of the famous ship has an irresistible appeal for many people. The present-day condition of the Titanic is another story. Several years before the movie, explorers located the remains of the ship. Some engineers had predicted that, even after 90 years, the ship's hull would be fairly well preserved because the oxygen level in the ocean at that depth was low enough to retard spontaneous rusting of iron. When the first videos of the sunken ship surfaced, many were surprised to see the hull festooned with long strands of iron oxides (rust). The hull had been undermined so extensively that it now seems likely that within the next decade or two it will collapse from its own weight. Microbes have been at work.

Some bacteria and archaea can obtain energy by combining oxygen and iron to form iron oxides. These iron-oxidizing microorganisms prefer a low oxygen environment, because at normal oxygen concentrations they cannot compete well with spontaneous rust formation. The location of the Titanic placed it in an environment that was especially conducive to microbial attack. Most fans of the Titanic probably view the microbial destruction of the ship as a near sacrilege, with microorganisms playing the role of relentless despoilers of a memorial to the passengers and crew. In fact, the dissolution of the Titanic is a manifestation of the health of the planet, an example of resurrection, not death. As iron oxides fall from the ship into the ocean sediment, they will be processed by still other microorganisms in the sediment. Ultimately, at least some of the iron from the Titanic will end up as part of living creatures. The cycle of life and death continues.

timeline have plants and animals begun to make their own contribution, but microorganisms remain a dominant force in controlling the composition of the atmosphere. Every time you breathe, you interact with the microbial world.

Microorganisms could and did exist for billions of years without plants and animals. Plants and animals, however, are totally dependent on the microbial world for survival. This realization has prompted scientists to change the way they look for life on other planets. Now, instead of looking for little green men, astrobiologists are looking for evidence of microbial life.

Types of Microorganisms

Prokaryotes and eukaryotes

Early microscopic examination of cells from plants and animals, and of single-cell microorganisms revealed that the internal structure of plant and animal cells is more complex than that of most microorganisms (Fig. 1.1). Each plant and animal cell has a nucleus consisting of a tightly coiled **deoxyribonucleic acid (DNA) genome** enclosed in a membranous covering (the **nuclear membrane**). Each cell also has a set of membranes external to the nucleus (the **endoplasmic reticulum** and **Golgi bodies**), that are involved in the synthesis and processing of proteins. Finally, in addition to the nucleus, each cell contains other membrane-enclosed organelles, such as **mitochondria** (animals and plants) and **chloroplasts** (plants), in which energy is generated.

Close examination of the microorganisms we now call bacteria and archaea revealed a lack of such internal organelles, and the presence of a DNA genome that is not enclosed in a membrane. To distinguish between cells that lack internal compartmentalization and cells that exhibit such compartmentalization, the terms **prokaryote** and **eukaryote** were coined. A prokaryote is a cell that lacks a nuclear membrane; a eukaryote is a cell with a nuclear membrane. The definition focuses on the nuclear membrane

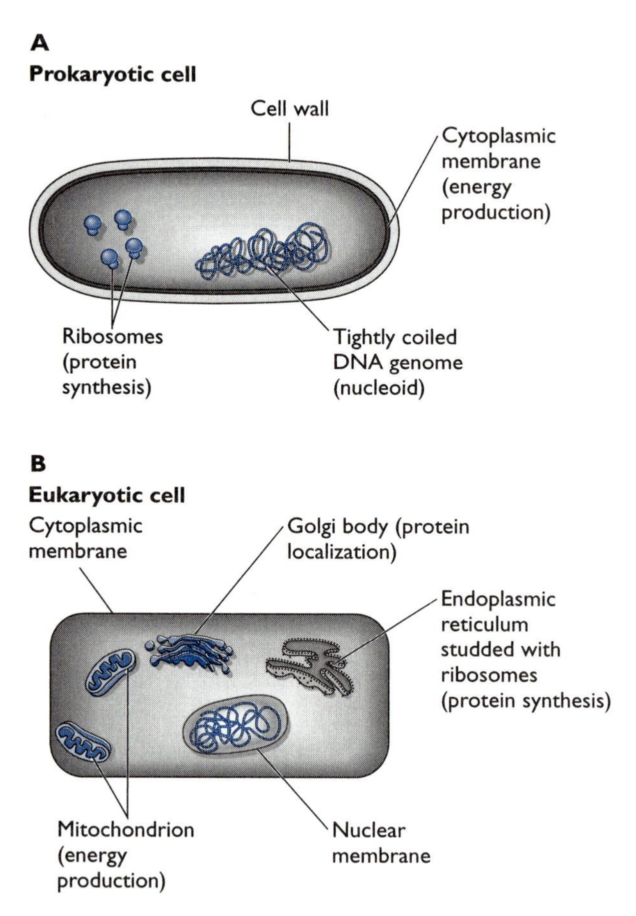

Figure 1.1 Internal structure of bacterial and archaeal cells (prokaryotes) as compared with plant and animal cells (eukaryotes). (A) The prokaryotic cell has a cytoplasmic membrane that is the site of energy production and is surrounded by a rigid cell wall. Protein synthesis is carried out by ribosomes free in the cytoplasm. The genome is not surrounded by a nuclear membrane. **(B)** The eukaryotic cell has a cytoplasmic membrane but is not always surrounded by a rigid cell wall. The cytoplasm contains a variety of organelles such as the mitochondrion, the site of energy production. Ribosomes are not free in the cytoplasm but are associated with the endoplasmic reticulum. A major distinction between prokaryotic and eukaryotic cells is the nuclear membrane that surrounds the genome in eukaryotic cells.

because some eukaryotic microorganisms lack intracellular organelles and some lack visible Golgi bodies. The microbial world has prokaryotic and eukaryotic members. The shapes and some of the characteristics of these free-living microorganisms are shown in Figures 1.2 and 1.3. Features of viruses, which are dependent on other organisms for their reproduction, and thus not free living, are summarized in Figure 1.4 (page 8). In other chapters, we will examine the different kinds of microorganisms more closely; however, it is useful at the outset to take a brief overview of their properties.

The prokaryotes: bacteria and archaea

The smallest free-living microorganisms are prokaryotes—**bacteria** and **archaea.** The archaea, which were recognized only recently as a separate microbial group, were so named because they were thought to be the oldest life forms on Earth. Scientists know now that bacteria are equally ancient, but the name archaea has stuck.

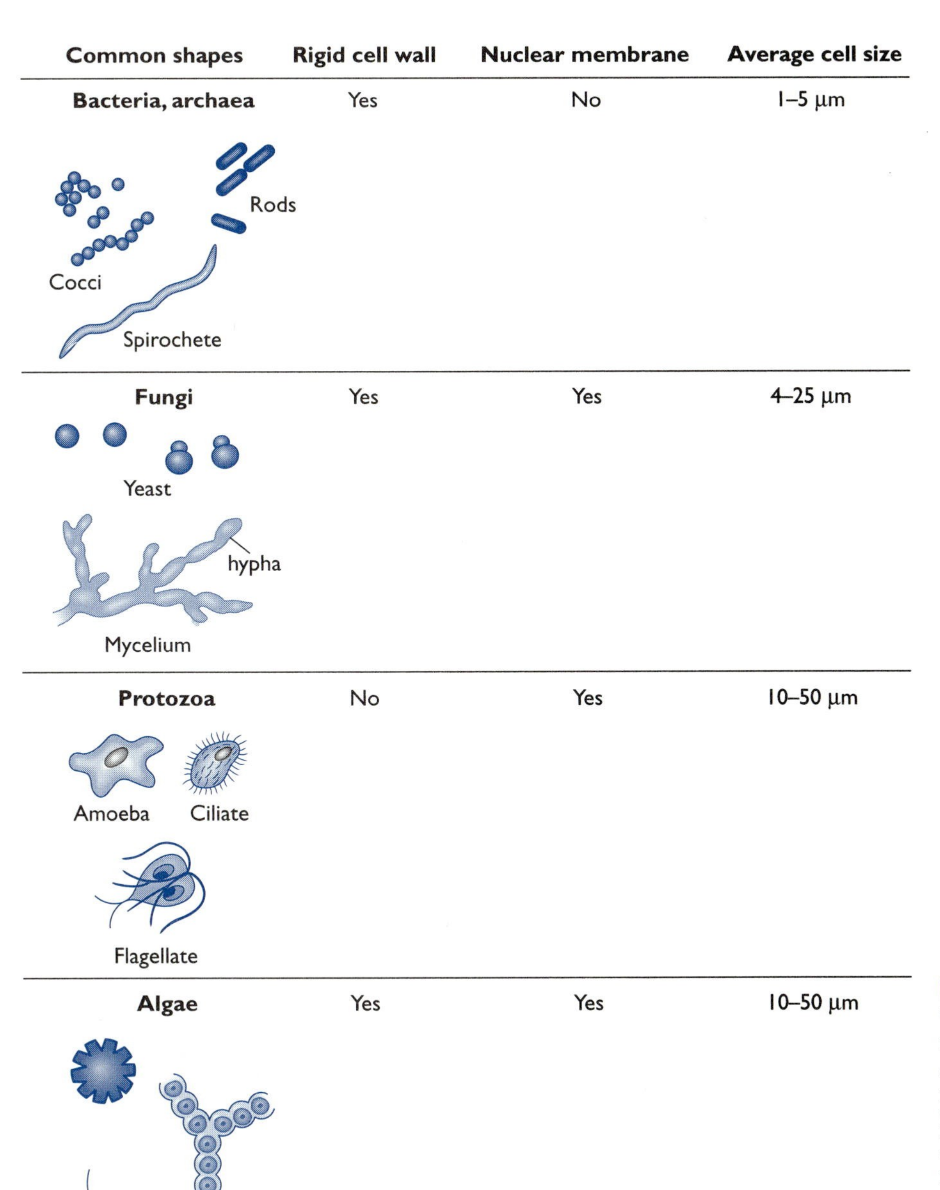

Common shapes	Rigid cell wall	Nuclear membrane	Average cell size
Bacteria, archaea	Yes	No	1–5 μm
Fungi	Yes	Yes	4–25 μm
Protozoa	No	Yes	10–50 μm
Algae	Yes	Yes	10–50 μm

Figure 1.2 Shapes and characteristics of free-living microorganisms. The major groups of microorganisms exhibit a variety of shapes. A rigid cell wall is found in all except the protozoa. The bacteria and archaea are the only free-living microorganisms that lack nuclear membranes. The ranges of size overlap, but the bacteria and archaea tend to be smaller than organisms in the other groups.

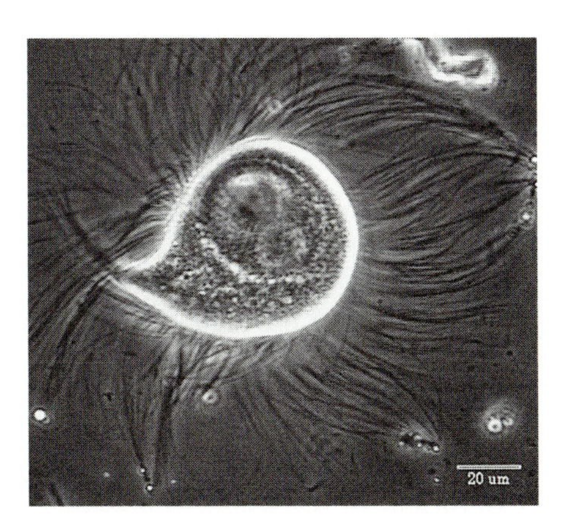

Figure 1.3 A ciliated protozoan. Courtesy of David Graham, Department of Microbiology, University of Illinois, Urbana, Illinois.

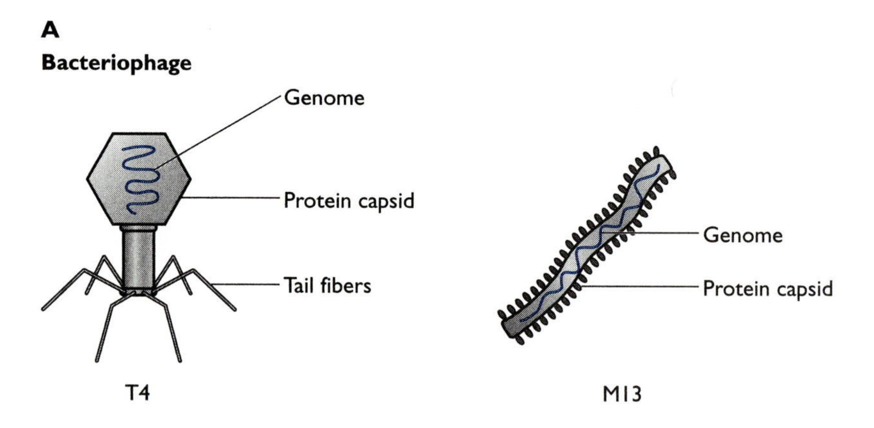

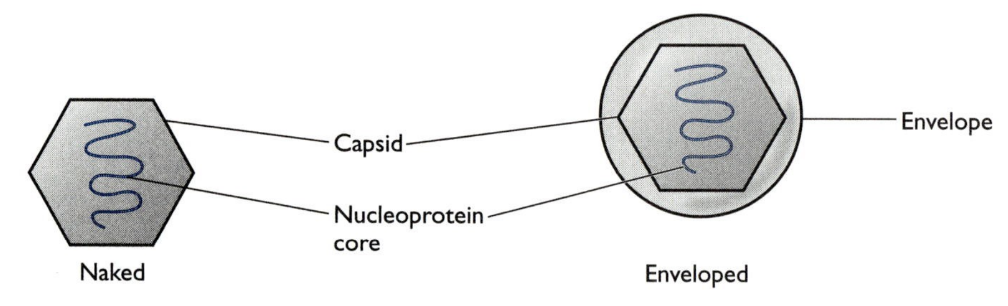

Figure 1.4 Features of viruses. (A) Bacteriophages are viruses that attack bacteria. **(B)** Eukaryotic viruses attack eukaryotic cells. Each type of virus includes a nucleic acid-protein core enclosed in a protein capsid. Enveloped viruses are surrounded by a phospholipid envelope, but naked (or nonenveloped) viruses are not.

Sizes and shapes. Most bacteria and archaea have simple shapes—rods, spheres (cocci) or spirals (spirochetes)—and divide by **binary fission** (Fig. 1.5 and Fig. 1.6). On average, bacteria and archaea are about 1–5 μm long and 1–2 μm in diameter. They can be seen with a light microscope set at a magnification of 100–200×. Not all bacteria are this small, however. The largest bacterium found was isolated from sediments off the coast of Namibia. This bacterium, *Thiomargarita namibiensis*, is a whopping 1 mm in diameter, about the size of the head of a fruit fly and easily visible to the unaided eye. Are there larger bacteria? *T. namibiensis* was not discovered until 1998. There may well be more bacterial behemoths waiting to be discovered.

Just as the upper limit of prokaryotic size has been called into question, the lower limit of bacterial or archaeal size has also become controversial. There have been reports, admittedly disputed, of organisms less than 1 μm in diameter identified as bacteria. This size is only marginally bigger than the largest viruses. Scientists have always assumed that bacteria could not be much smaller than 1 μm in diameter because the minimal amount of DNA needed to direct the survival and reproduction of a free-living organism takes up at least this minimal storage space. The controversy over whether these tiny organisms are really bacteria illustrates a major problem in microbiology: how to determine whether a microorganism seen with the microscope is actually alive. If the organism moves or is caught in the act of

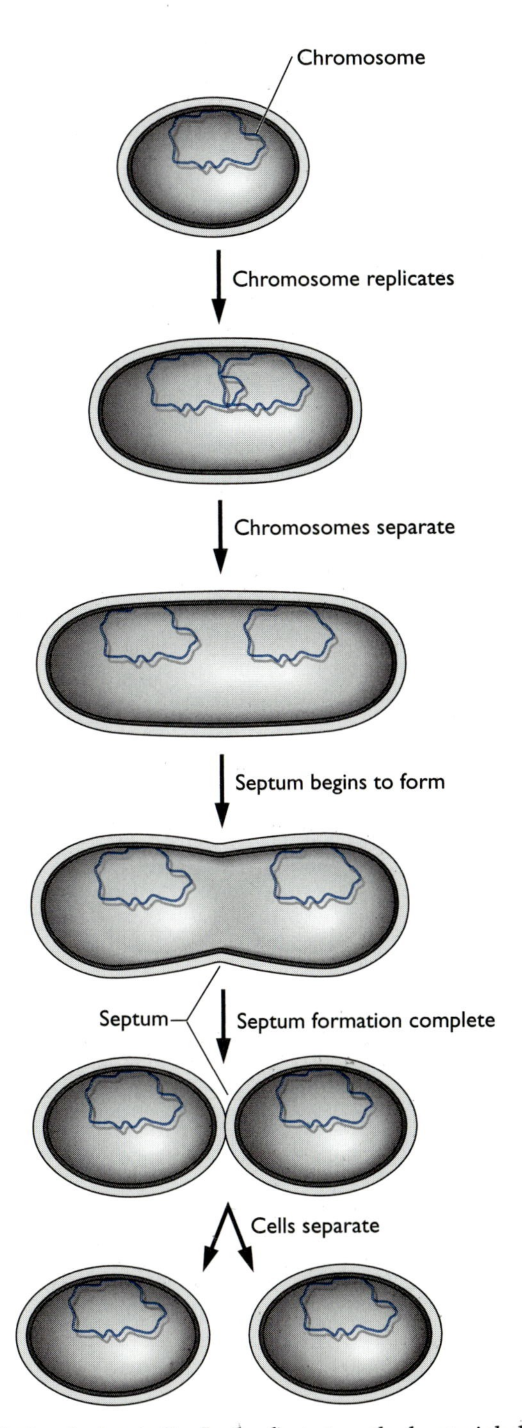

Figure 1.5 Binary fission in bacteria. In the first step, the bacterial chromosome replicates. Then the cell begins to elongate, and the two chromosomes separate. At the same time, a septum begins to form between the two chromosomes. After septum formation is complete, the new cells separate.

dividing, then clearly it is viable. However, if it is neither moving nor actively dividing, the only way to establish its viability is to demonstrate that it can be grown in the laboratory. It is now clear that most of the bacteria and archaea in nature have not been cultivated in the laboratory, so failure to cultivate is not proof that a microorganism is dead.

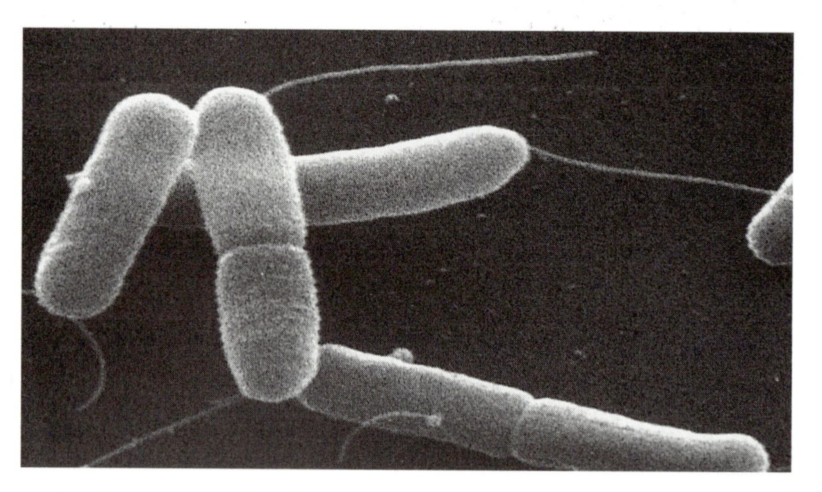

Figure 1.6 ***Escherichia coli* undergoing cell division.** The string-like appendages (flagella) are used for motility. © David M. Phillips / Visuals Unlimited.

Other properties of bacteria and archaea. Bacteria and archaea, with few exceptions, have rigid **cell walls.** A rigid cell wall is necessary because the high concentrations of salts and other molecules packed into the small cell interiors exert powerful outward pressure. Without strong walls, the cells would explode. The cell wall also gives a bacterium or archaean its characteristic shape. Although some prokaryotes can change their shapes under certain conditions, such changes are uncommon. For a long time, biologists believed that these simple shapes indicated that the bacteria and archaea were not very diverse, certainly not as diverse as insects or other animals. Microbiologists argue that this view of microorganisms is incorrect, because it assumes that shape and other visible features are the only indications of **diversity.** Diversity in the microbial world, especially that of prokaryotes, is more accurately reflected by metabolic capabilities. Some archaea, for example, grow best in boiling water and "freeze" at temperatures humans consider normal. Some bacteria can grow in polar ice or in sulfuric acid. Bacteria and archaea produce and consume compounds such as methane, nitrogen gas, and sulfides that are neither produced nor utilized by plant and animal cells.

Bacteria and archaea typically have only 1 or 2 chromosomes, whereas the genomes of eukaryotes are more complex and contain more DNA. Human cells, for example, have 23 pairs of chromosomes and about 1000 times the amount of DNA contained in the average prokaryotic genome. The prokaryotes make up for this relative lack of genomic complexity by their unparalleled ability to mutate their genomes rapidly and to add new DNA segments acquired from other microorganisms. They also have a much shorter generation time than animals visible to the naked eye, a trait that allows them to evolve more rapidly than larger animals. In recent years, some bacteria have reminded us of their superior ability to evolve rapidly by becoming resistant to the antibiotics we use against them, often within a few years after the introduction of a new antibiotic. Not all such genetic changes are the result of mutations. Bacteria and archaea can also acquire DNA from other microorganisms—even from members of different species—and incorporate it into their genomes. This process, which has been called **horizontal gene transfer** or "prokaryotic sex," has clearly played an important role in the evolution of prokaryotes.

Distinguishing bacteria from archaea. Bacteria and archaea share many traits. How do they differ from each other? If the answer were obvious, microbiologists would not have taken so long to realize that there are two distinct groups of prokaryotes. The fact that these two types of prokaryotes were different first became evident in the 1980s from analyses of gene sequences. One major difference between bacteria and archaea is in the composition of their cell walls. Bacteria have a cross-linked polysaccharide cell wall matrix called **peptidoglycan.** Archaea have no peptidoglycan in their cell walls. A more striking contrast is in the structure of the lipids that make up their cytoplasmic membranes. Bacteria, like eukaryotic cells, have membranes composed of a bilayer of phospholipids (Fig. 1.7). Archaea have a membrane composed of a different type of lipid, with some lipids spanning the membrane that make it partially a monolayer. Microbes that can grow at the highest extremes of temperature have so far all proven to be archaea. It is possible that their unusual membrane structure gives archaeal cells greater stability under such conditions. All of the prokaryotes capable of chlorophyll-based photosynthesis are bacteria. Photosynthetic archaea exist but do not have chlorophyll. Another metabolic distinction is that all of the prokaryotes capable of degrading complex polymers are bacteria. Bacteria are major contributors to the recycling of biomass. Archaea seem to be limited to the use of simple molecules as sources of carbon and energy.

There is one apparent distinction between bacteria and archaea which—if true—is difficult to understand. Whereas bacteria cause many diseases of animals and plants, no disease-causing archaea have been identified. All other groups of microorganisms have disease-causing members, so why should archaea be any different? There are many archaea that can grow at 37°C, the temperature of the human body. In fact, there are archaea that live in the human lower intestine. It is possible that, because archaea have been recognized as a distinct category of microorganism comparatively recently, the disease-causing ones simply have been missed so far.

Eukaryotic microbes: protozoa, algae, and fungi

Eukaryotic microorganisms can be divided into three major categories: **protozoa, algae,** and **fungi** (Fig. 1.2). Generally, they are about ten times the size of bacteria, although some small protozoa are similar in size to the larger bacteria.

Properties of eukaryotic microbes. Eukaryotic microorganisms display a more impressive diversity in shape than prokaryotes. They also vary in the ways in which these shapes are determined. Protozoa do not have rigid cell walls, but algae and fungi do. The shapes of protozoa probably are maintained, instead, by an interior protein skeleton called the **cytoskeleton.** The cytoskeleton is a network of protein fibers that provides structural rigidity but also flexibility. The flexible cell surface allows protozoa to engulf and digest bacteria and smaller eukaryotic microorganisms. Algae do not need this structural flexibility because they get their energy from **photosynthesis,** the conversion of light energy into biomass, and thus do not need to eat other microorganisms to survive. Fungi secrete polymer-degrading enzymes that digest cellulose and other materials from

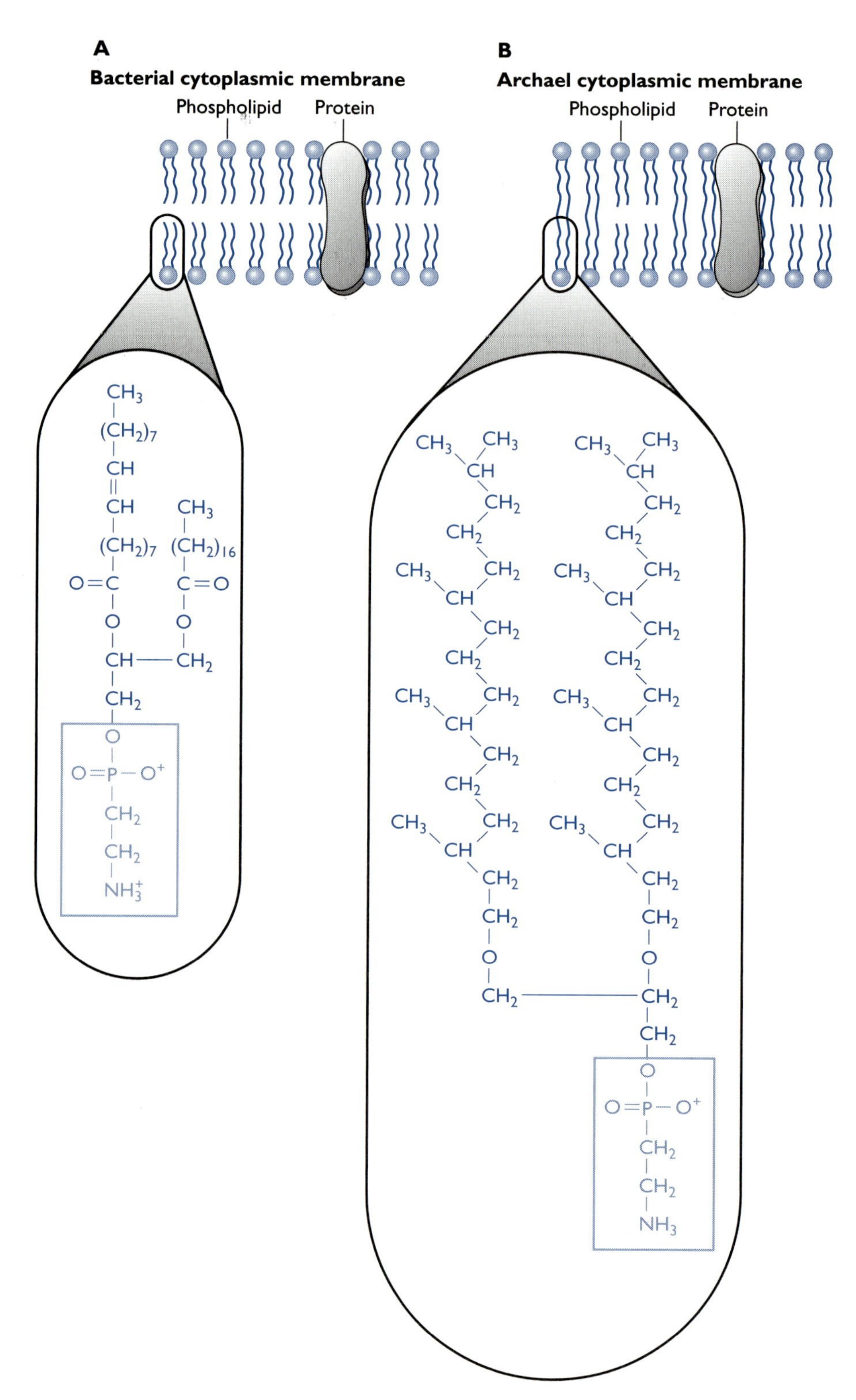

Figure 1.7 Structure of bacterial and archaeal cytoplasmic membranes. The major structural difference between the two types of cytoplasmic membranes is that the bacterial membrane **(A)** is a phospholipid bilayer, whereas the archaeal membrane **(B)** is a partial monolayer. Both types of membranes have proteins interspersed throughout.

the external environment into subunits small enough to be taken up without resorting to engulfing their energy sources.

Protozoa and algae normally grow as single cells that divide by binary fission. Fungi, however, can form large multicellular branches called **hyphae** (Fig. 1.2). In the hyphae, new cells are produced by binary fission but remain part of the growing hyphal structure. A mass of hyphae is called a **mycelium.** You have probably seen the mycelial form of fungus as the blue or green fuzz that appears on food stored too long in the refrigerator. There are also fungi called **yeasts,** that grow as single cells and reproduce by **budding.** Many fungi grow strictly as mycelia or as yeasts, but some can switch from yeast to mycelial form and back again (**dimorphic fungi**). Fungi that cause serious human infections are usually dimorphic fungi, living in the mycelial form outside the human body but switching to the yeast form when they infect. Both yeasts and mycelial cells have rigid cell walls. Many eukaryotic microorganisms have both asexual and sexual reproductive modes.

Ecological roles. Protozoa, algae, and fungi vary in their ecological roles. Protozoa graze on other life forms. Many environmental protozoa live by ingesting and killing prokaryotes or smaller eukaryotes. Others, such as the protozoa that cause malaria, live by parasitizing larger animals. Algae form the basis of the food chains of other animals, especially in marine environments where the algae are important members of marine **phytoplankton.** Bacteria are the other important component. Fungi, like bacteria, are major recyclers of dead biomass, especially biomass from plants. Fungi also infect plants, especially where there is wounded tissue.

All three groups of eukaryotic microorganisms contain members that cause human disease. Sometimes these diseases are due to microbial growth inside the body (**infection**). Examples of infections caused by eukaryotic microorganisms are the protozoal diseases malaria and sleeping sickness and the fungal diseases athlete's foot and vaginitis. Most of the life-threatening fungal diseases occur primarily in people who are immunocompromised.

Some fungi and algae cause disease without infecting, by producing toxic substances (**toxins**) that can harm animals. Algae produce low molecular weight toxins that can kill animals and fish. In 1997, a small alga, *Pfiesteria,* nearly destroyed the fishing industry in the eastern United States when a bloom filled the water with neurotoxin, harming fishermen and causing a panic among fish consumers. Fungi also produce low molecular weight toxins such as aflatoxin, which causes liver cancer. Most of these toxins probably are not designed to damage animals because they do not help the microorganism colonize the animal body. They must have some other function, as yet unknown.

Viruses

Viruses are smaller than bacteria, too small to be seen with an ordinary light microscope. An **electron microscope,** which has a greater resolving power than a light microscope and can achieve a 1000× magnification or higher, is needed to get a good look at viruses. One unusual feature of viruses compared with free-living microorganisms is that their genomes are not necessarily composed of DNA. Some viruses have genomes composed of **ribonucleic acid (RNA).** The genomes of viruses are much smaller than those of free-living microorganisms. Because viruses use the cells they infect as the source of most of the compounds they need to reproduce, they need only genes that enable them to take over the infected cell's biosynthetic machinery. Both prokaryotes and eukaryotes can be attacked by specific viruses. Some shapes of prokaryotic and eukaryotic viruses are shown in Figure 1.4.

Viruses have very simple structures. The viral RNA or DNA genome, which is usually associated tightly with proteins that stabilize the genome and perform essential functions during replication, is enclosed in a tightly packed protein coat called the **capsid.** The capsid protects the genome during those periods when the virus is outside the cell it invades. Some viruses have an additional coating called an **envelope,** consisting of a protein-containing phospholipid bilayer. Viruses with an envelope are called **enveloped viruses,** and viruses that lack an envelope are called **naked** or **nonenveloped viruses.**

A virus infects a cell by first attaching to the surface, then releasing its genome into the cell's interior. The attachment of a virus to its target cell is highly specific. A virus will only attack cells that have the right type of surface receptor for that virus. Thus, a virus that attacks cells of one bacterial species will not necessarily attack cells of another species. Similarly, viruses that attack one type of animal cell may not attack other cell types. The second step in the infection process is the takeover of the biosynthetic machinery of the infected cell by proteins produced from viral genes. The virus uses the biosynthetic machinery of the infected cell to make many copies of its genome and its proteins. Finally, these components assemble into intact viruses, and the viruses exit the infected cell.

Viruses cause a number of human diseases, including acquired immunodeficiency syndrome (AIDS), hepatitis, polio, and influenza. Viruses also cause disease in many animals and plants. Viruses that infect bacteria are abundant in some environments and may contribute to keeping the numbers of bacteria in check.

The Activities of Macrobes Affect Microbes

The central theme of this chapter has been that microorganisms affect animals and plants in a variety of important ways. This interaction is not unidirectional, however. It is appropriate to end this chapter with a brief look at how the activities of plants and animals affect microorganisms.

Direct effects of plants and animals on microorganisms

Microorganisms are everywhere and dominate most aspects of life on Earth, but they are not impervious to the activities of the larger life forms. The evolution of plants and animals has provided microorganisms with new places to live and new sources of food. Every animal and plant carries large populations of microorganisms in or on their bodies. These microbial populations stay with them throughout life. Dead plants and animals provide food for the microorganisms that recycle dead biomass. These effects of macrobes on microbes are unintentional, but humans have taken intentional actions to control microbes. Humans produce disinfectants and antimicrobials to kill microbes. These efforts, however, have had little overall effect on the microbial community as a whole. In only two cases have humans managed to eradicate or even come close to eradicating a microbial species: the viruses that cause polio and smallpox. Both of these were human-specific viruses that could be forced out of the human population by vaccination. There are no known cases, so far, in which a free-living microbe has been eradicated.

Indirect effects

More than any other animal, humans have had profound effects on the environment. Sometimes human activities produce unintended—and quite unwelcome—effects on the microbial world. Two examples of this phenomenon are the rise of Legionnaire's disease (a type of bacterial pneumonia) and the development of a zone in the Gulf of Mexico where oxygen concentrations have fallen too low to support marine animal life (hypoxic zone). These examples are worth considering in some detail because they illustrate the strong connections between the microbial and human worlds.

Legionnaire's disease. Throughout most of the 20th century, scientists believed that no new infectious diseases were likely to emerge because every microorganism capable of causing human disease surely would have done so already. Surprisingly, however, a number of apparently new diseases have appeared in recent years. These diseases, which have been called **emerging infectious diseases,** are usually the result of increased human contact with a microorganism that has been around for many years without the opportunity to infect humans. In 1976, a form of pneumonia, afterward named **Legionnaire's disease,** appeared suddenly among attendees at an American Legion convention. Since then, numerous outbreaks of the disease have occurred in hospitals, hotels, and nursing homes—locations with large populations of people whose immune systems were somewhat impaired by advanced age, smoking, or other conditions. The culprit responsible for the sudden appearance of Legionnaire's disease was the air-conditioning system. Prior to the 1970s, air-conditioned buildings were not as common as they are today. As hotels and hospitals began to install cooling systems, an unintentional window of opportunity was opened for a bacterium, later named *Legionella pneumophila,* to encounter susceptible humans. *Legionella* is normally found in freshwater ponds and lakes, but it adapted readily to growth in the water of air-conditioner cooling towers. Contaminated water from these towers was circulated through the air-conditioning coils and escaped through leaks in the tubing into the interior of the building. Older people, especially those with impaired lung defenses, inhaled the bacteria and acquired a new form of pneumonia.

Hypoxic zone in the Gulf of Mexico. Fishermen call it the **dead zone,** a growing area of hypoxic water in the Gulf of Mexico where oxygen concentrations have fallen so low that marine animals that require oxygen cannot survive. Similar dead zones have appeared in other coastal waters where rivers empty into the ocean (Fig. 1.8). The hypoxic zone in the Gulf of Mexico is one of the largest examples of this phenomenon. How did the dead zone come into existence? The answer starts far up the Mississippi River, where the river runs past midwestern farms and picks up nitrogen fertilizer in runoff. Human and animal wastes also contribute to the nitrogen load of the river. Ultimately this nitrogen-rich water reaches the Gulf of Mexico. In the past, the growth of phytoplankton in the gulf was limited by the low availability of some important nutrients, such as nitrogen. Nitrogen runoff into the Mississippi River brought increased nitrogen into the gulf, upsetting this equilibrium. An explosive increase in phytoplankton biomass resulted. Part of the phytoplankton population died and sank to the bottom of the gulf. There, it became a source of nutrients for

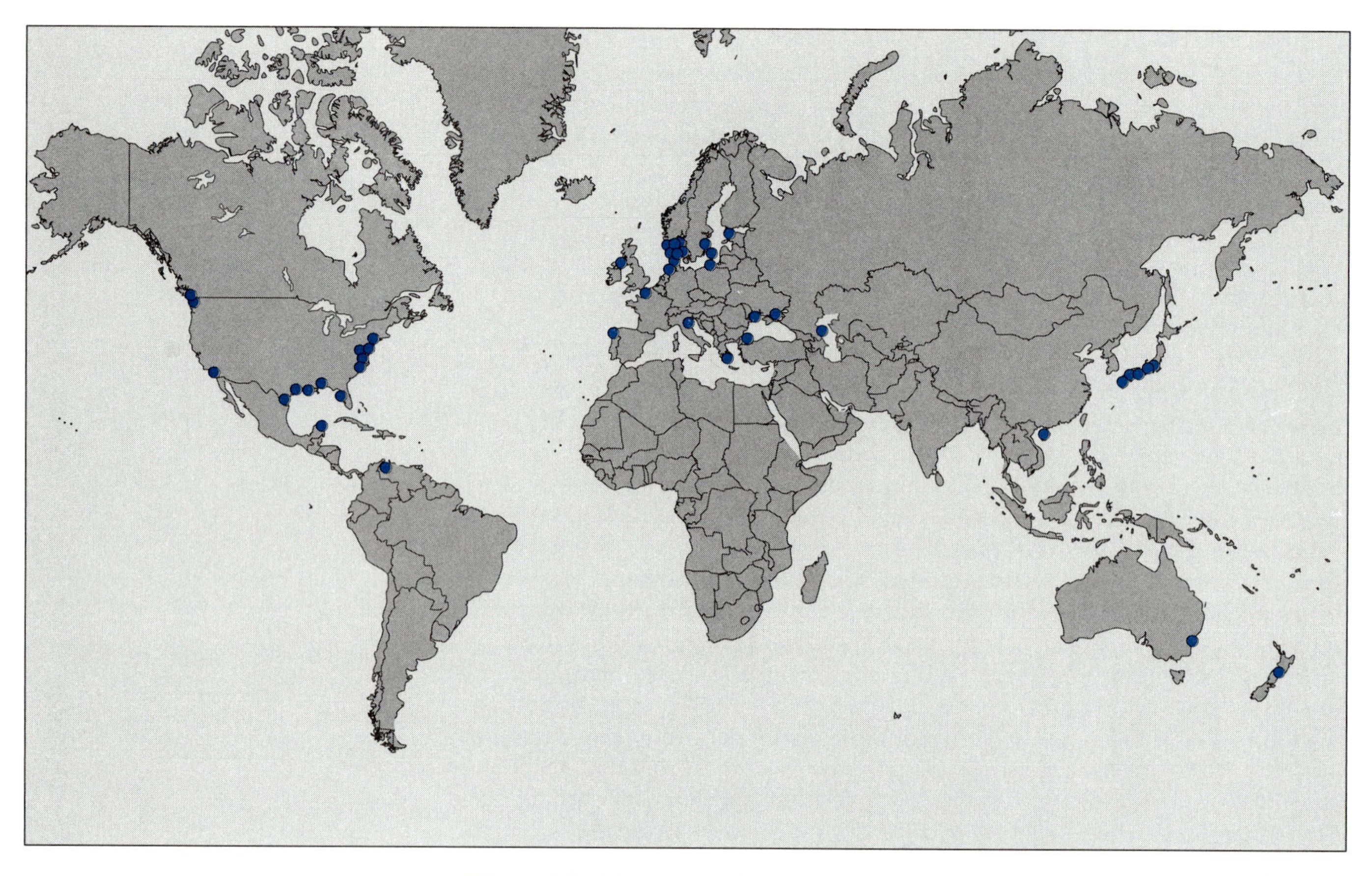

Figure 1.8 Major hypoxic zones of the world. Most of the hypoxic zones (●) are due to nutrient discharge from rivers into the marine environment. Source: Gulf of Mexico Hypoxia: Land and Sea Interactions. 1992. Report from Council for Agricultural Science and Technology, Ames, Iowa.

microorganisms in the sediment. As these microorganisms consumed the dead biomass, they also consumed oxygen. Oxygen consumption was too rapid to allow the replacement of oxygen by diffusion from the surface of the water. A new equilibrium had been established, one that unfortunately was inconsistent with the survival of marine animals.

Exploring microbe-macrobe interactions. To live comfortably with microorganisms and to gain as much benefit from their activities as possible, we need to understand better the many interactions microorganisms have with each other and with the macrobial world. This book will explore these interactions and the metabolic capabilities of microorganisms that make them possible.

chapter 1 at a glance

Comparison of the main types of microorganisms

Microorganisms	Characteristics	Beneficial roles	Role in disease
Prokaryotes			
Bacteria	Rigid cell wall Divide by binary fission Some capable of photosynthesis	Recycle biomass Control atmospheric composition Component of phytoplankton and soil microbial populations	Cause of many human diseases Some produce protein toxins
Archaea	Rigid cell wall Unusual membrane structure Photosynthetic members lack chlorophyll	Produce and consume low molecular weight compounds Aid bacteria in recycling dead biomass Some are extremophiles	No known role in human disease
Eukaryotes			
Protozoa	No rigid cell wall Many live by ingesting other microbes or chunks of biomass No photosynthetic members	Minor role in recycling Keep populations of bacteria, archae, and smaller eukaryotes in check	Cause of many human diseases
Fungi	Rigid cell wall Single-cell form (yeast) reproduces by budding Multicellular form (hyphae, mycelium) No photosynthetic members	Recycling biomass Stimulate plant growth	Cause of many human infections (mainly in immunocompromised people) Some produce toxic substances
Algae	Rigid cell wall Photosynthetic	Important component of phytoplankton	None known to cause infection but some produce toxic substances during algal blooms

How plants and animals affect microbes:

Living plants and animals provide new colonization sites and nutrients for microbes.

Dead plant and animal biomass provides nutrients for microbes.

Humans perturb the environment, changing conditions that affect microorganisms.

- Air-conditioning and *Legionella* pneumonia
- Fertilizer-intensive agriculture and the Gulf of Mexico hypoxic zone

key terms

algae photosynthetic eukaryotic microbes with rigid cell walls

archaea prokaryotic microbes with a cytoplasmic membrane structure that differs from that of bacteria and eukaryotes; includes microorganisms that can grow at very high temperatures and microorganisms that make methane

bacteria prokaryotic microbes with rigid cell walls

eukaryote cell with its DNA genome enclosed in a membrane (nuclear membrane)

fungi eukaryotic microbes with rigid cell walls; includes the yeast form (single cells that reproduce by budding) and the mycelial form (with multicellular branches called hyphae)

prokaryote cell lacking a nuclear membrane

protozoa eukaryotic microbes without rigid cell walls; many live by ingesting other microorganisms or fragments of biomass

viruses microorganisms that are not free-living but replicate by using the biosynthetic machinery of free-living organisms

Questions for Review and Reflection

1. In what sense is Earth the "planet of the microbes"? How would you explain this concept to family and friends?

2. If you accept the "planet of the microbes" idea, how would you look for life on other planets?

3. Microbiologists tend to stress the differences between prokaryotes and eukaryotes. In the case of microorganisms, what are some similarities?

4. Plant biologists might well question the assertion that microorganisms are at the base of terrestrial food chains because that honor has traditionally been awarded to plants. What arguments could be made to support the contention that microbes are at the base of terrestrial food chains? What arguments can be made against such a contention?

5. If microorganisms are really in control of our planet, why are they affected at all by human activities?

6. Superficially, the emergence of Legionnaire's disease and the development of the hypoxic zone in the Gulf of Mexico seem to be very different stories, but they have some common themes. How are they similar?

7. The term "dead zone" has a negative connotation. How could a different perspective of the process at work there define the same area as a "live zone"?

2 Diversity and History of Microorganisms

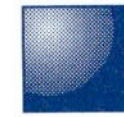

Don't underestimate microbes. There is nothing like a 3-billion-year evolutionary head start to even whatever odds humans may try to impose.

In Chapter 1, the assertion was made that, despite microorganisms' simple appearance, the microbial world is actually as diverse as the macroscopic world of plants and animals. This is a major shift from biologists' earlier view of microorganisms as relatively nondiverse. How did such a shift occur? Traditional measures of diversity, developed by scientists working on plants and animals, emphasized physical traits such as appendages, shape, and coloring. This type of classification system, however, is of limited utility for microorganisms. In particular, superficial physical traits cannot be used to assess relationships between microorganisms and macroorganisms, because there are no physical traits common to both groups to use as yardsticks for comparison.

Measuring Diversity in the Microbial World

A molecular yardstick for measuring relatedness

The scientific world needed a classification system based on some trait possessed by all living organisms. In the 1980s, scientist Carl Woese put forth a radically new suggestion that went straight to the center of the diversity issue. He suggested that the deoxyribonucleic acid (DNA) sequences of certain common genes could be used to determine the relatedness of different organisms. The example Woese picked was a gene that encoded an RNA molecule found in ribosomes, **ribosomal ribonucleic acid (rRNA). Ribosomes,** the protein-RNA complexes on which proteins are synthesized, are found in all prokaryotes and eukaryotes. Although there are differences in size between the ribosomes of prokaryotes and eukaryotes, the sequences of the rRNA molecules from all sources contain regions that are very similar (that is, highly conserved regions). These allow the sequences to be aligned and compared (Fig. 2.1). There are three rRNA molecules in each ribosome. Woese chose the intermediate-sized rRNA, **16S rRNA** (prokaryotes) or **18S rRNA** (eukaryotes). This molecule was large enough to contain enough information for genetic comparisons but small enough for the gene to be sequenced easily.

Figure 2.2 shows the way in which the sequences of rRNA genes from different organisms can be used to deduce relationships between creatures as diverse as bacteria and humans. The sequences used in Figure 2.2

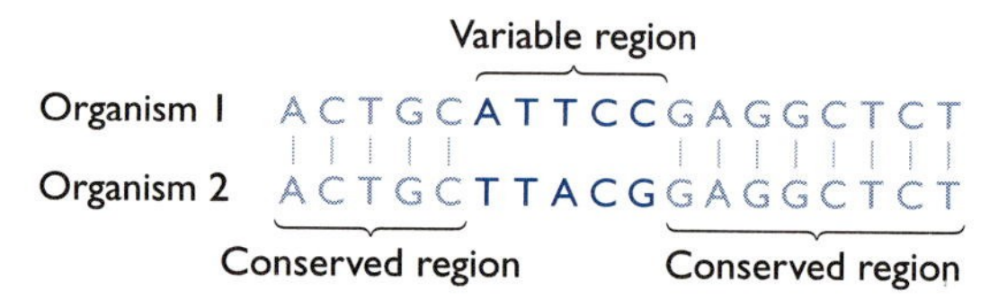

Figure 2.1 Comparison of sequences of rRNA genes from diverse organisms. If two regions of DNA sequence, such as the ones shown here, contain some regions that are similar, it is possible to align them and identify regions where there are sequence differences (variable regions). This is the way rRNA gene sequences are aligned and compared. The information used to determine how related two sequences are can be found in the region where the differences are, but the conserved regions are important for making sure that the same variable regions from different sequences are being compared. In this figure, for simplicity, only the sequence of one strand of each DNA molecule is shown.

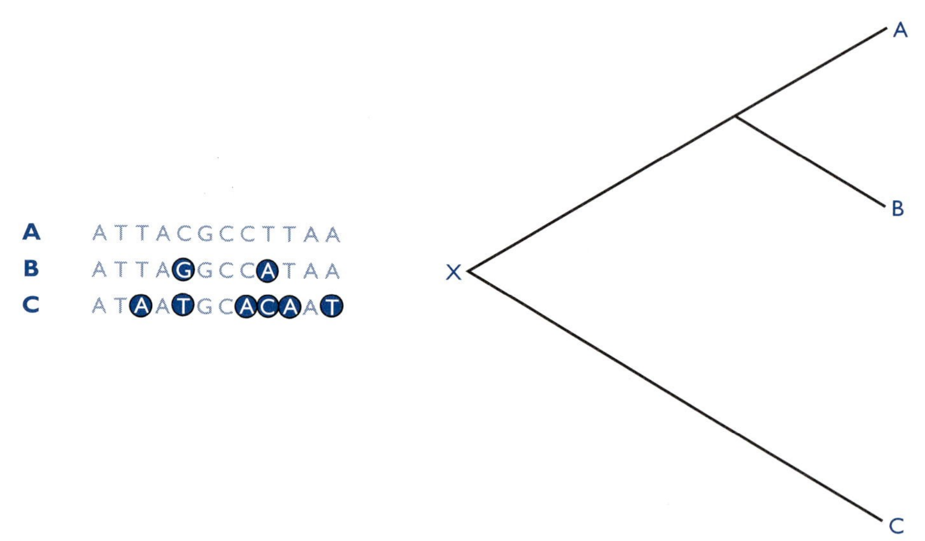

Figure 2.2 Relationship between diverse organisms based on sequences of rRNA genes. This figure does not show actual rRNA sequence data but illustrates how differences in sequence data can be used to establish relationships between different organisms. In this simple example, sequences from three microorganisms (A, B, and C) are compared. Again, for simplicity, only one strand of the DNA is shown. Note that the sequences from A and B are identical except for two bases (highlighted letters). By contrast, the sequence from A differs from that of C by six bases. The sequence from B also differs from that of C by six bases. Thus, A and B are more closely related to each other than to C. This information is depicted using lines with lengths proportional to the number of differences. Thus, whereas A and B are separated by two units of line length, each is separated from C by six units of line length. In the tree shown here, there is an implicit assumption that A, B, and C all descended from the same ancestor X. This type of tree is called a rooted tree. But the important information in the figure is in the total branch lengths separating the letters.

are not the sequences of actual rRNA genes. 16S rRNA genes are larger than 1000 base pairs (1 kbp) in size and thus are too large to display in a single figure. The principle of comparison is the same, however. Genes from different organisms will have similar but not identical rRNA gene sequences. If the sequence of the gene from organism A has only 2 differences from the sequence of the gene from organism B, but has 6 differences from the sequence of the gene from organism C, then A and B are assumed to be more closely related to each other than either is to C. This relatedness

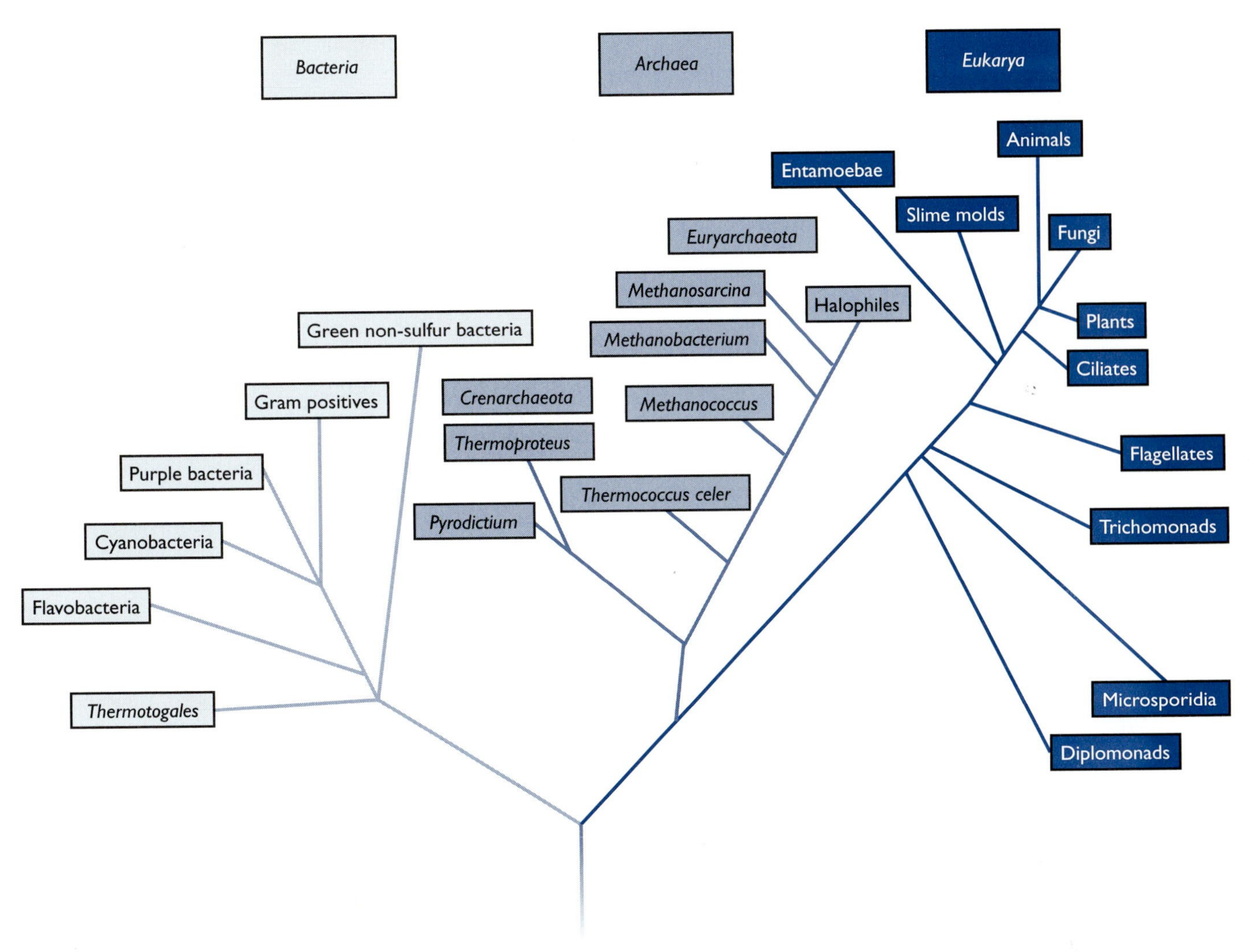

Figure 2.3 The three-domain view of life. When an analysis like that illustrated in Figure 2.2 was performed using the rRNA gene sequences of many different organisms, the tree shown in this figure was the result. The longer the branch lengths, the greater the diversity in the group. Note how much shorter the branch lengths are for newly evolved organisms such as fungi, plants, and animals than the branch lengths for prokaryotic and lower eukaryotic microorganisms.

is expressed in the length of lines that connect A, B, and C on a phylogenetic tree. The lines joining A and B will be shorter than the ones joining A and C or B and C.

The three-domain view of life

When scientists used rRNA genes as a yardstick to establish the relationships between prokaryotic and eukaryotic organisms, they came up with a surprising picture (Fig. 2.3). As expected, the eukaryotic organisms grouped together, although they exhibited a high degree of diversity (long branch lengths). Note, however, that plants, animals, and fungi are relatively nondiverse (short branch lengths), compared with the prokaryotic organisms. The surprising finding was that bacteria grouped separately from archaea. Thus, instead of the two-domain grouping of prokaryotes and eukaryotes, there were **three domains:** two prokaryotic and one eukaryotic. Moreover, the archaea appeared to be more closely related to the eukarya than to the bacteria. The three-domain classification scheme is now widely accepted by microbiologists, who have subsequently found other supporting evidence. The three-domain classification scheme remains controversial, however, among some plant and animal biologists who have difficulty accepting the idea that microorganisms are far more genetically diverse than plants or animals. The three-domain model fits quite well, however, with what would be predicted from a comparison of evolutionary time scales of microorganisms and macroorganisms. To see why this is so, consider the history of life on Earth and the relative evolutionary time scales of microbes and macrobes.

The Origins of Microbial Diversity

Early Earth

An easy explanation for the diversity found in the microbial world is that microorganisms, especially prokaryotes, have a much longer evolutionary history than plants or animals and thus have had more time to evolve into diverse forms. Also, because they were here first, they enjoyed unchallenged access to all of the sites available in or on Earth. Finally, microorganisms have been exposed to and survived cataclysmic conditions unknown by higher animals and plants. Plants and animals are relative newcomers and have only had to prove their adaptive capacity for several hundred million years, a fairly short span of years in evolutionary time. During this time, conditions on the earth's surface have been conducive to survival of plants and animals. Certainly, there have been many examples of species extinction, but, by and large, the temperature has remained fairly stable, there have been few collisions with really large meteors, volcanic activity has been moderate, and the oceans have remained well mixed and oxygenated.

These conditions are by no means typical of Earth's past, marked by ice ages, periods of heavy volcanic activity, and periods when entire oceans became anoxic. Microorganisms have proven their ability to face challenges unimaginable to us today. Moreover, microorganisms did not simply occupy various niches offered by the earth. Through their chemical activities, they transformed the earth and its atmosphere in a number of ways. Some of these changes actually contributed to making the earth habitable for the plants and animals that appeared much later.

The earth is about 4.5 billion years old. Scientists estimate that the first living creatures appeared about 4 billion years ago, shortly after the earth's surface had cooled enough to allow liquid water to form (Fig. 2.4). These creatures were most similar to modern day prokaryotes—bacteria and archaea. Because some microorganisms living on Earth today are capable of growing in boiling water, life could clearly have begun while the earth's surface was still very hot (the aptly named **Hadean period**). Also, because the sun was only about two-thirds as bright as it is today, the earth's surface would have become habitable faster than if the sun had been brighter.

Early Earth was far from being an Eden. It was as inhospitable as any planet ever conjured up by science fiction writers. Earth's early history was punctuated by a succession of catastrophic impacts, as large chunks of space rock crashed into the earth's surface. Such a bombardment created the craters on the moon. Most evidence of these impacts on the surface of the earth have now been effaced by the effects of plate tectonics and erosion, but Earth sustained even more impacts than the moon. Life during the high-impact period would not have been easy. Some impacts were powerful enough to vaporize oceans, creating clouds of steam that would have sterilized the earth's surface. These events may not have completely obliterated emerging life forms. Microorganisms could have survived this period

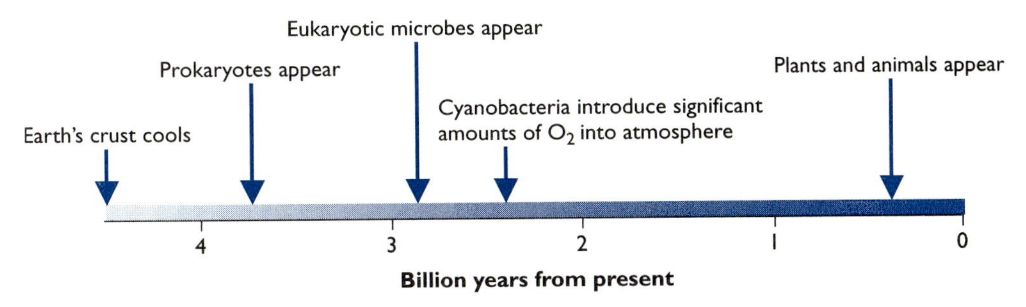

Figure 2.4 Approximate timing of major events in the history of life on Earth.

deep underground. Some may have had the capacity of modern microorganisms to produce tough survival forms called **spores.** Although direct exposure to steam would have killed them, some spores could have survived slightly cooler conditions that would still have been hot enough to kill an actively growing microorganism. Fortunately for life on Earth, this massive bombardment stopped early in the planet's history.

Even when large chunks of extraterrestrial rock were not bombarding the earth's surface, conditions were harsh. Because the ozone layer had not yet formed, the earth's surface was exposed to strong ultraviolet (UV) radiation. Despite these unpromising conditions, however, prokaryotic life began to develop. Some microorganisms may actually have been able to live on the earth's surface. One bacterium, *Deinococcus radiodurans,* can survive doses of radiation 3000 times greater than the lethal dose for humans. Most organisms, however, probably developed in the subsurface of land masses or beneath the ocean surface, where they were protected to some degree from the UV radiation.

After Earth's traumatic beginning, things settled down a bit. However, volcanic activity was much greater than today. This activity actually helped keep the earth's surface from freezing, by spewing carbon dioxide into the atmosphere and creating a **greenhouse effect.** During this period, microorganisms multiplied and began to occupy all the microenvironments the earth offered, from the bottom of the ocean to its surface, from temperate regions to polar ice, from the surface of land masses to the deep subsurface. Adaptation to these places, each of which offered different potential nutrients and different conditions of pH and temperature, required that microorganisms develop very diverse traits.

The panspermia theory

Geologists cannot pinpoint precisely the date when life first appeared, because the small size of microorganisms means they leave no clear and convincing geological record. Yet all of the available evidence supports the hypothesis that life appeared soon after the earth's crust cooled. Many scientists have found it difficult to believe that something as complex as a free-living cell could evolve in only a few hundred million years. This has led some to theorize that life may not have arisen on Earth but rather came to this planet from elsewhere, most probably Mars (the **panspermia theory).** The rationale for the panspermia theory begins with the observation that when the planets first formed, some cooled faster than others. Mars, being both smaller and farther from the sun than Earth, probably cooled much more quickly. Recent evidence suggests that Mars may also have had water in those times. During the bombardment period, Mars, too, was hit repeatedly by large chunks of space rock, and these huge impacts would have knocked fragments of the planet's surface into space. Material ejected from Mars would have been drawn earthward by the gravitational pull of the sun. It is clear from the properties of modern prokaryotes that some could have survived in the frigid vacuum of space during a long journey to Earth. If life did arise originally on Mars, some of this life could have landed on the newly cooled Earth and inoculated the planet with living cells. As future expeditions to Mars bring back subsurface materials, scientists may find evidence of these ancient life forms.

The oxygen revolution

A revolutionary development occurred between 2.5 and 2 billion years ago, changing the earth and its atmosphere completely. Oxygen gas began to appear in significant amounts in the earth's atmosphere as a result of a microbial metabolic process called **oxygenic photosynthesis.** Although many compounds such as water contained bound oxygen, there had been no oxygen gas in the atmosphere. Oxygenic photosynthesis differed from earlier forms of photosynthesis, in that it split water and released oxygen. The bacteria responsible for this new type of photosynthesis are called **cyanobacteria.** This first appearance of oxygen left a tangible geological record: banded iron formations in rock. Iron in the earth's crust reacted with oxygen to form black iron oxides, producing dark bands. Cyanobacteria left a fossil record, too. Some cyanobacteria accumulated to form large mounds called **stromatolites** (Fig. 2.5). Geologists have found fossilized stromatolites dating back 3 billion years and microfossils of individual cyanobacteria cells that date to 3.5 billion years ago. Cyanobacteria brought the oxygen level of the earth's atmosphere up to about 10% of today's level, high enough to create conditions that favored evolution of oxygen-utilizing organisms.

Oxygen changed everything. One major effect was to expand the simple cycles of carbon, nitrogen, and other compounds to include more oxidized products. Thus, for example, sulfides could now be oxidized to sulfate, and iron could be oxidized to iron oxides. A second major effect was

Figure 2.5 Fossilized stromatolite. Cyanobacteria accumulate into large mounds (stromatolites) which may become fossilized. © Albert Copley / Visuals Unlimited.

the formation of the **ozone layer.** Once the ozone layer formed, UV-sensitive microorganisms could begin to move out of the ocean, where waters had provided some protection from UV light, and into terrestrial habitats. A third major effect occurred somewhat later, when the cyanobacteria were incorporated by eukaryotic cells to form the **chloroplasts** of algae and green plants.

How did scientists determine that chloroplasts, which are now found only inside plant or algal cells, were once bacteria? The first clue came from finding that cyanobacteria and plants perform photosynthesis in exactly the same way, using the same type of chlorophyll. A second clue came from analysis of the rRNA gene sequences of chloroplasts, which revealed that the chloroplast sequence was most closely related to that of cyanobacteria. In a similar development, which probably occurred at about the same time that cyanobacteria were incorporated by eukaryotic cells to form chloroplasts, other eukaryotic cells acquired nonphotosynthetic bacteria, which became the mitochondria of today's eukaryotic cells.

The first eukaryotes

Pinpointing the date of appearance of the first eukaryotes has also proven difficult. Because many eukaryotes are oxygen dependent, scientists had theorized that protozoa first appeared about 2 billion years ago. However, there are modern protozoa that live in anoxic environments, so protozoa could have emerged before the appearance of oxygen in the atmosphere. Recent evidence has pushed the likely date of the first protozoa back to about 3 billion years ago. Bacteria and archaea are clearly older than protozoa, but the first eukaryotes may have appeared very early in evolution. An early appearance of the first eukaryotes is consistent with the high degree of diversity in this group. Algae presumably appeared after cyanobacteria, because their chloroplasts were derived from cyanobacteria. They probably evolved within the last 2 billion years.

The fungi appeared only comparatively recently, during the last several hundred million years. This late appearance of a type of microorganism is difficult to understand, given that all other types have such ancient lineages. Scientists have speculated that because terrestrial fungi are closely associated with plants, fungi might have co-evolved with plants. Another possibility is that there are much older lines of fungal descent that remain to be discovered. Until recently, fungi were thought to be exclusively terrestrial. Now scientists are finding marine fungi in ocean sediments and other locations far from land. Additional studies of marine fungi may answer the question of whether fungi are truly the newcomers that they appear to be or whether they have much more ancient roots.

Viruses

Where do viruses fit into this picture? The short answer is that they don't. Viruses have no ribosomes, so the rRNA yardstick does not apply. Moreover, viruses have such small genomes and are so diverse in their gene sequences that no highly conserved molecular yardstick has been identified for viruses. A crude way of estimating the age of viruses is to assume that they did not appear earlier in evolution than their current host animal or microbe. Thus, for example, according to such reasoning, a virus that infects only humans was unlikely to have appeared before the evolution of humans. Or a bacterial virus specific for *Escherichia coli* probably appeared after *E. coli* first evolved. A problem with linking the evolution of a virus with its current host is that we know such associations can change. As more is learned about viruses, scientists may discover new ways to deduce the sequence of events in the evolution of viruses.

Mining Diversity

The domestication of microorganisms

The domestication of microorganisms began many thousands of years ago with the discovery of alcohol fermentation and the effects of yeast in bread. Once scientists acquired the ability to identify and study the microorganisms responsible for such previously mysterious processes, the pace of discovery of new microbial processes accelerated. By the mid-1900s, bacteria and fungi were being used to produce a variety of products ranging from amino acids and vitamins to antibiotics. Two developments since the 1970s have reopened the frontier of possibilities for mining the microbial world for new products and processes.

Genetic engineering and PCR: the new bonanza. Scientists learned how to introduce segments of foreign DNA into bacteria, plants, and animals to provide the recipient organism with new traits. Bacteria have been sharing DNA segments with each other naturally for billions of years, but, with the advent of genetic engineering, scientists could control this process and extend it to plants and animals. (Genetic engineering and its applications in biotechnology will be the subject of later chapters in this book.) So, genetic engineering created another way to exploit microbial diversity: the directed creation of organisms with desirable new traits. This discovery created a new phenomenon, the **biotechnology industry.** An associated development, **polymerase chain reaction (PCR)**, accelerated the diversity gold rush.

PCR is a process that makes many copies of (amplifies) a specific portion of a larger DNA molecule (Fig. 2.6). Only minute amounts of DNA are

Figure 2.6 What the polymerase chain reaction (PCR) accomplishes. The details of the process, including an explanation of how the ends of the amplified region are set, will be provided in Chapter 5. It is important to understand first, however, what PCR seeks to do. Starting from a large segment of DNA isolated from an organism or even an environmental site, the DNA-copying enzyme DNA polymerase makes many copies of a single specific segment of the larger DNA segment. In the process, the mixture containing the DNA polymerase and the DNA is cycled between 95°C (to separate the strands of newly synthesized DNA) and 55°C (to allow DNA polymerase to make yet another copy of each separated strand).

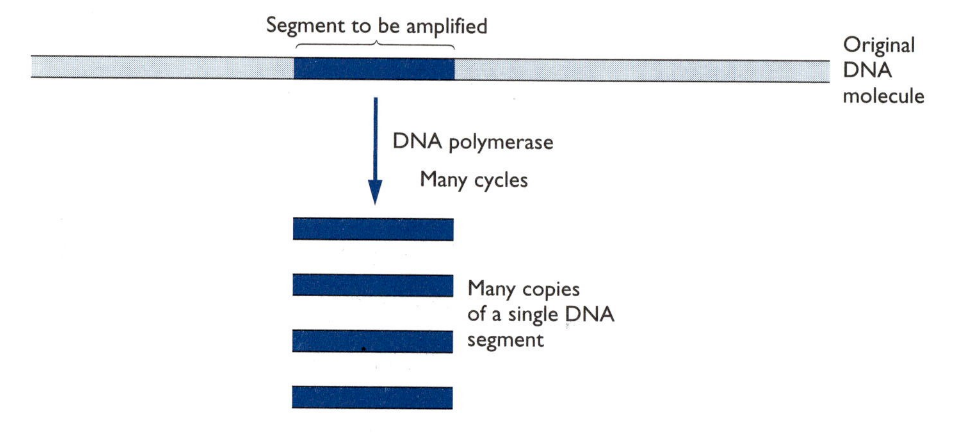

box 2–1 *Some Current Uses of PCR*

PCR is changing modern biology.

- Cancer biologists can identify a gene believed to be involved in causing a particular type of human cancer and amplify it both from healthy people and people with the cancer. The DNA sequences of the genes can then be compared to determine whether the people with cancer have an aberrant form of the gene and, if so, how the genes differ. This approach may be the basis for discovering new types of anticancer therapies that target genes or the proteins they encode.
- Forensic scientists selectively amplify specific regions of human DNA to obtain a "molecular fingerprint" that is person-specific. Amplification can be done from a spot of blood or a dried sperm sample. Thus, it is now possible to state with great certainty whether a blood sample or a sperm sample came from a specific suspect.
- Microbiologists with a new microorganism in hand amplify certain DNA segments that act like fingerprints for microorganisms. The DNA sequence of the amplified segments gives a tentative identification of the microorganism within weeks, replacing uncertain identification processes that once took years. The DNA sequence of the fingerprint molecule also tells scientists if this organism is a new species, genus, or kingdom of microbial life.
- Microbiologists interested in learning what microorganisms might be living in a Yellowstone hot spring amplify a mixture of DNA segments that provide a fingerprint for the whole community of microorganisms. This process is more complicated than the ones previously mentioned, but it allows scientists to detect uncultivated as well as cultivated microorganisms and thus provides an indication of the true diversity of microbes at the site.
- PCR has created a new field: **molecular archaeology.** One example is provided by scientists who wanted to know why the influenza pandemic of 1918 was so much more lethal than any flu epidemic since that time. Influenza is caused by a virus. PCR was used to amplify segments of the influenza virus genome from lung biopsy specimens taken from people who died in the 1918 epidemic. The DNA sequences of these segments were compared to those from modern influenza viruses to learn how the 1918 virus differed. The 1918 virus was probably a hybrid of a swine influenza virus and a human influenza virus. This knowledge helps health care officials look for potential recurrences of such forms of influenza virus.

necessary as a starting point. DNA genomes of even simple microorganisms can contain thousands of genes. Those of humans and other animals contain tens of thousands of genes. Just as a person might choose to make many copies of a favorite photograph to share with friends, that same person can now choose to make many DNA copies of a small segment of a much larger DNA genome. This simple-sounding process has proven to have hundreds of applications (Box 2–1) and has created a rapidly growing industry that currently grosses about $100 billion a year. The magnitude of the revenues generated by PCR drew substantial public and scientific attention, but what focused that attention on microbial diversity and its potential for future exploitation was the **DNA polymerase** that made PCR possible.

PCR amplification requires that the reaction mixture, which contains DNA polymerase, be exposed to temperatures ranging from 55–95°C on a recurring basis (Fig. 2.6). Normally, enzymes would be inactivated rapidly by exposure to such high temperature. The DNA polymerase that could take this kind of punishment was obtained from a bacterium growing in hot springs of Yellowstone National Park. Previously, microbiologists had focused their attention primarily on a few well-studied microbial species, such as the bacterium *E. coli* or the yeast *Saccharomyces cerevisiae.* Suddenly scientists rediscovered what the earliest microbiologists had known years before—that there is a great deal of unexplored diversity in the microbial world.

Extent of microbial diversity remaining to be exploited. A second development that renewed interest in microbial diversity was the realization that only a tiny fraction (<1%) of the microorganisms in nature have been cultivated and studied. By contrast, more than half of the species of plants and mammals and at least a fourth of the species of arthropods are known. To date, only a few thousand species of bacteria have been named. Even fewer species of archaea and eukaryotic microorganisms are known. Estimates suggest that there may be hundreds of millions of microbial species left to discover. The number of species does not give a full sense of the diversity of microorganisms, however, because scientists who study plants and animals use a different standard for creating a species from the one used by microbiologists. Most microbiologists now use the **"70% rule"** as a guide for designating a species. That is, two microorganisms are in the same species if they are 70% or more identical at the DNA sequence level. Applying this criterion to higher organisms would place all arthropods in a single species and all primates in a single species. This would certainly simplify matters for biology students, but is not likely to be accepted by the scientists who work on either arthropods or mammals. How species should be defined remains a major question in microbiology. Scientists still argue over whether to use metabolic activities, DNA sequence, or a combination of both to define a species. And what does it mean to have a species that contains such a diverse mixture of organisms?

Although arguments over definition of species may seem arcane, the diversity that gives rise to this argument has a very practical implication. It means that the microbial world has been virtually untouched with respect to human exploitation. Given the numerous advances the domestication of certain microbes has already produced, imagine the richness of new possibilities just waiting to be found. Put this together with the fact that PCR and DNA sequencing now make it much easier, cheaper, and faster to identify and characterize new microorganisms than ever before, and the microbial world becomes a real treasure trove. The lure of this vast potential treasure has created a group of scientists called **bioprospectors,** who are sampling many environments for new microorganisms. The products of their research already are beginning to appear. New enzymes have been identified for the PCR industry. For instance, your stone-washed jeans are not pounded on river rocks; they are treated with enzymes found by bioprospectors. The detergent industry is developing new enzymes to improve the cleaning properties of laundry soaps. This is only the beginning.

Bioprospectors in hot water: literally and figuratively

Because many bioprospectors assume that exotic sites are most likely to yield the most unusual microorganisms (a common misperception), their zeal to uncover new organisms and enzymes has triggered a veritable assault on places with hot springs and other unique or extreme habitats, like Yellowstone National Park. Until recently, bioprospectors could be seen dipping their sampling devices into hot springs—and sometimes falling into them. (Two scientists have already been killed in such accidents.)

The people who administer parks and protected habitats like Yellowstone are now trying to restrict access to such sites. Eager bioprospectors can do considerable damage to a site. Moreover, bioprospectors have not been especially eager to share the profits of their discoveries with the owners (federal, state, or private) of the land where the original material

box 2–2 *Who Owns the Microbes of Yellowstone National Park?*

U.S. National Park Service officials became frustrated by bioprospectors who were reaping lucrative profits by harvesting microorganisms found in national parks. None of the companies voluntarily shared any profits, so the Park Service decided to make a deal with Diversa Corporation, a biotechnology company. Diversa had been particularly aggressive in finding and marketing new microbial products.

To understand why park officials might have been frustrated, consider the bacterium *Thermus aquaticus*, which was originally isolated from a Yellowstone hot spring. The Taq polymerase, a DNA polymerase obtained from *T. aquaticus*, was the first of a number of similar enzymes that led to the creation of the lucrative PCR industry. Hoffman LaRoche Inc., which owns the patent for Taq, currently makes at least $100 million a year on the polymerase, a figure that could rise to $1 billion a year within the next decade. Manufacturers of PCR thermocycling machines have made similarly spectacular profits from the availability of enzymes like Taq polymerase. More recently, a DNA ligase enzyme that forms the basis for another type of amplification called ligase chain reaction (LCR), which is becoming increasingly important for the diagnosis of serious illnesses such as chlamydial disease, was isolated from a Yellowstone hot spring. Once again, none of the profits went to the park.

What complicates this debate is the fact that, in the past, the taking of small amounts of sample from an environmental site without accounting to anyone for it was considered to be a legitimate and unrestricted activity. The pharmaceutical industry has screened hundreds of thousands of soil samples from all over the world in its search for new antibiotic-producing organisms. No royalties were paid or expected.

The situation in Yellowstone brought the issue of ownership to the fore, largely because of the number of bioprospectors who descended on the park and the enormous profits being generated. Given that Park Service officials are not lawyers or financiers, the deal they struck so quickly and so quietly with Diversa makes a certain amount of sense. Diversa is a company that was created primarily by scientists interested in microbial diversity. When seeking permission from park officials to look for unusual microbes, Diversa had agreed readily to give Yellowstone Park some portion of the proceeds of any microbial discoveries made on park land.

Yet some people were worried that ceding ownership of organisms in a publicly-owned park to a single company might be setting a bad precedent. Two environmental advocacy groups brought a lawsuit against the National Park Service to negate the agreement with Diversa and to institute a careful public review of alternative options. The lawsuit was filed in December 1997. In April 2000, a judge upheld the original agreement between the National Park Service and Diversa. The environmental groups are going to appeal the decision. There is much to be said on both sides of this argument. Stay tuned for future developments.

was obtained. Yellowstone and other parks are now restricting the access of scientists—especially those from industry—to vulnerable and potentially lucrative sampling sites. Additional attempts by U.S. National Park Service administrators to harvest some of the profits of bioprospectors have created a new level of controversy and raised unprecedented legal issues (Box 2–2).

Genome sequences of microorganisms

We know, based on the rRNA gene yardstick, that microbes are diverse; however, this does not provide much information about what these diverse microorganisms are capable of doing. Scientists suspect that, even in the case of those microorganisms that have been cultivated and intensively studied, there are still many things that have not been learned. A direct way to learn all there is to know about a microorganism is to examine the sequence of its genome. The sequence of bases (**nucleotide sequence**) in the genome of an organism contains all the information about what that organism is and does. The problem lies in how to translate the DNA sequence into information about physical traits. Until recently, most of the focus of DNA sequencers was on obtaining the basic sequence information.

A number of viral genomes have been sequenced. Because of their small size, obtaining their sequence was not as much of a challenge as obtaining those of free-living microorganisms. The first complete genome sequences for bacteria and archaea began to appear in the 1990s. By 2010, there should be at least 100 complete prokaryotic genome sequences. A few genome sequences for eukaryotes are beginning to appear. The genome sequence for the yeast *S. cerevisciae* (genome size about 11 Mbp) and for the protozoal parasite *Giardia lamblia* (genome size about 12 Mbp) have been completed. The genomes of important plants and animals and, of course, the human genome are being sequenced. Because of the sizes of these genomes, such projects will take much longer than the sequencing of the genomes of microorganisms. Attention now is turned to the question of what can be learned from the raw sequence data.

In the case of the well-studied viruses, most of the genes were already known from biochemical studies of viral proteins. In this case, the main benefit of having the gene sequences is that the genes can be cloned more easily and used to produce large quantities of viral proteins. These proteins are being used in vaccine preparations and to search for new antiviral compounds that will bind to and inactivate viral proteins. Genome sequencing has another potential application in the field of virology. Today, if a new viral disease appeared, the first step would be to obtain its genome sequence. Thus, even before the virus could be cultivated or identified by conventional means, there would be extensive genetic information available, which, when compared to knowledge of known viruses, would quickly reveal the virus type and possibly suggest antiviral compounds that might be effective.

Scientists also have discovered from genome sequences that a lot remains to be learned, even about well-studied bacteria like *E. coli*. About one-third of the genes on the *E. coli* chromosome have no known function. This does not mean they have no function, but that scientists have not discovered the traits encoded by the genes. Scientists are now attempting to figure out what these genes encode. Comparisons of genome sequences may explain why some microorganisms cause disease and others do not. Researchers are currently comparing the genome sequence of the diarrhea-causing bacterium *Salmonella typhimurium* with that of a closely-related strain of *E. coli* that does not cause disease. Genes found in the *S. typhimurium* genome but not in the *E. coli* genome may be the key to understanding what traits of *S. typhimurium* account for its ability to cause disease.

How will genome sequence information help with microorganisms that have not yet been cultivated? Once scientists know the DNA sequence of an interesting gene, they can use PCR to amplify this gene from DNA isolated from a microbial community. (More information about how this is done and how DNA sequencing works will be provided in Chapter 5.) The important point to be made, however, is that it is now possible to isolate genes from environmental sites. This capability may allow scientists to characterize at least parts of the genomes of organisms that have not been cultivated. In medicine, scientists are using such approaches to identify disease-causing microorganisms that have not yet been cultivated. The availability of gene sequences is revolutionizing microbiology and is providing new ways to mine diversity.

chapter 2 at a glance

A molecular yardstick for assessing diversity

Sequences of 16S (prokaryote) or 18S (eukaryote) rRNA genes from two different organisms compared; greater sequence similarity means closer relatedness

- Why 16S/18S rRNA genes?
  - All free-living organisms have them
  - Small enough to sequence easily, but large enough to contain sufficient information to make comparisons
- Conclusion from rRNA gene comparison: three-domain model of life
  - Bacteria
  - Archaea
  - Eukarya

Origins of diversity in the ancient age of microorganisms

Assumption: longer evolutionary history and more challenges creates more diversity

History of life on Earth

- Early Earth (4.5–2 billion years ago)
  - Absence of oxygen in atmosphere
  - Absence of ozone layer
  - Cataclysmic bombardments, volcanic activity
  - Ice ages and ages in which the ocean become hypoxic
  - Prokaryotes evolve early
- Oxygen revolution
  - Cyanobacteria invent oxygenic photosynthesis; put oxygen in the atmosphere
  - Ozone layer forms
  - Cyanobacteria later become chloroplasts of algae, green plants
  - Phytoplankton begins to form
- First eukaryotes
  - Probably protozoa
  - Time of appearance perhaps before the oxygen revolution
  - Algae evolve later but are also ancient
  - Fungi evolve very recently, along with plants and animals
- Viruses: currently no way to place them in the larger evolutionary picture

Mining diversity

New opportunities for domesticating microorganisms

- Genetic engineering provides new tools of study
- New appreciation of microbial diversity provides new subjects for study
- Example of PCR industry shows possibilities

Bioprospectors use these insights to find new products and processes

(continued)

chapter 2 at a glance *(continued)*

Genome sequences: a new way to study microbial diversity

More than 100 genome sequences of bacterial and archaeal genomes may soon be complete

Only a few sequences of eukaryotic microbe genomes

New challenge to convert DNA sequence information into information about physical traits and activities of the microbes

key terms

bioprospectors scientists searching for new products and processes in diverse microorganisms

biotechnology use of genetic engineering and related DNA-based techniques to develop new products and processes

cyanobacteria bacteria that were first to practice oxygenic photosynthesis, a process that splits water and releases O_2, and introduced O_2 into the earth's atmosphere; later became the chloroplasts of plants

genome sequence the arrangement of bases in an organism's genome; contains information about the organism's traits

molecular yardstick use of sequences of highly conserved genes, such as the gene for 16S or 18S rRNA, to deduce relatedness between different organisms and to estimate diversity in a particular group of organisms

panspermia theory hypothesis that life on Earth may actually have arisen on other planets and arrived on material blasted free by impacts with large asteroids.

polymerase chain reaction (PCR) process for selectively making many copies of a single segment of an organism's genome; example of a large new area of biotechnology made possible by mining microbial diversity

three-domain model model based on rRNA gene sequence analysis that divides life into three domains: bacteria, archaea, and eukarya

Questions for Review and Reflection

1. In Chapters 1 and 2, mention was made of the fact that prokaryotes (and possibly some eukaryotes) can transfer segments of their DNA to members of other species. How could this activity make it more difficult to use a molecular yardstick to determine the relatedness of microorganisms?

2. Suppose you discovered a new organism that looks like a bacterium. You are able to grow it in the laboratory and have obtained the sequence of its 16S rRNA gene. How could you use this information to determine tentatively the species of the microbe? How might you run into trouble doing this?

3. What does the history of life on Earth have to do with microbial diversity? What are some reasons that prokaryotes and eukaryotic microorganisms should be more diverse than plants and animals?

4. How could scientists prove that the panspermia theory is correct, and what problems would they encounter attempting to do so?

5. In what ways did the cyanobacteria make possible the later evolution of plants and animals?

6. Many microorganisms are able to grow in the absence of oxygen, but no plants or animals are known to have this capacity. How can you explain this?

7. Why did the discovery of the DNA polymerase that made PCR possible, and the consequences of this discovery, excite the bioprospectors?

8. Do you think that Diversa should "own" the microbes collected from Yellowstone National Park? If so, why? If not, how would you solve the problem of compensating the park for profits made from its microbial denizens?

9. What sorts of things could you learn from a genome sequence of a microorganism? This topic will come up again in later chapters, but this is a good time to start thinking about the question.

3 The Challenge of Cultivating and Identifying Microorganisms

Microbes are like exotic plants. They grow in profusion in nature but many wilt when brought into the laboratory garden.

Scientists interested in zebras or ferns can simply go into the wild, capture the animal or dig up the plant, and bring it back. It is easier to study an animal or plant if it can be successfully bred or propagated in captivity, but this is not necessary. For microbiologists, the situation is different. In some cases, a single species of microorganism grows in abundance in a natural setting and large numbers of organisms can be obtained for study, even if the microorganism cannot yet be grown in the laboratory. Such cases are rare. In most cases, microbiologists need to induce microorganisms to reproduce in the laboratory to obtain large numbers of a single microorganism. This allows researchers to study enzymes or other proteins made by the microbe, to follow the chemical reactions it is performing, or to characterize its cell wall. In a clinical setting, cultivation of the suspected disease-causing microbe is necessary for identification and to determine its susceptibility to antimicrobial drugs. Scientists generally want many copies of a single organism (**pure culture** of the organism), because interpreting the results of experiments involving mixtures of microorganisms (**mixed culture**) can be quite difficult.

Why Cultivation of Microorganisms Is Important

Microbiologists are faced with two problems when they set out to study a new microorganism. First, they must identify the laboratory conditions that induce the microorganism to divide and reach high concentrations (**cultivation**). Second, they must be able to **isolate** the organism from all the other microorganisms that may be present in the same site. Later, they will need to confront a third problem, how to **identify** the organism. That is, they will need to find traits of the organism that allow them to distinguish it from others that may look the same and act very similarly. Every new microorganism presents its own special challenges. To introduce you to the types of problems microbiologists face, we will describe in detail two examples of the use of cultivation, isolation, and identification.

An Example from Clinical Microbiology: A Case of Suspected Meningitis

Clinical background

By far the most intensive efforts to cultivate and identify microorganisms have focused on bacteria that cause disease. Accordingly, the first example comes from clinical microbiology: a microbiologist's response to a possible case of meningitis in a college student. Meningitis is an inflammation of the membranes that cover and protect the brain (**meninges**). Many microbes produce inflammation of the meninges if they reach the brain. Inflammation can cause increased pressure on the brain, which may result in brain damage and death. Outbreaks of meningitis on campuses have become a regular feature of college life in the U.S. midwest. Outbreaks have occurred in other locales as well. Most outbreaks occur when people (some of whom carry meningitis-causing bacteria in their throats) are crowded together in close environments. Meningitis is a medical nightmare, because it can be fatal unless treated quickly and appropriately. The symptoms are often nonspecific—headache and other flu-like symptoms. Sometimes, but not always, a rash appears. Meningitis cases are rare, except during an outbreak, so it is easy for physicians to fail to diagnose meningitis in the first cases of an outbreak.

In this example, for simplicity, we will assume that there is an outbreak in progress, so physicians and nurses are already watching closely for possible meningitis cases. Most cases of meningitis in the college age group are caused by the bacterium *Neisseria meningitidis.* In the course of the disease, the bacteria enter the spinal fluid, a site that normally contains no microorganisms. Thus, the spinal fluid should contain a pure culture of *N. meningitidis.* The physician takes a sample of spinal fluid with a syringe and sends it to the microbiology laboratory. The physician probably will not wait for the lab results but will start treatment immediately. Laboratory test results are a good backup if the initial diagnosis proves wrong. Also, this information can help physicians diagnose the next cases of meningitis more accurately.

First step: microscopic examination

The microbiologist in the hospital laboratory needs information about the probable cause of the patient's meningitis as rapidly as possible. A quick, low-tech test is the examination of spinal fluid under a microscope. Many years ago, a scientist named Hans Christian Gram invented a stain that has proven very useful in the microscopic diagnosis of bacterial infections, called the Gram stain. The steps in the Gram stain procedure are illustrated in Figure 3.1. Some bacteria stain **gram-positive** (blue), whereas others stain **gram-negative** (red). In Chapter 4, the physical basis for the different Gram staining reactions will be explained. For now, it is enough to say that characteristics of the cell wall account for the different outcomes of the staining procedure. Under the microscope, cells of *N. meningitidis* appear as pairs of spheres stuck together (diplococci), and they stain gram-negative (Fig. 3.2). Most of the clinically significant cocci stain gram-positive. Gram-negative cocci are uncommon. Thus, a simple Gram stain gives a quick confirmation of the physician's initial diagnosis. This information can also help the physician decide which is the best antimicrobial compound to use for treatment.

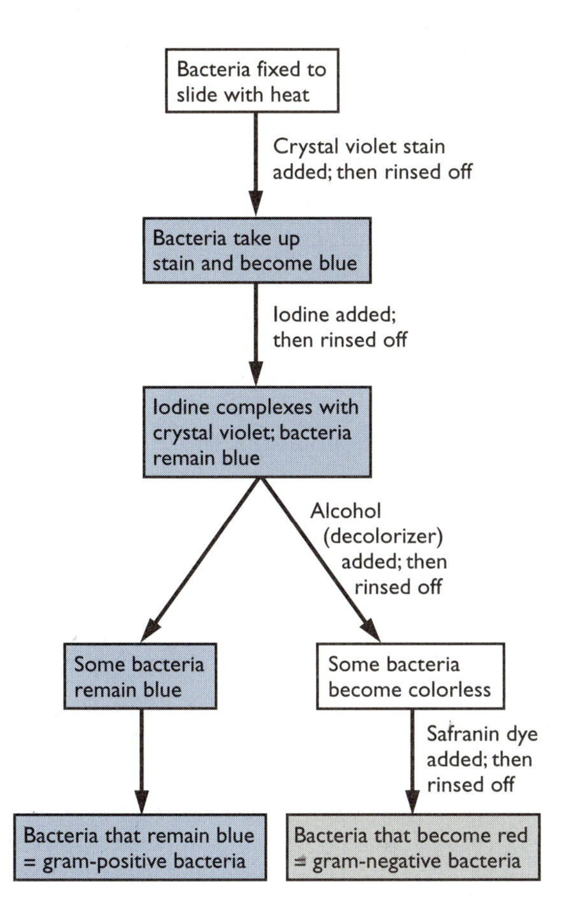

Figure 3.1 Gram stain procedure. A drop of bacterial culture is placed on a glass slide and allowed to dry. The slide is passed over a gas flame to "fix" the bacteria to the slide. The bacteria are covered with crystal violet dye for about a minute, and then the slide is rinsed. Iodine is added next to the slide. It enters the bacterial cells and forms a complex with the crystal violet. The slide is rinsed again. This is followed by the decolorizing step, in which the slide is rinsed with alcohol. This causes the crystal violet–iodine complex to be released from gram-negative bacteria but has no effect on the gram-positive cells. After rinsing with water, the slide is covered with safranin dye (a red counterstain). The safranin stains the gram-negative bacteria red, but the gram-positive cells remain blue. See Color Plates 1–4.

Second step: cultivation

Choice of medium and atmosphere. The laboratory microbiologist next needs to cultivate the bacteria to confirm the tentative identification and to determine whether this strain is the same one that has infected other patients in the same outbreak. Fortunately, the microbiologist will not have to discover anew how to cultivate *N. meningitidis.* Years ago, scientists figured out how to induce this organism to reproduce itself in the laboratory. For growth of microorganisms, scientists can choose to use either liquid or agar-solidified media (agar plates). In this case, the microbiologist chooses to start with the agar medium, for reasons that soon will become apparent. *N. meningitidis* is a **fastidious** microorganism. That is, it requires a rich mixture of different sugars, vitamins, and amino acids. Accordingly, the medium contains blood (a good source of many vitamins and nutrients) and a number of other additives. Such a complex medium is called a **rich medium.** Many organisms, however, will grow on a medium that contains very few components. Why some microorganisms need a rich medium and others do well on a much more restricted medium will be explained in a later discussion of bacterial energy metabolism (Chapter 8).

The solid or liquid phase of the medium is only part of what is needed to induce a microorganism to reproduce. Many microorganisms require an atmosphere different from normal air. *N. meningitidis,* for example, requires oxygen and can grow with an air atmosphere, but it also grows faster if the normal atmosphere is supplemented with additional carbon dioxide (CO_2).

Figure 3.2 *Neisseria meningitidis* in spinal fluid. The gram-negative bacteria occur as pairs of cocci (small dark spots). This photo also shows neutrophils that are attracted to the site of the infection. The neutrophils can be identified by their lobed nuclei (large dark ovals). Neutrophils ingest and kill bacteria. © A. M. Siegelman / Visuals Unlimited.

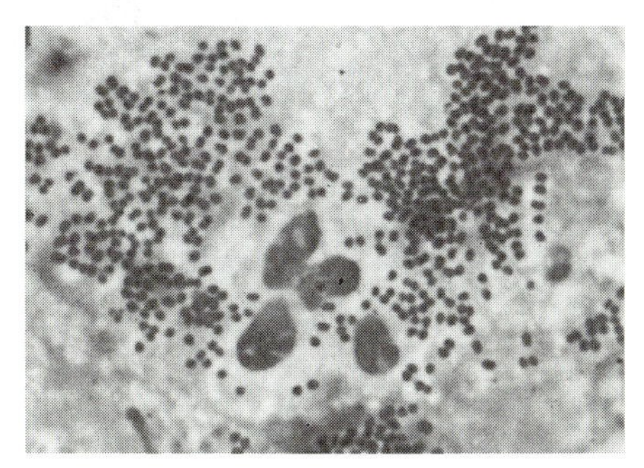

Obtaining a pure culture. In the spinal fluid, bacteria are present as a pure culture, but when a spinal fluid sample is taken with a syringe, bacteria from the patient's skin may contaminate the sample. Many skin bacteria grow faster than *N. meningitidis,* so if the microbiologist inoculates the spinal fluid specimen directly into liquid medium, a contaminant could overgrow the organisms of interest. Thus, the microbiologist's first step in cultivation will be to use agar medium to separate physically the different bacteria that may be in the sample. To separate bacteria, the microbiologist uses a technique called streaking, illustrated in Figure 3.3. The streaking process dilutes the number of bacteria originally applied to the plate, so that individual bacterial cells are separated spatially at some point in the streak.

Wherever a cell is deposited, a visible mound of cells, called a **colony,** will form after the plate has been incubated. In the case of *N. meningitidis,* it will take a day or two at 37°C for a visible colony to form. Some bacteria form colonies overnight. Others, such as the bacterium that causes tuberculosis, may take weeks to form visible colonies. The time spent waiting for bacteria to grow can mean days will pass before laboratory results are available. If two cells on the plate are close together, as they are early in the

Figure 3.3 Streaking plates to obtain isolated colonies. (A) Petri plates containing solid medium are marked into four quadrants. A loopful of a bacterial culture is spread across one quadrant. The loop is sterilized in a flame to kill any bacteria remaining on it. The plate is rotated. **(B)** The sterile loop is used to pick up a small amount of the material that was streaked in quadrant I; the material is then streaked in quadrant II. **(C, D)** This procedure is repeated two more times to inoculate quadrants III and IV. This is a dilution process that results in a smaller number of bacteria being spread on each succeeding quadrant. After the plate has been inoculated, it is incubated. **(E)** After the appropriate amount of time, isolated colonies of bacteria should appear in quadrant IV.

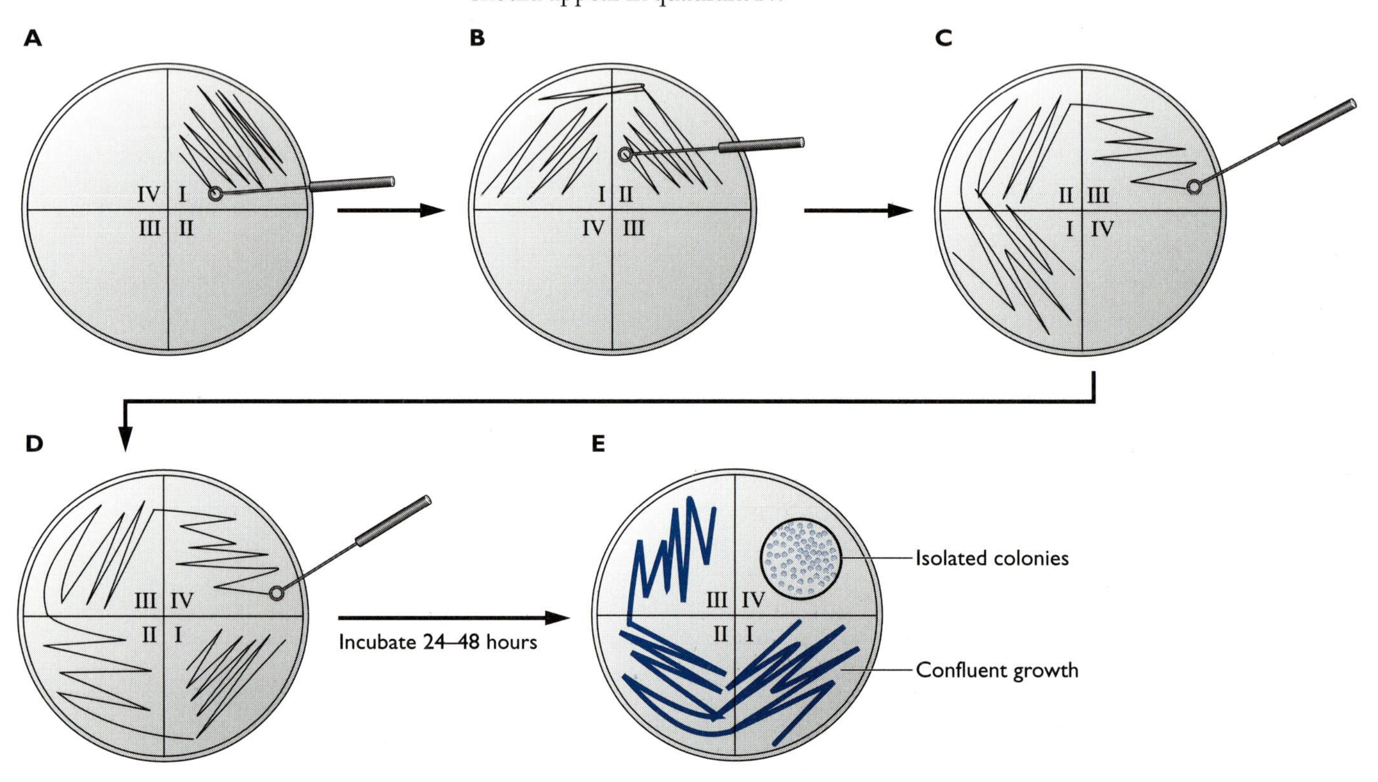

streak, the colonies will grow together to produce **confluent** growth. In a later part of the streak, individual organisms are separated far enough from each other to allow well-separated single colonies to form. The isolated colonies are the ones the microbiologist wants because each of these should contain a pure culture. See Color Plate 5.

Another approach to obtaining isolated colonies is shown in Figure 3.4. The specimen is diluted sequentially, and a portion of each dilution is spread on an agar plate. At some point in the dilution series, the bacteria will be dilute enough to form isolated colonies on the agar plate. The dilution method can also be used to quantify the number of bacteria in the original specimen (Fig. 3.4). Because the streaking procedure is faster and requires fewer plates, a microbiologist in a clinical laboratory will use that method unless there is a need to quantify the number of bacteria.

Colonies formed by different types of bacteria can have different appearances. Colonies of *N. meningitidis* have a somewhat wet appearance, whereas colonies of a common bacterial contaminant from skin, *Staphylococcus epidermidis,* have a chalky white appearance. The visible features of colonies on the agar plate can guide the microbiologist in deciding whether the specimen was contaminated and, if so, which colony to select for future study. See Color Plate 6.

Confirming the initial identification. To confirm the tentative identification of the bacterium as *N. meningitidis,* the microbiologist will perform additional tests. Some of the criteria used to identify *N. meningitidis* are: growth at 30–37°C but not at room temperature, a positive oxidase test, and the ability to grow on medium containing glucose or maltose as the only

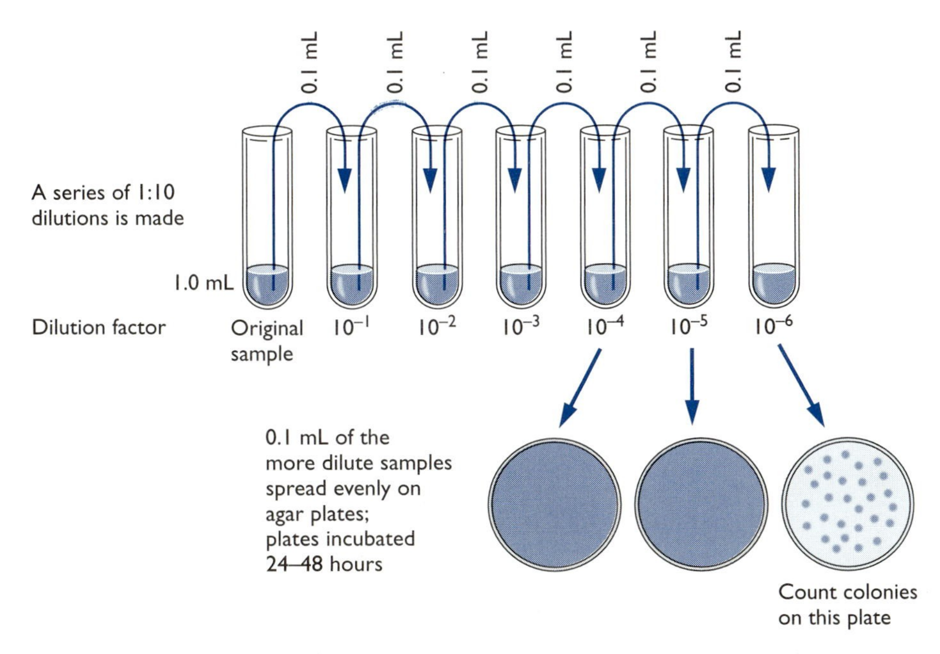

Calculation of number of viable bacteria in the original sample:

$$\frac{\text{Number of colonies}}{\text{Volume plated}} \times \frac{1}{\text{Dilution factor}}$$

Figure 3.4 Serial dilution of cultures to obtain isolated colonies. A series of 1:10 dilutions is made from a culture of bacteria. Samples are taken from several of the tubes that are most dilute and are spread evenly on plates containing solid medium. After the appropriate incubation, some of the plates should have individual colonies that can be counted. From these data it is possible to determine how many bacteria were present in the original culture.

source of carbon, but not on medium containing sucrose or lactose. In the **oxidase test,** an enzyme produced by the bacteria converts a colorless chemical to a purple color (oxidase-positive). These tests are designed to separate *N. meningitidis* from other bacteria that cause the same disease or have similar traits. Very few cocci stain gram-negative, but bacteria that have this Gram-staining trait but do not cause meningitis are found in the throats of some people. The ability to use the different sugars listed separates *N. meningitidis* from these bacteria. Few bacteria are oxidase-positive, so this test is good for differentiating *N. meningitidis* from other bacteria that cause meningitis.

The clinical tests needed to identify a bacterial species are performed in a specific order, with each group of tests dictated by the outcome of previous ones. Thus, finding that a bacterium is a gram-negative diplococcus and is oxidase-positive leads the microbiologist to proceed to the sugar utilization and other tests. There is no single regimen of tests that identifies all disease-causing bacteria. Instead, microbiologists use tests like the Gram stain and information about the disease the patient is experiencing to choose the specific set of tests to be used. Large notebooks contain tests appropriate for isolation and identification of different disease-causing bacteria. Often, in addition to tests used to identify the microorganism, the microbiologist will perform other tests to determine whether the organism is susceptible to different antibiotics. In a case of meningitis, the results of such tests may be available too late to be useful to the person from whom the spinal fluid was drawn. Yet the results of such tests may provide a useful guide for treating the next case of meningitis.

Third step: using DNA fingerprinting to determine whether the isolate is responsible for other recent infections

Metabolic tests to establish a species identification are useful for confirming the identification of the disease-causing microbe and can help the physician decide on the appropriate antimicrobial agent to use. However, these tests will not answer the crucial question of whether this is the same strain of *N. meningitidis* that has infected other students during the outbreak. It is important to know this. If an outbreak is emanating from a single source, steps can be taken to screen people exposed to the single source to find out whether they are carrying *N. meningitidis* in their throats. If so, they can be treated to eliminate the bacteria. The alternative is to wait for new cases to turn up at a clinic or hospital. Because some of these cases may arrive at medical facilities too late for effective treatment, this is an unacceptable alternative.

Investigators can determine whether the strain that caused the most recent case of meningitis is the same as the one that infected other students by comparing the **deoxyribonucleic acid (DNA) "fingerprints"** of the different strains of *N. meningitidis* isolated during the outbreak. A current method of choice for doing this is **pulsed field gel electrophoresis (PFGE).** To explain how PFGE works, it is first necessary to introduce a type of DNA-degrading enzyme called a **restriction enzyme.** Restriction enzymes cut at specific sequences in the DNA. Some examples are shown in Figure 3.5. Incubating chromosomal DNA from a bacterial isolate with a restriction enzyme generates a mixture of fragments of different sizes. These can be separated in a gel matrix that has been placed in an electric field. DNA is negatively charged, so the fragments will migrate through the gel toward the positive pole. The larger fragments will move more slowly than the smaller ones (Fig. 3.6). The gel is then stained with **ethidium bromide,** a

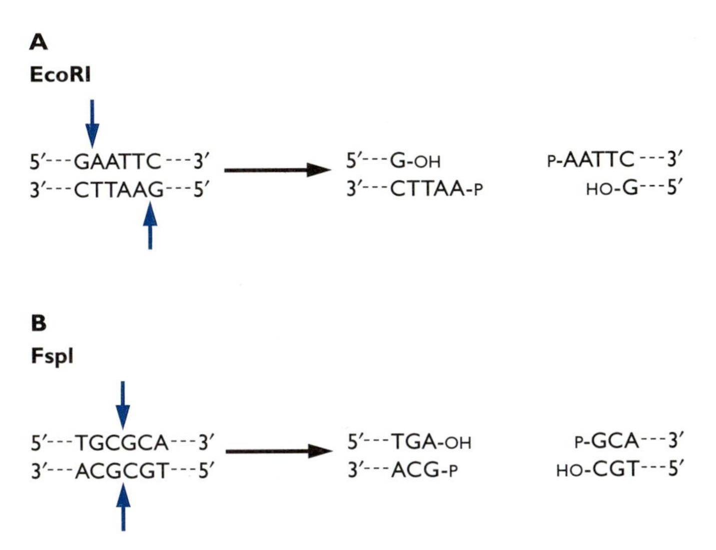

Figure 3.5 Examples of the types of cuts made by restriction enzymes. Each restriction enzyme recognizes a unique sequence in DNA. **(A)** Some restriction enzymes, such as *Eco*RI, cleave DNA so as to leave staggered ends. **(B)** Others, such as *Fsp*I, make cuts that leave blunt ends.

Figure 3.6 Electrophoresis of DNA cut by restriction enzymes. After digestion by restriction enzymes, DNA is placed in wells in agarose gels and subjected to an electric field (electrophoresis). DNA migrates to the positive pole. Fragments are separated on the basis of size, with the smaller segments moving farther than larger fragments. After electrophoresis is complete, the gel is stained with ethidium bromide, a dye that intercalates into double-stranded DNA. The stained bands can be seen under ultraviolet light.

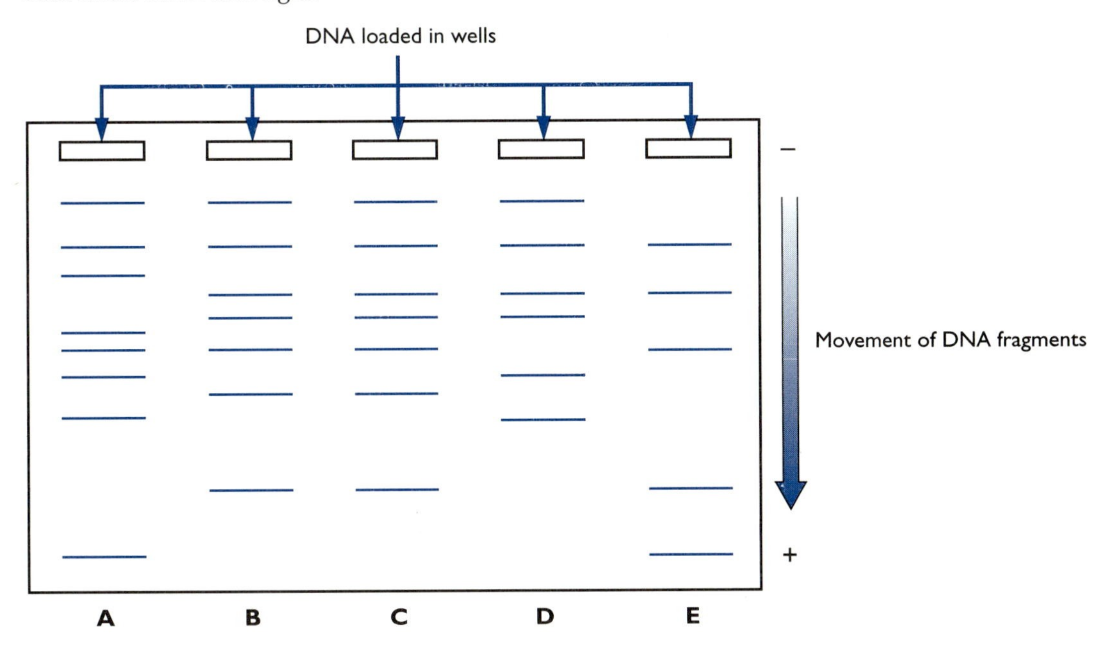

chemical that binds to double-stranded DNA and fluoresces under ultraviolet light.

Two different strains of *N. meningitidis* will have chromosomes with very similar DNA sequences. But, as stated in the previous chapter, two strains of a bacterial species will usually have enough sequence differences to give different patterns when their DNA is cut with the appropriate restriction enzyme. The chromosome of a bacterium is large, ranging from 1–6 Mbp of DNA. To generate simple patterns on a gel, scientists choose restriction enzymes that recognize sequences of at least 8 bp in length. Such enzymes will cut a large genome into a relatively small number of segments, but all these segments will be quite large. DNA fragments in an electric field move in a snake-like fashion from pore to pore in the gel matrix. In a unidirectional electric field like that shown in Figure 3.6, large DNA segments get stuck early in the process of electrophoresis and do not migrate very far. Scientists have discovered, however, that pulsing the electric field in different directions (Fig. 3.7) allows large pieces of DNA to move far enough into the gel to produce a clear pattern. This is called pulsed field gel electrophoresis. A PFGE analysis of strains

Figure 3.7 Pulsed field gel electrophoresis (PFGE). DNA is cleaved into large fragments that would not enter the gel under the usual electrophoretic conditions. However, by applying the current first in one direction, then in the opposite direction (pulsing), the large pieces of DNA make their way through the gel matrix. The bands are stained with ethidium bromide.

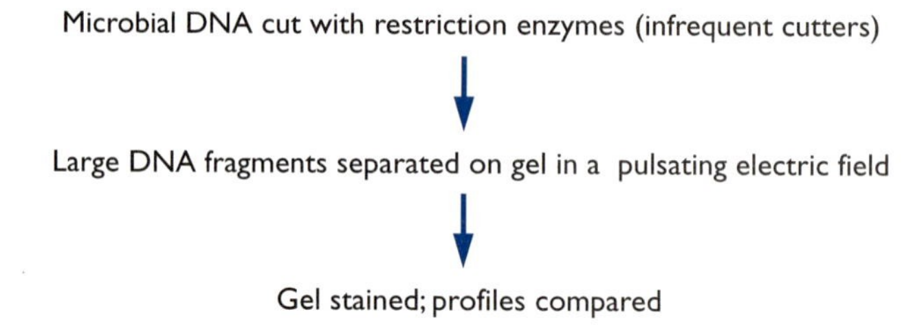

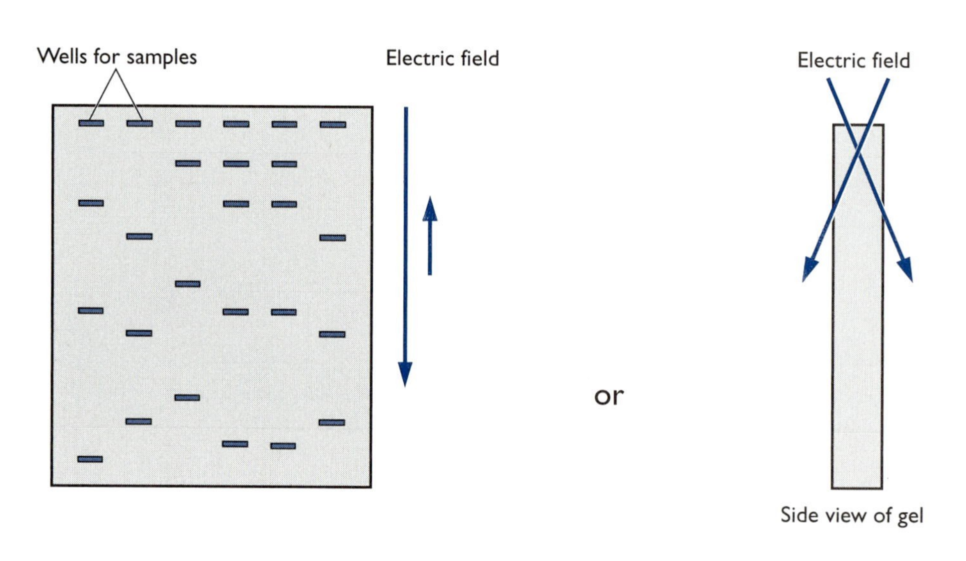

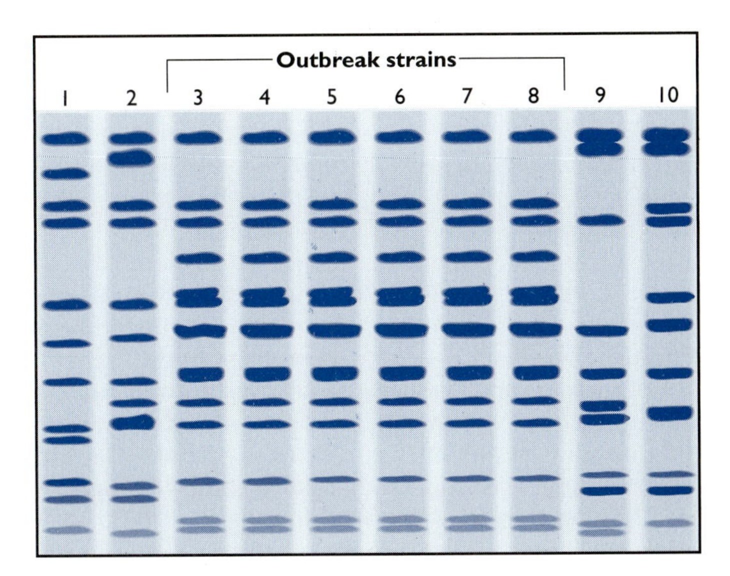

Figure 3.8 PFGE analysis of an outbreak of meningitis. Comparison of migration of DNA fragments from different control strains of *N. meningitidis* not associated with the outbreak (lanes 1, 2, 9, and 10) with DNA fragments from different isolates associated with the outbreak (lanes 3, 4, 5, 6, 7, and 8). Note that the strains associated with the outbreak have an identical pattern that is different from those of the control strains.

from an actual meningitis outbreak is shown in Figure 3.8. This outbreak clearly emanated from a single source.

Considering alternative causes of meningitis

Yeasts. If the patient with suspected meningitis had been immunocompromised (for example, because of human immunodeficiency virus [HIV] infection) and if the case had occurred in the absence of any other recent cases, the physician might consider the possibility that meningitis was caused by the yeast *Candida albicans*. *C. albicans* is a common occupant of the human mouth. It rarely causes disease in healthy people but is a serious threat to people who are immunocompromised. The first clue to a *C. albicans* infection is the appearance of ovoid forms larger than bacteria in the spinal fluid specimen. Some of these ovals may have buds or longer projections called pseudohyphae. These features provide evidence that a yeast is responsible for the infection. The Gram stain is not useful in the case of yeasts, but morphology is. If the microbiologist decides to confirm the diagnosis by isolation of the yeast, the procedure is much the same as for bacteria. Agar medium containing carbohydrates and other nutrients required by yeasts is streaked and incubated until isolated colonies form. Yeasts are mostly oxygen consumers, so ordinary air is a suitable atmosphere. Yeast colonies generally form more slowly than most bacterial colonies and may take days to appear. In the meantime, bacteria may overgrow the yeast colonies before they become large enough to see. To prevent this, scientists use **selective media,** which have acidic pH or contain antimicrobials or other compounds that prevent the growth of organisms other than the one of interest.

Selective media are also useful in cases in which the area being sampled contains many types of bacteria. For example, a fecal specimen contains hundreds of different species of bacteria, many of which can grow on ordinary laboratory medium. Thus, if a scientist is interested in cultivating the diarrheal pathogen, *Salmonella typhimurium,* he or she will use media and atmosphere that select for the growth of *Salmonella* and against the growth of the many other bacteria found in a fecal specimen.

Viruses. Viruses can cause meningitis, although viral meningitis is generally not as serious as bacterial or fungal meningitis. If the patient has viral meningitis, examination of the specimen using a light microscope will reveal no microorganisms, because viruses are too small to be seen with a light microscope. Most hospitals cannot afford to maintain an electron microscope for visualizing viruses. Thus, different approaches are used to diagnose viral infection. One of these is the serological test. Serological tests use blood proteins called antibodies that bind specifically to certain organisms, either to detect these organisms in the specimen being examined or to detect the patient's response to infection. Serological tests will be explained in later chapters, where we describe the properties of antibodies.

Normally, the laboratory microbiologist would not cultivate the virus as part of routine analysis of a spinal fluid specimen because cultivation of viruses is expensive, time consuming, and does not help with treatment. In most cases, the disease resolves spontaneously without permanent damage to the brain (except in meningitis caused by HIV). Moreover, no antiviral therapy is yet available for the most common causes of viral meningitis. A scientist might, however, want to cultivate the virus for research purposes, especially if there is some reason to think it may be a new kind of virus.

To cultivate viruses that cause human disease, tissue culture cells provide a source of cells to support viral replication. **Tissue culture cells** are human or animal cells that have been immortalized so that they continue to divide. Normal animal cells become terminally differentiated and stop dividing. Cultured cells are animal cells that have been treated to make them return to the undifferentiated state. This is done by plating large numbers of cells in plastic tissue culture dishes that contain liquid medium. A few cells will have undergone DNA rearrangements that allow them to continue dividing. The cells that descend from such cells are called a line of tissue culture cells. Cancer cells are often used as the starting point for developing a line of tissue culture cells because they are already partially de-differentiated. For example, HeLa cells, a commonly used type of cultured human cells, were derived from cervical cancer tissue.

For routine use, cultured cells inoculated into liquid medium in a plastic dish will attach to the bottom surface of the dish and divide to form a monolayer of cells. The sample containing the virus is diluted, and each dilution is added to a petri dish containing a monolayer of cultured cells. Viruses kill the cultured cells, so, instead of a colony, the virus shows itself by forming a hole, called a plaque, in the monolayer. Usually, it takes days for plaques to form. Staining the monolayer with a purple dye that is taken up by living cells makes the plaques more obvious as clear zones against a darkly colored background (Fig. 3.9). Because different viruses target different cell types, it is necessary to have some idea of which virus might be the culprit before trying to cultivate it. Many important viruses still have not been cultivated in tissue culture cells.

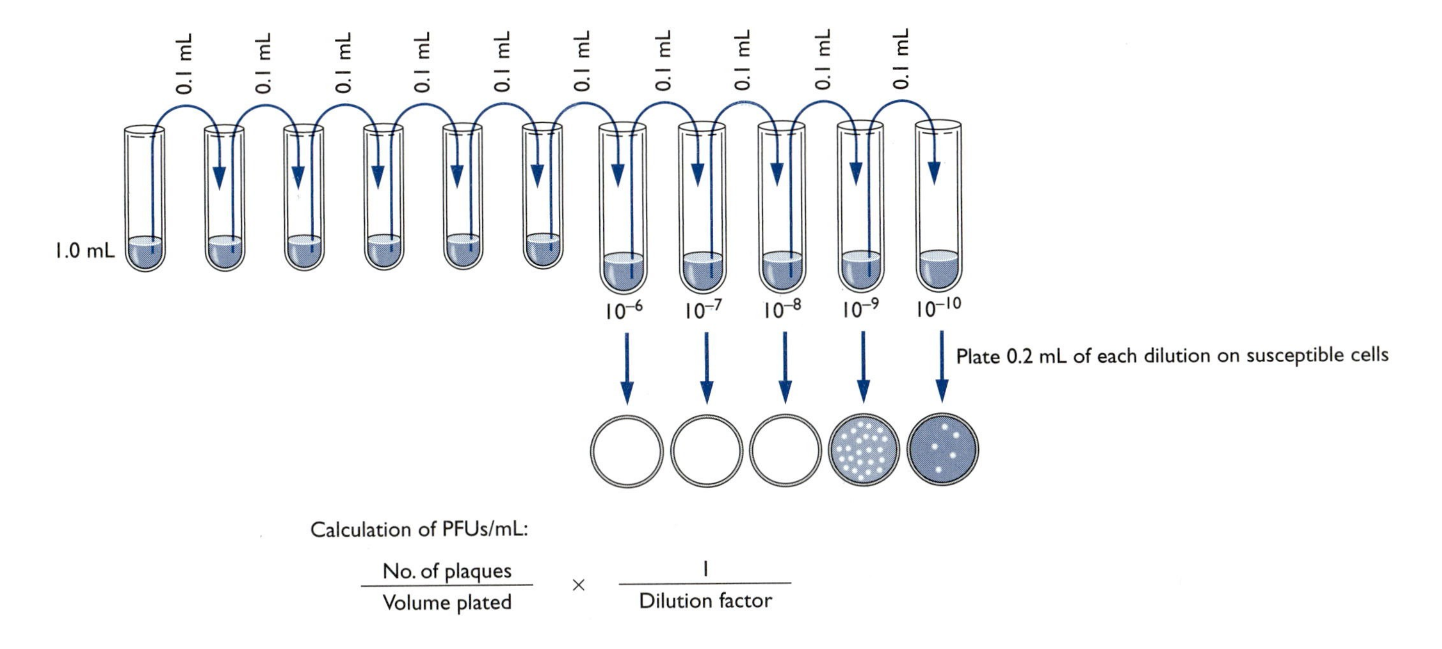

Figure 3.9 Plaque assay for determining number of infectious viral particles in a sample. A series of tenfold dilutions of a sample containing viruses is made. A small amount of appropriately diluted samples is added to petri dishes containing tissue culture cells. A single virus will bind to a cell, replicate, kill the cell, and spread to other cells in the immediate vicinity, causing a small area of dead cells (a plaque). The plaques can be detected when a vital stain (one that stains only living cells) is added to the plate. The living (uninfected) cells take up the dye and form a blue background. Because they are dead, the cells in the plaques do not take up the dye and so appear as clear areas in the blue field. Plaques can be counted, and the number of infectious virus particles (plaque forming units; PFUs) in the original sample can be calculated. This is analogous to the method of determining the number of bacterial cells by plating serial dilutions.

Safety in the Clinical Laboratory

Clinical microbiologists routinely grow cultures of microorganisms that cause human disease. Because these bacteria can be a danger to others as well as to the clinical microbiologist, stringent safety precautions are observed. The primary precautions depend on the characteristics of the organism being tested. *N. meningitidis,* for example, does not survive very long outside the body. Although a bit of culture spilled on a piece of paper covering the lab bench will be autoclaved just in case, this is not as great a hazard as a hardier organism might present in the same setting. A microbe's route of entry into the body is also important in safety considerations. HIV, for example does not penetrate intact skin, so scientists working with it are far more concerned about a needle stick incurred while working with the virus than with virus spilled on plastic gloves worn by the investigator.

Yet another factor that must be considered is the treatability of the infection if an accident occurs. For example, a strain of the bacterium that causes tuberculosis and is resistant to all known antituberculosis drugs will be handled more cautiously than a strain that is susceptible to the drugs. The most dangerous microbes are usually handled only in special high-containment facilities, not in hospital general laboratories. The efficacy of these safety procedures is evident from the fact that, even among scientists working with highly virulent organisms such as HIV, infections acquired during experiments are extremely rare. Knowledge, coupled with common sense, is a microbiologist's best protection.

Analyzing a Complex Microbial Community: The Microbiota of the Human Colon

Extensive research and testing has gone into the development of methods to grow and identify disease-causing bacteria and fungi. Methods for cultivating members of natural microbial communities are much less highly developed. In fact, as mentioned in Chapter 2, only a tiny fraction of the microorganisms in most natural communities have been cultivated. One major problem is that the variety of microorganisms in a natural setting is much greater than that encountered in the clinical setting.

Assessing the success of cultivation attempts

How does an investigator know how many organisms have been missed in initial attempts to isolate and identify members of a microbial community? One commonly used way to estimate the degree to which all the organisms in a site have been cultivated is to determine the total count of microorganisms by microscopy. The sample is diluted sequentially, and a calibrated portion of each dilution is introduced into a **Petroff-Hauser counter.** This device displays microorganisms against a grid background that aids in counting the microorganisms. Counts obtained by microscopy can be somewhat inaccurate, because it is often difficult to distinguish between microorganisms and pieces of debris shaped like microorganisms. Also, the implicit assumption is made that all of the organisms seen in the microscope are alive. Nonetheless, this approach to estimating population size is a useful way of determining how successful cultivation efforts have been. In a later chapter, you will be introduced to a new approach to analyzing a microbial community, based on 16S/18S rRNA genes. First, however, it is important to understand the classic cultivation-based approach to characterizing a microbial community.

The microbial population that normally resides in the human colon (the colon microbiota) provides a good example of problems encountered in analyzing a microbial community. It is also one of the few naturally occurring microbial communities from which microbiologists have been able to cultivate most of the microorganisms present. Microbiologists were interested in describing this community because it is likely to be involved either directly or indirectly in diseases such as colon cancer and inflammatory bowel disease.

Initial microscopic analysis

The first step in analyzing a microbial community is microscopic examination of material from the site. Examination of colon contents from a person living in the developed world would reveal many rods and cocci whose small sizes suggest that they are bacteria. A Gram stain reveals gram-positive rods and cocci and gram-negative rods. There are methane-producing archaea in the human colon, but their numbers are far lower than the number of bacteria, so they do not show up in the microscopic analysis, which only detects the numerically major populations. The colon contents of someone living in a developing country might contain protozoa as well as bacteria.

Some protozoa cause diarrhea, but they can also be carried by people who have no obvious symptoms. The protozoa stand out, because they are larger and have more diverse and complex morphologies than the bacteria.

Cultivation of colonic bacteria

When a scientist first begins to explore the microbial population of a particular site, there is usually little or no information about what types of media or conditions to use. The site itself may offer some clues. The colon, for example, is rich in polysaccharides and proteins but deficient in simple sugars and amino acids. Thus, adding polysaccharides and proteins to the medium may help to cultivate the organisms. Initially, however, microbiologists did not consider the features of the colonic environment and instead started with media that were known to grow *Escherichia coli* and many other types of bacteria. These investigators used a rich medium replete with sugars, amino acids, and vitamins, and they incubated their plates aerobically. Not surprisingly, because the medium used was one that supported the growth of *E. coli,* and because *E. coli* was originally isolated from a sewer, one of the most prominent bacteria cultivated under these conditions was *E. coli.* On the basis of their results, scientists concluded that *E. coli* was the main species of bacteria in the colon.

Other scientists pointed out, however, that the concentration of *E. coli* determined by dilution and plate counting was only about 10^8 bacteria per gram wet weight of contents, whereas a microscopic estimation of the number of bacteria put the total concentration of bacteria at 10^{11} per gram wet weight of contents. Clearly the numerically predominant bacteria had not yet been cultivated. The trick that allowed these numerically predominant species of bacteria to be cultivated was to exclude oxygen from the atmosphere in which the plates were incubated. Instead, an atmosphere that consisted of nitrogen, CO_2, and hydrogen was used. The agar medium was also heated to expel oxygen, then cooled under this oxygen-free atmosphere to make it oxygen-free as well.

In retrospect, excluding oxygen was an obvious step toward cultivating colon microorganisms, because the colon is a very anoxic environment. Most of the oxygen that is swallowed is taken up by cells of the small intestine or oxygen-utilizing bacteria in the colon. The combination of an oxygen-free atmosphere and the addition of some special vitamins and minerals to the medium eventually allowed scientists to cultivate the numerically predominant gram-negative and gram-positive bacteria. Bacteria that grow in the absence of oxygen are called **anaerobes.** Bacteria that can *only* grow in the absence of oxygen are called **obligate anaerobes.** The predominant colon bacteria are all obligate anaerobes. Cultivating the predominant bacteria led to the discovery of a number of new genera of bacteria, such as *Bacteroides* species (gram-negative rods) and *Peptostreptococcus* species (gram-positive cocci). Identification of new species in a nonclinical setting is still performed in much the same way as identification in clinical settings. That is, the patterns of carbon source utilization, staining characteristics, motility or lack of it, and other features are compared to those of other organisms. Such comparisons can be misleading because, for example, many bacteria use the same carbohydrates as carbon sources. Increasingly, therefore, scientists are relying on the rRNA gene sequence to corroborate the results of metabolic tests. rRNA gene sequences of *Bacteroides,* for example, gave the first indication that this genus was not closely related to *E. coli* but represented a different group of bacteria.

Probiotics. Disruption of the colon microbiota by antibiotics or cancer chemotherapy can cause diarrhea and other intestinal discomforts. Physicians have long sought an effective way to repopulate the bowel with a healthy microbiota after such treatments. The identification of the major bacterial species in the colonic microbiota has provided a rational basis for designing "cocktails" of bacteria that might solve this problem in the future. The large probiotics industry that already exists is based on the notion that regular ingestion of freeze-dried preparations containing live bacteria can prevent or cure all kinds of ills. These preparations are readily available in most health food stores. Unfortunately, there is little credible evidence to support the claims made for these preparations.

Most of the currently available probiotics contain species of *Lactobacillus* or *Bifidobacterium.* At one time these were thought to be major colonic bacteria. Because they were among the few colonic genera that had not been found to cause disease, they were declared "good" bacteria—bacteria that could counter the effects of the "bad" bacteria. More recent efforts to determine the composition of the colonic microbiota have revealed that, although these genera are present, they are not major members of the microbiota. Moreover, some of the *Lactobacillus* preparations contain bacteria isolated from yogurt, not from the human colon, and are not able to colonize the human colon. For a more successful probiotic story, see Box 3–1, which describes bioremediation of the microbiota of cattle.

Cultivation of colonic archaea

About 20% of the human population produces enough methane in their colons for that methane to be detectable in their breath. This methane is produced by the **methanogenic archaea.** The most common colonic species of methanogen is *Methanobrevibacter smithii.* Cultivation of methanogens, which long eluded scientists, required another atmospheric trick. Many methanogens produce methane by using H_2 to reduce CO_2 to methane (CH_4). That is, the substrates used for growth are gases rather than sugars or other dissolved compounds. In addition, methanogens are very sensitive to oxygen. Simply providing an atmosphere that contained H_2 and CO_2 but no oxygen did not support good growth of the methanogens, however. It was necessary to provide the gases at pressures higher than atmospheric pressure to facilitate their entry into the liquid or solid phase of the medium, and it was necessary to replenish them from time to time as they were used up by the archaea.

How would a scientist go about isolating a numerically minor population such as the colonic archaea? Streaking is unlikely to be successful, because many bacteria can grow on the type of medium that might support methanogens and, because of their greater numbers, these bacteria would overgrow the slow-growing and less numerous archaea. A common strategy is to use a type of cultivation called **enrichment culture.** A selective medium, which allows growth of the organism of interest but discourages the growth of other microbes, is inoculated with colon contents. After several transfers, if the attempt is successful, the desired microbes will be the numerically predominant population in the medium. In the case of the methanogenic archaea, a simple medium with a H_2-CO_2 atmosphere is the base medium. Antibiotics are added to discourage the growth of bacteria.

Can a growth medium be too rich?

A rich medium was appropriate in the two examples just described. Such success stories have led microbiologists to resort automatically to relatively rich media. It seems reasonable enough to offer the microorganism a buffet

box 3–1 *Bovine Bioremediation*

Cattle have a digestive organ called the rumen, which contains a high concentration of bacteria and other microbes. The rumen is located before the true stomach in the digestive tract of the animal. In the rumen, bacteria break down polysaccharides in grass and other forage, ultimately releasing as end products the short-chain fatty acids acetate, propionate, and butyrate:

CH_3-COOH (Acetate) CH_3-CH_2-COOH (Propionate)

$CH_3-CH_2-CH_2-COOH$ (Butyrate)

These simple compounds are absorbed by the ruminant and used as sources of carbon and energy. A ruminant depends absolutely on the rumen fermentation for its nutrition. Disruptions in the microbiota of the rumen can be deadly. Even before the discovery of microorganisms, peasant farmers in some parts of the world knew to give a cow that was not digesting its food properly part of the cud from a healthy animal, thus inoculating the sick animal with the microbiota of a healthy one. Apparently, this worked often enough to make the practice part of veterinary folk medicine.

In recent years, microbiologists have gone one step further in solving ruminant health problems with bacteria. In Australia, cattle thrive on the abundant grazing ranges, but poisonous plants also grow along with the grass. Cattle sometimes eat these plants and sicken and die. The problem seemed intractable until Australian microbiologists came to the rescue. One of the biggest toxic plant problems was the legume *Leucaenia*. To protect itself from predators, *Leucaenia* produces mimosine, a toxic compound.

HO, N, NH_2, COOH, O

Mimosine

Scientists had noticed that some ruminants in Hawaii and Indonesia could eat mimosine-producing plants with impunity and wondered if some component of the ruminal microbiota might be detoxifying the compound. Bacteria can degrade so many different types of compounds that this seemed a reasonable hypothesis. The scientists set up enrichment cultures in which mimosine was the sole source of carbon and energy, inoculated the cultures with ruminal contents from the Hawaiian and Indonesian ruminants, and eventually isolated several different types of bacteria that could degrade mimosine to nontoxic components. These bacteria were fed to Australian cattle and established themselves in the rumen. The treatment was successful, and now Australian cattle can tolerate an occasional meal of *Leucaenia*. Scientists are now trying to isolate bacteria that degrade other plant toxins that present problems for cattle ranchers. This success story is an excellent example of the many potential benefits of cultivating microorganisms. Bioremediation is not just for toxic waste dumps.

Source: C. S. Stewart. 1994. Plant–Animal and Microbial Interactions in Ruminant Fibre Degradation. pp. 13–28. In: *Microorganisms in Ruminant Nutrition*. (R. A. Prins and C. S. Stewart, eds.) Nottingham University Press, Nottingham, U.K. 249 p.

of different menu choices, and this approach works well in certain cases. There are some cases, however, in which a rich medium can be lethal for microorganisms. This has proven true in the case of bacteria from the open oceans.

Two culprits that make media toxic for some microorganisms are metals and agar. Metals such as copper are common contaminants of commercially available amino acid mixes or salts used to make laboratory media. Microorganisms adapted to grow in areas such as the open oceans, where copper is present in very low concentrations, may find the concentrations of copper in standard laboratory media toxic. Scientists have found that removing metals like copper from ingredients used to make laboratory media has allowed them to cultivate microbes they previously were unable to grow.

Agar is a polysaccharide obtained from algae. The crude form usually used to make solid media has a tan color. Purified agar is white. The tan color comes from contaminants that bind to the white agar polysaccharides. The contaminants contain many extraneous compounds. No one has analyzed these contaminants in detail, but some are toxic for some microorganisms. Washing agar with water to remove the contaminants and produce a white form of the agar has proven to be a key step for cultivating many types of environmental bacteria.

Some of these same bacteria have another quirk: they prefer to grow inside an agar matrix rather than on the agar surface. One technique for growing such bacteria is the agar shake method. Bacteria are inoculated into molten agar, and the agar is shaken to disperse the bacteria. Still other microorganisms find agar so unacceptable that other surfaces, such as cellulose filter paper, are needed.

Consortia

During the past century and a half, microbiologists have been very successful at isolating microorganisms from complex communities and growing them in pure culture. Yet most microorganisms still elude efforts to cultivate them. Some possible explanations for this failure were outlined in the previous section. But the medium and the atmosphere may not always be the problems. Microbiologists are beginning to suspect that some—perhaps many—microorganisms may be unable to grow in pure culture. In their natural settings, microbes are most commonly found in complex communites that contain eukaryotes, bacteria, and archaea. Many microbiologists have now had the experience of being able to simplify a community but finding themselves unable to move to the pure culture stage.

Consider the example of a microbiologist interested in a polluted site where microorganisms are degrading jet fuel under anoxic conditions. Microbes that can carry out such a reaction could be extremely valuable for bioremediating other similarly-polluted sites. The microbiologist takes an enrichment culture approach, by designing a simple medium in which jet fuel is the main source of carbon. The medium is inoculated with soil from the site. After a period of time, microorganisms grow in this medium, and some of the jet fuel components disappear. The microbial population has been simplified, compared with that contained in the original soil sample. Microscopic analysis of the microbes in the medium suggests that there may be only three or four types of microorganisms, whereas the original soil sample contained hundreds of species.

Normally, the enrichment step would be followed by isolation on an agar-solidified version of the same medium. In at least some cases, the research effort fails at this step. Although the mixture of organisms continues to grow when diluted into more liquid medium, no colonies form on the solid version of the medium. Such an outcome suggests that more than one microorganism is participating in the desired reaction and that separating the organisms makes them unable to grow on jet fuel medium. Such cooperative associations of microorganisms are called **consortia.** A later chapter (Chapter 8) will outline some reasons why certain pairs or multiples of microbial species must rely on each other to carry out difficult reactions. This remains a research area of considerable mystery, however.

chapter 3 at a glance

Steps in clinical idenfication of *Neisseria meningitidis* in a suspected case of meningitis

- Gram stain results: gram-negative diplococcus
- Streak on blood agar medium to obtain a pure culture (single colony)
- Test pure culture for metabolic traits such as ability to utilize certain sugars
- DNA fingerprinting of the isolate and isolates taken from other patients to ascertain whether all infections in the current outbreak were caused by the same strain of *N. meningitidis*
    - Restriction enzymes used to cut chromosomal DNA at specific sites
    - Pulsed field gel electrophoresis separates different-sized DNA fragments to generate a strain-specific pattern of bands

Other microorganisms that can cause meningitis

- Yeast (*Candida albicans*)
    - Isolated similarly to bacteria, but usually grow more slowly
    - Use selective media to discourage overgrowth by bacteria
- Viruses (many different kinds)
    - Inoculate into a plate containing tissue culture cells
    - Viral replication kills cells, creating a hole (plaque) in the monolayer

Analysis of a complex microbial community: the human colonic microbiota

- Microscopic analysis: gram-positive cocci and rods and gram-negative rods predominate
- Cultivation of bacteria by dilution and plating
- Comparison to microscopic counts done with a Petroff-Hauser counter: check what proportion of the bacterial population is being cultivated
- Isolation of colonic archaea: methanogens
    - Concentration much lower than concentration of bacteria
    - Perform an enrichment using selective media containing antibiotics to prevent overgrowth of archaea by bacteria
    - Supply growth substrates, H_2, and CO_2 at above atmospheric pressure to facilitate entry of gases into the liquid medium
    - Appearance of the product, CH_4, indicates that archaea are growing
    - Gram stain is not appropriate, but many methanogenic archaea have a blue fluorescence when viewed with a fluorescence microscope

Why have microbiologists failed to cultivate many of the microbes in the environment?

- Rich media may contain toxic components, such as metals
- Need to use washed agar or some other surface
- May need a consortium of different species to carry out the desired reaction

key terms

anaerobe microorganism that grows without oxygen; obligate anaerobes grow only in an environment free of oxygen

atmosphere often overlooked but critical component of efforts to cultivate microorganisms

dilution and plating different dilutions of a mixture are plated on solid medium to obtain isolated colonies at some dilution; used to enumerate microorganisms in a mixture

enrichment procedure use of selective liquid media to allow a subset of microorganisms with the desired trait to proliferate; usually followed by isolation of colonies on the same medium solidified with agar

Gram stain staining procedure used to classify bacteria; gram-positive bacteria stain blue, gram-negative bacteria stain red

Petroff-Hauser counter device used to enumerate microorganisms by microscope count instead of cultivation

pure culture culture that consists of only one strain of a microorganism

rich medium medium that contains many different components

selective medium medium that permits growth of only certain microorganisms and discourages the growth of others

streaking one method for spatial separation of microorganisms on a solid surface to obtain a pure culture; cannot be used to enumerate microorganisms

Questions for Review and Reflection

1. Suppose the microbiologist working on the meningitis case decided to take a short cut and bypass the streaking step, going straight to the sugar utilization steps. If the culture was not pure, how could this affect the test outcome? How could the microbiologist check to make sure that the culture was not mixed?

2. Traditionally, mixed cultures have been the anathema of microbiologists, because they can change in composition and give confusing test results. Yet in the case of the jet fuel enrichment, it appears that the scientist is going to have to live with a mixed culture. Why might it be easier to get reproducible results from a mixed culture in this case than it would have in the meningitis case? [Hint: Consider the nature of the medium.] What could the scientist working on the jet fuel project do to make sure the mixed culture was stable and not changing in composition?

3. The metabolic tests (sugar fermentation tests, oxidase test) used to identify *N. meningitidis* were designed specifically for that species. Identifying different species usually requires different batteries of tests. Yet the DNA fingerprinting analysis has been applied to a variety of different bacteria. Explain why.

4. In the section on safety in clinical laboratories, different types of accidents (for example, a spill on gloves as opposed to a needle stick) were discussed with respect to their seriousness. Why not simply treat every spill or release as an equally serious event?

5. When scientists first attempted to cultivate colon bacteria, they thought they were using nonselective conditions because they used a rich medium. Yet this medium was selective, because it cultivated only a portion of the colonic bacteria. How was it selective?

6. A scientist has a liquid containing a mixture of bacteria from a soil sample. When the sample is diluted 1:1000 and 0.1 mL is plated on an agar plate and incubated, there are 50 colonies on the plate. What was the concentration of bacteria that grew on this medium in the original mixture? How many colonies would you expect to see on a plate on which 0.1 mL from the 1:10,000 dilution was plated?

7. What kinds of behavior by microbes could make it difficult to get an accurate Petroff-Hauser count? What remedial actions could be taken?

8. Why will viruses grow in some cultured cell lines and not in others? [Hint: Reread the brief description of viral replication in Chapter 1.]

4 Structural Features of Prokaryotes

Microbes are the ultimate handymen. With a very simple genomic toolbox, they can construct almost anything.

The surface of a bacterium is the portion of the microbe that encounters the external environment most directly. Thus, it is not surprising that bacteria deploy components of their surfaces in a variety of ways: for motion, for adherence to surfaces, for sensing the environment, and for acquiring nutrients.

The Surface of Bacteria

Swimming and sticking

Motility from flagella. Bacteria in a liquid environment may need to move through the liquid to find a more congenial location. Motility is commonly provided by **flagella,** long, spiral-shaped rods that stick out from the surface of the cell. The filament of the flagellum is built up from many copies of the protein flagellin. Where the filament enters the surface of the bacterium, there is a hook in the flagellum and the flagellum is attached to the cell surface by a complex of proteins called the flagellar motor (Fig. 4.1). The flagellar motor rotates the flagellum, causing the bacterium to move through the environment. Some flagellated bacteria have a single flagellum, and others have multiple flagella. In the case of *Escherichia coli,* rotation of the flagellar motor in one direction causes all the flagella to point in the same direction, so that the cell moves forward. Rotation in the other direction causes the flagella to point in different directions, and the cell tumbles in place. Tumbling reorients it, so that when flagellar rotation is reversed once again, the cell sets off in a new direction (Fig. 4.2).

By periodically reversing the direction of flagellar rotation, *E. coli* follows a **random walk** that allows it to sample its environment for various nutrients. If a change of location causes it to sense a higher level of a particular nutrient, the flagellar motor can turn primarily in the direction that causes forward motion, and the bacterium moves up the gradient of the desired nutrient. This process of converting environmental signals into this type of directed motility is called **chemotaxis** (Fig. 4.3).

Some strains of *E. coli* cause diarrhea. These strains prefer the surface of the small intestine to the interior of the intestine. Motility and chemotaxis presumably help *E. coli* to move to the intestinal surface. Recently, scientists have discovered that bacteria in the environment move much more

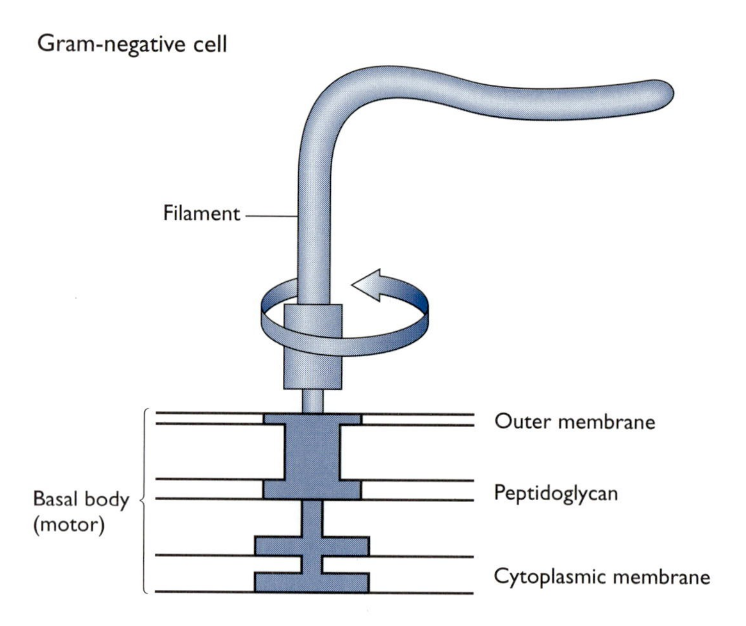

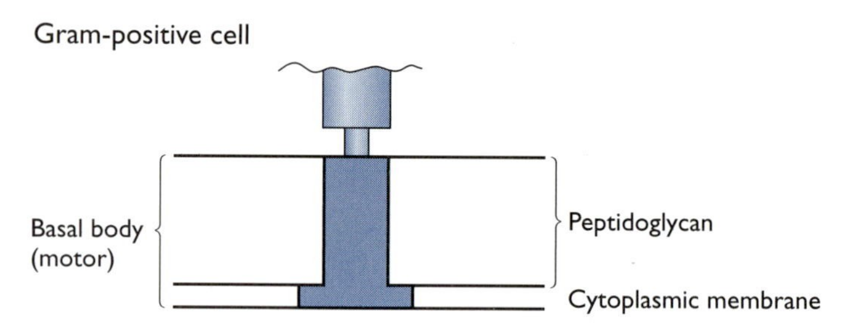

Figure 4.1 Structure of flagella. One end of the flagellum is embedded in the cell wall and cytoplasmic membrane. A protein complex, the flagellar motor, rotates the flagellum. The filament of the flagellum, which is composed of a single protein subunit flagellin, extends outward from the cell.

A Forward movement

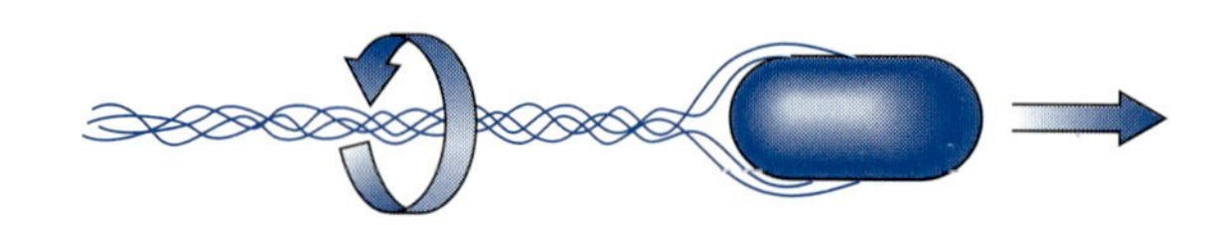

B Tumbling

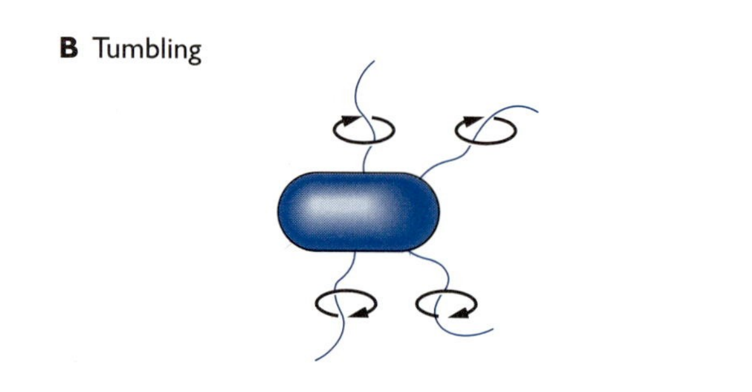

C Reverse direction

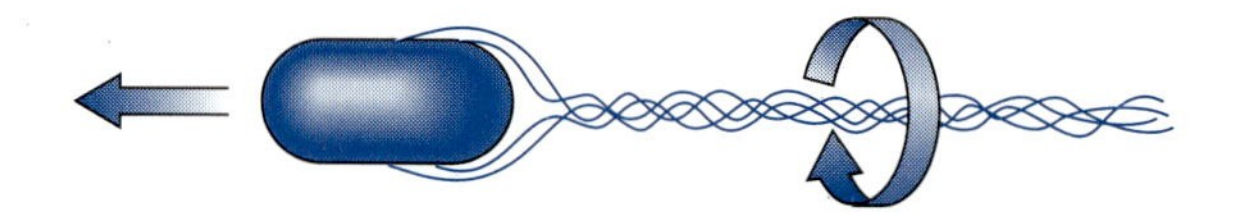

Figure 4.2 Bacterial motion mediated by flagella. (A) Flagella rotating in one direction cause the bacterium to move in a straight swimming motion. **(B)** When the direction of rotation of the flagella is changed, the bacterium tumbles. **(C)** Once all of the flagella are rotating in the same, opposite direction, the bacterium swims in the reverse direction.

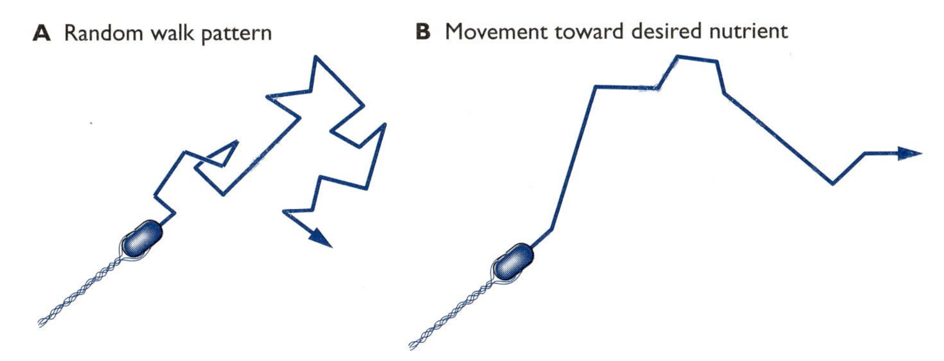

Figure 4.3 Chemotaxis. **(A)** If a bacterium does not detect something either desirable or undesirable, it moves in different directions (random walk pattern). **(B)** Some bacteria can sense a gradient of desirable or undesirable compounds. Prolonging the swimming mode when the bacterium is moving into areas of increasing concentrations of a desirable compound allows the bacterium to move in the general direction of higher concentrations of that compound. Similarly, bacteria can move away from undesirable compounds. Proteins associated with the flagellar motor bind the compound and alter the times of swimming and tumbling.

rapidly than those kept in laboratory medium for prolonged periods. Apparently one effect of laboratory medium is to make bacteria lazy and flabby.

Other types of motility. The spiral-shaped **spirochetes** use flagella for motility, but their flagella are located between an outer membrane and the cytoplasmic membrane. The flagella are wrapped around the cell body, and when they contract they move the cell forward in a corkscrew motion. Spirochetes move best in viscous environments, such as the mucin that lines the intestinal tract. They also appear to bore through or between mammalian cells. Lyme disease and syphilis are among the diseases caused by spirochetes (Fig. 4.4). In the course of the disease, the bacteria move from the bloodstream into the tissue. Their particular type of motility may contribute to their ability to make this transfer.

Gliding motility, which allows some bacteria to glide over surfaces, is common in environmental settings. Although scientists have studied this type of motility for years, they are still unsure exactly how gliding works. The process does not involve flagella. Other protein projections called pili (see next section) help some bacteria move along surfaces. The flagella bind the bacterium to a surface, and the bacteria move along by a **twitching motility.** *Neisseria meningitidis* exhibits this type of motility.

Some bacteria are able to move within mammalian cells after they invade them. *Shigella dysenteriae,* a gram-negative, rod-shaped bacterium that causes bloody diarrhea (dysentery), possesses this capability. The bacterium invades a cell, and actin fibrils begin to condense around one end of the rod-shaped bacterium, propelling it through the cell. Actin fibrils are normally the major component in the mammalian cell cytoskeleton, but the bacteria divert some of this actin to move them within the cell (Fig. 4.5). The actin tails that form on the bacteria also propel them from one cell to the other. This may be the real purpose of actin-based motility—to infect adjacent cells. Inside mammalian cells, the bacteria are protected from the immune system, so moving from cell to cell is a way to avoid attack by the

Figure 4.4 ***Borrelia burgdorferia.*** This spirochete is the cause of Lyme disease. © Charles W. Stratton / Visuals Unlimited.

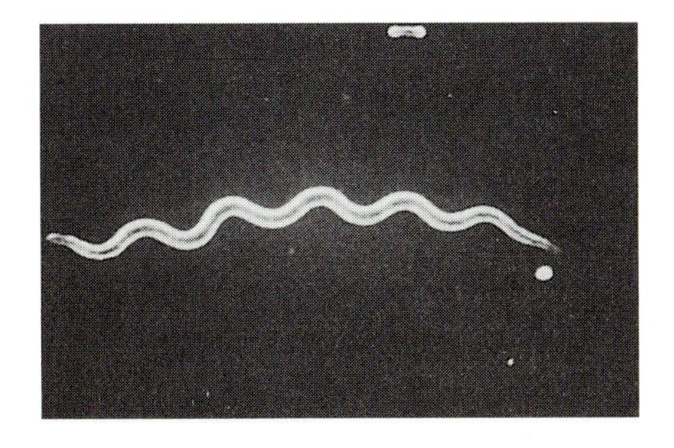

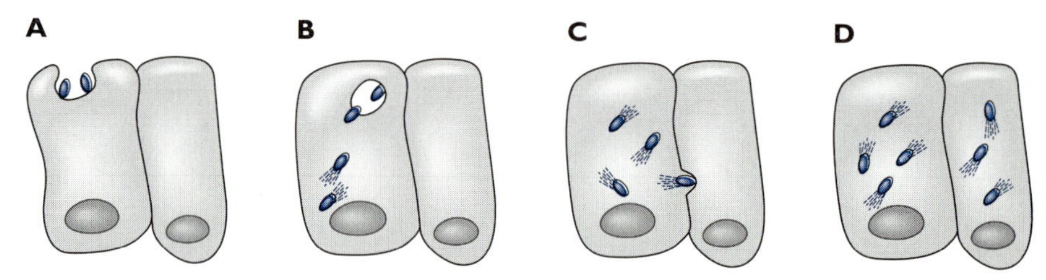

Figure 4.5 Movement of *Shigella* within and between cells. **(A)** Initially, *Shigella* is taken up in a vesicle by an epithelial cell. **(B)** *Shigella* is released from the vesicle and recruits human cell actin from human cell cytoskeleton (small hair-like segments) to build an actin "tail" on one end of the cell. **(C)** This tail pushes the bacteria through the human cell cytoplasm and can propel it into adjacent cells as well. **(D)** *Shigella* with actin tails moving within epithelial cells.

immune system. Medical textbooks classify *S. dysenteriae* as "nonmotile," because these bacteria do not move when placed in a drop of liquid (the classical definition of motility). Within cells, however, the bacteria are clearly motile.

Adherence. A close examination of the surfaces of most *E. coli* cells reveals, in addition to the long spiral-shaped flagella, many shorter, straight projections from the cell surface. These structures, like flagella, are constructed from a single type of protein subunit, pilin, and are called **pili.** Some shorter, thinner projections have been given the name **fimbriae,** but because the role and structure of pili and fimbriae are the same, these terms will be used interchangeably. The purpose of pili or fimbriae is to allow the bacterium to attach to surfaces (Fig. 4.6). Attachment to the surface of the

Figure 4.6 Binding of pili to host cells. **(A)** Pili are hairlike projections, shorter than flagella, on the cell surface, that allow a bacterium to attach to a surface. **(B)** The attachment of the end of the pilus to a molecule on the surface is very specific.

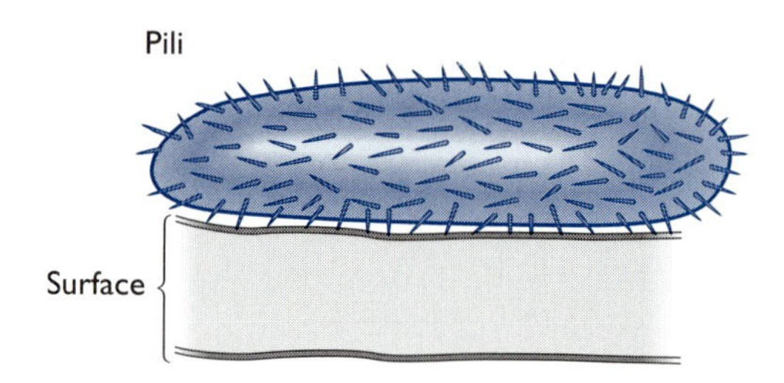

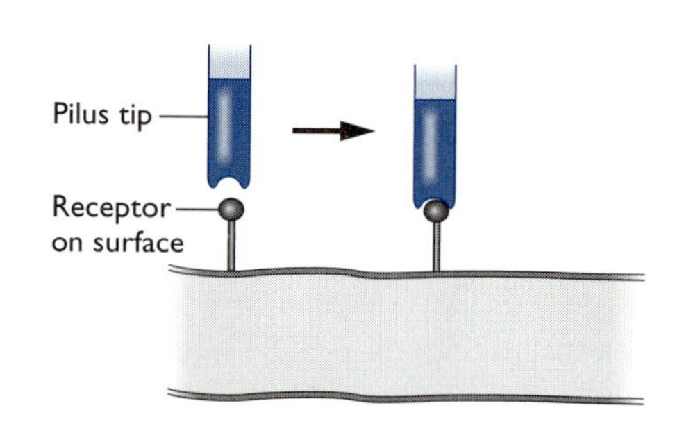

intestine could be very important for an *E. coli* that has used flagella to reach it, because unless it attaches, the fast flow of contents through the small intestine would wash the bacterium out of the site. Continuing to stay in the site by using flagella to swim against the current would consume a lot of energy.

The attachment of the tip of the pilus to a surface is usually quite specific. That is, the tip of the pilus, which is made of proteins different from pilin, is designed to bind a specific molecule on the surface of the cell to which the bacteria attach (Fig. 4.6).

Surface polysaccharides

Capsules. The surface of colonies of *N. meningitis* (discussed in Chapter 3) has a shiny, wet appearance. This is because the surface of *N. meningitidis* is covered by a layer of polysaccharides called a **capsule.** Capsules are usually composed of polysaccharides, although a few bacteria have protein capsules. The main role of capsules for disease-causing bacteria is protection from one of the human body's most important defense mechanisms, the phagocytic cell. **Phagocytic cells** constantly patrol the blood and tissue. If they encounter a bacterium, they try to ingest it and kill it. Some bacteria, such as *N. meningitidis,* use capsules to prevent themselves from being ingested by phagocytic cells. (This protection mechanism will be explained in a later chapter.)

Biofilms. Surface polysaccharides also are used by bacteria as a means of adherence. In many natural settings, where the flow of liquid past an area would wash bacteria away, bacteria form multilayer communities called biofilms. **Biofilms** usually contain more than one species of bacteria. The first layer of the biofilm binds to a surface with pili or some other attachment mechanism, then succeeding layers adhere to the first layer, using a polysaccharide slime to cement them together. Bacterial biofilms cause a variety of problems. A biofilm on a ship's hull slows down forward progress, because water flows past the hull more sluggishly. Biofilms inside pipes produce acid that undermines the integrity of the pipes. Dental plaque begins as a bacterial biofilm that later becomes calcified. Bacterial biofilms that form on contaminated plastic medical implants (heart valve replacements, for example) are almost impossible to eliminate with antibiotics, necessitating the surgical removal of the implants. Bacteria in a biofilm are probably less susceptible to antibiotics than free-swimming (planktonic) bacteria because the antibiotic does not diffuse readily through the polysaccharide layer.

The bacteria in dental plaque or in a pipe biofilm may not be dividing very often but are metabolically active. They can produce enough acid to damage the surface to which the biofilm adheres. Although biofilms cause problems for us, they are the natural way of life for many bacteria.

The bacterial cell wall

Electron microscope examination of cross-sections of gram-positive and gram-negative bacteria has revealed the difference between these staining types. This difference lies in how their cell walls are organized. The rigid cell wall of bacteria consists of a highly cross-linked polysaccharide-peptide matrix called **peptidoglycan** (Fig. 4.7). The backbone of peptidoglycan is a polysaccharide consisting of repeating N-acetyl-glucosamine-N-acetyl-muramic acid units. Attached to most of the N-acetyl-muramic residues are

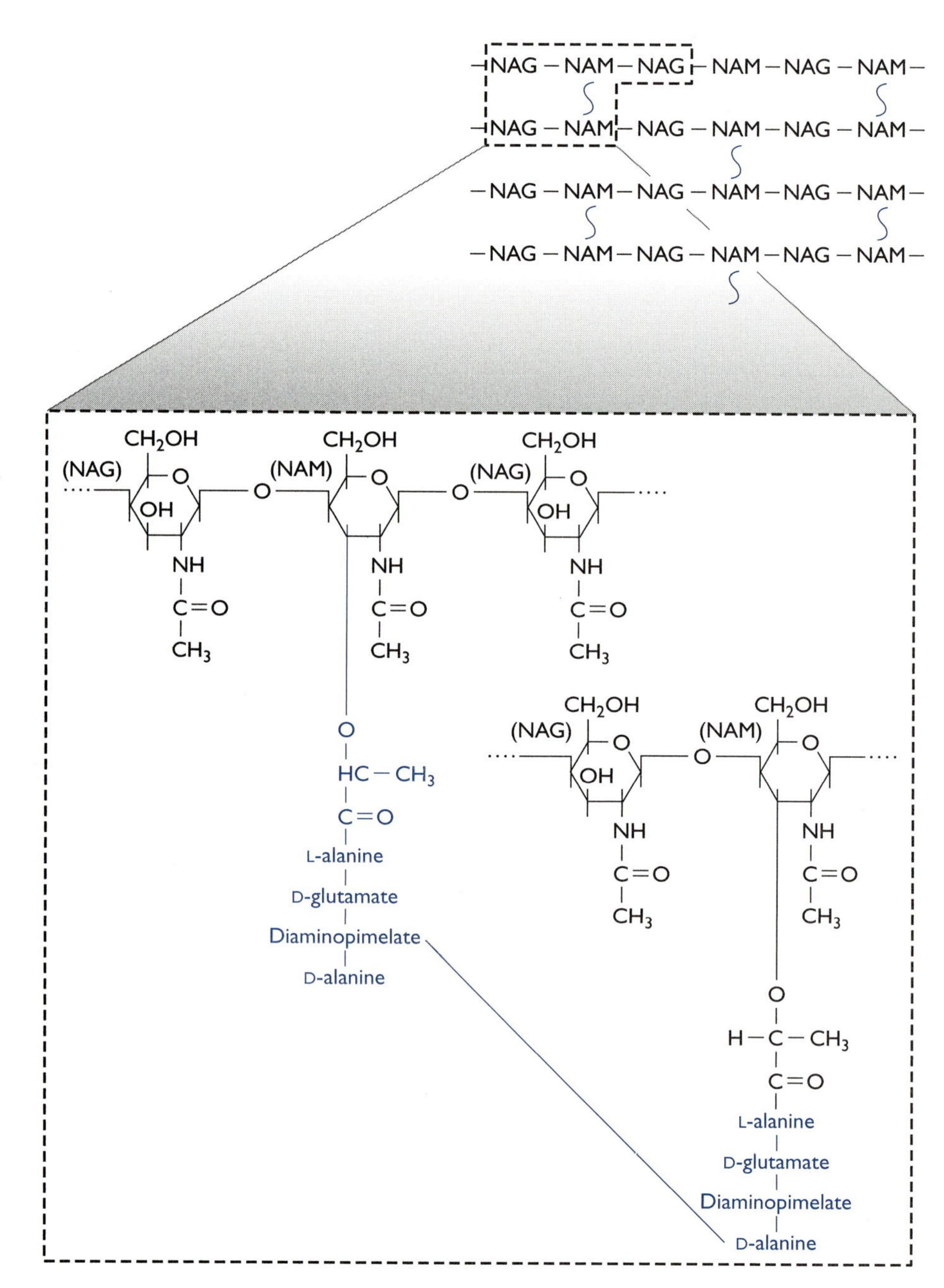

Figure 4.7 Structure of *E. coli* peptidoglycan. Peptidoglycan consists of a polysaccharide backbone (N-acetyl-muramic acid and N-acetyl-glucosamine) that is wrapped around the cell. Adjacent portions of the backbone are crosslinked by peptides attached to N-acetyl-muramic acid residues. NAM, N-acetyl-muramic acid; NAG, N-acetyl-glucosamine.

peptides, which, in the final form of peptidoglycan, form cross-links between N-acetyl-muramic acid residues on adjacent segments of the backbone. Peptidoglycan is also called **murein.**

Gram-positive cell wall. Gram-positive bacteria have a thick peptidoglycan layer covering their surfaces (Fig. 4.8). This layer may be covered with a coat of tightly packed proteins, called an S-layer, but it is exposed to

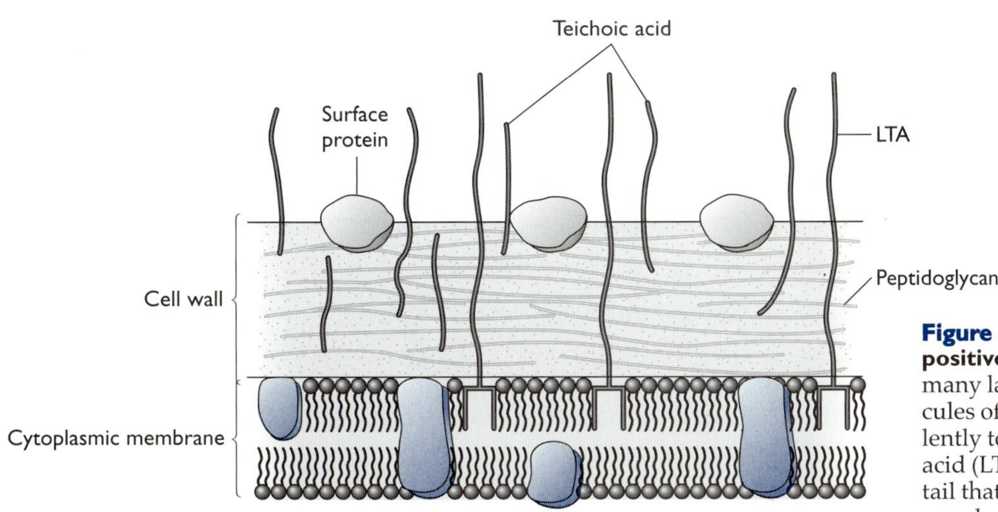

Figure 4.8 Structure of the gram-positive cell wall. The cell wall contains many layers of peptidoglycan. Molecules of teichoic acid are attached covalently to peptidoglycan. Lipoteichoic acid (LTA) is teichoic acid with a lipid tail that is inserted into the cytoplasmic membrane. The cell wall may have a protein coat on its surface. These proteins may be widely spaced, as shown here, or tightly packed in an S-layer.

the environment in the sense that proteins and other molecules diffuse freely into and through the peptidoglycan. Woven through the peptidoglycan layer is a charged polysaccharide called **teichoic acid.** Some molecules of teichoic acid are anchored in the cytoplasmic membrane, which lies under the peptidoglycan, by a lipid anchor. These molecules are called **lipoteichoic acid (LTA)** molecules. During an infection, LTA molecules, released by bacteria killed in their battle with the phagocytic cells, trigger an inflammatory response. (This response is complex and will be explained in detail later in a chapter on the defenses of the human body.) For now, it is enough to say that many of the symptoms of an infection caused by gram-positive bacteria are the result of the body's alarm response to LTA. The body apparently recognizes LTA as a signature for gram-positive bacteria.

Gram-negative cell wall. Gram-negative bacteria have a different type of cell wall (Fig. 4.9). It consists of a thin layer of peptidoglycan that is covered by an **outer membrane.** For additional stability, the outer membrane is attached to the peptidoglycan by a lipoprotein, one end of which is covalently attached to peptidoglycan and the other end is embedded in the outer membrane. The membrane is not a phospholipid bilayer, although it does contain phospholipids. The inner layer of the outer membrane consists of phospholipids. The outer layer is composed of a polysaccharide-lipid molecule called **lipopolysaccharide** (**LPS**). The lipid portion of LPS forms the outer leaflet of the membrane. The polysaccharide portion is exposed on the cell surface. LPS plays the same role during an infection as the LTA of gram-positive bacteria. LPS, released from lysing bacteria, triggers an inflammatory response that is responsible for many of the symptoms of an infection.

The outer membrane of gram-negative bacteria would be impermeable to the small molecules that bacteria need for food were it not studded with **pores** formed by proteins called **porins.** These pores allow nutrients from the external environment to diffuse into the space between the outer and cytoplasmic membranes, a space called the **periplasm.** The outer membrane also contains proteins that bind and internalize vitamins or iron. Flagella and pili protrude through the outer membrane. Gram-negative bacteria use their outer membrane in a variety of ways.

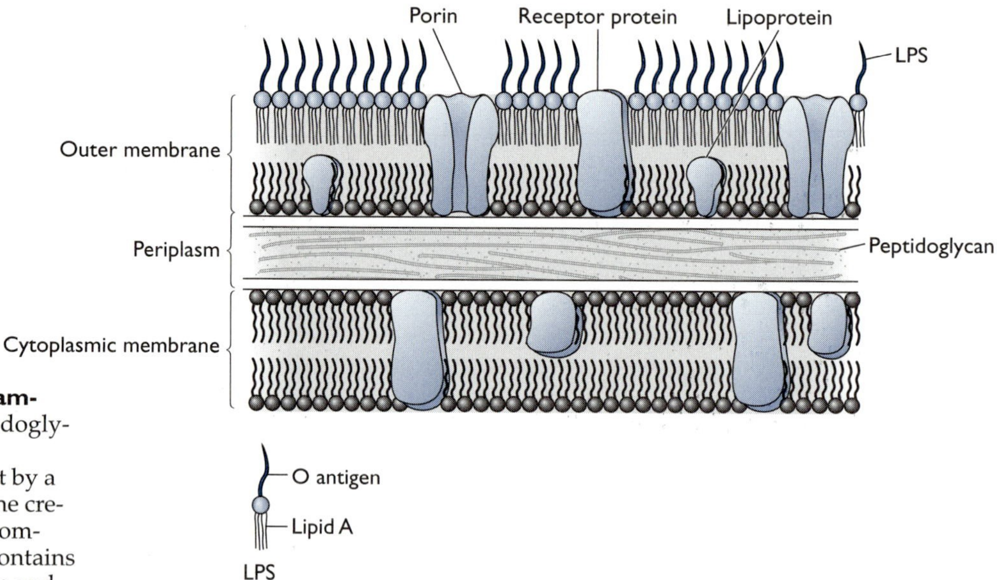

Figure 4.9 Structure of the gram-negative cell wall. The thin peptidoglycan layer is covered by an outer membrane, which is attached to it by a lipoprotein. The second membrane creates the periplasm as a separate compartment. The outer membrane contains phospholipids on its inner surface and lipopolysaccharide (LPS) on its outer surface. LPS has a lipid portion, which is embedded in the outer membrane, and a polysaccharide portion, which is exposed on the surface of the outer membrane.

Because the pores do not allow molecules as large as proteins to cross, the outer membrane excludes external proteins from the periplasm and keeps proteins secreted by the bacterium into the periplasm in this intermembrane space. One of the human body's defenses against bacteria is an enzyme called lysozyme. **Lysozyme** attacks the sugar backbone of peptidoglycan. The outer membrane of gram-negative bacteria protects peptidoglycan from lysozyme attack by excluding it from the periplasm.

The outer membrane also prevents some bulky antibiotics, such as vancomycin, from entering the periplasm. This is the reason most gram-negative bacteria are naturally resistant to vancomycin, whereas gram-positive bacteria are generally susceptible. Most other antibiotics are small enough to pass through outer membrane pores, but, under antibiotic pressure, bacteria can mutate their porins to be even more exclusive, keeping the antibiotics out that normally would have entered the periplasm. In the periplasm, the bacteria store enzymes like alkaline phosphatase, which probably helps to degrade incoming DNA fragments from other bacteria. The periplasm also contains oligosaccharides that help the bacteria adjust to changes in the osmolarity of the medium.

So why does one type of cell wall allow bacteria to retain the crystal violet–iodine stain, even when the bacteria are washed with the ethanol solution that removes the dye from other bacteria? This question still has not been answered. The trait presumably has something to do with the thick peptidoglycan walls of gram-positive bacteria because, from the cytoplasmic membrane on into the interior, gram-negative and gram-positive bacteria are quite similar.

The cell surface of archaea

Some archaea have cross-linked cell walls that resemble peptidoglycan in structure, but with a different amino sugar that replaces muramic acid. This type of cell wall is called **pseudomurein.** Other archaea have cell walls that consist of tightly packed hexagonal protein subunits (the **S-layer**). A few genera of archaea have a sheath that covers the S-layer. Because archaea do not have peptidoglycan, they are not susceptible to antibiotics, such as penicillin, that interfere with peptidoglycan synthesis. Scientists interested in isolating archaea take advantage of this by adding penicillin or other antibiotics to their enrichment and isolation media. The antibiotics prevent the faster-growing bacteria from overgrowing the slow-growing archaea.

The Cytoplasmic Membrane and the Cytoplasm

Composition and function of the cytoplasmic membrane

Both gram-positive and gram-negative bacteria have phospholipid bilayer cytoplasmic membranes, and both employ these membranes in the same way. The cytoplasmic membrane serves many functions (Fig. 4.10). The unusual membrane lipids of the archaea were described in Chapter 2. Although the lipid composition of archaeal membranes differs from that of bacterial membranes, the cytoplasmic membrane proteins of archaea play much the same roles as those of bacteria.

Transport of nutrients. Most molecules in the environment cannot diffuse through the lipid portion of the cytoplasmic membrane and must be admitted through proteins called **transport proteins.** Transport proteins convey sugars and other essential nutrients into the cytoplasm, where they are used as sources of carbon and energy. Transport proteins are very specific for selected substrates. This permits a bacterium to allow entrance of beneficial compounds and to exclude others that may be detrimental to the cell. Because transport proteins are specific for one or a few compounds, a bacterium needs many different transport proteins to transport the mixture of nutrients it needs. Usually, the transport of nutrients requires energy because the concentration of nutrients is much lower outside the cell than in the cytoplasm. Because molecules would normally diffuse from a higher to a lower concentration, it takes energy to make them move in the other direction.

Energy generation: the electron transport system. The cytoplasmic membrane of bacteria is an important part of the energy machinery of the cell. Enzymes and other molecules in the cytoplasmic membrane pump protons out of the cell. The proteins and other molecules that do this are called, collectively, the **electron transport system** because they separate

Figure 4.10 The cytoplasmic membrane of bacteria. The cytoplasmic membrane is a phospholipid bilayer that contains many different types of proteins involved in transport of nutrients, regulation of gene expression, energy generation, and other activities.

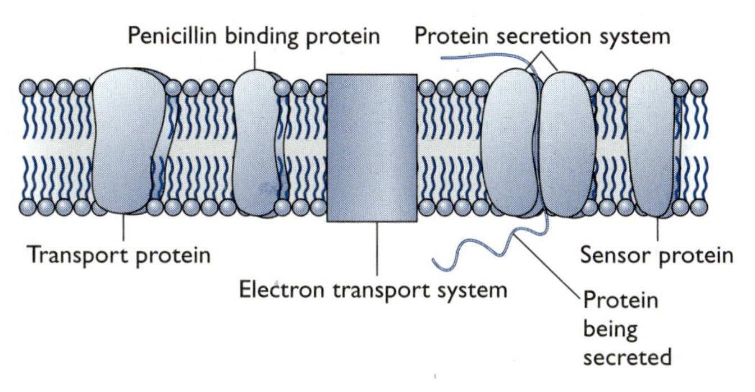

electrons from protons, expel the protons to the outer surface of the cytoplasmic membrane, and then dump the electrons on a molecule called the terminal electron acceptor. The expelled protons generate energy by reentering the cell through a cytoplasmic membrane protein complex called adenosine triphosphate (ATP) synthase. ATP is the main energy currency of the cell. Protons reentering the cell can also interact with the flagellar motor directly to provide energy for rotation of the motor or with transport proteins to provide energy for transport of nutrients into the cell. The electron transport system will be explained in detail in Chapter 8.

Proteins that synthesize peptidoglycan. The cytoplasmic membrane also contains proteins that play a role in the synthesis and turnover of peptidoglycan. In order for the bacteria to elongate and divide, they need to break down peptidoglycan in the region being elongated and build it up again. The proteins that help to synthesize or break down peptidoglycan are the target of the antibiotic penicillin and are called **penicillin-binding proteins.** Proteins involved in the synthesis of LTA or LPS are also located in the cytoplasmic membrane.

Secretion system. The cytoplasmic membrane contains proteins that secrete proteins, made in the cytoplasm, into the periplasm. This complex of proteins is called the **secretion system.** Obviously, bacteria do not want to secrete all the proteins they make, only those destined for the periplasm or extracellular fluid. The proteins destined to be secreted are identified by a short sequence of amino acids at the amino terminal end of the protein. During the secretion process, this signal is clipped off by a peptidase to produce the final secreted form of the protein.

Regulatory proteins. The cytoplasmic membrane is also part of the "brain" of the bacterium because it contains proteins that sense changes in the external environment and convey those signals to proteins in the interior. In turn, these proteins respond by changing the proteins currently being made into a more appropriate set. If, for example, a bacterium in water that is capable of infecting the human small intestine is ingested, it will need to adapt rapidly in order to survive passage through the stomach and to attach to the surface of the small intestine. Among other adaptations, it will need to produce pili that are appropriate to the human intestine. This type of response to external cues is called **regulation,** and the proteins that register and transmit the environmental signal are called **sensor** proteins. Not all proteins involved in regulation are located in the membrane. In fact, many are located in the cytoplasm of the cell, and bacteria often use both cytoplasmic membrane proteins and cytoplasmic proteins to coordinate their response to environmental changes.

The cytoplasm

The cytoplasm of a bacterial cell contains enzymes that generate ATP directly by oxidizing glucose or other carbon sources (Chapter 8). Some of these enzymes interact with the electron transport system in the cytoplasmic membrane to produce ATP indirectly. The cytoplasm also contains some of the enzymes involved in synthesis of the subunits of peptidoglycan. These enzymes cooperate with cytoplasmic membrane proteins that carry out the final assembly of peptidoglycan. Ribosomes, the site of protein synthesis, are located throughout the cytoplasm. The mass of DNA that forms the genome of the cell (**nucleoid**) and the proteins that interact with it are also found in the cytoplasm. If the double-stranded DNA molecule that makes up the nucleoid were to be stretched out in linear form, it would be many times longer than the bacterial cell. To pack the DNA into a manageable-sized nucleoid, bacteria use two strategies. First, the double-stranded helix of the DNA is wound very tightly (**supercoiled**). This causes the DNA molecule to collapse into itself, forming a compact mass. Bacteria also have **histone-like proteins** that bind to the DNA and help maintain the structure of the nucleoid. Enzymes involved in the synthesis of DNA and ribonucleic acid (RNA) are located in the cytoplasm. Finally, many of the proteins that help to orchestrate the response of the bacterium to environmental changes (regulatory proteins) are located in the cytoplasm. Some of these interact with cytoplasmic membrane proteins.

The Synthesis of Peptidoglycan: An Important Target for Antibiotics

Steps in peptidoglycan synthesis

The description of the various features of the bacterial cell given in the previous sections does not do justice to the dynamic interplay of the different cell components. As an example of how bacteria coordinate the activities of the different cell compartments, consider the synthesis of peptidoglycan. Peptidoglycan synthesis is also an excellent example of how antibiotics work and the way in which bacteria become resistant to them. Both of these aspects of peptidoglycan synthesis will be explored in this section.

Synthesizing and exporting the subunits of peptidoglycan. The first steps in peptidoglycan synthesis occur in the cytoplasm and result in the construction of the subunits that will later be assembled to produce peptidoglycan. Enzymes in the cytoplasm construct a unit that consists of N-acetyl-muramic acid linked to a pentapeptide (Fig. 4.11). This unit then interacts with a complex lipid in the cytoplasmic membrane, **bactoprenol.** It not only exports the peptidoglycan subunit from the inner face of the cytoplasmic membrane to the outer face but also serves as the matrix on which an enzyme adds the second backbone sugar, N-acetyl-glucosamine, to form the peptidoglycan disaccharide-peptide repeating subunit. This same bactoprenol is also involved in synthesis of LPS and LTA. The early steps in peptidoglycan synthesis must be carried out in the cytoplasm because some of them require ATP or other energy-rich molecules such as uridine diphosphate (UDP) available only there.

Assembling peptidoglycan. The enzymes that assemble the subunits into intact peptidoglycan are located in the cytoplasmic membrane. As the subunits are passed through the membrane, the disaccharide portion is first attached to the terminal sugar molecule on the growing backbone of the peptidoglycan chain. Other enzymes then cleave the D-alanine-D-alanine at the end of the peptide and use the energy in this bond to attach this peptide to another peptide on an adjacent part of the sugar backbone. This cross-linking step is critical for the strength of the peptidoglycan layer. It is also the target for two important classes of antibiotics, the β-lactam antibiotics and the glycopeptides.

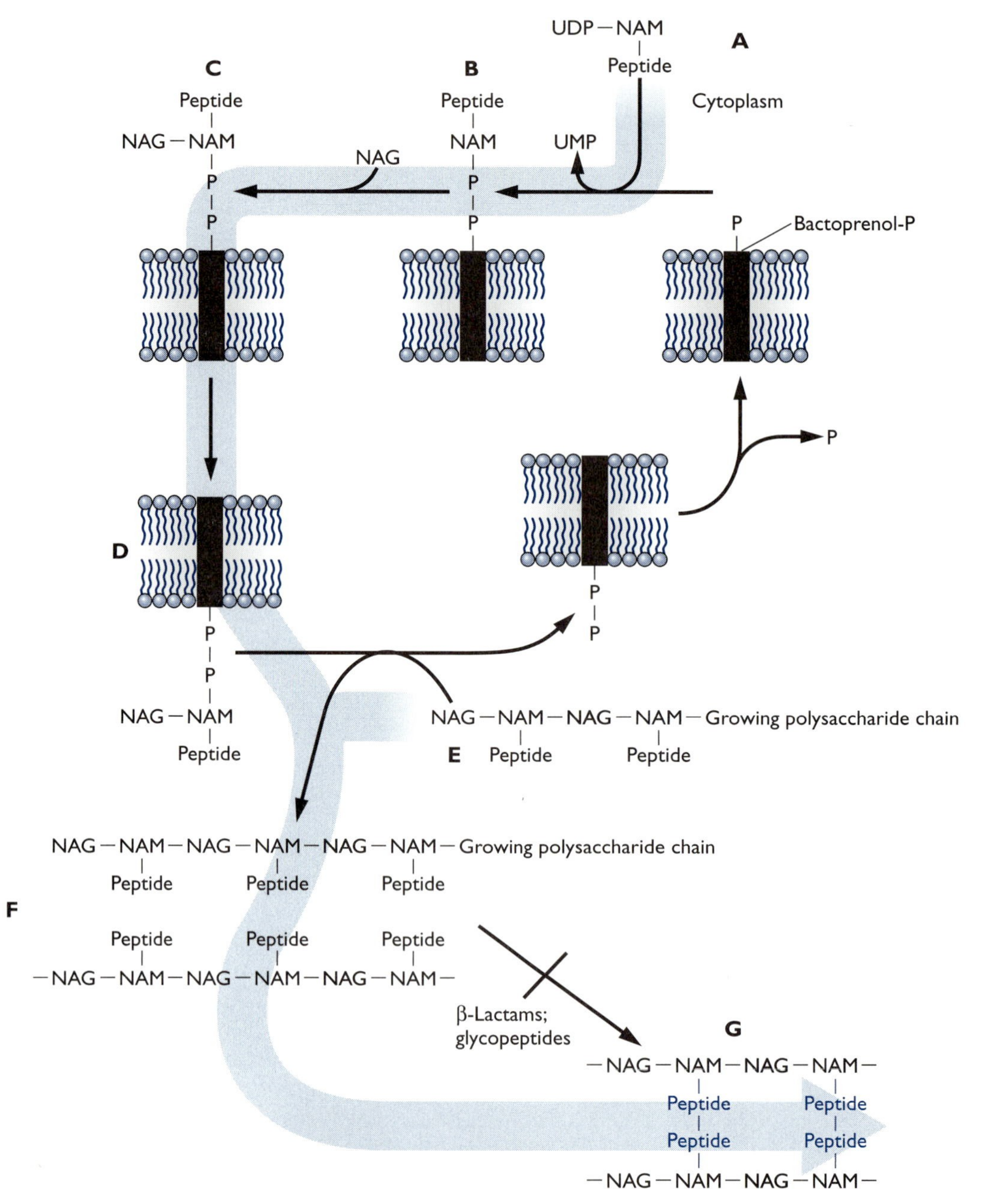

Figure 4.11 Synthesis of peptidoglycan. Steps in peptidoglycan synthesis and the locations where they occur are shown. **(A)** N-acetyl-muramic acid (NAM) is linked to a peptide in the cytoplasm. **(B)** The NAM-peptide subunit binds to the phosphorylated bactoprenol in the cytoplasmic membrane. UDP, an energy rich compound, donates a phosphate group. **(C)** N-acetyl-glucosamine (NAG) is added to the NAM-peptide subunit. **(D)** Bactoprenol exports this new unit across the cytoplasmic membrane. **(E)** These units are linked together to form a growing polysaccharide chain. **(F)** The assembly of peptidoglycan occurs on the surface of the outer membrane and is catalyzed by enzymes embedded in the cytoplasmic membrane. **(G)** The cross-linking step is the one inhibited by β-lactam antibiotics and the glycopeptide vancomycin.

Antibiotics that inhibit the cross-linking of peptidoglycan. Antibiotics are chemical compounds that kill or inhibit the growth of bacteria. Most of them are specific for bacteria and do not affect other types of microorganisms. Antibiotics exert their effects by binding to some important bacterial molecule and impairing its function. The β-lactam antibiotics bind to and inactivate the proteins that catalyze the cross-linking reaction (Fig. 4.11). The glycopeptides bind to the D-alanine-D-alanine part of the peptide itself. Because these molecules are large and bulky, they block the cross-linking enzymes from attaching to the peptide and thus stop the cross-linking reaction.

The β-lactam family of antibiotics includes the penicillins and cephalosporins, two very important groups of antibiotics. The β-lactam ring, which is the active part of the molecule, is shown in Figure 4.12, along with the structures of some of these antibiotics. The carbapenems and monobactams are the most recent additions to the family. Although it may not be obvious from the flat projection of the β-lactam ring shown in Figure 4.12, in three dimensions this ring resembles the D-alanine-D-alanine region closely enough to fool the cross-linking enzymes trying to cleave it. Instead, the enzymes attach the antibiotic covalently to themselves, rendering them inactive. Failure to cross-link peptidoglycan ultimately undermines the integrity of the cell wall and the bacteria lyse.

The most widely used member of the glycopeptide family of antibiotics is vancomycin, currently the last drug available for treating diseases caused by strains of *Staphylococcus aureus* and other gram-positive cocci that have become resistant to all other antibiotics. As mentioned earlier, vancomycin is too large to diffuse easily through the outer membrane pores of gram-negative bacteria and is thus ineffective against most of them.

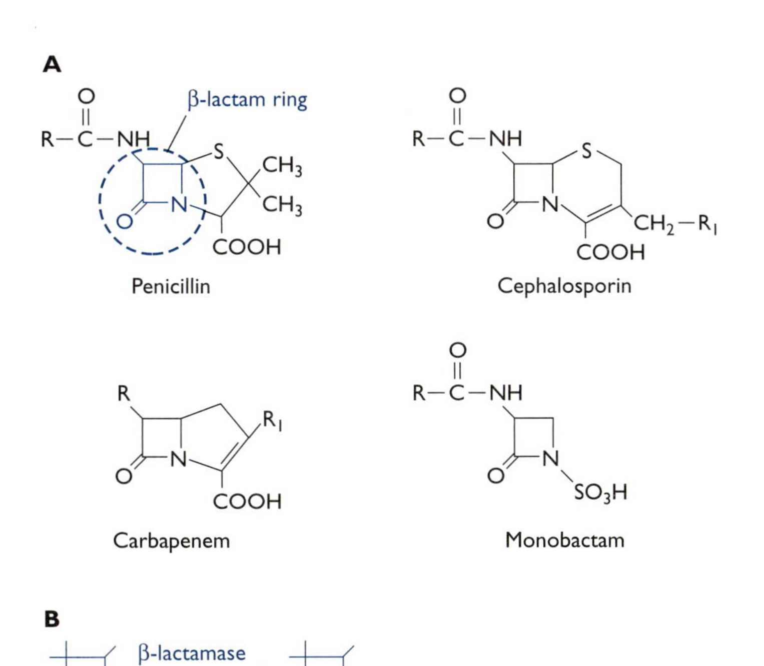

Figure 4.12 Structure of the β-lactam ring and some β-lactam antibiotics. (A) Structure of the major families of β-lactam antibiotics. **(B)** Bacterial β-lactamases hydrolyze the C-N bond of the β-lactam ring, thus inactivating the antibiotic.

How bacteria become resistant to β-lactams and vancomycin. There are three main strategies bacteria can use to become resistant: inactivate the antibiotic, change the antibiotic target so that it still works but no longer binds the antibiotic, or limit access of the antibiotic to its target. An example of resistance resulting from inactivation of the antibiotic is the enzyme β-lactamase. This enzyme hydrolyzes the C-N bond in the β-lactam ring. Because this ring is essential for activity of the antibiotic, hydrolysis of this bond inactivates the antibiotic. Many gram-positive and gram-negative bacteria now produce β-lactamases. But scientists are fighting back (Box 4–1).

The gram-negative bacteria secrete their β-lactamases into the periplasm, where they are available to protect the enzymes that cross-link peptidoglycan. Because the pores limit the diffusion of the antibiotic somewhat, the enzyme has a better chance to inactivate all of the antibiotic molecules that enter the periplasm. In fact, some bacteria have mutated their porins so that they limit even more the diffusion of the antibiotic into the periplasm. Gram-positive bacteria do not have this option. They secrete their β-lactamase, but because the secreted enzymes can diffuse away from the cell, they usually have to secrete much more enzyme than the gram-negative bacteria. Some gram-positive bacteria have developed a different resistance strategy. They mutate their cross-linking enzymes so that they no longer bind the antibiotic.

So far, the only known mechanism of resistance to vancomycin is changing the D-alanine-D-alanine of the peptide to D-alanine-D-lactate or some other variation on this theme. Vancomycin does not bind these substitute peptides. Of course, the bacteria also must provide a new enzyme to create the new dipeptide and new cross-linking enzymes that recognize it. Vancomycin resistance among the disease-causing gram-positive bacteria is still fairly uncommon. But the genes needed to confer this type of resistance are now present in some bacterial species. These genes, as well as those needed for resistance to β-lactam antibiotics, are currently being spread among different species of bacteria by horizontal gene transfer (Chapter 7).

box 4–1 *Battling β-Lactamases*

As more and more reports of bacteria that produced β-lactamases appeared in the literature, the first response of scientists was to make new β-lactam antibiotics that were resistant to hydrolysis by bacterial β-lactamases. One of the reasons for the large variety of β-lactam antibiotics now on the market was this effort to counter the rising tide of β-lactamase–producing bacteria. Unfortunately, the bacteria responded by mutating their β-lactamases so that they could inactivate the new antibiotics. Another strategy to counter these insidious bacterial enzymes was needed. Scientists discovered variants of the β-lactam antibiotics that were not active as antibiotics but were very effective at inhibiting β-lactamases. One such compound is clavulanic acid.

Clavulanic acid

By combining clavulinic acid or some other β-lactamase inhibitor with a β-lactam antibiotic, antibiotics that once would have been inactivated by the bacterial enzymes could be recycled. Clavulanic acid eliminated the bacterial enzyme while the antibiotic proceeded to do its job of killing the bacteria. The popular antibiotic preparations Augmentin and Timentin are combinations of clavulanic acid and ampicillin.

chapter 4 at a glance

Features of the bacterial surface

Organelle or molecule	Composition	Location	Function
Flagella	Protein subunits	Extends outward from surface except in spirochetes where flagella are wrapped around cell, inside outer membrane	Motility
Pili	Protein subunits; tips bind to specific molecule	Extends outward from cell	Attachment; twitching motility
Capsule	Usually loose network of polysaccharides	Covers surface of cell	Protection from phagocytes; biofilm formation
Lipopolysaccharide	Lipid, polysaccharide	Outer layer of gram-negative outer membrane; lipid portion embedded in membrane; polysaccharide exposed on surface	Stabilizes membrane; elicits an inflammatory response in the human body
Porins	Proteins	Embedded in gram-negative outer membrane	Form pores that allow diffusion of nutrients through outer membrane
Lipoprotein	Lipid, protein	Attached to peptidoglycan and embedded in outer membrane	Stabilizes cell
Lipoteichoic acid	Lipid, polysaccharide	Found in peptidoglycan layer of gram-positive bacteria	Unknown; elicits an inflammatory response in the human body
Peptidoglycan	Polysaccharide backbone cross-linked with peptides	Gram-positive bacteria (thick layer), usually exposed to environment but may be covered by an S-layer; gram-negative bacteria (thin layer), covered by the outer membrane	Maintain shape of cell

Functions and location of important components of the cytoplasmic membrane and cytoplasm

Component	Location(s)	Function
DNA nucleoid	Cytoplasm	Genome
Histone-like proteins	Cytoplasm, attached to DNA	Stabilize nucleoid
Enzymes involved in synthesis of DNA, RNA	Cytoplasm	Replication of the genome, transcription of RNA.
Ribosomes	Cytoplasm	Protein synthesis
Enzymes involved in breaking down substrates	Cytoplasm, some interact with the electron transport system.	Energy generation.
Electron transport system	Cytoplasmic membrane	Energy generation.
ATP synthase	Cytoplasmic membrane	Uses protons pumped out of the cell by the electron transport system to make ATP
Secretion system	Cytoplasm (associated with membrane), cytoplasmic membrane	Secrete certain proteins from cytoplasm to external environment
Peptidoglycan synthetic machinery	Cytoplasm, cytoplasmic membrane	Synthesize and assemble peptidoglycan

(continued)

chapter 4 at a glance (continued)

Antibiotics that inhibit peptidoglycan synthesis

Types of antibiotics
- β-lactams (for example, penicillin, cephalosporin)—inactivate enzymes that cross-link peptidoglycan.
- Glycopeptides (for example, vancomycin)—bind the D-alanine-D-alanine part of the peptide portion of peptidoglycan and block cross-linking enzymes from acting on it.

Mechanisms of resistance
- Inactivate antibiotics
    - β-lactamases
- Modify target so antibiotic no longer binds
    - Mutant cross-linking enzymes (β-lactam resistance)
    - Modify D-alanine-D-alanine (vancomycin resistance)
- Prevent access to target
    - Porins of gram-negative bacteria limit diffusion of antibiotic into periplasm

key terms

adherence attachment to surfaces; can be mediated by pili (rod-like projections) or capsular polysaccharides

antibiotics chemical compounds that inhibit the growth of or kill bacteria by binding to specific target molecules (for example, proteins involved in peptidoglycan cross-linking) and inhibiting their function

biofilms community of microorganisms attached to a surface and held together by a polysaccharide matrix

chemotaxis sensing the concentration of a particular chemical and responding by moving toward or away from it

cytoplasm interior of the cell; contains DNA nucleoid, ribosomes, and enzymes that generate energy and perform biosynthetic reactions

cytoplasmic membrane phospholipid bilayer containing proteins involved in substrate transport, energy generation, peptidoglycan synthesis, secretion of proteins, and sensing of environmental signals

gram-negative cell wall thin layer of peptidoglycan covered by a membrane (outer membrane); outer membrane contains protein pores and lipopolysaccharide (LPS)

gram-positive cell wall thick peptidoglycan layer containing lipoteichoic acid (LTA) with no outer membrane; sometimes has an S-layer overlying the peptidoglycan

motility ability to move through the environment; different types include flagella-driven, gliding, corkscrew movement of spirochetes, twitching, and intracellular actin polymerization

peptidoglycan rigid cell wall of bacteria; consists of a polysaccharide backbone wrapped around the cell and stabilized by peptide cross-links

S-layer, pseudomurein rigid cell walls of archaea; S-layers are composed of a dense layer of protein subunits; pseudomurein resembles peptidoglycan but has no muramic acid

Questions for Review and Reflection

1. Why do bacteria have pili? Why not just use flagella for swimming *and* attachment?

2. Many antibiotics work on both gram-positive and gram-negative bacteria. Explain why this is the case. Explain why this is not true of vancomycin, which works best on gram-positive bacteria.

3. The U.S. Navy has spent a lot of money trying to figure out how to keep microbes from forming biofilms on a ship's hull. Why would microbes form a biofilm on a ship? What advantage does it give them?

4. Why is chemotaxis seen only in motile bacteria?

5. Why don't bacteria have pores in their cytoplasmic membrane like those in the outer membranes of gram-negative bacteria?

6. Give at least two reasons why the outer membrane and periplasm of gram-negative bacteria gives them an advantage over gram-positive bacteria.

7. How does the outer membrane of a gram-negative bacterium differ from the cytoplasmic membrane in composition and function?

8. Why do bacteria make pieces of peptidoglycan in the cytoplasm, then export the units to the surface of the cytoplasmic membrane for assembly? Why not carry out all these steps on the surface of the cytoplasmic membrane?

9. How does the action of vancomycin resemble that of penicillin? How does it differ?

10. Penicillin and vancomycin inhibit the same step, cross-linking of peptidoglycan. Why then is it that a bacterium that has become resistant to penicillin and other β-lactam antibiotics does not become resistant to vancomycin at the same time?

5 Bacterial Genetics I: DNA Replication and Genetic Engineering

Critics of genetic engineering portray the process as "unnatural." Yet, scientists learned genetic engineering by watching bacteria perform this life-saving and adaptive process on themselves—a process that sometimes works to our detriment. Genetic engineering by bacteria makes it possible for them to become resistant to antibiotics and capable of causing new diseases. For humans, learning to perform genetic engineering better and faster than bacteria may well prove to be a matter of life or death.

The original motivation for studying the way in which bacteria make copies of their genomes was the hope that such information would shed light on the way that human cells carry out what was believed to be a similar process. As it turned out, this expectation was justified, because many of the basic features of DNA replication are conserved in all free-living creatures. However, these studies had an unexpected extra dividend: the molecular revolution that has transformed biology. Molecular techniques that emerged from the studies of DNA structure and replication include DNA hybridization, DNA sequencing, polymerase chain reaction (PCR), and cloning. Another unanticipated dividend was the discovery of a new mechanism of antibiotic action and the development of a new class of antibiotics, the fluoroquinolones, which now are used widely to treat human disease. These examples show how basic science investigations that initially may seem abstruse can turn out to have significant real-world impacts.

Steps in Bacterial DNA Replication

Opening up the supercoiled double helix

DNA is a **double-stranded helix,** held together by hydrogen bonds between the bases, which are attached to a **deoxyribose-phosphate backbone** (Fig. 5.1). These bases pair very specifically, **adenine** (**A**) with **thymine** (**T**) and **cytosine** (**C**) with **guanine** (**G**), a feature that helps to assure accurate copying of the DNA molecule. In the helical form of DNA, an even more important source of stability is the interaction between adjacent bases stacked on each other. The structure of DNA is so stable that it takes temperatures near 100°C to separate the two strands. The tight and

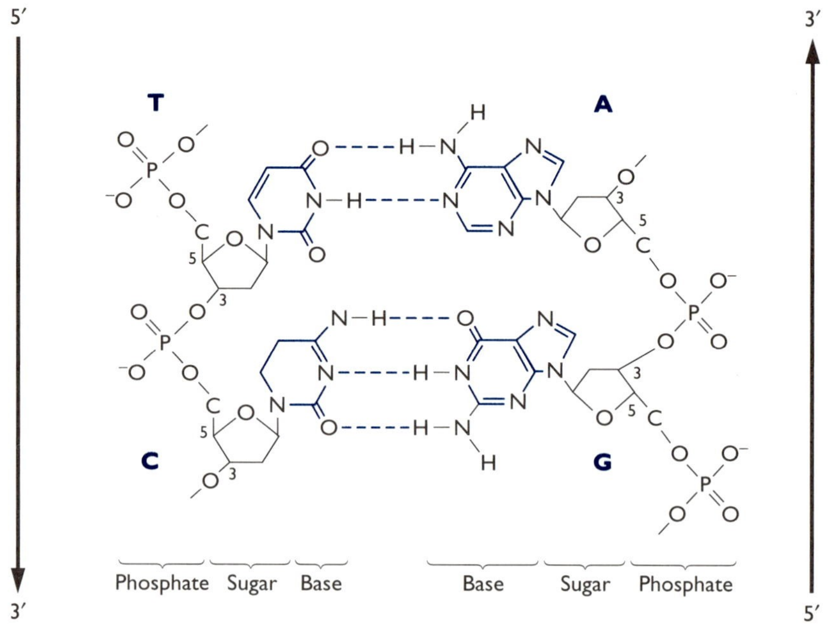

Figure 5.1 Structure of DNA. Each strand of DNA has a deoxyribose-phosphate backbone to which different bases (adenine [A], thymine [T], cytosine [C], and guanine [G]) are attached. The two strands interact with each other through hydrogen bonds that form between the bases on different strands. The specificity of the interaction is determined by two hydrogen bonds that restrict binding of A to T and three hydrogen bonds that restrict binding of C to G. A and G are purines; C and T are pyrimidines. Additional stability comes from interactions between adjacent bases in the stack of bases that forms when the DNA is in its helical form. The structure of RNA is similar, except that the nucleotides in the backbone contain ribose instead of deoxyribose. RNA contains the base uracil (U) instead of thymine.

specific interactions between the strands of a DNA molecule are the basis for DNA hybridization, a widely used molecular biology technique that will be described later.

The DNA in an average bacterial genome, if laid out linearly, would be much longer than the bacterium. To pack this much DNA into the interior of a bacterial cell, the DNA must be coiled so tightly (**supercoiled**) that it collapses on itself to form the **nucleoid.** To copy the strands of such a tightly coiled molecule, it is first necessary to relieve the supercoiling and open up sections of the DNA helix. This allows **DNA polymerase,** the enzyme that will copy each DNA strand, to gain access to the strands. Local uncoiling is begun by enzymes called **topoisomerases.** These enzymes nick one strand of the DNA, allowing the DNA to become uncoiled in the region of the nick. Enzymes called **helicases** separate the strands.

Copying both strands

DNA polymerase has two features that are critical to genetic engineering applications. First, although DNA polymerase can bind to a single-stranded DNA segment, it binds more tightly and copies DNA more efficiently and accurately if it starts at a single-stranded region of DNA adjacent to a double-stranded region. In bacterial DNA synthesis, this double-stranded region is formed by a short segment of RNA that associates with the DNA strand (Fig. 5.2A). This RNA segment is called an **RNA primer.** In later descriptions of PCR and DNA sequencing, the primer will be a DNA segment, but the principle is the same: a double-stranded region on an otherwise single-stranded DNA molecule directs efficient and accurate binding of DNA polymerase and orients subsequent copying of the single DNA strand. A second key feature of DNA polymerase is that, once it binds the DNA strand, it synthesizes new DNA only in a **5' → 3' direction**. That is, if a small region of double-stranded nucleic acid occurs in a larger, single-stranded region, DNA polymerase will not proceed to synthesize DNA in both directions from the

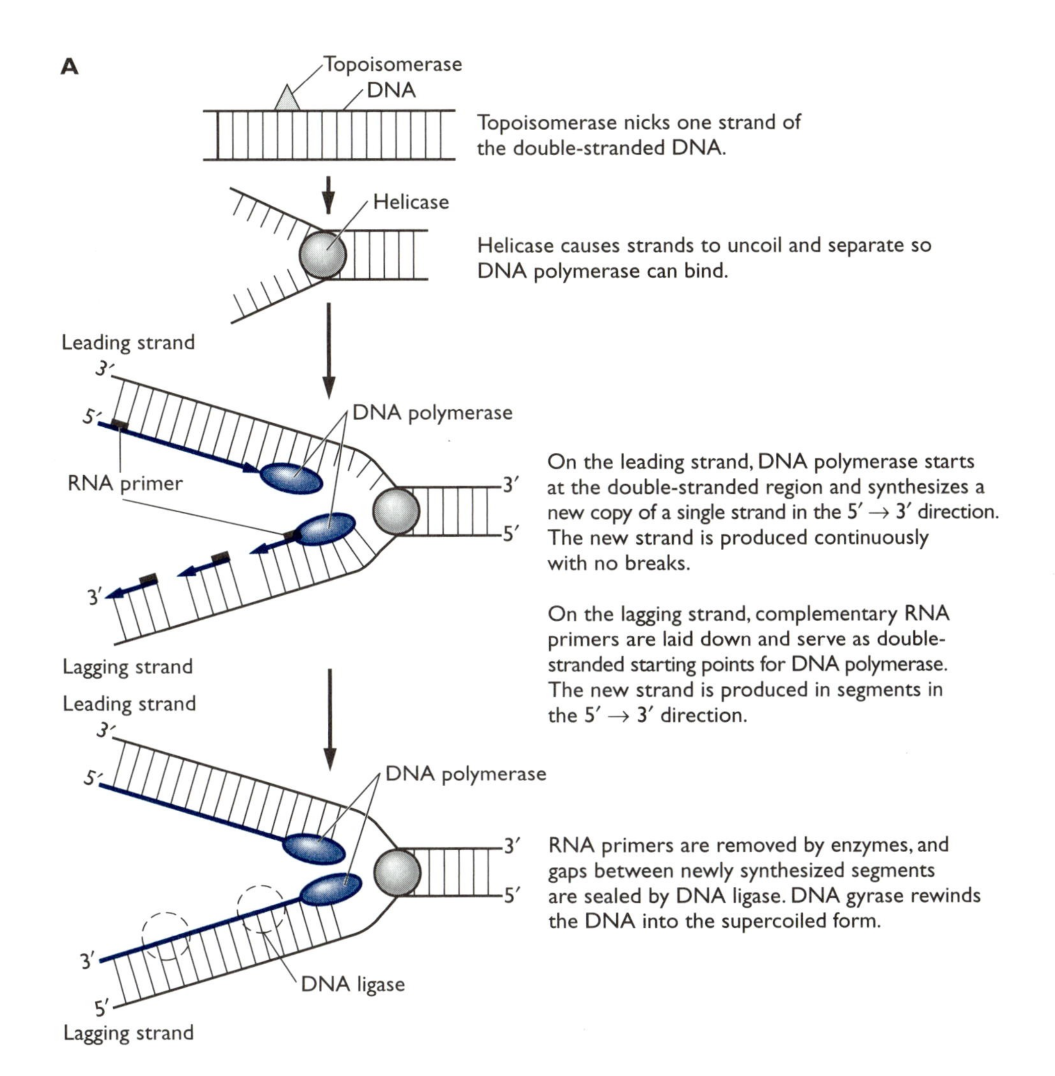

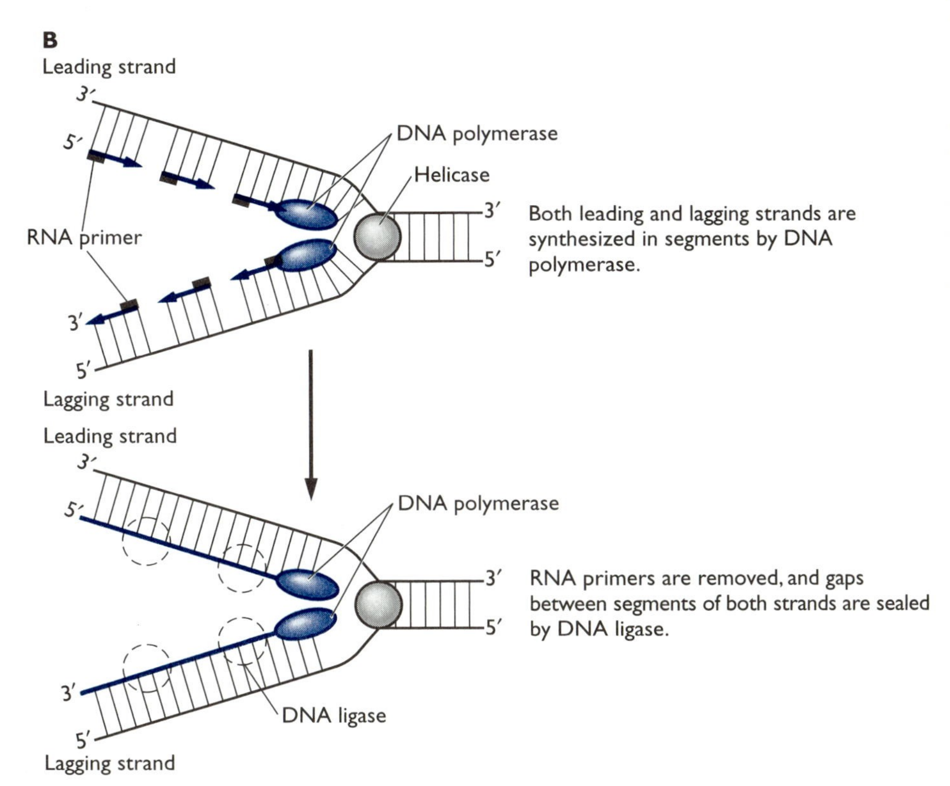

Figure 5.2 DNA replication. (A) The original version of the DNA replication process. For simplicity, only one direction of replication is shown here. In reality the same process is occurring in the other direction as well, so that the replication region resembles a bubble instead of a single fork. Enzymes nick one strand of the DNA to open up the supercoiled structure. Then, helicase causes the DNA strands to uncoil and separate. On the leading strand, DNA polymerase begins at a double-stranded region and continuously synthesizes a new copy of the single strand in a 5' → 3' direction. On the lagging strand, complementary RNA primers are laid down and serve as a double-stranded region from which the DNA polymerase can synthesize short segments in a 5' → 3' direction. RNA primers are removed and the gaps between the segments are sealed by DNA ligase. After the new DNA is formed, DNA gyrase rewinds it into the supercoiled form. **(B)** The newer—but still controversial—understanding of DNA replication. The process is the same except that both the leading and lagging strands are copied in segments.

double-stranded region but only in the 5' → 3' direction. DNA polymerase attaches the 3'-OH of the last deoxyribose backbone sugar of the most recently incorporated nucleotide to the 5'-phosphate group of the deoxyribose on the incoming nucleotide.

On the **leading strand,** the copying procedure appears to be relatively straightforward. That is, DNA polymerase starts from a double-stranded region formed by an RNA primer bound to the single DNA strand and synthesizes a copy of the strand, moving 5' → 3' as the double-stranded DNA is pulled apart. The moving region where the strands are being separated and copied is called the **replication fork.** Copying the other strand is a bit trickier because of the requirement for movement of DNA polymerase only in the 5' → 3' direction. On that strand, called the **lagging strand,** small RNA primers anneal to newly exposed segments of the DNA to serve as a starting point for 5' → 3' synthesis of a DNA copy. The result is that the synthesis of the lagging strand occurs in segments. After synthesis, enzymes remove the RNA primers and replace them with DNA copies. An enzyme called **DNA ligase** seals the gaps between the newly synthesized DNA segments.

This account of DNA synthesis is one that has been accepted for decades and appears in most textbooks. Recently, however, it has become controversial. Many scientists now believe that synthesis of the copy of the leading strand, like that of the lagging strand, is discontinuous, although the segments on the leading strand are probably longer than those on the lagging strand. That is, although synthesis of a copy of the leading strand could proceed continuously, it does not do so. Instead, RNA primers are used, and synthesis proceeds in discontinuous segments, which are then ligated, after removal of the primer, by DNA ligase (Fig. 5.2B). This view of leading-strand synthesis makes sense when one considers that otherwise, DNA polymerase would have to synthesize DNA continuously over a long region—half the chromosome. Synthesis in segments seems more likely.

How can scientists differ on such a fundamental aspect of DNA synthesis, a process that has been studied intensively for decades? The version shown in Figure 5.2A was derived from *in vitro* experiments in which DNA synthesis was reconstituted in a test tube. In these experiments, synthesis of the leading strand had to proceed over relatively short distances compared with those that must be traversed in a chromosome. The newer version (Fig. 5.2B) was deduced from studies of bacterial replication using living bacteria, where the leading strand must be copied over long distances. This reassessment of the elements of DNA synthesis provides an excellent example of the dynamic nature of scientific investigation, with current dogma always up for reevaluation and revision.

Why do bacteria use RNA primers for discontinuous synthesis of a new DNA strand? Why not use DNA primers, making it unnecessary to remove and replace the RNA primers before ligating the DNA segments? The most appealing answer to this question takes into consideration the fact that when DNA polymerase tries to synthesize a new strand of DNA from an entirely single-stranded template, DNA polymerase makes mistakes until it is firmly seated on the DNA strand and is proceeding actively. Accordingly, the early part of the newly synthesized DNA is very likely to carry errors. Providing a primer increases substantially the efficiency of binding and the accuracy of copying by DNA polymerase. The RNA nature of the primer identifies it as an initially synthesized nucleic acid segment, which is prone to errors. These primers are easily distinguished from double-stranded DNA by enzymes that remove them. The gap resulting from removal of the

RNA primer has a double-stranded DNA segment to act as a primer for filling in the gap. The overall accuracy of copying is thus preserved.

Restoration of supercoiling

After the two strands are copied, the supercoiling must be restored. An enzyme that helps to rewind the DNA into its supercoiled form is DNA gyrase, the main target of a class of antibiotics called fluoroquinolones (Fig. 5.3). Fluoroquinolones bind to DNA gyrase and prevent it from restoring the right degree of supercoiling. This kills the bacterium by preventing orderly completion of DNA synthesis. Bacteria become resistant to fluoroquinolones by mutating their DNA gyrase so that it no longer binds the fluoroquinolone.

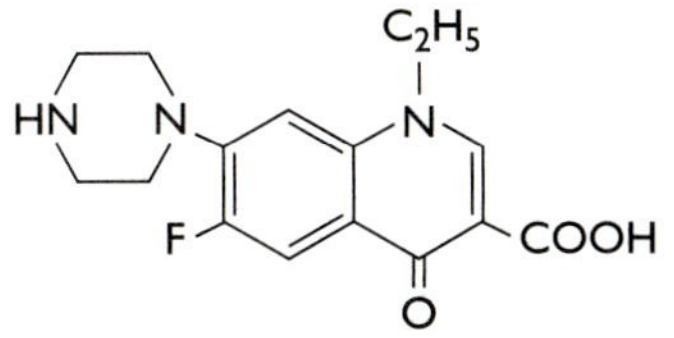

C_2H_5 HN N N F COOH O

Figure 5.3 Structure of norfloxacin, a quinolone. Quinolones bind to DNA gyrase and prevent it from restoring the correct amount of supercoiling to newly synthesized DNA, thus killing the bacterium.

Bacterial cell division

Synthesis of the chromosome begins at a specific site, the **replication origin,** and requires several specific proteins to get the process started. In the case of a circular bacterial chromosome, replication occurs in both directions away from the replication origin. About halfway around the chromosome are a series of **termination sites,** where the two newly synthesized copies of the halves of the chromosome are joined and the two chromosomes separate. Because linear bacterial chromosomes have been discovered only recently, nothing is known about their synthesis. During cell division, DNA replication occurs at specific sites on the cytoplasmic membrane, where the chromosome is tethered. As the cell elongates, the two copies are pulled apart. The attachment of the chromosome copies to specific sites on the cytoplasmic membrane helps to insure that each daughter cell gets one copy of the chromosome. Recent studies indicate that this separation may involve an active, motor-driven process that is a primitive version of the process seen in eukaryotic cell division. Division by binary fission allows bacterial populations to expand very rapidly if conditions for division are favorable. *Escherichia coli,* for example, can complete a division in 20 minutes if sufficient nutrients are available. The time it takes a bacterium to divide is called the **generation time.**

A bacterium coming into a new environment—a flask of laboratory medium, for instance—must first adapt to the new conditions. Chapter 6 describes this process of adaptation. In this phase called the **lag phase** (Fig. 5.4), bacteria do not divide, and their numbers stay constant. Once they have adapted, assuming that abundant nutrients are provided, they enter a phase of rapid division by binary fission, and their numbers increase quickly. You could say that bacteria are creatures that multiply by dividing. This phase of rapid division is called the **exponential phase,** because the number of bacteria in the site rises exponentially in powers of two. How quickly bacterial numbers can rise to high levels during this type of growth is illustrated by the calculation shown in Figure 5.4. A clinical example of this type of growth is a urinary tract infection, in which bacteria that have entered the bladder can grow exponentially with short generation times on the nutrient-rich urine stored in the bladder. People who contract urinary tract infections experience very rapid onset and worsening of symptoms, which are the direct consequences of rapid bacterial division.

At some point, even in the rich medium provided by generous scientists, one or more nutrients run out, or end products rise to inhibitory levels and cell division ceases. This stage, when bacterial numbers remain relatively constant, is called **stationary phase.** After a time without sufficient nutrients, bacteria may begin to lyse and the number of bacteria will begin to decrease. This stage is called **death phase.** Recently, scientists studying the action of the penicillin family of antibiotics have become quite interested in death phase. Little is known about what goes on during death phase, but one feature is the activation of enzymes called **autolysins.** Autolysins, as the name suggests, degrade peptidoglycan and thus contribute to lysing the bacterial cell. Recent findings indicate that one effect of the penicillin family of antibiotics, in addition to preventing the formation of peptidoglycan cross-links (Chapter 4), is to stimulate, either directly or indirectly, the activity of autolysins and thus hurry the bacteria into death phase.

Scientists see the type of growth illustrated in Figure 5.4 in the laboratory, but this pattern of growth is probably uncommon in nature, especially in the external environment. In most natural settings, low nutrients or other adverse conditions limit cell division. The water of the open ocean, for example, is poor in nutrients. In terrestrial environments and in sediments of rivers or oceans, nutrients may come and go, leading to short spurts of growth resembling exponential phase, separated by long periods

Figure 5.4 Growth curve of bacteria in laboratory medium. When bacteria find themselves in a new environment, they must first adapt to the conditions before they start dividing (lag phase). After they have adapted, bacteria divide rapidly by binary fission, which results in an exponential rise in the number of bacteria (exponential phase). Once nutrients become limiting, the bacteria stop dividing (stationary phase). After a period of starvation, the bacteria may begin to lyse, and the number of cells in the culture declines (death phase).

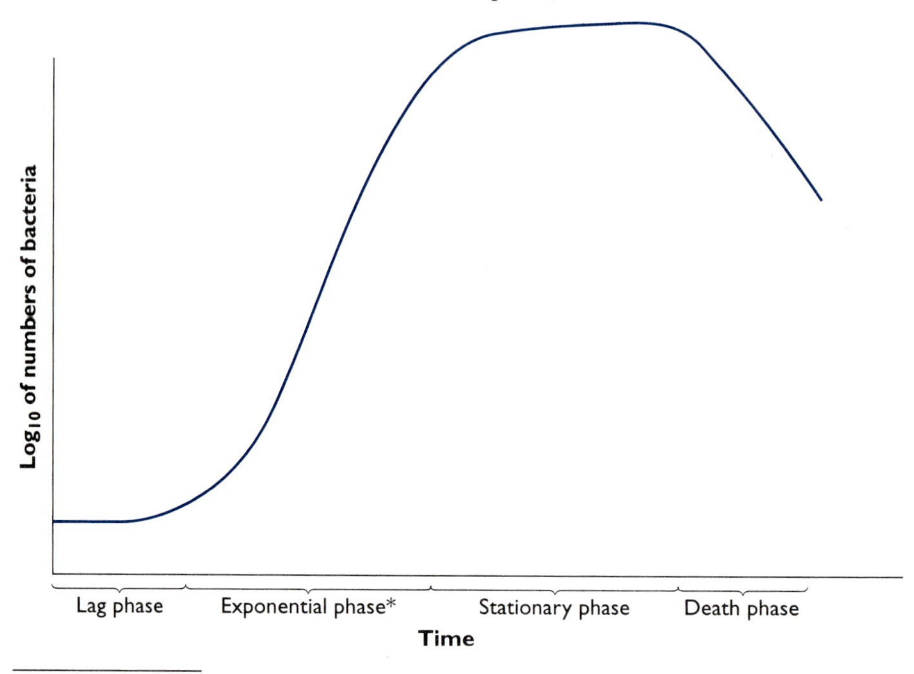

*The increase in cell number that occurs during exponential growth is defined by the formula $N = N_0 2^n$, where N = current number of bacteria, N_0 = initial number of bacteria, and n = number of divisions. For example, if $N_0 = 10$ bacteria, after 20 divisions the culture will contain 1.05×10^7 bacteria ($N = 10 \times 2^{20}$). If bacteria divide every 20 minutes, the process of going from 10 to 10 million bacteria will take only 20 × 20 minutes, or 6.6 hours.

of starvation. Scientists have seen this phenomenon at the bottom of the sea, when a dead whale or other large marine animal has died and sunk to the bottom. Away from coastal regions, the sea bottom is deficient in macrobial and microbial life because of the lack of nutrients. On and around a dead whale, however, communities of microbes and larger animals bloom until the carcass is completely degraded. The microbes obviously were there to begin with but were not actively growing until transient nutrients became available.

The type of growth shown in Figure 5.4 has been useful to scientists studying bacteria, because high population densities can be generated in a relatively short period of time. It is important to keep in mind, however, that the growth conditions used in laboratory studies are not necessarily similar to those that the bacterium evolved to cope with in a natural setting.

Asymmetrical cell division during sporulation

Normal division of actively growing bacterial cells is a symmetrical process. Some bacteria, in response to adverse conditions, can also carry out a form of asymmetrical cell division that results in production of a spore. **Spores** are very tough survival forms that can later develop into **vegetative cells** when conditions improve. Spores survive boiling; for this reason, laboratory media must be sterilized at 220°C to kill spores that might be in media components. When canning foods, it is also necessary to achieve this level of heat to eliminate such dangerous spore-forming bacteria as *Clostridium botulinum,* the cause of botulism. Although spore-formers are common in soil and water environments, only a few of the medically important bacteria form spores. Species of *Clostridium* cause **gas gangrene, botulism,** and **tetanus.** *Bacillus anthracis,* another spore-former, is the cause of **anthrax.** Anthrax, a rapidly fatal disease in people who develop the pulmonary version of the disease, has recently been appropriated by terrorist groups. The fact that spores survive for years and can be stored dry makes this bacterium particularly attractive as a potential weapon. Fortunately, despite several attempts, bioterrorists have yet to succeed in using anthrax spores to kill people. Anthrax was once an important disease of farm animals, causing high mortality. One of Pasteur's first vaccines protected animals against anthrax. Today, antianthrax vaccines have virtually eliminated anthrax from animal populations in developed countries.

The process of sporulation is illustrated in Figure 5.5. The chromosome is duplicated, and the copies separated. Only one copy will end up in the spore. Some spore-forming bacteria produce spores at one end of the cell (as shown in Fig. 5.5). Others produce them in the middle of the cell. Because the spore is created inside the bacterium, which then lyses and releases the spore, such spores are called **endospores.** Fungi produce spores externally (exospores). The spore consists of the normal contents of a bacterial cell, stabilized with a high concentration of calcium dipicolinate and covered with layers of densely packed peptidoglycan and protein (the spore coat) that can withstand many environmental challenges, such as high heat, drying, and ultraviolet (UV) light. Spores are smaller and more dense than vegetative cells. They are also metabolically inactive. The process by which a spore is made has been well studied, but little is known about the process by which the spore is converted into a vegetative cell or what signals tell the spore that it is time to begin this process.

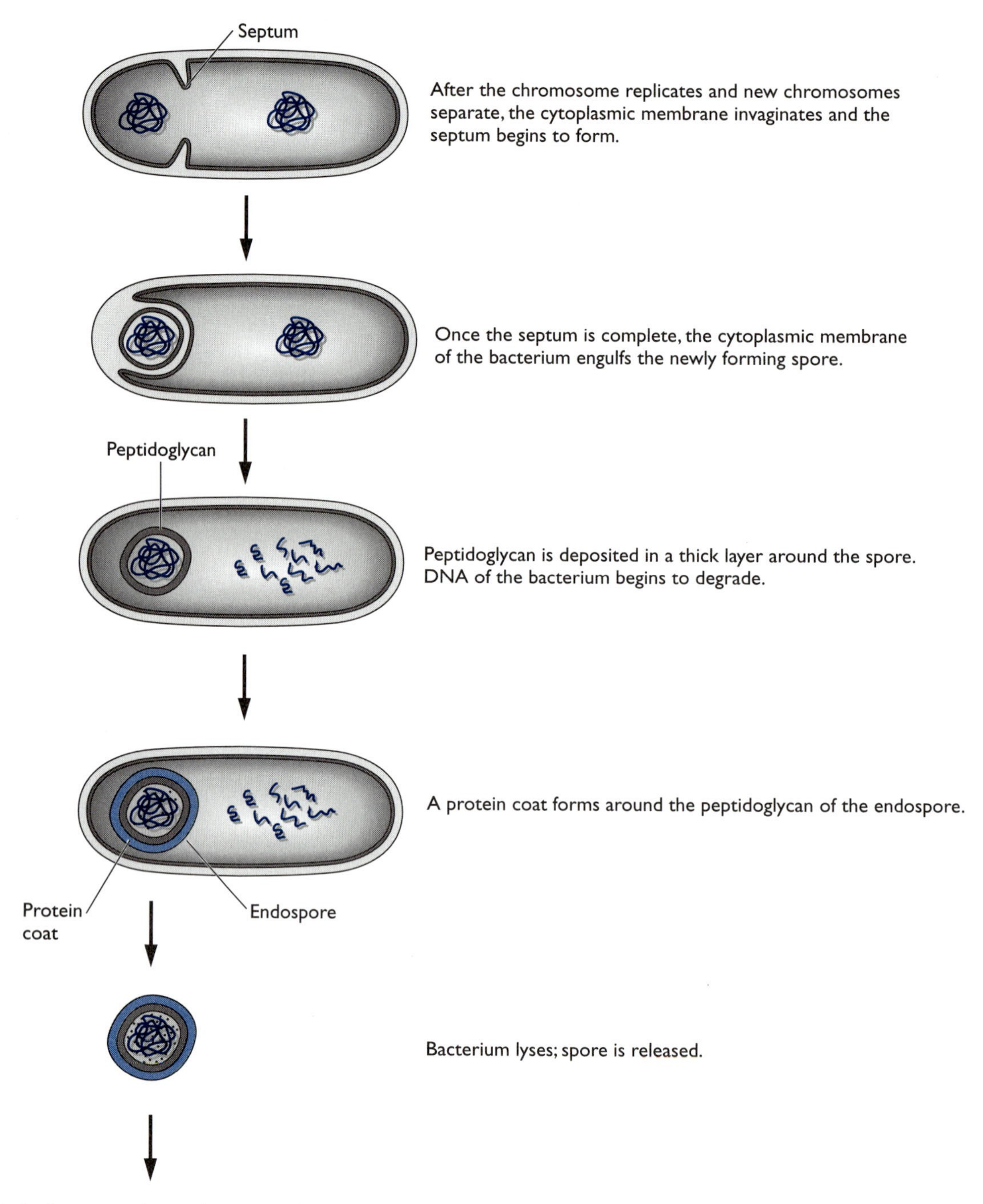

Figure 5.5 Sporulation in bacteria. The chromosome is replicated, and the daughter chromosomes separate. A septum begins to form around one chromosome, eventually walling off a portion of the cell. Next, a thick layer of peptidoglycan and a layer of protein are laid down around the septum, forming an endospore. The cytoplasm inside the endospore contains the normal contents of a bacterial cell plus a high concentration of calcium dipicolinate, which probably contributes to its resistance to heat and drying. The spore is then released from the cell by lysis of the cell.

Correcting Mistakes and Healing Lesions

Damage to DNA

Microbes that lived near the surface of the ocean or on the surfaces of land masses of early Earth had a serious problem: damage to DNA as a result of punishing radiation from sunlight bombarding the planet's surface. Even after the ozone layer formed, radiation still presented a problem, especially for microbes that lived in regions where intense sunlight was the rule. UV radiation kills bacteria by causing breaks in DNA. UV radiation chemically cross-links pyrimidine bases to produce **pyrimidine dimers** (Fig. 5.6). Because these dimers distort the structure of DNA, they can cause DNA polymerase to insert the wrong base when it passes through the region. In fact, scientists use UV radiation to induce the formation of mutations in bacteria. To avoid adverse consequences of dimer formation, bacteria remove pyrimidine dimers and surrounding bases from the affected DNA strand. Unless the bacteria repair such gaps rapidly, each gap will become a double-stranded break when the next replication fork moves through the region (Fig. 5.7). A fragmented chromosome can no longer replicate, and the bacterium will die. Other types of radiation, such as ionizing radiation, also cause breaks in DNA but by a different mechanism. The result, however, is the same: a fragmented chromosome that makes the bacteria nonviable. Today, the food industry is putting this knowledge to use by irradiating food to kill bacteria. In surgical theatres, UV lamps are sometimes suspended over the operating table during long procedures to reduce the bacterial load in the air and on the patient's skin.

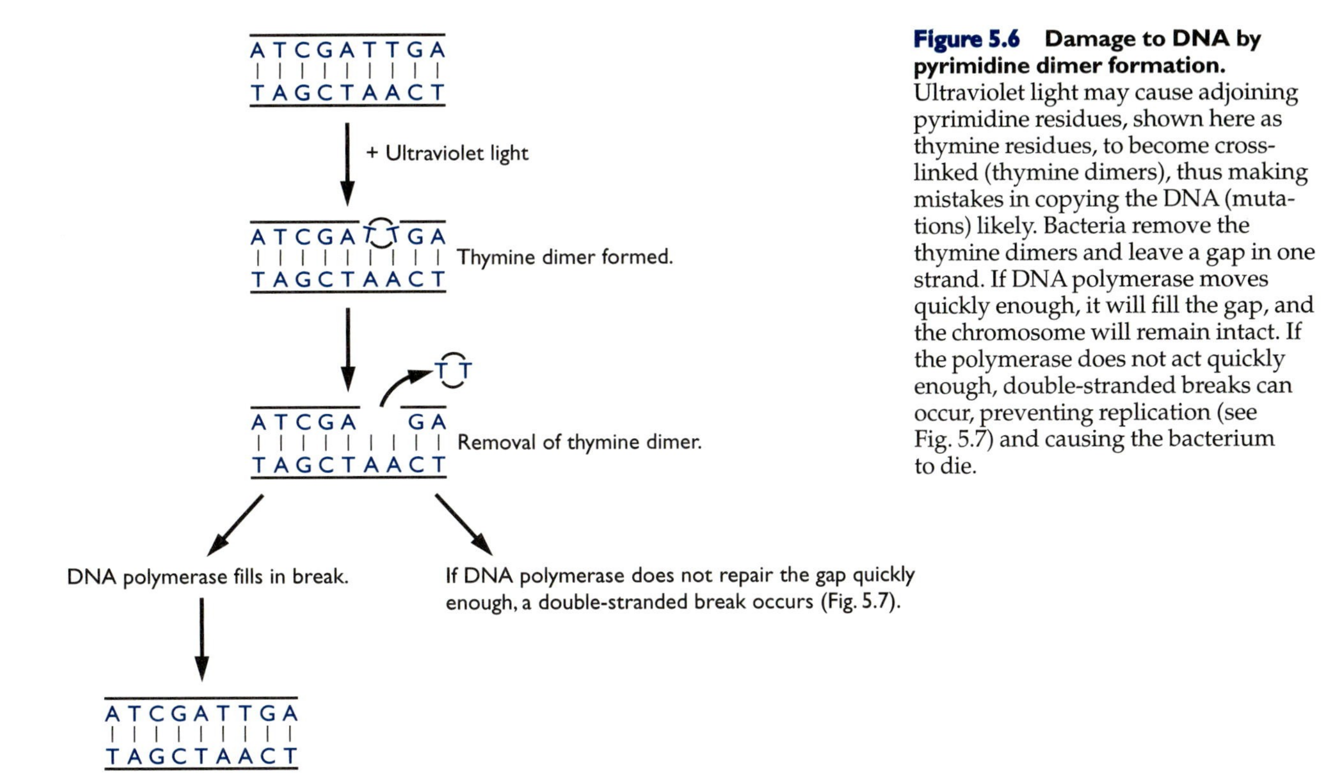

Figure 5.6 Damage to DNA by pyrimidine dimer formation. Ultraviolet light may cause adjoining pyrimidine residues, shown here as thymine residues, to become cross-linked (thymine dimers), thus making mistakes in copying the DNA (mutations) likely. Bacteria remove the thymine dimers and leave a gap in one strand. If DNA polymerase moves quickly enough, it will fill the gap, and the chromosome will remain intact. If the polymerase does not act quickly enough, double-stranded breaks can occur, preventing replication (see Fig. 5.7) and causing the bacterium to die.

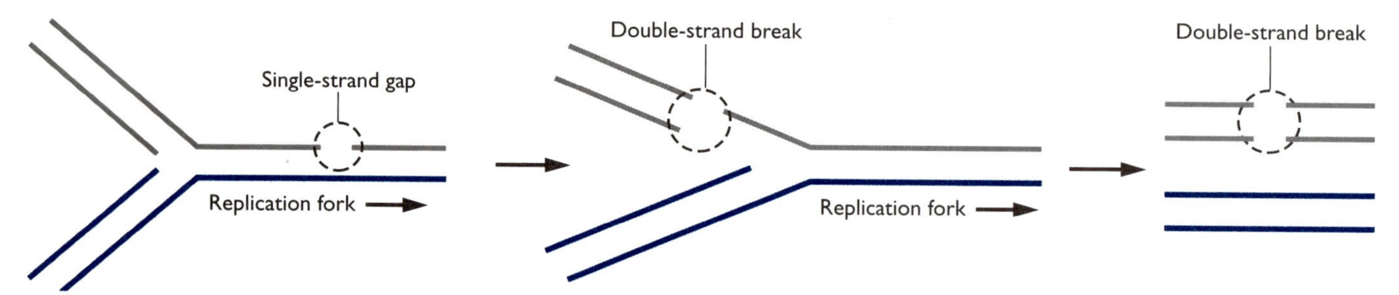

Figure 5.7 How a single-strand gap can become a double-stranded break in the DNA. As a replication fork moves through an area that contains a gap in one strand of the DNA, the DNA polymerase will not synthesize new DNA opposite the gap. Even though replication may resume on the other side of the gap, a double-stranded break has been introduced into one copy of the chromosome. This phenomenon can occur in either the leading or lagging strand.

Homologous recombination

Bacteria can repair double-stranded breaks in their DNA through a process called **homologous recombination.** Homologous recombination is only a part of a more complex DNA repair system, but it plays a major role in DNA repair, especially in repair of lesions created by radiation. Homologous recombination also plays a role in the acquisition of new traits by bacteria and is used by scientists in genetic engineering. Accordingly, the focus here will be on homologous recombination to the exclusion of other DNA repair mechanisms.

Homologous recombination is illustrated in Figure 5.8. Two segments of DNA, identical in sequence, can interact in a crossover event so that one DNA segment replaces the other. Rec proteins mediate the annealing and crossover reactions shown in Figure 5.8A. The protein RecA is responsible for the initial annealing step, in which identical DNA segments are paired. An event in which a DNA segment enters the chromosome, but which does not involve DNA similarity or RecA, is called a RecA-independent process. Such processes will be described in Chapter 7. Although bacteria can live in the laboratory without this enzyme, they become exquisitely sensitive to UV light and to chemicals that cause breaks in DNA.

The mechanism by which homologous recombination can heal double-stranded breaks in DNA is shown in Figure 5.8B. Homologous recombination replaces the double-stranded break with two single-stranded gaps. Note, however, that if a replication fork sweeps through the region before the single-stranded gaps can be repaired, a bacterium using this strategy risks loss of the other copy of the chromosome. The more radiation experienced by a bacterium, the less the chance that the repair system can repair new breaks. Some bacteria cope with radiation much better than others (Box 5–1).

An important role of homologous recombination is to help bacteria acquire new genes. Although the homologous recombination system requires virtually identical segments of DNA to make crossovers, these identical regions may flank regions containing different sequences from that of the bacterium's original DNA. If so, these sequence differences may be incorporated when crossovers occur on either side of this region. An example of this type of adaptation is provided by *Streptococcus pneumoniae,*

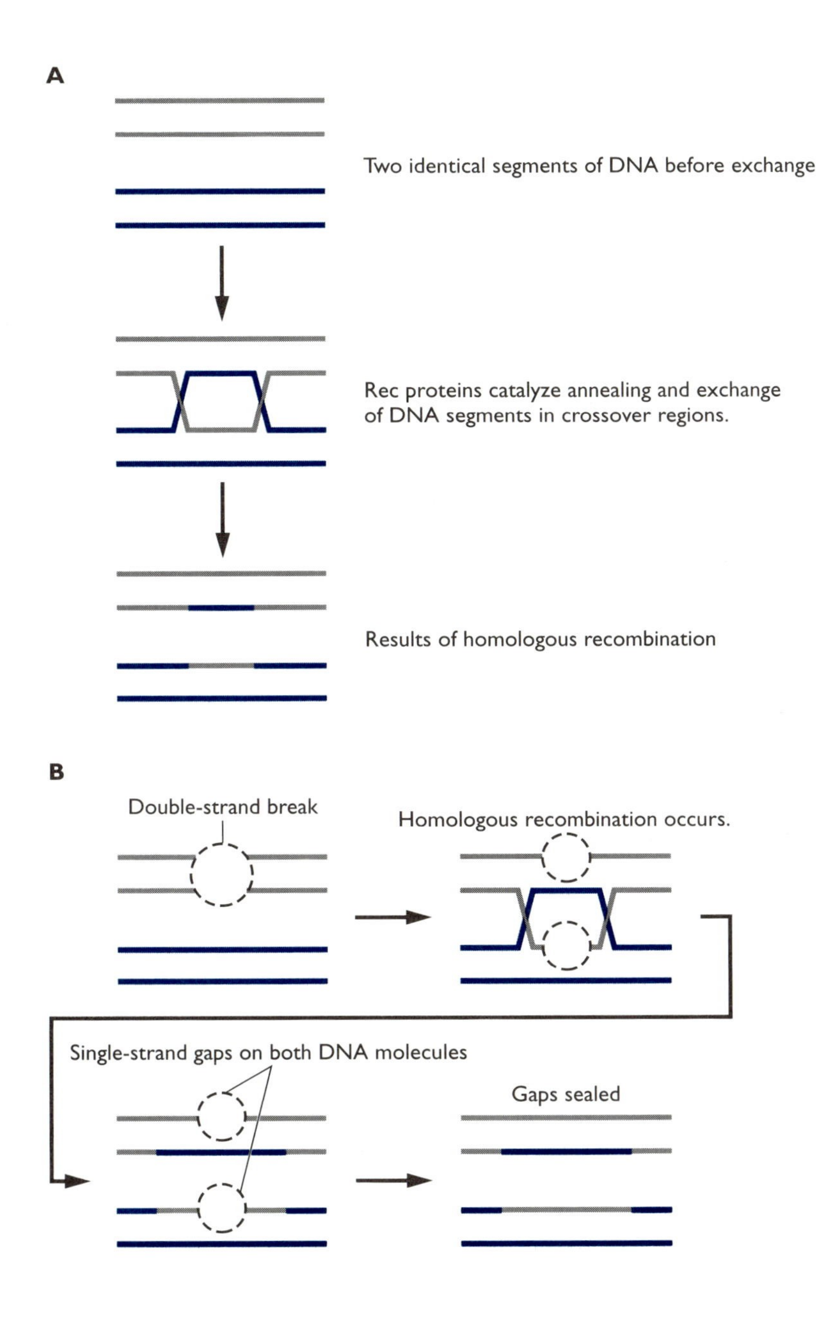

Figure 5.8 Homologous recombination. (A) By the process of homologous recombination, two segments of DNA with identical sequences can exchange regions. This exchange is initiated when part of one strand of a DNA molecule invades a second DNA strand and vice versa. This realignment of strands forms a crossover. The Rec proteins catalyze the annealing and exchange of portions of the two strands of DNA. The final result is that each of the two original DNA segments now has a new segment that it obtained from the other DNA segment. If the segments are completely identical, there will be no change in the two DNA strands. But if the two DNA segments contain regions with sequence differences flanked by regions of identity, where the crossovers occur, the two resulting DNA molecules will have exchanged the variable region. **(B)** The double-stranded break can be repaired by homologous recombination. After the recombination event occurs, each double-stranded DNA molecule will have a single-stranded gap that can be sealed by DNA polymerase.

the most common cause of bacterial pneumonia. Some strains of *S. pneumoniae* have become resistant to penicillin by mutating their penicillin-binding proteins, proteins involved in cell wall synthesis that are inactivated by penicillin (Chapter 4). Strains of *S. pneumoniae* that are susceptible to penicillin can acquire DNA from resistant strains by taking it up from the environment, a process called transformation, which will be described in Chapter 7. Once in possession of DNA from a resistant strain that has a mutant penicillin-binding protein, the susceptible strain uses homologous recombination to replace its own penicillin-binding protein with the mutant, resistant form and thus becomes resistant itself (Fig. 5.9A). The DNA sequences of the different penicillin-binding proteins are identical enough in some regions to allow homologous recombination to occur, leading to incorporation of dissimilar regions that lie between the identical

box 5–1 Extreme Radiation Resistance

Some bacteria are capable of winning the race to repair DNA strand breaks, even in the face of very high levels of radiation. In fact, bacterial biofilms have been found growing on the surfaces of spent nuclear fuel rods stored in nuclear reactors. An example of a bacterium that is unusually resistant to ultraviolet and ionizing radiation is *Deinococcus radiodurans*. This hardy, gram-positive bacterium can survive doses of radiation that would kill other living things. Its ability to do this arises from an unusually active DNA repair system that includes an extremely effective homologous recombination system. *D. radiodurans* is of interest to scientists developing new bioremediation strategies because it is one of the few microbes that can survive the radiation found in sites contaminated with high levels of radioactive substances. Cleaning up pollutants at such sites manually is quite expensive, because of the protective gear needed for clean-up workers. Seeding a site with bacteria, which then proceed to carry out the work of decontamination, has both economic and safety advantages over the current process of removing large amounts of contaminated soil by hand or treating soil on the site with repeated applications of chemicals. Scientists now are using cloning to introduce genes into *D. radiodurans* that would allow this bacterium to degrade toxic pollutants, such as trichloroethylene and chlorobenzene.

Source: Lange, C. C., L. P. Wackett, K. W. Minton, and M. J. Daly. 1998. Engineering a recombinant *Deinococcus radiodurans* for organopollutant degradation in radioactive mixed waste environments. Nature Biotechnol. 16:9229–9233.

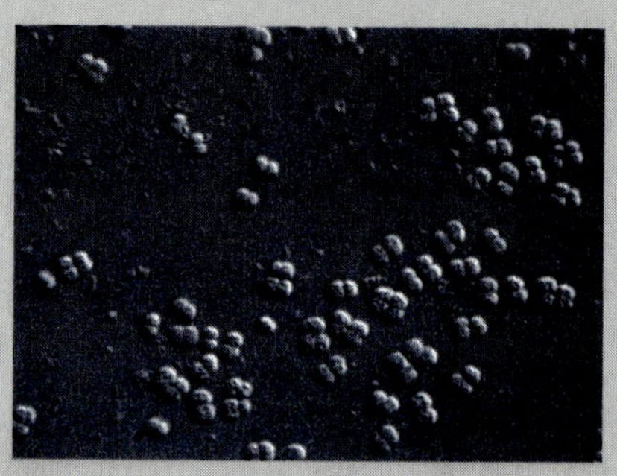

Deinococcus radiodurans. This bacterium can survive heavy doses of radiation, because it has a very active DNA repair system.

regions. The new hybrid gene encodes a penicillin-binding protein that no longer binds penicillin. Figure 5.9B illustrates the ways in which genes encoding the same penicillin-binding proteins can vary in different *S. pneumoniae* strains. Scientists use similar strategies to change genes in a microbe under study (Chapter 7).

In Chapter 18, yet another example of the ways in which bacteria use homologous recombination to their advantage will be described. In this case, the bacterium that causes gonorrhea uses homologous recombination to change its surface proteins so that it can evade the killing action of the immune system.

Plasmids and Plasmid Replication

Characteristics of plasmids

Plasmids are DNA segments that replicate autonomously. Although plasmids use DNA polymerase and some other host cell proteins to replicate themselves, they have their own origin of replication and sometimes provide proteins that participate in replication. For some plasmids, the process of replication is **bidirectional,** like that of chromosomal replication. That is, replication begins at the replication origin and proceeds in both directions to form an expanding bubble. Other plasmids use a **rolling circle** type of replication, in which DNA polymerase proceeds unidirectionally from the origin of replication around the plasmid to terminate DNA synthesis at the replication origin.

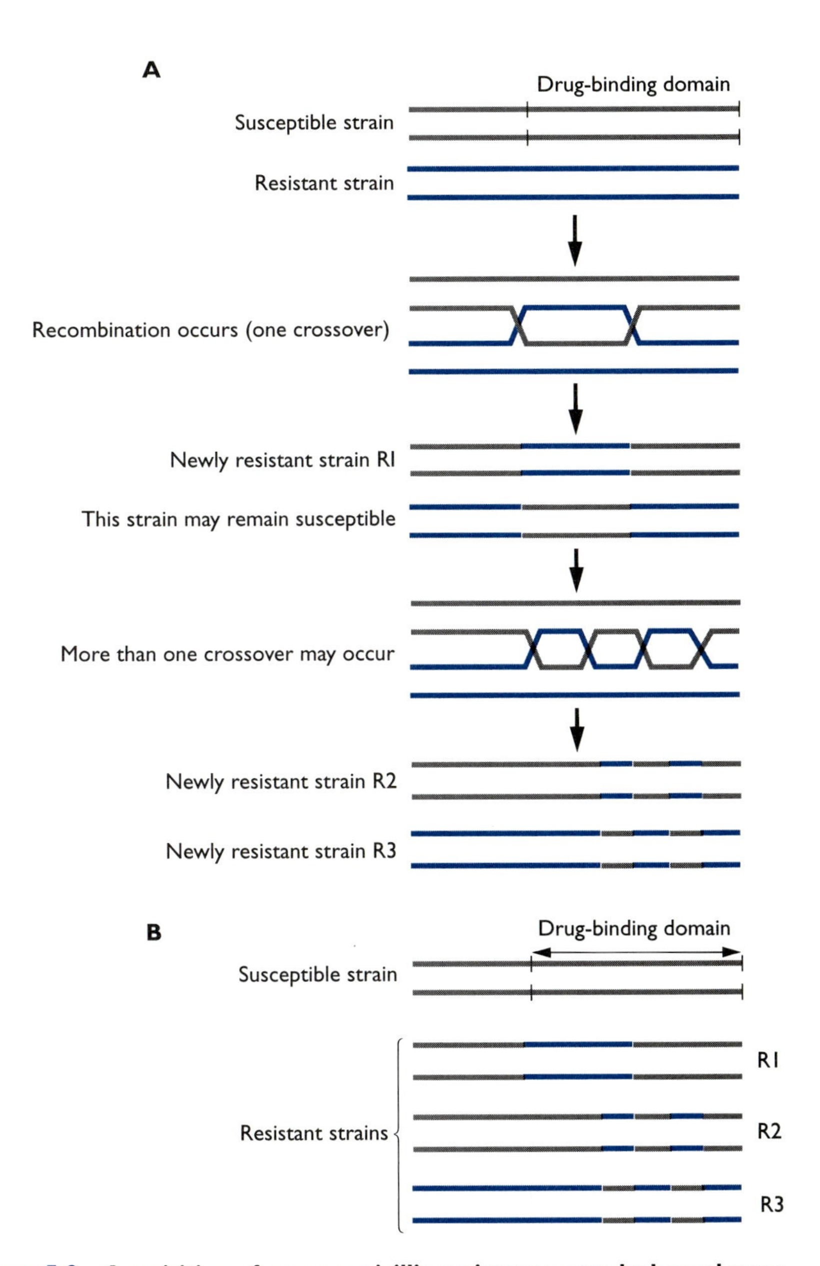

Figure 5.9 Acquisition of a new penicillin resistance gene by homologous recombination in *Streptococcus pneumoniae*. *S. pneumoniae* can become resistant to penicillin by acquiring a gene for a mutant penicillin-binding protein. Different resistant strains of *S. pneumoniae* have genes for penicillin-binding proteins that differ slightly in sequence from susceptible strains. **(A)** Homologous recombination between DNA of a susceptible strain and that of a resistant strain can introduce mutations into the gene in the susceptible strain. This may not result in the susceptible strain becoming resistant. It may require several homologous recombination events to make the susceptible strain resistant. For the sake of simplicity, multiple crossovers are shown together, but they probably occurred as separate events. **(B)** Analysis of a series of resistant strains shows that a variety of homologous recombination events have occurred to produce new proteins that resist binding by penicillin.

Because plasmid replication is usually not tied to replication of the chromosome, there can be multiple copies of the same plasmid in a cell. The number of plasmid copies per cell is called the **plasmid copy number.** In bacteria isolated from nature, plasmid copy numbers are generally low (1–10 copies per cell). Plasmids have a variety of ways of controlling their copy numbers. Control of copy number is probably a way to prevent the process of plasmid maintenance and replication from imposing too much of an energy burden on the bacterial cell. For use in the laboratory, scientists have obtained mutant plasmids that have very high copy numbers (hundreds of copies per cell). Such plasmids are lost easily unless there is a selective pressure that maintains them in the cell.

Gene cloning: an application of plasmid characteristics

Cloning vectors. Plasmids provide the basis for **gene cloning.** A plasmid used in cloning is called a **cloning vector.** The process of cloning introduces genes of interest into the cloning vector. The cloning vector can then be used to make many copies of the cloned gene in *E. coli* or some other bacterial host. Cloning vectors usually are high-copy-number plasmids. Plasmid cloning vectors are small, about 5–15 kbp in size. This trait makes it easy to separate them from chromosomal DNA when the bacteria are lysed to release the plasmid. Isolation of the cloning vector is the first step in a cloning experiment.

Isolation of plasmid DNA from chromosomal DNA is called a plasmid preparation procedure. One such procedure is shown in Figure 5.10. Bacteria are lysed with detergent, releasing the DNA. Addition of salts precipitates chromosomal DNA, which is much larger and thus precipitates at a lower salt concentration than the plasmid DNA. The still-soluble plasmid DNA is collected. The solution containing the plasmid is then passed through a column or otherwise treated to concentrate the plasmid and separate it from cell proteins, some of which can degrade the plasmid. A widely used cloning vector is shown in Figure 5.11. The cloning vector has a region containing many restriction sites that are recognized by different restriction enzymes. This is the region into which the DNA to be cloned will be

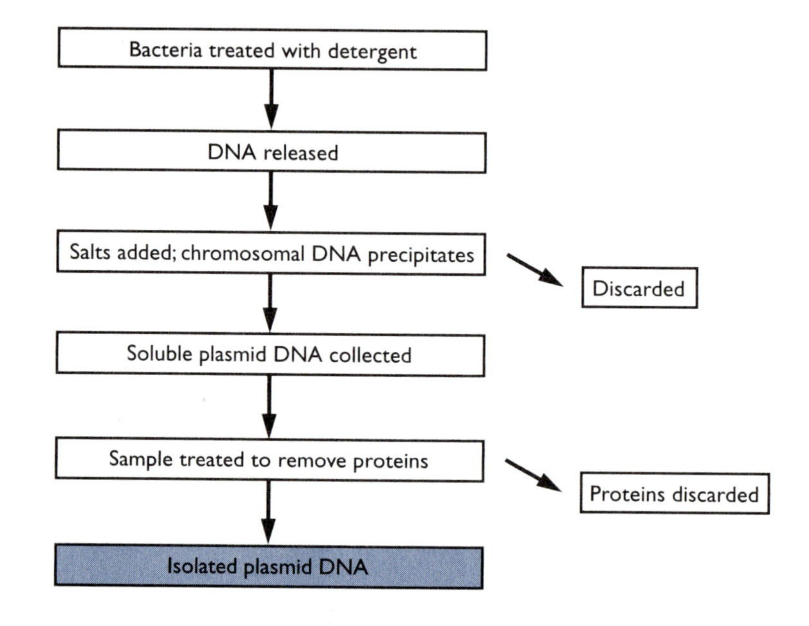

Figure 5.10 Isolation of plasmid DNA. Bacteria that have been harvested from a culture are treated with detergent to lyse them. The material that is released contains proteins, chromosomal DNA, and plasmid DNA. The next step is to increase the salt concentration, which will cause the much larger chromosomal DNA molecules to precipitate, but the plasmid DNA and proteins will remain soluble. The soluble portion is treated to remove the proteins, leaving the plasmid DNA in solution.

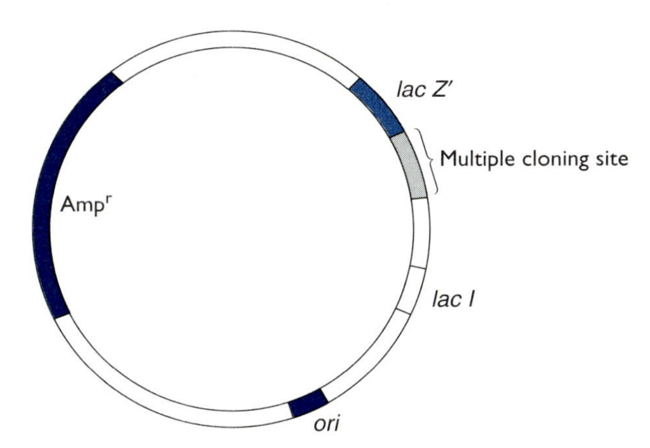

Figure 5.11 pUC19, a widely used cloning vector. The multiple cloning site contains a variety of restriction sites. The origin of replication (*ori*) and the gene encoding ampicillin resistance (Amp^r) also are indicated. The *lac Z'* gene encodes part of an enzyme that makes a colorless compound in the medium turn blue. If DNA has been inserted in the multiple cloning site, this enzyme is not made. This blue/colorless screen helps to identify vectors that contain inserted DNA (colorless colonies).

inserted. Note that the cloning vector also contains a gene that confers resistance to ampicillin. Such selectable markers are necessary in later steps of cloning to determine which cells, in this example *E. coli,* received a copy of the cloning vector.

The cloning process. The cloning process is illustrated in Figure 5.12. The cloning vector is first cut by a restriction enzyme that cuts once within the plasmid. The restriction enzyme step produces a linearized form of the cloning vector. The DNA to be cloned is also digested with the same restriction enzyme. In the example shown in Figure 5.12, the object is to clone gene X from the chromosomal DNA of a bacterium. For the scientists attempting to make *D. radiodurans* capable of degrading trichloroethylene (Box 5–1), gene X would be the gene or genes that encode enzymes capable of detoxifying trichloroethylene. Digestion of chromosomal DNA with the same restriction enzyme used to digest the plasmid produces a mixture of linear chromosomal DNA fragments. The restriction enzyme is inactivated, and the linear vector is mixed with the chromosomal fragments. In this case, because the restriction enzyme used leaves three-base single-stranded segments at the ends of each fragment, these "**sticky ends**" will hybridize with complementary ends of other fragments. The desired outcome is for the sticky ends of the chromosomal DNA fragments to hybridize with the sticky ends of the cut plasmid DNA but, of course, this is not the only possible outcome. Some of the chromosomal fragments will hybridize with each other to make circles of chromosomal DNA. Similarly, plasmid sticky ends can hybridize with each other, restoring the original plasmid. After hybridization is complete, DNA ligase is added to seal the nicks, producing intact double-stranded circular molecules.

This ligated mixture is then introduced into *E. coli* by a process called **transformation.** Laboratory transformation is a process by which DNA is forced into *E. coli* by high salt conditions and heat shock (**chemical transformation**) or by a transient high electric field (**electroporation**). In Chapter 7, a natural form of transformation will be described, but, because most bacteria are not naturally transformable, they must be forced to take up DNA. The scientist doing the cloning wants the plasmid and anything that has been ligated

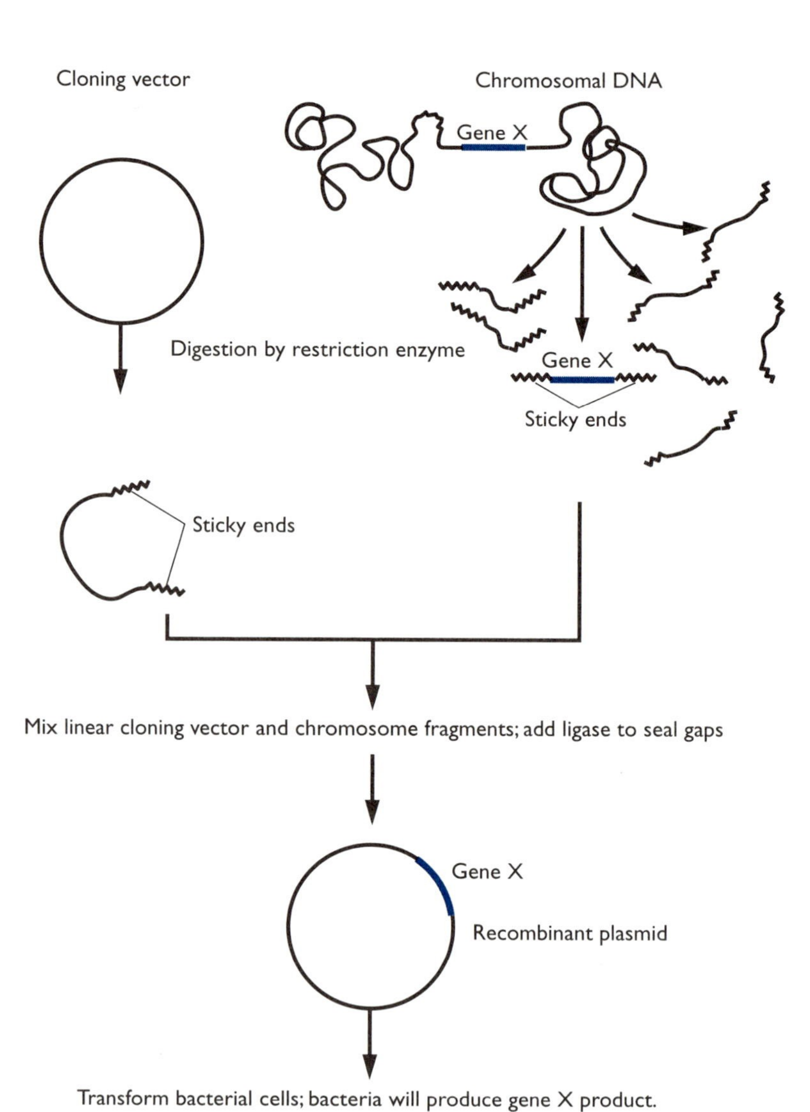

Figure 5.12 The cloning process. The plasmid cloning vector and chromosomal DNA (which contains the gene to be cloned, gene X) are digested by the same restriction enzyme. This causes the cloning vector to become linear. Chromosomal DNA, containing gene X, is also cleaved. The restriction enzymes are inactivated, and the linear vector and chromosomal fragments are mixed. If the restriction enzyme used produces sticky ends, the sticky ends of some of the chromosomal fragments will hybridize with complementary sticky ends of the vector. DNA ligase is added to seal the gaps. The intended result is a recombinant plasmid containing gene X that can be used to transform a bacterial cell. Selection for the antibiotic resistance gene by including antibiotic in the medium yields only cells that contain the plasmid. In reality, a large percentage of the linear cloning vectors re-anneal to form a circular plasmid that does not contain the gene of interest. This is why it is necessary to select or screen transformants for clones that have the desired gene.

into it, not ligated chromosomal fragments that are not in a plasmid. Because only the plasmid can replicate in *E. coli,* recipient bacteria that pick up chromosomal fragments will quickly lose these nonreplicating fragments during division.

Selection and screening for the desired clones. Chemical transformation and electroporation are not 100% efficient. Thus, only a minority of the *E. coli* cells exposed to the transformation process will receive a plasmid. To select for this minority, another feature of the cloning vector is brought into play: the **selectable marker.** The selectable marker usually is an antibiotic resistance gene. Cells that inherited the plasmid become resistant to the antibiotic and can be selected by plating the bacteria on agar plates containing the antibiotic that selects for bacteria carrying the resistance gene. Not all selected cells will contain a plasmid that has picked up a cloned insert. If the desired gene X has a selectable phenotype (for example, antibiotic resistance), it is easy to find the cells in which the plasmid that contains gene X resides by selecting for the phenotype conferred by gene X. In the case of genes encoding enzymes that allow a bacterium to degrade trichloroethylene, having these genes might allow *D. radiodurans* to grow on medium that contains trichloroethylene. If not, it is necessary to screen a number of colonies to find the one with the cloned X gene. A color reaction that detects trichloroethylene degradation, allowing colonies containing the cloned genes to be differentiated from colonies that contain only the cloning vector, is an example of a screening strategy. The cloning vector shown in Figure 5.11 allows a similar sort of screening. If the plasmid contains no cloned DNA, the enzyme encoded by *lac* Z' makes the colonies blue when the enzyme's substrate (X-gal) is provided in the medium. If DNA has been inserted in the plasmid, *lac* Z' product is not made and the colonies are colorless. Many tricks are used in screening, but these are beyond the scope of this text. The gene X ligated into the plasmid cloning vector is said to be cloned, because now many copies are made by the bacterium that contains it.

Uses of cloning. Why would a scientist want to clone a gene? Providing a new trait for a bacterium is a common use of cloning. Another common goal is to induce bacteria to produce large amounts of the protein encoded by the gene. An example is human insulin. Human insulin is obviously safer and more likely to be effective for human diabetics than insulin from another animal species. Getting large quantities of insulin from humans is quite difficult, making the product very expensive. Once the human insulin gene was cloned, scientists were able to induce bacteria to make the protein, and human insulin became abundant and inexpensive enough for regular use by diabetics. Proteins used in vaccine preparations are now being produced in a similar way. This process has safety implications as well. Growing large amounts of a strain of *E. coli* that produces surface proteins of the influenza virus is obviously much safer than growing and processing large quantities of the virus itself. Scientists who work on various aspects of HIV biology often use cloned segments of the viral genome rather than live viruses.

The importance of plasmids for bacteria

Plasmids clearly are important to humans; they have made many scientific advances possible. But why are plasmids important to bacteria? Plasmids are a useful vehicle for acquisition of new traits. For example, soil bacteria acquire plasmids that allow them to utilize a new carbon source. Bacteria that cause human disease acquire plasmids that make them resistant to antibiotics or other toxic substances.

Plasmids sometimes persist, not because they are doing the bacterium good, but because they have a strategy for insuring that they will be maintained in the bacterium. For example, some plasmids produce both a toxic protein that can kill the bacterium and a protein that acts as an antidote to the toxin. If one of the daughter cells in a division event does not receive a copy of the plasmid, it still receives copies of the toxic protein and the antidote protein. The antidote protein, however, is shorter-lived than the toxic protein, so in the absence of constant production of antidote from the plasmid-encoded gene, the daughter cell is killed. Such plasmid strategies for maintaining themselves have been called **plasmid addiction systems.**

Plasmids can be transferred from one bacterium to another, by a process that will be described in Chapter 7. This type of DNA exchange among bacteria is called **horizontal gene transfer.** Horizontal transfer of plasmids and other types of DNA is proving to be much more common in nature than scientists had suspected previously. Because these transfer events can occur between bacteria belonging to different species and genera, bacteria can sample a huge range of traits, keeping the ones that are beneficial.

Other Applications of Information About DNA Replication

DNA hybridization and Southern blotting

The fact that two strands of DNA bind tightly and specifically to each other if they share complementary sequences has made possible **DNA hybridization,** a powerful technique now used widely in all areas of biology. To see how DNA hybridization can be used, consider a scientist who has a collection of bacterial isolates and wants to know which one carries a particular antibiotic resistance gene, gene A. Testing the strains for growth on a medium that contains the antibiotic does not necessarily answer the question, because resistance to the antibiotic could be the result of any of a number of different genes. Also, some bacterial isolates may not yet be making enough of the resistance protein to cause them to be resistant to the level of antibiotic being tested.

Spot blot. To determine which bacterial isolates contain gene A, the scientist first obtains DNA from each of the isolates in the collection, heats the DNA to boiling to separate the DNA strands, and attaches the single-stranded DNA to special paper, which will hold the DNA in place even when the paper is washed (Fig. 5.13). Such paper is called a membrane, and the DNA applied to a small section of the membrane is called a **spot** or **dot blot.** The scientist then takes a portion of gene A and labels it with a radioactive tag or dye to create a **DNA probe.** The DNA probe is made

Figure 5.13 Spot (or dot) blot. DNA from members of a group of bacterial isolates is tested to determine whether each contains gene A. The DNA from each isolate is heated to separate the DNA strands. A sample of each of the treated DNA samples is added as a small spot to a membrane (spot blot). The resulting blot is covered with a solution that contains a labeled single-stranded DNA probe that is complementary to gene A. The blot is washed to remove any unbound probe. The membrane is treated so that the probe can be visualized. Spots that are positive (that bound the probe) carry gene A.

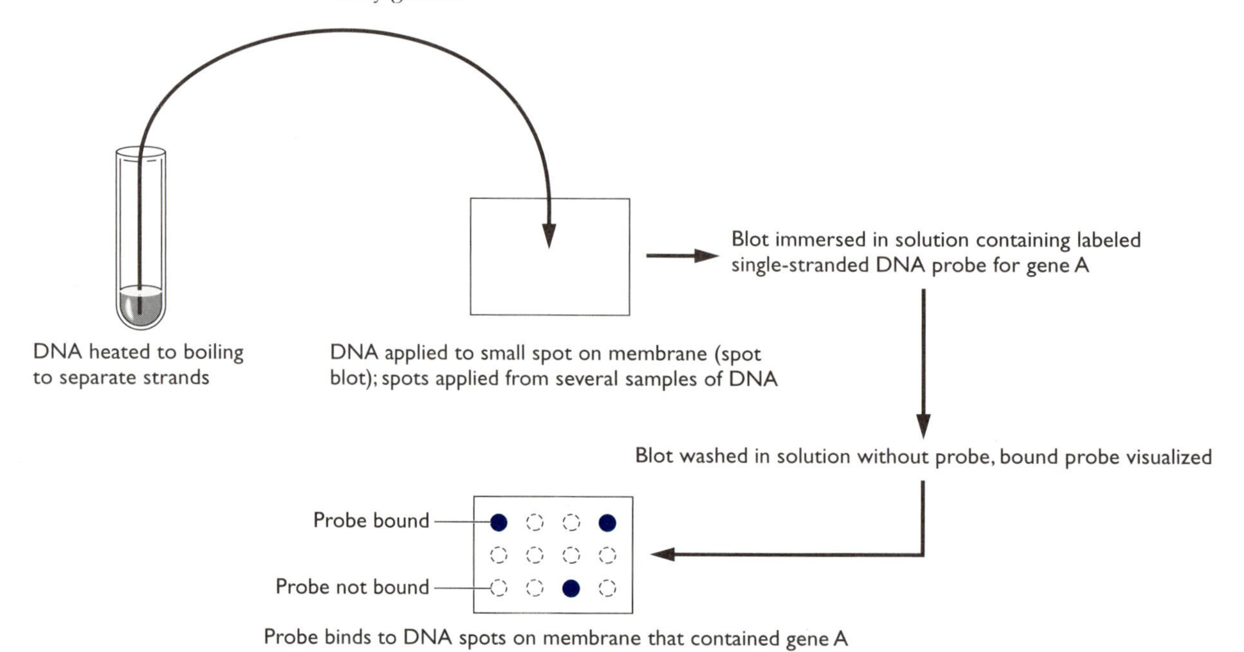

single-stranded by boiling. Once it is single-stranded, the probe is placed on the blot. Single-stranded DNA from a strain that contains gene A will bind the labeled single-stranded probe that hybridizes to gene A. Single-stranded DNA from a strain that does not have gene A will not bind the probe. The blot is then washed to remove unbound probe, and the bound, labeled probe is detected. Spots that are labeled are from strains that carry gene A. This approach could be used equally effectively to determine which bacterial isolates belong to a particular species. In this case, the DNA probe would not be the resistance gene A but a DNA segment found in that species and not in others.

Whether a DNA segment hybridizes to DNA that has a complementary sequence depends on temperature and salt concentration. If the probe DNA is identical to the DNA on the blot, the probe DNA will hybridize at higher temperatures and lower salt concentrations than if the probe DNA had some base changes that made it slightly different from the target gene. The researcher can decide how much difference in sequence between probe DNA and target DNA is acceptable and, to obtain the desired degree of specificity, adjust the temperature and salt concentration at which the hybridization reaction occurs. An example of the use of spot blots is the initial test for human immunodeficiency virus (HIV) in blood. A labeled probe detects HIV genes. A complication that will be addressed in Chapter 18 is that HIV and some other viruses have an RNA genome. RNA is naturally single-stranded and will hybridize to a complementary single-stranded DNA probe. In some applications, a DNA copy of the RNA is made using an HIV enzyme called reverse transcriptase.

Southern blot. A more complicated version of this type of hybridization test, but one that provides more information, is the **Southern blot.** The Southern blot is named after scientist E. M. Southern, who developed the technique. Since then, another hybridization test that detects RNA rather than DNA has been called a Northern blot as a play on Southern's name. The Southern blot process is shown in Figure 5.14. DNA from a bacterial strain is first digested with a restriction enzyme. The fragments are separated on an agarose gel placed in an electric field. Because the fragments are usually much smaller than those separated by pulsed field gel electrophoresis (PFGE; Chapter 3), a simple, linear electric field is used. So many fragments are separated in the gel at this point that, if the gel were stained with a stain that detects all DNA, the resulting staining pattern would resemble a long, vertical smear.

To look for one or a few specific genes in this mixture, DNA fragments from the gel are transferred to a membrane to generate a blot. In the process, the DNA fragments also are treated to make them single-stranded, so that all the DNA on the membrane is single-stranded. Now the labeled single-stranded probe is applied to the membrane. It will only bind to a fragment on the membrane that contains the complementary DNA sequence. Thus, when the bound probe is visualized, one or more bands will appear in those lanes where DNA from a strain carrying the gene of interest was loaded.

DNA sequencing

The process of DNA sequencing. DNA sequencing takes advantage of the fact that DNA polymerase starts from a double-stranded region and proceeds in the 5' $\rightarrow$ 3' direction to copy the single-stranded region from that point. A second feature of DNA polymerase that is important for DNA

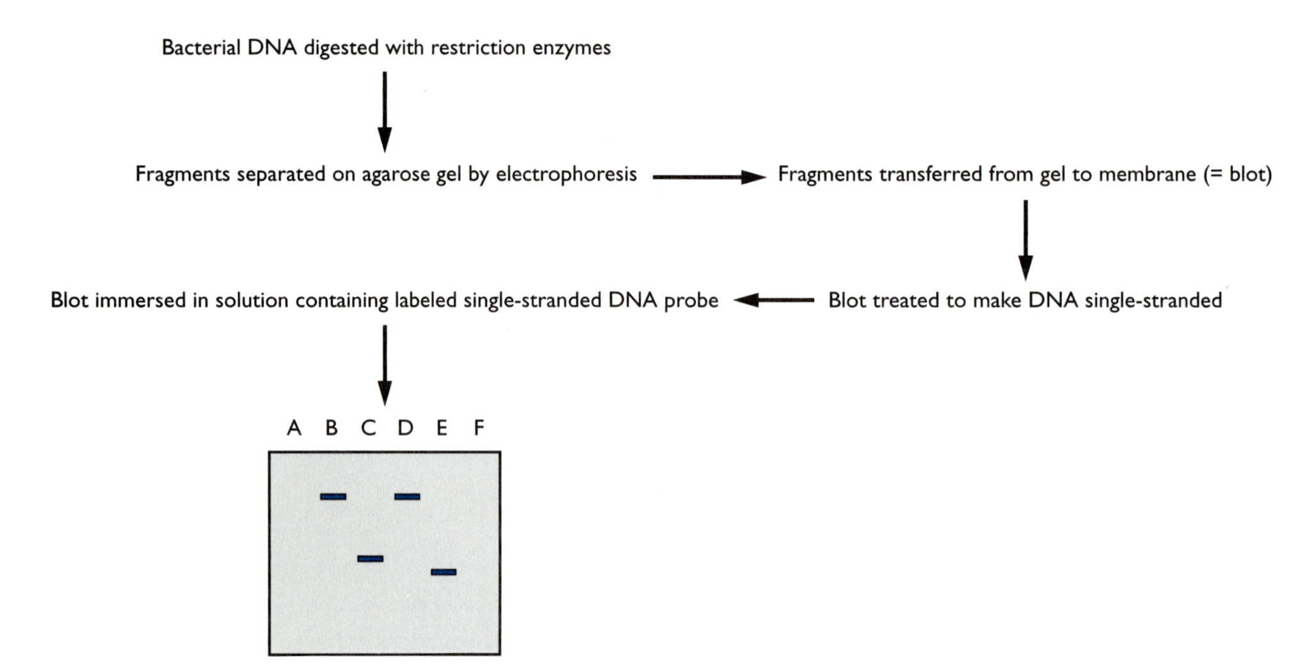

Figure 5.14 Southern blot. DNA from a bacterial strain is digested with restriction enzymes. The fragments that result are separated on an agarose gel by electrophoresis. Once separation is complete, the gel is treated to make the DNA single-stranded and is then placed on a membrane. The DNA fragments are transferred from the gel to the membrane (blot). The blot is then covered by a solution containing a labeled, single-stranded DNA probe for a specific gene. Any unbound probe will be washed from the gel, and the gel will be treated so that bound probe can be visualized. In the gel shown here, strains B through E all have the gene of interest, but the genes are in different locations on the chromosome or in a region whose sequence varies from strain to strain (except in strains B and D), because the restriction fragments differ in size.

sequencing is that when DNA polymerase integrates an altered base that lacks a 3'-OH group (**dideoxynucleotide**; Fig. 5.15) into the growing DNA copy, DNA synthesis ceases. The dideoxynucleotide cannot be attached to the next base because of the missing 3'-OH. DNA polymerase thus stops synthesizing DNA and releases the incomplete fragment.

The way in which these features of DNA polymerase are put to work to determine the sequence of bases in a segment of DNA is shown in Figure 5.15A. The DNA to be sequenced is first made single-stranded. To direct DNA polymerase to the correct starting place, a single-stranded segment of DNA called the **primer** is synthesized. The primer is complementary to the DNA at the point where the sequencing is to begin. Bacteria use RNA primers—so why don't scientists do the same? The reason is that it is technically easier to synthesize and check the sequences of single-stranded DNA segments than RNA segments. In the reaction mix are the primer, DNA polymerase, the DNA strand to be sequenced, each of the four normal bases (A, T, C, and G), and one of the four dideoxynucleotides (**dideoxyadenosine triphosphate [ddA], dideoxythymidine triphosphate [ddT], dideoxycytidine triphosphate [ddC]**, or **dideoxyguanosine triphosphate [ddG]**). The dideoxynucleotides are labeled with radioisotopes or a dye.

A Reactants

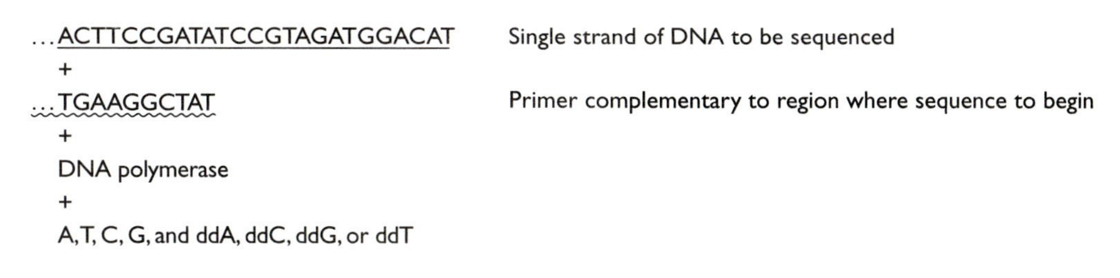

B Results of reaction that contained A, T, C, G, and ddA

Three other reactions each containing A, T, C, G, and one dideoxynucleotide would be carried out as well.

C Results obtained when all four reaction mixtures are electrophoresed on the same gel are shown below. Lanes are labeled with the base replaced by the dideoxynucleotide. (Numbers in parentheses in the "A" lane correspond to the four DNA segments generated in the reaction diagrammed above.)

Autoradiogram of sequencing gel (labeled nucleotides expose X-ray film)

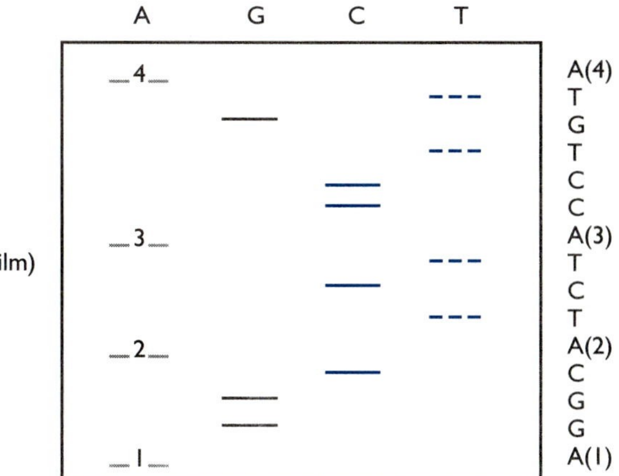

Figure 5.15 Original DNA sequencing procedure. (A) The segment of DNA to be sequenced is made single-stranded and then mixed with a primer (which is complementary to the region where sequencing is to begin), DNA polymerase, the four nucleotides, and one of four labeled dideoxynucleotides. **(B)** Results of the reaction when dideoxyadenosine triphosphate (ddA) was used. ddA was inserted randomly at the appropriate places in the sequence. This resulted in fragments of four different lengths. Other reactions containing one of each of the other three dideoxynucleotides were carried out also. **(C)** The four reaction mixtures were placed on a gel and subjected to electrophoresis. The gel was placed on a sheet of X-ray film. Because the dideoxynucleotides are labeled with radioisotopes, the X-ray film will be exposed wherever a dideoxynucleotide occurs. The sequence can then be read by starting with the smallest fragment (in this gel, labeled 1), which happens to be in the A lane. This is followed by the next smallest fragment, which is in the G lane. By moving up the gel to increasingly larger fragments, the sequence of bases is determined. Usually a DNA-sequencing gel will separate hundreds of bands instead of the small number shown here.

The dideoxynucleotides will be incorporated at random, thus stopping DNA synthesis wherever the complementary base is encountered. A series of DNA fragments of differing lengths will be generated. In the original form of DNA sequencing, the labeled fragments were separated by electrophoresis, and the radioisotopes were detected on X-ray film (Fig. 5.15B).

More recently the entire process has been automated. Each of the dideoxynucleotides is labeled with a different colored dye instead of radioisotopes. The fragments of DNA, labeled with the dye of the dideoxynucleotide that terminates the DNA synthesis, are passed through a column and separated on the basis of length. A detector records the color of the dye. A typical read-out from a DNA sequencing machine is shown in Figure 5.16, along with the sequencing machine's identification of the base to which each peak corresponds.

Polymerase chain reaction (PCR)

PCR was introduced in an earlier chapter (Chapter 2) as a method for selectively amplifying a particular segment of DNA. Here, the method for performing such an amplification is explained in more detail. Once again, the key feature of DNA polymerase for this application is its penchant for synthesizing a DNA copy of a single strand of DNA starting from a double-stranded region of the DNA. The steps in a PCR amplification reaction are illustrated in Figure 5.17. Suppose the scientist in the earlier hybridization section wants to use PCR instead of hybridization to determine which bacterial isolates contain resistance gene A. The first step would be to design and synthesize primers, segments of single-stranded DNA about 20–30 bases long, that bind to the ends of the region to be amplified, in this case gene A. In DNA sequencing, there was only one primer, but two primers are

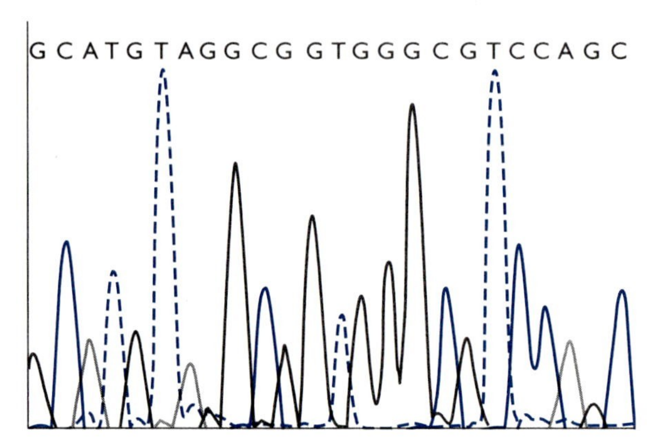

Figure 5.16 Sequencing using an automated DNA sequencing machine. Each of the dideoxynucleotides is labeled with a colored dye. Once the fragments have been synthesized by DNA polymerase, they are separated in a DNA sequencing machine column on the basis of length. The machine also records the color of the dye as labeled fragments come off the column. The results that are printed out show the location of each fragment and identify the base to which it corresponds. The printout comes in four colors, with adenine in green, thymine in red, cytosine in blue, and guanosine in black. (These are represented in blues and blacks on this figure, but the information about the sequence is the same.) DNA sequencing machines can read as many as 500–700 bases in a single run.

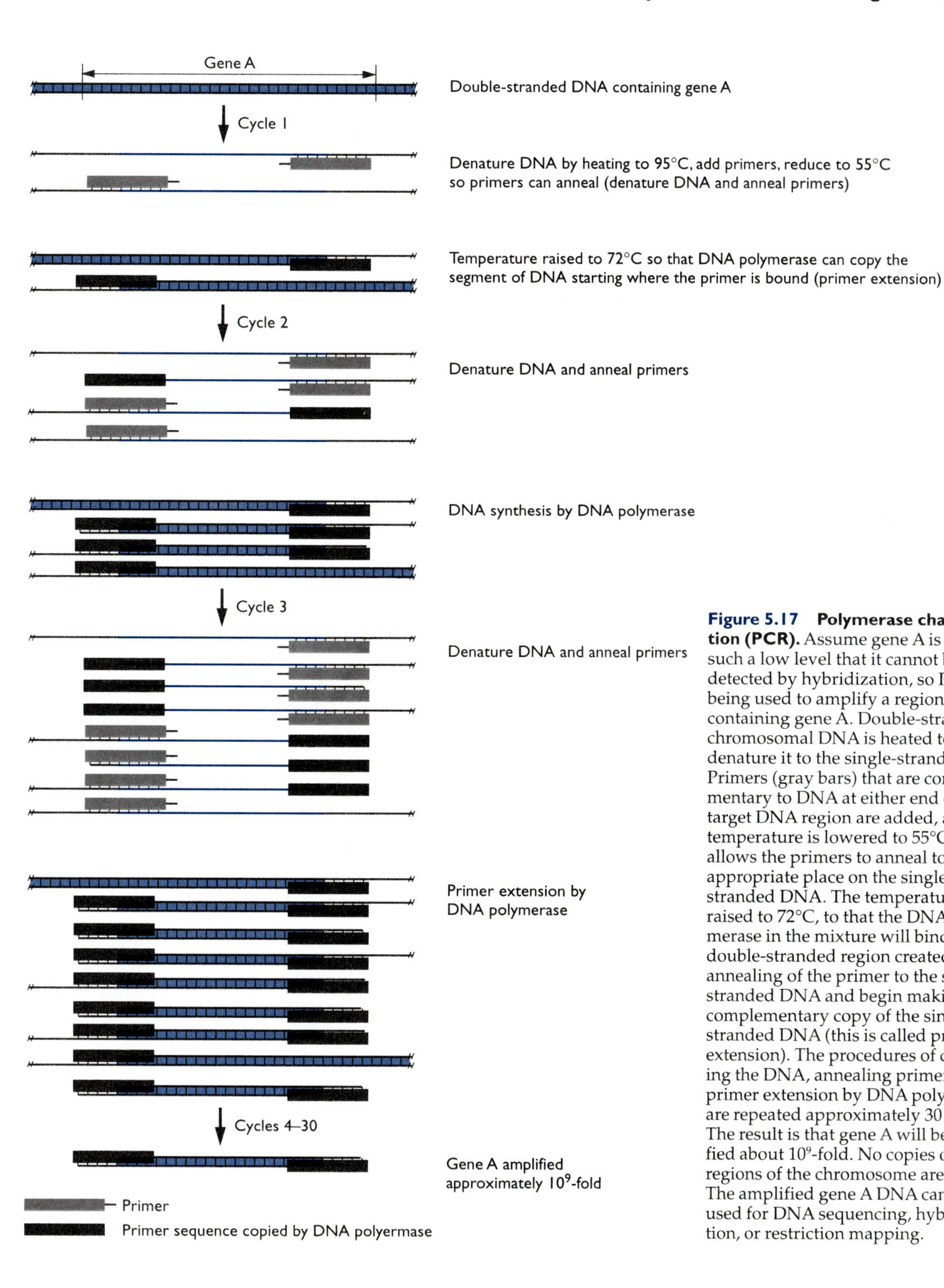

Figure 5.17 Polymerase chain reaction (PCR). Assume gene A is present at such a low level that it cannot be detected by hybridization, so PCR is being used to amplify a region of DNA containing gene A. Double-stranded chromosomal DNA is heated to 95°C to denature it to the single-stranded form. Primers (gray bars) that are complementary to DNA at either end of the target DNA region are added, and the temperature is lowered to 55°C, which allows the primers to anneal to the appropriate place on the single-stranded DNA. The temperature is raised to 72°C, to that the DNA polymerase in the mixture will bind to the double-stranded region created by the annealing of the primer to the single-stranded DNA and begin making a complementary copy of the single-stranded DNA (this is called primer extension). The procedures of denaturing the DNA, annealing primers, and primer extension by DNA polymerase are repeated approximately 30 times. The result is that gene A will be amplified about 10^9-fold. No copies of other regions of the chromosome are made. The amplified gene A DNA can now be used for DNA sequencing, hybridization, or restriction mapping.

used in PCR. The places at which these two primers bind dictate the endpoints of the amplified fragment. Thus, the sequences of the two primers determine the size as well as the particular DNA fragment to be amplified.

Steps in the PCR amplification process. If the scientist has the DNA sequence of gene A, he or she can design primers that bind to either end of the gene (Fig. 5.17). Primers are synthesized chemically to the specifications of the scientist who orders them. At the beginning of the reaction, the sample is heated to 95°C to separate the DNA strands of the test material, which in this case is an entire bacterial chromosome. The primers are added and the temperature reduced to 55°C, so that the primers can anneal with their complementary sequences, if these are present in the DNA being tested. Of course, the strands of the chromosomal DNA also begin to re-anneal with each other. However, because the concentration of primers is high and because it takes longer for strands of very large DNA molecules to line up with each other than for short segments like the primers to anneal to their complementary sequences on the chromosomal DNA, the primers win most of the time.

The temperature of the reaction mixture is raised to 72°C so that the thermophilic DNA polymerase can copy the segment of DNA, beginning from the region of double-stranded DNA created by the binding of the primer. DNA synthesis of both strands continues until the temperature is raised once again to 95°C to separate the DNA strands and cause DNA polymerase to stop synthesis. On the next round, DNA segments, limited in size to the sequence between and including the primers, begin to appear. These fragments quickly predominate and, after 20–30 cycles, become the predominant form. Thus, after amplification, if the DNA is electrophoresed on a gel, only a single band corresponding to the desired amplified region appears if the specimen contained the gene being targeted. If the gene was not in that sample, no amplified product results. The chromosomal DNA is still there but at such a low level that it does not show up on a stained gel.

Uses of PCR. PCR has been used to solve an amazing number of problems in many different areas of inquiry. Some of these applications were described in Chapter 2. In microbiology, a common application of PCR is to use it to identify the species of a microbial isolate. Before PCR, it could take months or years to determine the identity of a microbe isolated from nature. Identification took so long because it was based on phenotypic traits, such as sugar utilization, Gram stain, or ability to tolerate oxygen. Although this approach sounds simple, in practice it is anything but simple. Many traits must be scored, and the more a scientist learned about the microbe, the more traits there were to test. In addition, scientists were faced with the nagging question of which traits were most important for distinguishing one microbe from another and whether important traits had been overlooked.

The PCR amplification and sequencing of ribosomal RNA (rRNA) genes has dramatically changed the process of species identification. Today, if a scientist wants to know the probable genus and species of a microbial isolate, PCR amplification is used to amplify the 16S rRNA gene (prokaryotes) or 18S rRNA gene (eukaryotes). The PCR primers bind to conserved regions of the rRNA gene. The amplified region is sequenced, and, in a process that takes minutes, the DNA sequence of the gene is submitted to a computer program that compares the sequence with that of millions of known rRNA gene sequences. The program determines what microbes are the closest relatives of the isolate in question. In some cases, the rRNA gene sequence is not closely related to any known microbe. In this case, the scientist knows that the microbe may represent a new species or genus. The tentative identity of a microbial isolate can now be established within a week.

It is important to check such tentative identifications by analysis of traits of the isolate, because PCR can amplify DNA contaminants along with the isolate of interest. Also, the alignment programs used to match the sequence with known rRNA gene sequences can make mistakes by choosing a less than optimal alignment. For these and other reasons, it is wise to back up a tentative identification with other observations. Nevertheless, having an idea of the type of microbe one is dealing with saves much time and effort and can suggest new traits to test. A later chapter (Chapter 21) will describe the application of PCR sequence analysis to characterizing complex microbial communities.

Genetic Engineering and the Race for Human Survival

At the beginning of this chapter, the quest to learn more about the ways in which bacteria genetically engineer themselves was portrayed as a matter of life or death. What is the argument for such a dramatic perspective? Bacterial genetic engineering affects us because bacteria are genetically engineering themselves to become more resistant to antibiotics and to acquire traits that enable them to cause new kinds of disease. *E. coli* O157:H7, the "killer *E. coli*" of media fame, is an example of the latter sort of bacterial genetic engineering.

How can we use genetic engineering to combat these wily microbes? First, clinical tests based on hybridization or PCR will soon provide rapid methods for identifying the cause of an infection and offer early indications of its susceptibility to antibiotics. Genetic engineering helps in the war against viral diseases, too. If a new form of influenza virus emerges, for example, its genome can be amplified rapidly by PCR and sequenced in a very short time to reveal its properties. Even if the virus is unknown, PCR amplification and DNA sequencing are possible and can provide even more valuable information. In this case, PCR is trickier, because there is no previous sequence information available to help in the design of PCR primers. One way to do this is to use as primers mixtures of short DNA segments (6–10 bp long) with randomized sequences. Of course, this generates a variety of different-sized DNA fragments that must be sorted out.

Cloning of microbial genes—both viral and bacterial—has already provided a cheaper and safer source of vaccines. DNA sequence analysis of the genomes of disease-causing bacteria is now being used to find new targets for antibiotics and vaccines. Genetic engineering of plants, to be described in a later chapter on biotechnology, not only promises to provide more nutritious foods and plants that have less need for pesticides and herbicides but also may produce novel forms of vaccines that can be administered through cheap and commonly consumed foods. Genetic engineering in animals could reduce the reliance of farmers on antibiotics to prevent disease and increase growth. Through genetic engineering, scientists are turning against bacteria the very traits that have stood the bacteria in such good stead for billions of years. It is a war we cannot hope to win, but genetic engineering is leveling the playing field.

chapter 5 at a glance

DNA Structure and DNA Synthesis and their applications

Structure of DNA/step in DNA synthesis	Application
Structure of DNA: complementarity of the two strands	DNA hybridization Southern blot
Opening up the double helix (topoisomerases relieve supercoiling, helicases unwind the DNA)	
DNA polymerase copies each DNA strand, starting from a double-stranded region created by an RNA primer	Primers used for DNA sequencing and polymerase chain reaction (PCR)
RNA primers are removed and gaps filled	
DNA ligase seals nicks in newly synthesized strand	Used in cloning to seal cloned segment into the cloning vector
DNA gyrase restores supercoiling	Target of fluoroquinolone antibiotics

Growth phases of bacteria in laboratory medium

Lag phase: bacteria adapt to new conditions, numbers remain constant.

Exponential phase: bacteria divide exponentially ($N = N_0 2^n$, where N_0 is the initial number of bacteria, n is the number of divisions, and N is the number of bacteria after n divisions).

Stationary phase: one or more nutrients become limiting, increase in number of bacteria ceases.

Death phase: bacteria begin to lyse, probably as a result of activity of autolysins (proteins that break down the cell wall); number of viable bacteria decreases.

Homologous recombination

Important part of the DNA repair system of all organisms

Helps to repair double-stranded breaks induced by radiation

Allows bacteria to acquire new forms of a gene, e.g., mutant penicillin-binding proteins

chapter 5 at a glance *(continued)*

Comparison of different molecular techniques

Technique	Primers/DNA polymerase	Size fractionation by gel electrophoresis	Hybridization	Restriction enzymes
Spot blot	No	No	Yes (probe)	No
Southern blot	No	Yes	Yes (probe)	Yes
DNA sequencing, PCR*	Yes	Yes	Yes (primer)	No
Cloning	No	Sometimes†	Yes (sticky ends anneal)	Yes

*Difference is that DNA sequencing uses dideoxynucleotides that terminate the synthesis of the DNA copy.
†The cloning procedure described here did not employ a size fractionation step, but sometimes scientists will size-fractionate the DNA to be cloned if they know the approximate size of the DNA fragment they want.

Comparison of DNA replication by vegetative cells and DNA replication during sporulation

Steps in division by vegetative cells	Differences in division associated with sporogenesis
Copies of chromosome are separated as cell elongates prior to division. Division is symmetric.	Division is asymmetric, no elongation.
Both daughter cells are the same size; both are viable.	One daughter cell is smaller than the other; only the one destined to be a spore will survive.
Both daughter cells contain all the molecules necessary for viability.	The spore contains all the molecules necessary for viability; can later become a vegetative cell.
Both daughter cells have normal peptidoglycan cell walls.	Spore has a thick covering (spore coat) that includes layers of protein as well as peptidoglycan.

Questions for Review and Reflection

1. How does each of the following differ from or resemble DNA replication in bacteria: DNA sequencing, polymerase chain reaction (PCR), cloning?

2. Suppose someone is trying to sabotage your experiments. The villain tries putting a fluoroquinolone antibiotic in tubes containing your reactions. Which of the following, if any, would be affected by this dastardly action: DNA hybridization, DNA sequencing, PCR, and/or cloning?

3. Figure 5.2A shows only one replication fork when, in fact, there are two, moving in opposite directions away from the replication origin. Draw the real-life DNA replication process with two replication forks forming an expanding bubble.

4. In the debate over safety of genetically engineered plants, antibiotechnology activists alleged that, in cloning, antibiotic resistance genes that ended up in the plant cells might leave the plant and enter bacteria, making the latter resistant to antibiotics. Later, we will assess the first part of this allegation, entry into the bacteria, but from what you have learned in this chapter, what would be the limitations on entry of the resistance genes (if they got into a bacterium) into the chromosome?

5. What is the payoff in the strategy of using an intact region of one chromosome to save a broken copy? If this strategy risks the loss of both chromosomes, why not just forget homologous recombination and write off the broken chromosome as a loss and proceed with the intact one?

6. The villain of question 2 is back. Having failed to disrupt your PCR reactions with antibiotics, the diabolical fiend now thinks he/she has a better idea for sabotaging your PCR reactions. Suggest some possibilities. (Don't try this at home—or in the lab!)

7. PCR and DNA sequencing both rely on primers and DNA polymerase.

Questions for Review and Reflection *(continued)*

How do they differ, and what is the importance of these differences? Why are the primers for both processes usually at least 20 bp in length?

8. Why does it make sense that DNA replication and cell division during sporulation are asymmetric? Why is asymmetric cell division limited to sporulation and not employed for general cell division?

9. In what sense could a plasmid be called a mini-chromosome? Chromosomes usually are considered to encode functions essential for the life of the cell. Does the fact that plasmids carry genes that are not essential under all conditions rule them out as chromosomes? [If you have difficulty answering this, you are in good company. Scientists differ in their answer to this question, especially in the case of very large plasmids that are close to a chromosome in size.] What types of genes on the "plasmid" would cause you to consider it a second chromosome?

10. Look at Figure 5.17 and explain why the fragments of the desired size increase in number exponentially (2^n) whereas the longer fragments increase linearly (2n).

11. You have cloned gene X. A critic says that gene X did not come from your organism but from a contaminant in your culture. How would you use Southern blot analysis to prove your critic wrong? (Hint: After making sure your culture is pure, obtain chromosomal DNA from your isolate and use your plasmid clone as a hybridization probe.)

key terms

cloning of a DNA segment the process by which multiple copies of a piece of DNA are made by inserting the segment into a cloning vector

cloning vector circular plasmid designed to receive DNA segments for cloning; most have a high copy number

DNA gyrase enzyme that reintroduces supercoils into DNA that has been uncoiled during replication; a target of the fluoroquinolone antibiotics

DNA hybridization process in which a sample of DNA is heated to separate the strands; the single-stranded DNA is bound to a membrane; the membrane is incubated with a solution containing labeled single-stranded DNA from a known gene (the probe); if the probe binds, this shows that the DNA in the original sample contained the gene encoded in the probe DNA

DNA ligase enzyme normally used by bacteria to connect segments of DNA produced during the replication process; in genetic engineering, the enzyme that covalently connects the DNA segment to be introduced into a cloning vector to the ends of the linearized cloning vector

DNA polymerase multiprotein complex that produces a single-stranded DNA copy of a single-stranded DNA template.

DNA sequencing process in which the nucleotide sequence of a segment of DNA is determined; dideoxynucleotides are used

genetic engineering complex variety of techniques, ranging from DNA hybridization and DNA sequencing to polymerase chain reaction and cloning, that use bacterial enzymes and processes to manipulate the composition of DNA molecules or detect specific DNA sequences

homologous recombination process by which identical or nearly identical segments of DNA pair with each other and cross over, so that one strand of DNA from one double-stranded molecule now serves as the complementary strand to the single strand from the other DNA molecule; used to repair breaks in the DNA or introduce new DNA

lagging strand DNA strand that must be copied in segments from RNA primers as the replication fork moves along the DNA molecule

leading strand DNA strand that could be copied as a single 5' → 3' segment, but probably is copied in segments

molecular biology term effectively interchangeable with genetic engineering, except that molecular biology encompasses a somewhat wider range of activities, including the process of microbial DNA replication and other macromolecular synthetic processes (described in Chapters 6 and 7)

plasmid circular or linear DNA segment that replicates independently of the bacterial chromosome

polymerase chain reaction (PCR) a method for amplifying DNA in vitro; involves oligonucleotide primers complementary to nucleotide sequences flanking a target sequence with subsequent replication of the target sequence

replication fork edge of the unwound portion of DNA being copied; has a fork-like appearance

restriction enzyme enzyme that cleaves double-stranded DNA at specific sites that usually consist of 4–8 base pairs; some leave "sticky ends," others leave blunt ends

6 Bacterial Genetics II: Transcription, Translation, and Regulation

Bacteria are quick-change artists. They have to be. Without warning, a bacterium chilling out in a pot of dressing on a salad bar is dumped into a vat of stomach acid. Or a bacterium about to be injected into a human arm by an insect will soon encounter an environment very different from the one inside the insect. Only the adaptable survive, and bacteria are nothing if not survivors.

Bacteria have two strategies for adaptation: a long-term strategy and a short-term strategy. The long-term strategy is mutation. The short-term strategy involves controlling which proteins are produced under any specific set of conditions. Mutation of DNA sequences can produce desirable new traits, as is the case with the mutant penicillin-binding proteins that confer penicillin resistance (Chapter 4). Yet mutation as an adaptive strategy has some drawbacks. First, a random mutation can be lethal when the change in DNA sequence makes a vital protein inactive. Many bacteria must die so that enough mutations can be tested to find the ones that are successful. Second, because acquisition of a desirable new trait may require multiple mutations, adaptation by mutation takes time, possibly years. Finally, an adaptation achieved by mutation is reversible only by additional rounds of mutation. It is not an easy on/off phenomenon. A bacterium living in the real world must be able to alter multiple activities within minutes, then alter them back again if conditions change once more. Thus, in addition to mutation, bacteria also need a rapid, reversible strategy for change.

Short-Term Adaptation

The short-term strategy that allows bacteria to make fast, easily reversible changes in their activities takes advantage of the way bacteria make a protein using information encoded in their DNA. First, an RNA copy of the gene is made (**transcription of DNA**), then this RNA copy (**messenger RNA [mRNA]** or transcript) is used as a template from which ribosomes produce a protein (**translation of the mRNA;** Fig. 6.1). Within minutes, bacteria can change the number of mRNA copies made of the DNA, the

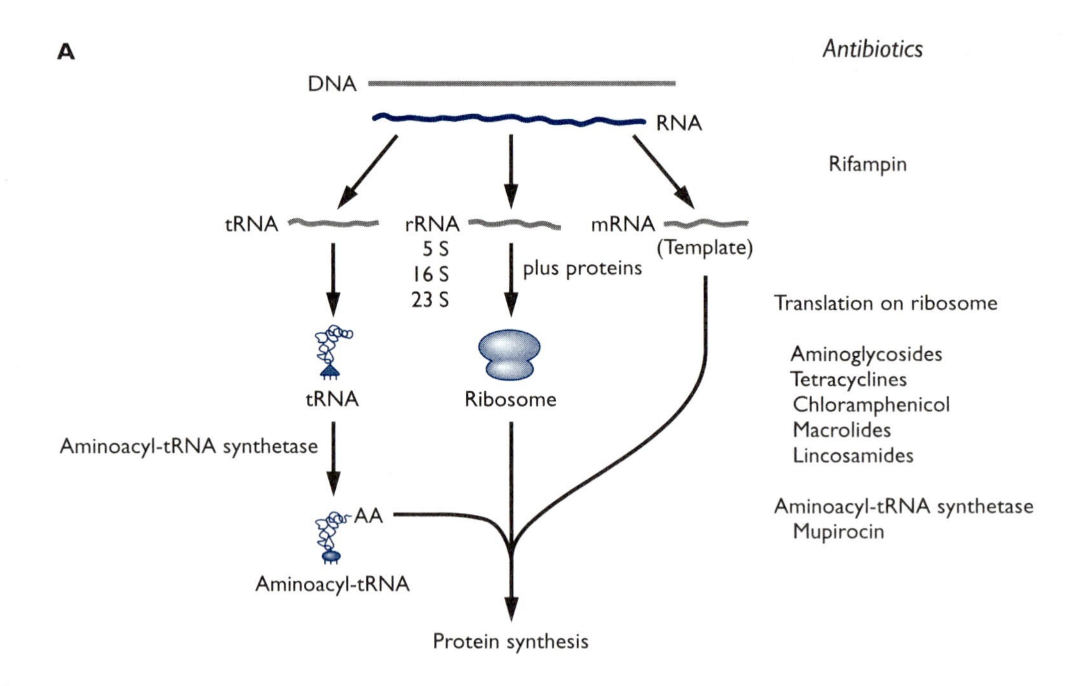

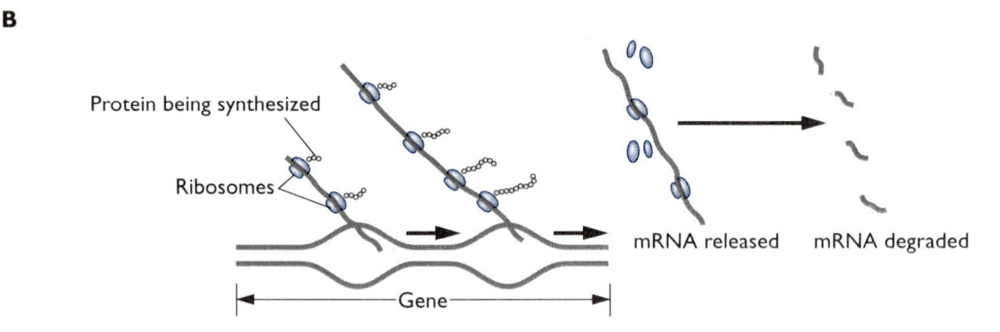

Figure 6.1 Overview of transcription and translation and antibiotics that inhibit these processes at different levels. (A) In the first step, an mRNA copy of the gene is made (transcription). This RNA can encode a protein (mRNA) or a ribosome-associated RNA (rRNA, tRNA). The antibiotic rifampin inhibits transcription. The next step is to translate the mRNAs into the amino acid sequence of a protein. tRNAs carry the individual amino acids to the mRNA, which is being translated by a ribosome (composed of rRNA and proteins). A number of antibiotics can inhibit translation by binding to a variety of sites on the ribosome. **(B)** In bacteria, transcription and translation occur simultaneously, with ribosomes translating the mRNA as it is being transcribed from the DNA.

number of proteins translated from an mRNA, or both, to increase or decrease the amount of that particular protein in the cell. The process of changing the number of transcripts made is called **regulation of gene expression.** Changing the number of proteins translated from the mRNA is called **translational regulation.** Because existing mRNA molecules and proteins are being degraded constantly, a decrease in transcription and/or translation has as rapid an impact on the protein content of a cell as an increase in transcription or translation. Another important feature of regulation is that it can change the levels of many proteins at the same time. Thus, in response to a change in environmental conditions, it is not unusual to see many proteins rise in abundance and many other proteins decrease in abundance, all within a matter of 5–10 minutes.

Some antibiotics inhibit the processes of transcription or translation (Fig. 6.1). Note that rifampin, an antibiotic that interferes with transcription, is different from those that interfere with translation, because they have different targets: the enzyme that makes the mRNA and the ribosomes. Translation has been an especially popular target for antibiotics, largely because so many different steps in translation can be inhibited. Now that the importance of regulation of transcription and translation for bacterial survival has been appreciated, scientists are also looking for antibiotics that interfere with regulation.

Scientists also are interested in understanding better the DNA sequence signals that bacteria use to direct transcription and translation. This knowledge will help scientists manipulate the expression and properties of important genes. The ability to manipulate gene expression has been responsible for many advances in the biotechnology industry, such as overproduction of amino acids and production of human insulin by bacteria. Another important application of this information, to be described in more detail later in this chapter, is that it allows scientists to "read" genome sequence data by recognizing DNA segments that might encode proteins (**open reading frames**) or RNA species. Furthermore, the amino acid sequence of the protein, which can be deduced from the DNA sequence of the open reading frame, provides valuable clues to how many and what type of proteins are being produced by the microbe.

Transcription and Its Regulation

Steps in transcription

The steps in transcription are shown in Figure 6.2. Transcription of DNA is mediated by **RNA polymerase.** Like DNA polymerase, RNA polymerase is a multiprotein complex that uses DNA as a template and proceeds in a 5' → 3' direction as it copies the DNA strand. Instead of synthesizing a DNA copy of the strand, it synthesizes an RNA copy. RNA polymerase consists of a **core enzyme** containing four protein subunits (α, α, β, β'), which catalyze the addition of ribonucleotides to the growing mRNA chain but cannot recognize signals that tell it where to start and stop transcribing. The core enzyme interacts with other proteins that help it to initiate or terminate the transcript. RNA polymerase copies only one or a few genes, not the entire chromosome. DNA sequences that tell RNA polymerase where to start transcribing an mRNA copy of a gene are called **promoters** (Fig. 6.2). The promoter sequence also directs RNA polymerase to the correct DNA strand. For accurate recognition of the promoter sequence, the core enzyme must bind to an additional protein called **sigma factor** (Fig. 6.2). The core enzyme with a sigma factor attached binds to a promoter sequence very specifically and begins to transcribe the DNA in a 5' → 3' direction from that point. Shortly after initiation of transcription, sigma factor leaves the RNA polymerase complex, because its role of initiating transcription is finished.

RNA polymerase stops transcribing a DNA segment when it detects another type of sequence signature in the DNA, a **terminator sequence.** Some terminator sequences have **stem-loop structures** and are thus easy to recognize in a DNA sequence (Fig. 6.3). Because DNA is double stranded, it probably does not itself form the stem-loop structure shown in Figure 6.3. Instead, the single-stranded mRNA copy being made by RNA polymerase forms this

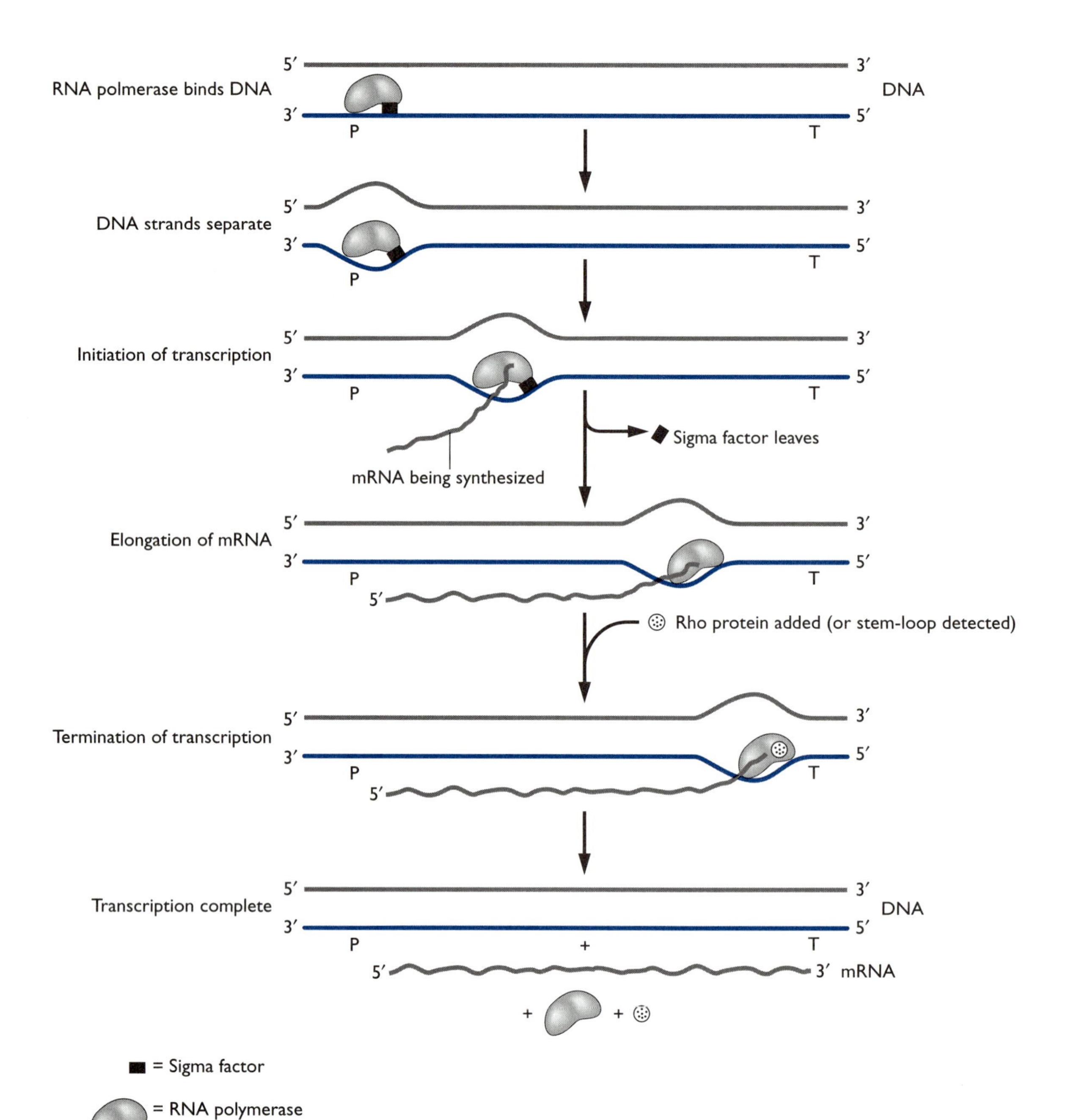

Figure 6.2 Steps in transcription. RNA polymerase is a multiprotein complex that transcribes in a 5' $\rightarrow$ 3' direction along a single DNA strand, forming an RNA copy as it moves. The RNA polymerase core enzyme is directed to the correct strand of DNA and the correct place to start transcribing by the promoter sequence on the DNA. Sigma factor is required for binding of the core enzyme (plus sigma factor) to the promoter sequence. Once the RNA polymerase is bound, RNA synthesis begins. This is called initiation of transcription. After this point, sigma factor has done its job and leaves the RNA polymerase. The RNA transcript (mRNA) continues to grow until the RNA polymerase recognizes a terminator sequence. The terminator sequence may be a stem-loop structure in the newly synthesized RNA. Other terminator sequences do not form stem-loop structures, so RNA polymerase requires the help of the rho protein to recognize them (rho-dependent terminators). Once the RNA transcript has been completed, the rho protein separates from the RNA polymerase, which leaves the DNA.

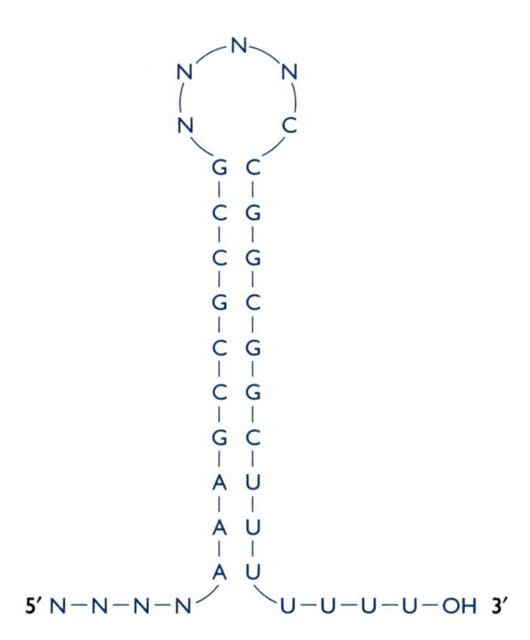

Figure 6.3 Stem-loop structure of some terminators. The stem-loop structure forms when complementary sequences of RNA occur near each other. These sequences come together and form a stable structure by intrastrand base pairing. If these loops are followed by a string of U residues, they act as transcription terminators. In cases where no such structures form, rho is required for recognition of the termination signal. N indicates bases that can be A, U, G, or C.

structure, causing RNA polymerase to stop transcription. Not all terminators have such a stem-loop structure. Some terminators have no obvious conserved sequence features, so the signal must lie in some structure (not the stem-loop) formed by the RNA. To recognize these regions, the core enzyme needs to bind a protein, **rho** (Fig. 6.2). Such terminators are called rho-dependent terminators. The promoter and termination signals ensure that RNA polymerase makes the same size transcript every time and that the transcript contains the entire region needed for translation of an intact protein.

Regulation of transcription

Promoters. Some genes are transcribed more often than others, and a gene may be transcribed more often under some conditions than others. For example, genes that are transcribed to make the RNAs involved in translation, ribosomal RNA (rRNA) and transfer RNA (tRNA), tend to be transcribed at the same level under most conditions, except at very slow growth rates when the cell is trying to save energy by shutting down transcription and translation of most genes. This is also true of genes encoding proteins that are needed all the time. Genes that are expressed at the same level under different conditions are said to be expressed **constitutively.** Genes encoding proteins that are needed only under some conditions and that are transcribed only under those conditions are said to have **regulated expression.** To understand how regulated genes work, it is first necessary to understand the features that make a particular DNA sequence a promoter.

Strong promoters are bound efficiently by RNA polymerase and are thus expressed at a high level; **weak promoters** bind RNA polymerase much less efficiently and are thus expressed at a low level. In the case of

Escherichia coli, most promoters share a two-part consensus sequence, shown in Figure 6.4. One sequence, **TATAAT,** is located about 10 bases upstream from the point where transcription begins (**transcription start site**). The second sequence, **TTGACA,** is located about 35 bases upstream from the transcription start site. Because the site at which transcription starts is arbitrarily labeled +1 and bases upstream from this point are given negative numbers, the TATAAT sequence is called the **–10 region** and the TTGACA sequence is called the **–35 region** of the promoter. Strong promoters can vary somewhat in their –10 and –35 sequences, but most have this recognizable motif. Weak promoters have fewer matches with the conserved bases in this consensus region and thus do not bind RNA polymerase very effectively. Genes with weak promoters are either meant to be expressed at a very low level all the time or, if they are to be expressed at a higher level under some conditions, require accessory proteins called **activators**, which help RNA polymerase bind to the promoter when more active expression of the gene is needed. Strong promoters of gram-positive bacteria and other bacteria that are not closely related to *E. coli* have different consensus sequences from those shown in Figure 6.4, but their consensus sequences generally have the same arrangement: conserved sequences near the –10 and –35 positions upstream from the transcription start site.

Operons. Not all genes have their own promoters. Sometimes, several genes that are transcribed in the same direction share the same promoter (**operon**; Fig. 6.5). In such cases, the transcript originates from the promoter of the first gene and continues through the genes immediately downstream from the first gene until a termination signal is encountered. This produces a transcript that encodes more than one protein. Operons are rare in eukaryotes and in eukaryotic viruses. They are found in

Figure 6.4 Structure of a strong promoter. In *E. coli* and closely related bacteria, most strong promoters have two consensus sequences: one at the –10 region and one at the –35 region. Transcription starts at a site arbitrarily labeled +1, so the –10 and –35 regions are located about 10 bp and 35 bp, respectively, upstream from the transcription start site. The consensus sequence that starts approximately 10 bases upstream from the start site is called the –10 region. The –35 region logically occurs approximately 35 bases upstream from the start site. These are the sequences recognized by a sigma factor called sigma-70. Promoters found in other species or promoters recognized by other sigma factors can have different consensus sequences and even different spacing of the consensus regions, but the overall organization of the promoter sequences is similar.

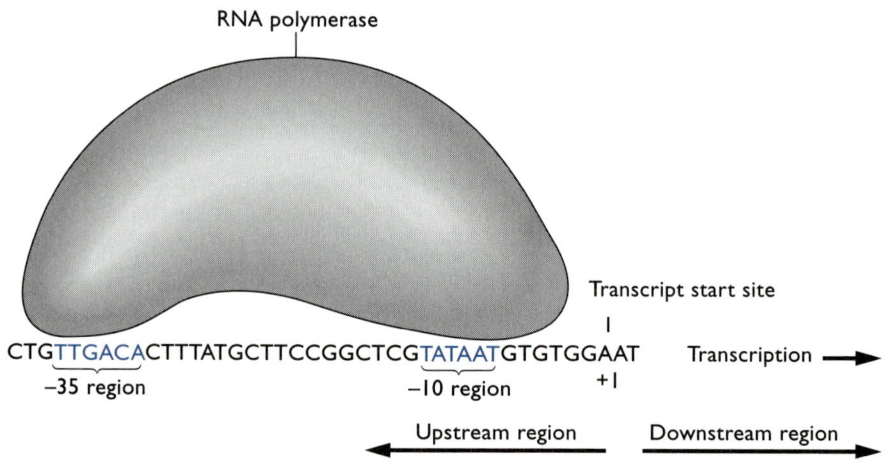

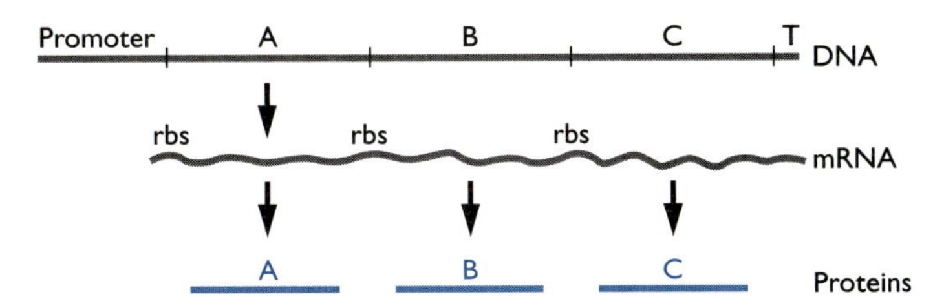

Figure 6.5 **Organization of an operon.** An operon consists of two or more genes transcribed from a single promoter. The promoter at the beginning of the first gene controls the transcription of the subsequent genes. The mRNA transcript that results has multiple ribosome binding sites and encodes more than one protein.

archaea, but not as frequently as in bacteria. It is not clear why bacteria so often have their genes organized in operons, whereas archaeans and eukaryotes make little or no use of the operon strategy.

Role of sigma factors in controlling gene expression. A sigma factor is required for efficient initiation of transcription. The strong promoter sequence shown in Figure 6.4 is recognized by RNA polymerase with a sigma factor called **sigma-70.** Sigma-70 recognizes most of the genes expressed in *E. coli. E. coli* also uses other sigma factors, which recognize different consensus sequences. For example, genes transcribed in response to nitrogen limitation have promoters that are recognized by a sigma factor called **sigma-55.** In spore-forming bacteria, special sigma factors recognize promoters of genes involved in sporulation and are brought into play only when the bacteria need to form spores. By using different sigma factors, bacteria gain an additional level of control over what types of genes are expressed under different conditions. These special sigma factors are only made under conditions in which the set of genes they control would be useful to them, thus effectively keeping these genes unexpressed when not needed.

A more finely tuned control is conferred by repressors and activators, accessory proteins that alter the ability of RNA polymerase (with its sigma factor) to bind to and transcribe from a specific promoter. Thus, two promoters that are recognized by the same sigma factor may have different levels of expression, depending on the conditions the bacterium is experiencing. This dual control is important, because bacteria experiencing nitrogen starvation, for example, may also be experiencing other limiting conditions, such as iron starvation or phosphate starvation. The appropriate response to a combination of nitrogen and iron limitation will be somewhat different from the appropriate response to a combination of nitrogen and phosphate limitation. Repressors and activators provide an additional level of control that allows the bacterium to modulate its response appropriately.

Repressors. Repressors are proteins that bind to the DNA near the promoter in such a way that transcription of a gene is prevented. Binding sites for repressor proteins are called **operator sites**. Repressors block expression of a gene either by binding to a sequence that lies between the promoter and the transcription start site of the gene or by binding to a sequence that overlaps the promoter (Fig. 6.6). When a repressor protein is bound to the DNA,

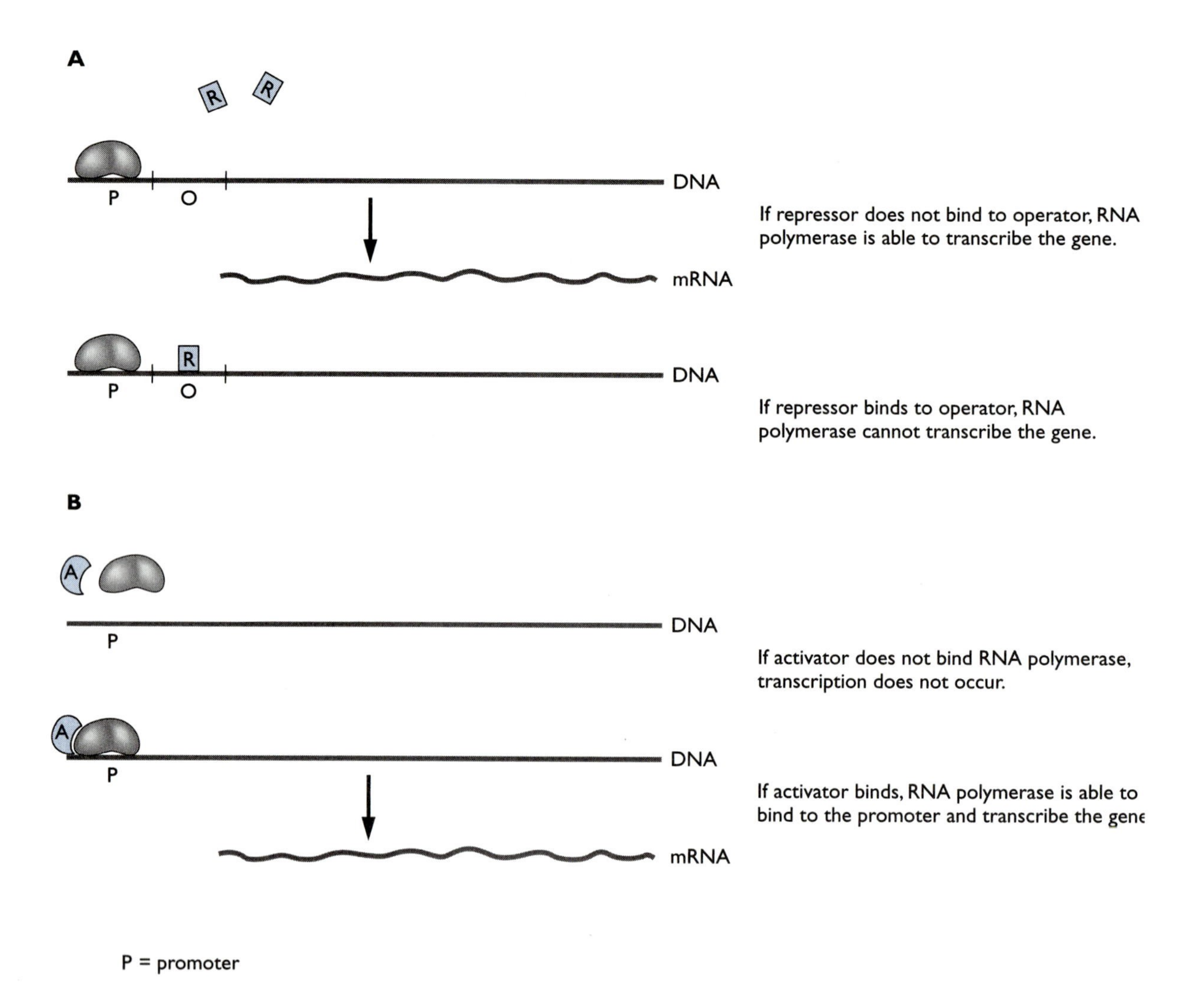

Figure 6.6 Control of transcription by repressors and activators. (A) The repressor binds to a site on the DNA called the operator, which is located between the promoter and the transcription start site of the gene. If the repressor is not bound to the operator, the RNA polymerase is able to transcribe the gene. If the repressor binds to the operator, the RNA polymerase does not begin transcription. Whether a repressor binds or not depends on some signal it receives from the environment. **(B)** Activators help RNA polymerase bind to weak promoters and to begin transcribing. If the activator does not bind, the RNA polymerase will not bind well and transcription will not occur. Some genes are controlled by both repressors and activators.

RNA polymerase either cannot bind the promoter region or, if it binds, cannot proceed with transcription.

To regulate gene expression, a repressor must be able to respond to a chemical signal (usually a small molecule) that reflects a change in the environment of the bacterium. This can be accomplished in two ways. First, a repressor protein may bind to its operator site in the absence of a chemical signal that triggers expression of the gene. But under conditions in which expression of the gene is needed, the chemical signal binds the repressor, causing the repressor to change conformation and cease binding the operator region (Fig. 6.6). Transcription can then proceed. When the chemical signal molecule is no longer present, the repressor resumes its original conformation and binds once more to its operator site. Transcription stops. An example of this type of repressor is a protein that controls the expression of genes that allow *E. coli* to utilize the disaccharide lactose. On entering the cell, lactose is modified to a form that binds the repressor protein, causing the protein to leave the operator site. When lactose is no longer available, the repressor can return to its normal conformation and bind to the operator site, stopping transcription.

Another type of repressor works in the opposite direction. That is, the repressor must bind its chemical signal in order to achieve a conformational change that allows it to bind its operator site and stop transcription. An example is the repressor that regulates the synthesis of a toxic protein produced by *Corynebacterium diphtheriae,* the bacterium that causes diphtheria, (Fig. 6.6). *C. diphtheriae* growing in the throat produce this toxin only when levels of iron in this environment are low. As long as iron levels are high, iron enters the bacterium and binds the repressor protein, changing its conformation so that the repressor-iron complex binds the operator and prevents transcription. Low iron levels lead to a reduction in the number of iron-repressor complexes, and the ironless repressor protein takes on a conformation that no longer binds to the operator site. Transcription occurs. Why would production of a bacterial toxin be controlled by iron concentrations? Diphtheria toxin kills human cells somewhat indiscriminately, releasing stores of iron for the bacteria. A good way for the bacterium to respond to low iron levels is to use iron concentration in the environment to control diphtheria toxin production. The bacterium produces a toxin that will raise iron levels in its vicinity only when it is starving for iron.

Activators. Activators are proteins that help RNA polymerase plus sigma bind to weak promoters (Fig. 6.6). Weak promoters have such poor consensus sequences that they bind RNA polymerase poorly, if at all, unless an activator protein is present. Activators not only help RNA polymerase bind to a weak promoter but also help it separate the strands so that transcription can begin. They do this by binding DNA sequences upstream from the promoter and interacting with RNA polymerase. Some activators, like repressors, change from inactive to active conformation, depending on whether or not they are binding a chemical signal. In other cases the activator is changed from the inactive to the active form when it is modified covalently by another protein—the one that actually senses the chemical signal. A common form of this latter type of activation is called a **two-component regulatory system.** Two-component regulatory systems control many bacterial genes, including some that encode functions important for adaptation of disease-causing bacteria to the human body. Two-component systems are so widely used by bacteria that scientists are considering them as possible targets for antibiotics.

The two parts of a two-component regulatory system are a cytoplasmic membrane protein (the **sensor**) and a cytoplasmic protein (the **activator)** that binds to DNA near the promoter sequence and stimulates RNA polymerase to begin transcription (Fig. 6.7). The regulatory signal (for example, a change in substrate concentration, temperature, or osmotic strength) causes a conformational change in the sensor protein. This conformational change causes the sensor protein to phosphorylate the activator protein. The phosphorylated form of the activator protein is the form that stimulates transcription. The phosphate group is constantly being removed from the phosphorylated form of the activator protein by a phosphatase. This is

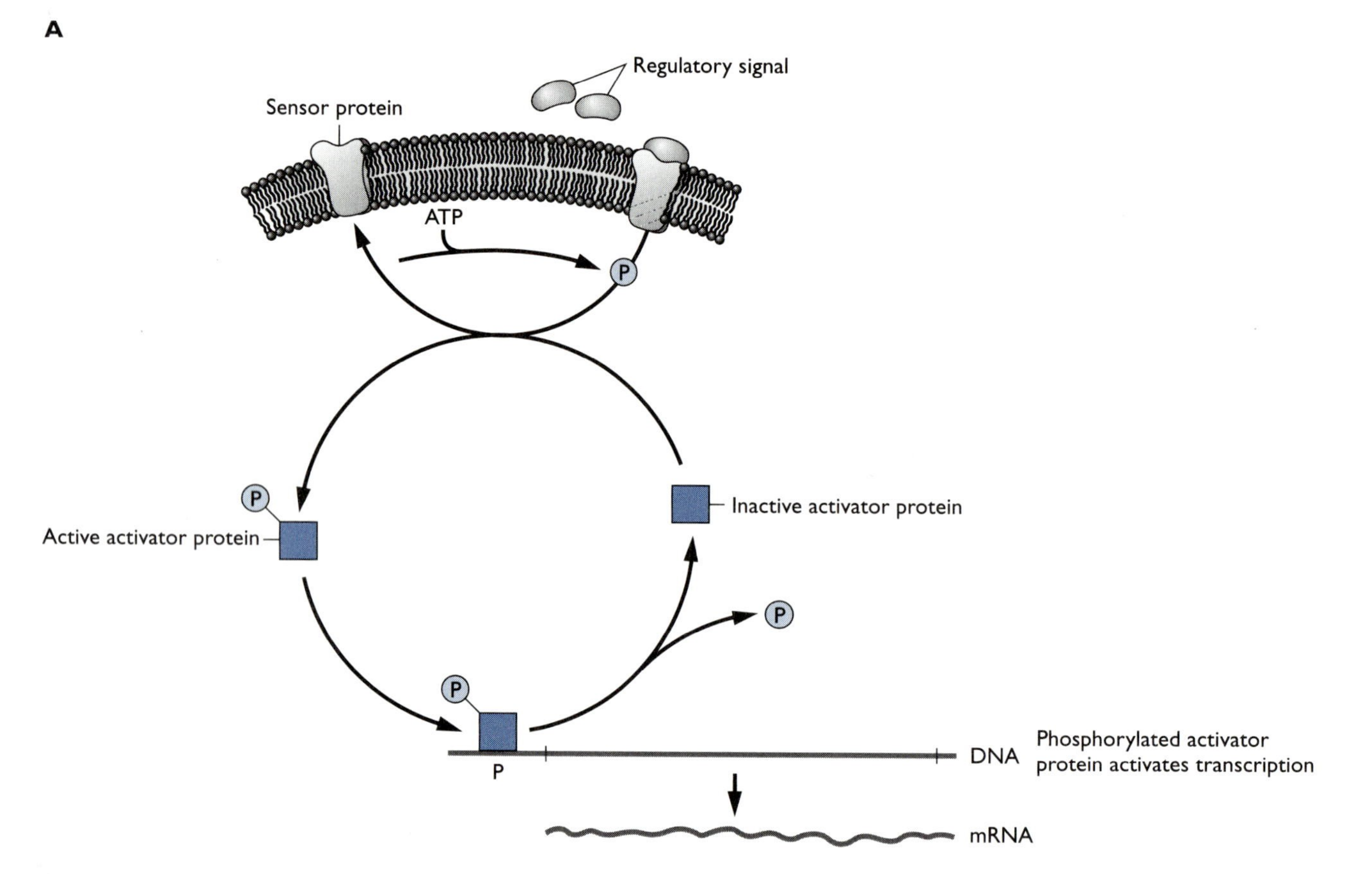

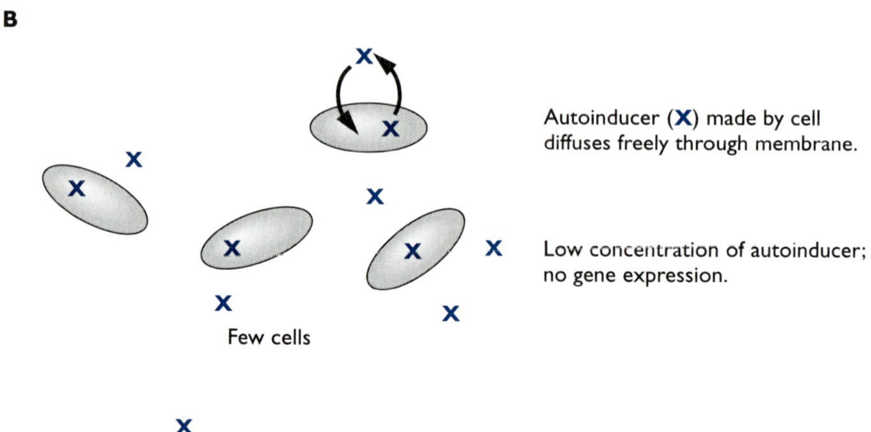

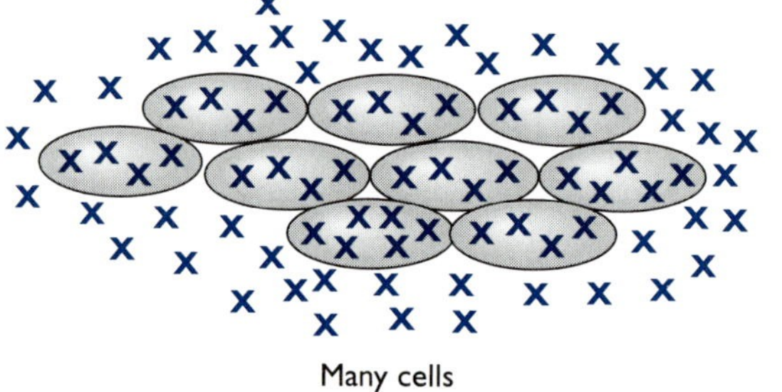

Figure 6.7 Regulation of transcription by two-component regulatory systems and quorum-sensing systems. **(A)** In a two-component regulatory system, a sensor protein in the cytoplasmic membrane of the bacterium undergoes a conformational change in response to a regulatory signal in the environment. The conformational change stimulates the sensor protein to phosphorylate an activator protein. The phosphorylated activator protein activates transcription. The phosphate group is regularly removed from the activator protein as a control measure, to prevent the bacterium from constantly transcribing the gene unless the regulatory signal is present in the environment. **(B)** Quorum-sensing systems recognize signals produced by bacteria. The signals are autoinducers that diffuse through the bacterial membrane. When bacterial numbers are low, the concentration of autoinducers is low. When bacterial numbers increase, the concentration of autoinducers increases. When the level of autoinducer is high enough in the bacterial cell, it will bind the appropriate repressor or activator and allow expression of the gene controlled by the autoinducer.

important for making the switch reversible, because the sensor protein must constantly rephosphorylate the activator. If the sensor stops detecting the signal, it stops phosphorylating the activator. The activator protein becomes de-phosphorylated and gene expression is turned off.

Gram-positive bacteria: peptides

EMRLSKFFRDFILQRKK

SKNSQIGKSTSSKCVFSFFKKC

MAGNSSNFIHKIKQIFTHR

Gram-negative bacteria: homoserine lactones

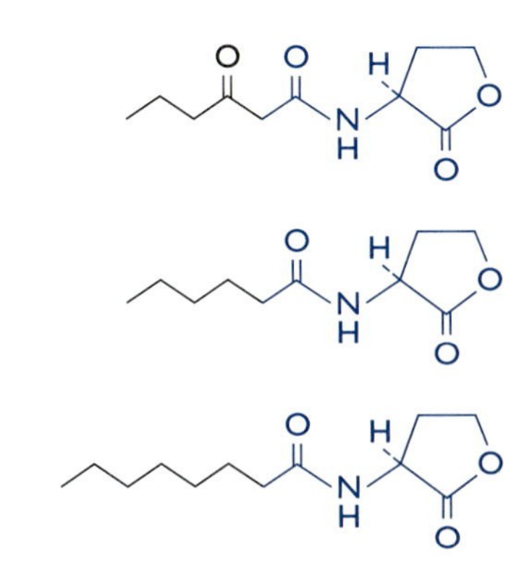

Figure 6.8 Structures of some autoinducers in quorum-sensing systems. Autoinducers produced by gram-positive bacteria are peptides. The amino acid sequence of three of these peptides is shown. (The letter code for the amino acids is given in Fig. 6.10.) Autoinducers produced by some gram-negative bacteria are homoserine lactones. Three examples from the bacterium *Vibrio fischeri* are given.

Quorum-sensing systems. A quorum-sensing system is a type of regulatory system that recognizes a signal produced by the bacterium itself instead of a chemical signal from the environment. **Quorum-sensing** systems can control either repressors or activators. The name "quorum sensing" is derived from the fact that this type of regulatory system senses the concentration of a particular species of bacteria in a site. The signaling molecule in a quorum-sensing system is called an **autoinducer.** Some structures of autoinducers are shown in Figure 6.8. The bacterium continuously produces the autoinducer, which diffuses readily through the bacterial membrane. As long as the concentration of that particular type of bacterium remains low, the concentration of autoinducer outside and inside the bacterium remains low. But if the concentration of bacteria reaches a high enough level, the concentration of autoinducer outside and inside the bacterium increases and finally achieves a level inside the bacterium that allows it to bind the repressor or activator it controls. Binding to the activator or repressor changes the conformation of the protein, allowing it to interact with the RNA polymerase in a way that stimulates gene expression. In the case of an activator, the altered conformation allows the activator to assist RNA polymerase to start transcribing the gene. In the case of a repressor, the altered conformation causes the repressor to leave its operator site, allowing RNA polymerase to initiate transcription.

Two examples illustrating reasons why bacteria might want to monitor the levels of members of their own species are plant pathogens that produce rot and bacteria that form biofilms. The gram-negative plant pathogen *Erwinia* produces rot by excreting enzymes that degrade polysaccharides in the plant cell wall. This provides food in the form of sugars released from the plant wall and liquefies the area, making it easier for the bacterium to spread through plant tissue. Because the degradative enzymes are excreted, other bacteria could harvest the yield from enzymatic breakdown of plant tissue. Also, if an insufficient number of *Erwinia* cells are present, not enough enzyme will be excreted to liquefy the plant tissue. The bacteria solve both problems by waiting to produce the degradative enzymes until the concentration of their species is high enough to make enzyme production worthwhile. Similarly, bacteria that form biofilms must monitor concentrations of like bacteria in the region before taking the trouble to produce the polysaccharide slime and making other changes appropriate to the biofilm form of life. A later chapter on the interactions between bacteria and invertebrates (Chapter 23) includes an example of a quorum-sensing system that turns on light production in a bacterial population that protects a small squid from predators.

Rifampin, an antibiotic that inhibits transcription. RNA polymerase is the target for rifampin (also called rifampicin; Fig. 6.9), an antibiotic currently used to treat or prevent bacterial infections such as tuberculosis and meningitis. **Rifampin** binds to the β subunit of RNA polymerase, distorting it so that it cannot perform transcription. As with the fluoroquinolones, which inhibit DNA gyrase by a similar mechanism (Chapter 5), bacteria can become resistant to rifampin by mutating their RNA polymerase β subunit

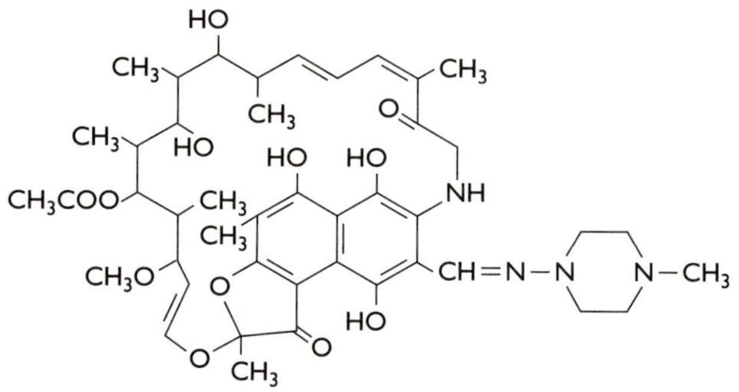

Figure 6.9 Structure of rifampin. Rifampin binds to one of the four subunits (β subunit) of the RNA polymerase core enzyme. Binding of the antibiotic prevents the RNA polymerase from transcribing the gene. Bacteria can acquire mutations in the β subunit so that the enzyme no longer binds rifampin, making the bacteria resistant to the drug.

so that it continues to function in transcription but no longer binds rifampin. Unfortunately, such mutations occur rather easily and resistance arises readily. This is one of the reasons rifampin is used to prevent or treat only a limited number of human diseases. Rifampin is important, however, because it shows that RNA polymerase could be a good target for future antibiotics.

Translation and Its Regulation

Translation, the production of proteins from mRNA

Amino acids, the building blocks of proteins. Proteins are made of 20 primary amino acids (Fig. 6.10). Amino acids differ in chemical characteristics. Some interact best with an aqueous environment (**hydrophilic amino acids**). In Figure 6.10, the hydrophilic amino acids are divided into subgroups of polar (hydrophilic but not charged) and charged amino acids. These amino acids tend to occur in parts of a protein that are exposed to aqueous environments, such as proteins in the cell cytoplasm (Fig. 6.11). The amino acids that do not interact well with an aqueous environment but prefer a hydrophobic environment (**nonpolar amino acids**) interact optimally with each other or with lipids. Such amino acids are most likely to be found inside a protein with its surface exposed to an aqueous environment or embedded in a membrane (Fig. 6.11). In proteins that form pores in membranes, such as the outer membrane porins of gram-negative bacteria, polar amino acids line the channel that is exposed to the aqueous environment and the nonpolar amino acids interact with the membrane lipids. Proteins destined to be transported across the cytoplasmic membrane (secreted proteins) are recognized by the proteins of the secretion system that transports them, on the basis of a short stretch of hydrophobic amino acids at the beginning of the protein. These and other amino acid traits are being used by scientists to predict features of a protein by using computer programs to "translate" the DNA sequence into the string of amino acids it encodes. These predictions are not always accurate but are widely used to provide initial clues about the structure of a protein and its cellular location (cytoplasm, membrane, or secreted through the cytoplasmic membrane).

Hydrophilic amino acids

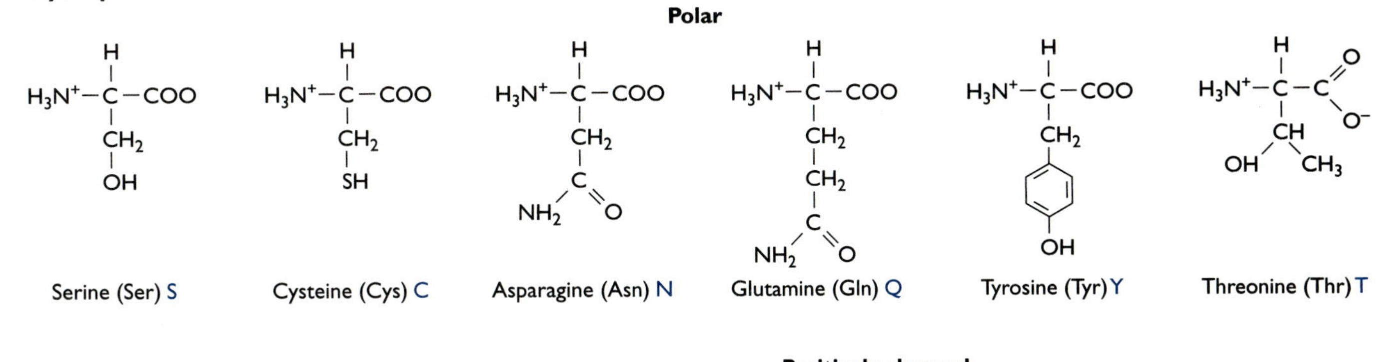

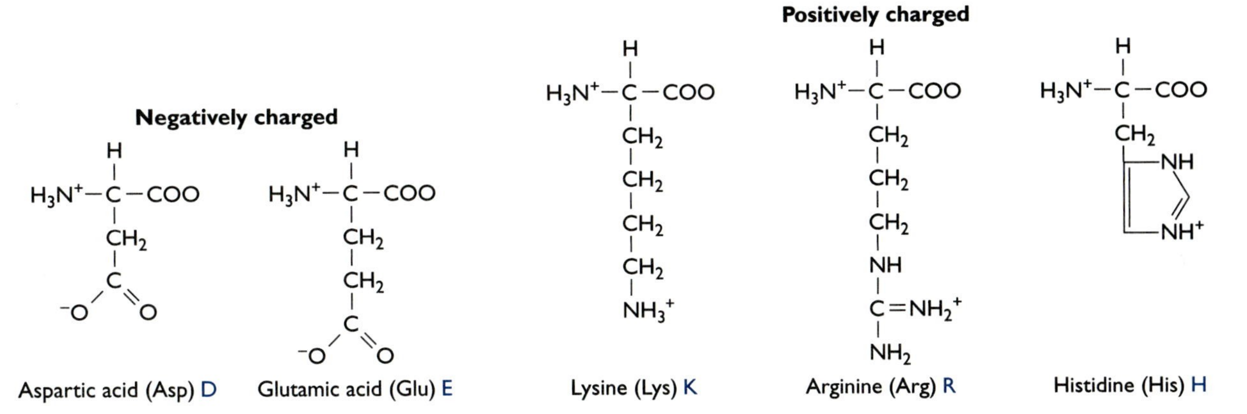

Hydrophobic amino acids (nonpolar)

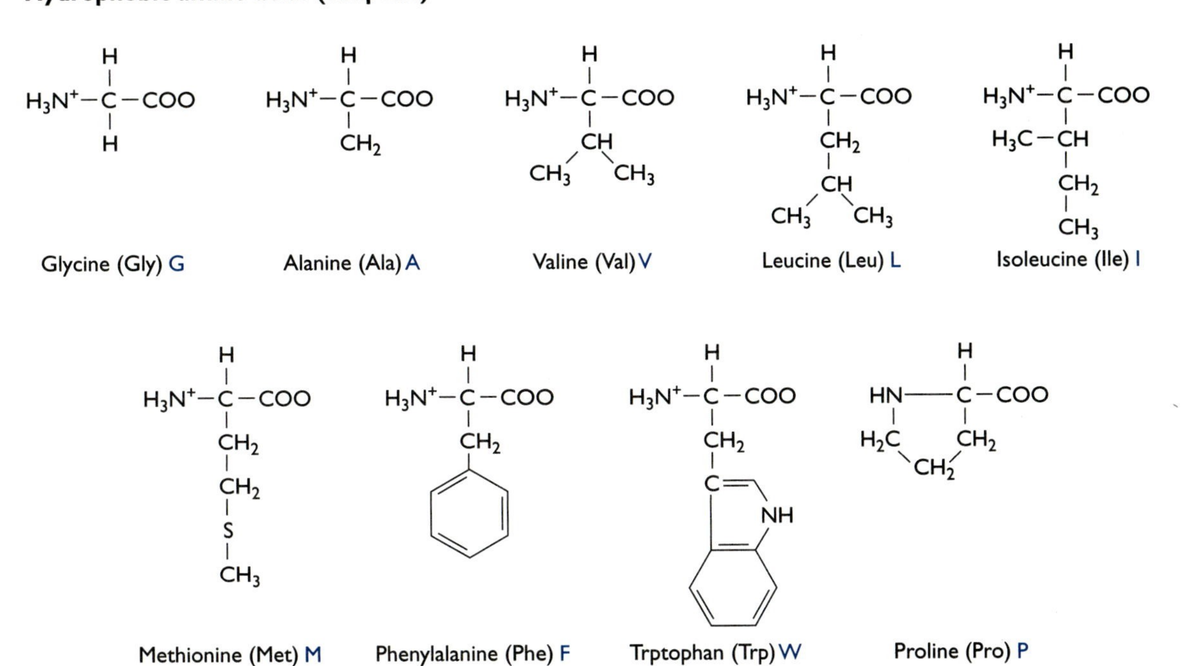

Figure 6.10 Structures of amino acids commonly found in proteins. The hydrophilic amino acids may be neutral, negatively charged, or positively charged. The hydrophobic amino acids are all neutral. The three-letter abbreviation is given in parentheses, and the single-letter designation is in blue.

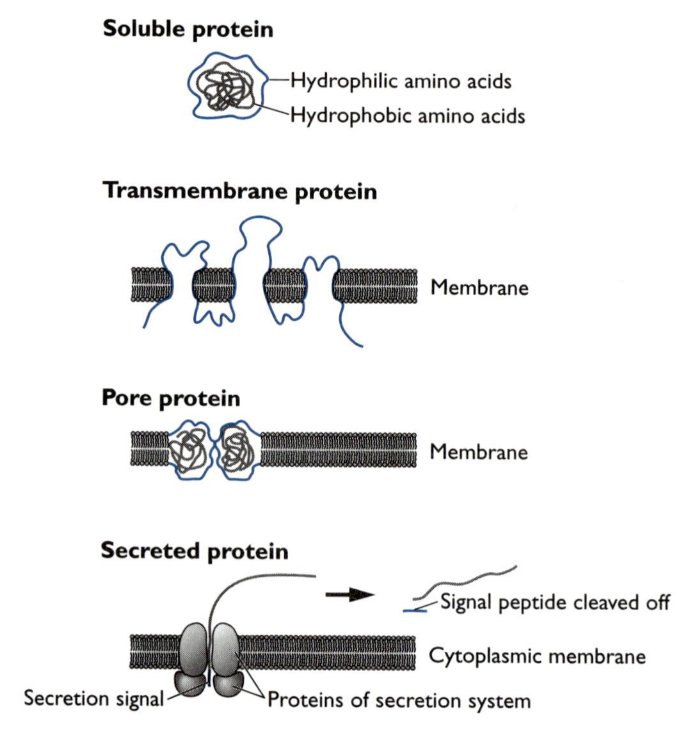

Figure 6.11 Arrangement of hydrophilic and hydrophobic amino acids in proteins. Hydrophilic amino acids usually occur in parts of the protein exposed to the aqueous environment, whereas hydrophobic acids tend to be located on the inside of proteins exposed to an aqueous environment or in regions of a protein embedded in a membrane. In soluble proteins (e.g., cytoplasmic proteins), the hydrophilic amino acids are exposed on the outer surface of the protein or in regions such as the active site of an enzyme that is in contact with the aqueous phase. In membrane-spanning proteins, the portion of the protein embedded in the cytoplasmic or outer membrane is composed of hydrophobic amino acids, whereas the rest of the protein consists of mostly hydrophilic amino acids. Proteins that form channels, such as outer membrane porin proteins, have hydrophilic amino acids on the portions of the protein exposed to the aqueous environment, and the internal portion consists of hydrophilic amino acids. Secretion signals on secreted proteins are hydrophobic. During secretion, the secretion signal is removed.

When mRNA is translated into protein, each of the three-base codons on the mRNA codes for a particular amino acid (Fig. 6.12). Note that some amino acids are specified by more than one codon. Frequently, these codons differ in the third position. Where there is more than one codon for an amino acid, one codon is used more frequently than the others. The correspondence between the codons in the mRNA sequence and the amino acids that will be incorporated into a protein made from that mRNA template allows scientists to deduce the amino acid sequence of a protein from the DNA sequence of the gene. In this way, scientists can translate DNA sequences from an organism to obtain information about its properties. There are now a number of computer programs available for translating DNA sequences into protein sequences. An obvious problem with this approach is that, depending on which base one starts with, three different amino acid sequences can be obtained by translating different reading frames on each strand of the DNA. Scientists (and bacteria) solve the prob-

A Codons for amino acids found in proteins

UUU Phenylalanine	UCU Serine	UAU Tyrosine	UGU Cysteine
UUC Phenylalanine	UCC Serine	UAC Tyrosine	UGC Cysteine
UUA Leucine	UCA Serine	UAA Stop	UGA Stop
UUG Leucine	UCG Serine	UAG Stop	UGG Tryptophan
CUU Leucine	CCU Proline	CAU Histidine	CGU Arginine
CUC Leucine	CCC Proline	CAC Histidine	CGC Arginine
CUA Leucine	CCA Proline	CAA Glutamine	CGA Arginine
CUG Leucine	CCG Proline	CAG Glutamine	CGG Arginine
AUU Isoleucine	ACU Threonine	AAU Asparagine	AGU Serine
AUC Isoleucine	ACC Threonine	AAC Asparagine	AGC Serine
AUA Isoleucine	ACA Threonine	AAA Lysine	AGA Arginine
AUG Methionine (start)	ACG Threonine	AAG Lysine	AGG Arginine
GUU Valine	GCU Alanine	GAU Aspartic acid	GGU Glycine
GUC Valine	GCC Alanine	GAC Aspartic acid	GGC Glycine
GUA Valine	GCA Alanine	GAA Glutamic acid	GGA Glycine
GUG Valine	GCG Alanine	GAG Glutamic acid	GGG Glycine

B Structure of N-formylmethionine

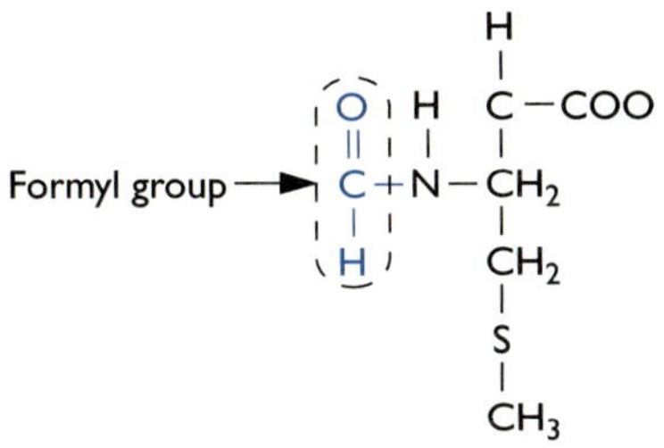

Formyl group → O=C(H)–N(H)–CH(COO)–CH_2–CH_2–S–CH_3

Figure 6.12 Three-base codons for the 20 amino acids found in proteins. (A) Some amino acids (such as trp and met) are specified by only one codon, whereas others are specified by several codons. **(B)** Structure of f-Met, the amino acid that serves as the N-terminal amino acid in proteins.

lem of which amino acid sequence is really the protein encoded by the genes of interest by making use of the signals that dictate which is the first codon (**start codon**) and which is the last to be used in translating a protein (**stop codon**). Start codons and stop codons are easily recognized by their sequence (AUG, start; UCA, UAA, and UAG, stop; Fig. 16.2). The DNA sequence between a start codon and a stop codon is called an open reading frame, indicating that the sequence between the two codons is likely to encode a protein (Fig. 6.13). Often more than one start codon is near the beginning of an open reading frame. In such cases, deciding exactly where

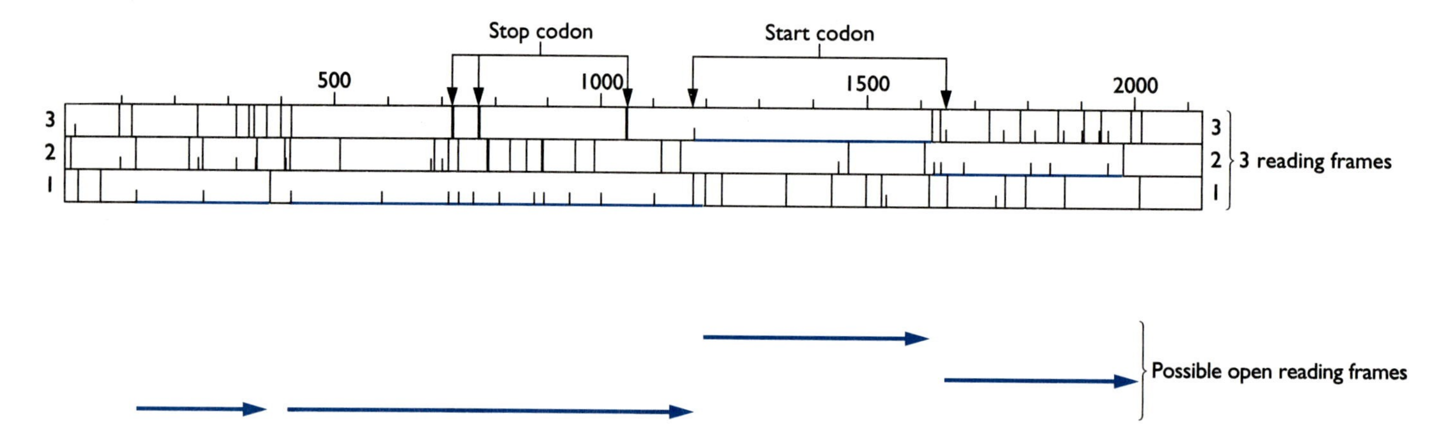

Figure 6.13 Determination of possible open reading frames in a DNA sequence. Stop codons and start codons are easily detected because of their unique sequences. Using this information, it is possible to find open reading frames and, from the DNA sequence, to determine the amino acid sequence of the protein encoded by this region of DNA. Possible start codons are indicated by short vertical bars. Stop codons are indicated by longer vertical bars. The horizontal arrows at the bottom of the figure indicate the length and direction of transcription of possible *orfs* deduced from the DNA sequence. They are displaced vertically from each other to indicate that they are in different reading frames.

the amino acid sequence of the putative protein begins requires a more detailed analysis of the actual protein encoded by the open reading frame. One example of how this type of analysis, coupled with information that can be deduced from the amino acid sequence of a protein, is being used to hunt for a vaccine against *Neisseria meningitidis* type B, a common cause of epidemic meningitis, can be found in Box 6.1.

Components of the translation apparatus. Ribosomes, the protein-RNA complexes on which mRNA is translated into proteins, have been very popular targets for antibiotics, and many of the antibiotics currently in use are protein synthesis inhibitors. Most of the newest antibiotics, including those recently approved by the U.S. Food and Drug Administration and those currently in clinical trials, inhibit bacterial protein synthesis by binding to ribosomes.

Bacterial ribosomes are composed of two subunits, a large (**50S**) subunit and a small (**30S**) subunit. The S is a sedimentation unit. The size of a large molecule or complex can be estimated by determining how rapidly it travels through a gradient from the top of a centrifuge tube to the bottom when the tube is subjected to high *g* forces in a centrifuge. The smaller the protein or complex, the more rapidly it travels. (The reason that the whole ribosome is **70S,** whereas the two subunits are 50S and 30S, is that the intact ribosome sediments differently than the sum of the two subunits because of its more compact structure.) The 50S and 30S subunits are composed of three different types of rRNA (5S rRNA, 16S rRNA, and 23S rRNA) and a number of ribosomal proteins.

Another important component of the translation machinery is **tRNA.** tRNA carries each amino acid to the ribosome, where it associates the amino acid with a particular codon on the mRNA being translated by the ribosome. Covalent attachment of an amino acid to a tRNA molecule is shown in Figure 6.14. Energy in the form of adenosine triphosphate (ATP) is required for attachment, a process called **charging.** The tRNA with its amino acid attached retains enough energy to enable the ribosome to trans-

box

6–1 *Putting Genome Sequences To Work: the Search for a More Effective Meningitis Vaccine*

Neisseria meningitidis, a cause of epidemic bacterial meningitis, was introduced in Chapter 3. Because outbreaks can arise so unexpectedly and can kill before effective preventive action is taken, vaccination of populations at high risk is an solution to preventing such outbreaks. In the developed world, institutionalized populations, such as college students and military recruits, would be targets for vaccination. In the developing world, especially Africa, where outbreaks of epidemic *N. meningitidis* kill or neurologically impair many thousands of people each year, widespread vaccination of the entire population would be highly desirable.
A vaccine is available against two types of *N. meningitidis,* types A and C. Different types have differences in their capsular polysaccharides that make it necessary for the immune system to respond to each type separately. Unfortunately, no vaccine exists for type B, the most common cause of epidemic meningitis. In part, this is because type B meningococcus mimics human cells by covering itself with sialic residues, sugars commonly found on the surfaces of human cells. To foil this bacterial strategy for evading the immune system, it is important to find bacterial molecules that, unlike sialic acid, identify the bacterium unambiguously for immune system attack. As will be explained in Chapter 14, the best vaccine against a disease like that caused by *N. meningitidis* consists of one or more proteins that are located on the bacterial surface.

Recently, the genome sequence for *N. meningitidis* type B has become available, and scientists set out to use this information to find a vaccine against type B disease. Immediately, the investigators began a hunt through all the possible proteins encoded by open reading frames for proteins that might be localized to the surface of the bacteria. *N. meningitidis* is a gram-negative bacterium, so a protein on the surface of the outer membrane would be most likely to elicit a protective immune response. Pili, which extend outward from the cell surface (Chapter 4), would be obvious vaccine candidates. But an additional characteristic of *N. meningitidis* complicates this approach. The bacteria change the amino acid sequences of their pili from time to time, so that the immune system does not recognize them. Because of this, scientists looked for other proteins with a similar location that were not so variable in sequence At this point, identification of outer membrane proteins from DNA sequence information is still an elusive goal. Yet, there are some steps that can be taken to sort through all the possible genes to find ones that might qualify as surface-exposed proteins. The strategy used by the genome sequencers includes the following steps:

- Identify all possible open reading frames and obtain the amino acid sequence of the protein.
- Identify possible secreted proteins by looking for a signal sequence at the amino terminus of the translated protein.
- Among possible secreted proteins, look for outer membrane proteins by looking for the amino acid sequence motifs that are known, e.g., transmembrane regions (Fig. 6.11) or a hydrophobic region that anchors the protein in the outer membrane.
- Clone and express promising-looking proteins in *E. coli* to obtain large quantities of protein (Chapter 5).
- Test the proteins for the ability to elicit a protective immune response.

Scientists used these and similar criteria to narrow the field of 570 possible secreted proteins in the genome sequence to 350 proteins. Although this number seems large, modern technology makes it a manageable number for further testing to determine whether the proteins elicited the desired immune response. As this is written, the proteins identified are still being tested as vaccine candidates. Although the results are incomplete at this point, the approach shows how genome sequence data can be put directly to work to develop important pharmaceutical compounds. If this strategy or some variation of it is successful, the development of vaccines could be speeded up significantly. This example also illustrates why it is so important to refine further our ability to deduce features of protein structure and localization from the amino acid sequence data.

Source: **Pizza, M., et al.** 2000. Identification of vaccine candidates against serogroup B meningococcus by whole-genome sequencing. *Science* 287:1816–1820.

fer that amino acid to the growing protein chain. Each amino acid is attached by a specific tRNA synthetase enzyme to the tRNA with an **anticodon** segment complimentary to the triplet code for that amino acid.

Not much attention was paid at first to the charging of tRNA as a target for antibiotics. One antibiotic, **mupirocin,** inhibited **tRNA synthetases** but had very limited use because it could be administered only topically. Because of the need for new antibiotics, however, pharmaceutical companies are taking another look at the tRNA synthetases as possible drug targets. They hope to find new nontoxic versions of mupirocin or other compounds that inhibit tRNA synthetases.

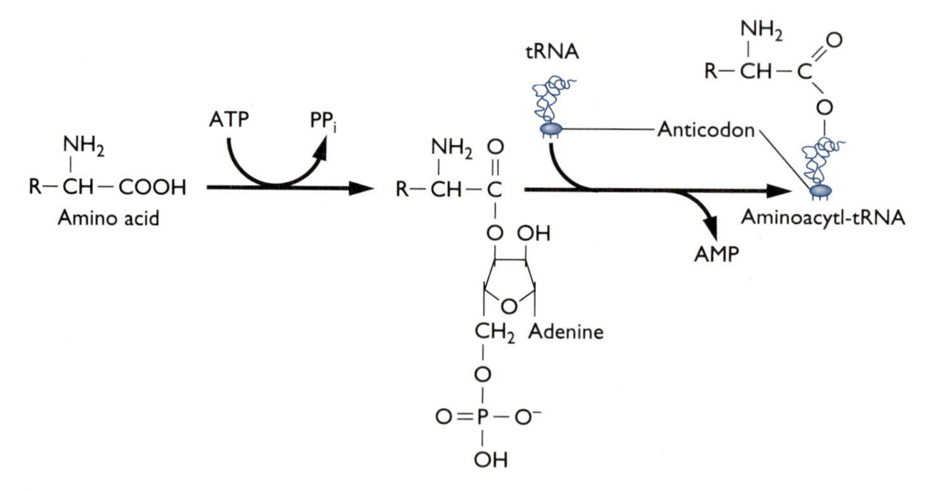

Figure 6.14 Attachment of an amino acid to tRNA. Amino acids are attached to the appropriate tRNA by an enzyme, aminoacyl-tRNA synthetase. Adenosine triphosphate (ATP) provides the energy for the reaction. The tRNA has an anticodon (made up of three bases) for the amino acid it carries that is complementary to the codon on the mRNA.

Steps in translation. The steps in translation are shown in Figure 6.15. The ribosome binds to the 5' end of the mRNA at a special recognition sequence, called the **ribosome binding site** (**rbs**). The assembled ribosome (70S) has two **aminoacyl-tRNA (AA-tRNA) binding sites,** the **A (aminoacyl) site** and the **P (peptide) site.** These sites bind the tRNA so that the anticodon part of the tRNA structure aligns with the triplet codon on the mRNA and so that adjacent AA-tRNAs are close enough together that the peptide bond can be formed between the amino acids. Translation starts when the first AA-tRNA binds the ribosome-mRNA complex (**initiation complex**). This first AA-tRNA is usually a formyl-methionine **(fMet)-tRNA,** the codon for which is AUG. Thus, **AUG** is called the start codon. This end of the protein is called the **amino terminus** or N-terminus of the protein (Fig. 6.15). Methionines that occur after the start codon are not formylated. The formyl group may help the bacterium identify the amino terminus of the protein. The use of fMet as the first amino acid in a protein is unique to bacteria and is not found in archaea or eukaryotes

As the ribosome moves along the mRNA, the tRNA carrying the growing peptide chain is located in the P site. The next incoming AA-tRNA binds to the A site. The peptide bond is formed, shifting the peptide to the AA-tRNA in the A site. The empty tRNA leaves the P site, and the tRNA carrying the peptide chain moves to the P site as the ribosome moves (**translocates**), so that the A site is now over the next codon. In addition to the P and A sites, a third tRNA binding site, the **E (exit) site,** receives the empty tRNA before it is released. Eventually, the ribosome encounters a stop codon, which does not code for any amino acid (**UAA, UAG, UGA**). At the stop codon, the protein is released, and the ribosome leaves the mRNA and dissociates into the 30S and 50S subunits (Fig. 6.15). The last amino acid is called the **carboxy terminus** or C-terminus of the protein.

Regulation of translation. The sequence of the mRNA at the ribosome binding site, the site where the ribosome binds and begins translating an

A

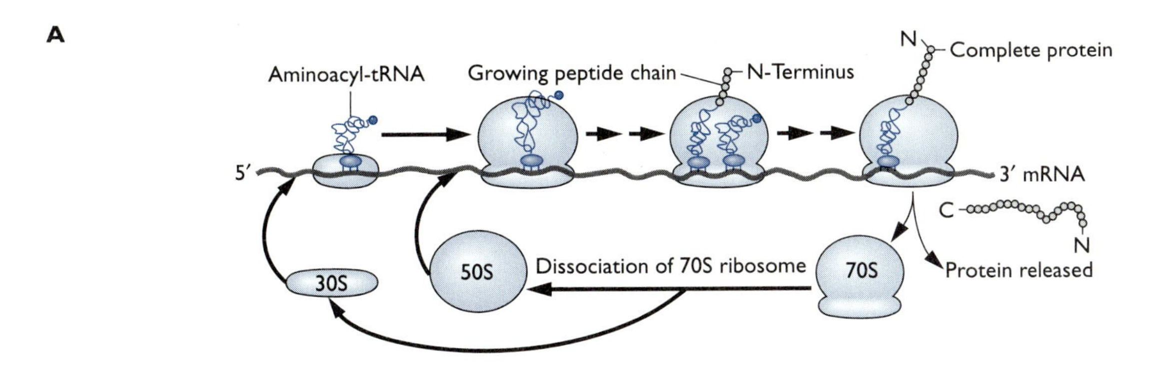

B

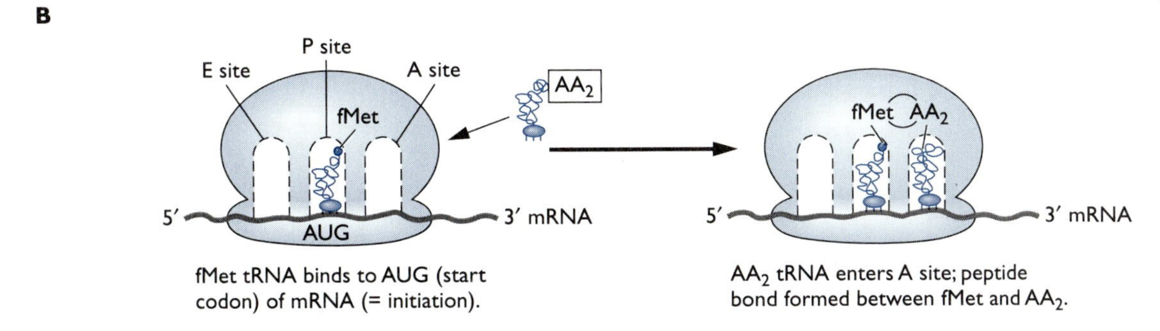

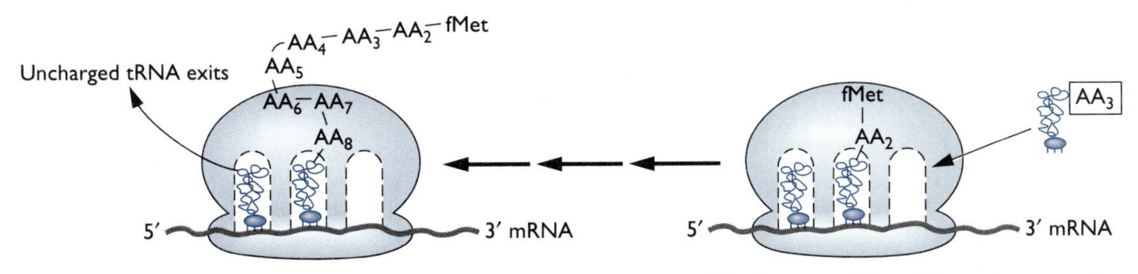

C Features of proteins

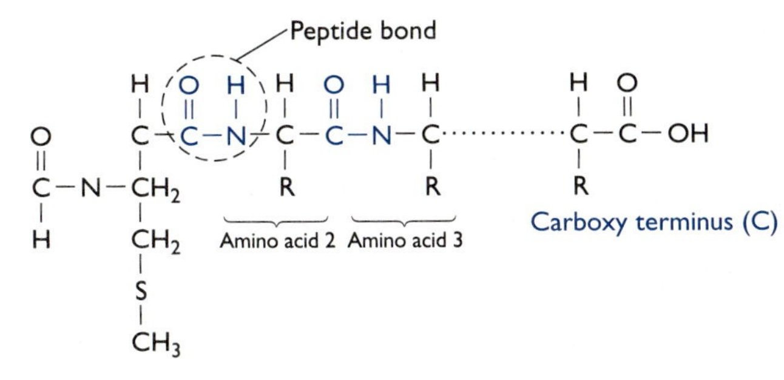

N-formyl-methionine
Amino terminus (N)

R = side group unique to a specific amino acid.

Figure 6.15 Translation of mRNA. (A) Overview of translation. An aminoacyl-tRNA binds to the 30S ribosomal subunit. This complex recognizes the ribosomal binding site on the mRNA. The 50S subunit joins this complex to form the 70S ribosome, which carries out translation. The first amino acid in the protein is called the amino-terminal end of the protein. Once the protein is synthesized, it is released from the ribosome and the ribosome dissociates again into 30S and 50S subunits. **(B)** Details of translation. The 70S ribosome has three sites. The tRNA carrying the growing peptide chain is located at the P site. Incoming aminoacyl-tRNAs (AA-tRNAs) bind at the A site. The E site receives the empty tRNA before it exits the ribosome. Initiation of translation occurs when tRNA carrying formyl methionine (fMet-tRNA) binds to the start codon (AUG) of mRNA. This occurs in the P site of the ribosome. fMet is the amino-terminal amino acid. As the ribosome moves along the mRNA, an AA-tRNA enters the A site. A peptide bond is formed between fMet and the new amino acid (AA1). fMet is released from the tRNA, which moves to the E site. The tRNA, which is bound to the growing chain, is translocated to the P site. This process continues until the protein is completely synthesized and released from the ribosome. The final amino acid added to the protein serves as the carboxy-terminal end. **(C)** Structural features of a protein. The amino-terminal amino acid is fMet, which is linked to a second amino acid by a peptide bond. Subsequent amino acids also are linked to each other by peptide bonds. The final amino acid added to the protein serves as the carboxy terminus. After the protein is completed, the formyl group is removed from fMet and the amino group is restored.

mRNA, also determines how often ribosomes bind that site and thus how many proteins are produced from that mRNA. An example of how bacteria use this type of regulation is provided by some bacterial protein toxins that have two protein subunits, a B subunit that binds the target human cell and an A subunit that enters the cell and damages it. Such bacterial toxins, which are important in a number of bacterial diseases, will be discussed in more detail in Chapter 15. For now, however, consider the following problem a bacterium must solve. Suppose, as is the case with a number of bacterial toxins, that although several copies of the B subunit protein are needed, only one copy of the A subunit protein is necessary. Moreover, the bacteria want to make both components of the toxin at the same time. A solution to this problem is shown in Figure 6.16. The genes for subunit A and subunit B are organized in an operon. Thus, the same promoter responding to the same conditions controls the expression of both genes. Yet the toxin needs five copies of B and one copy of A to be fully active. This ratio can be attained if the ribosome binding site of the gene encoding B is five times as effective at initiating translation as the ribosome binding site of the gene encoding A. Thus, from a single transcript of the operon, one A protein is produced for every five B proteins. Because the proteins are being synthesized at the same time in close proximity to each other, it is easier for the toxin complex (1 A, 5 B) to be assembled.

Effects of mutations

Knowing the signals that direct RNA polymerase to initiate or terminate transcription and direct ribosomes to initiate or terminate protein synthesis makes it possible to understand how mutations in different parts of a gene affect its expression and the protein it encodes. Mutations in the promoter region can increase expression of a gene, decrease expression of a gene, or eliminate regulation of a gene whose expression is normally regulated, causing it to be expressed constitutively. Alteration of the ribosome binding site can eliminate or increase production of a protein from an mRNA transcript. Within the sequence that is translated into a protein, changing a base

Figure 6.16 Regulation of translation. Two genes in a single operon may be translated in different amounts, depending on the strength of their ribosome binding sites (rbs). In this figure, which shows an operon encoding a toxin that is composed of one A subunit and five B subunits, gene A has much weaker rbs than gene B. This means that ribosomes will bind to gene B much more frequently (five times more frequently in this figure) than to gene A. The result is that gene B will be translated five times as frequently as gene A. This allows the bacterium to produce the appropriate number of each type of subunit from a single transcript.

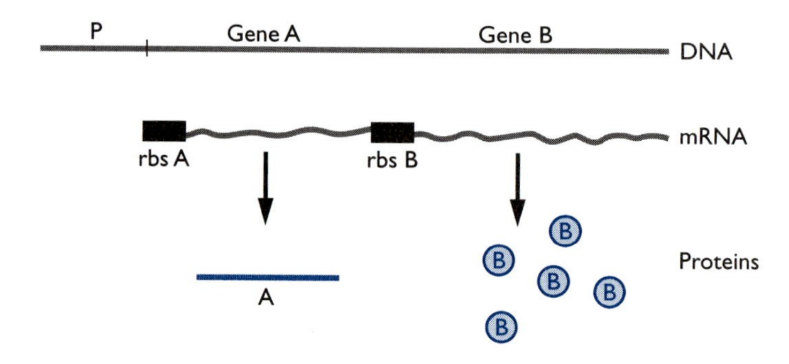

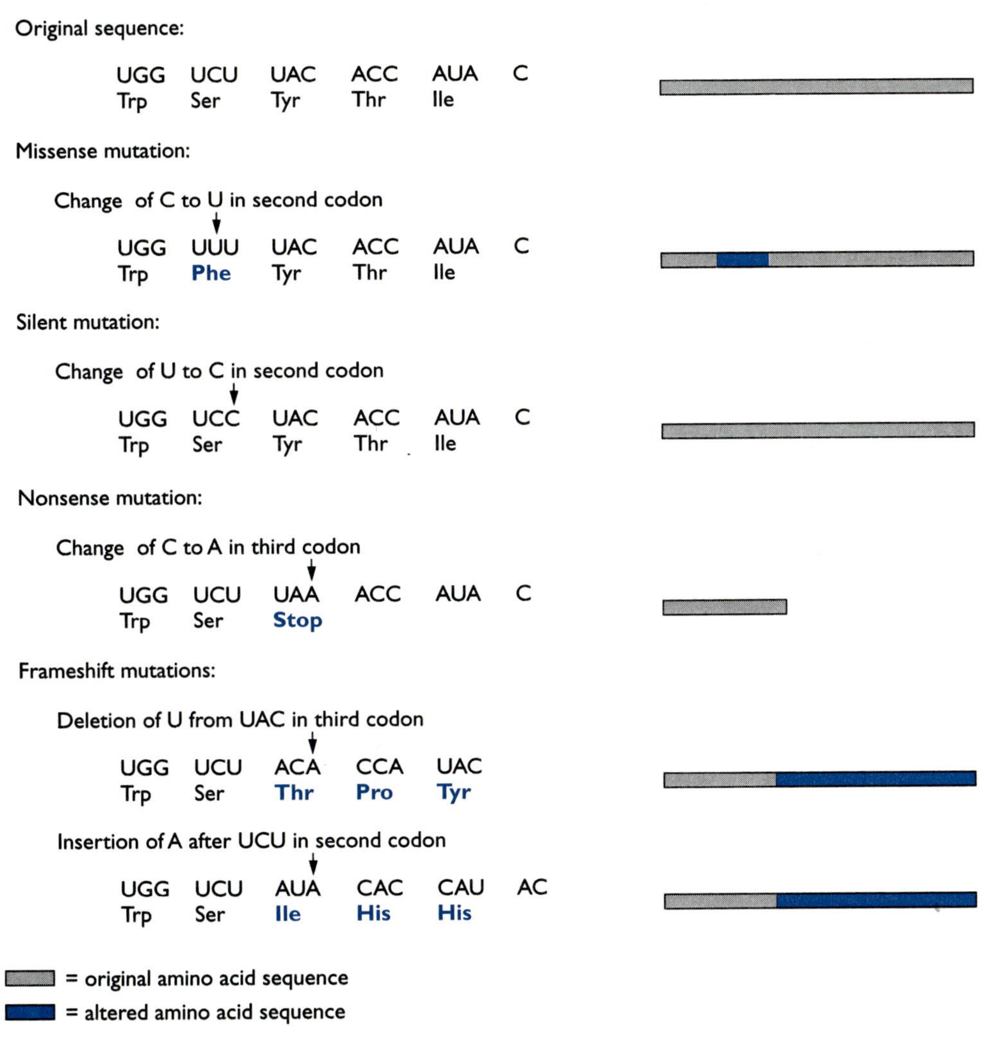

Figure 6.17 Consequences of mutations occurring at different sites in the DNA. Different types of mutations have different effects on the results of translation. Missense mutations change only one amino acid in the protein. Silent mutations will produce no changes in the sequence of amino acids. Nonsense mutations create a stop codon, which results in a truncated protein. Frameshift mutations cause a variety of effects, some more serious than others.

in a codon may change the amino acid inserted at that point. Such mutations are called **missense mutations** (Fig. 6.17). Changing an amino acid often makes a protein less active or less stable, but sometimes such mutations can make the protein work better. Missense mutations were responsible for the changes in penicillin-binding proteins that made some of them resistant to penicillin. Not all missense mutations affect the protein. Because some amino acids are encoded by more than one triplet of bases, changing a base may change the DNA sequence but not the amino acid. Or the new amino acid may have no effect on protein activity or stability. Such mutations are called **silent mutations.**

Single-base changes can have far more drastic effects on a protein (Fig. 6.17). A mutation that changes a codon from one that encodes an amino acid to a stop codon will terminate the protein prematurely (**nonsense mutation**). Insertion or deletion of a base can change the reading frame (**frameshift mutation**). That is, as the ribosome moves along the mRNA, reading three bases at a time, it will move into another reading frame when it hits the added or deleted base. The remainder of the protein after this point will have a different amino acid sequence from the original protein and will terminate at a different point. Finally, if the stop codon of a protein is changed to one that codes for an amino acid, the length of the protein will be extended. Nonsense mutations, frameshift mutations, and mutations that eliminate stop codons are usually deleterious to the cell, because the protein is no longer functional or is rapidly degraded. Such mutations are clearly harmful for bacteria. Thus, random mutation of a bacterium's genome results in the death of many bacteria, with only an occasional survivor carrying a beneficial or silent mutation.

Antibiotics that inhibit translation

Mechanisms of action. The translation machinery of bacteria is a popular target for antibiotics, largely because the complex process of translation can be inhibited in so many ways. An antibiotic could inhibit initiation or one of the steps in elongation. Several different classes of antibiotics inhibit protein synthesis, each with a different target. These antibiotic classes include the **aminoglycosides** (e.g., streptomycin, spectinomycin, gentamicin, and kanamycin), the **macrolides** (e.g., erythromycin), the **tetracyclines** (e.g., tetracycline and doxycycline), the **lincosamides** (e.g., lincomycin and clindamycin), and **chloramphenicol.** All of these prevent protein synthesis by binding to ribosomes and inhibiting translation. Structures of antibiotics representative of these classes are shown in Figure 6.18. Note the variety of structures, an indication that the different antibiotics bind to different sites on the ribosome.

Aminoglycosides and tetracyclines bind to the 30S ribosomal subunit. Binding of aminoglycosides affects the interaction between the ribosome and the mRNA. In some cases (for example, streptomycin), the effect is to freeze the ribosome on the mRNA so that it cannot continue the translation process. In other cases (for example, spectinomycin), a later stage of translation is affected. Binding of tetracycline distorts the A site so that the AA-tRNA cannot bind to this site, thus stopping the elongation of the protein. The other classes of protein synthesis inhibitors also prevent steps in elongation but accomplish this by binding to the 50S subunit.

Resistance to protein synthesis inhibitors. Bacteria have developed a variety of mechanisms for resisting protein synthesis inhibitors. Bacteria become resistant to aminoglycosides by producing an enzyme that modifies the aminoglycoside and inactivates it. The **aminoglycoside-modifying enzymes** include one that takes the acetyl group from acetyl-S-CoA and joins it covalently to the antibiotic and others that take the adenyl group or phosphate group from ATP and join them to the antibiotic (Fig. 6.19). These modifications prevent the antibiotic from binding the ribosome. One mechanism of resistance to chloramphenicol also involves inactivation of the antibiotic. An enzyme, **chloramphenicol transacetylase,** takes the acetyl group from acetyl-S-CoA and uses it to produce acetylchloramphenicol, which is inactive (Fig. 6.19).

Bacteria become resistant to erythromycin and lincosamides by acquiring a gene that encodes an rRNA methylase. This methylase adds a methyl

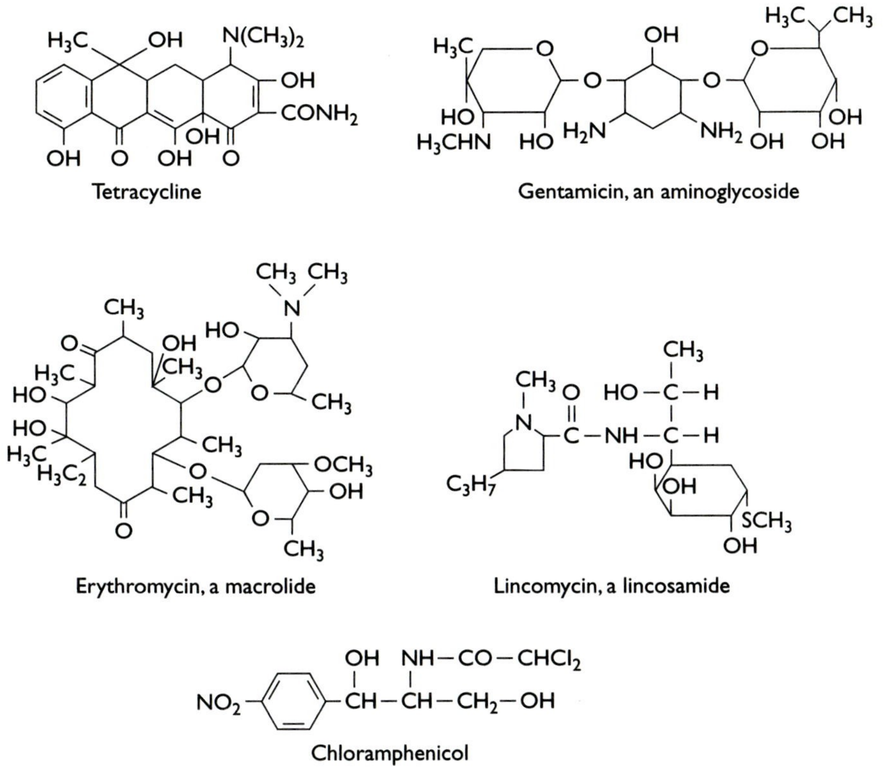

Tetracycline

Gentamicin, an aminoglycoside

Erythromycin, a macrolide

Lincomycin, a lincosamide

Chloramphenicol

Figure 6.18 Structures of some antibiotics that inhibit translation. Aminoglycosides and tetracyclines bind to the 30S ribosomal subunits. The other drugs prevent elongation of the growing protein chain by binding to different sites on the ribosome.

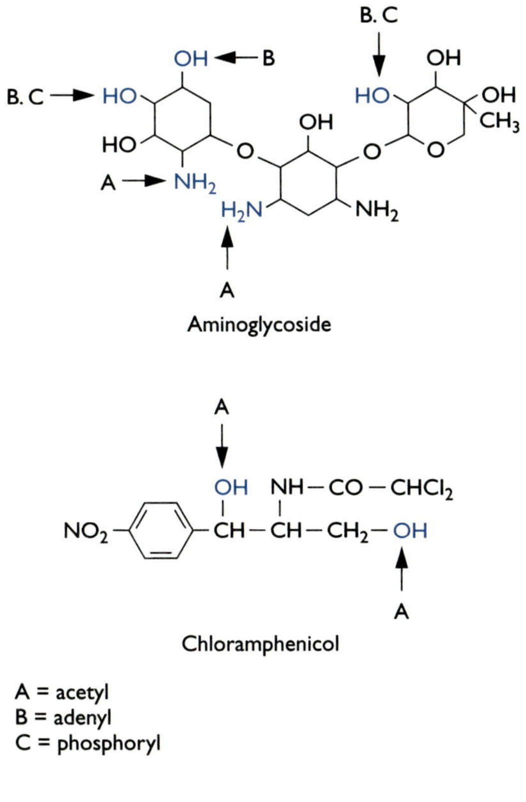

Aminoglycoside

Chloramphenicol

A = acetyl
B = adenyl
C = phosphoryl

Figure 6.19 Mechanisms of resistance to protein synthesis inhibitors. A variety of enzymes can modify the aminoglycosides by adding a group to the antibiotic. Some groups that may be added are acetyls, adenyls, and phosphates. Chloramphenicol can be modified by chloramphenicol transacetylase, which adds the acetyl group from acetyl-S-CoA to form acetyl-chloramphenicol. The modified forms of the antibiotics are inactive because they no longer bind the ribosome.

group to a specific adenine residue in ribosomal RNA. Ribosomes modified in this way no longer bind erythromycin and lincosamide. The two most common mechanisms of resistance to tetracycline are ribosome protection and tetracycline efflux. The **ribosome protection** mechanism is similar to the mechanism of resistance to erythromycin and lincosamides. The resistance protein binds to the ribosome and somehow prevents tetracycline from binding and stopping translation. The way in which the resistance protein accomplishes this is not known. **Tetracycline efflux** is mediated by a cytoplasmic membrane protein. This protein forms a pump that pumps tetracycline out of the cell as fast as it is transported in. Thus, the intracellular concentration of tetracycline is kept too low for effective binding to the ribosomes.

Until recently, tetracycline efflux was thought to be the only example of the efflux type of resistance, but recent studies have shown that efflux pumps exist for other antibiotics, such as chloramphenicol and macrolides. Some of these efflux pumps confer resistance to more than one class of antibiotic and are similar to the multidrug-resistance pumps seen in some human tumor cells, which protect the tumor from antitumor drugs.

Post-Translational Control

For some proteins, transcription and translation occur at more or less constant levels, but the activity associated with the protein varies with the conditions the bacterium is experiencing. This type of regulation can occur in several ways, but only two examples will be given here to illustrate the principle. Generally, **post-translational control** involves enzymes. Enzymes are proteins that bind a substrate, which they proceed to modify. They can also bind other compounds, called inhibitors, that reduce or eliminate their ability to modify their substrate. One way bacteria use this type of inhibition is illustrated by amino acid biosynthetic pathways. The enzymes involved in producing an amino acid form a pathway that involves many different steps. Some of these steps consume energy. If the bacterium itself has made an excess of the amino acid or if the amino acid suddenly becomes available in the external environment, it is no longer expedient for the bacterium to expend the effort to synthesize the amino acid. A way to switch off the biosynthetic pathway when excess amino acid accumulates is to have the amino acid bind an enzyme in the biosynthetic pathway, often at a site different from its active site. Binding of the amino acid inhibits the activity of the enzyme by distorting its shape. When the level of amino acid decreases, the amino acid is no longer available to bind the enzyme and the enzyme becomes active again (Fig. 6.20). This type of regulation has been called **feedback inhibition,** because the bacterium is using amino acid concentrations as a feedback indicator that it is time to shut off synthesis of the amino acid.

Feedback inhibition is a problem for the biotechnology industry. For example, amino acids used as supplements in certain types of foods are produced by bacteria. Suppose that a research team has gone to considerable lengths to increase the activities of one or more enzymes in the pathway for synthesis of the amino acid, but their efforts then are foiled by feedback inhibition, which shuts down the pathway as soon as the amino acid begins to increase in the growth medium. Feedback inhibition becomes the enemy of the production team, and ways to overcome it must be found. One way to overcome feedback inhibition is to have the bacteria produce much higher levels of the inhibited protein so that a much higher

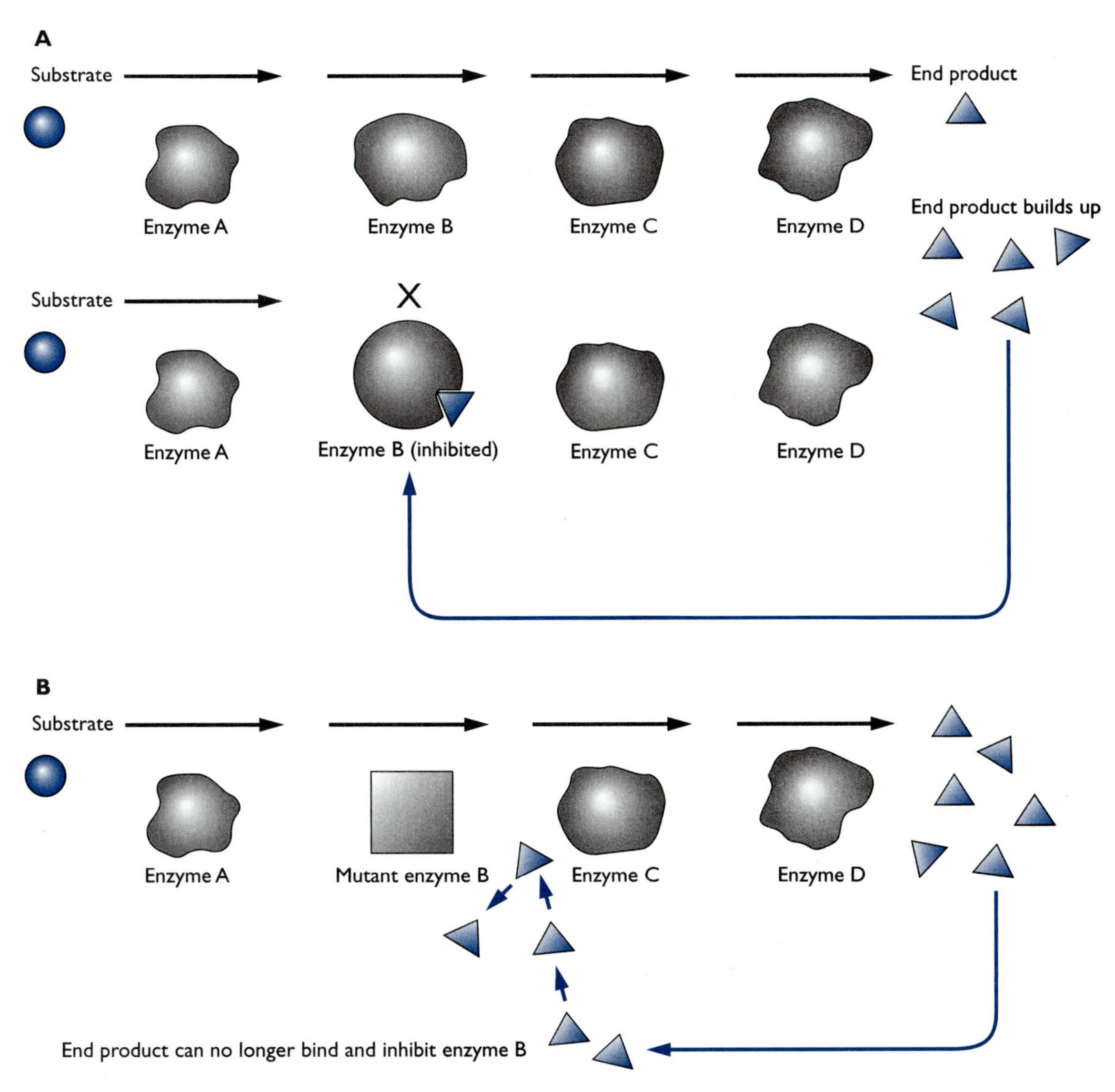

Figure 6.20 Example of feedback inhibition. A substrate is converted to an end product in a series of reactions catalyzed by different enzymes. Arrows show the flow of substrate through the pathway. **(A)** If a bacterium has produced a sufficient amount of end product, it is not energy efficient to continue to produce more of that end product. In this case the end product will bind to one of the enzymes in the pathway and alter it so that it is no longer active. Once the end product is used, the inhibition of the enzyme will be reversed and the process of converting the substrate to product will continue. **(B)** In industrial applications in which large amounts of end product are desired, feedback inhibition creates a problem. One way to avoid this is to select for mutant enzymes that will no longer bind the end product, and thus the process can continue uninterrupted.

concentration of amino acid is needed to shut off the system. Another way is to obtain a mutant form of the inhibited enzyme that is no longer inhibited by the amino acid.

Another way to regulate the activity of an enzyme is to attach a chemical group that inactivates the enzyme but that can be removed later to restore activity. An example of this type of regulation is provided by nitrogenase, a complex of enzymes that converts atmospheric nitrogen to ammonia, which is then used by the organism for synthesis of nitrogen-containing compounds. The nitrogenase complex consumes substantial amounts of energy. Thus, if exogenous ammonia becomes available, it is highly desirable to shut down this energy-expensive complex. This is done by covalently modifying one of the enzymes in the complex, shutting off its activity. The covalent modification is reversible so that at a later time, when ammonia is no longer available, the bacteria can remove the covalently bound group and reactivate the nitrogenase complex. Nitrogenase will be described in more detail in a later chapter on some important environmental bacteria that are part of marine phytoplankton (Chapter 22).

Genome Sequences

Reading genome sequences

Much has been written in the press about how genome sequences will produce a wealth of new information, but these accounts rarely explain how this wealth of information will be obtained and put to use. The first complete genome sequences of free-living organisms were those of prokaryotes, the genomes of which usually are smaller than 6000 kbp in size. Once the bases are assembled into a continuous genome sequence and stored in a computer, scientists must then decode all those As, Cs, Ts, and Gs. The first step is to use a computer program to identify open reading frames (*orf*s) as shown in Figure 6.13. The computer translates the DNA sequence into amino acids. Scientists look for start codons and stop codons to identify open reading frames. Some *orf*s will be encoded by one strand of the DNA and others by the other strand of the DNA. The direction that leads from start codon to stop codon is called the direction of transcription and is indicated by an arrow. Arrows indicating *orfs* encoded by one DNA strand point in one direction and those encoded by the other strand point in the opposite direction. In the map of an entire genome, there are too many *orfs* to allow this sort of representation, and *orfs* are represented by bars whose lengths indicate their sizes (Fig. 6.21). Scientists assume that the promoter is going to be located upstream from the start codon in the sequence. In some bacteria, scientists are beginning to be able to identify promoters by their sequence, but this type of analysis can be unreliable. This is especially true in the case of promoters that are regulated by activators and thus may not have a strong promoter consensus sequence. Also, if two open reading frames read in the same direction and the stop codon of one is very close to the start codon of the next one, the possibility that these open reading frames are in an operon must be considered.

Once a possible *orf* is identified, scientists take the amino acid sequence of the translated *orf* and search databases for known proteins with a similar sequence. For example, if the amino acid sequence of a possible *orf* from a new bacterium is most closely related to penicillin-binding proteins of other bacteria, the *orf* may encode a penicillin-binding protein that confers penicillin resistance. At this point, the identity of the gene is still hypothetical. Scientists must make mutations in the gene or otherwise show that a protein with the expected properties is actually made. Only about two-thirds of *orf*s in the genomes of bacteria for which genome sequences are available are recognizable as encoding proteins of known function and some of these identifications are tenuous. It is somewhat daunting that after decades of intensive study of these bacteria, a large number of *orf*s still have no known function.

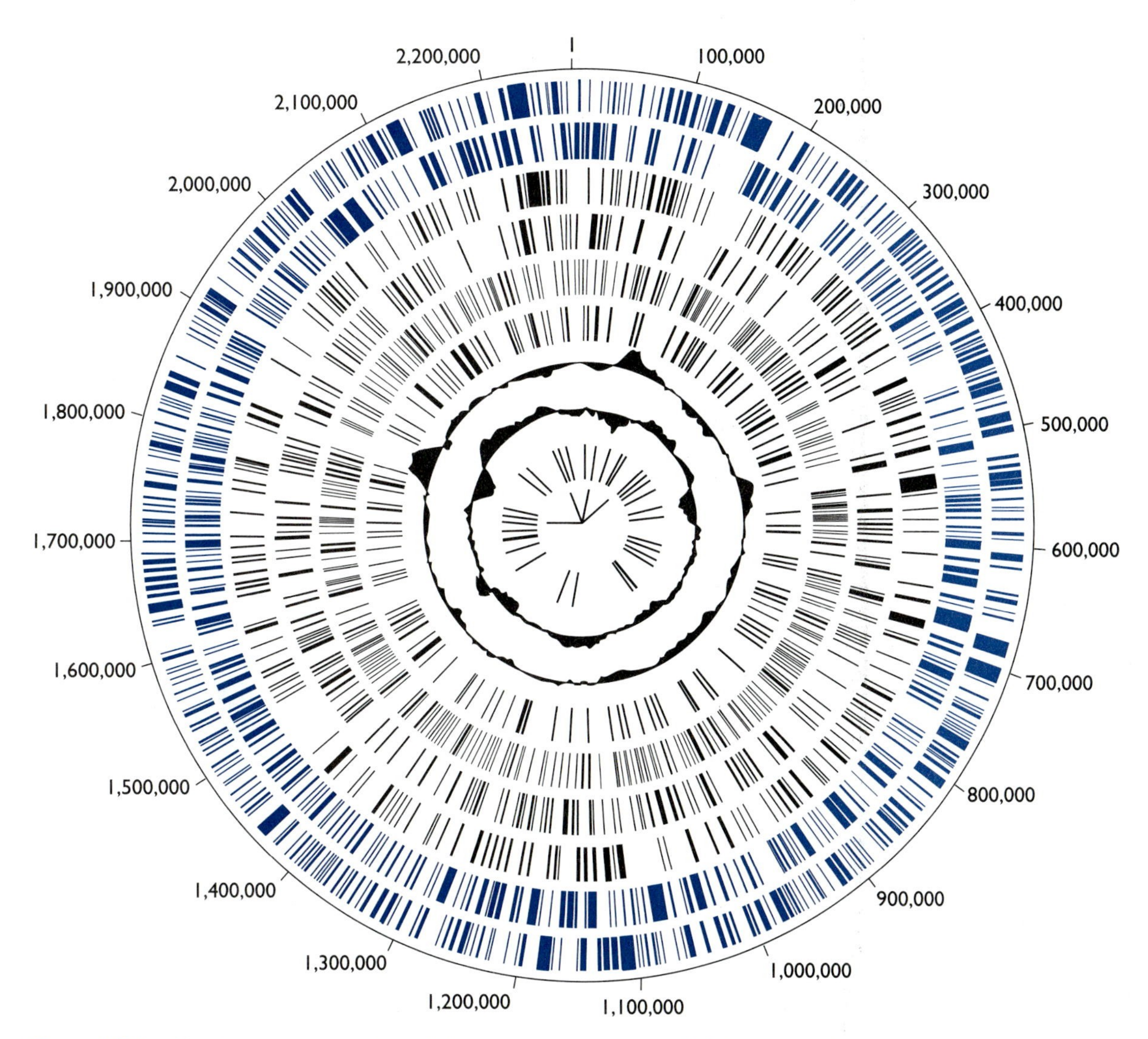

Figure 6.21 Circular representation of the genome sequence of *Neisseria meningitidis serogroup B*. Numbers on the circumference indicate the number of bases. The bars in the outer two circles show predicted open reading frames. The open reading frames of the outer circle are transcribed in the opposite direction from the open reading frames shown in the second circle. There are more than two circles of bars to allow the display of genes encoding a particular type of function(s) such as rRNA and tRNA genes. Reprinted with permission from Tettelin, H., et al. 2000. Complete genome sequence of *Neisseria meningitidis* serogroup B strain MC58. Science 287:1809–1815. © American Association for the Advancement of Science, 2000.

What about RNAs that are not translated, such as rRNAs or tRNAs? These usually are easy to find, because rRNA and tRNA sequences are highly conserved in bacteria. Thus, they can be identified by using the DNA sequence to search the databases for such genes. Some special databases contain only rRNA or tRNA sequences.

Deducing gene function from gene regulation

Measuring gene regulation with reporter group fusions. If you are trying to extract information from a gene's sequence about the possible role of the gene, one way to guess its function is to determine what conditions regulate its expression. For example, if a gene is expressed only when the concentration of iron in the medium is very low, but not when the concentration of iron is high, the gene product may well have some role in acquisition of iron. This type of information narrows the possible identity of a gene and helps the investigator develop new and more detailed hypotheses about the gene's function. If the gene in question encodes an enzyme that is detectable in enzyme assays, the activity of the enzyme produced can be used as an indicator of what conditions induce expression of the gene. Many genes, however, encode proteins that are not easy to assay. To get around this problem, scientists created **reporter gene fusions** (still called operon fusions by some).

In a reporter group fusion, the promoter of the gene of interest is cloned, so that it is upstream from a gene (the reporter gene or **reporter group**) that encodes an easily assayable enzyme. In other words, the promoter that once controlled expression of one gene now controls expression of the reporter gene. The *E. coli* gene, ***lacZ,*** which encodes the enzyme **β-galactosidase,** an enzyme that hydrolyzes lactose to glucose and galactose, is commonly used as a reporter gene. Fusions made using this reporter gene are called *lacZ* fusions. β-galactosidase will also hydrolyze an artificial substrate, called X-gal, which consists of galactose attached to an indoyl compound "X" that turns blue when it is released from the galactose. Thus, β-galactosidase action on this artificial substrate causes a change in color of X-gal from colorless to blue. This makes it very easy to assay β-galactosidase activity on plates (where the colonies turn a blue color). In wild-type *E. coli, lacZ* has its own promoter, but scientists have removed the natural promoter from the form of *lacZ* used in promoter fusions, so that *lacZ* expression now reflects the activity of whatever promoter is cloned upstream from it.

Construction of a gene fusion is illustrated in Figure 6.22. The promoter region of the gene of interest (gene Q) is cloned, so that it becomes the promoter of *lacZ.* This construct is introduced into the bacterium from which the promoter sequence came, and the bacterium is exposed to a variety of conditions. The conditions that cause β-galactosidase to be produced, as indicated by a color change in X-gal, are the conditions that cause the original form of the gene to be expressed. If low iron in the medium leads to high expression of gene Q, colonies plated on low iron medium will turn blue.

Reporter gene fusions, although extremely useful, have some shortcomings. One is that only a single gene can be studied at a time. A scientist confronting an entire genome sequence and wanting to know which genes have their expression regulated by low iron is not going to be able to screen every possible promoter region in the genome sequence by fusing each of them to β-galactosidase. Another drawback of the reporter gene fusion approach is that the investigator must be able to introduce the cloned gene fusion into the bacterium of interest. This is sometimes difficult or impossible. These two problems are solved by a new technology called microarrays, developed for screening genes from the entire genome sequence of an organism to determine which genes are expressed under different conditions.

Microarrays. Suppose a scientist has the complete genome sequence of *E. coli* and wants to know how many *E. coli* genes respond to low iron conditions by changing their levels of expression. This scientist will turn to a **microarray** for an answer. Microarrays take advantage of the fact that if a gene is being expressed, RNA copies of it will be present in the cell. The higher the level of expression, the more RNA copies there will be. If the level

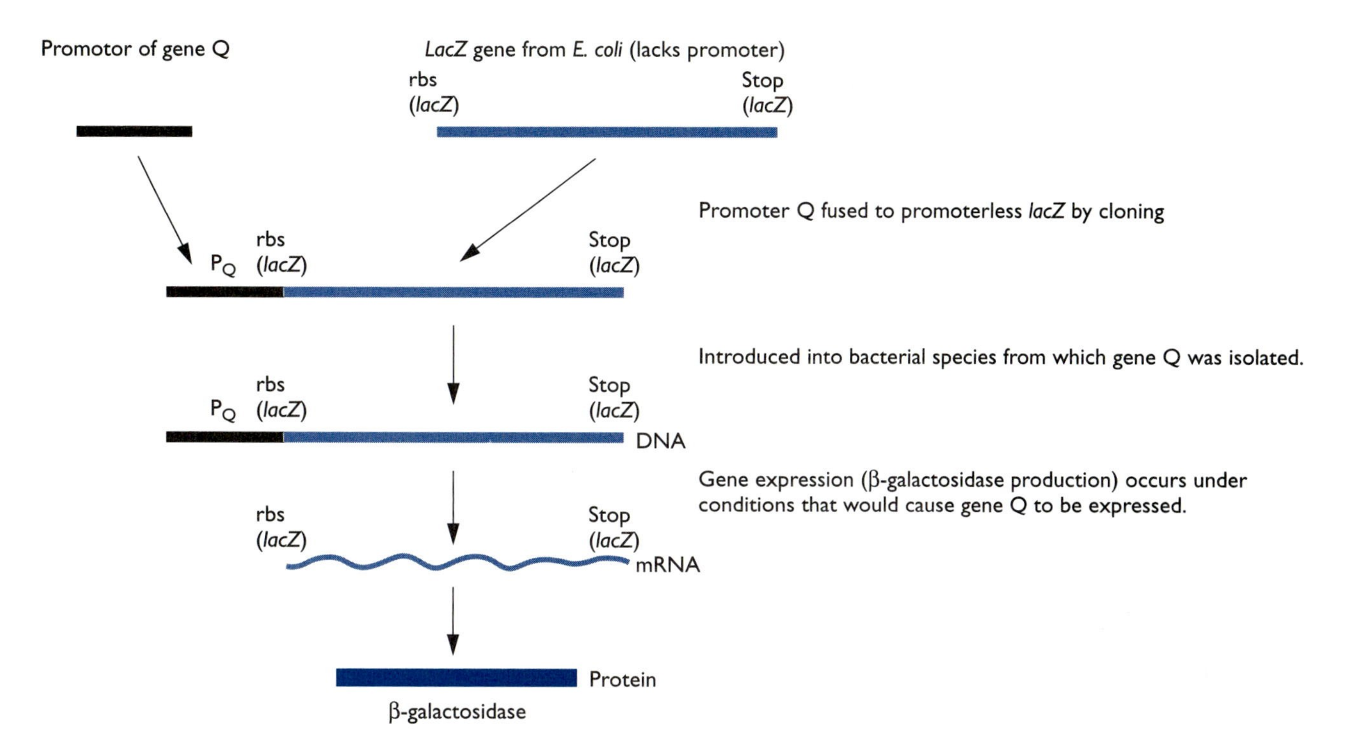

Figure 6.22 Construction of a reporter gene fusion. The promoter from the gene of interest (gene Q) is cloned so that it becomes the promoter of the reporter group, in this case the *lac*Z gene from *E. coli.* The promoter normally associated with the *lac*Z has been removed, so the only way that β-galactosidase (the enzyme encoded by *lac*Z) can be produced is if the promoter of gene Q allows the *lacZ* gene to be expressed. This new construct is introduced into the bacterium from which gene Q came, and the bacterium is exposed to different conditions. The conditions that cause β-galactosidase to be produced are assumed to be the conditions that cause the original form of gene Q to be expressed.

of expression changes, so will the number of RNA copies. A microarray is a chip divided into thousands of tiny squares. In each of the squares, portions of the single-stranded DNA from a specific gene have been attached (Fig. 6.23). Thus, each square represents a different gene on the *E. coli* chromosome. Such chips are now becoming available for a number of microorganisms.

To determine how many of these genes are expressed under low iron conditions compared with high iron conditions, *E. coli* is grown in medium containing a low iron concentration. For comparison, a parallel culture is grown in medium containing a high iron concentration. mRNA is isolated from each of these cultures and labeled with fluorescent nucleotides. The mRNA from low iron medium is incubated with one copy of the chip, and the mRNA from high iron medium is incubated with a second copy of the chip. The mRNA will bind specifically by RNA-DNA hybridization to DNA representing the gene from which it is transcribed. Thus, some squares will bind labeled mRNA and will thus become fluorescent. Other squares, with DNA segments that do not hybridize to any mRNA molecule in the cell, will not bind labeled mRNA and thus will not become fluorescent. Unbound mRNA is washed away, and fluorescence from bound mRNA is visualized. The squares in which DNA binds labeled mRNA are the ones containing a gene that is expressed under the condition being tested. Some squares on the two chips will bind mRNA from cells grown

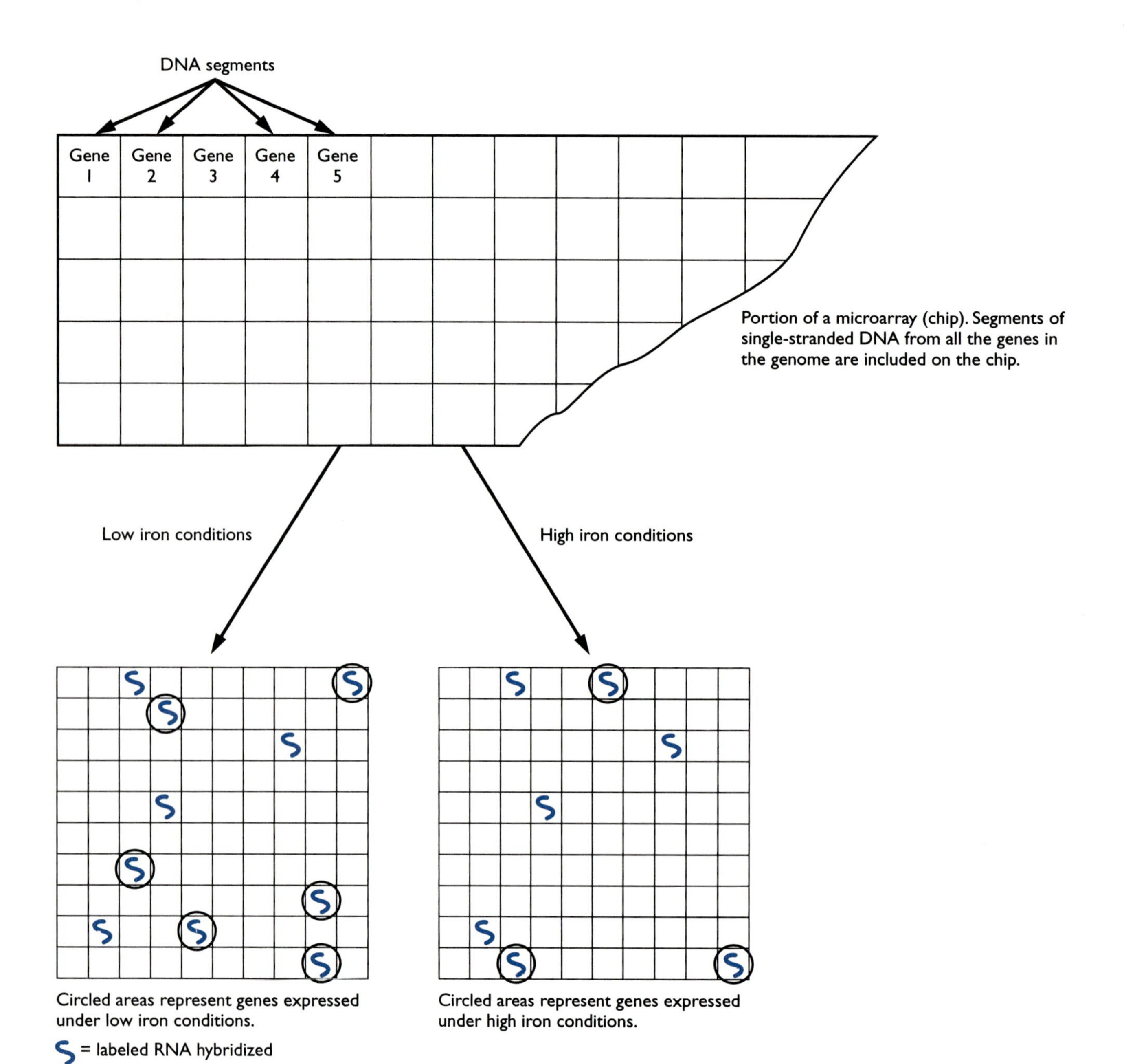

Figure 6.23 Microarrays. Microarrays are chips divided into thousands of tiny squares. Each square contains portions of single-stranded DNA from a specific gene. In the experiment in this figure, we are asking which genes respond to low iron conditions and which to high iron conditions. One batch of bacteria were grown under low iron conditions and the second under high iron conditions. mRNA isolated from the culture grown under low iron concentrations is labeled and incubated with one chip, and the mRNA from the culture grown under high iron concentrations is labeled and incubated with the second chip. The mRNA will bind to DNA of the gene from which it was transcribed. By looking at the squares on the chips where the label can be detected, it is possible to determine which gene was expressed under low iron conditions, which under high iron conditions, which under both conditions, and finally, which under neither condition.

under both conditions, and some squares will bind mRNA under neither condition. These contain genes in which expression is not controlled by iron concentration.

The scientist in this example is interested in those squares that bind mRNA under one of these growth conditions but not the other. These are the iron-regulated genes (Fig. 6.23). The genes labeled by the mRNA can be identified by consulting a key that lists which gene is represented by each square. Microarrays allow a scientist to screen an entire genome sequence for expression of the genes it contains. This approach is currently being used to identify genes involved in responses to a variety of conditions.

The next big step. Analysis of DNA sequence data and gene regulation are only a very small first step toward unlocking the information stored in a microbial genome sequence. Knowing that expression of a gene is regulated by iron levels does not help much if you have no idea how a bacterium acquires and utilizes iron. The reason scientists have been able to deduce as much as they have to date from genome sequences and their regulation is that there is information from biochemical and genetic studies about how bacteria use various enzymes and structural proteins to carry out essential life processes. If a search of the databases for a protein with an amino acid sequence similar to the one translated by computer from a DNA sequence finds a matching protein from another bacterium, this information is useful only if prior studies have identified the function of the matching protein. Information on the various physiological processes carried out by bacteria and how these processes interact is far from being complete enough to serve as an adequate basis for gaining all the information about the capabilities and activities of a bacterium from raw genome sequence data.

Chapter 8 will introduce the kinds of physiological studies that will allow genome sequencers to jump to the next level of understanding. An important point to consider is that a great step forward from bacterial genome sequences into deducing bacterial activities is also potentially a great leap forward toward interpreting genome sequences of plants and animals. Many of the basic physiological processes of a plant or animal cell are similar to those found in bacteria and archaea–a point that should not be surprising, given that prokaryotes were our ancestors. As more is learned about how bacteria and archaea carry out basic physiological tasks and coordinate these functions, new insights into the functions of plants and animals may well emerge.

chapter 6 at a glance

Summary of signals that control transcription and translation of a gene

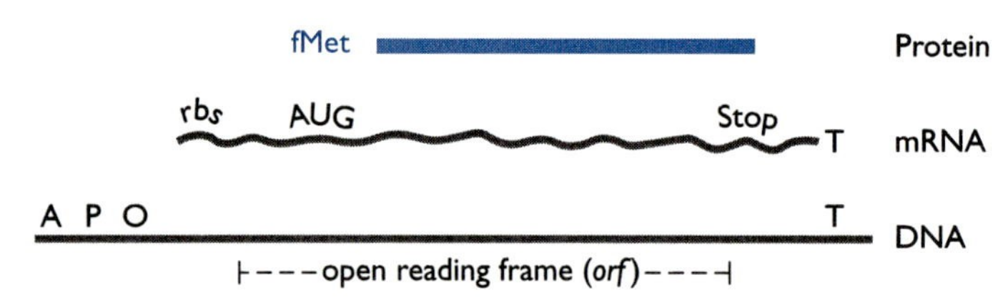

A = activator site; P = promoter; O = operator (repressor binding site); T = terminator; rbs = ribosome binding site; AUG = start codon; stop = stop codon (UAG, UGA, UAA); fMet = formylmethionine (amino terminus of protein)

Signals	Determine	Mutation can affect
Promoter, terminator	Length of transcript	Length of transcript Strength of promoter
Operator, activator binding site	Whether transcript is made and how many are made (level of gene expression)	Regulation of expression
Ribosome binding site (rbs)	Where ribosome binds Sequence affects amount of protein made	Whether protein is made and how much is made
Start codon, stop codon	Where protein starts, stops Length of protein	Whether protein starts, stops in right places
DNA sequence of *orf*	Amino acid sequence of protein	Premature termination (nonsense) Change of amino acid (missense) Different amino acid sequence after a certain point (frameshift)

Mechanisms for gene regulation

Type of regulation	Level at which regulation occurs
Transcriptional control	
Sigma factor	Ability of RNA polymerase to bind a certain set of promoters
Repressor, activator Two-component Quorum sensing	Ability of RNA polymerase to bind promoter and start transcribing gene under different conditions
Translational control	
Ribosome binding site	Ability of ribosome to bind mRNA and start translating
Post-translational control Feedback inhibition, covalent modification	Activity of protein affected

chapter 6 at a glance

Chemical signals

Origin and type of signal	Type of regulation	Effect of signal
Environment	Sigma factor	Determine whether special sigma factor is produced
	Repressors, activators	Binding of signal to repressor or activator changes conformation, determines if it will bind DNA and affect transcription
	Two-component system	Sensor responds to signal by phosphorylating activator or repressor, enabling it to aid RNA polymerase to transcribe gene
Internal	Quorum sensing	Bacteria produce autoinducer molecule, use it to assess their concentration; autoinducer binds repressor or activator, changing its conformation so that the repressor or activator binds to or ceases to bind to DNA
Internal or external	Feedback inhibition	Amino acid or other product of pathway (exogenous or endogenous) inhibits an enzyme in the pathway
	Covalent modification	Modification activates or inactivates enzyme

Analysis of genome sequences

Identify possible open reading frames by looking for start and stop codons, machine-translate to obtain amino acid sequence of protein

Use comparisons to sequences of known gene or proteins (computer database searches) to identify possible function of each gene

Use amino acid composition of proteins to guess protein structure and cellular localization

- Amino-terminal hydrophobic stretch: signal sequence
- Proteins with membrane-spanning regions or channel structure: membrane proteins
- Proteins with hydrophilic amino acids on surface: cytoplasm or periplasm

Use reporter gene fusions or microarrays to classify genes by the conditions under which they are expressed

Information from studies of microbial physiology helps to identify roles of proteins.

key terms

activator protein that enhances binding of RNA polymerase to the promoter and helps it to initiate transcription

gene segment of DNA that encodes a protein or an RNA species such as rRNA or tRNA

gene expression the process of converting DNA sequence (the gene or open reading frame) to RNA sequence (transcription) and from there to amino acid sequence of protein (translation); process can be regulated at each level (regulation of gene expression); constitutive expression, gene expressed under all conditions; regulated expression, gene expressed only under some conditions

messenger RNA (mRNA) RNA molecule transcribed from DNA that contains the coding sequence for at least one protein

mutations changes in DNA sequence of genes or regulatory regions can lead to altered expression of a gene, altered amino acid sequence, or altered protein length; may result in nonfunctional protein

open reading frame (*orf*) region of the DNA sequence that starts with a start codon and ends with stop codon

operator site region of DNA to which a repressor binds and prevents mRNA synthesis

operon a series of genes controlled by a single operator; single transcript encodes multiple proteins

promoter site on DNA where RNA polymerase binds to initiate transcription; promoter consensus sequences identify strong promoters

regulation of gene expression use by bacteria of DNA sequences and proteins that bind them (sigma factors, activators, repressors) to alter the number of transcripts or proteins made from a particular gene

repressor protein that binds close to operator and prevents transcription

ribosomes protein-rRNA complexes on which mRNA is translated into protein; bind mRNA at ribosome binding site, leave mRNA at stop codon

RNA polymerase an enzyme that moves along selected regions of the DNA, forming a complementary strand of mRNA; requires sigma factor for initiation of transcription and rho factor for termination of some transcripts (rho-dependent terminators)

transcription synthesis of an mRNA strand complementary to a single strand of DNA; resulting RNA can be mRNA (encodes a protein) or ribosome-associated RNA (rRNA, tRNA)

transfer RNA (tRNA) the form of RNA that carries an amino acid to the ribosome; at the ribosome the anticodon region of the tRNA binds to a codon on the mRNA specific for the amino acid that the tRNA carries and transfers the amino acid to the growing peptide chain

translation synthesis of a protein by a ribosome as it moves along mRNA, converting mRNA sequence into amino acid sequence of the protein

Questions for Review and Reflection

1. Make an educated guess as to why the structure of RNA differs from that of DNA. Why doesn't RNA polymerase start from a DNA primer the way DNA polymerase does?

2. Scientists analyzing DNA sequence data speak of "translating" DNA sequence (by computer) into the amino acid sequence of a protein, thus appearing to ignore mRNA. Why can they do this? In what senses are "open reading frame" and "gene" two terms for the same thing?

3. The sequences of promoters and ribosome binding sites can differ considerably from one gene to another, although there are often common motifs. Why don't all genes have the same promoters and the same ribosome binding site sequences? In what sense are sigma factors regulatory proteins?

4. What are the characteristics of promoters controlled by repressors versus those controlled by activators? Why are activator binding sites often located upstream of the promoter site, whereas those for repressors are located downstream of or overlapping the promoter region?

5. How does a quorum-sensing system differ from a two-component regulatory system? Which one of these would be most appropriate for controlling expression of genes involved in biofilm formation? For controlling genes that encode flagella proteins?

6. In the next chapter, horizontal transfer of genes from one bacterium to another will be described. If a bacterium of one species acquires a gene from a bacterium of another species, the gene may not be expressed at first. However, if selective pressure is great enough, the gene may ultimately be expressed in its new bacterial host. Explain this phenomenon.

7. Can you think of reasons why cells use mRNA as an intermediary instead of translating proteins directly from DNA?

8. Some amino acids are associated with two or more codons. One of these is usually used far more often than the other. How could bacteria use this feature as a crude way to control the levels of some proteins?

9. In the tRNA molecule, why is the codon that interacts with mRNA separated spatially from the site where

Questions for Review and Reflection *(continued)*

the amino acid is attached? What characteristics must a tRNA synthetase have to charge such a bifunctional molecule?

10. What is the function of the A, P, and E sites on the ribosome? Why is it necessary to have more than one binding site on the ribosome for the tRNA-amino acid and tRNA peptide?

11. Why does it make sense that most of the protein synthesis-inhibiting antibiotics affect translocation instead of blocking initiation? In what way would an antibiotic that interfered with charging of the tRNA by tRNA synthetases affect protein synthesis? What does the fact that a single antibiotic interferes with most tRNA synthetases tell you about tRNA synthetases?

12. What is the advantage for the bacterium of using the ribosome binding site sequence to vary the amount of production of different proteins?

13. Suppose you use ultraviolet light to generate random mutations in the genome of a bacterium that produces the enzyme β-galactosidase. You then screen colonies on a plate containing X-gal for mutants that are white not blue (mutants that no longer produce active β-galactosidase). You find several colonies that no longer produce the active enyme. What are some ways the loss of enzyme activity could have occurred?

14. Why might post-translational regulation (feedback inhibition, covalent modification) be advantageous to a bacterium?

15. In the section on genome sequence analysis, the statement was made that using DNA sequence motifs to identify promoters can be unreliable. Explain this statement. Why can the same thing be said about predicting the structure of a protein or its location in the cell from DNA sequence data?

16. In microarray experiments it is necessary to compare at least two conditions. Why? Is this also true for promoter fusion experiments? Suppose you fuse β-galactosidase to a region you think contains a promoter, but detect no expression of the gene in *E. coli.* Does this mean no promoter is in that region? Are there other possible explanations?

7 Bacterial Genetics III: Horizontal Gene Transfer Among Bacteria

Bacteria evolve not only by mutating their genomes but also by receiving and incorporating DNA segments from members of other species. How can one define an individual bacterium if large portions of that bacterium's genome are cobbled-together gene segments acquired from a mixture of other bacterial species that are not its ancestors but its contemporaries? Blithely indifferent to maintaining their own identity, bacteria are using this capacity for acquiring genes from other organisms to allow them to become resistant to antibiotics, to become able to cause new diseases, or otherwise to increase their options for survival.

Bacterial reproduction is usually described as asexual, because bacteria have no equivalent of the genetic fusion of two different cells that is characteristic of sexual reproduction in eukaryotes. Nonetheless, bacteria do have the ability to exchange segments of DNA with other bacteria. Because these segments can become fixed in a bacterium's genome and confer new traits, gene exchange among bacteria could be considered to be a form of bacterial sex. DNA exchange between bacteria is called **horizontal gene transfer** to differentiate it from vertical gene transfer, the inheritance of a gene from a progenitor. Virtually all of the information available on horizontal gene transfer among microbes is limited to bacteria, although new studies indicate the same will be true for the archaea. Accordingly, bacteria will be the main focus of this chapter.

For decades, bacteria have been known to be capable of horizontal gene transfer in the laboratory. In fact, information about horizontal gene transfer was the foundation of the genetic engineering revolution. Only when scientists could introduce foreign DNA into bacteria in the laboratory did cloning become possible (Chapter 5). Until recently, however, many scientists have downplayed the importance of horizontal gene transfer as a form of bacterial evolution in nature. It is easy to understand this skepticism. Bacteria from different species, and especially bacteria from different genera, are likely to have promoters and other regulatory signals different enough to make it doubtful that these signals would function in a distantly related bacterium. Thus, the argument went, even if physical transfer of the

DNA occurred and the DNA segment was fixed in the recipient's genome, such events would have no practical consequences, because no new traits would be acquired by the recipient of the DNA. If the incoming DNA segment integrated in such a way as to disrupt a gene, the acquisition of the DNA segment might even be deleterious. Reasonable as such arguments may seem, the conclusion they point to—that acquisition of beneficial traits by horizontal gene transfer is a rare event in nature—is being proved wrong. Evidence is accumulating to support the contention that not only does horizontal gene transfer occur frequently in nature but that it plays a significant role in bacterial evolution.

Evolutionary and Medical Significance of Horizontal Gene Transfer

Evolutionary significance

One line of evidence for horizontal gene transfer as an evolutionary force is coming from examination of bacterial genome sequences. Finding genes with two very similar DNA sequences in two distantly related bacteria suggests that these sequences might have been acquired by horizontal gene transfer. The number of such examples is increasing. Our growing appreciation of how much horizontal gene transfer may have contributed to the evolution of bacteria has caused scientists to question whether the standard evolutionary tree, characterized by well-isolated branches (Chapter 2), should in fact be replaced by interwoven nets of branches (Fig. 7.1). These nets may transcend the prokaryotic world, because known examples of introduction of bacterial DNA into eukaryotic cells, in the form of chloroplasts and mitochondria, raise the possibility that horizontal gene transfer may have occurred between prokaryotes and eukaryotes. Some genes from mitochondria, which were once free-living bacteria, have migrated to the nucleus of the eukaryotic cell, making the horizontal transfer of these genes complete. Horizontal transfer of genes among bacteria also raises questions about the integrity of species classification schemes based on phenotypic traits, such as the ability to use a particular substrate. If two species have been differentiated on the basis of differences in their ability to utilize lactose and if the genes allowing lactose utilization can be transferred by horizontal gene transfer, how firm then is this definition of species?

A solution of sorts to such quandaries may be that some types of genes are acquired and retained more readily than are others. That is, a gene encoding resistance to an antibiotic may be acquired without much danger to its recipient, because expression of the gene does not affect basic cellular functions. The acquisition of a gene encoding an RNA polymerase subunit from a distantly related bacterium, however, could be quite dangerous. The problem with nets could well disappear if scientists can define a subset of genes that are almost never acquired and kept by the recipient bacterium. These genes could then serve as a stable definition of what a species is and return the integrity of the original evolutionary tree concept.

Medical significance

Effects of horizontal gene transfer, seen from the viewpoint of evolutionary analysis of bacterial lineages, may seem a rather abstruse topic. Yet, horizontal gene transfer among bacteria is having immediate, practical effects on human health. Scientific studies have shown that not only are antibiotic-resistance genes being spread by horizontal gene transfer but that horizontal gene transfer is contributing to the evolution of disease-causing bacteria

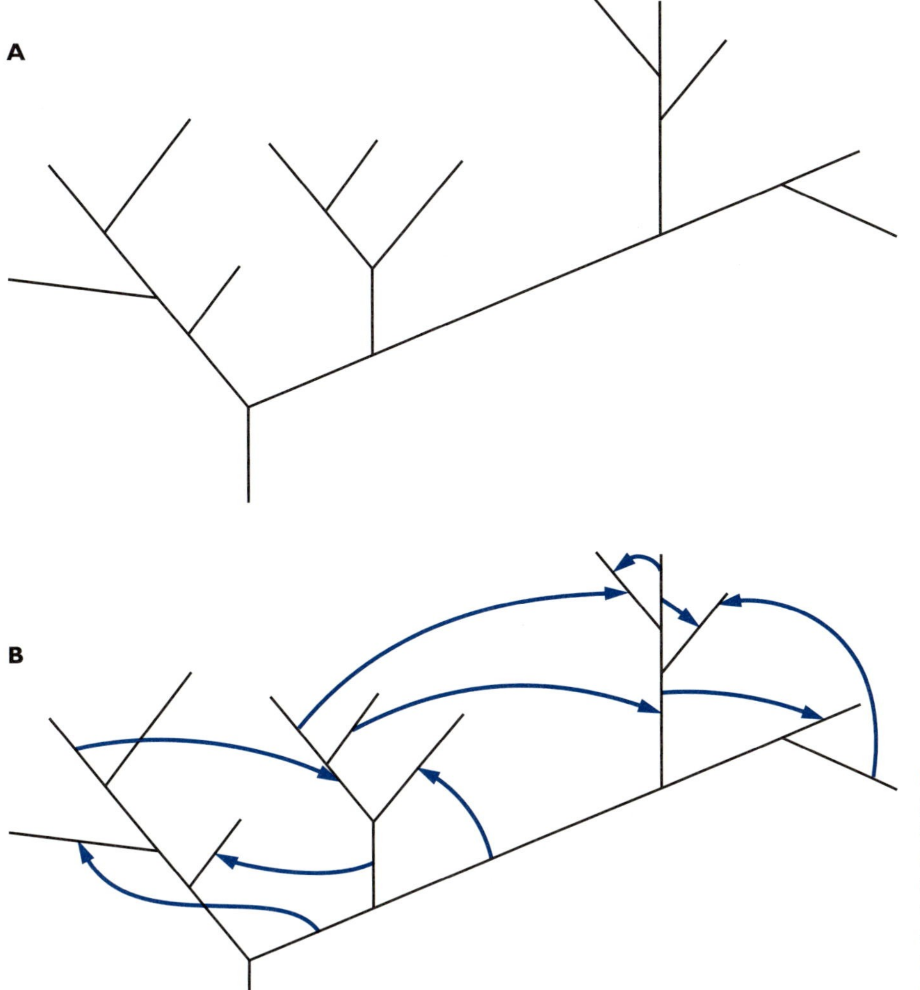

Figure 7.1 Effect of horizontal gene transfer on attempts to deduce the evolutionary ancestry of bacteria. **(A)** A simple phylogenetic tree based on rRNA sequences (see Chapter 2), which takes into account only vertical transmission of genetic information. **(B)** Horizontal gene transfer events (indicated by arrows) can obscure the lines of descent and make it difficult to trace a bacterium's evolutionary history.

(Fig. 7.2). An example of the latter phenomenon is "killer *Escherichia coli*," (*E. coli* O157:H7; Chapter 17), which causes bloody diarrhea and kidney failure in children. Such strains appear to have been created by a horizontal gene transfer event, when a less virulent strain of *E. coli* acquired genes that allowed it to make a new toxin. The toxin is thought to be responsible for kidney failure, a symptom not previously associated with *E. coli* infections. On a more positive note, horizontal gene transfer among soil bacteria growing in polluted sites has created bacteria able to degrade new pollutants, bacteria that may be useful in the bioremediation of such sites.

Types of Horizontal Gene Transfer

Gene transfer events that involve members of different species or different genera are called **broad host range** gene transfers. Those that occur between very closely related bacteria, such as members of the same species, are called **narrow host range transfers.** Broad and narrow gene transfer events are mediated by the same mechanisms of DNA transfer, although

A

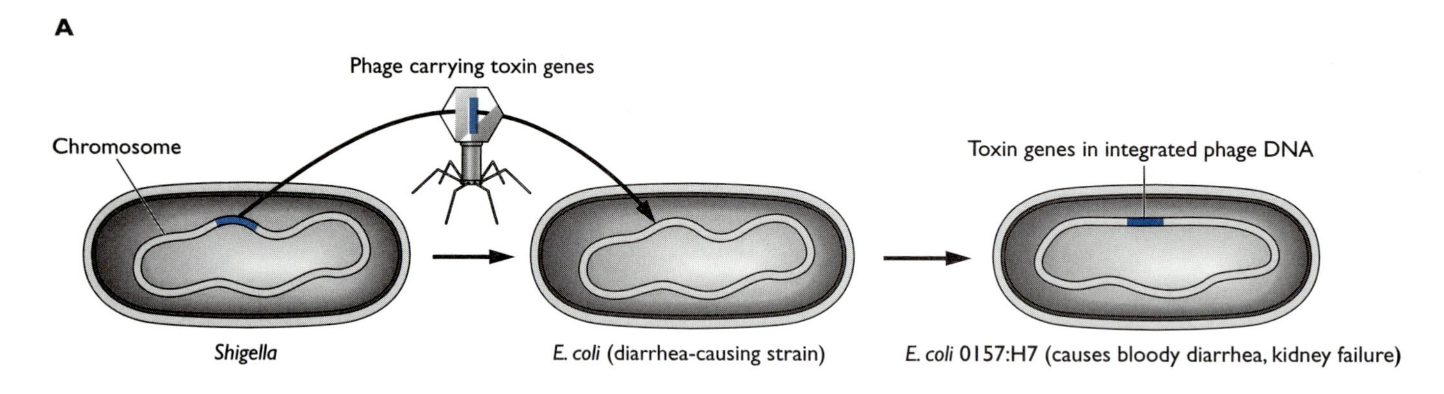

B

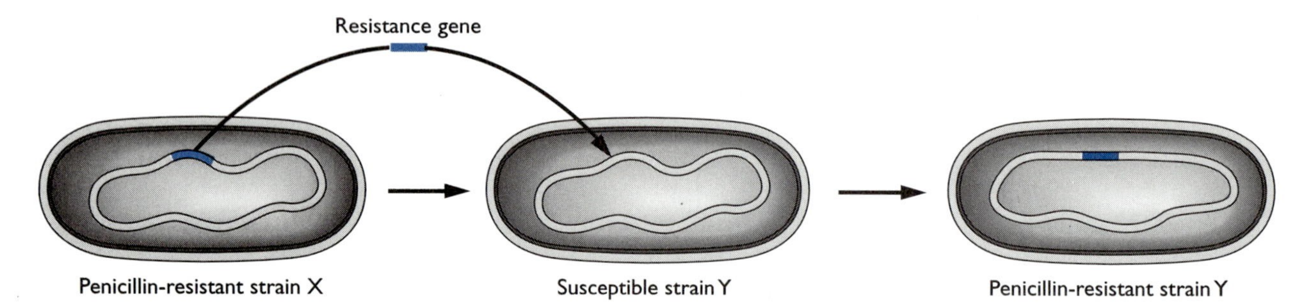

Figure 7.2 Horizontal gene transfers that have made some bacteria more dangerous to humans. (A) Horizontal gene transfer event that is thought to have created *Escherichia coli* O157:H7 from an *E. coli* strain that caused less serious disease. A bacteriophage from *Shigella* entered this strain and integrated itself into the bacterial genome (lysogeny). Genes on the phage encoded a toxin that enabled the strain to cause a more serious disease. **(B)** Horizontal gene transfer introduces antibiotic resistance genes into a previously susceptible bacterial strain.

some mechanisms favor narrow over broad host range transfer for reasons that will soon become evident.

Three types of horizontal gene transfer have been studied extensively: uptake of free DNA from the environment (**transformation**), DNA transfer mediated by bacteriophages (**transduction**), and direct cell-to-cell transfer of DNA (**conjugation**). Although these mechanisms have received most of the research attention, it is worth noting that there is a fourth mechanism of DNA transfer that may have far-reaching practical consequences: the international transportation system, which conveys travelers and goods all over the world. Although this chapter will focus on the first three types of horizontal gene transfer, it is important to realize that once a strain of bacteria with new characteristics has arisen in one part of the world, it can be rapidly disseminated.

Transformation

Scientists who work with eukaryotic cells use the word "transformation" to describe what happens to cells that lose the tight control of growth rate that is characteristic of normal cells and begin to divide actively (**immortalization**). This type of transformation can occur as a result of mutation or uptake of DNA. Bacteriologists use the term transformation to mean simply the uptake by bacteria of DNA from the environment. Because bacteria normally divide by binary fission, they are naturally immortalized. There are two types of transformation: **natural transformation,** the kind that occurs in nature, and **artificial transformation** by chemical shock or electric shock, the kind practiced in laboratories to make recalcitrant bacteria take up DNA they would not normally accept. These two types of transformation have different properties (Fig. 7.3).

In natural transformation, DNA from the environment first attaches to the surface of the bacterial cell (Fig. 7.3A). Then, one strand is nicked and digested away by bacterial enzymes to make the DNA single-stranded. The single-stranded form is taken up by an active DNA transport system, a system in which specialized proteins internalize the DNA by an energy-dependent process. Segments of single-stranded DNA enter the cell cytoplasm.

If the incoming DNA has regions of sequence identity with the chromosome of the recipient (for example, if the incoming DNA comes from another strain of the same species), the incoming DNA can enter the chromosome by homologous recombination (see Chapter 5). Homologous recombination can replace one copy of a gene with another. This is probably the way that mutant penicillin-binding proteins of *S. pneumoniae,* one of the few human-associated bacteria capable of natural transformation, are being passed around in this species. Another type of recombination, called illegitimate recombination, does not require homology between the incoming DNA and its chromosomal target. This process is not well understood but is usually quite inefficient compared with homologous recombination.

Natural transformation systems seem to be uncommon among bacteria that normally occupy the bodies of humans or animals, although some naturally transformable bacteria are important disease-causing organisms, such as *S. pneumoniae* and the bacterium that causes gonorrhea. Because most bacteria associated with humans, such as *E. coli,* are not naturally transformable, it is necessary to use artificial conditions to force them to take up DNA if one wants to introduce DNA in a cloning experiment. As described in Chapter 5, one step in cloning DNA in *E. coli* is to subject the bacteria to alternate salt shock and heat shock or to a high-voltage electric field (electroporation) to force the uptake of DNA. These measures bypass the need for specialized protein complexes that introduce DNA into the cell. This type of transformation introduces intact double-stranded DNA into the cell cytoplasm (Fig. 7.3B).

Results of recent studies suggest that, unlike human and animal microbes, soil and water microbes often have natural transformation systems. Naturally transformable soil bacteria can account for as much as 5% of the cultivatable bacterial population. This interesting observation remains unexplained. One possible explanation arises from observations that nutrients are often more limiting in soil than in animal intestines. DNA released from dying microbes may be an important food source for soil bacteria. In contrast to the intestinal tract, where naked DNA is rapidly degraded by nucleases, naked DNA is very stable in soil, because it is bound to soil particles. This binding protects it from nucleases. Any bacterium that can remove this bound DNA and internalize it would have access to a unique food source.

Bacteriophage transduction

Bacteriophages, also called phages, are viruses that attack bacteria (Chapter 3). When a bacteriophage attacks a bacterial cell, it first attaches to a receptor on the bacterial surface, then injects the DNA in its capsid into the bacterial cell (Fig. 7.4). At this point, one of two things can happen. First,

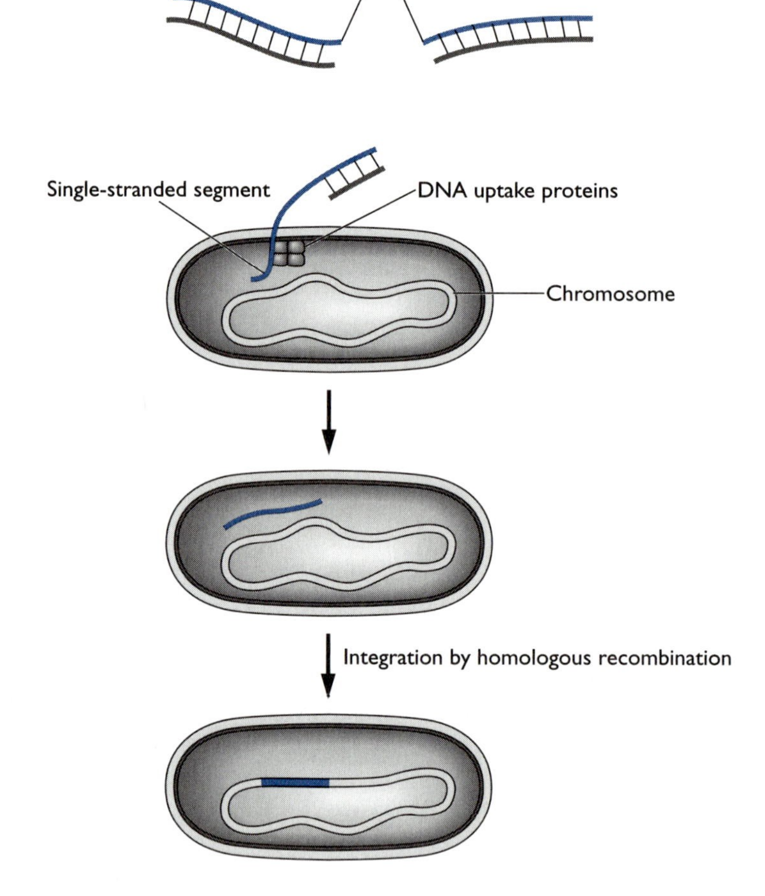

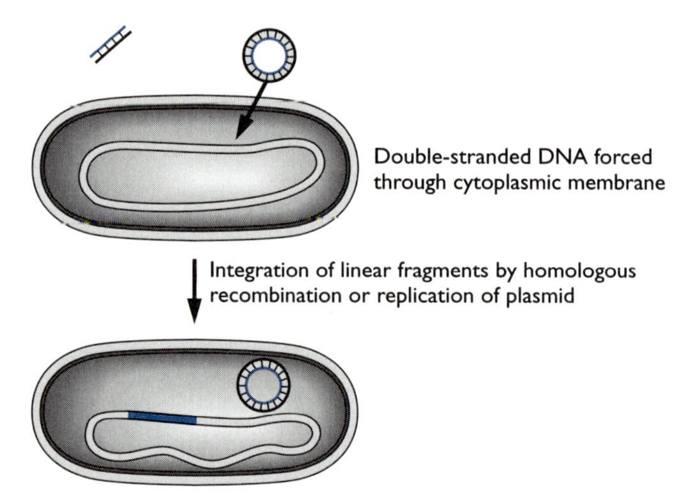

Figure 7.3 Transformation mechanisms. (A) Natural transformation. Double-stranded DNA from the environment is nicked in one strand, which is then digested away, leaving the second strand. This single-stranded segment is taken up by a protein complex into the bacterial cytoplasm, where it may be further degraded or enter into the bacterial genome. **(B)** Artificial transformation. Chemical or electrical field stresses force double-stranded DNA through the cytoplasmic membrane into the cytoplasm. The DNA can be a linear segment, which integrates by homologous recombination, or a plasmid.

the phage can proceed to reproduce itself, eventually killing the bacterial cell (Fig. 7.4), a process called **lytic infection.** Alternatively, the phage DNA can integrate into the bacterial chromosome, a process called **lysogeny** (Fig. 7.2A). Not all phages can perform lysogeny. The ones that can may later become activated by environmental stresses to enter the lytic growth mode. An interesting feature of lysogeny is that many bacterial toxin genes are carried as part of phage genomes. No one knows why a self-respecting bacteriophage would be lugging around genes encoding proteins that damage human cells, but at least some of them do. Some examples of phage-encoded toxins are botulinum toxin (botulism), diphtheria toxin (diphtheria), cholera toxin (cholera), and the toxin acquired by *E. coli* O157:H7 (bloody diarrhea and kidney failure).

Phages producing a lytic infection can also transfer DNA from the lysed bacterium to other bacteria. This process, called **generalized transduction,** is illustrated in Figure 7.4. A phage that is replicating lytically chops up the

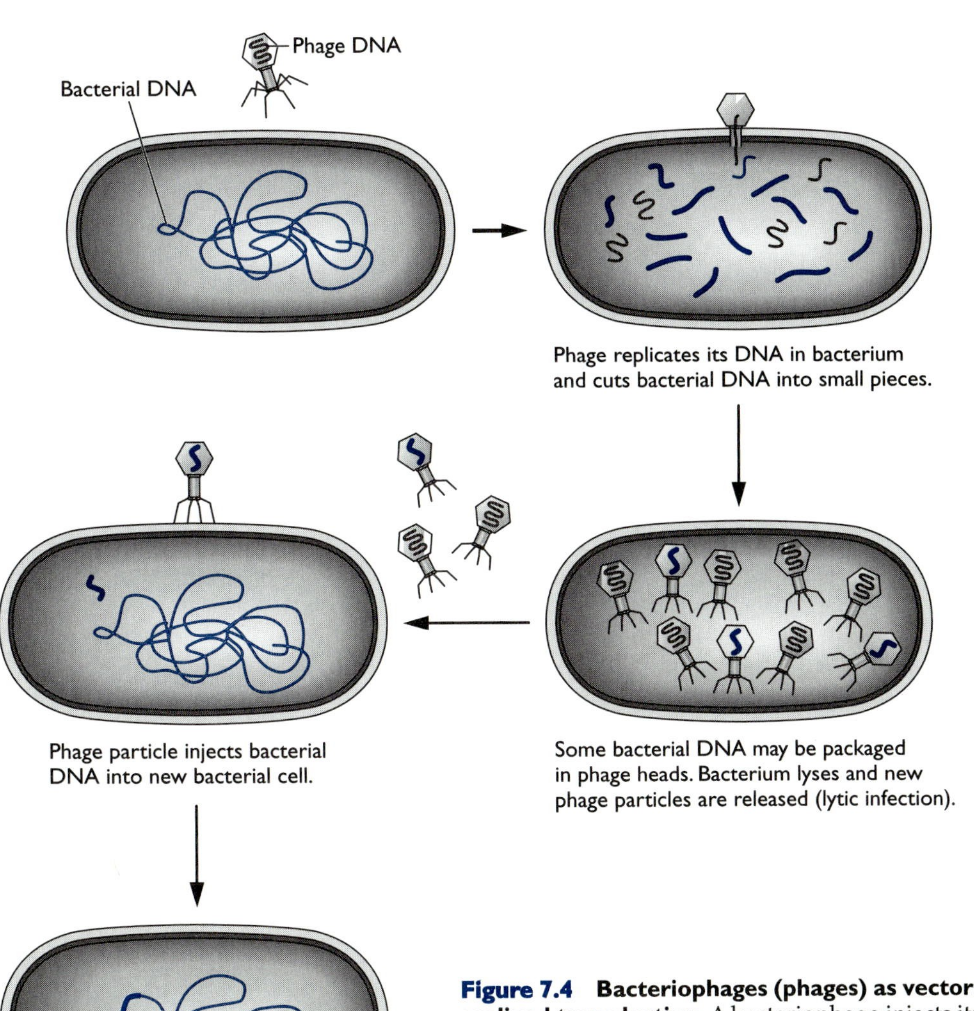

Figure 7.4 Bacteriophages (phages) as vectors for the transfer of DNA via generalized transduction. A bacteriophage injects its DNA into a bacterium and begins to replicate in a lytic cycle that will eventually kill the bacterium by causing it to lyse. During lytic growth, the phage accidentally packages a double-stranded segment of bacterial DNA in its capsid, then transfers this bacterial DNA segment to another bacterium as it tries to infect. The DNA segment will either integrate into the bacterial genome of the cell receiving it or be lost during subsequent replication events.

chromosome of the bacterium it infects. In the frenzy to assemble new phage particles, a piece of bacterial chromosomal DNA (instead of the phage's own genome) may be inadvertently packaged in the phage head. The phage particle carrying bacterial DNA can later attach to another bacterium and inject the bacterial DNA into it. Because this phage head does not contain the viral genome, introduction of DNA by the phage does not cause either lysogeny or lytic growth but serves only as a means of introducing foreign DNA into the bacterial cell. The incoming double-stranded DNA segment will either integrate into the chromosome by homologous recombination or be lost.

Before more advanced molecular techniques became available, scientists used generalized phage transduction in the laboratory to move DNA segments from one bacterial strain to another. Generalized transduction may have a role in natural settings, too. Microscopic examination of water, soil, and rumen contents by electron microscopy has revealed that phages are found very commonly in nature. Phages may also be transferring DNA from bacterium to bacterium in these environmental settings. Only recently have scientists begun to look into this possibility.

Conjugation

Conjugation is the term used to describe direct cell-to-cell transfer of DNA. During conjugation, two bacteria (a donor and a recipient) make tight contact with each other (Fig. 7.5). Cell-to-cell contact is mediated by a complex set of proteins that form the connection between the mating cells (the **mating bridge**). The proteins of the mating bridge also mediate the transfer of DNA from the donor to the recipient. Although chromosomal DNA can be transferred by conjugation under some conditions, DNA transferred by conjugation is most commonly carried on specialized gene transfer elements that are capable of forming and utilizing the mating bridge. Two such specialized gene transfer elements are self-transmissible plasmids and conjugative transposons. Both are composed of double-stranded DNA and both can carry other genes besides those needed for the actual process of conjugation.

Self-transmissible plasmids. Plasmids were introduced in Chapter 5, primarily as cloning vectors. Here they are considered as conjugal gene transfer elements. Only a fraction of plasmids, however, carry genes that encode the proteins that form the mating bridge and carry out transfer of the DNA segment. Such plasmids are called **self-transmissible plasmids.** Because of the large number of genes needed for conjugation, self-transmissible plasmids are usually quite large, at least 30 kbp in size.

The process of conjugal plasmid transfer is illustrated in Figure 7.5. After tight contact has been made between donor and recipient cells by means of proteins encoded on the plasmid, the mating bridge forms. In the donor cell, enzymes associated with the mating bridge make a single-stranded nick at a position on the plasmid called the **transfer origin** (***oriT***). One end of the nicked DNA strand is attached to a protein of the mating bridge, and the remainder of the strand is gradually passed to the cytoplasm of the recipient, until one entire single-stranded copy of the plasmid is in the recipient cell. While this is happening, the strand being removed from the plasmid in the donor is duplicated by the donor's DNA polymerase. In the recipient, the recipient's DNA polymerase copies the incoming single-stranded plasmid. Then, the end of the plasmid in both the donor and the recipient is released from the mating bridge and attached to the other end of the plasmid, regenerating the circular form of the plasmid in both cells.

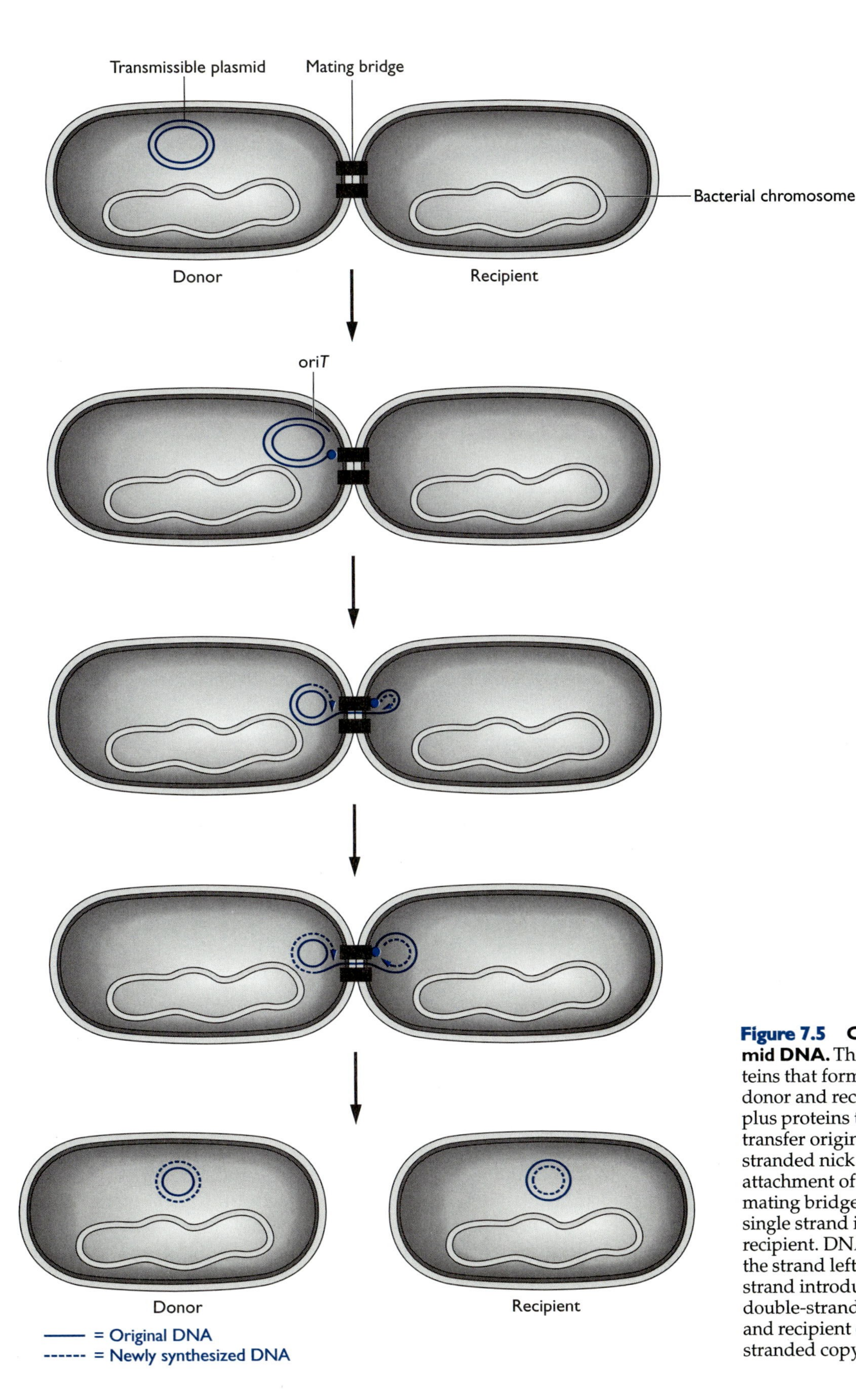

Figure 7.5 Conjugal transfer of plasmid DNA. The plasmid encodes proteins that form a channel between the donor and recipient (mating bridge) plus proteins that nick the plasmid at its transfer origin (*oriT*). The single-stranded nick at the *oriT* is followed by attachment of the single strand to the mating bridge, which then transfers the single strand into the cytoplasm of the recipient. DNA polymerase makes both the strand left in the donor and the strand introduced into the recipient double-stranded, so that both donor and recipient end up with a double-stranded copy of the plasmid.

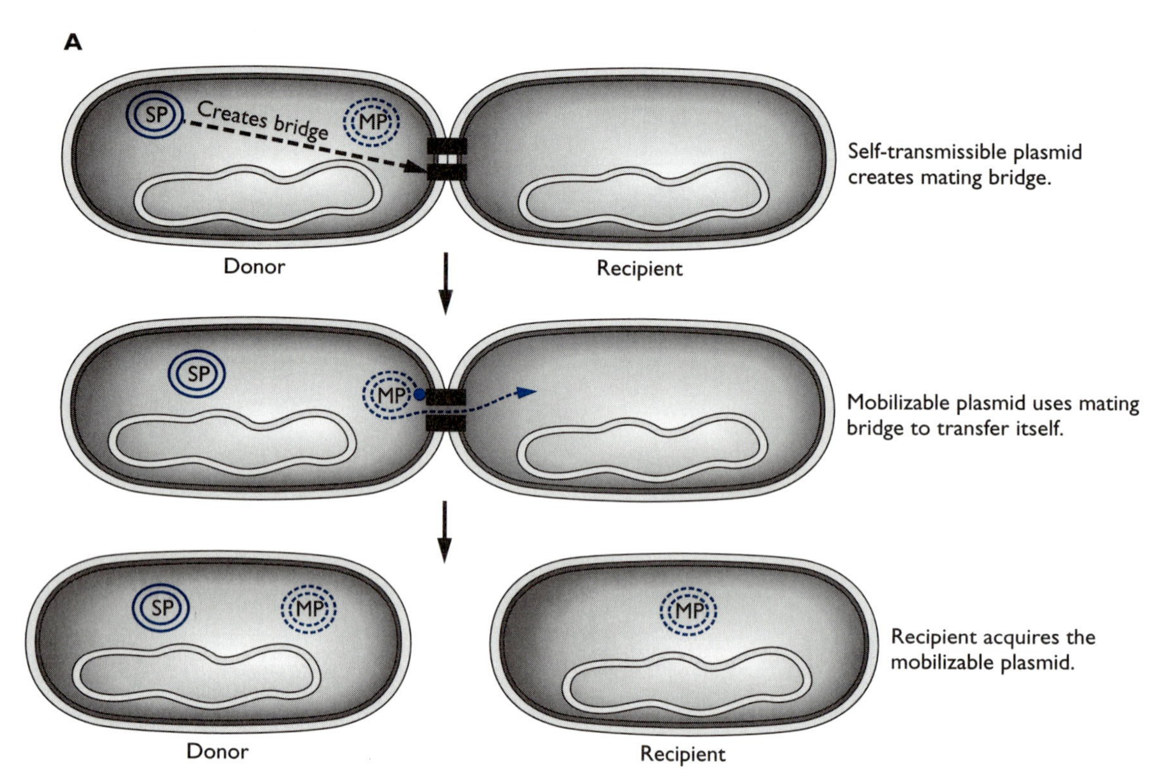

Figure 7.6 Mobilization of plasmids. (A) Mobilization of a mobilizable plasmid (MP), which has an *oriT* and genes encoding proteins that make the single-stranded nick at the *oriT* but lacks the genes encoding the mating bridge. If a self-transmissible plasmid (SP) in the same strain as the mobilizable plasmid provides the mating bridge, the mobilizable plasmid can use this bridge to transfer itself.

Mobilizable plasmids. Some plasmids, called **mobilizable plasmids,** cannot transfer themselves, because they lack the genes needed to create the mating bridge. However, they carry genes encoding the proteins that nick the plasmid at its *oriT.* Such plasmids can make use of the mating bridge furnished by a separate self-transmissible plasmid to complete their own transfer (Fig. 7.6A). Thus, self-transmissible plasmids are said to mobilize the mobilizable plasmid (transfer *in trans*). Both mobilizable and self-transmissible plasmids have played major roles in transferring antibiotic-resistance genes to a variety of bacteria. Genes encoding toxins and other traits that contribute to the ability of a bacterium to cause disease can be carried on plasmids. Plasmids may also carry genes encoding metabolic traits, such as the ability to utilize a sugar or a pollutant as a source of carbon and energy.

In the type of conjugal transfer just described, the recipient acquires a plasmid, but the donor gets nothing extra. In a related type of conjugal transfer, called **retrotransfer,** the donor can acquire a new plasmid, too. Retrotransfer is illustrated in Figure 7.6B. The donor carries a self-transmissible plasmid, and the recipient carries a mobilizable plasmid. The self-transmissible plasmid transfers to the recipient, where it mobilizes the recipient's plasmid back into the donor. When the dust clears, both recipient and donor have two plasmids instead of one. Retrotransfer could be a way for bacteria carrying self-transmissible plasmids to shop around for useful genes that may be carried on mobilizable plasmids in other bacteria.

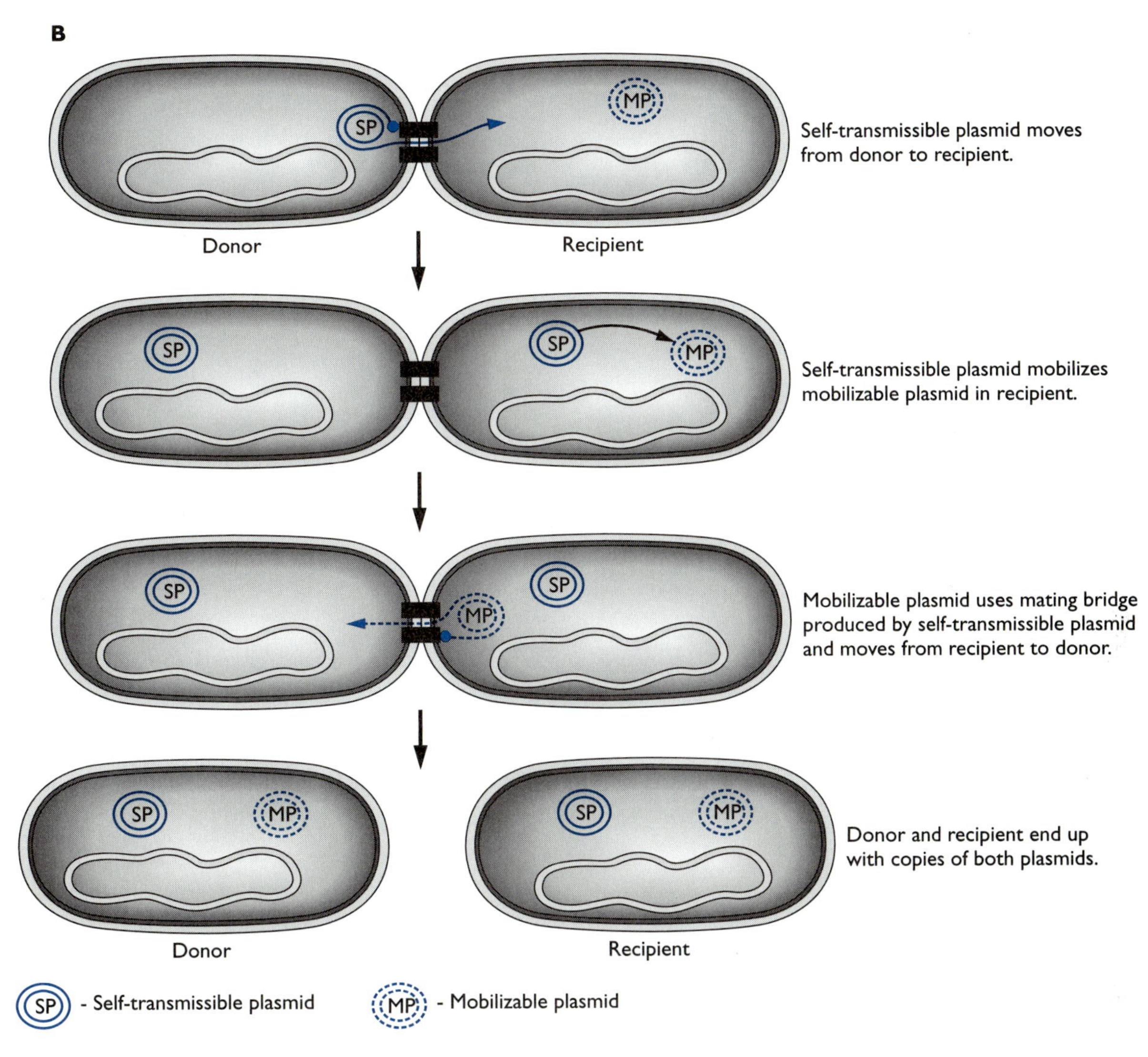

Figure 7.6 *(continued)* **(B)** Retrotransfer is a mobilization event involving the recipient as well as the donor. An SP in the donor enters the recipient and mobilizes an MP so that it moves "backwards" from the recipient to the donor. Both donor and recipient end up with both plasmids.

Conjugative transposons. A second type of self-transmissible gene transfer element is the **conjugative transposon**. Conjugative transposons are normally integrated into the bacterial chromosome. In response to some signal, they excise themselves from the chromosome to form a covalently closed circle, which looks like a plasmid but does not replicate (Fig. 7.7). This circular intermediate is then nicked at an internal *oriT* and transferred to a recipient, much as a plasmid is transferred (Fig. 7.5). In the recipient, the circular form of the conjugative transposon integrates itself into the recipient genome. Conjugative transposons have now been found in a number of bacterial genera and are capable of carrying a variety of types of genes. Conjugative transposons, like self-transmissible plasmids, can mobilize plasmids that cannot transfer themselves.

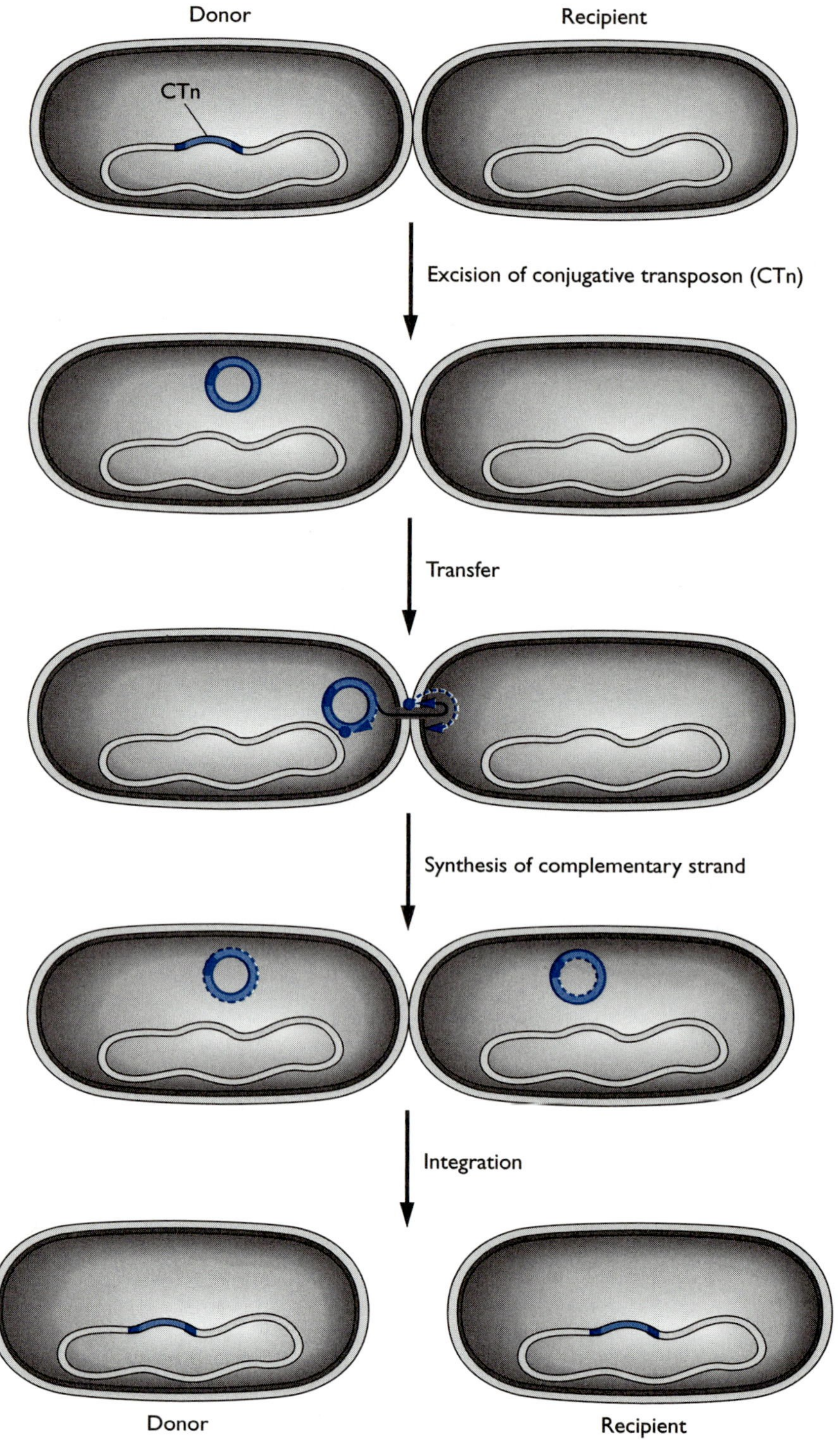

Figure 7.7 Steps in the transfer of a conjugative transposon. The conjugative transposon is integrated into the donor cell's genome. It excises itself to form a circular intermediate, which then transfers by conjugation similarly to a self-transmissible plasmid. The transposon circular form then integrates into the recipient's genome.

box 7–1 *The Archeology of Horizontal Gene Transfer: Finding Traces of Previous Gene Transfer Events*

In the laboratory, the frequency of gene transfer events such as tranformation, transduction, and conjugation is usually rather low, especially when bacteria freshly isolated from nature are tested. Only about one in 10,000 recipients or fewer obtains a copy of the transferred DNA, even under optimized conditions. This observation leads to the question of how frequently horizontal gene transfer events actually occur in nature. Earlier in the chapter, we implied that such transfers might occur often. What is the reason for thinking this is the case?

A way to look for evidence of horizontal transfer events is to find virtually identical copies of a gene in bacteria that are not closely related. That is, if a gene has been transferred recently from bacterium A to bacterium B, the gene will not have had time to mutate appreciably and will have virtually the same sequence in B as in A. Another indication that a gene in bacterium B has been acquired from another bacterium is that the **%G+C** content of the gene is different from that of other genes in B. Bacteria vary in the frequency with which G and C are used compared with A and T. These differences are expressed as %G+C, the percent of total bases in a bacterium's genome that are G or C instead of A or T. If organism B has a gene with a sequence that is virtually identical to that of a gene found in A, and if the gene in B has a %G+C content similar to that of A's genes but different from that of B's genes, the gene was probably transferred horizontally from A to B.

What does analysis of natural bacterial isolates suggest about the amount of horizontal gene transfer that occurs among bacteria in nature? Virtually identical copies of antibiotic-resistance genes (>95% identical DNA sequence) have been found in members of distantly related genera and in both gram-positive and gram-negative bacteria. Perhaps even more surprising, identical genes have been found in bacterial species normally found in soil and species normally found in the human intestinal tract, indicating that some genetic conduit is open between bacteria that normally occupy different sites.

The search for evidence of horizontal gene transfer events in nature is not just an exercise in data collecting. The results of such searches can have profound economic and political implications. For example, reports suggesting that gene transfers had occurred between genera of bacteria normally found in the intestinal tract of animals and those normally found in the human intestine have fueled controversy over the use of antibiotics in agriculture. Nearly half the antibiotics produced each year in the United States are used in agriculture to treat sick animals, prevent infections, and stimulate growth of animals. Concerns have been raised about the possibility that antibiotic-resistant bacteria in an animal's intestine, selected by antibiotic use on the farm, might enter the food supply and make their way into the human intestine. There, they might be able to transfer their resistance genes to human intestinal bacteria or even to human pathogens passing through the intestine. Evidence that such transfers might be occurring is coming from studies that find virtually identical copies of resistance genes in animal intestinal and human intestinal bacteria. This evidence is controversial, because it is based only on DNA sequence identity. The %G+C data do not indicate clearly the direction of transfer, so the transfers could have occurred from human intestinal bacteria to animal intestinal bacteria rather than the other way. This is a good example of a way in which scientists must sometimes grapple with incomplete but suggestive data when asked to advise the regulatory agencies on such politically-charged issues as whether antibiotic use in agriculture should be curtailed.

Source: Alliance for the Prudent Use of Antibiotics. Reservoirs of Antibiotic Resistance Network. Available at: http://www.roar.antibiotic.org. Accessed April 26, 2000.

Horizontal gene transfer events may occur fairly often in nature. Mechanisms for detecting traces of such events that occurred at some point in the past are described in Box 7–1.

Insertion Sequences, Transposons, and Integrons

Plasmids and conjugative transposons often undergo changes that involve the insertion or deletion of segments of DNA. Similar rearrangements are seen in genes carried on the chromosome. Such events are mediated by DNA segments called insertion sequences and transposons that move about within a bacterial cell, promoting various changes in the DNA. The changes they make are larger than those caused by changes of one or a few bases in a DNA sequence. They make a contribution to evolution of bacterial traits that can be as significant as the contribution of single base mutations.

Insertion sequences and transposons

Insertion sequences (**ISs**) are DNA segments, usually about 1–2 kbp in size, which can move from one site in the DNA of a bacterium and integrate into another site on the DNA. Most of them integrate randomly and integration occurs independently of homologous recombination. That is, no sequence identity is required between the IS and the DNA segment into which it integrates. An IS usually contains only a single open reading frame, which encodes an enzyme that mediates the excision and integration of the IS. Because the sole activity of an IS is to move itself from one place to another in the genome, ISs have been called "jumping genes" or "selfish DNA." An

Figure 7.8 Comparison of an insertion sequence (IS) and a transposon. (A) The IS contains the gene that encodes an enzyme (transposase) that allows it to excise itself from and enter DNA randomly. Many ISs also have promoters that point outward from the ends. This allows an IS to provide a new promoter for a gene if it integrates into the promoter region of that gene. **(B)** A transposon is composed of two ISs and intervening DNA, which can carry antibiotic resistance genes or genes conferring other traits. The ISs and the DNA they flank now move as a single unit. **(C)** Consequences of IS or transposon insertion for genes in the region where the element inserts.

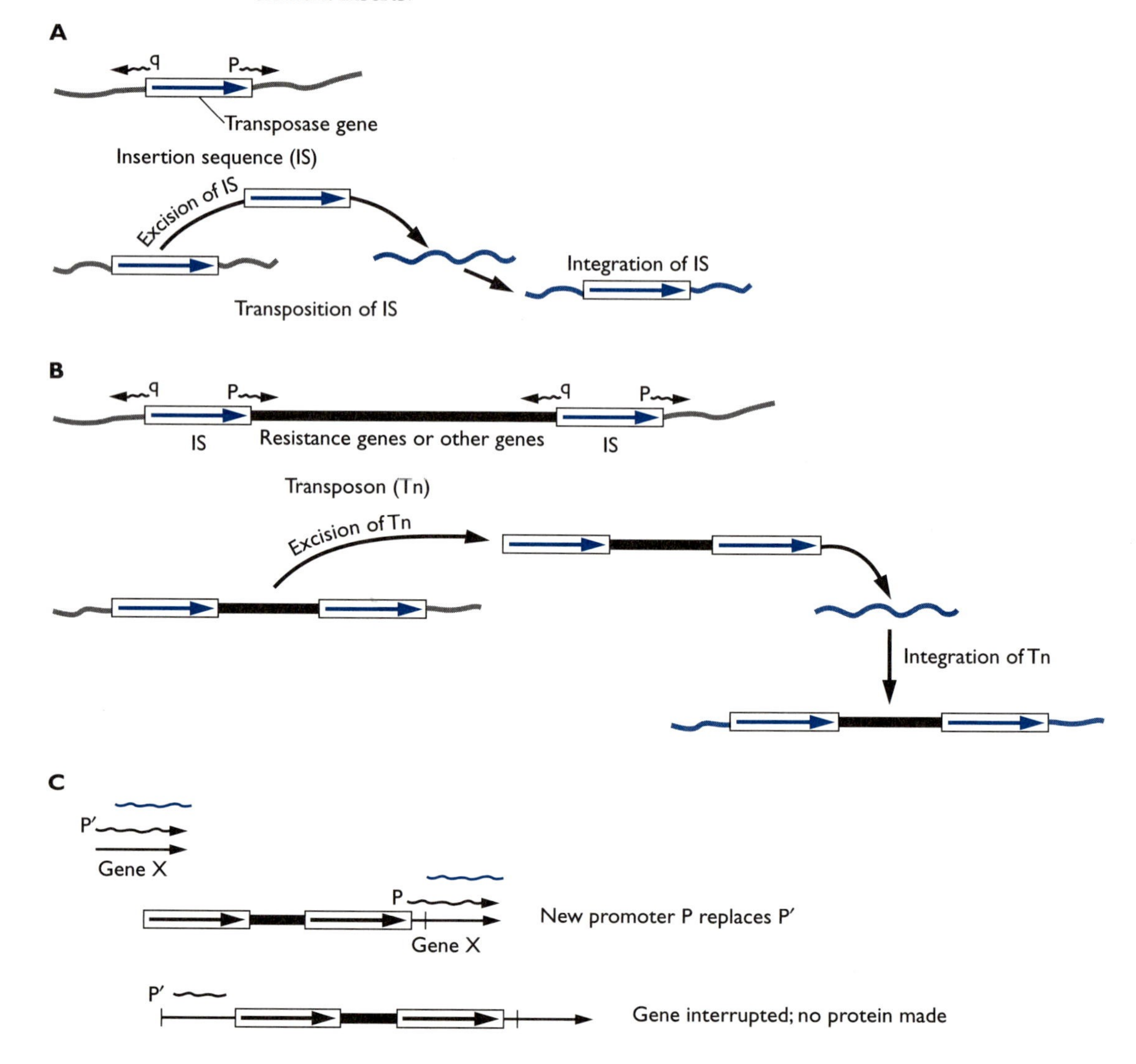

important property of ISs is that near their ends they often have promoters that are pointing outward (Fig. 7.8A), a trait that allows them to activate expression of adjacent genes (Fig. 7.8C). ISs move within cells, unless they can hitch a ride on a transmissible element (such as a plasmid or conjugative transposon) and use that vehicle to move from one cell to another. Sometimes, two ISs flank a chromosomal DNA segment and the whole assembly, called a transposon, begins to move as a single unit (Fig. 7.8B). **Transposons** play an important role in moving antibiotic-resistance genes and other genes from place to place in the genome and from genome to plasmid or the reverse.

An example of what ISs can do is provided by the evolutionary odyssey of the ***ampC* gene** of *Klebsiella,* a bacterium that is closely related to *E. coli* and causes bacterial pneumonia (Fig. 7.9). The *ampC* gene encodes a β-lactamase that inactivates many β-lactam antibiotics (see Chapter 4), including some of the newest ones. Initially, the *ampC* gene was in the *Klebsiella* chromosome and was very poorly expressed because it had an extremely weak promoter. Because expression was poor, the amount of protein produced was low. The

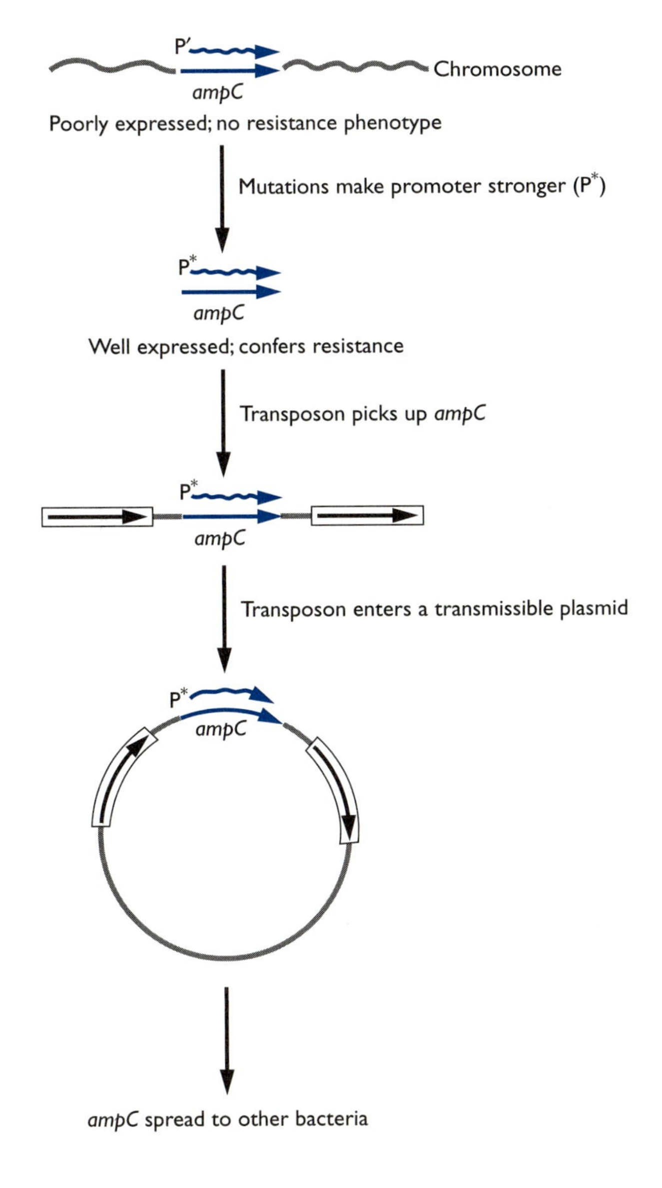

Figure 7.9 The evolution of the *ampC* gene of *Klebsiella* species from a gene that conferred little resistance to β-lactam antibiotics to a significant clinical threat to human health. The *ampC* gene encodes a β-lactamase that inactivates many of the more advanced β-lactam antibiotics, but this gene was clinically irrelevant as long as its expression was too low to confer significant levels of resistance. Mutations in the promoter region, which increased expression of the gene, were followed by incorporation of the gene into a transposon. The transposon then integrated into a plasmid that is now transferring the *ampC* gene into different *Klebsiella* strains.

gene thus conferred such a weak level of resistance that only someone looking very hard would have noticed it. *ampC* was certainly not conferring a level of resistance that was clinically significant. Before long, however, under antibiotic selection, the gene was being expressed at much higher levels, as a result of mutations in its promoter. Soon after, the *ampC* gene acquired a couple of ISs, and the resulting transposon was able to move *ampC* onto a transmissible plasmid, which is now spreading to many strains of *Klebsiella*. What was once a clinically irrelevant resistance gene has become the terror of hospitals—all thanks to ISs. In other cases, expression of resistance genes and other genes has been activated when an IS is integrated into the gene's promoter region.

Integrons

Integrons are IS elements or transposons with an unusual feature: they create large clusters of genes that move as single units. Integrons have developed a highly efficient strategy for creating gene clusters (Fig. 7.10). In addition to a **transposase** gene, which allows the integron to move from one site in the DNA to another, integrons contain a gene that encodes an **integrase,** which mediates a form of site-specific integration that incorporates the DNA segments that will form the gene cluster into a specific site on the integron. Adjacent to the integrase gene is a promoter, which controls expression of the genes accumulated by the integron. Incoming DNA circles (**gene cassettes**), which carry a sequence that allows them to recognize and

Figure 7.10 Structure of an integron. (A) Integrons are IS elements or transposons with an additional capacity for forming multiple gene clusters. The integron contains an integrase gene (*Int*), the protein product of which mediates the integration of gene cassettes into a specialized target site on the integron. All of the gene cassettes integrated into this site are transcribed in the same direction. A promoter (P) adjacent to the site where the gene cassettes are integrated controls the expression of the operon created by integration of multiple gene cassettes. **(B)** Example of an operon created by insertion of multiple cassettes into an integron. The operon contains various resistance genes.

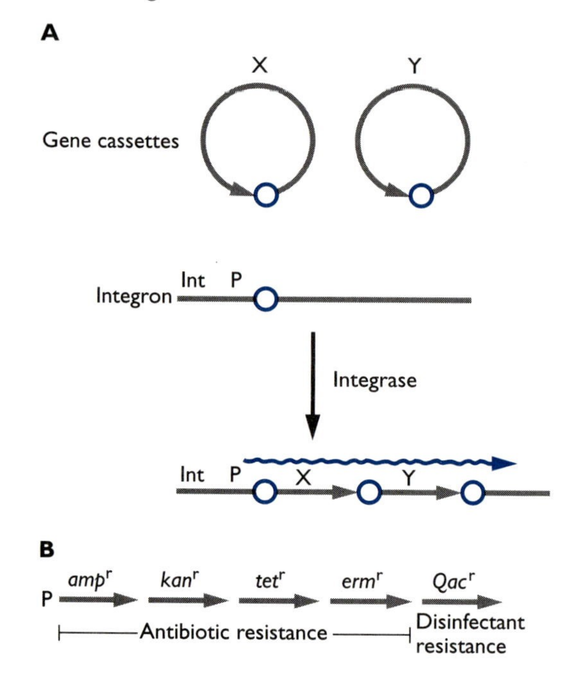

integrate specifically into the integron's integration site. All the incoming genes integrate so that they are transcribed in the same direction, thus creating an operon that is expressed from the integron promoter. No one knows where the gene cassettes come from, but they must be abundant because there are many examples of gene clusters formed in integrons from them (Fig. 7.10). Scientists studying resistance genes were the first to discover integrons, so most of the examples of integron-mediated gene cluster formation involve antibiotic-resistance genes. Some scientists suspect, however, that integrons have had a much more pervasive influence on creation of operons than was previously appreciated and may have played a major role in the evolution of operons.

The ominous thing about the antibiotic-resistance gene clusters created by integrons is that, because they are all controlled by a single promoter, loss of one gene as a result of deletion is likely to eliminate expression of other genes in the cluster. This means that selective pressure exerted by a single antibiotic tends to maintain the entire gene cluster intact in the strain and prevent loss of any of the genes in the gene cluster. That is, a single antibiotic of one type can select for maintenance of resistance genes that confer resistance to antibiotics of other types. Also, if the integron-assembled cluster enters a plasmid or conjugative transposon, the whole cluster can now be transferred to other bacteria en masse.

Bacterial Gene Transfer as a Source of Genetic Tools

Gene disruptions

Scientists who want to identify the function of a particular gene can use gene expression as a guide to function (Chapter 6), but a more certain guide to function is to stop the gene from being expressed and then determine the consequences. One way to stop a gene from being expressed is illustrated in Figure 7.11A. An internal segment of the gene is cloned into a plasmid that does not replicate in the target organism. This plasmid, along with its gene segment, is then transformed into the target organism by transformation or conjugation, with selection for an antibiotic resistance gene carried on the plasmid. The only way the plasmid can rescue itself is to recombine by homologous recombination with the gene in the chromosome. Because the plasmid contains only a portion of the gene, this recombination event disrupts the gene, ensuring that the protein encoded by the gene will not be made intact.

An alternative strategy is to use transformation to introduce linear DNA carrying the gene, into which has been cloned an antibiotic-resistance gene (Fig. 7.11B). In this case, two crossovers are necessary to introduce the resistance gene into the chromosome, replacing the gene with a disrupted copy. The effect is the same as that of the single crossover event: the disruption of the gene. Scientists interested in putting genome sequences to work are using both these strategies to interrupt genes so that the effects of their disruption can be determined. This is a very powerful way to ascertain the function of a gene.

Transposon Mutagenesis

In the case of the gene disruptions just described, a specific gene is targeted. But what if the researcher wanted to find genes that were previously unknown? Transposons come to the rescue here, because they integrate randomly into DNA. Often, integration of a transposon disrupts a gene, because

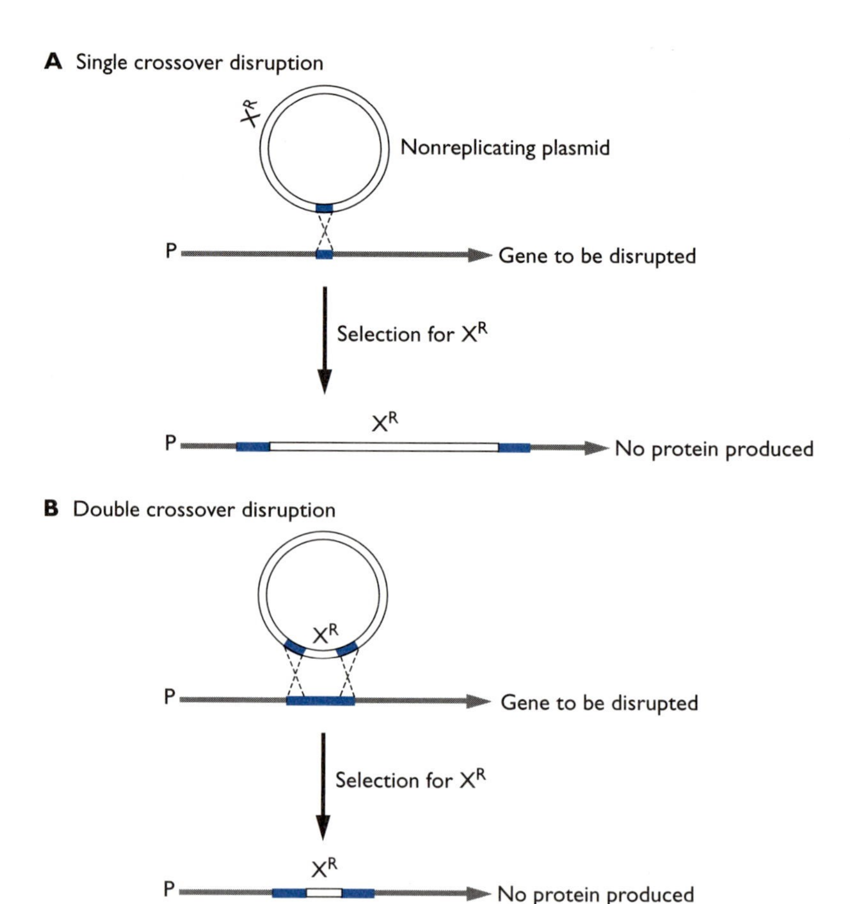

Figure 7.11 Two methods for disrupting a gene. (A) Homologous recombination integrates an internal segment of the gene via a single crossover. The plasmid carrying the cloned internal segment does not replicate in the recipient in which the gene disruption is being made. **(B)** A linear DNA segment is introduced that consists of the gene interrupted by some selectable marker gene such as an antibiotic resistance gene. A homologous recombination event involving two crossovers replaces the wild-type gene with the interrupted gene.

the transposon inserts into the coding region of the gene. One strategy for using **transposon mutagenesis** to find new genes is illustrated in Figure 7.12. The transposon is introduced into the bacterial strain on a plasmid that does not replicate in that strain. Selection for the antibiotic-resistance gene on the transposon ensures that only cells in which the transposon left the plasmid and entered the genome will survive the selection. Continuing the example of Chapter 6, in which a scientist was interested in genes needed for growth at low iron concentrations, the scientist would screen a collection of bacterial cells each with its own transposon insertion for those no longer able to grow on agar medium containing a low iron concentration (Fig. 7.12). Such mutants presumably had a transposon inserted into a gene necessary for growth under low iron conditions. Once interesting transposon-generated mutants are identified, it is then possible to find where the insertion occurred by cloning the gene, using the transposon as a marker.

Why is transposon mutagenesis so useful? To understand why, let us return to the example of the search for information about how a bacterium survives in low iron conditions. Studies of other bacteria suggest that the bacterium under study might have a special surface receptor for iron and proteins that internalizes the iron bound to the receptor. But what if the bacterium of interest does not use the standard mechanisms for obtaining iron when iron concentrations are low? Analyzing genes, the loss of which makes the bacterium unable to grow in low iron conditions, helps to determine whether those genes are like known iron-acquisition genes or are part

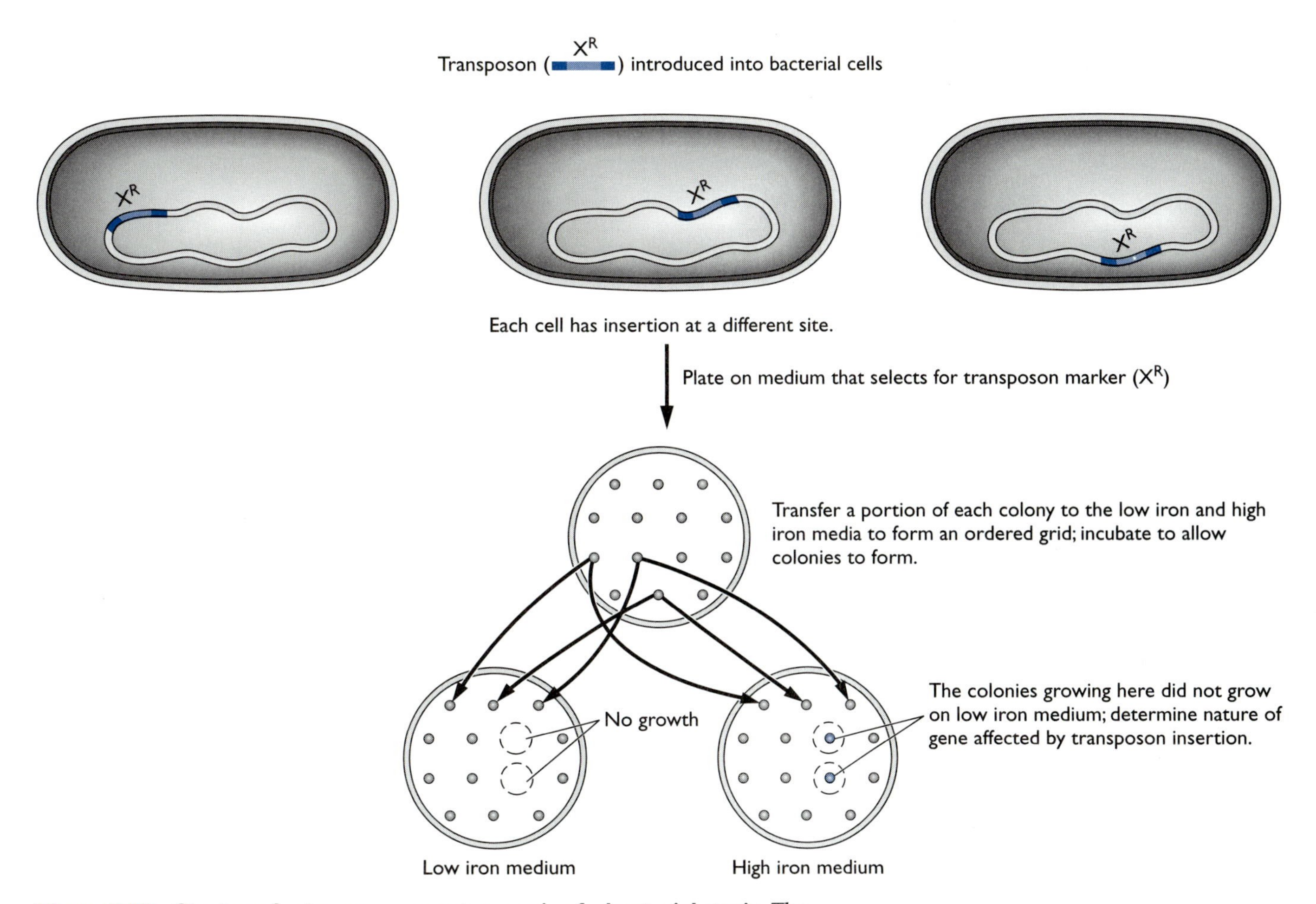

Figure 7.12 Strategy for transposon mutagenesis of a bacterial strain. The transposon, carrying a selectable marker, such as an antibiotic-resistance gene, is introduced into the bacterial strain of interest with selection for the marker on the transposon. Only bacterial cells with the transposon integrated somewhere in their genomes will survive the selection. Transposon-generated mutant strains are screened for the phenotype of interest, in this case inability to grow under low iron conditions. Mutants unable to grow under these conditions are subjected to further study.

of some new type of pathway. The power of transposon mutagenesis is that it enables scientists to look into the unknown and ask the question: how does this bacterium cope with low external iron levels? The answer may be unexpected.

Beyond Bacteria: Transposons and Horizontal Gene Transfer in Higher Forms of Life

Insertion sequences and transposons give bacteria a way to make rather crude changes in their activities by inactivating genes (if the IS or transposon lands inside a gene) or activating genes (if the IS or transposon lands in a promoter region). Is this strategy of DNA manipulation limited to simple organisms such as bacteria? Not at all. Transposons were first noted in plants. Those ears of corn with the multicolored kernels that are used as fall decorations provide an example of the activities of the plant equivalents of bacterial transposons. More recently, studies of the human genome have revealed that a significant portion of the human genome contains transposable elements that are in a constant state of movement. The effects of these "jumping genes" will be muted in differentiated cells, because these cells have only a limited lifetime. Their effects on the germ line cells will take a long time to become apparent because of the long generation time of humans. Still, animal and plant genomes are clearly fluid constructs. It is a natural, involuntary function, and its role in human evolution remains unknown. It can't be all bad, or we would not be here to study the phenomenon.

Transposable elements in plants and the human genome seem to move only within the cell in which they are found. The possibility that horizontal gene transfer of such transposable elements occurs and that they spread from organism to organism has become more believable as a result of new studies of the phenomenon in insects. Insects, like plants and humans, have transposon-like elements of their own. Two types of insect transposable elements are the mariner elements, which move similarly to bacterial transposons; and retrotransposons, which move via an RNA intermediate. Mariner elements were first found in the genome of the fruit fly *Drosophila melanogaster.* Scientists have now used polymerase chain reaction (Chapter 5) to screen the genomes of other insects for copies of the *Drosophila* mariner elements. They found evidence for mariner elements in the genome of the horn fly, *Haematobia irritans,* the mosquito, *Anopheles gambiae,* and the golden lacewing, *Chrysoperla plorabunda.* Sequence analysis of these mariner-like elements revealed that in some cases the mariner elements in these other insect genera differed from those in *Drosophila* by only a few bp out of the 1044 bp mariner element. Because the insect species involved last had a common ancestor more than 200 million years ago, the high degree of DNA sequence identity between copies of the transposons in different insects clearly indicates that the elements were transferred recently from one species to another. Evidence has also been found for the movement of retrotransposons between different species of *Drosophila.*

In the case of bacterial horizontal gene transfer, the mechanisms of transfer have been identified: transformation, phage transduction, and conjugation. So far, no such active systems for transferring DNA have been found in higher eukaryotes. So, how are the transposons of insects being transferred? One possibility is that biting insects, such as mites that prey on fruit flies and other insects, could be the mechanism of spread. It remains to be seen whether specialized forms of horizontal gene transfer analogous to natural transformation or conjugation occur between plant and animal species.

chapter 7 at a glance

Comparison of transformation, phage-mediated transfer, and conjugation

Type of transfer	Type of DNA transferred	Mechanism of transfer
Transformation		
Natural transformation	Linear single-stranded DNA	Protein complex in cytoplasmic membrane takes up DNA into cytoplasm
Artificial transformation	Linear or plasmid double-stranded DNA	No specialized uptake mechanism DNA forced through membrane by chemical shock or an electric field
Phage-mediated transfer		
Lysogeny	Linear double-stranded phage DNA	Phage genome integrates into bacterial genome
Generalized transduction	Linear double-stranded bacterial DNA segment	Phage accidentally incorporates bacterial DNA segment into its capsid during lytic growth; transfers bacterial DNA segment to another bacterium
Conjugation	Single-stranded copy of a plasmid or conjugative transposon circular form	Plasmid or conjugative transposon circular form transfers from cell to cell through the mating bridge formed between a donor bacterial cell and a recipient bacterial cell

Insertion sequences, transposons, and integrons

Type of element	Features
Insertion sequence	Encodes transposase (catalyzes movement from one site on the DNA to another) Often has promoters pointing outward from its ends Can provide a new promoter for a gene (integrates in promoter region) Can inactivate a gene (integrates inside the open reading frame)
Transposon	Usually consists of two insertion sequences flanking genes such as antibiotic resistance genes or other genes unrelated to transposon movement Moves as a single unit; can provide a new promoter or inactivate a gene, similar to IS
Integron	Specialized IS or transposon containing an integration site and an integrase that creates clusters of genes (operon), all expressed from the same integron promoter

Genetic tools

Gene disruptions—use homologous recombination to disrupt specific genes

Transposon mutagenesis—uses random transposon insertions (independent of homologous recombination) to disrupt and identify unknown genes

key terms

Key terms

bacteriophages viruses that attack bacteria; also called phages; may introduce new genes into bacteria by lysogeny (integration of viral genome) or generalized transduction (accidental transfer of bacterial genes)

conjugation direct cell-to-cell transfer of bacterial DNA via a multiprotein mating bridge

conjugative transposon self-transmissible chromosomal element that excises itself from the chromosome to form a circular intermediate, which is transferred to a recipient, then integrates into the recipient genome

gene disruption mutants mutants generated by using single-crossover or double-crossover homologous recombination to interrupt a gene so that the protein it encodes is no longer produced

horizontal gene transfer transfer of DNA from one organism to another; in bacteria, mediated by transformation, phage transduction, or conjugation

insertion sequence (IS element) DNA segment (usually 1–2 kpb in size) that integrates randomly into DNA; contains a gene encoding transposase, the enzyme that allows the IS element to move from one site to another, plus promoters pointing outward from each of its ends; can activate expression of genes or disrupt genes

integron transposon that also contains an integrase gene, the protein product of which facilitates the integration of gene cassettes into a special site on the integron, creating gene clusters that are under the control of a promoter provided by the integron

plasmids DNA segments (often circular) that replicate autonomously; some carry genes that allow them to transfer themselves by conjugation from a bacterial donor to a bacterial recipient (self-transmissible plasmid); some carry genes that allow self-transmissible plasmids to mobilize them from a donor to a recipient cell (mobilizable plasmid)

transformation uptake of DNA from the extracellular fluid; natural transformation system consists of cytoplasmic membrane complex that imports single-stranded DNA into the bacterial cytoplasm; artificial transformation employs conditions that force bacteria to take up DNA from the extracellular environment

transposon DNA segment flanked by two insertion sequence elements, which now move as a unit; may contain antibiotic resistance genes or genes for a variety of other metabolic traits

transposon mutagenesis use of transposons to create a collection of mutants in which most mutants will have the transposon interrupting a gene; employed to generate mutations that identify genes important for a particular trait

Questions for Review and Reflection

1. Explain why it is usually much more advantageous for a bacterium to obtain new genes by horizontal gene transfer than by mutation. In what cases could it be harmful?

2. The possibility was mentioned that some types of genes are transferred much less frequently than others. Does this mean that transfer of the DNA is less frequent or that maintenance of the DNA in the recipient is less likely, or both?

3. Natural bacterial gene transfer systems make use of a protein complex to move DNA across the cytoplasmic membrane. Why would such a complex system be needed? In what ways does the DNA transfer process involved in phage transduction and conjugation resemble or differ from that involved in natural transformation?

4. Could *E. coli* O157:H7 have acquired toxin genes by generalized phage transduction instead of lysogeny? How? Which process is likely to be more efficient and why?

5. Why does DNA transferred by conjugation have to be in a circular form? Why is this not a requirement for phage transduction and natural transformation?

6. Transformation and phage transduction are usually associated with narrow host range transfers, whereas plasmids and conjugative transposons can mediate broad host range as well as narrow host range transfers. Explain this difference.

7. How are conjugative transposons similar to plasmids? What advantage might their mode of spread have over that of plasmids?

8. If you were studying a bacterial strain that could transfer certain traits to other bacteria, how would you decide whether transfer was by conjugation, transformation, or phage-mediated processes?

9. A bacterial strain contains a small plasmid that carries gene X. This small plasmid is neither mobilizable nor self-transmissible. A scientist finds, however, that a few cells of the bacterial strain carrying this plasmid have now become capable of transferring it by conjugation to other bacteria. Now the plasmid, which still carries gene X, has become much larger than it was before. What might have happened? (Hint: Consider the possibility that the strain carried a conjugative transposon in its chromosome.)

10. Scientists using virtually identical gene sequences as a criterion for horizontal gene transfers are often asked how they can rule out a phenomenon called convergent evolution, in which the amino acid sequence of a protein is so highly conserved by selection that the gene encoding the protein cannot mutate freely to change its amino acid sequence. How many DNA sequence differences could occur without changing the amino acid sequence of the protein? Could this be a way of differentiating between horizontal gene transfer and convergent evolution? (Hint: Review material on codons in Chapter 6.)

11. You have cloned a gene that may encode a β-lactamase. That is, its sequence is similar to that of known β-lactamases. Explain how you would use gene disruption analysis to determine whether the gene is conferring resistance to penicillins on the strain from which you cloned it.

12. Suppose you tried but were unable to clone the gene from the strain mentioned in question 11. How could you use transposon mutagenesis to find the elusive gene?

13. Suppose you are working with a strain of bacterium that is deficient in homologous recombination. Which of the horizontal gene transfer mechanisms—transformation, phage lysogeny or transduction, or conjugation would be unable to transfer genes to this strain?

14. Why use a transposon instead of an insertion sequence for transposon mutagenesis? Both integrate randomly, so why the preference for the transposon?

15. How would you clone a gene into which a transposon had been inserted? (Hint: Review Chapter 5 on cloning.) Assume that the transposon carries an antibiotic-resistance gene.

8 Bacterial Energetics

You can have the entire DNA sequence of a microbe's genome in your computer, but the microbe will not come to life in your understanding until you begin to factor in the life force that puts all the gene products into motion in a living cell. That life force is energetics.

To increase in numbers, bacteria need carbon, nitrogen, phosphorous, sulfur, and other compounds to make nucleic acids, proteins, peptidoglycan, and other cell components (Chapter 3). Acquiring these compounds and assembling them into macromolecules requires energy. In this chapter, we introduce one way bacteria extract energy from compounds they encounter in the environment: by oxidizing sugars such as glucose. In later chapters, other types of energy-yielding strategies, such as photosynthesis, will be described. The bacteria that utilize sugars as a source of carbon and energy play an important role in recycling biomass, the bulk of which consists of polysaccharides and proteins. Understanding how bacteria extract energy from different sources is also important to bioprospectors, who are looking for microorganisms that carry out novel reactions. Whether or not a particular reaction is possible often depends on the conditions the microbe experiences and the concentrations of substrates and products.

How Bacteria Make a Living

Recycling biomass: from the dead zone to the human colon

In Chapter 1, the hypoxic zone in the Gulf of Mexico (called the "dead zone" by fishermen) was used as an example of one way in which human activities can lead to changes in microbial activities. Nitrogen runoff from farms and other sources flows into the Gulf, initiating blooms of phytoplankton. Dead phytoplankton then settles to the bottom, where it is digested by other microbes. In the process, microbes consume dissolved oxygen from the water faster than it can be replaced. In this chapter, we will explore the microbial digestion of dead biomass, the process that has led to a drastic decrease in the amount of dissolved oxygen in the Gulf waters. This phenomenon occurs everywhere and is a necessary part of the recycling carried out by microbes. In fact, from the microbial point of view, the hypoxic area of the Gulf is very much a live zone, where bacteria extract carbon and energy from the biomass and convert it into more microorganisms.

No one has examined in detail the microorganisms at work in the Gulf sediments, but this type of process has been studied in the case of the human colon and in other sites where biomass is being degraded. Based on what is known about other settings, we can construct a picture of what happens to the dead biomass in the Gulf sediments and explain why oxygen is being consumed in the process.

Breakdown of polymers

Biomass is a complex mixture of polysaccharides, proteins, and lipids. Because polysaccharides make up the bulk of biomass, especially that of plant origin, and because many bacteria prefer to utilize sugars as sources of carbon and energy, we will focus on polysaccharide digestion. The enzymes that degrade polysaccharides hydrolyze the bonds that connect the sugar molecules to each other, releasing individual sugar molecules or short fragments of the polysaccharide, which are then further digested to individual sugars (Fig. 8.1A). Because polysaccharides are large molecules and are usually part of an even larger matrix of other polymers, taking them intact into the cytoplasm is not an option. Some bacteria solve this problem by secreting polysaccharide-degrading enzymes into the extracellular fluid, where the enzymes can attack the polysaccharide (Fig. 8.1B). The bacteria then take up the sugars released by enzymatic digestion.

In an environment where the polysaccharide-degrading bacteria have many competitors that could use the products of the polysaccharide-degrading enzymes, this strategy is not advisable. A somewhat different strategy used by bacteria in competitive environments is to tether the polysaccharide-degrading enzymes to their surfaces and adhere to the polysaccharide matrix they are degrading. This makes it easier to trap the sugars their polysaccharide-degrading enzymes have released. *Clostridium* species, a group of gram-positive bacteria that are important polysaccharide degraders in many environments, use this strategy. A variant on this strategy is used by polysaccharide-utilizing colonic *Bacteroides* species and some related gram-negative bacteria found in the rumen of cattle. These bacteria bind the polysaccharide to proteins in their outer membranes, digesting the polysaccharide as it is translocated across the outer membrane.

Although we will emphasize sugar metabolism in this chapter, it is worth noting that bacteria use a similar strategy for attacking proteins. Proteases, the enzymes that hydrolyze proteins to amino acids, are sometimes extracellular and sometimes cell-associated. The amino acids generated by protease action are then taken into the cell cytoplasm, where they are used directly to synthesize proteins or are oxidized further to serve as sources of carbon, energy, and nitrogen. Although bacterial digestion of polymers is usually a good thing, it can be associated with disease (Box 8–1).

Transport across the cytoplasmic membrane

Sugars that reach the cytoplasmic membrane are transported across the membrane by an energy-consuming process. Although many different sugars are used by bacteria as sources of carbon and energy, the focus here will be on utilization of glucose. Glucose is the main component of the plant polysaccharide cellulose and is a major constituent of many other polysaccharides. Many bacteria that utilize glucose transport it into the cell using the **phosphotransferase system** (Fig. 8.2). Glucose is transported into the cell

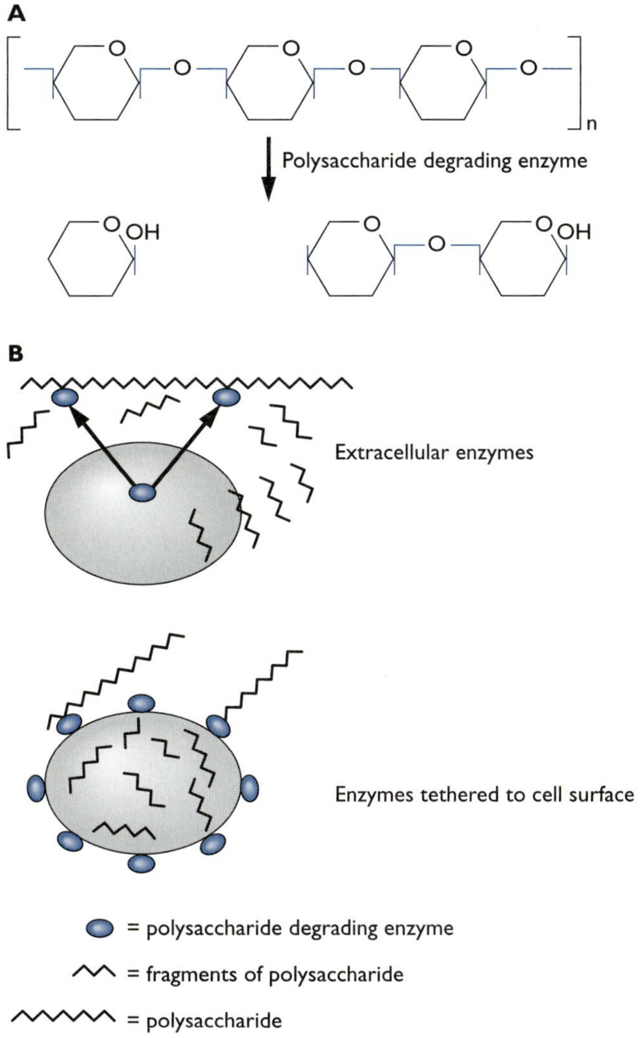

A

Polysaccharide degrading enzyme

B

Extracellular enzymes

Enzymes tethered to cell surface

= polysaccharide degrading enzyme

= fragments of polysaccharide

= polysaccharide

Figure 8.1 Hydrolysis of polysaccharides. (A) Large polysaccharides are degraded into individual sugars or small polysaccharide fragments by enzymes. **(B)** Some bacteria secrete enzymes into the extracellular cellular environment, where they degrade the polysaccharides. Others keep the enzymes attached to their surfaces.

box 8–1 *The Dark Side (for Us) of Bacterial Digestion of Polymers*

Breakdown of the polymers that compose biomass is generally a very important microbial activity, one that is essential for the continuation of life on this planet. In some cases, however, polymer digestion by bacteria has adverse consequences. The consumption of oxygen in Gulf of Mexico waters is one example. Two others are rot in fruits and gas gangrene. Bacteria that live on fruits such as apples usually cannot get through the waxy coating with which the fruit protects itself. But small wounds caused by wind-propelled particles or insect damage give the bacteria access to the plant tissue beneath the surface. The bacteria secrete enzymes that digest the polysaccharides responsible for structural integrity of plant tissues. The result is a rotten spot on the surface of the fruit. Growers prevent this bacterial activity by spraying fruit trees with antibiotics to inhibit the growth of the bacteria.

A human equivalent of plant rot is gas gangrene, a serious disease caused by *Clostridium perfringens*. *C. perfringens* is an obligate anaerobe but can survive in deep wounds, which can become quite anoxic. Not only do these clostridia produce proteases and other enzymes that digest tissue, but, in the process of utilizing the digestion products, they produce copious amounts of CO_2. The physical pressure of the trapped gas causes additional tissue damage. A large region of dead tissue can develop. The cells of the human body's defense system cannot penetrate this damaged area once it gets large enough. As a result of the destruction of blood vessels in the area, antibiotics do not penetrate either. Thus, either the dead tissue must be removed surgically or the entire limb must be amputated.

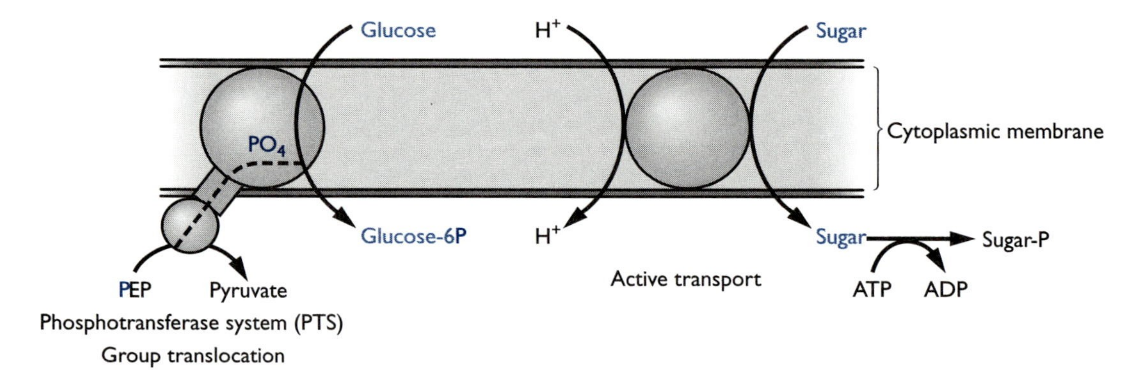

Figure 8.2 Two types of glucose transport. Glucose is phosphorylated by the phosphotransferase system (PTS) as it enters the cell. Certain other sugars are taken up by active transport and are phosphorylated only after they enter the cytoplasm.

and phosphorylated in the process to generate glucose-6-phosphate. The source of the phosphoryl group is phosphoenolpyruvate (PEP). PEP is one of the **high-energy phosphorylated compounds** used by bacteria (and by other living creatures as well) to store or provide energy (Fig. 8.3). The high-energy phosphate compound used most extensively as an energy donor or for energy storage is adenosine triphosphate (ATP). PEP and other phosphorylated compounds are used in more limited types of reactions (Fig. 8.3).

Figure 8.3 High-energy phosphate compounds and some reactions in which they play a role.

Compound	Reactions
Adenosine triphosphate (ATP)	Many catabolic and biosynthetic reactions
Guanosine triphosphate (GTP)	Protein synthesis
Uracil triphosphate (UTP)	Polysaccharide synthesis
Phosphoenol pyruvate (PEP)	Peptidoglycan synthesis, phosphotransferase system

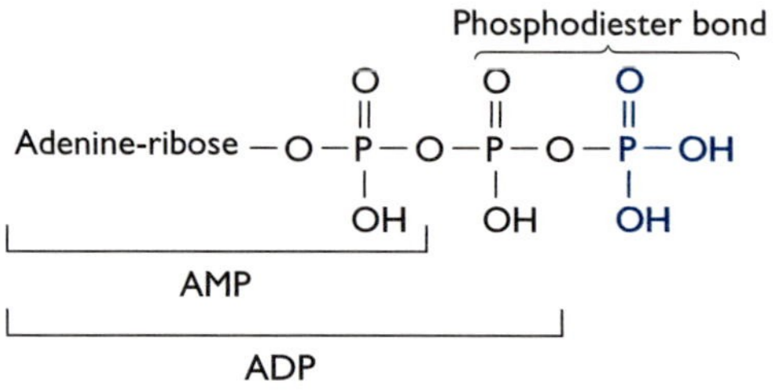

Phosphodiester bond

Adenine-ribose —O—P(=O)(OH)—O—P(=O)(OH)—O—P(=O)(OH)—OH

AMP

ADP

HOOC—CH(—O—P(=O)(OH)—OH)—CH_2

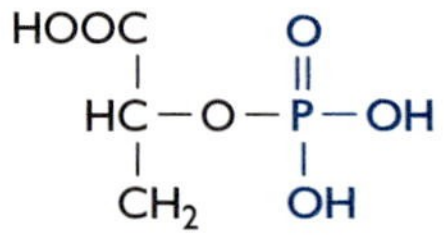

Bacteria can also transport sugars into the cytoplasm by **active transport,** a process that requires energy but translocates the sugar without phosphorylating it. One type of active transport system, which co-transports a proton with the substrate, is shown in Figure 8.2. ATP can also be used as a source of energy. Sugar molecules transported into the cell by active transport are then phosphorylated by enzymes called hexokinases, which transfer a phosphate group from ATP to the sugar to form a phosphorylated sugar.

The fate of glucose-6-phosphate

Different bacteria make different uses of glucose-6-phosphate. A commonly used pathway is the Embden-Meyerhof-Parnas (EMP) pathway, also known as **glycolysis.** Glycolysis is used not only by many sugar-utilizing bacteria but is also the pathway used by human cells to extract carbon and energy from glucose. This pathway first isomerizes glucose-6-phosphate to fructose-6-phosphate, then uses ATP to add a second phosphate group, producing fructose-1,6-bisphosphate (Fig. 8.4A). The next steps take fructose-1,6-bisphosphate, a 6-carbon (C6) compound, and break it into two 3-carbon (C3) compounds, which are ultimately oxidized to two molecules of pyruvate, also a C3 compound. In the process, four ADP molecules are phosphorylated to make four ATP molecules and two molecules of nicotinamide adenine dinucleotide (NAD^+) are reduced to reduced nicotinamide adenine dinucleotide (NADH). The role of compounds such as NAD^+/NADH, that are

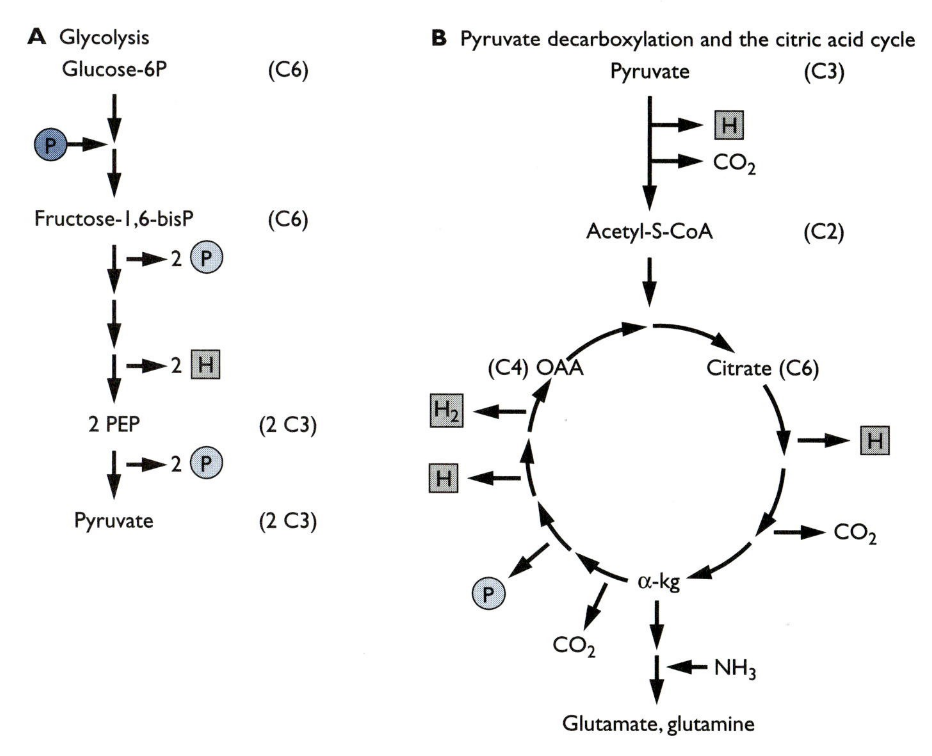

Figure 8.4 Overview of glycolysis and the citric acid cycle. (A) Glycolysis, the process by which glucose-6-phosphate, a C6 compound, is broken down into two molecules of pyruvate (C3), results in a net gain of two ATP molecules and two NADH molecules. **(B)** Some bacteria further oxidize pyruvate in the citric acid cycle, thus generating more ATP. One of the products of the citric acid cycle is α-ketoglutarate, which is combined with ammonia to produce glutamate and glutamine.

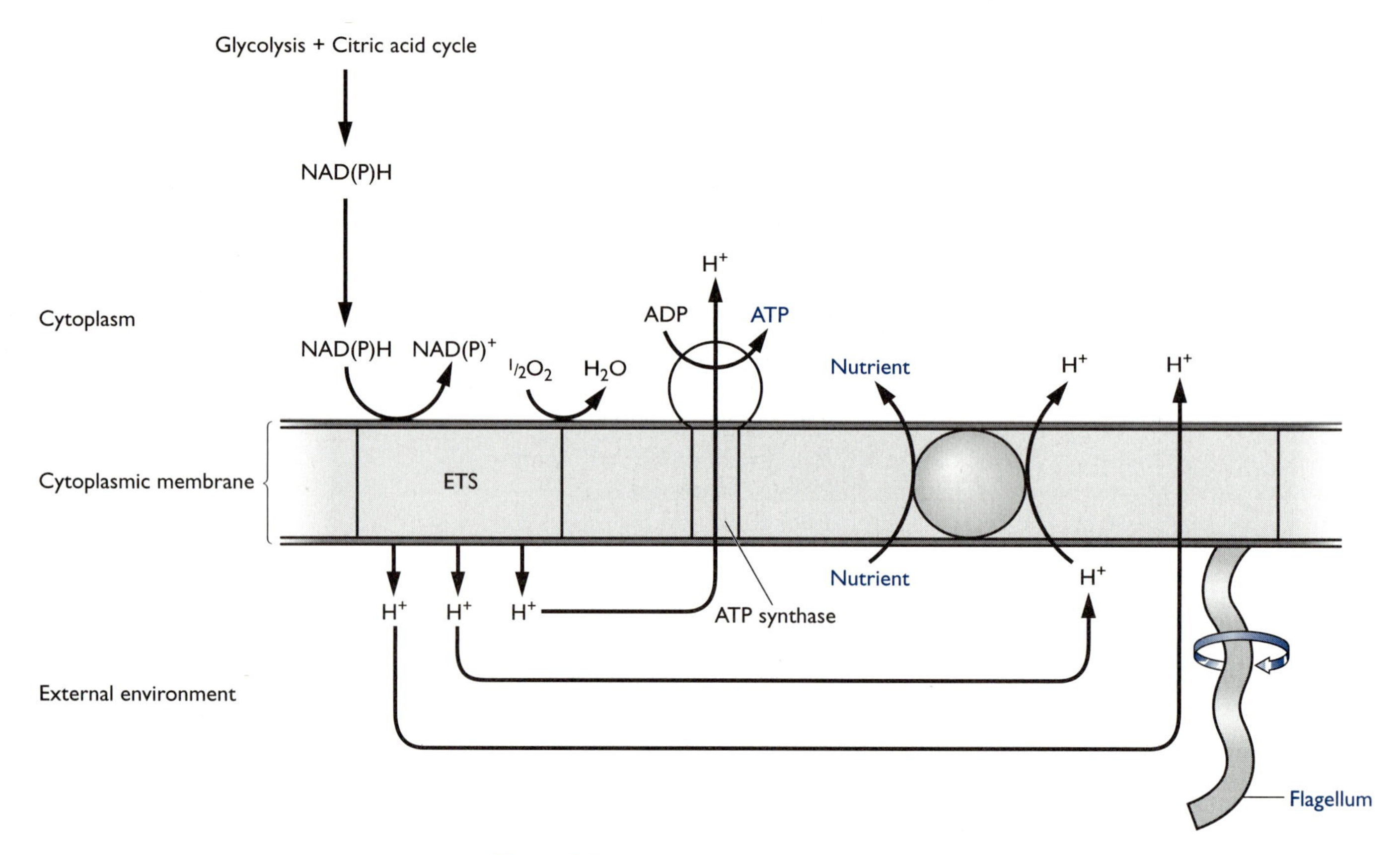

Figure 8.5 Overview of how the electron transport system works. NAD(P)H is used indirectly to yield adenosine triphosphate (ATP) and perform other kinds of work. The electron transport system (ETS) pumps protons across the cytoplasmic membrane and, in the process, transfers electrons to a terminal electron acceptor. The proton gradient that is formed can do work, such as rotating flagella or generating ATP.

involved in oxidation and reduction reactions, are illustrated in Figure 8.5. NAD^+ and NADH acquire and donate protons and electrons. An acquisition of protons and electrons by a compound is called a **reduction** of that compound, and a loss of protons and electrons is called an **oxidation.** The pathway from glucose-6-phosphate to pyruvate represents an oxidation of glucose and a transfer of protons and electrons to NAD^+. To this point, the bacterium has dephosphorylated two high-energy compounds (PEP/ATP) and has produced four ATPs for a net gain of two ATPs. NADH can be used to drive the synthesis of more ATP, as will be described shortly.

Some bacteria stop at this point, using NADH to reduce pyruvate to lactate and then excrete the lactate. Others decarboxylate pyruvate (C3) to form an acetyl (C2) group, then use a pathway called the citric acid cycle to oxidize the acetyl group to carbon dioxide and water (**citric acid cycle**; Fig. 8.4B). In the process, more molecules of NAD(P) are reduced and a molecule of ATP is generated from ADP.

Glycolysis is not the only pathway from glucose-6-phosphate to C3 compounds. Other pathways that accomplish this have in common with glycolysis that they oxidize the glucose to produce NADH or NADPH. Some also generate at least one ATP.

The electron transport system

How is NAD(P)H used to produce ATP from ADP? The answer is a combination of the **electron transport system** (ETS) and a protein complex called **ATP synthase.** An overview of this process is shown in Figure 8.5. The role of the ETS is to pump protons from the cytoplasm to the outside of the cytoplasmic membrane and dispose of the electrons left over from this process by transferring them to a terminal electron acceptor. The ETS depicted in Figure 8.5 uses oxygen as the terminal electron acceptor. When oxygen acts as a terminal electron acceptor (aerobic respiration), the reduction of oxygen produces water. This explains why oxygen consumption is often associated with biomass recycling.

The cytoplasmic membrane is impermeable to protons, so a proton gradient is generated by the ETS, yielding an excess of protons outside the cytoplasmic membrane. These protons can perform several different kinds of work (Fig. 8.5). They can reenter the cytoplasm through a protein complex called ATP synthase, driving the phosphorylation of an ADP to an ATP. It takes about three protons moving through the ATP synthase to drive the phosphorylation of an ADP to an ATP. Protons can also interact with transport proteins to provide energy for active transport of substrates. Protons provide energy to drive rotation of the flagellar motor.

The Sugar Utilization Pathway in More Detail

Energetics

In the preceding overview of glycolysis and the citric acid cycle, ATP was used to contribute or store energy. How does hydrolysis of compounds like ATP and PEP drive some reactions, whereas production of these compounds is used to store energy released in other reactions? Chemists have calculated the energy stored in various compounds by determining how much energy would be required to form the compound (the free energy of formation). To determine whether a reaction is possible, the sum of the energies of the initial reactants is subtracted from the sum of the energies of the products to determine $\mathbf{\Delta G^{0'}}$. ($\Delta G^{0'}$ is the energy difference at pH7, 25°C, and 1 molar concentrations of all reactants.) If $\Delta G^{0'}$ is negative, the reaction is exergonic. That is, it is thermodynamically feasible and the reaction will go in the direction drawn in the equation. If $\Delta G^{0'}$ is positive, the reaction is endergonic and is not thermodynamically feasible.

Consider, for example, the reaction:

$$\text{glucose} + P_i \rightarrow \text{glucose-6-P},$$

where P_i is inorganic phosphate. The $\Delta G^{0'}$ value for this reaction is positive ($\Delta G^{0'}$ = +14kJ/mol), so the reaction cannot proceed in this direction. If, however, this reaction is coupled with the hydrolysis of PEP to pyruvate, a reaction with a large negative $\Delta G^{0'}$ ($\Delta G^{0'}$ = –62kJ/mol), the combined reaction is energetically favorable:

$$\text{glucose} + P_i \rightarrow \text{glucose-6-P} \qquad \Delta G^{0'} = +14 \text{ kJ/mol, and}$$
$$\text{PEP} + H_2O \rightarrow \text{pyruvate} + P_i \qquad \Delta G^{0'} = -62 \text{ kJ/mol.}$$

The combined reaction is:

$$\text{glucose} + \text{PEP} + H_2O \rightarrow \text{glucose-6-P} + \text{pyruvate} \qquad \Delta G^{0'} = -48 \text{ kJ/mol.}$$

This reaction is feasible. If, on the other hand, a reaction has a large enough negative $\Delta G^{0'}$, it can drive the phosphorylation of ADP to make ATP. Thus:

$$PEP + H_2O \rightarrow pyruvate + P_i \qquad \Delta G^{0'} = -62 \text{ kJ/mol, and}$$
$$ADP + P_i \rightarrow ATP + H_2O \qquad \Delta G^{0'} = +35 \text{ kJ/mol.}$$

The combined reaction is:

$$PEP + ADP \rightarrow pyruvate + ATP \qquad \Delta G^{0'} = -27 \text{ kJ/mol.}$$

Thus, the energy in PEP is used to add a high-energy phosphate group to ADP to form ATP. A reaction in which an enzyme in the cytoplasm uses energy from one high-energy organic compound (in this case PEP) to convert ADP into ATP is called **substrate level phosphorylation** to distinguish it from ATP production via the ETS and ATP synthase.

The use of $\Delta G^{0'}$ values has one important limitation. These values are computed for conditions in which both substrates and products are all available at 1 M concentrations. In nature, such high concentrations are seldom encountered. If the concentrations are not 1 M, corrections must be made. This has considerable practical importance for scientists interested in finding organisms that carry out novel reactions. In some cases, having the right ratio between products and substrates can actually make a reaction favorable that is unfavorable at 1 M concentrations. Scientists use this information to suggest how to grow the microbe to optimize the likelihood that the reaction will occur.

Oxidation-reduction reactions

The oxidation-reduction reaction is another important type of reaction and can be expressed as:

$$A_{red} + B_{ox} \rightarrow A_{ox} + B_{red}.$$

That is, one compound (A_{red}) is oxidized to A_{ox}, and another compound (B_{ox}) is reduced to B_{red}. Oxidation removes electrons from a compound and reduction adds them. NAD^+ and NADH are often involved in such reactions. An example is the reduction of pyruvate to lactate, a reaction carried out by bacteria that ferment glucose to lactate:

$$NADH + H^+ + \underset{\text{pyruvate}}{CH_3\text{-}\mathbf{CO}\text{-}COOH} \rightarrow NAD^+ + \underset{\text{lactate}}{CH_3\text{-}\mathbf{CHOH}\text{-}COOH}$$

The structures of NADH and NAD^+ are shown in Figure 8.6. NAD^+ can accept one proton and two electrons. This is the reason for writing NAD^+ for the oxidized form and NADH for the reduced form. The structures of two other molecules involved in oxidation-reduction reactions, flavin adenine dinucleotide (FAD) and reduced flavin adenine dinucleotide ($FADH_2$), are also shown in Figure 8.6. FAD can accept two electrons and two protons (or two H atoms). Flavins, like FAD and flavin mononucleotide (FMN), and lipids called quinones (which also accept two protons and two electrons), will figure as important groups in the explanation of the way the ETS works.

To predict the direction of different oxidation-reduction reactions, a measure somewhat different from $\Delta G^{0'}$ is used. For these reactions, scientists use $\mathbf{E_0}$ (in millivolts, mV), which is the electrical potential of a particu-

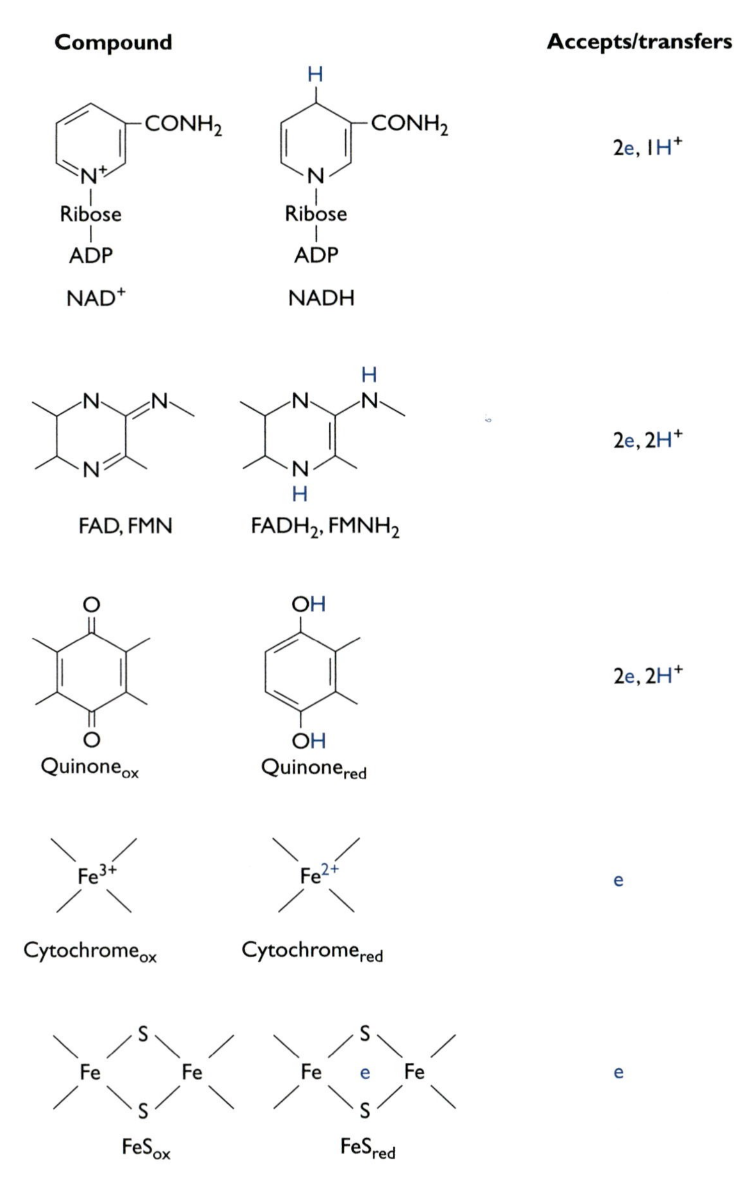

Figure 8.6 **Structures of molecules involved in oxidation-reduction reactions.**

lar redox couple, such as NAD^+ and NADH. Electrons will flow from a redox pair with a more negative potential to one with a more positive potential, so the difference in E_0 indicates the direction in which an oxidation-reduction reaction will proceed. (The change in E_0 values in a reaction is proportional to $-\Delta G^{0'}$, so E_0' is just another way of keeping track of whether a reaction is feasible.)

Some examples of E_0' values are provided in Table 8.1. For example, NADH (E_0' = –320 mV) can reduce the oxidized form of FAD (E_0' = –220 mV), and the reduced form of FADH can reduce the oxidized form of a quinone (E_0' = –75 mV). Oxygen has one of the highest positive E_0' values (+820 mV). E_0' values are important for determining the order of the reactions that comprise an ETS.

TABLE 8.1 Some E_0' Values for Redox Reactions

Redox couples	E_0' (mV)
$NAD^+/NADH$	–320
$FAD/FADH_2$	–220
Pyruvate/lactate	–140
$Quinone_{ox}/quinone^{a}_{red}$	–75 – +100*
$Cytochrome_{ox}/cytochrome^{a}_{red}$	+46 – +350*
O_2/H_20	+820

*Range indicates values for different quinones and cytochromes.

Oxidation of glucose-6-phosphate to pyruvate

Examples of steps in glycolysis are shown in Figure 8.7. Starting with glucose-6-phosphate, an ATP-dependent phosphorylation step produces fructose-1,6-bisphosphate, which is then cleaved into two phosphorylated C3 compounds, 3-phosphoglyceraldehyde and dihydroxyacetone phosphate. The dihydroxyacetone phosphate is converted enzymatically into 3-phosphoglyceraldehyde, so only the 3-phosphoglyceraldehyde is shown in Figure 8.7. An oxidation step is coupled to the addition of an inorganic phosphate (P_i) to 3-phosphoglyceraldehyde to form 1,3-bisphosphoglycerate. This compound contains enough energy to participate in a substrate level phosphorylation reaction to convert an ADP to an ATP. A second substrate level phosphorylation reaction occurs when PEP is converted to pyruvate. Most of the steps in glycolysis can proceed in either direction, except for the step that produces fructose-1,6-bisphosphate and the conversion of PEP to pyruvate. These steps are irreversible, because production of fructose-1,6-bisphosphate is driven by ATP hydrolysis and the dephosphorylation of PEP is associated with a very large negative $\Delta G_0'$. These unidirectional steps are important to ensure that the pathway moves in the right direction.

In a living cell, carbon does not always flow through the glycolytic pathway at the same rate. In *Escherichia coli,* for example, if the concentration of ATP falls and the concentration of ADP rises—an indication that more ATP needs to be generated—ADP binds to and activates the enzyme that synthesizes fructose-1,6-bisphophate. This ensures that carbon flow from glucose-6-phosphate to pyruvate will occur at the highest possible rate. If the level of PEP rises, however, indicating that the cell has plenty of energy, PEP binds to the same enzyme and inhibits its action, thus slowing the rate of glycolysis until additional energy is needed.

Glycolysis and the citric acid cycle do more than generate energy. Enzymes in these pathways also produce intermediates that serve as substrates for enzymes in biosynthetic pathways that generate amino acids, sugars, lipids, and other important cell components. These are listed in Table 8.2. During cell growth, most of the carbon flows through the energy-yielding pathway. However, small amounts of these intermediates are diverted into biosynthetic reactions as needed.

Pyruvate decarboxylation and the citric acid cycle

Up to this point in the pathway, a C6 compound has been converted into two C3 compounds. The decarboxylation of pyruvate is the first step in which a molecule of CO_2 is released. The decarboxylation of pyruvate is accompanied by an oxidation **(oxidative decarboxylation)** reaction. In this

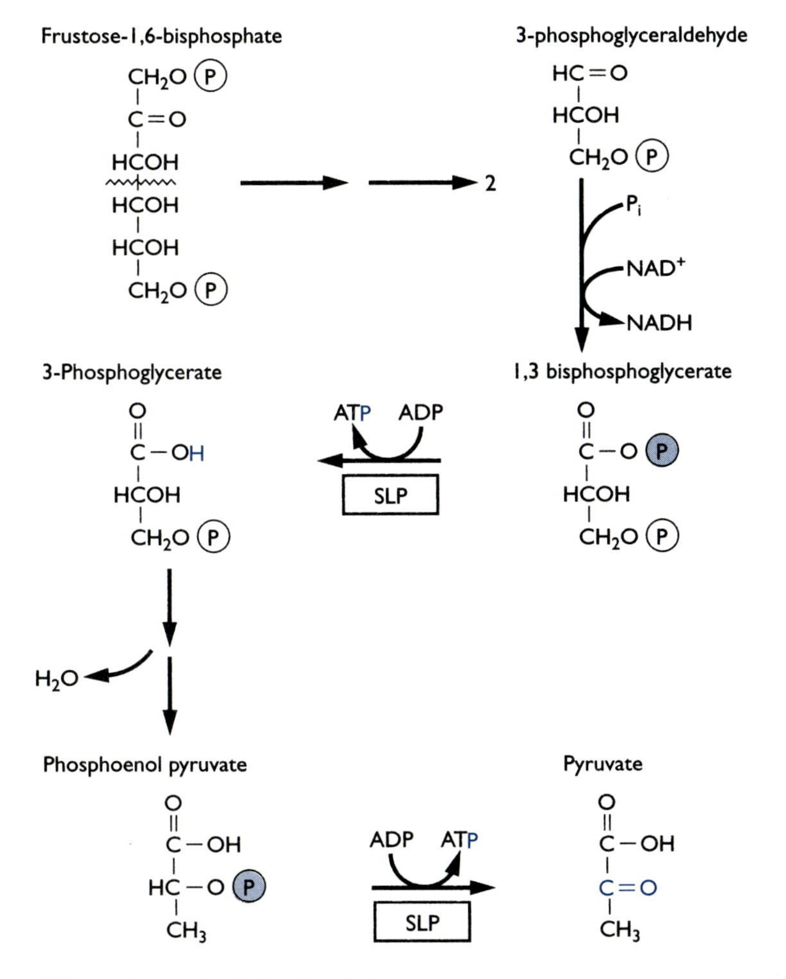

Figure 8.7 Some important reactions in glycolysis. Fructose-1, 6-bisphosphate is cleaved to two 3-phosphoglyceraldehydes. P_i is added to 3-phosphoglyceraldehyde to yield 1,3-bisphosphoglycerate. This compound can participate in substrate level phosphorylation (SLP) to generate adenosine triphosphate (ATP) and 3-phosphoglycerate. The end result is pyruvate. Multiple arrows indicate that more than one reaction occurs. Blue highlight represents molecules involved in the reaction.

TABLE 8.2 Glycolytic and Citric Acid Cycle Intermediates: Starting Points for Biosynthesis

Intermediate	Products synthesized
Glucose-6-P	Polysaccharides, nucleic acids, amino acids
Fructose-6-P	Peptidoglycan sugars
Phosphoglyceraldehyde/ dihydroxyacetone phosphate	Phospholipids
3-phosphoglycerate	Amino acids
Phosphoenol pyruvate	Amino acids, muramic acid
Oxaloacetate	Amino acids
α-ketoglutarate	Amino acids; entry point for NH_3

reaction, an important cofactor in bacterial catabolic and biosynthetic reactions makes its first appearance: HS-coenzyme A (HS-CoA). The structure of HS-CoA is shown in Figure 8.8A. HS-CoA can form high-energy thioesters (through its HS group) with compounds like the acetyl group of pyruvate (Fig. 8.8B). The activated form of acetyl, acetyl-S-CoA, can now transfer the acetyl group to the 3-carbon compound oxaloacetate to form citric acid, the entry point into the citric acid cycle. Alternatively, bacteria can use the energy stored in acetyl-S-CoA to drive the synthesis of an ATP by converting acetyl-S-CoA to acetyl-phosphate, then to acetate, converting an ADP to an ATP in the process.

The citric acid cycle performs three functions (Fig. 8.4). First, it removes sequentially the two carbons of the acetyl group, completing the oxidation of glucose to CO_2. Second, it generates two NAD(P)H, one $FADH_2$, and one ATP. Third, it provides carbon compounds (α-ketoglutarate and oxaloacetate) that are key substrates in the synthesis of 10 amino acids. Moreover, α-ketoglutarate is an entry point for ammonia into the biosynthetic reactions. From α-ketoglutarate and ammonia, bacteria can produce glutamate and glutamine, the amino acids that donate amino groups in the synthesis of other amino acids.

The electron transport system

To extract energy from NADH and its equivalents, bacteria use an ETS. A typical ETS is shown in Figure 8.9. NADH is oxidized to NAD^+, transferring a proton and two electrons to a flavoprotein. Flavoproteins are proteins with flavin molecules like $FADH_2$/FAD as parts of their structure. Note that on the scale of E_0' values, NADH/NAD^+ has a more negative value than $FADH_2$/FAD. Because flavin molecules can carry two protons

Figure 8.8 Structures of (A) HS-coenzyme A and (B) acetyl-S-coenzyme A.

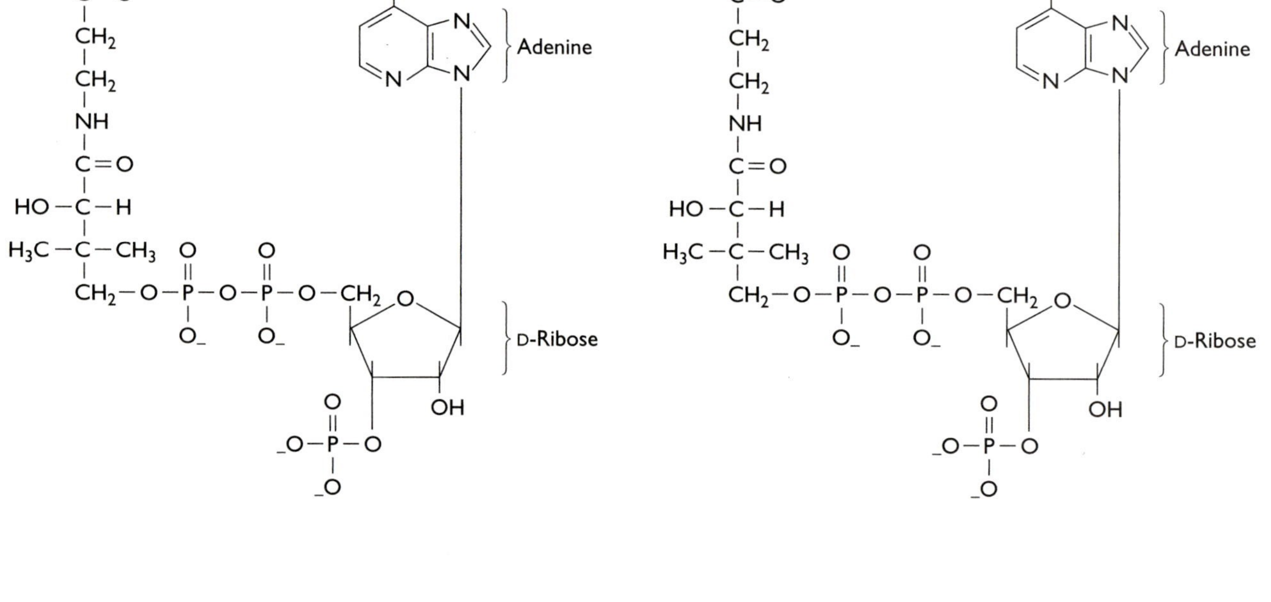

and two electrons (Fig. 8.6), the flavin group can pick up a proton from the cytoplasm. The flavoprotein transfers two electrons to the next protein in the system, a protein with an iron-sulfur cluster (iron-sulfur protein; Fig. 8.6). The iron-sulfur protein cannot accept protons, because its iron-sulfur cluster only carries electrons. So the two protons are released to the outer surface of the cytoplasmic membrane, thus in effect pumping them out of the cell. To prepare to receive more electrons, the iron-sulfur protein transfers the electrons it currently holds to a lipid quinone. A quinone that accepts two electrons can now accept two protons as well. The quinone picks up the protons from the cytoplasmic side of the membrane.

Once again, a molecule carrying protons and electrons (the quinone) interacts with a protein that can only accept electrons (cytochrome), and two more protons are extruded from the cell. The cytochrome must get rid of these electrons so that it can accept the next pair of electrons from the quinone. To do this, the cytochrome interacts with an enzyme, cytochrome oxidase, which oxidizes the cytochrome, relieving it of its two electrons and transferring them to a terminal electron acceptor. In some cases, the cytochrome oxidase itself extrudes two more protons. Some electron transfer systems have more components than the simple system shown in Figure 8.9, but the basic principle is the same: molecules such as quinones that can carry both protons and electrons extrude the protons to the outside of the cytoplasmic membrane and pass their electrons on to molecules that can only accept electrons. The molecules involved in the electron transport process just described are present in relatively limited amounts in the cell. They compensate by having the capacity to turn over massive amounts of protons and electrons in a short time.

In Figure 8.9 the terminal electron acceptor is oxygen, which is converted to water by cytochrome oxidase. Other electron acceptors, such as

Figure 8.9 Example of an electron transport system. Some systems are more complex, but this one illustrates the types of components normally found in such systems. Note that the order of these reactions is consistent with the E_0' values in Table 8.1. In the lower panel the reactions are drawn to emphasize the fact that all the molecules are recycled, except the terminal electron acceptor. CO, cytochrome oxidase; FP, flavoprotein; Q, quinone; cyt, cytochrome.

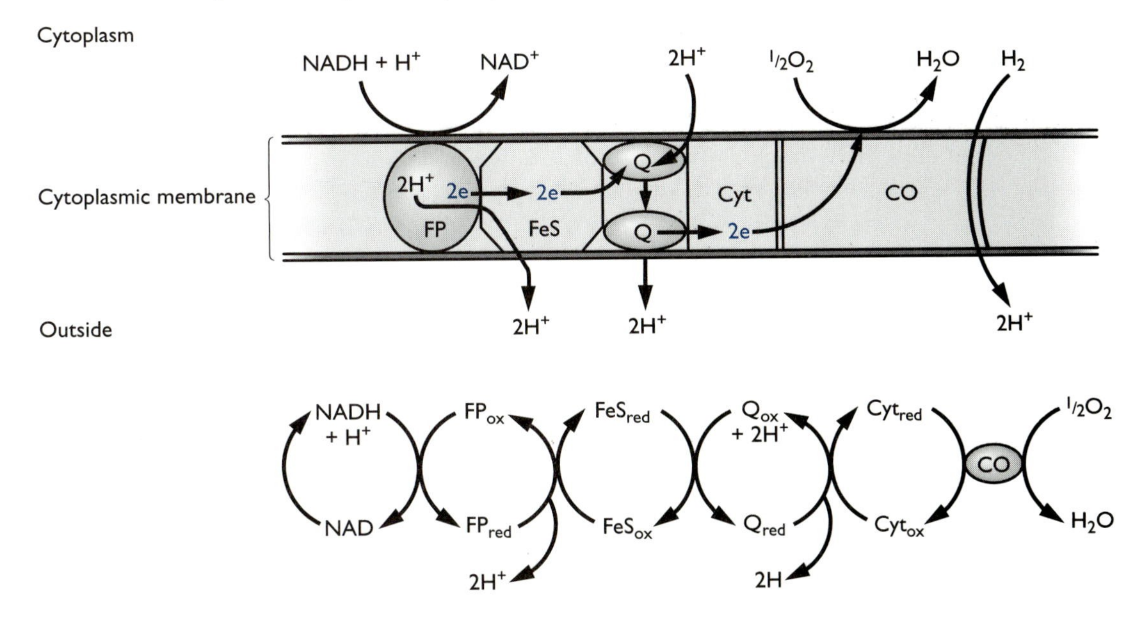

nitrate, sulfate, or even an organic compound, can also be used as terminal electron acceptors. The use of an inorganic compound as an electron acceptor is called **respiration.**

How Energy-Yielding Pathways Affect the Ecology of Bacteria

The dead zone bacteria

In the previous sections, we explained why biomass-utilizing bacteria would consume oxygen if they have an ETS that can use oxygen as an electron acceptor. Note that because oxygen consumption is associated with energy metabolism and because many electrons are being passed through the ETS to generate energy, many molecules of molecular oxygen are consumed. Bacteria are capable of running sugars very rapidly through the process described here and generating a large amount of NADH. Each molecule of NADH is responsible for the consumption of one molecule of oxygen.

Although bacterial activities in the dead zone have caused the deaths of most of the crabs and other marine animals that once lived there, the bacteria probably have not killed themselves. Instead, when oxygen is depleted, some will switch to other terminal electron acceptors. In addition, the reduced level of oxygen may provide opportunities for other groups of bacteria to increase in numbers. For example, some bacteria can use sulfate as an electron acceptor. Sulfate is present in high concentrations in sea water, but sulfate-reducing bacteria cannot grow in the presence of oxygen and thus are normally found in sediments where oxygen levels are very low and not in the water above the sediments. The disappearance of oxygen from the water might well allow the sulfate-reducers to move up into the water column. A bacterium's type of energy metabolism is a major factor in determining where it will be found in nature.

Lactic acid bacteria

Oxygen level is not the only factor that influences the activities of bacteria. Substrate availability is also a major factor. Consider, for example, the **lactic acid bacteria** located in the human mouth. These bacteria do not consume oxygen and are, in a sense, indifferent to it, because they can grow in the presence or absence of oxygen. Such bacteria are called **aerotolerant anaerobes.** Lactic acid bacteria are major microbial culprits in the formation of dental caries. To see why, consider their metabolism. They oxidize glucose to pyruvate, generating a net of two ATP and two NADH. Because they do not have an ETS, they cannot use the NADH to produce ATP. In fact, they have a second problem. An ETS oxidizes NADH to NAD^+, making the NAD^+ available to participate in more redox reactions. The lactic acid bacteria use their two NADH molecules to reduce two pyruvates to lactate as a means of regenerating NAD^+. Lactate is excreted into the medium.

The glucose oxidation system of the lactic acid bacteria does not generate as much ATP as that of bacteria with a citric acid cycle. The lactic acid bacteria make up for this by processing more glucose molecules per unit time than they would need to if they had a citric acid cycle. The lactic acid bacteria as a group are among the fastest growing bacteria known, with division times as low as 15–30 min. An experiment performed in many introductory microbiology laboratories is to inoculate samples from the surface of teeth into a medium containing a high concentration of glucose.

The lactic acid bacteria grow so rapidly and produce so much lactic acid, which inhibits the growth of other bacteria, that the lactic acid bacteria soon come to dominate the enrichment culture.

Lactic acid bacteria, such as *Streptococcus mutans*, grow in biofilms on the surface of teeth (**plaque**). In a person who does not consume much sucrose (a disaccharide composed of fructose and glucose), these bacteria usually do not produce enough lactic acid in a short enough period of time to damage the tooth enamel. A high sugar diet, however, allows them to produce high concentrations of lactic acid, which undermines the integrity of the tooth enamel and results in a cavity. (Some of the lactic acid bacteria produce other products besides lactate, such as acetate and ethanol. These pathways are beyond the scope of this text, however, and will not be covered.)

Another feature of the lactic acid bacteria deserves comment. Lactic acid bacteria generally require many amino acids and vitamins, because they are unable to make their own. One reason is that because they do not have the citric acid cycle, the citric acid cycle intermediates used as substrates for synthesis of important amino acids are not available. In particular, glutamate and glutamine, the entry points for ammonia nitrogen, are not made. In the mouth, there is an abundant supply of amino acids and vitamins from food. Lactic acid bacteria would not fare nearly as well in an environment where free amino acids and vitamins are not so freely available. This is another example of how the metabolic traits of a bacterium are an important factor in determining what niches the bacterium can occupy.

Escherichia coli

E. coli can grow on a much simpler medium than that required to support growth of the lactic acid bacteria. Because *E. coli* has a citric acid cycle, it can synthesize all of the amino acids it needs. Because it has the enzymes that convert α-ketoglutarate to glutamate and glutamine, it can use ammonia as a nitrogen source. *E. coli* can also utilize a variety of electron acceptors, including organic as well as inorganic ones. In the absence of oxygen or nitrate, *E. coli* switches to a fermentative metabolism. Organisms capable of switching from one type of metabolism to another are called **facultative bacteria.** The metabolic flexibility of *E. coli* allows it to live in a variety of settings, from water to the human bladder. In the human body, *E. coli* normally lives in the lower part of the small intestine and in the colon. Most of the carbohydrate that reaches this area is in the form of polysaccharides, because the small intestinal cells have removed virtually all of the easily digested sugars and amino acids before that point. *E. coli* can utilize a variety of sugars and disaccharides but does not utilize polysaccharides. So what is its natural substrate in the intestine? This question has never been answered, but this limitation of *E. coli* explains why the bacterium is not nearly as numerous in the intestine as polysaccharide-degrading bacteria, such as *Bacteroides* species.

Syntrophy

A different example of the way in which microbial energetics affect a microorganism's activities and ecology is a phenomenon known as **syntrophy.** Syntrophy is an interaction between two microorganisms that makes possible a reaction not energetically feasible for one of the microorganisms acting alone. That is, syntrophy allows one of the partners to perform a reaction that has a positive $\Delta G^{0\prime}$. How is that possible? The number that dictates whether a reaction will be feasible is not $\Delta G^{0'}$ alone. The important number is $\Delta G'$, the change in free energy in a reaction that takes into account the actual concentrations of subtrates and products.

For the reaction $aA + bB \rightarrow cC + dD$,

$$\Delta G' = \Delta G^{0'} + 2.3RT\log\{[C]^c[D]^d/[A]^a[B]^b\},$$

where R is a constant, T is the absolute temperature, and [A] is the concentration of reactant A. This equation may look complicated, but the idea it embodies is actually quite simple. The log of 1 is 0. The log of fractions less than 1 is negative. Thus, if the concentration of C or D is kept much lower than that of A and B, the second term becomes negative. Even if the reaction has a positive $\Delta G^{0'}$, the $\Delta G'$ value can become negative if the second term acquires a large enough negative value. Thus, if the $\Delta G^{0'}$ for a reaction is +20 kJ/mol, the reaction would be thermodynamically unfavorable if all substrates were present at concentrations of 1 M. Yet, if the actual ratio of products to substrates in the second term became much lower than 1, so that the value of the second term reached –32 kJ/mol, the overall $\Delta G'$ would be +20 – 32 kJ/mol = –12 kJ/mol, and the reaction would become thermodynamically feasible under those conditions. A way to make the concentration of products much lower than the concentration of substrates is for the microbe using the substrates to be in close association with a second microbe that consumes the products of the first microbe. The second microbe, in effect, makes the $\Delta G'$ value negative, even if the $\Delta G^{0'}$ value for the reaction carried out by the first microbe was positive.

As an example, consider the oxidation of butyrate to acetate under anoxic conditions. The equation for this reaction is:

$$\underset{\text{Butyrate}}{CH_3\text{-}CH_2\text{-}CH_2\text{-}COOH} + H_20 \rightarrow \underset{\text{Acetate}}{CH_3\text{-}COOH} + CO_2 + H_2 + H^+$$

The $\Delta G_0'$ value for this reaction is positive. Yet this reaction occurs in some natural settings. The explanation is that a second microorganism is consuming the hydrogen as fast as it is produced, keeping the net concentration of H_2 low enough to make $\Delta G'$ negative. This is an example of syntrophy in action. Syntrophy is proving to be important for finding microorganisms that degrade toxic wastes, because some pollutants are quite difficult to degrade. Syntrophic pairs appear to be able to accomplish reactions that might not otherwise be possible.

chapter 8 at a glance

Importance of energetics

Bacteria use energy to obtain carbon, nitrogen, phosphorus, and other elements to make cell components.

Bacteria need energy to carry out biosynthetic reactions.

How bacteria make a living: recycling biomass

- Polysaccharide-degrading enzymes
  - Cleave polysaccharides to sugars
  - May be free in extracellular fluid or tethered to cell wall
- Glucose transport across cytoplasmic membrane
  - Phosphotransferase system transports and phosphorylates glucose in one process.
    - Forms glucose-6-phosphate
    - Uses phosphoenol pyruvate (PEP) for energy
  - Active transport
    - Requires energy; does not phosphorylate glucose
    - May co-transport proton
    - Glucose phosphorylated in cytoplasm; uses ATP
- Glycolysis
  - One molecule of glucose-6-phosphate (C6 compound) converted through series of steps to two molecules of pyruvate (C3 compound)
  - Two molecules of NAD^+ reduced; 2 ATP molecules used; 4 ATP molecules generated; net gain of 2 ATP, 2 NADH
- Citric acid cycle
  - Acetyl group of acetyl-S-CoA oxidized to CO_2 and H_2O
  - Three molecules $NAD(P)^+$ reduced; one molecule of ATP generated
- Electron transport system
  - Pumps protons from cytoplasm to outside of cytoplasmic membrane
  - Transfers electrons to terminal acceptor
  - Proton gradient generated, does work
    - Protons moving through ATP synthase form ATP from ADP
    - Drives active transport of nutrients
    - Drives flagellar rotation

(continued)

chapter 8 at a glance *(continued)*

Energetics

- $\Delta G^{0'}$ = differences in energy of formation of products and substrates
  - Calculated at 1M concentration of substrates and products (unusual in nature)
  - Negative value means reaction is thermodynamically favorable
- Reactions can be coupled so that overall $\Delta G^{0'}$ is negative
- Substrate level phosphorylation, transfer of a phosphoryl group from an organic compound to ADP

Oxidation-reduction reactions

- Electrons and protons removed from one compound (oxidation) and added to another (reduction)
- E_0' is the electrical potential of oxidation-reduction; electrons move from more negative to less negative E_0'

Effects of energy-yielding pathways on bacteria

Type of energy metabolism determines where bacterium will be found in nature.

Dead-zone bacteria may shift from oxygen to other compounds as terminal electron acceptors.

Lactic acid bacteria

- Aerotolerant anaerobes
- Ferment glucose, produce lactic acid
- Need exogenous amino acids due to lack of a citric acid cycle

Facultative bacteria

- Without electron acceptors switch to fermentation
- Allows versatility in places to grow

Syntrophy

- Interaction between microbes allows reactions to go that would not be energetically feasible for one microbe alone.
- Possible because $\Delta G'$ can become negative despite a positive $\Delta G^{0'}$ if product concentrations low enough.

key terms

active transport process that translocates sugars across the cytoplasmic membrane without phosphorylating them; requires energy

citric acid cycle oxidation of acetate to CO_2 and H_2O; NAD^+ and FAD reduced, one molecule of ATP produced

$\Delta G^{0'}$ free energy change; sum of the energies of products minus sum of energies of substrate; if negative, reaction will proceed; if positive, reaction will not proceed or will go in reverse

E_0' oxidation-reduction potential of a redox couple; electrons flow from redox pair with more negative potential to one with more positive potential

electron transport system complex of proteins and other compounds that pump protons out of cytoplasm and transfer left-over electrons to a terminal electron acceptor such as oxygen

facultative bacteria bacteria that can change from one type of metabolism to another depending on growth conditions

fermentation use of organic compounds as electron acceptors

glycolysis oxidation of glucose to pyruvate; results in a net gain of two ATP and two NADH molecules

oxidation loss of protons and electrons by a compound

oxidation-reduction reactions reactions in which one compound is oxidized and another is reduced

phosphotransferase system protein complex that transports glucose into the cell and phosphorylates it in the process

reduction gain of protons and electrons by a compound

respiration use of an inorganic compound as an electron acceptor

syntrophy interaction between two microbes that allows a reaction that is not energetically possible for one organism acting alone

Questions for Review and Reflection

1. Why might gram-positive and gram-negative bacteria have slightly different strategies for degrading large polymers like polysaccharides?

2. How would you compare and contrast the roles of NAD(P)H and ATP in the metabolic reactions described in this chapter?

3. In some texts, the statement is made that lactic acid bacteria have a very inefficient metabolism compared to bacteria with a respiratory metabolism, because the lactic acid bacteria "throw away" the carbon and energy still stored in lactic acid. Yet, in a competitive situation, expecially under anoxic conditions, the lactic acid bacteria usually win the race to take over an enrichment culture. How can you explain this apparent anomaly?

4. You have been assigned to attend a seminar given by one of the authors of this book. One of the first sentences out of the speaker's mouth is "*Bacteroides* is an obligately anaerobic bacterium that lives in the colon." Mercifully, you drift off to sleep. But you have to write something in class about the subject of the seminar. What could you say about the probable features of *Bacteroides* from this statement, given that you at least recall that a major product in the colon is acetate and that *Bacteroides* is a major contributor to the colonic fermentation?

5. Why do bacteria need an energy intermediary like ATP? And, for that matter, why use NAD(P)H as an intermediary?

6. The reactions that couple catabolism (glycolysis) and anabolism (synthesis of amino acids and proteins) do not show this same pattern. Instead, the early substrates in important biosynthetic reactions such as synthesis of amino acids are actually intermediates of glycolysis or the TCA cycle. Why is this the case? Why doesn't this strategy deplete the energy-yielding pathways and cause a dangerous energy shortage?

7. Microbiologists who study bacterial oxidation and reduction of toxic inorganic compounds such as arsenate sometimes say that the bacteria they study "breathe" arsenate. Make an educated guess as to what they mean by such a statement.

8. Explain, using a numerical example, how the $\Delta G'$ value of a reaction can be negative despite the fact that the $\Delta G^{0'}$ value is positive. Assume that 2.3 RT = +60 mV and $\Delta G^{0'}$ = + 20 mV.

9. How do bacteria extract the energy needed to pump protons from the "electron tower" shown in Table 8.1?

10. In many enrichments for microbes that degrade difficult-to-degrade (energetically) compounds such as jet fuel, the end result is three or four different microbes, instead of the two in the syntrophic reaction described in this chapter. Could a syntrophic interaction have three members? Under what conditions? (You are at the forefront of research here, because no one has shown that three-way syntrophy can occur, although many scientists suspect it probably can.)

9 Introduction to Eukaryotic Microorganisms

The real missing link in the evolution of higher animals is not some dinosaur fossil but the earliest eukaryotes. Studies of these early eukaryotes may reveal how they evolved from prokaryotes. Whatever happened in that long-ago evolutionary transition resulted in an explosion of morphological diversity that makes the eukaryotic microbes so fascinating to watch. It also contained the seeds of what later blossomed into the visible plants and animals of today.

Views on the diversity of eukaryotic microbes, often called **protists,** are undergoing a massive shift, as scientists begin to realize the extent to which the diversity of this important group of microorganisms has been underestimated. Because no single chapter (or for that matter no single book) could do justice to this rapidly expanding field, this chapter will not attempt to provide complete coverage of protists. Instead, we will provide an overview of this exciting and important part of microbiology by describing some of its members, their activities, and their relationships to each other. Of necessity, the bulk of the coverage will focus on medically important eukaryotic microbes, because these have received the most research attention. However, because eukaryotic microbes play a number of important roles in the environment, where they are often found in close association with bacteria and archaea, a few examples of environmental protists also will be provided.

How Did Eukaryotes Arise?

Although scientists are far from having a definitive answer to this question, lack of certainty has not stemmed the flow of imaginative speculation swirling around this issue. An early theory of the origin of eukaryotes was proposed by Lynn Margulis. Working from the fact that such eukaryotic organelles as mitochondria and chloroplasts looked very much like bacteria, Margulis suggested that a bacterium with an anaerobic metabolism had engulfed a bacterium with an aerobic metabolism (later to become the mitochondrion). The original bacterium formed the nucleus of the new cell type. Later, as it became clear from phylogenetic studies that archaea were more closely related to eukaryotes than bacteria (Chapter 2), Martin and Muller proposed an alternative scenario, in which an archaean engulfed a bacterium to form what eventually became the eukaryotic cell. These two theories share a couple of fundamental problems. First, bacteria and archaeans do not ingest other organisms by engulfing them. Engulfment is a distinctively eukaryotic cell trait. It would be necessary to posit a prokaryotic ancestor, now no longer extant, that did the engulfing. Of course, as pointed out in earlier chapters, so little is known about the true diversity of prokaryotes that we cannot rule out the existence of prokaryotes capable of engulfing each other. A second problem with these theories is that genome sequences of eukaryotes are revealing that not all the "bacterial" genes of eukaryotes are found in mitochondria or chloroplasts but are found in the nucleus of the eukaryotic cell. Does this mean that Margulis was right after all, or were horizontal gene transfer events that occurred long after the emergence of the first eukaryote responsible for these bacterial genes being found in the genome of eukaryotic cells? Additional light may be shed on such fascinating questions as more genome sequences of lower eukaryotic cells become available. In the absence of concrete information about the origin of eukaryotic cells, it is necessary to fast-forward in evolutionary time to a period when bona fide eukaryotic cells, as we define them today, were present.

Categories of Eukaryotic Microbes

Importance of morphology in classification

Until recently, eukaryotic microbes were classified primarily on the basis of their appearance and such fundamental traits as photosynthesis or presence of a rigid cell wall. Classification based on morphological features was much more feasible in the case of the eukaryotic microbes than in the case of prokaryotes because of the rich diversity of structure and internal organelles found in eukaryotes. Electron microscopy made images of internal organelles even clearer, giving additional power to the morphology-based classification scheme. As the molecular yardstick approach is applied to supplement morphological analyses, a better view of the eukaryotic microbes is emerging—one that exposes their truly amazing diversity and brings into focus their evolutionary relationships. The terms used in this chapter reflect the vocabulary favored by microbiologists who study eukaryotic microbes of environmental as well as medical importance. Microbiologists who focus on eukaryotic microbes of medical importance use terminology that at first seems different, because it takes into account the site of infection (e.g., blood and tissue hemoflagellates). Nonetheless, the groupings that emerge from the newer phylogenetic analysis accord for the most part very well with the earlier classification schemes used to define the disease-causing eukaryotes. There are also differences in terminology among those who classify environmental eukaryotes. Again, the names may differ somewhat, but the basic groupings remain essentially the same.

The archaezoa

The first eukaryotes probably appeared more than 2 billion years ago. The microbes of today that are thought to resemble these early eukaryotes most closely are a group of organisms called the **archaezoa.** Examples of this group are **diplomonads** and **parabasalians.** The diplomonads and parabasalians lack such internal organelles as mitochondria that are commonly found in eukaryotes. Because they lack mitochondria, many of them live an anaerobic lifestyle and do not require oxygen to generate energy. The ability to live under anaerobic conditions would have been a requirement for microbes that were around before cyanobacteria first put molecular oxygen into the air and before plants came along to give the atmospheric oxygen concentration a further boost. Although this argument makes sense, it is important to note that some evidence suggests that the mitochondria-less archaezoa may once have had mitochondria but lost them. Genes normally found in mitochondria are still found in the nucleus of what are now amitochondrial microbes. In the modern world, many environmental niches that are rich in nutrients are either completely anoxic or hypoxic (e.g., sediments, the human and insect intestine). Eukaryotes that do not require oxygen would be able to colonize such niches. Mitochondria may have been lost because they were not particularly useful in such locations. Such events illustrate that just as evolution of a group of organisms can go forward, it also can go back (reductive or degenerative evolution). It is interesting to note that many of the so-called archaezoa today not only lack mitochondria but also have become obligate parasites of insects and animals. Yet their ribosomal RNA (rRNA) phylogeny places their first appearance long before modern animals appeared. Are these organisms degenerate or merely highly successful adapters to emerging opportunities? Early eukaryotes almost certainly possessed the genomic and metabolic flexibility of modern prokaryotes—the kind of flexibility that has allowed prokaryotes to adapt to plastic and antibiotics in mere decades.

Disease-causing diplomonads and parabasalians. Two examples of archaezoa that will be encountered in later chapters on human disease are ***Giardia intestinalis*** and ***Trichomonas vaginalis,*** pathogens of the small intestine (Chapter 17) and the vaginal tract (Chapter 18), respectively. *Giardia* lacks not only mitochondria but also **Golgi bodies,** organelles in which proteins are processed and directed to different locations in the cell. *G. intestinalis* also appears to reproduce only by asexual division and has no sexual phase. This and its other characteristics make it seem more similar to bacteria and archaea than to other protists. *G. intestinalis* is called a diplomonad because it has two nuclei and two sets of four flagella, one set associated with each nucleus (Fig. 9.1). To those who lean toward anthropomorphizing

Figure 9.1 ***Giardia intestinalis.*** *Giardia intestinalis* is an example of a diplomonad, with two nuclei and eight flagella. *G. intestinalis* has no mitochondria. In the human or animal intestine, it adheres to the intestinal lining (Chapter 17).

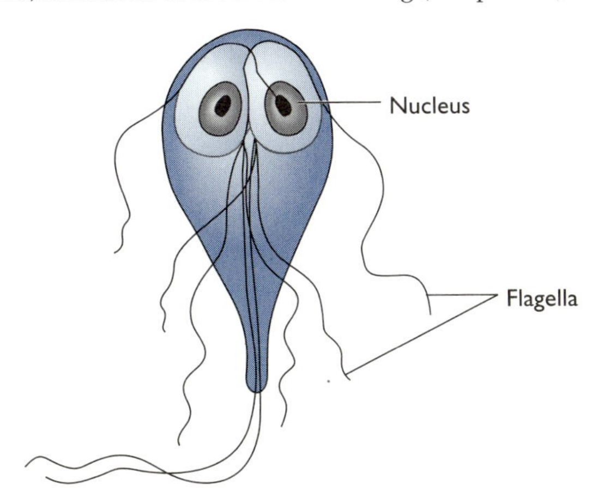

microbes, the two nuclei look like eyes, giving the organism a comic appearance. It is anything but funny, however, if it infects your small intestine. *G. intestinalis* can cause a long-term, debilitating form of diarrhea.

The sequence of the *G. intestinalis* genome has now been completed and reveals another prokaryote-like feature of this diplomonad: few, if any, introns have been found in the *G. intestinalis* genes examined so far. **Introns** are segments of DNA that encode a portion of RNA missing from the final form of the messenger RNA (mRNA) transcript. After transcription of the initial form of the RNA, introns are removed from the transcript to produce the final mRNA that will be translated into a protein. Some bacteria and archaea have introns in transfer RNA (tRNA) genes but not in genes encoding proteins.

T. vaginalis is a **trichomonad,** a division within the parabasalian group. The word parabasalian derives from what was first called a parabasal body but is now known to be an oddly shaped Golgi body. Parabasalians do not have mitochondria, but some are capable of meiosis as well as mitosis, making them more like higher eukaryotes than the diplomonads. Trichomonads have a single nucleus associated with 4–6 flagella (Fig. 9.2). In a wetmount, these flagella give *T. vaginalis* a lovely, falling-leaf type of motility. Presumably, this motility helps *T. vaginalis* move to its preferred location in the vaginal tract.

Beneficial archaezoa. Not all diplomonads and parabasalians cause disease. An interesting subgroup of the parabasalians is actually beneficial for its hosts. These are the parabasalians found in the guts of termites and wood-eating cockroaches. The termite *Reticulitermes,* found in many parts of the United States, has a number of protists in its hindgut. The function of one of these, ***Trichonympha,*** is understood best. *Trichonympha* is a stunning-looking protist because of its hundreds of flagella that swirl as the organism moves (Fig. 9.3). For the anthropomorphizers, *Trichonympha* could be considered the Rapunzel of the protist world. For *Reticulitermes,* however, *Trichonympha* has a much more prosaic but important function: it feeds the

Figure 9.2 ***Trichomonas vaginalis.*** *Tricomonas vaginalis* is an example of a parabasalian trichomonad, with one nucleus, several free flagella, and a flagellum attached to an undulating membrane. The parabasal body, a large Golgi apparatus, gives the group its name. Hydrogenosomes are organelles much like mitochondria, that convert pyruvate to acetate instead of carrying out respiration.

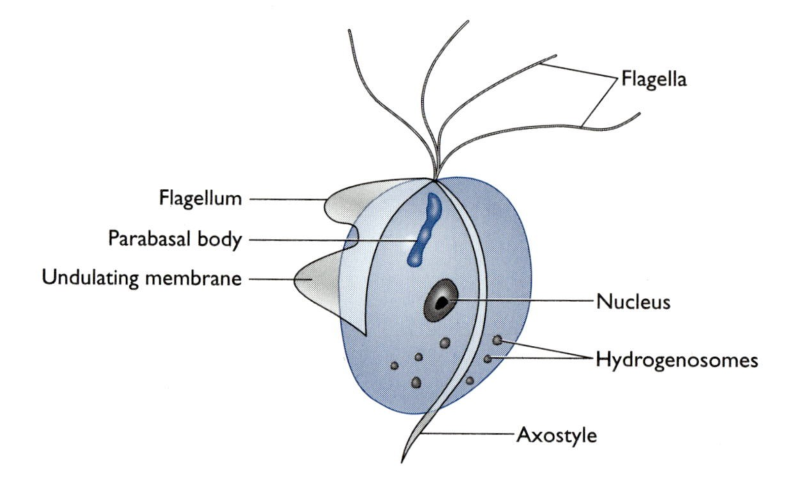

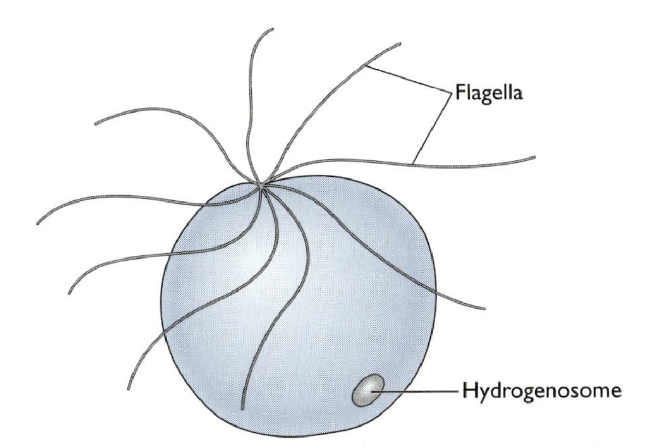

Figure 9.3 A termite hindgut parabasalian. *Trichonympha,* which lives in the hindgut of termites, converts wood chips into acetate, a compound that can be used as food by the termite. This type of parabasalian commonly has many nuclei and parabasal bodies (not shown).

termite. Termites can chew wood into small particles but cannot digest it. *Trichonympha* ingests the wood fragments, breaks them down into glucose units, and ferments the glucose to produce acetate. Acetate is absorbed and utilized as food by the termite (Fig. 9.4).

This brings us to an unusual parabasalian organelle, the **hydrogenosome.** Hydrogenosomes look like mitochondria, but they do not carry out oxygen-based respiration as mitochondria do. Instead, hydrogenosomes make acetate, carbon dioxide, and hydrogen from pyruvate, a set of reactions that can yield an extra ATP for an organism growing anaerobically. This is the source of the acetate that is so important for the survival of the termite. Moreover, the hydrogen and carbon dioxide are converted by hindgut bacteria called **acetogens** to produce more acetate for the insect (Fig. 9.4). So *Trichonympha* makes an indirect as well as a direct contribution to acetate production. Other wood-eating insects such as cockroaches have similar parabasalians that perform similar essential nutritional functions.

It is curious that diplomonads and parabasalians, which are supposed to be the modern versions of ancient protists, are today almost all obligate parasites or symbionts of insects or mammals. When these protists first evolved, such hosts or anything remotely resembling them were billions of years away in the evolutionary future. Two possible explanations could

Figure 9.4 Interactions between the termite *Reticulitermes* and its gut protists and bacteria. *Trichonympha* produces acetate, hydrogen, and carbon dioxide from cellulose. Acetogenic bacteria make more acetate from hydrogen and carbon dioxide. The acetate feeds the termite.

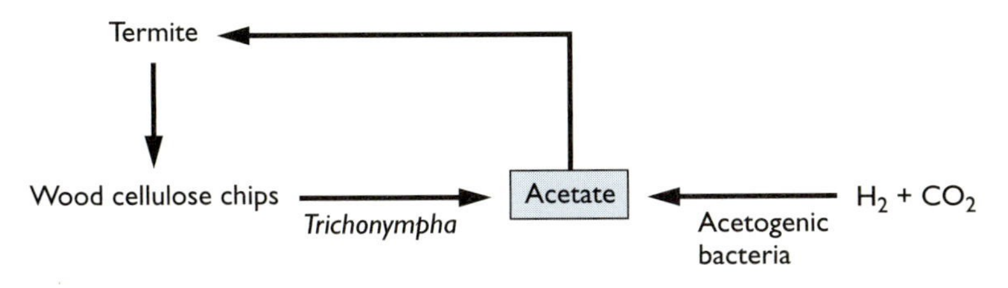

account for this apparent anomaly. First, early protists were likely to be every bit as adaptable as prokaryotes. The genetic diversity of modern protists supports this contention. If bacteria could adapt within decades to become resistant to antibiotics or to bind plastic, surely protists could develop a complex association with insects and mammals over a few hundred million years. A second explanation is that scientists have inadvertently adopted a biased view of this group by focusing on mammalian pathogens and insect symbionts. Far more free-living members of this group may exist than current research suggests. As scientists focus more on environments outside the mammalian and insect body, many more members of the archaezoa will be discovered.

More recently evolved protists

Another early-evolving group of protists is the **amoeboflagellates,** including the **kinetoplastids** and the **euglenoids.** These protists have mitochondria and most have an anaerobic lifestyle.

Amoeboflagellates. An example of an amoeboflagellate is ***Naegleria fowleri,*** one of the few members of this group capable of infecting humans. This organism causes a rare but rapidly developing form of meningitis that can kill within hours. Initially, cases of *N. fowleri* infection were seen mostly in fresh-water ponds in parts of the United States, where freezing winter temperatures did not destroy the organism. Cases of *N. fowleri* meningitis have also been associated with inadequately disinfected swimming pools. More recently, *N. fowleri* has been spotted in lakes around nuclear powerplants, such as the one in Clinton, Illinois. These lakes, which perform a cooling function for the nuclear reactors, provide water warm enough to support the growth of *N. fowleri* year-round. So far, no cases of *N. fowleri* meningitis associated with recreational use of lakes around nuclear power plants have been reported. Once again, *N. fowleri* provides an example of how eukaryotic microbes, like prokaryotic ones, are quick to take advantage of changes humans introduce into their environment.

Amoeboflagellates feed on bacteria and can reproduce very rapidly under the right conditions, with generation times as low as 2 hours. Their name derives from the fact that the organism has both an amoeboid form and a flagellated form (Fig 9.5). The amoeboid form is probably the predominant form in nature, allowing the organism to move over surfaces in search of prey. The flagellated form allows the organism to move from one location to another when food becomes scarce. The amoeboid form can also form cysts under adverse conditions. Cysts have a function similar to that of bacterial spores, in that they allow the organism to survive under conditions that would kill the actively replicating form.

Euglenoids. Euglenoids, like other amoeboflagellates, are found in fresh-water ponds, but none are known to cause disease in humans or animals. If, as a high-school student, you used a microscope to view pond water to which some grass had been added and let the mixture sit in a sunny location for several days, you probably would have seen *Euglena* species. The best studied of these is ***Euglena gracilis,*** a photosynthetic euglenoid (Fig. 9.6). *E. gracilis* has a single nucleus and a single flagellum. It also contains many **chloroplasts,** organelles that allow it to convert sunlight into chemical energy. An organelle called an **eyespot** allows it to migrate toward the level of light where its photosynthetic machinery

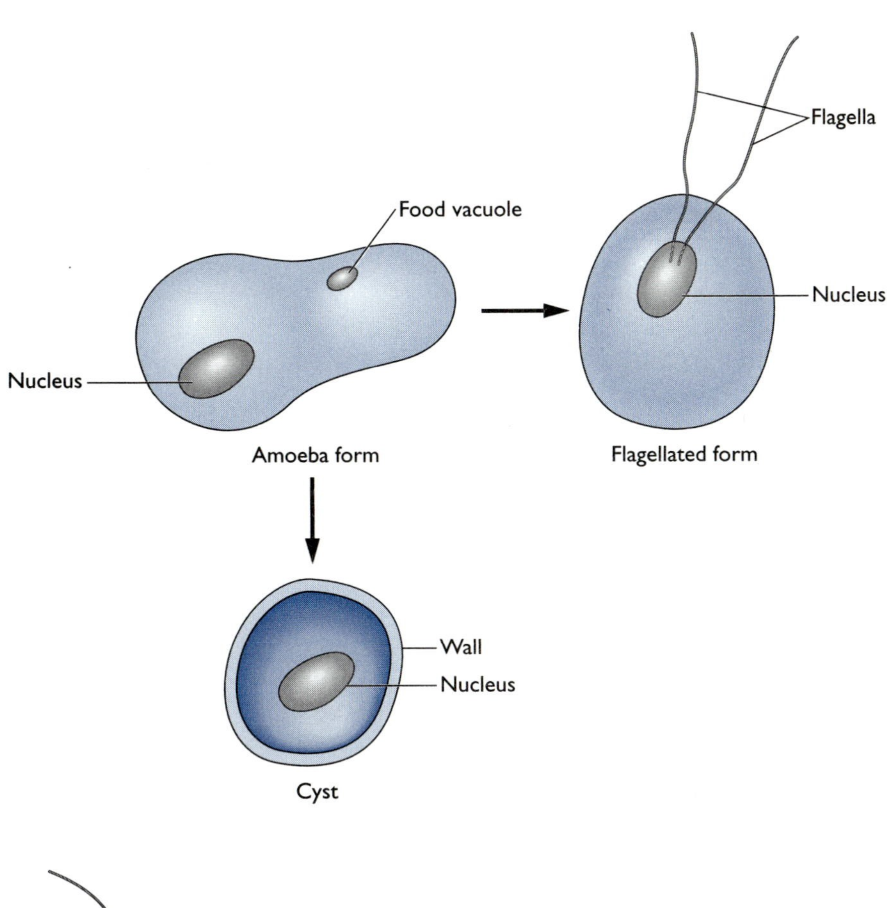

Figure 9.5 Stages of the amoeboflagellate, *Naegleria*. The amoeba form moves by gliding over surfaces. The flagellated form, which develops if nutrients are limited, swims through the environment in search of new food sources. The cyst form develops in very adverse conditions, in which this tough survival form can survive but the amoeba form or flagellated form cannot.

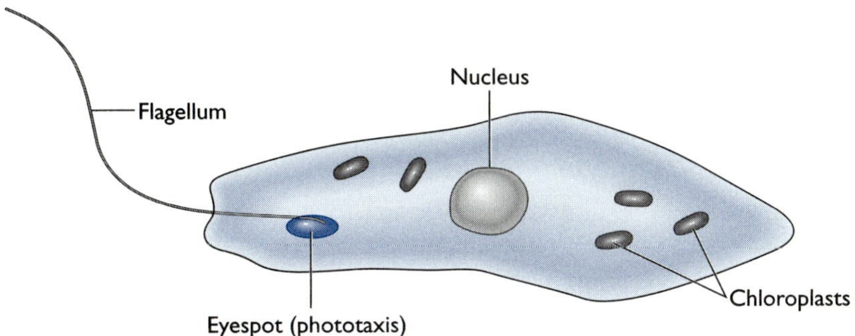

Figure 9.6 The euglenoid, *Euglena gracilis*. Euglenoids have an elongated shape, a single nucleus, and a single flagellum. This euglenoid is photosynthetic, but many euglenoids are not and live by preying on other microbes.

works best, a process called **phototaxis.** Many members of the euglenoid group are nonphotosynthetic. These so-called colorless euglenoids lack the chloroplasts that give the photosynthetic euglenoids their greenish hue. Colorless euglenoids prey on other organisms, including photosynthetic euglenoids. The colorless and photosynthetic euglenoids do not represent two distinct groups. Under some conditions, photosynthetic (green) euglenoids can be converted to colorless euglenoids. This suggests that at least some euglenoids carry the genetic information to allow them to have either lifestyle. If so, the mechanisms by which they switch from one metabolism to another are not understood at all and present a fascinating area of future research.

Kinetoplastids. Many kinetoplastids are free living and are found widely in soil and water. They live by ingesting bacteria. Some members of this group are parasitic, however, and cause serious human diseases. Examples of these diseases are sleeping sickness (*Trypanosoma brucei*) and leishmaniasis (*Leishmania* species). **Leishmaniasis** can take several forms. One form, mucocutaneous leishmaniasis, infects the nose and surrounding areas and destroys the cartilage, sometimes making a hole in the soft palate of the mouth. The disease is usually not fatal but is terribly disfiguring, because the nose appears to have melted on the face. Another form, visceral leishmaniasis, is often fatal. Leishmaniasis, like sleeping sickness, is primarily found in tropical areas of the world.

Both *T. brucei* and *Leishmania* species are transmitted by insects, the tsetse fly and the sandfly, respectively. The associations between the kinetoplastids and their insect host are not well understood, but lately scientists have begun to look into these associations in more detail. They hope to design more effective insect control without having to resort to expensive and potentially dangerous insecticides.

The kinetoplastids have a single nucleus and a single flagellum. The flagellum is attached to a membrane that undulates as the organism moves (**undulating membrane**; Fig. 9.7). Movement of the flagellum causes the membrane to undulate. At the base of the flagellum is an organelle called a **kinetoplast,** which gives this group its name. The kinetoplast is a mitochondrion with an unusual feature. Whereas most mitochondria have a single circular genome, the genome of the kinetoplast consists of a few large chromosomes and thousands of small DNA circles. These small circles contain genes that might encode proteins, except for the fact that the genes are missing key residues that would place a U in the mRNA. The transcripts are translated by a process called RNA editing. The RNA copy of the DNA is edited by having U residues added to it, so that the final mRNA can be translated into a protein. The reason for this unusual form of RNA processing is not known, but the phenomenon is being studied intensively by scientists interested in the flow of information from DNA through RNA and into proteins.

Apicomplexa. The apicomplexa contain some important human pathogens: ***Plasmodium*** species, the cause of **malaria** (Chapter 20), and ***Cryptosporidium parvum,*** a cause of diarrhea (Chapter 17). The apicomplexa have very complex life cycles, which include the **sporozoite** (the infective form for humans), the **merozoite** (the form that invades and lyses human cells), and the **gametes** (the sexual forms that appear late in the infection). *Plasmodium* species are transmitted by mosquitoes and seem to be obligate pathogens of mammals. *C. parvum,* a free-living apicomplexan, is transmitted to humans in water and milk. The disease-causing apicomplexa,

Figure 9.7 The kinetoplastid *Trypanosoma brucei.* A single flagellum is attached to a long, undulating membrane, which allows the microbe to move through the human bloodstream. The kinetoplast is a type of mitochondria that has a few large chromosomes and many small DNA circles.

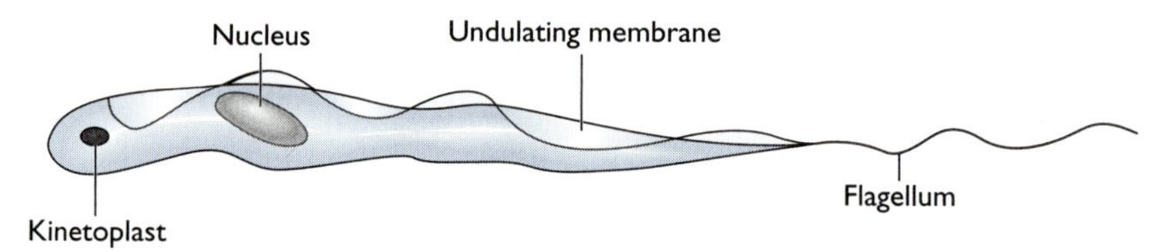

like the disease-causing kinetoplastids, have adapted to live in human tissues and blood. As will be seen in Chapter 20, the apicomplexan that causes malaria interacts with cells in the liver and bloodstream as it carries out its complex developmental cycle. A similarly complex life cycle occurs in the insect that transmits *Plasmodium* species.

Dinoflagellates. The dinoflagellates are characterized by two flagella, and most are photosynthetic (Fig. 9.8). Photosynthetic dinoflagellates are sometimes called algae but, in reality, belong in their own separate group. One of the photosynthetic dinoflagellates, ***Pfiesteria,*** has some unusual features, including a life cycle that consists of many stages. It becomes photosynthetic not by making its own chloroplasts but by using a hose-like appendage to remove chloroplasts from true algae. Because of their mode of acquisition, the chloroplasts have been called **kleptochloroplasts**. Finally, *Pfiesteria* produces low-molecular-weight toxic compounds, which can kill fish and produce neurological symptoms in humans.

Pfiesteria, subsequently dubbed "the cell from hell," made the news in 1997, when a bloom of it nearly destroyed the fishing industry on the U. S. eastern coast near the Chesapeake Bay. The *Pfiesteria* bloom put toxins into the water that caused lesions in fish. Fishermen exposed to the toxins developed neurological symptoms, such as confusion and staggering. Fortunately the symptoms were of short duration, and none of the affected fishermen experienced long-term neurological damage. Reports of the toxins, however, caused a wave of panic among consumers of seafood from this area. Although no cases of disease from consuming fish or shellfish from the affected area were reported, consumers stayed away from Chesapeake Bay seafood, driving some fishermen into bankruptcy and bringing many others close to financial ruin. Now consumer confidence in Chesapeake Bay produce has been restored, but local fishermen will not soon forget this traumatic period.

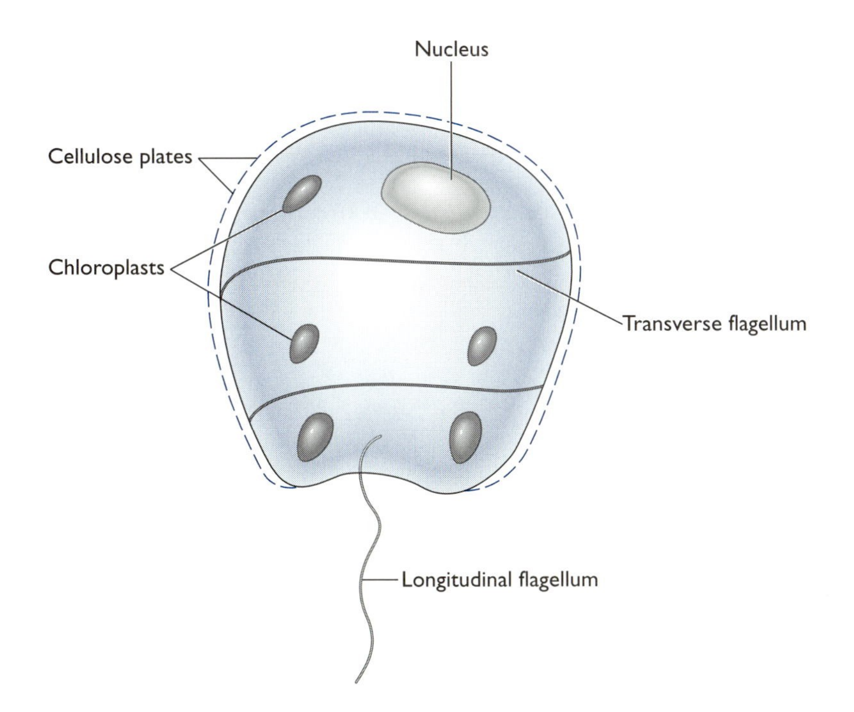

Figure 9.8 Features of a dinoflagellate. The dinoflagellate has two flagella, one wrapped around the cell. Many dinoflagellates, including *Pfiesteria,* are covered by cellulose plates.

Why did *Pfiesteria* bloom at that specific time? The answer is still not clear. Fishermen blamed nitrogen-rich runoffs from poultry farms, an allegation stoutly denied (with some justification) by the poultry producers. Human sewage was also a suspected cause. Unless and until such a bloom occurs again, questions about its origins probably will not be answered. Dinoflagellates are also responsible for the so-called **red tides**. The name derives from the fact that blooms of the photosynthetic organisms stain the water red. Toxins released from the dinoflagellates kill fish and can harm humans who eat fish or shellfish contaminated with water from the bloom area.

Last but by no means least: algae, diatoms, and fungi

Algae and diatoms. Among the latest arrivals on the eukaryotic microbial evolutionary scene are algae, diatoms, and fungi. Diatoms and algae are important parts of marine and freshwater phytoplankton. Diatoms and algae convert sunlight to energy, fix carbon dioxide to form complex organic molecules, and fix atmospheric nitrogen into more complex nitrogenous compounds. These organisms are a source of food for many marine animals, which in turn provide food for other animals. Diatoms and algae play a similar role in freshwater environments. They have left a formidable geological record. Algae have rigid cell walls, composed of calcite, which survive for long periods of time after the algae die. Over centuries, deposits of algal cell walls form large, white geological deposits. Tourists sailing off the coast of England often pause to admire the legendary White Cliffs of Dover. Few realize that these beautiful cliffs are the aggregated "skeletons" of billions of algae. Diatoms, too, have left their mark. Their skeletons, composed of silica, precipitate calcium carbonate when the organisms die. Under the ocean are many large geological deposits formed from diatom remains. For a much newer role for diatoms, see Box 9–1.

Tourists may see the White Cliffs of Dover as a scenic attraction, but biologists view it in another way. Evidence suggests that the carbon dioxide concentration of the atmosphere is rising, largely as a result of industrial emissions. If you are one of those who believe that this increase may lead to global warming and that global warming may have adverse consequences for animals and plants, your mind naturally turns to strategies for lowering the concentration of atmospheric carbon dioxide. The skeletons of algae and diatoms point to a possible solution. These skeletons are, in effect, "buried" carbon dioxide—carbon dioxide that had been incorporated into living things but not rereleased into the environment after their deaths. The increasing use of carbon dioxide by algae and diatoms could have the effect of converting carbon dioxide (greenhouse gas) into inert carbon. We will return to this topic in Chapter 22.

For the most part, the activities of algae and diatoms are beneficial. Algae and diatoms are important components of plankton, the basis of marine and terrestrial food chains. Algae are used to produce agar and a polysaccharide called **carrageenan,** which is used as a thickening agent in puddings and ice cream. Nonetheless, algae and diatoms can occasionally cause death and destruction. Algal blooms contributed to the formation of the hypoxic zone in the Gulf of Mexico, which was described in Chapter 1. In lakes and ponds, algal blooms wreak similar havoc when dead algal biomass sinks to the bottom and stimulates the activities of microbes that consume oxygen, causing major fish die-offs and upsetting the normal ecology of the site. Algae and diatoms cause no known infections in humans and animals.

box 9–1 *Diatoms Help Police with Their Investigations*

According to forensic pathologists, drowning can be an especially difficult diagnosis to verify, especially if the victim has been moved after death. Movement can occur in the water, as the body is washed away from the original site of drowning, or can occur if the body is physically removed from the water and taken to another site. Diatoms have two features that make them useful diagnostic indicators as to where the drowning occurred. First, the silica skeletons of diatoms give them very distinctive appearances, and there is a vast range of diatom morphologies. Second, different sites in a river or lake tend to have different populations of diatoms. Thus, a look at the variety of diatom shapes in a particular location is almost like taking the "fingerprint" of that location. When a person drowns, there is a sudden intake of water into the lungs and sinuses. After that, exchange of water between the external environment and tissues is much slower. So, for a reasonable length of time after drowning, the drowned person's lungs and other tissues are marked with the diatom population of the site. The lack of diatoms may even help to identify bodies that were killed on land, then dumped into the water.

Sources: Ludes, B., et al. 1999. Diatom analysis in victim's tissues as an indicator of site of drowning. Int. J. Legal Med. **112:**163–166.

Pollanen, M. S. 1998. Diatoms and homicide. Forensic Sci. Int. **91**:29–34.

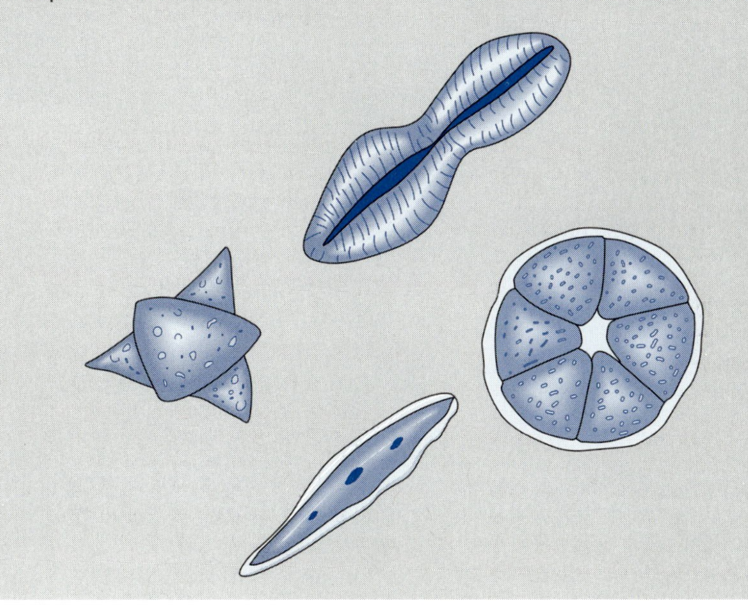

Fungi. Fungi are important recyclers of dead biomass. They are instrumental in breaking down dead trees by adhering to the dead plant tissue and secreting enzymes that degrade the cellulose and **lignin** (a polyphenolic compound that gives wood its tough, rigid property). Fungi also associate with the roots of trees and other plants, where they somehow stimulate the plant roots to proliferate, thus increasing the absorptive capacity of the plant's root structure. Fungi grow in large, multicellular strands called **hyphae,** which can become very large. White rot fungi on dead trees are an example of this phenomenon. **Lichens** are another example of visible microbial assemblages. Lichens are composed of fungi and a photosynthetic component, either algae or cyanobacteria. The fungus gives the lichen its macroscopic structure. The photosynthetic microbe fixes nitrogen from the atmosphere and converts it into forms that can be used by the fungus. Lichens play an important ecological role in the formation of soil. Lichens growing on rock erode the rock structure with acid end products from fermentation of organic compounds. Fissures in the rock are further widened by the physical penetration of the lichen structure into them. Lichens erode rocks in this way, producing soil that can support the growth of other lichens and plants.

As already described in Chapter 3, fungi can grow as single-celled yeasts, which reproduce asexually by budding, or as long multicellular filaments

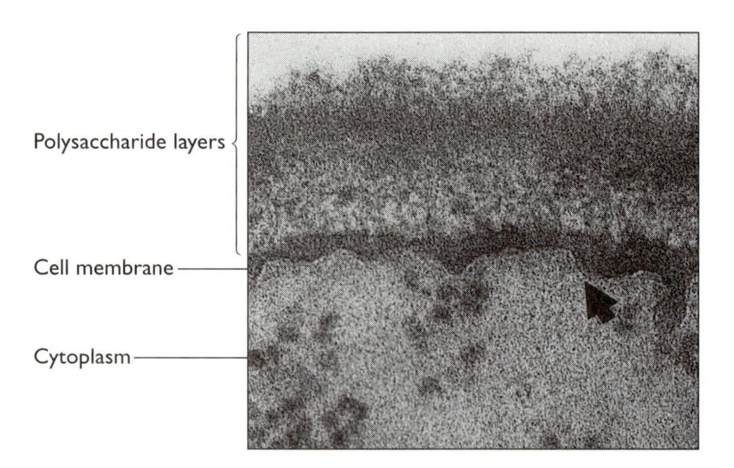

Figure 9.9 Structure of the fungal cell wall. A photomicrograph showing the layers of a fungal cell wall. The layers consist of various polysaccharides. © M. A. Persi, Visuals Unlimited

called hyphae. Along hyphae, spores form outside fungal cells (**exospores**) or in sacs that have different names, depending on where in the sac the spores are located. Some examples are conidiospores and sporangiospores. Yeasts can also produce spores. Yeasts and fungi have a sexual reproduction mode, as well. Some characteristics of the fungal cell wall and cytoplasmic membrane are illustrated in Figure 9.9. The cell wall consists of layers of polysaccharides, which form a rigid matrix. Cytoplasmic membranes of fungi contain a lipid called **ergosterol,** the fungal equivalent of the human lipid, cholesterol (Fig. 9.10).

Yeasts have long been a mainstay of the food and pharmaceutical industry. Yeast-based fermentation of glucose to ethanol is the foundation of the beer and wine industry and also is essential in the production of ethanol used as a fuel additive. In bread making, baker's yeast ferments the sugar in the dough, producing carbon dioxide gas that causes the bread to rise. The yeast itself contributes to the flavor of bread. Fungi have been an important source of antibacterial compounds. Penicillin is produced by ***Penicillium,*** the bread mold. Although far more antibiotics are now produced by bacteria themselves, especially the gram-positive ***Streptomyces*** species, fungi were the sources of some of the first antibiotics and have continued to provide important pharmaceutical compounds.

Some fungi cause human disease. The most common fungal infections are those of the skin, nails, and hair. These include athlete's foot, jock itch, ringworm, and nail infections that deform the nail. The fungi that cause superficial infections of skin, nails, and hair can do so because they are able to hydrolyze keratin. **Keratin** is a tough protein found in dead skin cells and in nails. Because it cannot be digested by most microorganisms, keratin serves as a defense of skin and nails. Even keratinolytic fungi, however, cannot cause infection if skin or nails are undamaged. Small cuts and fissures in the skin or damage to nails allows the fungi to colonize. Their growth in the area causes inflammation, the sources of the itching and other symptoms of athlete's foot, jock itch, and ringworm. A general feature of infections caused by fungi and yeasts is that they are most likely to occur in people whose defenses have been breached in some way. The fungi that cause **cutaneous infections** are unable to invade underlying tissues and establish infection.

Figure 9.10 The structure of ergosterol, a sterol component unique to fungal cell membranes.

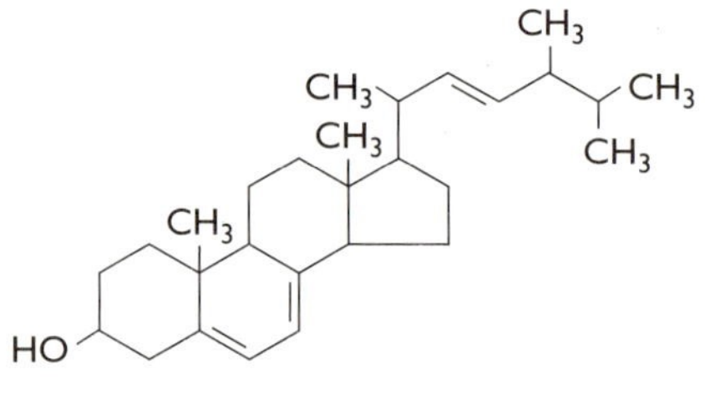

CH_3 CH_3 CH_3 CH_3 CH_3 CH_3 HO

The most common entry site for more invasive fungal infections is the lungs. Several fungal pathogens cause lung infections. Usually these fungi are capable of growing in the yeast form in the human body. Fungal pathogens that differentiate into the yeast form at 37°C in the human body but revert to the mycelial form in the environment are called **dimorphic fungi.** Characteristics of fungi that allow them to infect the lung and other tissues remain poorly characterized. One is the formation of pseudohyphae, mycelium-like projections from the yeast form, which exude hydrolytic, tissue-damaging enzymes from their tips. A common cause of fungal lung infections in the United States is the dimorphic fungus, ***Histoplasma capsulatum.*** *H. capsulatum* is found widely in soils in the Ohio River Valley, especially in soils contaminated with bird feces, which presumably serve as a source of nitrogen for the fungi. *H. capsulatum* has followed pigeons and starlings into the cities, expanding its range somewhat.

The relatively innocuous nature of *H. capsulatum* for otherwise healthy people is indicated by the finding that, in regions where this fungus is widespread, more than half of the people living there show immunological signs that they have experienced an infection. Yet symptomatic disease is relatively uncommon, indicating that most people with healthy immune systems who encounter *H. capsulatum* dispose of it readily. Among those not seriously immunocompromised, the people most likely to develop symptomatic *H. capsulatum* lung infections are those with somewhat depressed lung defenses as a result of old age, smoking, or other lung damage. Additional examples of fungal infections will be encountered in Chapters 16 and 18. Fungi and protists are not infectious disease problems for humans only, however, as described in Box 9–2.

Certain fungi, such as ***Aspergillus*** species (color plate 7), produce toxic compounds. *Aspergillus* produces **aflatoxin,** a low-molecular-weight toxin that causes liver cancer (Fig. 9.11), especially in people infected with hepatitis viruses or who have sustained liver damage in some other way. **Ergot,** the active ingredient in the hallucinogenic drug LSD, is also produced by fungi. During the Middle Ages, storage of grains under moist conditions allowed such fungi to grow on the grain and produce hallucinogenic toxins. These toxins may have been the origin of such legendary spectacles as St. Vitus's

box 9–2 *Protist-Caused Diseases: Important Contributors to the Extinction of Some Species?*

Fungi cause infections in many animals other than humans. Recent events have caused naturalists interested in species diversity and species extinction to take microbes more seriously. A type of fungus called **chytrid fungus** has been causing mass die-offs of frogs in the rain forests of Central America and Australia. Chytrid fungi differ from other fungi in that they have a flagellated form as well as hyphae and spores. The fungus, *Batachochytrium*, resembles the fungi that cause athlete's foot, in that it infects keratinized tissues and degrades the protein keratin. Infection of the frog's skin causes extensive destruction. Because a frog breathes through its skin, this suffocates the animal. *Batachochytrium* appears to have been responsible for at least two species extinctions.

In other areas, protozoal microbes, bacteria, and viruses are causing high mortality in populations of birds, kangaroos, and crayfish. Scientists interested in maintaining species diversity and preventing extinctions are now facing the fact that infectious disease may be nearly as important as loss of habitat as a factor contributing to extinction of species of wild animals.

Source: Daszak, P., A. Cunningham, and A. Hyatt. 2000. Emerging infectious diseases of wildlife: threats to biodiversity and human health. Science. **287**:443–449.

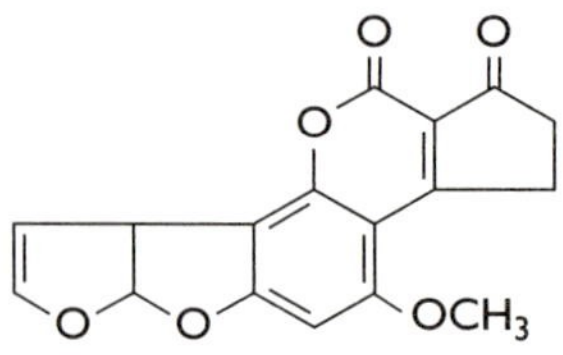

Figure 9.11 Structure of aflatoxin, the liver carcinogen produced by the fungus *Aspergillus niger.*

dance, when entire populations of some villages appeared to go mad, dancing wildly in the streets until they were felled by exhaustion. Other explanations assume that mass delusions or genetic disorders were responsible, but a more convincing explanation is fungal toxins. The neurotropic compounds produced by fungi are not broken down by heat and so would persist in bread made from contaminated flour.

Antimicrobial compounds directed against protists and fungi

Antiprotozoal compounds. A number of compounds inhibit the growth of or kill protists that cause human infections. Examples of these are provided in Table 9.1. Some are antimicrobial compounds that are also used to treat bacterial infections. The protists in which growth is inhibited by antibacterial compounds tend to be the more primitive protists, such as *Trichomonas* or *Giardia.* More evolutionarily advanced protists, such as *Trypanosoma* and *Plasmodium* species, tend to require antimicrobial drugs that target them specifically. Treatment of malaria will be described in Chapter 20 as an example of the way in which treatment can be complicated by a complex protist life cycle like that of the apicomplexa.

Antifungal compounds. A number of antifungal compounds are available. Many inhibit synthesis of ergosterol, the essential fungal membrane sterol, or interact with ergosterol to make fungal membranes leaky (Fig 9.12). Ergosterol is similar in structure to the human sterol cholesterol and probably serves a similar function in fungi. Because the enzymes that synthesize ergosterol differ enough from those that synthesize cholesterol to reduce human side effects, ergosterol-synthesizing enzymes are a good target for antifungal drugs. Skin infections caused by fungi are treated with topical ointments that contain antifungal drugs such as **tolnaftate.** Nail infections are more of a challenge. One drug that has been effective in treating nail infections is **griseofulvin,** an antifungal compound that disrupts the mitotic spindle of fungi. The fact that griseofulvin accumulates in keratinized tissues such as nails makes it particularly useful for this application. Scientists are now looking more closely at antifungal compounds that interfere with synthesis of the fungal cell wall, another unique fungal target.

Eukaryotes on the Edge: Eukaryotic Extremophiles

So much has been written about the ability of prokaryotes, especially the archaea, to live under conditions that seem extreme to us, such as temperatures near boiling or freezing and pH values as low as 0 or as high as 10,

Table 9.1 Examples of antimicrobial compounds that inhibit the growth of protists

Drug	Mechanism
Tetracycline*	Inhibits protein synthesis
Emetine	Inhibits protein synthesis
Suramin	Inhibits glycolysis in trypanosomes
Primaquine	Inhibits mitochondrial respiration in *Plasmodium* species (malaria; Chapter 20)
Norfloxacin*	Inhibits DNA replication
Pentamidine	Inhibits DNA synthesis in kinetoplast of trypanosomes

*Drug has antibacterial activity as well as antiprotist activity.

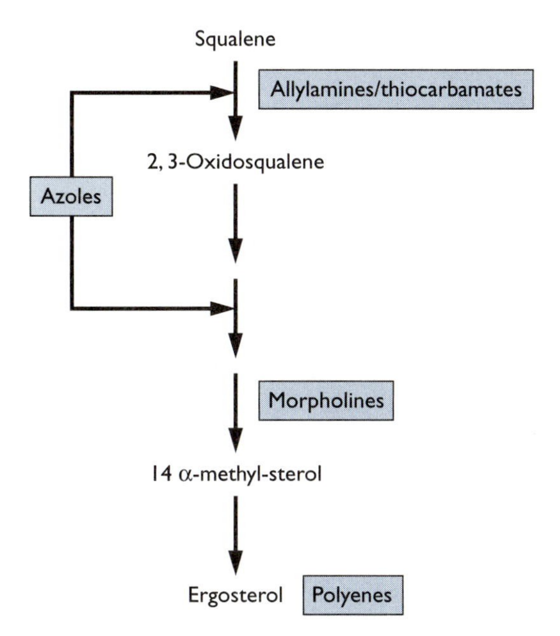

Figure 9.12 Synthesis of the fungal sterol ergosterol and some of the antifungal compounds that interfere with its synthesis.

that many people have gotten the impression that eukaryotic microbes are the wimps of the microbial world. This view is now being challenged, as scientists find these supposedly fragile organisms thriving at extremes of temperature and pH. Granted, the archaea are still the extremophile champions, but the eukaryotes are beginning to appear as serious contenders in the "extreme living conditions" sweepstakes.

What first attracted attention to the eukaryotes' ability to survive the extremes normally associated with prokaryotes was not a microbe but a 3-meter invertebrate, the tubeworm ***Alvinella,*** which lives on the sides of hydrothermal vents in a cloud of sulfide that wells up through these vents (Chapter 23). An environment less likely to support a eukaryote could hardly be imagined. Yet the worm routinely experiences temperatures as high as 70–80°C and sulfide concentrations that would kill most animals and plants. The ability to tolerate sulfides is attributed to a bacterial **endosymbiont** of *Alvinella* that oxidizes sulfide to sulfate, generating energy in the process that benefits its bacterial host and fixing carbon dioxide into forms that probably feed the worm. The ability to tolerate a temperature extreme, however, is a trait of the worm itself.

As scientists looked more closely at extreme environments, they began to find protists along with the bacteria and archaea. Although the ability of eukaryotic microbes to grow at high temperatures (above 80°C) is controversial, there is no question that eukaryotic microbes have been found growing in Anarctica, where temperatures hover close to or below 0°C, and in acid mine drainage sites, where the pH is close to 0. In Africa, Lake Nakuru, which, despite its high pH (pH 10), supports a large population of flamingos, contains not only the cyanobacteria on which the flamingos dine but also several species of eukaryotes that may be living on the fecal droppings of the flamingos. Eukaryotic microbes have also been found in sites characterized by very high salt concentrations and at the bottom of the ocean, where pressures are very high.

How eukaryotes, especially those without rigid cell walls and with phospholipid membranes that are supposed to be unable to withstand extremes of any kind, tolerate such extremes remains to be determined. The fact that, despite all received wisdom to the contrary, the eukaryotic microbes (and macrobes) are present and apparently thriving in locations once thought to be the sole domain of prokaryotes needs to be explained. For those science fiction buffs who may have been discouraged by pronouncements about the alleged inability of eukaryotic cells to withstand extremes of any kind, the new findings that eukaryotes are tougher than we thought should be a source of inspiration and hope for finding life that resembles us—or at least our closest microbial relatives—on other planets.

Eukaryotic Model Systems

Just as *E. coli* has served as a model system for studies of bacterial genetics and physiology, eukaryotic microbes serve as important model systems for the study of eukaryotic genetics and physiology. One of the first nonfungal eukaryotic microbes to be isolated in pure culture was ***Tetrahymena.*** Although fungi and yeast can be isolated in much the same way as bacteria, because they grow on similar substrates, protists such as *Tetrahymena* represented a larger challenge, because the protists normally grow by feeding on bacteria. But in culture, the bacteria had a disturbing tendency to overgrow the protist. Discovering how to grow *Tetrahymena* on defined medium was a major accomplishment. Studies of *Tetrahymena* produced, among many other landmark discoveries, the discovery that RNA could act like an enzyme. A *Tetrahymena* enzyme contained an RNA species that actually catalyzed the reaction—a cleavage of an RNA substrate—that was originally attributed to the protein. Now catalytic RNAs, called **ribozymes,** are being developed for a variety of applications, including treatment of diseases caused by RNA viruses such as human immunodeficiency virus.

The simple bakers' yeast, ***Saccharomyces cerevisciae,*** introduced a revolution in eukaryotic genetics in the 1980s. Armed with a microbe that could be grown and manipulated genetically almost as easily as *E. coli*, scientists proceeded to make a variety of discoveries about regulation of eukaryotic genes, splicing of RNA, and protein processing. Because *S. cerevisciae* and other yeasts were late evolutionary arrivals, it is not surprising that they served as good models for some aspects of mammalian cell physiology, such as protein processing and regulation of gene expression. Also, *S. cerevisciae* has both sexual and asexual replication modes, a trait that allowed scientists more leeway in manipulating the organisms and led to important discoveries about sexual mating of eukaryotic cells.

Another relatively late evolutionary arrival, the slime mold ***Dictyostelium,*** has served as a model of another sort. *Dictyostelium* is a social amoeba that, under appropriate conditions, aggregates to form a slug-shaped form consisting of many individual cells. The cells change their association with each other to develop into a **stalk** and finally into a **fruiting body** (Figure 9.13). *Dictyostelium* continues to serve as a model for social behavior of eukaryotic cells and for development of complex multicellular structures.

It is easy for those who are new to biology and still heavily influenced by appearances to smile at a designation once given *S. cerevesciae* by its

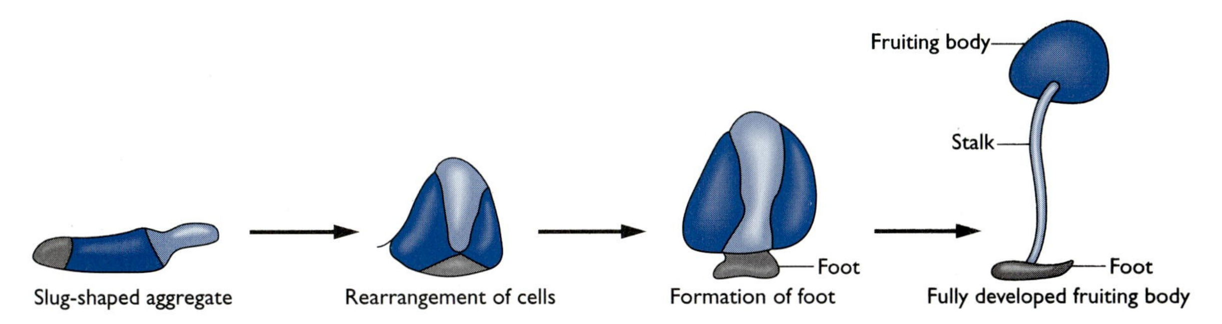

Figure 9.13 Stages in the life cycle of *Dictyostelium*. *Dictyostelium* is a social amoeba that forms a slug-shaped aggregate, which then progresses to a fruiting body.

supporters: a mouse in a test tube. Or to scoff at the idea that studies of *Dictyostelium* might reveal the secrets of human tissue development. Yet, these claims are not at all far fetched. The key to most basic processes of human cells is locked in the genomes of humans' earliest ancestors, the eukaryotic and prokaryotic microbes. Discoveries about how the bacterium *E. coli* replicated its DNA, transcribed it into RNA, and then translated the RNA into proteins proved to be paradigms, with a few modifications, for cells of much more advanced organisms, such as plants and animals. Likewise, processes carried out for the eukaryotic microbes are yielding additional paradigms that help us better understand our biology. In order to learn a language, it is first necessary to acquire a vocabulary and an understanding of the grammatical rules that string the words together in a comprehensible way. Simple model systems are the quickest and easiest ways to learn the vocabulary and grammar that make it possible for us to begin to interrogate the more complex organisms, including ourselves, that would otherwise appear completely mysterious.

chapter 9 at a glance

Overview of eukaryotic microbes covered

Category of microbe	Characteristics	Examples
Archaezoa	No mitochondria Some lack Golgi bodies Flagella Today, many are symbionts or parasites of animals, insects	Diplomonads (*Giardia intestinalis,* diarrhea) Parabasalians (*Trichomonas vaginalis,* vaginitis; *Trichonympha* species, symbiont of termites)
Amoeboflagellates	Feed on bacteria Amoeboid form, flagellated form, cyst form Flagella (on flagellated form) Most free-living in soil, water	*Naegleria fowleri,* meningitis in humans
Euglenoids	Some photosynthetic (chloroplasts), others feed on microbes Single flagellum Eyespot (photosynthetic members, for phototaxis) No known animal pathogens	*Euglena gracilis* (photosynthetic, found in ponds and lakes)
Kinetoplastids	Kinetoplast (mitochondria with few large genome segments, many tiny DNA circles) Flagella attached to undulating membrane Members include important human pathogens, many insect-transmitted	*Trypanosoma brucei* (sleeping sickness, tsetse fly) *Leishmania* species (leishmaniasis, sandfly)
Apicomplexa	Complex life cycles Some forms have flagella Members are important human pathogens	*Plasmodium* species (malaria, mosquito; Chapter 20) *Cryptosporidium parvum* (diarrhea; ingested; Chapter 17)
Dinoflagellates	Complex life cycles Two flagella Most are photosynthetic Some produce toxins	*Pfiesteria* (blooms; toxins kill fish, harm humans) Other dinoflagellates (red tides; toxins kill fish, harm humans)
Algae, diatoms	Rigid cell walls (algae, calcite; diatom, silica) Photosynthetic, important component of phytoplankton "Skeletons" form geological deposits	Phytoplankton Blooms that create hypoxic zones
Fungi	Rigid cell walls Mycelia, spores Some have yeast form Nonmotile Some members cause human disease Serious infections mainly in immuno-impaired people	Component of lichens (with algae) Athlete's foot, other skin and nail infections *Histoplasma capsulatum* (dimorphic fungus, lung infection)

chapter 9 at a glance

Antimicrobial compounds

Category of microbe	Target of antimicrobial
Protists	Many targets, including protein synthesis, DNA replication
	Some antibacterial compounds work on more primitive protists (e.g., tetracycline)
Fungi, yeasts	Membrane sterol ergosterol a target of many compounds
	Other targets, such as cell wall synthesis, DNA replication

Eukaryotics at the extremes

pH values 0–10

Temperature 0–80°C

Characteristics allowing growth at extremes unknown

Examples of model systems

Tetrahymena (catalytic RNA)

Saccharomyces cerevisciae (yeast; basic eukaryotic processes, such as gene regulation, protein processing)

Dictyostelium (slime mold; social interactions, development)

key terms

algae and diatoms protists that have recently evolved; important components of phytoplankton; photosynthetic; many beneficial effects but can cause major fish die-offs if they reach high numbers

amoeboflagellates protists with an amoeboid form, a flagellated form, and a cyst form; most are free-living, feed on prokaryotes

apicomplexa group of protists with complex life cycles; have both sexual forms (gametes) and asexual forms, the sporozoite (infective form) and merozoite (replicative form); life cycles may involve both humans and mosquitoes (*Plasmodium* spp., the cause of malaria)

archaezoa microbes that probably resemble earliest eukaryotes; some members such as the diplomonads (e.g., *Giardia*) and parabasalians (e.g., *Trichomonas*) lack mitochondria (as a result of secondary loss) and may live under anaerobic conditions

dinoflagellates group of protists that have two flagella; many are photosynthetic; an example is *Pfiesteria*, which produces a neurotoxin that kills fish and causes transient neurological problems in humans

euglenoids protists with a single flagellum; some are photosynthetic, others are "colorless" and feed on prokaryotes; no known human pathogens

fungi major recyclers of dead biomass; degrade cellulose and lignin in dead plants; many disease-causing species; not photosynthetic; produce spores; may exist as yeasts, hyphae, or both (dimorphic)

kinetoplastids protists with a single nucleus, flagellum and undulating membrane; possess an unusual mitochondrion called the kinetoplast, the genome of which consists of a few large chromosomes and many small DNA circles

lichens assemblages of fungi and photosynthetic microbes, either algae or cyanobacteria; fungi give macroscopic structure, photosynthetic microbes fix nitrogen, which is used by fungi; important in eroding rocks to produce soil

phototaxis movement of certain protists (e.g., euglenoids) to a level of light where photosynthetic machinery works best; mediated by an organelle called an eyespot

protist term often used to describe eukaryotic microbes as a group

Questions for Review and Reflection

1. Question everything! Are the so-called archaezoa really all that ancient? What are some arguments to challenge the notion that current representatives of this group are good representatives of ancestors of eukaryotic microbes? What arguments support the claim that they are similar to the earliest eukaryotes?

2. Evolution of the eukaryotic microbes preceded that of plants and animals by billions of years, so current interactions between these microbes and later life forms should not obscure our vision of their history. Review the environmental roles of the eukaryotic microbes described in this chapter and develop hypotheses about their role in the world before the larger life forms appeared.

3. What does the fact that termites and wood-eating roaches rely on microbes to feed them suggest about development of a new generation of insecticidal compounds? What advantages might chemicals have that attack microbes instead of insect cells? What are some disadvantages?

4. Algae and diatoms are not known to cause human infections, although they can sometimes produce toxic compounds harmful to humans. This makes sense at first glance, because these microbes are photosynthetic. Is it possible that a "killer diatom" exists and might one day manifest itself? If so, what would be the likely scenario?

5. Why is a fungus more likely to cause human disease than a diatom? Can you formulate a set of characteristics that would help to predict what eukaryotic microbes might cause human infections?

6. Why might antibacterial compounds be effective against *Giardia* or *Trichomonas* but not against *Plasmodium* species or fungi?

7. The chapter ended with a short description of some eukaryotic microbes that are used as model systems for studying processes in higher animals. Explain how a simple single-celled organism might unlock secrets of the human body. Will these simple model systems help to decode the human genome sequence and, if so, how?

10 Viruses of Mammalian Cells

So, Naturalists observe, a flea
has smaller fleas that on him prey;
and these have smaller still to bite 'em;
and so proceed ad infinitum.
—Jonathan Swift

Viruses are totally dependent on free-living cells. In fact, it is likely that all free-living cells have viruses capable of infecting them. The viruses of prokaryotic cells have been described in previous chapters. The viruses of eukaryotic microbes have been poorly characterized but clearly exist; plants and arthropods have a variety of viruses. In this chapter, the focus will be on viruses of mammalian cells, the best characterized of all viruses of eukaryotic cells. This does not mean that viruses of plants, insects, and eukaryotic microbes are unimportant. In fact, the study of these viruses may well be a major area of research in the future. But, bowing to the reality of the huge literature on mammalian viruses, which may prove to be a valuable guide to future investigations of viruses of other eukaryotic cells, this chapter will confine itself to mammalian cell viruses, with an emphasis on viruses that attack human cells. In later chapters, examples of specific human viruses, such as hepatitis virus, polio virus, and human immunodeficiency virus (HIV), will be covered in more detail. This chapter will serve as an introduction to the properties of such viruses.

Structure and Classification of Mammalian Viruses

Components of enveloped and naked viruses

An introduction to viral structures was provided in Chapter 1 and will be reviewed only briefly here. The portion of a virus that is common to all viruses consists of a nucleic acid-protein core covered by a protein coat (**capsid**; Fig. 10.1). The capsid consists of many copies of one or a few polypeptides (**capsomeres**), which are tightly packed in an array that often has a crystalline appearance. The capsid gives the virus its shape, helps to organize the viral genome (which is packed inside it), and protects the viral genome from enzymes that might degrade it in the external environment. As will be seen shortly, the capsid also plays a role in the uncoating of the virus, the process by which the virus releases its genetic material into the

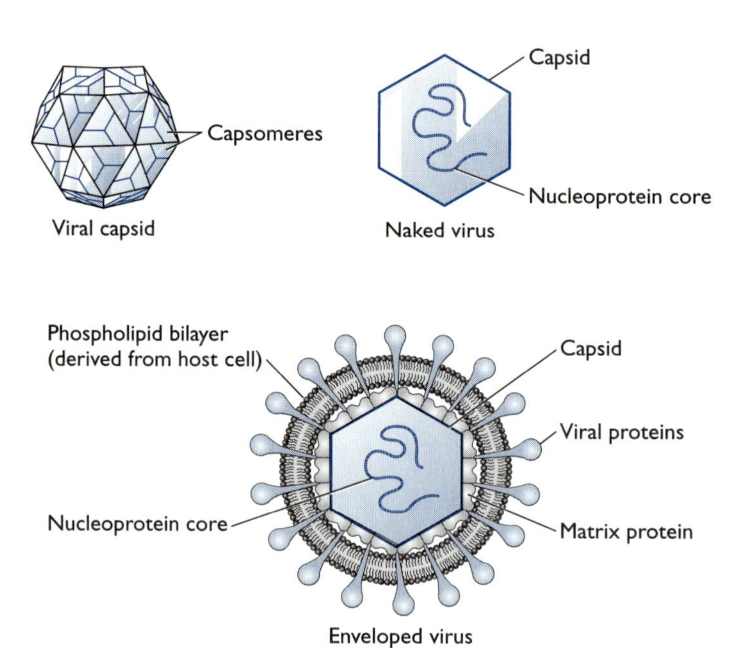

Figure 10.1 Components of naked and enveloped viruses. The nucleoprotein core contains the viral genome, together with proteins that play a structural role to organize the genome or perform an enzymatic function during replication. The capsid consists of tightly packed proteins (capsomeres) that give the virus its shape and protect the nucleoprotein core. Enveloped viruses have a phospholipid bilayer membrane that contains envelope proteins, which are used for attachment to host cells. Underlying the envelope are matrix proteins that stabilize the membrane and attach it to the capsid.

cell it is attacking. Later in the replication cycle, the capsid helps to organize copies of the viral genome that have been synthesized inside the infected cell.

Many viruses have an **envelope,** a protein-phospholipid layer that covers the capsid. Viral envelopes contain proteins that project outward from the viral surface, **envelope glycoproteins.** Envelope glycoproteins mediate attachment of an enveloped virus to the host cell surface. They do this by binding very specifically to carbohydrates or proteins on the surfaces of the cells the virus infects. The host cell proteins bound by the viral surface proteins are sometimes called **viral receptors. Matrix proteins,** which interact with the membrane-embedded portion of the envelope protein and with capsid proteins, link the envelope to the capsid and help to stabilize the viral particle. Matrix proteins also play an important role in the uncoating of enveloped viruses and in the assembly of viral particles in late stages of the viral replication process. Viruses without an envelope are called **unenveloped** or **naked** viruses. Because naked viruses have no envelopes, their capsid proteins are the ones that mediate attachment of the virus to the receptor molecules on the surface of the mammalian cell. As with enveloped viruses, the interaction between the capsid protein of the naked virus and the host cell viral receptor is very specific.

Both naked and enveloped viruses have **nucleoprotein cores** enclosed in their capsids. The proteins of the nucleoprotein core can have a structural function, such as stabilization of the viral nucleic acid during initial and late stages of viral replication, or an enzymatic function. Depending on the way a virus replicates, it may have to provide some proteins that are not found in human cells. For example, the replication cycle of HIV involves a step in which a DNA copy is made of the RNA genome of the virus. Because mammalian cells do not have enzymes that carry out such a reaction (**reverse transcriptase**), the virus must provide its own reverse transcriptase. Other examples of specialized replication enzymes provided by the virus will be given in a later section on strategies of viral replication.

In contrast to free-living microbes, which all have double-stranded DNA genomes, viruses can have genomes composed of single-stranded RNA, double-stranded RNA, single-stranded DNA, or double-stranded DNA. A similar range of genome types is also seen in bacterial viruses. Why viruses exhibit such a variety of genome types is not clear, but the fact that some of them have RNA genomes has a special significance for their evolution. RNA is less stable chemically than DNA. Moreover, enzymes that reproduce RNA tend to make more mistakes than those that reproduce DNA. As a result, viruses with RNA genomes tend to mutate more rapidly than organisms with DNA genomes. Two viruses with RNA genomes, HIV and influenza virus, are among the most rapidly mutating organisms known. Not all RNA viruses mutate rapidly, however, so more than the RNA composition of the viral genome is involved in viral mutation rates.

Viral genomes range in size from about 4 kbp, the size of a small bacterial plasmid, to about 200 kbp. The smallest bacterial genome is about 500 kbp in size. Thus, viruses range in genetic capacity from very simple, stripped-down genomes to complexities near that of a stripped-down bacterial genome. Viral genomes can be very small, because viruses have evolved to make use of many features of the cell they infect: the cell's ability to synthesize nucleotides and amino acids, its ability to generate energy, and its ability to synthesize proteins and nucleic acids. Enveloped viruses even get their phospholipid bilayer envelope from the nuclear membrane, the endoplasmic reticulum, or the cytoplasmic membrane of the infected cell. Thus, a virus needs only to supply genes encoding proteins of the envelope, capsid, and nucleoprotein core. The small genome is advantageous. A small genome can be replicated much more rapidly than a large one, allowing many copies of the viral genome to be made in a short period of time. In this way, viruses quickly outpace replication of the genome of the infected cell, competing successfully for components that should have been directed toward mammalian genome replication.

Nomenclature and classification of mammalian viruses

Virus classification is based on a number of characteristics (Table 10.1), including the presence or absence of an envelope, the shape of the virus, the characteristics of the virus's capsomeres, the composition of its genome, and the characteristics of its replication strategy. In some cases, the host range of the virus and its mode of spread (by insect, for example) play a role in classification. Viruses are organized into families, genera, and species. The family name ends in "viridae" and is not italicized. The genus name, which is italicized, ends in "*virus*." The species name, which is not italicized, also ends in "virus." For example, the virus that causes herpes cold sores has the species name herpes simplex 1 virus and is a member of the genus *Simplexvirus* and the family Herpesviridae. In this chapter, for

Table 10.1 Characteristics used to classify mammalian viruses

Presence or absence of envelope
Shape of virus (e.g., icosohedral, filamentous)
Characteristics of viral genome
RNA or DNA, single-stranded or double-stranded, linear or circular, segmented or nonsegmented
Mode of spread (e.g., insect vector)

simplicity, we will use primarily the family name and the species name. Examples of families of viruses covered in later chapters are provided in Figure 10.2. Evident relationships link virus families together taxonomically, as seen in Figure 10.2, but no single genetic yardstick, comparable to that for prokaryotic and eukaryotic ribosomal RNA sequences, exists for establishing relationships between mammalian viruses or between mammalian viruses and viruses of plants, arthropods, eukaryotic microbes, or prokaryotes.

Figure 10.2 Examples of viral classification. All of the viruses included in this figure are described in later sections of the text. Note the use of features listed in Table 10.1 to organize the various families and species of viruses. For simplicity, only the family name and the species name are given in this figure.

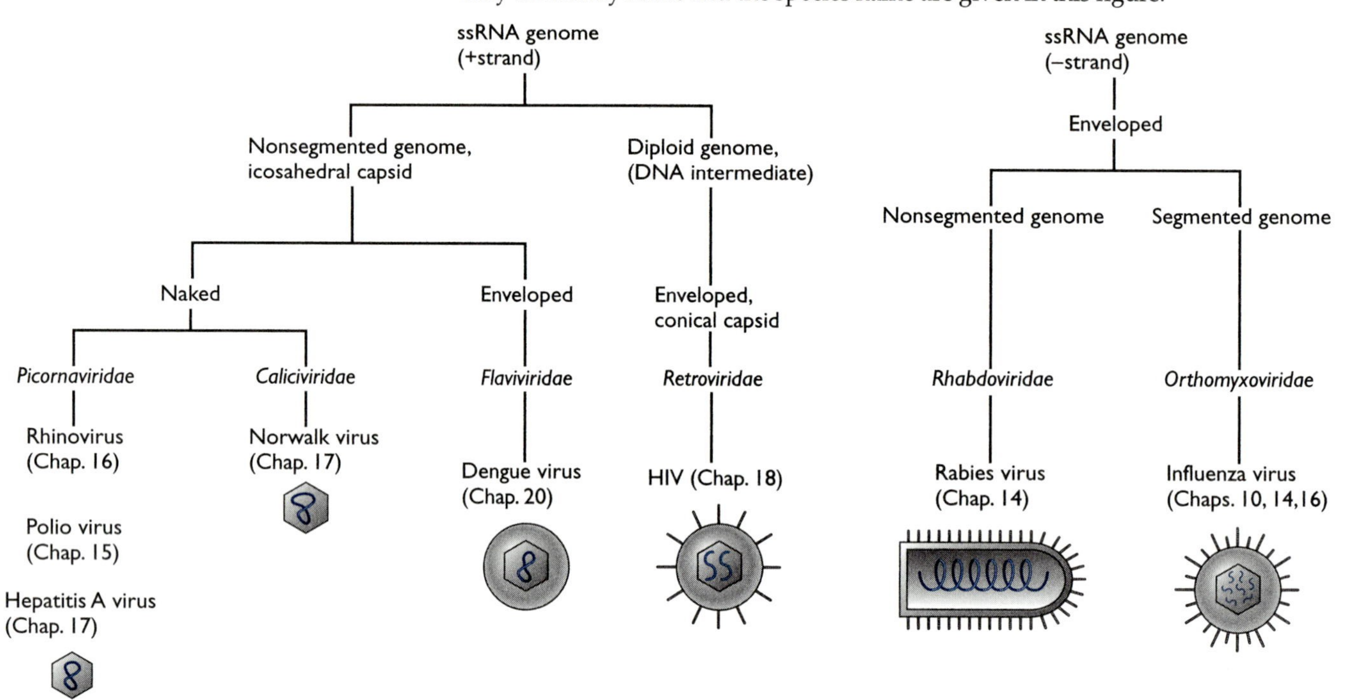

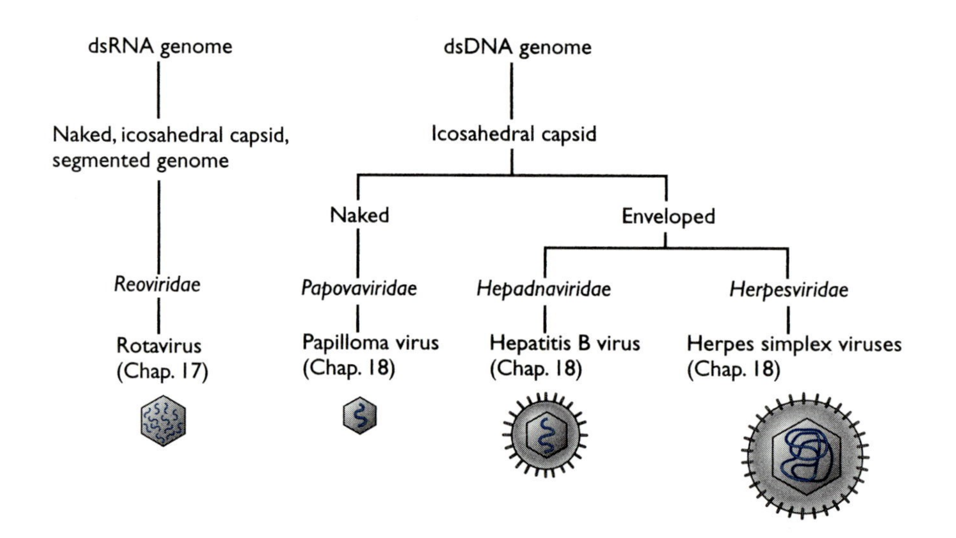

How Viruses Infect Mammalian Cells and Replicate Themselves

Overview of the steps in viral replication

The initial steps in the viral infection cycle are similar for all viruses. That is, the virus first binds to a specific molecule on the surface of the cell it will infect (**attachment**), then releases its nucleoprotein core into the host cell cytoplasm, a process called **uncoating.** Some viruses follow attachment with a step that fuses their surfaces with the membrane of the target cell and allows release of the nucleoprotein core into the cell cytoplasm (Fig. 10.3). Others undergo a more complex internalization and uncoating process, which involves endocytosis of the virus, acidification of the endocytic vesicle interior, fusion of the viral surface to the vesicle membrane, and, finally, release of the nucleoprotein core into the host cell cytoplasm.

After attachment and uncoating, copies of the viral genome and viral proteins are made. Viral genome replication can occur in the nucleus or in the cytoplasm, depending on the virus's replication strategy. Replication strategies vary with the nature of the viral genome, as will be explained later in more detail. Viral proteins are synthesized in the cytoplasm. Finally, new viral genomes and viral proteins are assembled and exit the cell. The two main routes of exit are **budding,** a process by which the virus pushes through a membrane acquiring an envelope in the process, and **lysis** of the host cell cytoplasmic membrane, allowing viral exit without budding. Enveloped viruses bud through membranes of the infected cell (nuclear membrane, endoplasmic reticulum membrane, or cytoplasmic membrane, depending on the virus). Simple as they seem at first glance, viruses exhibit an astonishing variety of strategies at each step of the replication cycle.

Influenza virus as an example of virus–host cell interactions

To convey a sense of the complex interaction between a virus and a mammalian cell, we will consider in detail a single type of virus, influenza virus. Influenza virus is not only a well-studied virus but it is also all too familiar to most people from personal experience with influenza, the disease caused by influenza virus. Thus, influenza virus provides a good illustration of virus–mammalian cell interaction. As we move through the stages of the influenza virus replication cycle, differences in the strategies used by other viruses will be noted. Another good reason to use influenza virus as an example is that, whereas influenza virus has some features in common with other viruses, it also has features that are unique. Its unique features underscore the fact that there is no such thing as a "typical" virus. Given their small sizes and relatively simple genomes, viruses display an impressive amount of variation in the ways they carry out different steps in their replication cycle,

Attachment and uncoating. Viruses seem so simple at first glance that it is hard to imagine them as the initiator in the virus–host cell interaction. Yet viruses have the capacity to manipulate the cell they are infecting into responding in ways that move the virus through its replication cycle. Often the action of the virus is subtle, consisting of tight binding of viral molecules to host cell molecules or production of an enzyme that alters the balance between host cell and viral replication processes. This is evident from the very beginning of the influenza virus replication cycle. The virus first attaches to its target mammalian cell by means of a protein called **hemagglutinin,** one of its envelope proteins. This protein was named for its ability

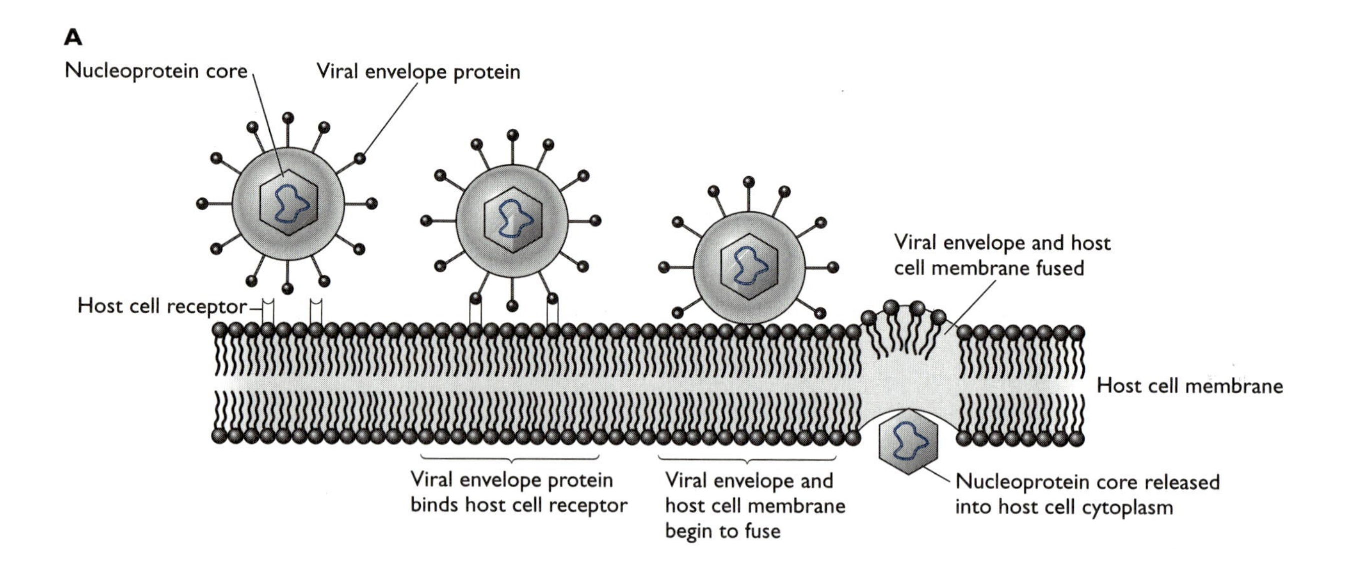

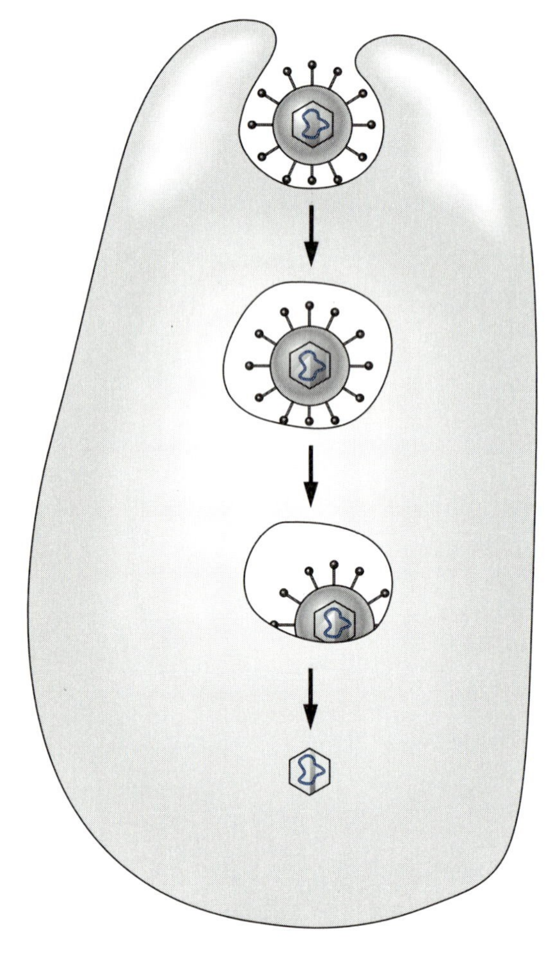

Figure 10.3 Entry and uncoating of viruses. The viruses in this figure have an envelope, but similar processes occur in the case of naked viruses. **(A)** Some viruses attach to the host cell receptor molecule they recognize, then fuse with the host cell cytoplasmic membrane and introduce their nucleoprotein core into the cytoplasm of the host cell. **(B)** Influenza virus and a number of other viruses use an internalization process in which the virus triggers its own uptake in an endocytic vesicle. After fusion of the viral surface with the vesicle membrane, the nucleoprotein core is released into the cytoplasm.

to agglutinate red blood cells. Hemagglutinin proteins form trimers that project out from the viral surface (Fig. 10.4), making the viral surface appear studded with spikes. These glycoprotein spikes bind very specifically to sialic residues on the surface of a mammalian cell. **Sialic acid** is a sugar found widely on mammalian cells (Fig. 10.5). Because many mammalian cells have sialic acid residues on their surfaces, influenza viruses can infect many different cell types. Other viruses that have surface proteins which recognize molecules found only on the surfaces of a limited number of cell types have a more restricted range of cells they can infect. Thus, for example, HIV infects only certain cells of the immune system, because HIV surface proteins recognize molecules found on the surfaces of these particular cells but not on other types of cells. Another level of specificity is the ability of the virus to replicate once it enters the cell. Some cells are permissive for replication of a particular virus but not for replication of others.

The model for binding of a viral surface protein to a molecule on the host cell (viral receptor) usually depicts a single viral protein interacting with a single receptor molecule on the host cell. Recently, in the case of HIV, it has become clear that some viruses may have more complex interactions with host cell surfaces. HIV surface proteins have a primary receptor, a protein called CD4, which is found on certain cells of the immune system, but, before the invasion process can proceed, stabilization of the interaction between the viral surface protein and CD4 requires interaction between HIV and secondary receptor proteins called coreceptors (Chapter 18). It is now evident that the use of both a receptor and a coreceptor may be much more common among viruses than was thought previously to be the case. In the case of influenza virus, the simpler model of the attachment process, in which a single viral protein binds to a single mammalian cell molecule, seems to be an accurate representation of the interaction. Binding of the viral envelope protein spike to sialic acid is very tight and is stable enough to trigger the mammalian cell to initiate endocytosis in the vicinity of the bound virus, forming an endocytic vesicle that contains the virus (Figure 10.3B).

Endocytosis is a normal pathway used by mammalian cells to ingest nutrients. The virus in effect subverts this normal pathway to facilitate

Figure 10.4 Envelope proteins of influenza virus. One of the envelope proteins (hemagglutinin) is involved in attachment of influenza virus to sialic acid residues on the host cell surface. Three hemagglutinin proteins are organized to form a protein spike, which actually mediates attachment to sialic acid residues. The second envelope protein (neuraminidase) cleaves the sialid acid residues from carbohydrate residues on the cell surface, thus releasing the virus from the cell. Four neuraminidase proteins form a unit in the envelope.

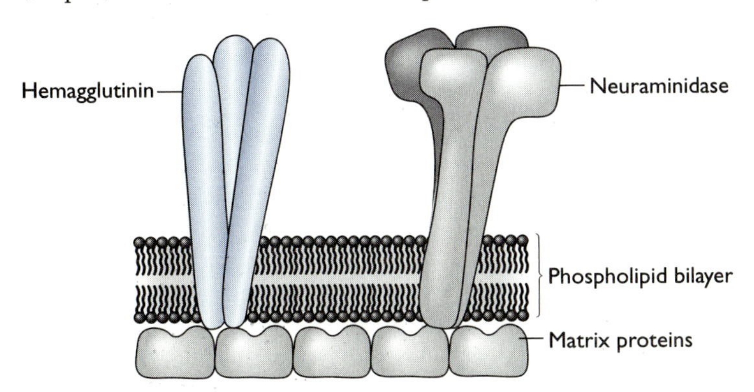

Figure 10.5 Structure of sialic acid. Sialic acid is a sugar found on the surface of most human cells, where it is attached covalently to other sugars.

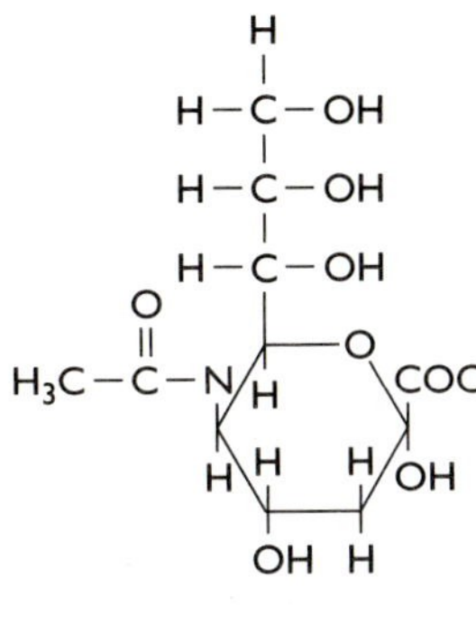

entry into the mammalian cell. After the endocytic vesicle is formed, it begins to acidify. Acidification leads to a conformational change in the matrix proteins that underlie the envelope, allowing the envelope to be removed and the nucleoprotein core to be released into the cell cytoplasm.

As already mentioned, not all viruses enter cells via the endocytic pathway. Some attach to and fuse directly with the cytoplasmic membrane of the mammalian cell, releasing the nucleoprotein core into the cytoplasm. A subset of viruses can use such a **fusogenic pathway** to move directly from an infected cell to a new cell without being released into the external environment. Such viruses have the advantage that, as long as they are inside cells, they are hidden from the immune system.

Replication of the viral genome and production of viral proteins. After release into the cytoplasm, the influenza viral core moves to the nucleus, where it gains access to the interior through pores in the nuclear membrane. Normally, proteins that transit nuclear membrane pores are restricted to those types of proteins that need to enter or leave the nucleus, but viral nucleoprotein core proteins are able to mimic these proteins sufficiently well to pass this cellular barrier. Once inside the nucleus, the viral genome begins to replicate. Most RNA viruses replicate in the cytoplasm, whereas DNA viruses replicate in the nucleus. Influenza virus is an exception to this rule. Influenza virus has a segmented genome made of single-stranded RNA segments. The sequence of these segments is the complement of the messenger RNA (mRNA) strand that will move into the cytoplasm to be translated into viral proteins. The mRNA strand is called the +strand, and the complement is called the –strand. Because the influenza virus genome is composed of –strands of RNA, these strands must first be copied into +strands, which will then be translated to produce viral proteins. This task is accomplished by an RNA-dependent RNA polymerase (**replicase**) encoded by the viral genome. The subunits of the replicase are proteins of the nucleoprotein core. Replicase makes a +stranded copy of the –strand. Some of these +strands become mRNAs, which are translated in the cytoplasm, and some remain in the nucleus, where the replicase uses them as a template to make new copies of the –stranded viral genome segments.

The replicase subunits are contained in the nucleoprotein core and are thus available from the start. In general, viruses with –strand RNA genomes must bring preformed RNA replicases with them to make +strands, which can be translated to provide more RNA polymerase subunits. A virus with a +strand RNA genome may be able to dispense with preformed replicase, because it can be translated directly, making its replicase on the spot. DNA viruses can use host enzymes for replication and transcription, although some provide their own replicases. Replication strategies for viruses with +strand RNA genomes, double-stranded RNA genomes, and DNA genomes are summarized in Figure 10.6.

Influenza virus must make some copies of its –strand genome that will serve as mRNA and some that will serve as templates for replicase to produce more –stranded copies of the genome. The virus solves this problem by a unique method of subverting normal host function. Normal mammalian cell mRNAs have an unusual structure on their 5' ends, called the 5' cap (Fig. 10.7A). A 5' cap signals the mammalian cell translation machinery that this mRNA is targeted for translation. Instead of making such a cap structure for its own +strand, the influenza virus replicase, once it has bound to the –strand viral RNA, acts as an endonuclease to cleave the 5' cap

Figure 10.6 Overview of viral replication strategies. **(A)** +strand RNA viruses, which are already in the messenger RNA (mRNA) form, can be translated to produce viral proteins, such as the replicase that will be used to reproduce the viral genome and make more viral mRNA for translation. **(B)** –strand RNA viruses require a preformed replicase, which was part of the nucleoprotein core, to make the mRNA copy of the –strand. Double-stranded RNA viruses use a replication strategy similar to that of the –stranded RNA viruses. **(C)** Retroviruses (such as HIV) make a DNA copy of their RNA genome using a special viral enzyme, reverse transcriptase, which is brought preformed into the infected cell as part of the nucleoprotein core. **(D)** DNA viruses usually use host cell enzymes for replication.

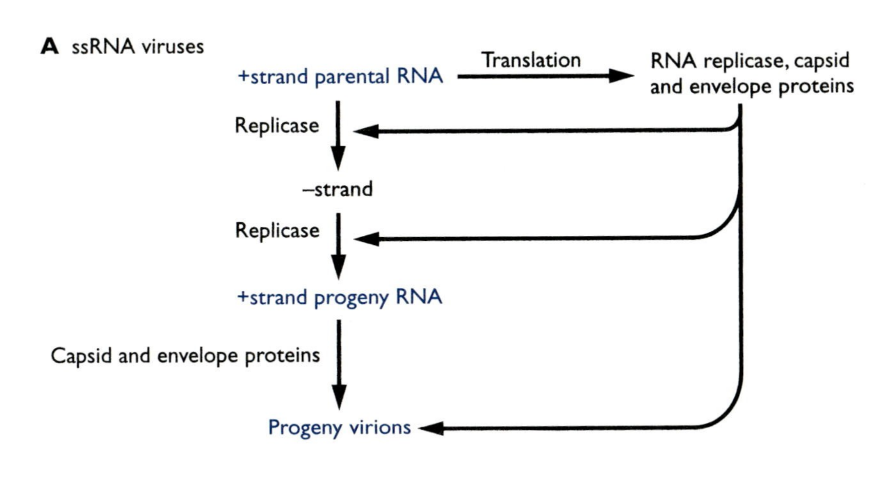

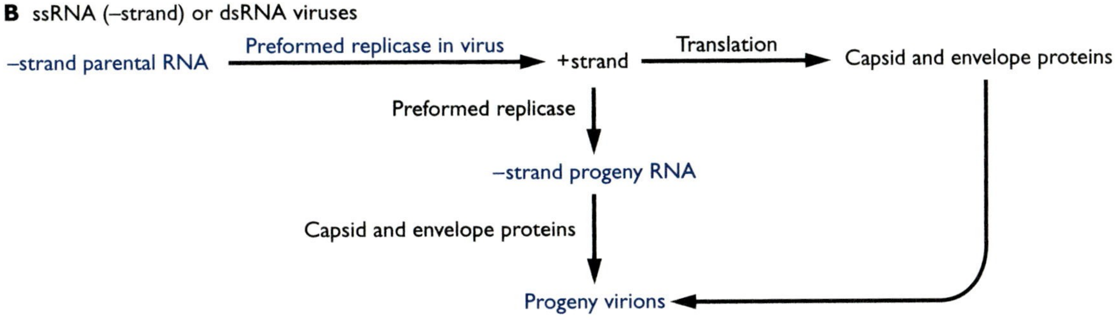

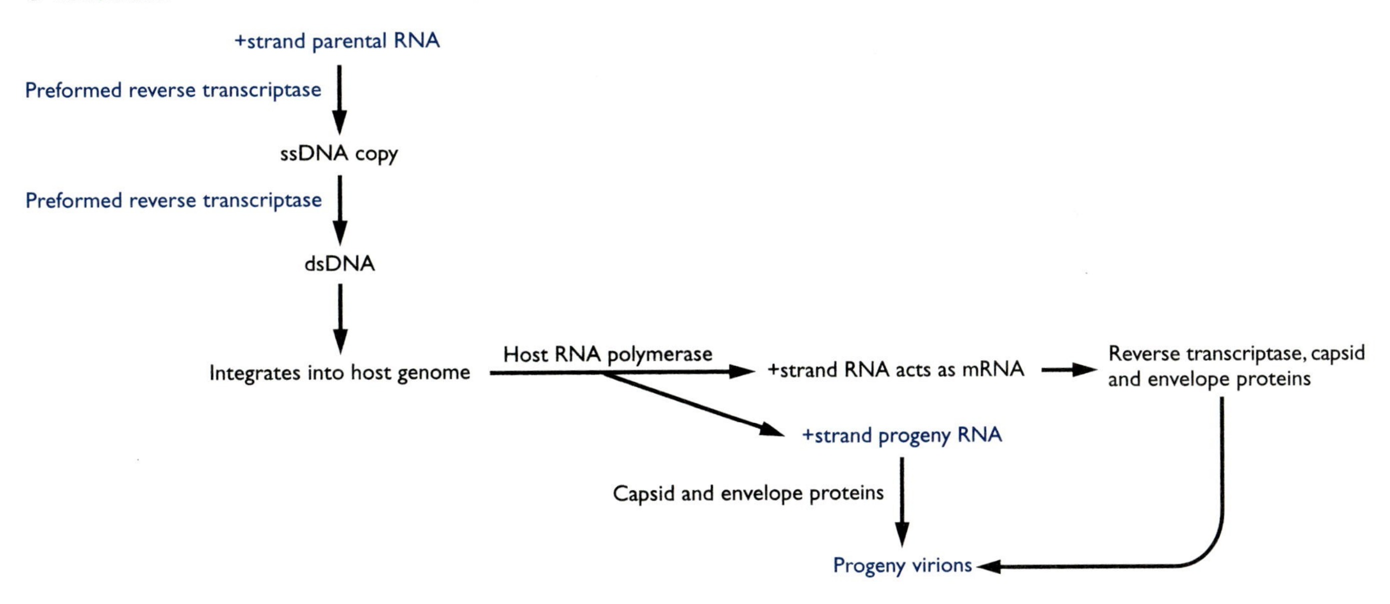

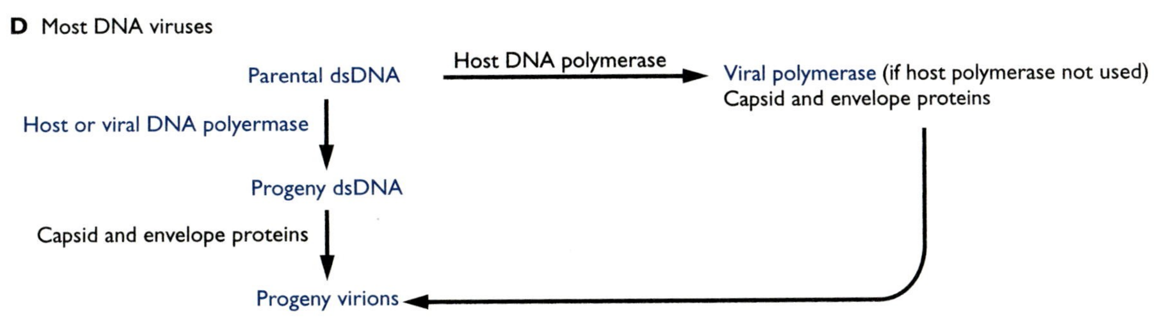

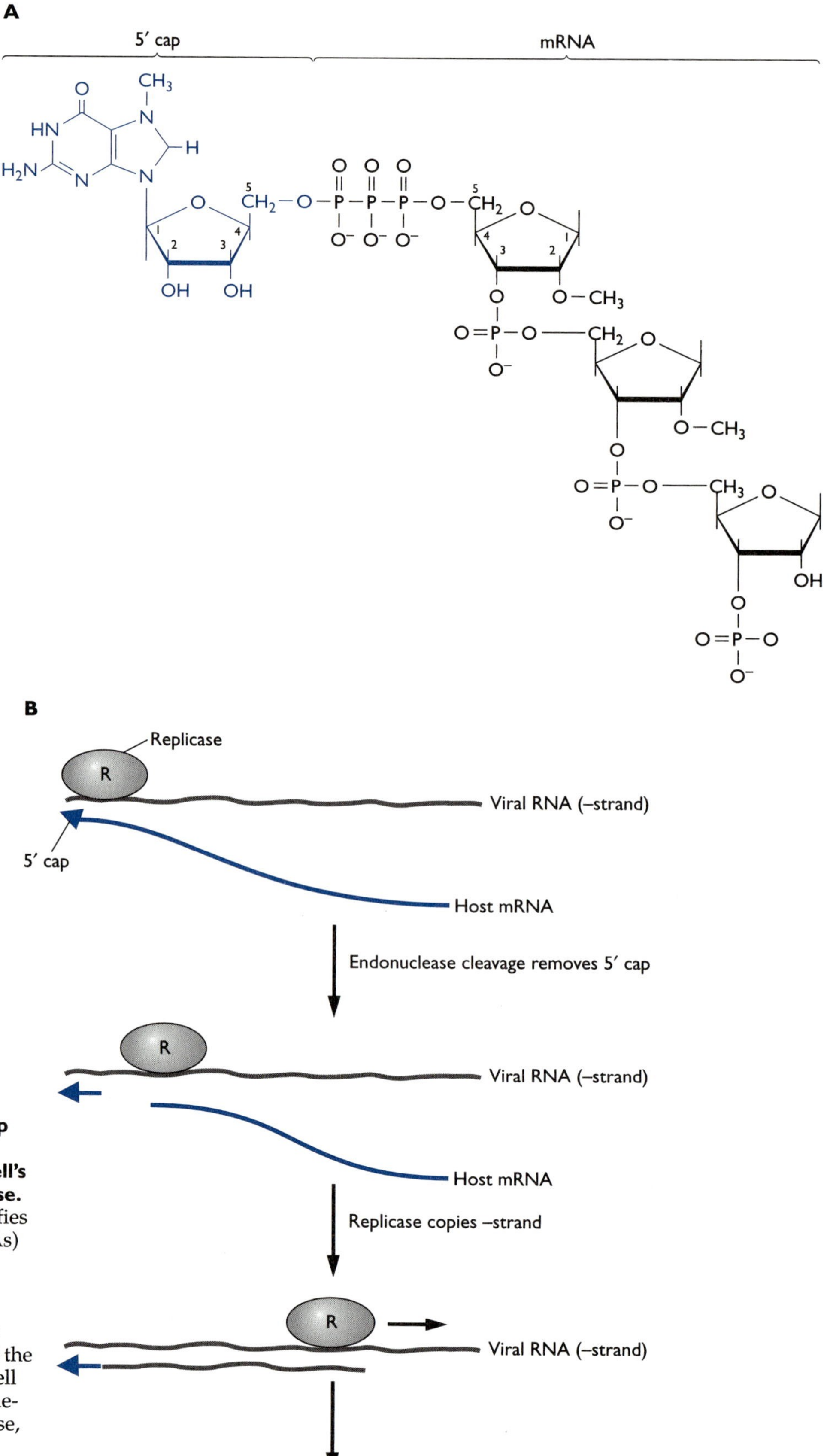

Figure 10.7 Structure of the 5' cap and the mechanism by which influenza virus subverts the host cell's translation machinery to its own use. **(A)** Structure of the 5' cap that identifies eukaryotic messenger RNAs (mRNAs) for translation. **(B)** The replicase of influenza virus binds to the viral –strand RNA, then attaches a 5' cap from a host mRNA. Endonucleolytic activity of the viral replicase clips off the 5' cap, releasing the rest of the host cell mRNA. The 5' cap then primes synthesis of the +strand by the viral replicase, resulting in a capped viral message. This strategy is unique to influenza virus.

structure plus 10–12 nucleotides from a host mRNA molecule (Fig. 10.7B). The captured cap region then primes synthesis of the +strand by the replicase, resulting in a capped +strand copy of the viral genome. This strategy accomplishes two goals. First, it targets certain viral +strands for translation. Second, it helps shut down translation of mammalian cell messages, thus leaving the translation machinery of the infected cell to devote itself to synthesizing viral proteins.

This "theft" strategy is one way to divert the host cell biosynthetic machinery to focus on translating viral transcripts and reproducing viral genomes instead of replicating the host cell genome and synthesizing host cell proteins, and all viruses must have some way of doing this. Some take advantage of their small genomes to replicate so rapidly compared to the host cell genome that they swamp the nucleic acid replication machinery of the host cell. In a similar way, their transcripts soon dominate in number, thus monopolizing the host cell translation machinery. Other viruses interfere with proteins that control host cell DNA or RNA synthesis, making it very inefficient compared to the more efficient synthesis of viral messages and genomes.

Whatever the strategy used by the virus to enhance its own replication at the expense of normal host cell functions, the usual effect is that host cell biosynthesis of its own nucleic acids and proteins declines drastically and synthesis of viral genome and viral proteins predominates. Such drastic effects on host cell biosynthetic functions can have an undesirable effect from the virus's point of view: **apoptosis,** a type of cell death, can be triggered, and the eukaryotic cell shuts down all its functions and dies. Some viruses counter apoptosis to keep the eukaryotic cell viable, from the biosynthetic point of view, long enough to give the virus time to complete its replication cycle.

Assembly and departure of virus particles from the infected mammalian cell. As replication of viral genomes and translation of viral proteins proceeds, copies of viral genomic segments and viral proteins begin to accumulate in the cytoplasm of the infected cell. When concentrations of these components are sufficiently high, the proteins of the nucleoprotein core interact with viral genome segments. Most RNA viruses have single genome segments, not 7–8 as in the case of influenza virus. The way in which influenza virus manages to gather and keep track of its 7–8 genome segments is not known. Sometimes the virus makes a mistake and packages a "wrong" segment. The significance of this mistake for influenza virus evolution and for human health is described in Chapter 16. For now, suffice it to say that this is the origin of the strains of influenza virus that have been called "killer influenza" because they are no longer recognized by the human immune system.

Once the nucleoprotein core has been assembled, the next step is to acquire the **matrix proteins** and envelope. While viral particles have been assembling in the cytoplasm, the influenza virus envelope proteins, the hemagglutinin and the neuraminidase, have been inserted into the cell membrane. The viral particles begin to bud out of the cell through the cytoplasmic membrane, acquiring their envelope in the process (Fig. 10.8).

At this point, the virus faces one final problem. The hemagglutinin spikes that allowed the virus to bind so tightly to the mammalian cell surface in the initial steps of infection now become a potential burden, because, as the virus buds through the membrane, the hemagglutinin spikes can bind

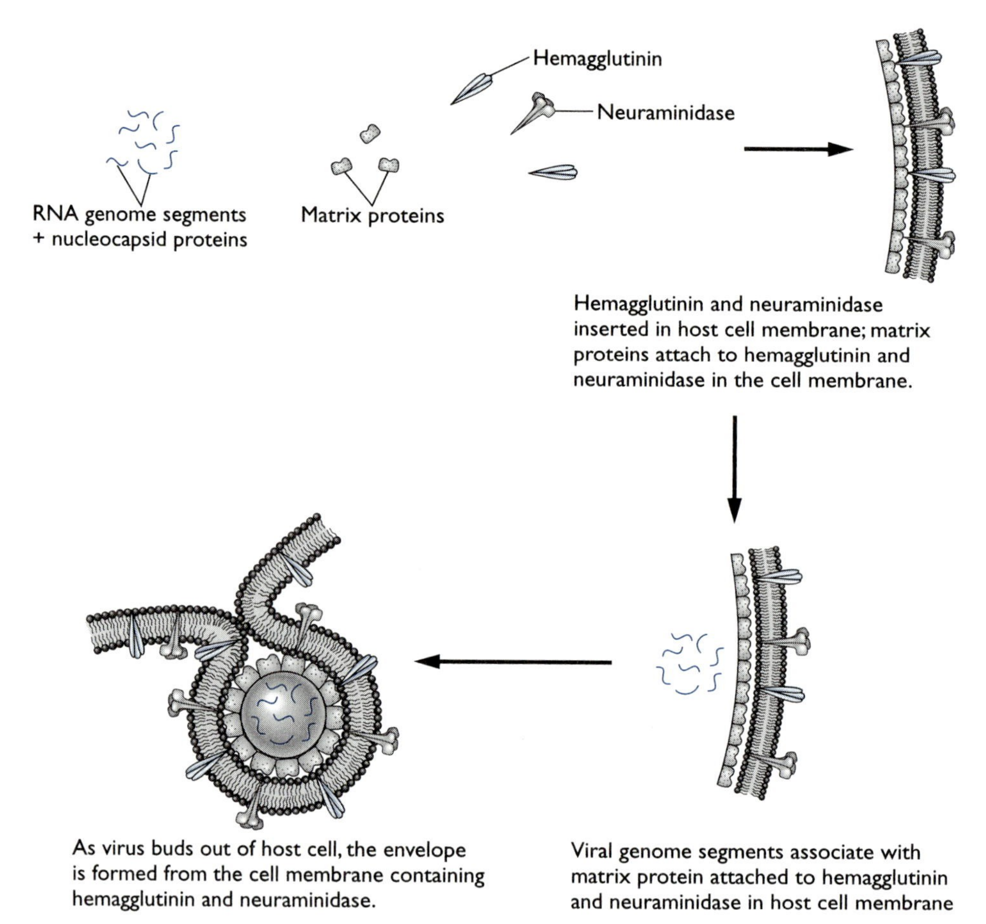

Figure 10.8 Budding of influenza virus through the host cell cytoplasmic membrane. Viral envelope proteins localize to the cytoplasmic membrane as they are made. Note that each genome segment has its own nucleocapsid proteins bound to the RNA segment rather than a standard capsid. These particles interact with viral matrix proteins and the envelope proteins as the virus begins to exit the cell. The virus acquires an envelope, complete with virally-produced envelope proteins, in the process of budding.

once again to sialic acid residues on the mammalian cell surface, tethering the virus to the cell it is trying to leave. The other envelope protein, neuraminidase, is thought to be the solution to this problem. **Neuraminidase** is an enzyme that cleaves the bond attaching sialic acid residues to other carbohydrate residues on the cell surface. This cleavage reaction helps to release the virus from the cell, so that the virus is now free to infect a fresh cell and repeat its replication cycle.

Antiviral Compounds

A review of the steps in viral replication shows that, although there are not many suitable targets for antiviral drugs, some do exist (Fig. 10.9). Stopping attachment of the virus to the host cell is a very attractive drug strategy, because if the virus never attaches to and enters the cell it will not be able to cause any damage. One way to stop attachment would be to provide mole-

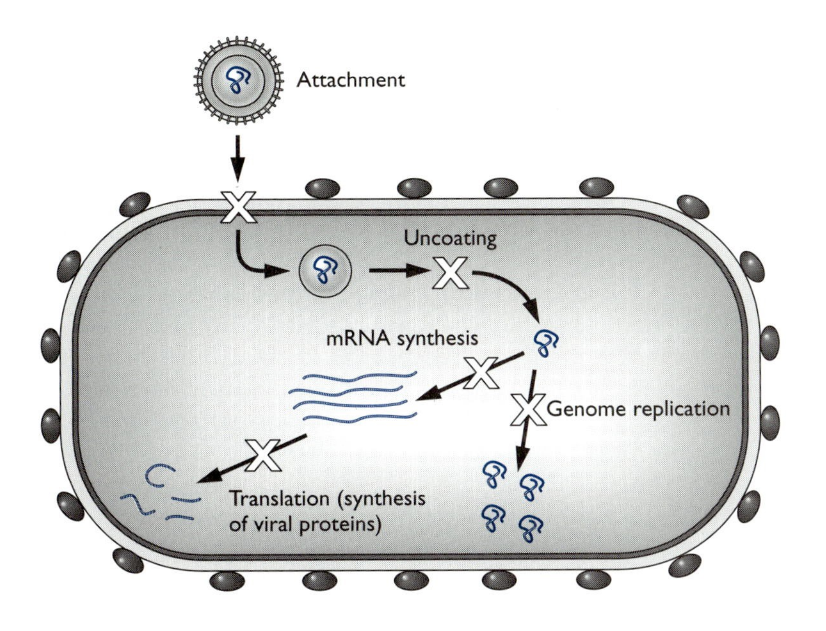

Figure 10.9 Targets for antiviral compounds. Xs mark the steps that are targeted by currently available viral compounds. Because viruses differ from each other in a variety of ways, an antiviral compound that works for one type usually does not work for other types of virus.

cules that resemble the viral receptor on the host cell surface enough to fool the virus into binding the receptor mimic instead of the cell itself. In the case of influenza virus, attempts have been made to use free sialic acid to prevent viral attachment. The problem with this approach is that the receptor mimic must be provided continuously and must have access to tissues the viruses are likely to infect. A more successful strategy for inhibiting attachment, vaccines that elicit an immune response against viral surface proteins, will be described in Chapter 14.

For viruses that are endocytosed and uncoat in endocytic vesicles, the uncoating process is a potential target. In fact, a successful anti-influenza drug, **amantidine,** inhibits uncoating of the virus. Amantidine does not interfere with the process of endocytosis—it would be toxic if it did—but instead binds to influenza virus matrix proteins and prevents the conformational change that is essential for uncoating.

For viruses with replication enzymes that are different from host cell enzymes, viral replication is a suitable drug target. Often the compounds used to inhibit viral replication enzymes are analogs of nucleotides, which prematurely terminate synthesis of viral genomes or viral messages or otherwise interfere with viral replication. Examples of such antiviral compounds and how they work are given in later chapters. A cornerstone of HIV therapy is a collection of such analogs that are preferentially incorporated by the HIV enzyme reverse transcriptase. Another unusual feature of the HIV replication process that serves as a target for a different type of antiviral drug is the formation of large polyproteins, which must be clipped apart by viral proteases to become activated (Chapter 18). The protease inhibitors used to treat HIV patients target this activity, not seen in human cells. Currently, viral attachment, uncoating, and replication are the main targets of antiviral compounds. It is conceivable that as more is learned about the replication cycles of viruses, new targets will be found.

Pathology of Viral Disease

Explaining symptoms

One factor that dictates the symptoms a virus causes is the type of host cell it infects. This specificity explains why hepatitis virus only infects liver cells and HIV targets certain cells of the immune system. In the case of influenza virus, the virus binds a receptor found on virtually all mammalian cells. Thus, some other factors must determine specificity because, in humans, influenza is normally a disease of the respiratory tract (Chapter 16). Birds can shed avian influenza virus from their intestinal cells, so clearly influenza is capable of attacking cells other than those of the respiratory tract. Why, then, is influenza usually localized to the cells of the human upper airway? One reason is that in humans the spread of influenza virus is by airborne transmission or by transmission of virus from hands to face and thence to the respiratory tract. Opportunity to encounter susceptible host cells is clearly a factor here. A second factor is probably the human immune response. Most people who encounter a strain of influenza virus have probably encountered a similar strain previously. Thus, the immune response kicks in early on, keeping the infection localized to the initial site of infection. It may not be completely effective at first if the new strain is not identical to the previously encountered one, but it may work early enough to allow the body to stem the infection. The symptoms of ordinary influenza are probably caused by the body's response to dying human cells (inflammation; Chapter 12). In an early response to infection, the human body mobilizes proteins called cytokines, which organize the defense responses of the body, together with cells that kill infected cells. This complex response to infection is the cause of generalized flu symptoms such as chills, fever, joint and muscle aches, and malaise. Thus, symptoms are not the result of viruses infecting the brain or the joints but result from the secondary effects of viral destruction of respiratory cells.

The importance of a timely immune response in limiting the spread of influenza virus is underscored by what happens when the human body encounters a strain of influenza virus that is completely new to the immune system. The influenza epidemic of 1918 and 1919, which killed millions of people worldwide is an example. Since that time, other outbreaks of "killer flu" have occurred with some regularity, but none quite so devastating or involving so many people. The cause of death in infected people is still somewhat controversial, because scientists and physicians at that time did not have the advanced diagnostic and analytical technologies we have today. Many of the deaths were undoubtedly caused by secondary bacterial infections made possible by the weakened defenses of the airway (Chapter 16). Nonetheless, some evidence suggests that the "killer flu" strains penetrated deeper into the lung than normal influenza infections and were thus more destructive. This could be explained by the delayed immune response that allowed the virus to gain access to susceptible cells it would not normally have had a chance to encounter.

Whatever the explanation for "killer flu," it is clear that the symptoms and severity of a viral disease depend on a constellation of factors. These include not only the specificity of a virus for a certain type of mammalian cell but also the route of transmission and the immune status of the person encountering the virus.

Outcomes of a viral infection

Symptoms of a viral infection are also influenced by the fact that there can be several possible outcomes of an initial infection event (Fig. 10.10). After

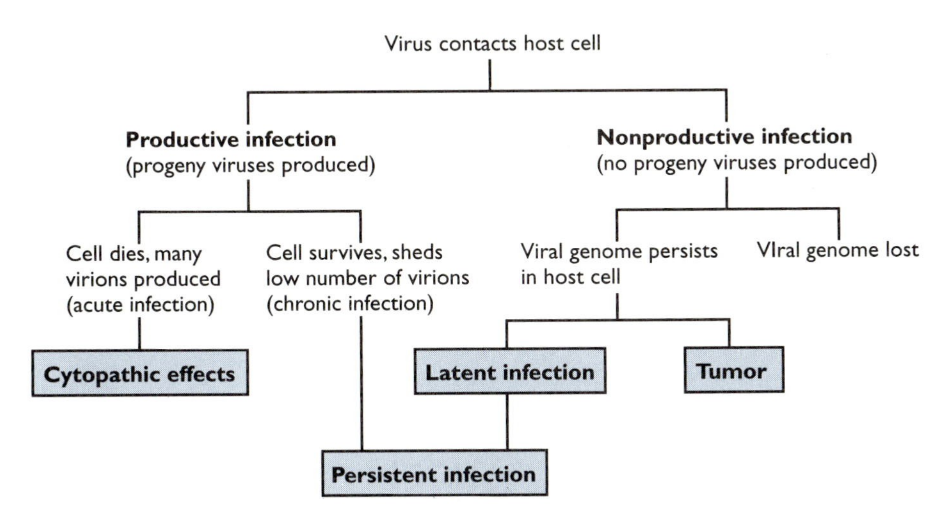

Figure 10.10 Possible outcomes of a viral infection.

initial entry of a virus into the body, the viral infection may remain localized, either because only certain cell types are affected (e.g., liver cells, gastrointestinal cells) or because the immune response quickly brings the infection under control. If the viral infection is cleared, symptoms will disappear. If viruses enter the bloodstream, the infection may spread throughout the body (systemic infection). Viruses can either spread as free viral particles in blood or by infecting the phagocytic cells that constantly move through the bloodstream (white cells; Chapter 12). These phagocytic cells are intended to be protective, but, if they become infected with viruses, they can actually help to spread the infection to distant parts of the body. This is true of HIV, which is capable of infecting a type of mobile phagocytic cell called the monocyte.

Both localized and systemic infections can become latent. A **latent infection** differs from clearance of viruses in the body in that, although symptoms have abated, the virus is still present. In the latent form, the virus is temporarily inactive and has been cleared from cells where it can replicate but is still capable of breaking out of the latent state to cause infection in the future. A good example of this latency is provided by herpes simplex virus, which causes lesions (cold sores, genital sores) when it is actively replicating in epithelial cells but has a latent stage in nearby neurons (Chapter 18). Emergence from this latent stage to reinfect epithelial cells prompts recurrences of herpes simplex symptoms.

The outcome of an infection caused by a particular virus can vary substantially from person to person (see Box 10–1 for some episodes that illustrate this dramatically). To some extent, this variation can be understood on the basis of the immune status of the person. A person who mounts a vigorous immune response is less likely to develop symptoms than a person who does not. Yet this cannot be the only explanation, because people with similar levels of immune competence can still exhibit great variations in the severity of symptoms. The number of viruses to which the person is exposed and the route of exposure are certainly important. Symptoms resulting from exposure to a lower number of viruses are more easily brought under control than are those from exposure to a higher number. Underlying conditions, such as concurrent infection with another microbe or conditions like diabetes that can weaken the immune system, can play an important role. The immune response may actually contribute to disease

box 10–1 *Two Episodes of Unintentional Inoculation with Live Viruses Illustrate the Range of Susceptibility to Viral Infections*

Case 1. The Cutter Incident. In the 1950s, before the development of a live polio vaccine, the polio vaccine consisted of formalin-killed viruses (Salk vaccine). A shipment of polio vaccine prepared by Cutter Laboratories in 1955 was not properly treated and contained live polio virus. Before this was discovered, nearly 120,000 schoolchildren were injected with the contaminated vaccine. Estimates based on a subsequent follow-up investigation and on information about the incidence of natural immunity in the U.S. population indicated that about half of the children inoculated with the vaccine already produced antibodies against the polio virus and were protected from contracting the disease. Approximately 25% of the remaining children were infected, as indicated by the appearance of symptoms or shedding of the virus in feces, but only 60 cases of paralytic polio (the most severe form of the disease) actually developed.

Case 2. Yellow fever episode. During World War II, yellow fever was a serious problem for U.S. troops sent to countries where the disease was endemic. Most military personnel were vaccinated to prevent yellow fever. In 1942, a shipment of vaccine that had been unintentionally contaminated with hepatitis B virus was administered to 401,535 military personnel. Of these, only 914 (about 0.2%) developed jaundice, and fewer than 50 developed the severe form of hepatitis.

When considering the figures quoted above, it should be kept in mind that the amount of virus actually injected in these cases was low. Exposure to higher levels of virus, as might happen during an epidemic, would result in a higher incidence of disease. What these cases illustrate, however, is the broad range of susceptibility even in populations that are relatively homogeneous with respect to age and standard of living.

Sources: Nathanson, N., and A. Langmuir. 1963. The Cutter incident. Poliomyelitis following formaldehyde-inactivated poliovirus vaccination in the United States during the spring of 1955. Amer. J. Hyg. **78**:16–81.

Sawyer, W., et al. 1944. Jaundice in army personnel in the western region of the United States and its relation to vaccination against yellow fever. Amer. J. Hyg. **39**: 337–430.

severity if the immune response mounted against an initial infection is inappropriate and exacerbates symptoms the second time the person is exposed. This seems to be the case with dengue virus infections, in which the second exposure produces worse disease than the first (Chapter 20). Such aberrant immune responses are uncommon, but they illustrate the complexity of the range of factors and conditions that mix together to determine the severity of symptoms.

Disease severity also may have a genetic component. For example, people with genetic defects in the gene encoding the coreceptor needed by HIV for effective attachment to human cells take longer to develop symptoms or may not develop symptoms at all (Chapter 18). Is this an HIV-specific phenomenon or are there genetic factors that affect the outcome of other viral infections? We need a better understanding of person-to-person variations in the severity of symptoms of viral diseases.

Why Aren't We Doing Better?

It is perhaps understandable that our attempts to understand disease-causing bacteria and eukaryotic microbes have foundered on the realization that these are formidable enemies—not only because of their adaptability but because they are so complex, both genetically and metabolically. Failures in this area are easy to rationalize. Yet, how can we rationalize similar failures to deal with microorganisms so simple that even the most primitive genome sequencing facility could sequence their genomes completely in a couple of weeks at the most? The answer is that their apparent simplicity is highly misleading. As isolated entities, these microorganisms are simple in the sense that they are composed, in most cases, of a very short shopping list of genes and proteins. What allows this apparent simplicity to balloon into complexity as great as that of the cells the viruses infect is the ability of viruses to take advantage of characteristics of eukaryotic cells in ways that can produce very complex responses. To revert to the words of Jonathan Swift at the beginning of the chapter, just as a flea biting a person can touch off all sorts of responses, from mental feelings of itching and irritation to results of infectious agents injected by the bite, something as simple as a virus has only to tweak a mammalian cell in the right way to elicit a cascade of responses, which, if understood, would lead to a far deeper comprehension of both viral and mammalian cell function.

chapter 10 at a glance

Components of mammalian viruses

Component	Composition and function
Nucleoprotein core	Contains viral genome, proteins needed for replication of viral genome (if genome cannot be translated directly) Examples of proteins: replicase that replicates –RNA strand, reverse transcriptase (HIV) that makes DNA copy of an RNA genome
Capsid	Tightly packed protein layer covering the nucleoprotein core Protects nucleoprotein core Mediates attachment to host cell receptor in case of naked virus
Matrix proteins	Proteins that attach envelope to capsid in enveloped viruses Role in uncoating
Envelope	Phospholipid bilayer containing envelope proteins Envelope proteins mediate attachment to host cell receptor

Viral classification (see Table 10.1)

Replication cycle

Step in replication	Influenza virus	Other viral strategies
Attachment to host cell	Envelope hemagglutinin protein spike attaches to sialic acid residues on host cell surface	Different viruses have different surface proteins that recognize specific host receptor molecules; interaction is a determinant of which cells are infected by virus
Uncoating, release of nucleoprotein core	Virus taken up in endocytic vesicle, fuses to vesicle membrane, nucleoprotein core released into cytoplasm	Some viruses fuse and uncoat without vesicle formation; others trigger envelope endocytosis like influenza virus
Replication of viral genome	Nucleoprotein core enters host cell nucleus Preformed replicase makes +strand (mRNA) copies of –strand genome Replicase acquires 5' cap structure from host mRNA	Most DNA viruses replicate their genomes in the nucleus; most RNA viruses replicate in the cytoplasm (influenza virus is an exception); Some use preformed replication enzymes, others do not (see Fig. 10.6)
Assembly of viral particles	Viral RNA genome segments and proteins assemble in cytoplasm; envelope proteins localize to patches of host cell cytoplasmic membrane	Some viruses assemble viral particles in nucleus
Exit of viruses from infected cell	Virus capsid buds through cytoplasmic membrane, acquiring an envelope Neuraminidase aids in release of budding viruses from cell	Some viruses bud through other membranes or do not bud at all (naked viruses), but lyse the cell to allow viral particles to escape

chapter 10 at a glance

Targets for antiviral compounds

Prevent attachment (Chapter 14)

Prevent uncoating (amantidine)

Interfere with replication of genome (nucleotide analogs)

Interfere with translation or activation of viral proteins (HIV; Chapter 18)

Pathology of viral infection

Most symptoms caused by host response to death of infected cells, which elicits an inflammatory response

Site of infection, extent of infection determined by:
- Types of host cells recognized by viral surface proteins
- Route of inoculation
- Rapidity and effectiveness of host immune responses

Outcomes of infection:
- Virus cleared without symptoms
- Localized symptoms
- Systemic infection
- Latent infection (virus still present but not actively replicating)

key terms

antiviral compounds not many targets for drugs; most suitable targets are the attachment step, the uncoating process in some viruses, and viral replication enzymes that are different from host enzymes

apoptosis type of cell death in which cell shuts down its biosynthetic functions; some viruses stimulate the process, others prevent it from occurring in order to complete their replication cycles

capsid protein coat that covers the nucleic acid-protein core of viruses; made up of many copies of one or a few polypeptides (capsomeres); gives viruses their shape and protects the genome from degradative enzymes in the environment

envelope protein-phospholipid layer that covers the capsid of many viruses; contains phospholipids from host cell membrane and viral glycoproteins that mediate attachment of the virus to host cell surface receptors; the envelope of influenza virus contains hemagglutinin and neuraminidase, two proteins that are the target of influenza vaccines

latent stage phase of viral replication in which the virus survives in cells but does not replicate; under certain circumstances latent viruses can become reactivated and start the infection cycle again

matrix proteins proteins that link the envelope to the capsid thus helping to stabilize the viral particle; interact with the viral proteins embedded in the envelope

minus (–) strand, plus (+) strand minus (–) strand is the complement of what will be the viral message; the plus (+) strand is the message

nucleoprotein core complex made up of the viral genome and proteins; covered by the capsid; the proteins may have a structural function or an enzymatic function; the genome consists of DNA or RNA and may be single-stranded or double-stranded, depending on the virus; the genome composition determines the mechanism by which the virus replicates

replication cycle basic steps in the viral replication cycle include: (1) attachment to the surface of the cell it will infect, (2) uncoating (release of the nucleoprotein core into the host cell cytoplasm), (3) replication of the genome and production of viral proteins, (4) assembly of new viral particles, and (5) release of new viral particles by budding or lysis of host cell

replicase preformed enzyme brought into the infected cell by influenza virus; duplicates the –strand to form viral transcripts and makes copies of the viral genome

reverse transcriptase an enzyme that catalyzes the formation of a DNA copy of an RNA genome of a virus; found, for example, in HIV

Questions for Review and Reflection

1. Viral attachment is obviously a key first step in a viral infection. How could this step be prevented? (You will be in a better position to answer this question after learning in Chapter 13 about the immune response to viral infections, but certain theoretical possibilities present themselves at this point.)

2. Based on what you know about the uncoating process, how might uncoating be prevented? Is it likely that a single antiviral drug would be able to prevent uncoating by more than one type of virus? Is there any step in common to most viruses that could be targeted to obtain broader spectrum antiviral compounds acting at this step?

3. Why do some viruses carry preformed enzymes in their nucleoprotein cores, whereas others do not? How does the viral replication strategy influence this aspect of virus design? Give specific examples for RNA viruses (+ strand, –strand) and DNA viruses.

4. If a successful virus can have a genome as small as 4 kbp, why are there viruses with genome sizes nearly 100-fold larger? Given that all viruses tap into the biosynthetic machinery of eukaryotic cells, why would some need genomes that encode many genes?

5. What are the possibilities for variation at each stage of the virus replication cycle? Can you explain from this analysis why there is no such thing as a "typical" virus?

6. What is latency and what is its practical significance? How might a virus in its latent stage differ from one in its actively replicating stage?

7. Many viruses do not have a latent stage. How do you explain this? What are the properties you would expect of a virus capable of latency? Does the answer to question 4 have any bearing on this issue?

8. Explain how a single simple virus could have the range of effects on different individuals seen in real life? How do differences in the virus and differences in the host affect this spectrum of effects?

9. The current mode of viral classification, based on viral structure, genome composition, and similar features, places viruses with such diverse disease effects as polio virus (neurological effects) and rhinovirus (common cold) in the same family. What does this say about viral classification schemes? (Hint: Don't jump to the conclusion that current classification schemes are useless. Look deeper.)

the delicate balance II between microbes and humans

By the time plants and animals began to evolve, microbes were already well in charge of planet Earth. Constant contact with millions of microbes was an inevitable fact of life from the start. Given this, it is not surprising that the plants and animals that managed to survive and reproduce reached some equilibrium with the microbial world. Many of the later-appearing life forms actually managed to gain some advantages from the microbes they encountered.

One successful strategy was to enlist neutral or beneficial microbes as defenses against microbes that might be harmful. This strategy, as played out in the human body, is described in Chapter 11. A second strategy for coping with microbes was to evolve a complex network of defenses to prevent microbes from gaining access to interior tissues and organs. These defenses are the subject of Chapters 12 and 13. Chapters 11–13 explain why, in a world teeming with microbes, microbial disease is a relatively uncommon experience for most of us. These chapters also explain why medical practices that impair these defenses transiently, in order to treat cancer or prevent rejection of a transplanted organ, may increase substantially the risk of microbial infection. Humans have learned how to bolster artificially the natural defenses of the human body. Chapter 14 explains how vaccines work to stimulate the immune system, providing immunity without the danger of acquiring an actual infection.

11 Microbial Populations of the Human Body

Philosophers and theologians regard humans as the crown of creation, but to the microbes we are no more—and no less—than the proverbial free lunch.

How would you characterize the human body? Most people would say that because human cells are eukaryotic cells, then humans must be eukaryotic organisms. This answer is not wrong, of course, but it is not entirely correct. Consider the fact that the bodies of humans and animals normally carry at least 10 times more prokaryotic cells than eukaryotic cells. Prokaryotes are smaller than human cells, so the human cells account for a far greater volume of tissue than the prokaryotic cells, but the prokaryotes are still a significant component of the body. These prokaryotes are not just contaminants; they play an important protective and nutritional role. The microbial populations that begin to develop in various body sites shortly after birth and remain in these sites throughout life are called the **microbiota** of the human body. (Some scientists use the term "microflora" rather than "microbiota," but, because prokaryotes are not flowers, the term "microbiota" is technically more correct.)

Getting to Know Your Prokaryotic Side: The Microbiota of the Human Body and Its Effects

Each part of the body that normally harbors microorganisms has a distinctive population of microorganisms adapted to the features of that specific site (Fig. 11.1). This will become evident as we consider the characteristics of different sites and of the microbes that normally reside in them. The microbiota of a site normally plays a protective role, in the sense that it occupies sites and consumes nutrients that might otherwise support the growth of disease-causing microbes. On occasion, however, members of the microbiota can cause disease. One way this can happen is for the composition of a microbial population to shift. For example, in **periodontal (gum) disease,** the population of bacteria occupying the gums shifts from being predominantly gram-positive, with some gram-negative members, to being predominantly gram-negative. The predominantly gram-negative population causes inflammation that can become severe enough to lead to tooth loss. The microbiota of a site can also cause disease if breaches in the defenses of the site allow microbes access to tissues and blood, sites that are normally

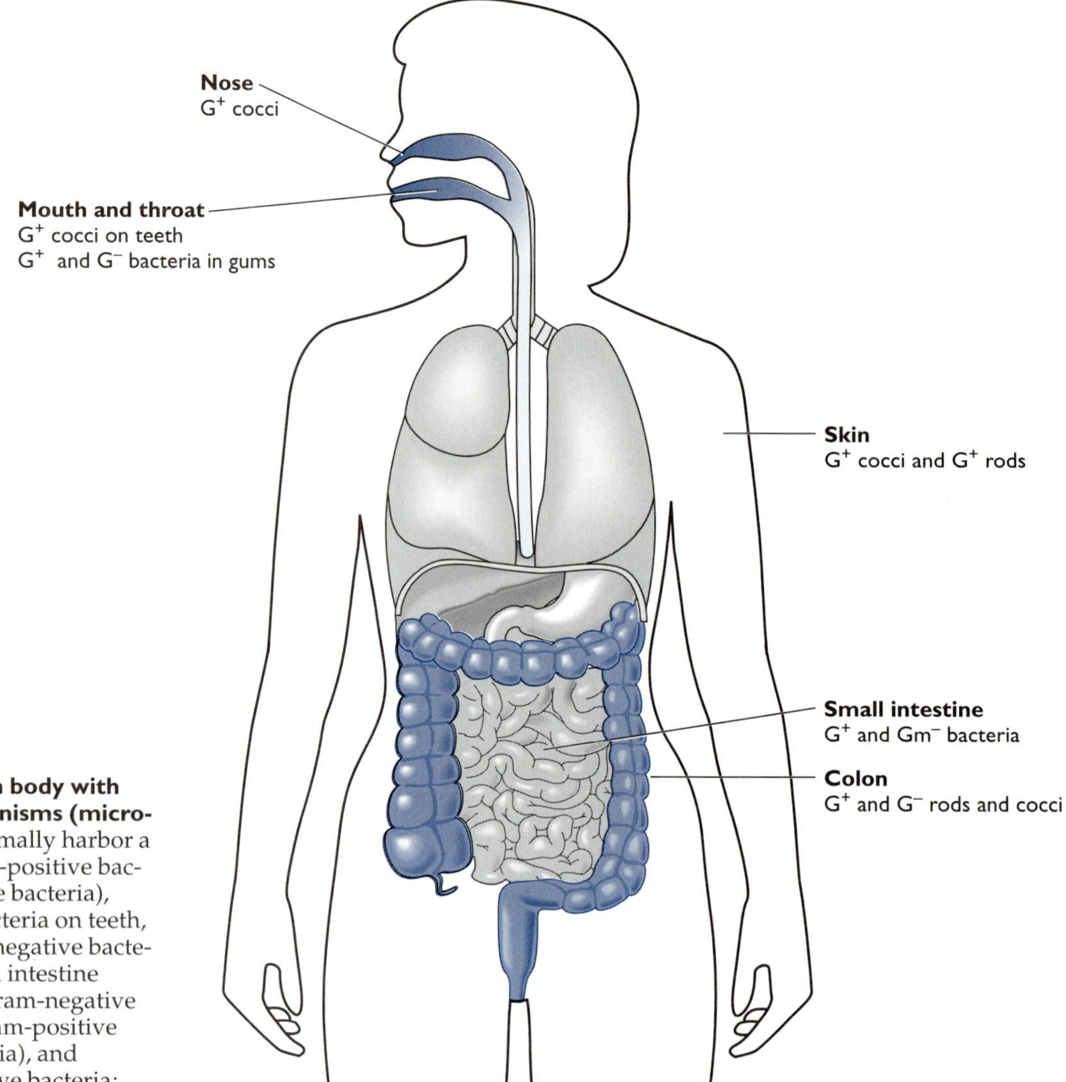

Figure 11.1 The human body with its associated microorganisms (microbiota). The areas that normally harbor a microbiota are skin (gram-positive bacteria), nose (gram-positive bacteria), mouth (gram-positive bacteria on teeth, gram-positive and gram-negative bacteria in gingival area), small intestine (few gram-positive and gram-negative bacteria), colon (many gram-positive and gram-negative bacteria), and vaginal tract (gram-positive bacteria; not shown).

free of microorganisms. For hospitalized patients, the microbiota of the body can be a major threat, in the form of postsurgical infections. These different aspects of the microbiota will be illustrated by the examples in this chapter.

The Microbiota of Human Skin

Features of skin

The surface of the skin does not provide a very hospitable environment for microorganisms. It consists of dead cells and is quite dry. Microorganisms prefer wetter environments with food sources that are easier to digest than dry dead cells. Also, the skin surface is normally slightly acidic, a condition that discourages the growth of many disease-causing microorganisms, which usually are adapted to growing at the pH found in most tissues (pH 7). Some microorganisms can colonize skin, however, and these tend to be benign or neutral organisms. From the body's point of view, it makes sense to limit the number of microorganisms on the skin without attempting to keep the skin completely free of microorganisms (even if this were possible). Sooner or later some microorganism would adapt to growth in this site, and it might not be benign. The protective strategy of encouraging certain benign microbial species to occupy a site is seen in all parts of the body that are accessible to microbes.

Features of the skin microbiota

The microbiota of the skin consists primarily of gram-positive bacteria. Two prominent examples are *Staphylococcus epidermidis*, a gram-positive coccus, and *Propionibacterium acnes*, a gram-positive rod. *P. acnes* is an obligate anaerobe, a fact that seems surprising at first. Yet the skin has sites, such as pores or glands, where oxygen concentrations are much lower than at the surface. *P. acnes* got its name from the suspicion that it plays a role in causing acne. Although *P. acnes* is found in acne lesions and antibiotic treatment can cure acne, the role of this particular species is still somewhat controversial.

Identifying *P. acnes* as a possible cause of acne might seem to raise questions about earlier assertions to the effect that the members of the normal microbiota are generally benign or neutral. Keep in mind, however, that acne is not a life-threatening infection of the kind that the body needs most to prevent. Also, *P. acnes* inhabits some sites that might be colonized by far more dangerous microorganisms. The lingering mystery is: why, if *P. acnes* does cause acne and if virtually everyone's skin carries *P. acnes*, do only some people develop acne?

S. epidermidis is more worrisome, because it can cause serious disease under certain conditions. *S. epidermidis* can cause life-threatening infections in hospitalized patients who have nutrients or pharmaceuticals introduced into their bloodstreams through needle-inserted plastic catheters. Biofilms of *S. epidermidis* form and grow along the catheter surface into the bloodstream. Careless surgeons can also contaminate plastic implants with *S. epidermidis* from their hands. On the implant, the bacteria form a biofilm that is virtually impossible to treat with antibiotics. Damage caused by the bacteria often necessitates a second surgical procedure to remove the contaminated implant and replace it with a sterile one.

S. epidermidis sounds like a pretty bad actor, until you realize that the human body did not evolve to prevent catheter-associated infections. Catheters are a recent innovation introduced by modern medicine. *S. epidermidis* had a clean bill of health until catheters and surgery came along. *S. epidermidis* is an excellent example of a theme introduced in Chapter 1: changes in human practices can provide new windows of opportunity for microbes. It is impressive that only a short time after plastic was invented and used in catheters and implants, *S. epidermidis* adapted to attach to plastic and use it as a vehicle for entry into the body. Manufacturers are developing catheters impregnated with antibacterial compounds, such as silver or antibiotics, in the hope of discouraging bacterial biofilm formation. Unfortunately, *S. epidermidis* is becoming increasingly resistant to many antibiotics.

The Microbiota of the Human Nose

Not much is known about the microbiota of the nose, except that it is predominantly gram-positive and contains some of the same organisms found on skin. The nose is an important site medically, because it is the place where *Staphylococcus aureus*, a very serious cause of infection, is most likely to be found. From the nose, *S. aureus* can be transferred to the skin, where it colonizes transiently before being displaced by skin microbiota. Or *S. aureus* can move from the nose of a food handler into food. About one-third of *S. aureus* strains are capable of producing a protein toxin, **enterotoxin,** which, if ingested, causes violent vomiting and intestinal cramps. This condition is rarely fatal but is highly unpleasant and is not the best way to remember a restaurant visit or a family gathering. The gloves now worn routinely by food handlers are designed to prevent transfer of bacteria from the body of the food handler into the food being prepared.

Carriage of *S. aureus* by medical personnel is far more serious, because *S. aureus* is a major cause of surgical wound and bloodstream infections in hospitals and other medical settings. This aspect of *S. aureus* has given rise to grim jokes about "hospital staph." About a third of the human population carries *S. aureus* at any given time. It would be a very useful preventive measure if microbiologists could figure out how to eliminate *S. aureus* from the nose, especially those strains that are resistant to many antibiotics. An antibiotic, mupirocin, is used to clear the nasal carriage of *S. aureus* by some hospital workers who work with highly susceptible patients. Monitoring such workers for carriage of *S. aureus,* however, is not standard practice in most hospitals. More information about the microorganisms normally found in the nose might help in the design of a strategy to use the microorganisms themselves to prevent *S. aureus* from colonizing.

The Microbiota of the Human Mouth

Development of the oral microbiota

The first microbes to colonize the mouth of an infant are gram-positive cocci, such as *Streptococcus salivarius. Lactobacillus* species (gram-positive rods) and some anaerobes are also present at low levels. The microbial population of the tongue and oral tissue is somewhat different from the population that develops on teeth, because of the different surfaces involved. Although the tissue of the mouth might seem like an aerobic site, obligate anaerobes can thrive there, too. When the teeth begin to appear in an infant, they provide new opportunities for microbial colonization, not only on the tooth surface itself but also in the **gingival crevice,** where the tooth emerges from the gum. Whereas the tooth surface is a relatively aerobic site, the periodontal pocket is mostly anoxic, and this difference influences the types of microbes that colonize each area. Gram-positive aerotolerant anaerobes, such as *Streptococcus sanguis, Streptococcus mutans,* and even a few obligate anaerobes, such as *Actinomyces* species, predominate on the teeth. The bacteria do not attach directly to the teeth but to the thin layer of salivary glycoproteins that covers the teeth. As the child grows older, the complexity of the biofilm communities on teeth increases.

Many of the bacteria normally found on teeth are lactic acid bacteria. Their role in dental caries was explained in Chapter 8. In the mouth, the numerically predominant lactic acid bacteria are *Streptococcus* species, such as *S. sanguis* and *S. salivarius.* These and other *Streptococcus* species colonize not only tooth surfaces but also the surfaces of the tongue and inner cheek. A much more threatening relative of these bacteria, *Streptococcus pneumoniae,* the most common cause of earache in children and pneumonia in adults, can colonize the mouth. It is not as successful at colonizing as the streptococci that normally occupy the mouth, but about one-fourth of the population carries *S. pneumoniae* in the mouth or throat. These colonized people are at risk for pneumonia and other diseases caused by *S. pneumoniae* because the first step in infection is colonization of the mouth and throat. This places the bacteria in a position to take advantage of breaches in the defense system that protects the lungs.

A serious case of the viral disease influenza can create such a breach, leading to the introduction of *S. pneumoniae* into the lung. This is why patients who develop pneumonia are often found to have recently had influenza. To the extent that the normal oral streptococci keep dangerous pathogens from colonizing the mouth and throat, they also help to protect us from lung disease.

Actinomyces species, the anaerobes found in the mouth, have an interesting shape. They are definitely prokaryotes and have a gram-positive cell wall, but they exhibit a branched appearance similar to that of fungi (Fig. 11.2). Relatives of the *Actinomyces, Streptomyces* species (which unlike the *Actinomyces* are obligate aerobes) live in the soil. Like fungi, the *Streptomyces* species are major producers of antibiotics.

Microbiota of the gingival area

The gingival area (where the tooth erupts from the gum) is normally colonized by a mixture of gram-positive and gram-negative bacteria, which are either aerotolerant or obligate anaerobes. Gingival bacteria form plaque on the root surfaces of teeth (subgingival plaque). In some people, if plaque growth continues, the microbial population begins to shift from a mixed gram-positive and gram-negative population to being exclusively gram-negative. The species composition shifts, and spirochetes begin to appear. This new

Figure 11.2 Morphology of *Actinomyces* species (A) and fungi (B). The filamentous morphology of *Actinomyces*, a gram-positive bacterium, is similar to that of the mycelial form of fungi.

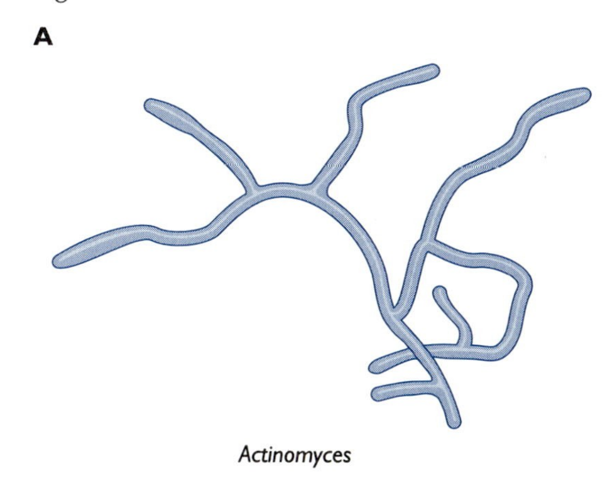

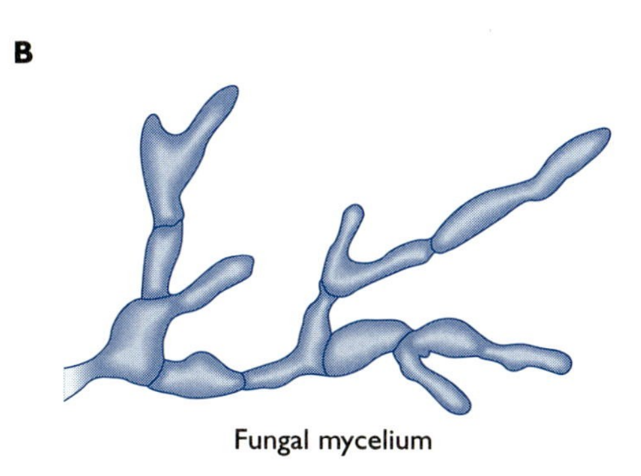

population of bacteria produces proteases that destroy gum tissue. The resulting inflammatory response gives rise to bleeding gums and eventually to receding gums and tooth loss. Some of the bacteria associated with periodontal disease are gram-negative anaerobic rods, such as *Porphyromonas gingivalis, Prevotella intermedia, Fusobacterium* species, or spirochetes, such as *Treponema denticola.*

It is not clear why periodontal disease predominantly develops in adults. Although there is a form of periodontal disease that occurs in children (juvenile periodontitis), the vast majority of periodontal disease cases occur in adults 40 years of age and older. Most people do not change their diet or living habits drastically between the ages of 20 and 40, so there may be factors associated with aging that contribute to the microbiota shift that characterizes the disease. It has been suggested that periodontal disease might be transmitted by kissing.

Although most of the microbes in the mouth are bacteria, some yeasts are also found there. One is *Candida albicans. C. albicans* is usually seen in the ovoid yeast form and reproduces by budding, but it can produce long appendages called pseudohyphae. Although *C. albicans* is usually benign, it can cause disease in people who are taking antibiotics or are immunocompromised. Because *C. albicans* is a normal denizen of the human body, it is not surprising that this yeast is a common cause of disease in people with human immunodeficiency virus and in cancer patients.

Thrush is a mild form of oral *C. albicans* infection that is sometimes seen in children whose microbiota is not fully developed and, less frequently, in adults who have taken many antibiotics. A whitish layer, consisting of dead human cells, appears on the tongue and throat. The most serious form of candidiasis, seen almost exclusively in people who are immunocompromised, is the systemic form, in which yeasts enter the bloodstream and spread throughout the body (systemic candidiasis). Systemic candidiasis can be deadly.

The Microbiota of the Stomach and Small Intestine

Is there a microbiota of the stomach?

For a long time, physicians believed that the human stomach was free of microbes, reasoning that no microbe could survive in such an acidic environment (pH 1–2). This is not necessarily true. The stomachs of mice contain a normal microbiota consisting of a bacterium, *Lactobacillus,* which colonizes one part of the stomach, and a yeast, *Torulopsis,* which colonizes the other part. Although the contents of the stomach have a very low pH, these microbes are taking advantage of the fact that the mucin layer, which covers the lining of the stomach and protects that lining from stomach acid, has a near-neutral pH. So far, no similar microbes have been identified in the human stomach. However, a bacterium has been discovered that could almost be called a member of the normal microbiota of the human stomach: the gram-negative bacterium, *Helicobacter pylori,* now known to be the cause of most gastric ulcers. In the developing world, most members of the human population carry *H. pylori* in their stomachs without developing disease. Although the carriage rate is lower in the developed world, many people carry this species by the time they are 40–50 years old. The fact that many people carry *H. pylori* without showing signs of disease has prompted microbiologists to suggest that it might be considered the normal microbiota of the human stomach. This fascinating microorganism will be described in Chapters 15 and 17.

The microbiota of the small intestine

For those microorganisms that prefer to grow at neutral pH values, conditions in the small intestine are definitely an improvement over those in the stomach. The pH of the small intestinal environment is much closer to neutrality than that of stomach contents. In the small intestine, however, microorganisms encounter another problem: the rapid flow of intestinal contents through the intestine. This rapid flow washes microbes out of the area, except for those able to adhere to the wall of the small intestine.

The body uses this strategy to give itself the first crack at the most digestible nutrients in the diet, such as simple sugars, amino acids, and fats. Microorganisms, especially bacteria, find these same nutrients very appealing. Human intestinal cells would be losers in a contest for sequestration of these nutrients if high numbers of microorganisms were allowed to build up in the lumen of the small intestine. The highest concentrations of bacteria in the small intestine are seen where the intestinal flow slows as intestinal contents reach the ileocecal valve that separates the small intestine from the colon. At this site, however, concentrations of bacteria are still relatively low ($<10^6$ per mL).

The Microbiota of the Human Colon

Characteristics of the colonic microbiota

Location of the bacteria. In the mouth and small intestine, most bacteria are attached to surfaces because otherwise salivary flow or rapid movement of contents would wash them out of the site. Attachment to the intestinal wall is not necessary in the colon, however, because the flow of contents through the lumen of the colon is slow; slow enough that microbes can easily divide rapidly enough to remain in the site. Moreover, nutrients are abundant. Not surprisingly, the concentrations of bacteria in the colon are very high, 10^{12}/g of contents. Microbes are packed in the lumen of the colon, accounting for about 30% of the contents of the colon. The colon is clearly designed to be a holding tank for bacteria, much like the rumen in cattle. (The microbiota of the human colon was described in Chapter 3.) Virtually all of the colon microorganisms are bacteria, predominantly gram-positive and gram-negative obligate anaerobes. Some methane-producing archaea are present, but at lower levels.

Although attachment to the wall of the colon is not essential for microorganisms to remain in the site, attachment to surfaces is still helpful to many of them. Incompletely digested plant fragments in the colon are covered with bacteria. There also may be bacteria that reside mainly in the mucin layer, which lubricates the intestine and prevents most microbes from reaching the surface of the colonic mucosa. Information about the distribution of microbes in the mucin layer and on the mucosal surface is incomplete, because of the difficulty in safely obtaining samples from healthy people without harming them. Thus, it is still unclear whether there is a special mucin-associated microbial community that differs from the microbial community in the lumen of the colon.

The colonic fermentation. In contrast to the mouth, where the main carbon source is simple sugars (for example, sucrose) or protein (in the case of the periodontal pathogens), the main source of carbon and energy for colonic bacteria is polysaccharides (Fig. 11.3). These are the polysaccharides in the human diet that cannot be digested with human intestinal enzymes.

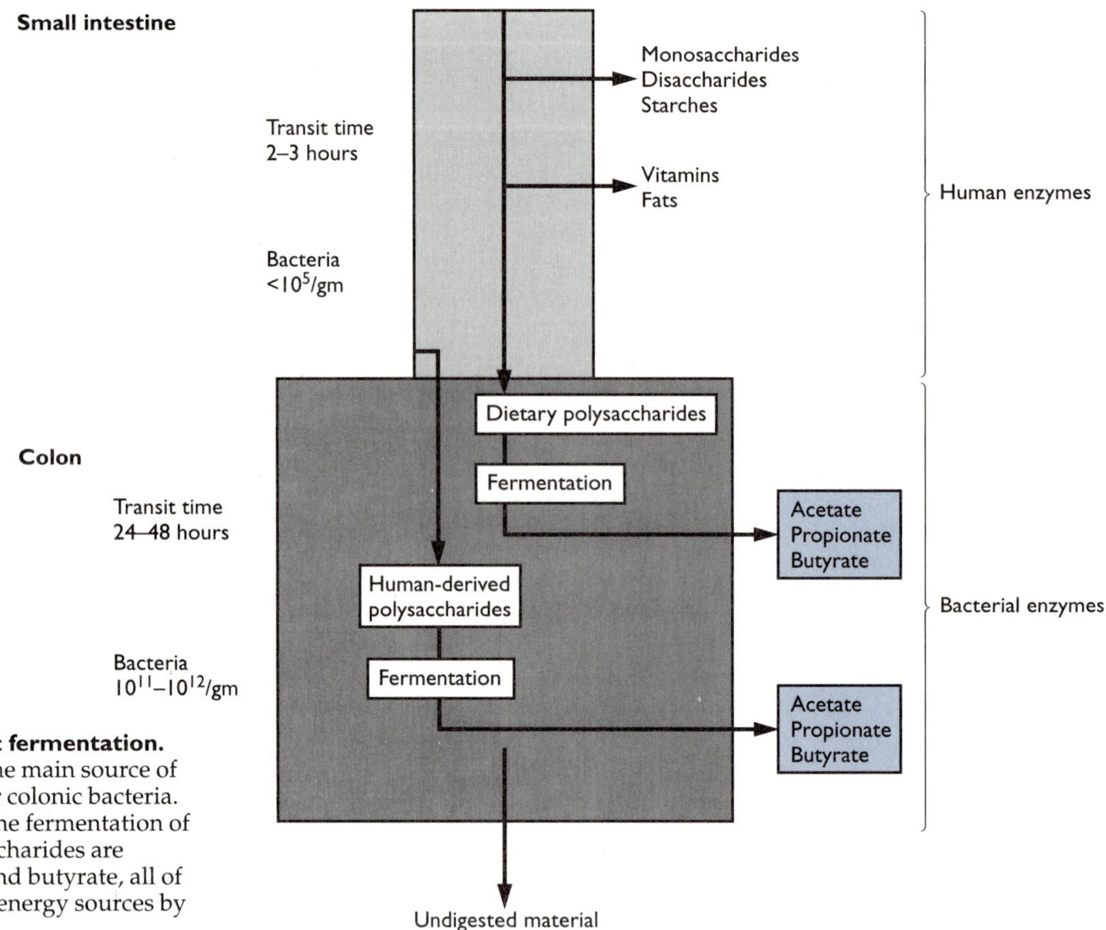

Figure 11.3 Colonic fermentation. Polysaccharides are the main source of carbon and energy for colonic bacteria. The end products of the fermentation of the degraded polysaccharides are acetate, propionate, and butyrate, all of which can be used as energy sources by colonic mucosal cells.

They include plant polysaccharides, such as **cellulose, xylan,** and **pectin.** Also entering the colon are human-produced polysaccharides, such as **mucins** and **mucopolysaccharides** (the polysaccharide glue that holds human cells together and is released when intestinal cells slough). These polysaccharides are readily degraded by colonic bacteria.

The end products of polysaccharide fermentation, mainly acetate, propionate, and butyrate, are absorbed by the cells of the colon mucosa and used as sources of carbon and energy. Thus, the colon can be considered to be a second organ of digestion, an organ that consists primarily of prokaryotic cells. The amount of energy the human body derives from the **colonic fermentation** has been estimated to be about 7%–10% of the total energy obtained from the diet. Interestingly, this is about the same amount of energy devoted to replacing intestinal mucosal cells, one of the most rapidly dividing cell populations in the human body. Because the rapid replacement of these cells is a defense against bacterial colonization of the intestinal wall, one might say that the bacteria are paying for the energy toll they take on the human colon.

For today's humans, especially those in developed countries, who, if anything, are over-nourished, the colonic fermentation is not of much importance. But in ancient times and in some countries where the available diet is high in plant polysaccharides and low in easily digestible substances, the colonic fermentation may represent the margin between survival and extinction. Moreover, there is some evidence that colon bacteria produce vitamins that contribute to human nutrition. The full story of the interaction between colonic bacteria and the human body is a long way from being told. The colon and its contents are a part of the human body about which little is known.

Composition and development of the colonic microbiota. Scientists have now established the identity of the major bacterial species in the colon. *Bacteroides* species are gram-negative anaerobes that account for about 25% of colon bacteria and are thought to play an important role in the colonic fermentation of polysaccharides. The other numerically predominant colonic bacteria are the gram-positive anaerobes, such as *Eubacterium, Peptostreptococcus,* and *Clostridium* species. As stated in Chapter 3, the colon is a very anoxic place, a fact that explains why more than 95% of colon bacteria are obligate anaerobes.

In infants, whose colons are free of bacteria at birth, the first bacteria to appear in the colon are oxygen-utilizing bacteria such as *Escherichia coli.* Once these bacteria are established, they render the colon anoxic enough to allow anaerobes like *Bacteroides* species to move in and take over. It takes about two years for the colonic microbiota of a child to reach its final adult state. Where do the bacteria that colonize the infant's colon come from? Most likely, the source is the parents and others who care for the child. Even the most fastidious people will have some colonic bacteria transiently on their hands. Although most of the colonic bacteria are obligate anaerobes, they can survive brief exposure to oxygen. Also, the infant's stomach is not as acidic as that of an adult, allowing more ingested bacteria to enter the intestine alive.

The period during which the microbiota is developing in an infant is a window of opportunity for pathogenic bacteria (Fig. 11.4). Spores of *Clostridium botulinum,* the cause of botulism in adults, are commonly found in such foods as honey. These spores pass harmlessly through the adult colon, because, even if the spores germinate, they could not compete with the colonic microbiota. In the infant colon, when there is less competition, the spores can sometimes germinate and produce botulism toxin, a potentially deadly neurotoxin. The toxin is absorbed from the colon and can cause a fatal paralytic disease called **infant botulism.** Fortunately, infant botulism remains a rare disease, but it illustrates the protective barrier provided by the mature form of the colonic microbiota.

Figure 11.4 Steps leading to infant botulism.

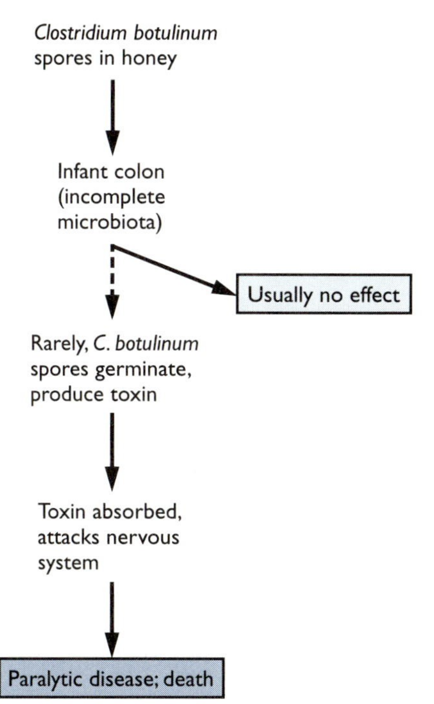

Another disease that illustrates the protective role of the normal microbiota is a disease of adults, **pseudomembranous colitis** (Fig. 11.5). People taking certain antibiotics that are particularly effective against colonic bacteria, can experience dramatic decreases in the numbers of the normally predominant anaerobes. Under such conditions, *Clostridium difficile,* which is normally present only in low numbers, can overgrow and produce two potent toxins that cause such severe damage to the lining of the colon that death can occur within days. Only about 5% of the adult population carries *C. difficile,* and in these people the colonic microbiota keeps the numbers of *C. difficile* low. Only with the advent of antibiotic therapy did pseudomembranous colitis become possible. This is yet another example of how changing human practices can produce, unintentionally, the conditions that make a new disease possible.

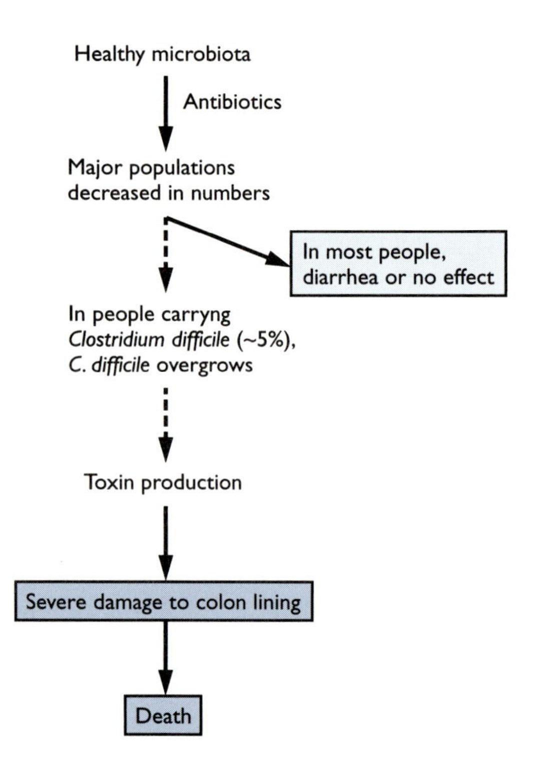

Figure 11.5 The process by which *Clostridium difficile* causes pseudomembranous colitis. Dashed arrows indicate that only a fraction of people continue to the next stage.

The colonic microbiota as a source of disease-causing bacteria

The colonic microbiota is protective if it stays in its natural site, but some members of the microbiota can cause serious infections when surgery or other trauma releases colon contents into the bloodstream. *E. coli* is the primary cause of the rapid death that sometimes results when an infected appendix bursts and releases intestinal bacteria into normally sterile areas of the body. Another numerically minor group of gram-positive cocci, *Enterococcus* species, is an increasingly common cause of post-surgical infections, especially in patients who are immunocompromised. Because they are among the most antibiotic-resistant bacteria known, *Enterococcus* species are much feared as a cause of disease in these patients. Some strains are now resistant to all known antibiotics.

If released from the colon, *Bacteroides fragilis* and other *Bacteroides* species can produce potentially lethal abscesses in nearly any organ of the body. Just as physicians once believed that no bacterium could live in the stomach, they long held the belief that obligate anaerobes could not infect human tissue, because tissue is aerobic. But tissue is not all that aerobic. Most of the oxygen is bound up in hemoglobin and is rapidly taken up and used by human cells. Moreover, areas where tissue damage has occurred become anoxic very rapidly because they no longer have a blood supply. As people age, their bodies acquire many such damaged areas as a result of infection or surgery.

The Human Vaginal Microbiota

Characteristics of the vaginal microbiota

The microbiota of the vaginal tract is most similar metabolically to the microbiota of the mouth. That is, it is normally dominated by lactic acid bacteria. In this case, however, the predominant bacteria are *Lactobacillus* species and not *Streptococcus* species. Less is known about the microbiota of the vaginal tract than about any of the other microbial populations of the human body. The sources of carbon and energy for these bacteria, for example, are unknown. Vaginal mucus and vaginal secretions are probably the main sources of nutrients. These contain an abundant source of sugars, amino acids, and vitamins that the lactic acid bacteria usually need. The concentration of lactic acid bacteria varies somewhat with the menstrual cycle, an observation that suggests an important role of vaginal secretions in the microbial ecology of the vagina.

The lactic acid bacteria of the vaginal tract clearly have a protective effect, because when antibiotic use reduces their numbers, disease-causing microorganisms can colonize the site. Women taking antibiotics for acne or urinary tract infections often develop **yeast infections,** as a result of overgrowth of the yeast *C. albicans*. At one time, scientists thought that the protective effect of lactobacilli was mediated mainly by their production of lactic acid, which makes the pH of the vaginal tract somewhat acidic. Now, however, another feature of vaginal *Lactobacillus* species has been identified that may make an even greater contribution to protection: production of hydrogen peroxide. The bacteria do not produce peroxide intentionally. It is a byproduct of their metabolism. A flavoprotein interacts with oxygen to produce a very reactive form of oxygen called superoxide radical, which is then converted to hydrogen peroxide. Hydrogen peroxide helps to keep yeasts and other potential pathogens under control.

Because yeast infections are so common, women often have resorted to folk cures, such as applying yogurt to the vagina. Yogurt contains lactobacilli, and the reasoning was that yogurt could thereby restore the normal microbiota and keep *C. albicans* under control. The problem with this reasoning is that pasteurized yogurt has few live bacteria and unpasteurized yogurt contains lactobacilli that are different from the ones normally found in the intestinal tract.

Bacterial vaginosis: a once-ignored disease assumes new importance

Just as periodontal disease is caused by a shift in the gum microbiota, the vaginal tract can experience a microbiota shift that produces a disease state (Fig. 11.6). **Bacterial vaginosis** is a disease caused by a shift from a predominantly gram-positive population to a predominantly gram-negative population. The main symptom of bacterial vaginosis is a discharge with a fishy odor. The fishy odor is the result of amines produced by bacteria in the shifted microbial population. No one took this disease very seriously (except afflicted women) until 1995, when two papers in the prestigious *New England Journal of Medicine* revealed a link between bacterial vaginosis and preterm birth.

Bacterial vaginosis is by no means the only or even the main cause of premature birth, but women with bacterial vaginosis have a significantly higher risk of preterm birth than do women with a normal microbiota. Moreover, administration of antibiotics that returned the vaginal microbiota to its normal state reduced the risk of preterm birth for women who had bacterial vaginosis to equal that of women who never had the disease. Scientists do not yet understand how bacterial vaginosis affects conditions in the uterus. One theory is that the slight inflammation that accompanies the disease and is responsible for the discharge might be interpreted by the body as a uterine infection. The body can respond to the threat of such an infection by initiating the birth process prematurely.

Figure 11.6 The steps leading to bacterial vaginosis.

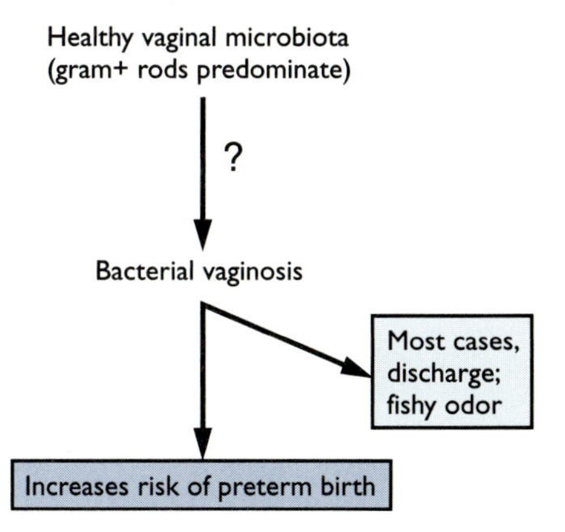

Pelvic inflammatory disease

The opening in the cervix allows sperm to enter the uterus and menstrual blood to flow out of the uterus. This opening can also admit vaginal bacteria to the uterus and fallopian tubes, areas that are usually free of microorganisms. Microorganisms entering the fallopian tubes can trigger an inflammatory response that damages the fallopian tubes. This condition is called **pelvic inflammatory disease (PID).** There is some uncertainty about the role played by the normal vaginal microbiota in PID. Women with PID may have members of the vaginal microbiota isolated from the inflamed area. The question is whether these bacteria are the cause of disease or simply are colonizing after damage caused by sexually-transmitted pathogens, such as *Neisseria gonorrhoeae,* the cause of gonorrhea. This issue sparked a huge controversy and a spate of lawsuits in the 1970s, when a popular intrauterine device called the Dalkon shield was alleged to be responsible for facilitating the entry of vaginal microbiota into the uterine and fallopian tubes and thus causing PID. These allegations are now thought to be untrue, but they illustrate the sorts of issues that can arise in connection with the microbiota.

Maintaining the Balance

The interactions between microbial populations of the human body and the body they inhabit are many and complex. Over evolutionary time, humans have developed an equilibrium with the microbial world, which consists of cloaking the body inside and out with microorganisms that are likelier to be friends than enemies. It should not come as much of a surprise that some of these microorganisms occasionally can cause disease when they manage to get into normally sterile areas of the body. In the next chapter, you will learn that certain molecules on the surface of most gram-positive and gram-negative bacteria can trigger a potentially devastating response if they get into the bloodstream. Thus, even microorganisms that do not normally cause disease can do so if given an opportunity to enter tissue and blood. Such incursions are called opportunistic infections.

Humans have no choice about whether to have a microbiota or not. We can, however, learn how to keep it in healthy balance and how to prevent entry of opportunistic microbes into the body. How to do this more effectively will be an important topic for future research.

box 11–1 *The Forgotten Colonic Eukaryotes*

Microbiological studies showing that the colonic microbiota consists solely of bacteria and archaea are somewhat misleading because they have focused on people living in developed countries. Throughout most of human history and in most parts of the world today, the colonic microbiota also includes protozoa and helminths (worms). In the developed world, these eukaryotes are associated with disease but most of the people who carry them have no symptoms. The loss of this component of the microbiota in developed countries—if indeed it has been lost entirely—has occurred within a relatively short time, over the past two centuries.

Is it possible that the loss of these populations has something to do with the rise in inflammatory bowel disease in the developed world? Some scientists think so and are looking into the possibility that loss of these eukaryotes may have been a mixed blessing. They argue that the human body evolved to deal with intestinal eukaryotes as well as prokaryotes. In fact, an entire arm of the immune system targets eukaryotic microorganisms, the same part of the immune response that is linked to allergic reactions. This being the case, the abrupt (in evolutionary time) loss of an ancient component of the colonic microbiota might have consequences.

chapter 11 at a glance

Roles of the microbiota

Protective—by competing with disease-causing microorganisms

Nutritional—providing nutrients for the human body

Potential source of infection—microbiota shift diseases and opportunistic pathogens

Features of the microbiota found in different sites

Site	Examples of microbes found in the site	Microbiota shift diseases	Examples of potential pathogens
Skin	Gram-positive cocci (*S. epidermidis*) Gram-positive rods (*Propionibacterium acnes*)	None known	*Staphylococcus epidermidis*
Nose	Gram-positive cocci, similar to skin	None known	*Staphylococcus aureus* (transient colonizer)
Mouth and throat	Teeth (*Streptococcus salivarius, Actinomyces* species) Gums (gram-positive cocci, some gram-negatives)	Periodontal disease	*S. pneumoniae* (transient colonizer)
Stomach	Controversial (e.g., *Helicobacter pylori*)	None known	*Helicobacter pylori*
Small intestine	Low numbers of bacteria (e.g., *E. coli*)	None known	*E. coli*
Colon	High numbers of bacteria (*Bacteroides* species, gram-positive rods and cocci; most are obligate anaerobes)	Infant botulism (*Clostridium botulinum*) Pseudomembranous colitis (*Clostridium difficile*)	*E. coli* *Enterococcus* *C. difficile* *Bacteroides*
Vagina	Most are gram-positive rods, cocci (e.g., *Lactobacillus* species)	Yeast infections after antibiotics Bacterial vaginosis	*Candida albicans*

Questions for Review and Reflection

1. Manufacturers of cosmetics advertise as a positive feature that their products neutralize the pH of the skin. If the acidic pH of skin is a protection from microbes, why hasn't this practice produced disastrous results for women using these products? [Hint: consider the microbial load of the skin today compared to centuries ago.]

2. If a surgeon is wearing gloves, how could he or she contaminate a plastic implant with skin bacteria?

3. In this chapter, the skin bacterium *Propionibacterium acnes* was not counted as an opportunist whereas *Staphylococcus epidermidis* was. Should it be? Could the status of a bacterium as an opportunist change with time? How?

4. What recent changes in human practices have made once innocuous members of the microbiota more dangerous? Does this mean that these changes should not have occurred?

5. Should dental caries be considered a microbiota shift disease? Why was it not included under that category?

6. Do you think *Helicobacter pylori* should be considered as the normal microbiota of the stomach? This is quite a controversial suggestion. Why?

7. Would you recommend treating infants with antibiotics to prevent infant botulism? Why or why not? Is there a better alternative?

8. At the level of genus and species, the microbiota of one person is very similar to that of another person, with some exceptions. Yet, some scientists have suggested that there might be person-to-person variations among the strains that colonize different people. How could you use pulsed field gel electrophoresis (Chapter 3) to settle this question? Could you also use this technique to answer the question of whether infants get their microbiota from their parents? How would you proceed?

9. Antibiotics are good for treating bacterial infections, but they can also cause disease. How is this possible?

10. Did opportunistic infections arise only after the advent of modern medicine? Could they have occurred in earlier times?

key terms

development of the microbiota change in the composition of microbial populations as an infant matures and finally develops an adult-type microbiota

microbiota microorganisms that are found at the same site on most people and are present in that site from childhood throughout life

microbiota shift disease a shift in the population of microbes that normally resides in a particular site, resulting in disease in that site or in immediately adjacent sites

nutritional role of the microbiota production by the colonic microbiota of fermentation products and vitamins that can be utilized by the human body

opportunists microorganisms that normally do not cause disease but can do so when the defenses of the body are undermined (for example, by catheters, surgery, or cancer chemotherapy)

protective role of the microbiota the ability of the microbiota to compete for colonization sites and nutrients with disease-causing microorganisms that could otherwise colonize the same site

transient colonizers microorganisms that occupy a site for limited periods of time but that in most cases do not become permanent colonizers of the site; examples are *Staphylococcus aureus* on skin and in the nose and *Streptococcus pneumoniae* in the mouth and throat

12 Fortress Human Body: The Nonspecific Defenses

Even the most modern police department, equipped with the most advanced technology, has trouble responding effectively to a 911 call without an address and the cooperation of those at the site of the emergency. Similarly, the neutrophils that cruise the bloodstream on the lookout for invading microbes also need the guidance and help provided by the proteins of the complement system.

Because we are creatures born into a world dominated by microorganisms, it would be surprising if our bodies did not reflect our long interaction with them. In fact, many features of the human body bear witness to this long process of coevolution. The overall result is a complex and very effective set of defenses that keeps the right kind of microorganisms in the right place and out of vulnerable areas of the body, such as internal organs and the bloodstream.

Defenses of the Human Body

Nonspecific defenses

It is customary to divide the defenses of the human body into two types: nonspecific and specific defenses. It is important to realize, however, that these two defense systems do not function independently. They communicate with each other on many levels, cooperating to vanquish would-be microbial invaders. These defenses are the reason why, for most of us, microbial disease is the exception rather than the rule. Conversely, disruption of these defenses increases considerably the risk of infection.

The first line of defense is the **nonspecific defense system,** which is always present. The system is called nonspecific because it is effective in eliminating most microbes. The defense systems cannot know in advance which infectious microbes might be tempted to invade, so the body has evolved defenses that react against generic categories of microorganisms, such as bacteria or viruses or protozoa. Rarely if ever included in discussions of this subject is the most important nonspecific defense: common sense. Avoiding contact with disease-causing microorganisms in the first place is clearly the best way to solve the disease problem. Prudent preventive measures, both on an individual and population-wide basis, constitute a vital barrier against disease.

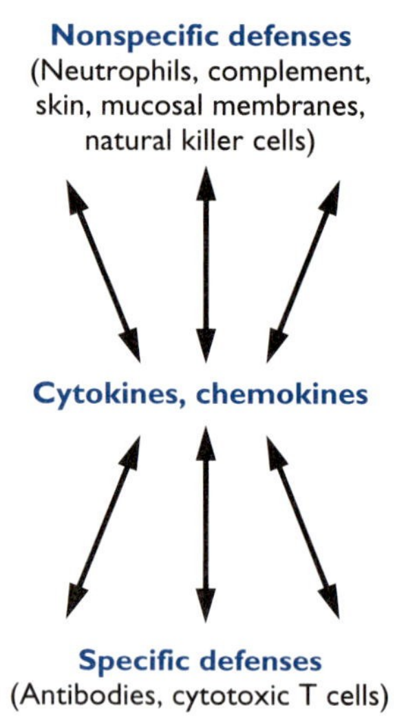

Figure 12.1 Components of the nonspecific and specific defenses. Activities of the nonspecific and specific defenses are coordinated by cytokines and chemokines.

Examples of nonspecific defenses include: the surfaces of the body (which act as physical and chemical barriers against infection), neutrophils (the white cells that patrol the bloodstream and rush to the site of an infection to do battle with invading microorganisms), complement and cytokines (which direct the activities of the neutrophils), and natural killer cells (which specialize in killing cells infected with viruses or other intracellular pathogens).

Specific defenses

The second line of defense is the **specific defense system.** Examples of specific defenses are antibodies and cytotoxic T cells, both of which target specific molecules of specific microorganisms. Specific defenses are the subject of the next chapter. Connecting these two defense systems is a complex network of proteins called cytokines and chemokines, which serve not only to stimulate the nonspecific and specific defenses to greater killing power, but also to administer and coordinate them (Fig. 12.1).

Body Surfaces: Skin and Mucosal Membranes

Body surfaces play an important part in the nonspecific defense system. One could argue that they are the most effective defenses we have, because if they succeed in preventing a potential invader from gaining access to underlying tissue or to the bloodstream, we experience no symptoms or ill effects. As will become evident, the internal defenses, especially the neutrophil defense system, can cause a great deal of collateral damage, including permanent damage to organs, in the process of trying to get rid of an invader that has gotten through the body's surfaces. Moreover, if the internal defenses overreact to an invader or mistakenly target the body's own tissues, these defenses can kill.

Skin and nails: a forbidding desert

Protective features of skin and nails. Human skin is an especially important barrier to disease. Its properties were described briefly in Chapter 11, but some salient features will be considered here in more detail (Fig. 12.2A). The surface layer of skin, the **epidermis,** consists of dead cells. These dead cells present a surface that is usually dry and slightly acidic, an environment that is not conducive to the growth of most microorganisms, which prefer wet environments. Viruses, which require living cells in which to replicate, are completely out of luck, because there are no live cells to keep them going. The dead cells of the skin are constantly being sloughed off so that any microbe that succeeds in attaching to skin is sloughed off as well. As the cells of the dermis are pushed outward toward the epidermis, they produce high levels of the protein **keratin,** which is difficult for most microbes to digest. Thus, as the skin cells die and reach the epidermis, they are especially resistant to microbial digestion, thus depriving skin-dwelling microorganisms of nutrients.

Nails are also composed of keratinized tissue. The fungi that cause athlete's foot or nail infections are able to utilize keratin as a source of carbon and energy. Nonetheless, even these fungi must find minute fissures or cracks in the skin to allow them to penetrate deeply enough into the surface layer to grow and cause the inflammation that people who are infected experience as pain and itching. Some microbes do manage to grow and persist on skin or in sweat glands or sebaceous glands (glands

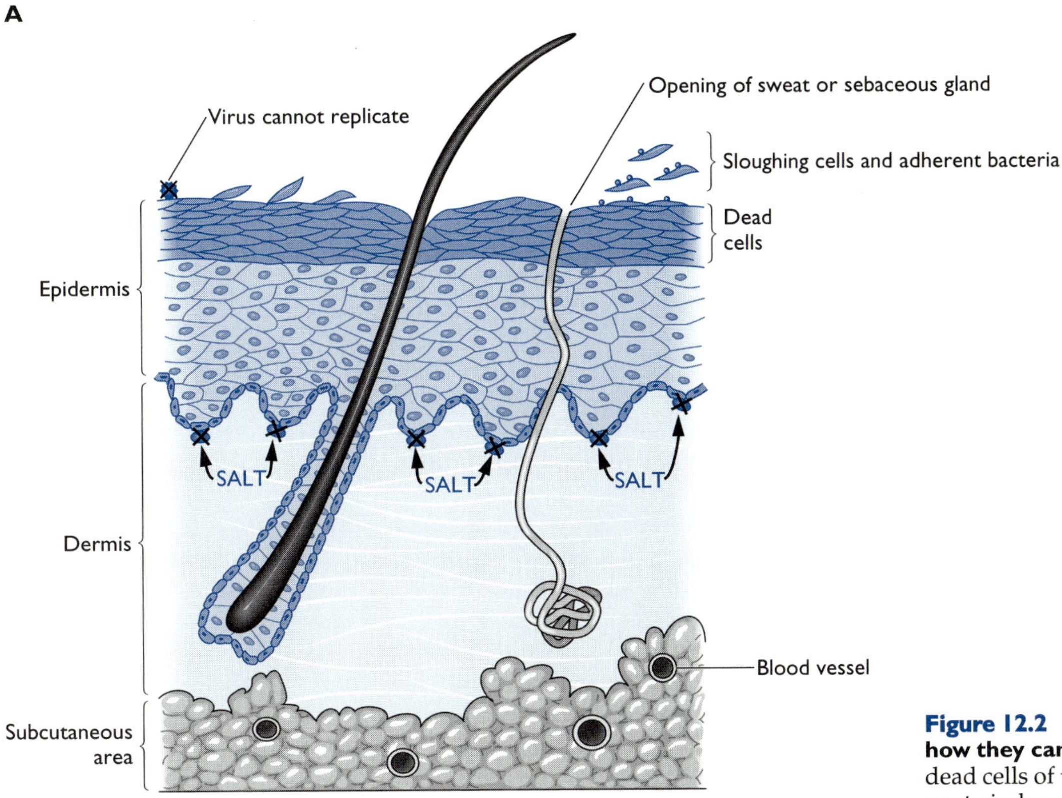

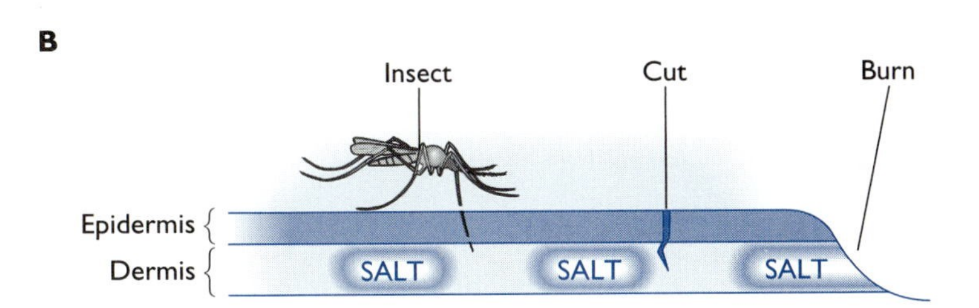

Figure 12.2 Defenses of the skin and how they can be breached. (A) The dead cells of the thick epidermis prevent viral replication and carry adherent bacteria with them when they slough off. Glands and hair follicles produce antimicrobial substances. The skin-associated lymphoid tissue (SALT) provides the specific defenses of the skin. The Langerhans cells are phagocytes associated with the SALT. **(B)** Defenses of the skin may be breached by insect bites, cuts, or burns.

that produce oil and other secretions). Usually these microbes, the normal microbiota of the skin, do not cause disease and, by competing for nutrients and colonization sites, may even prevent colonization of the skin by disease-causing bacteria.

Connection with the specific defenses. Underlying the skin is a specialized component of the immune system called the **skin-associated lymphoid tissue (SALT)**. This acronym designates a set of cells that responds to invaders by activating specific defense systems. The cell type that initiates the specific response is the **Langerhans cell,** a phagocytic cell that engulfs and destroys microorganisms. Langerhans cells also signal other cells of the immune system to respond to the new threat (Chapter 13).

Breaches in skin. The importance of skin as a nonspecific defense is evident from what happens when this barrier is breached. Cuts and burns allow bacteria on the surface of the skin or in the environment a chance to reach underlying tissue (Fig. 12.2B). In the case of burns, the specific and nonspecific defenses of skin are both impaired because they are destroyed by tissue damage resulting from the burn. Burn patients are thus especially susceptible to infection of the burned area. A person who survives the original trauma of serious burns must still survive the infections that often develop.

Microorganisms transmitted by biting arthropods also have a tailor-made vehicle for getting through intact skin: the insect itself. The mouth parts of blood-sucking arthropods, such as mosquitoes and ticks, penetrate into the dermis, where the blood supply of the skin is located, and convey the microorganisms they carry into the underlying live tissue.

Mucosal membranes

Mucosal membranes. Although technically inside the body, the intestinal tract, respiratory tract, vaginal tract, and bladder are all areas that are exposed constantly to material from the external environment. The lining of the intestinal tract, the lining of the airway of the lung, and part of the lining of the bladder consist of only a single layer of cells, called mucosal or epithelial cells. Even in areas where there is more than one layer of epithelial cells (e.g., mouth, vaginal tract, bladder), the barrier these cells provide to microbial invasion is not as thick as that of the skin. These barriers have to be thin and exposed to body fluids, because secretions pass through them. Moreover, in the intestinal tract and lung, fluids or gases from the outside world are absorbed across the membrane into the bloodstream.

Mucin. These fragile barriers would be breached easily by microorganisms were they not protected by a thick, sticky layer of a substance called mucin or mucus (Fig. 12.3). **Mucin** consists of a mixture of proteins and polysaccharides, the main role of which is to trap microbes and prevent them from reaching the epithelial cells. In the vaginal and intestinal tracts,

Figure 12.3 Defenses of mucosal membranes. The single layer of epithelial cells of the mucosa is covered by a thick layer of mucin, which traps most microbes before they reach the cells. It also contains antimicrobial compounds such as lysozyme and lactoferrin. Any microbes that penetrate the mucin and attach to epithelial cells will be eliminated when the cells slough off. The specific arm of the mucosal defense system is the mucosa-associated lymphoid tissue (MALT).

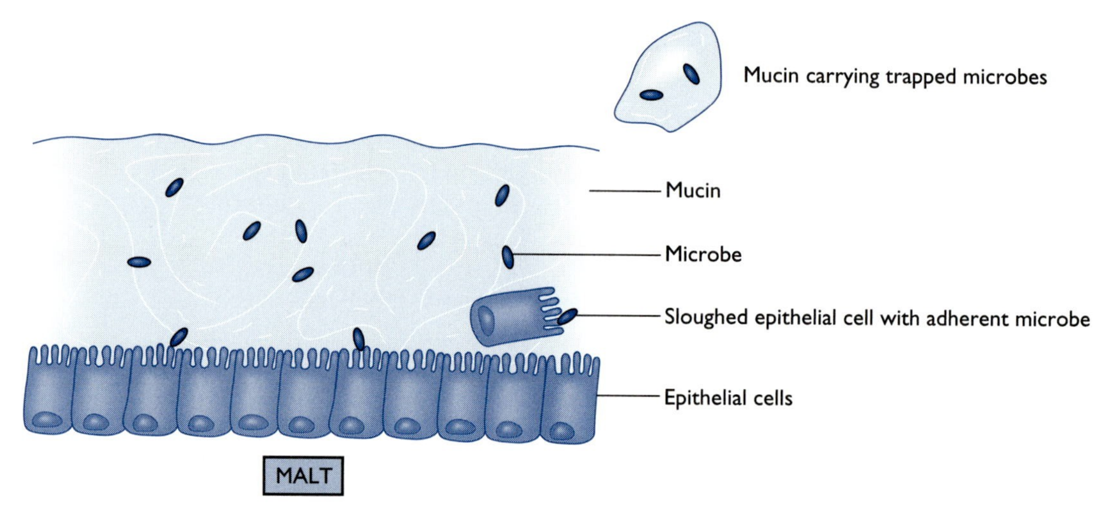

mucus also has a lubricating effect that prevents mechanical damage to the epithelial cells. Mucin not only traps microorganisms but also contains antimicrobial substances, such as **lactoferrin** (an iron-binding protein that deprives microbes of iron), **lysozyme** (an enzyme that digests the cell wall of bacteria), and **defensins** (small proteins that poke holes in microbial membranes).

Mucin is constantly being shed and replaced, so any microorganisms trapped in mucin are expelled eventually from the body. Epithelial cells are also replaced frequently. Thus, if a microorganism does succeed in penetrating the mucin layer and attaches to an epithelial cell, it is removed when the epithelial cell is shed (sloughed) and excreted. This constant replacement of the epithelial cells exacts a rather large energy toll. The other rapidly dividing populations of cells in the human body are the neutrophils and the cells of the immune system. This energy toll is the price we pay for peaceful coexistence with microorganisms. As described in Chapter 11, the energy provided to our intestinal cells by the fermentation activities of the intestinal microbiota is a partial payoff (intentional or not) by the bacteria for the toll they extract.

Association with the specific defenses. Like the skin, the mucosal membranes have their own special arm of the specific response system, the **mucosa-associated lymphoid tissue (MALT)**. The MALT associated with the intestinal tract is called the **gastrointestinal-associated lymphoid tissue (GALT)**. The first cell to act in the GALT is the macrophage, a cell that engulfs and degrades incoming microorganisms. This cell plays a role similar to that of the Langerhans cells of the SALT. The macrophage signals other cells of the specific defense system to mount a specific response against the invader (Chapter 13). From an examination of the SALT and GALT, a basic theme emerges that will be woven through the rest of the chapter: the nonspecific defenses repel most invaders, and the specific defense systems then cope with whatever gets through the nonspecific defenses.

The importance of an intact epithelial layer is evident from the fact that sores or other breaches of the mucosal membrane in the vagina or intestine increase the risk that exposure to human immunodeficiency virus (HIV) will result in the introduction of the virus into the bloodstream. Another example of what happens when the mucosal lining is breached is peritonitis. A burst appendix or accidental perforation of the bowel during surgery permits bacteria to gain access to the peritoneal cavity surrounding the bowel. If prompt antibiotic treatment is not initiated, infection can develop and bacteria can enter the bloodstream, a potentially fatal development.

Neutrophils and Complement

Characteristics of neutrophils and complement

Inflammation is the redness, pain, heat, and swelling seen around infected wounds. Inflammation can cause breaches in the epithelial membranes that line internal areas, such as the intestinal tract. Inflammation is a visible sign that the nonspecific defenses of tissues and blood are at work. It is also a sign that a successful defense effort by the nonspecific defenses of tissue and blood may produce some collateral damage to tissue. Many of the symptoms of infectious diseases are actually caused by the internal defenses of the body trying to fight off an invader, rather than by the microbes themselves.

When inflammation goes out of control, a potentially fatal process called **septic shock** can occur.

Role and features of neutrophils (PMNs). If microorganisms manage to bypass the skin and epithelial surfaces, the next line of defense is the phagocytic cells (neutrophils) and natural killer cells. **Neutrophils** specialize in killing extracellular microbes, whereas **natural killer cells** target microorganisms, such as viruses, growing inside human cells. Neutrophils have many names, including **polymorphonuclear leukocytes** (**PMNs**), and polys. The use of more general terms, such as "white cell" or leukocyte, sometimes used to indicate neutrophils, is not entirely correct. It is true that neutrophils are the predominant type of white cell or leukocyte, but they are not the only cells in this category. We will use the term neutrophil to designate these cells. The term "polymorphonuclear" refers to the fact that the nucleus of a neutrophil has a many-lobed structure that can actually appear to be multiple nuclei when the cell is viewed in cross-section (Fig. 12.4). Natural killer cells are also white cells but are more closely related to cells of the specific defense system, such as T and B cells, than to neutrophils. Neutrophils normally circulate in blood and are short-lived but numerous. Neutrophils are produced in the bone marrow and released into blood, where they mature rapidly and begin to circulate.

How neutrophils kill bacteria. The process by which neutrophils ingest and kill bacteria is illustrated in Figure 12.5. The microbe is first engulfed and enclosed in a membrane vesicle (**phagosome**). The vesicle interior begins to acidify as a result of the pumping of protons into the vesicle by an ATP synthase. The neutrophil also contains granules called **lysosomes,** which contain a number of substances that are toxic to microbes, including proteases and defensins (small peptides that integrate into and disrupt membranes). The most toxic weapon in the lysosome's armory is an enzyme called **myeloperoxidase.** Normally, this enzyme is inactive, but when the lysosome fuses with the phagosome to form the **phagolysosome,** myeloperoxidase is activated and produces **superoxide** by the following reaction:

$$\text{NADPH} + 2\text{O}_2 \rightarrow 2\text{O}_2^- \text{ (superoxide radical)} + \text{NADP}^+.$$

Superoxide is a very toxic form of oxygen in its own right, because it can oxidize and inactivate proteins and other molecules on the bacterium's

Figure 12.4 Structure of a neutrophil. The term "polymorphonuclear" leucocyte (an alternative name for neutrophil) derives from the multilobed structure of its nucleus.

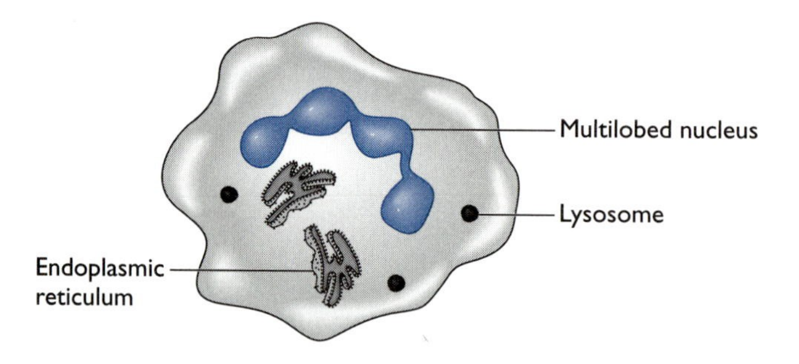

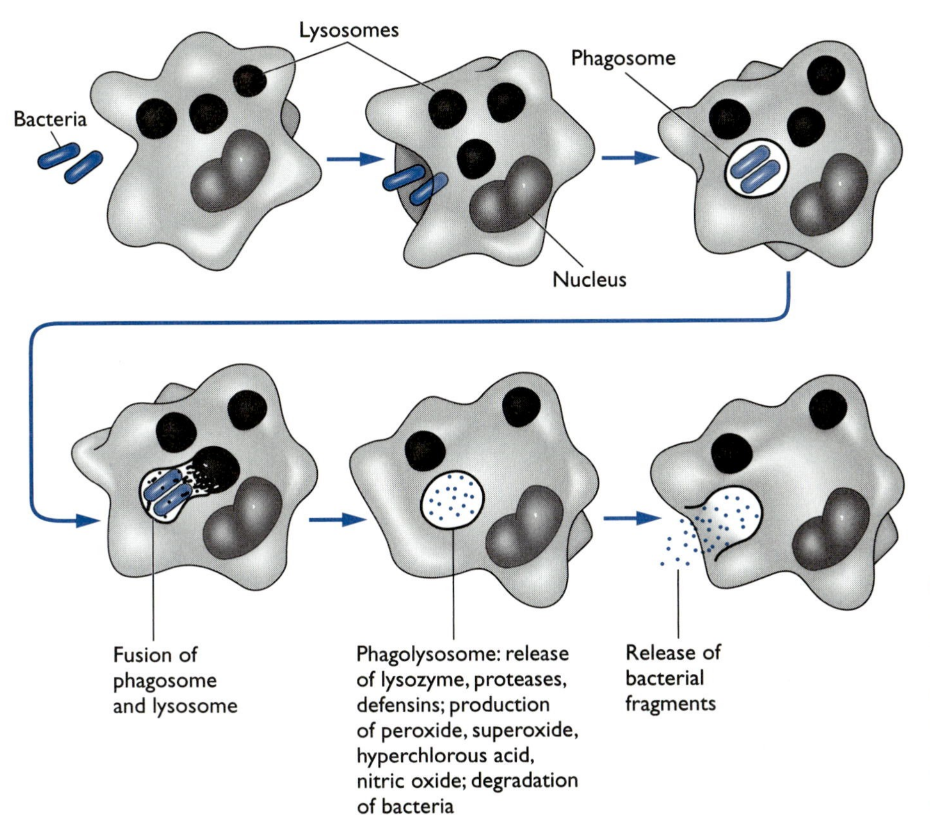

Figure 12.5 Ingestion and killing of microbes by a neutrophil. Bacteria are engulfed in a phagosome. The phagosome fuses with a lysosome, which contains numerous compounds toxic to microbes. Most bacteria are killed and degraded by these compounds. The fragments are released to the external environment.

surface. It becomes even more toxic when it combines with chlorine in a spontaneous reaction to form **hypochlorous acid** (household bleach). Few bacteria can withstand the toxic effects of superoxide and hypochlorous acid. If this assault is successful, the neutrophil ejects the fragments of the killed bacterium. Another toxic inorganic compound that probably helps to control invasive microorganisms is **nitric oxide (NO)**. This molecule is either produced enzymatically by human cells or forms spontaneously when nitrite encounters acidic conditions. NO, like superoxide, is a very reactive molecule that inactivates bacterial proteins. Because superoxide and NO are toxic to human cells, they must be carefully confined. This is the reason they are produced primarily in the phagolysosome.

During a microbial attack, when neutrophils become activated to produce a greater burst of superoxide and NO, the phagolysosomes sometimes fuse with the neutrophil membrane, releasing their toxic contents into the external environment. This causes destruction of nearby human cells, thereby contributing to the cell damage associated with the inflammatory process.

Microbial evasion of killing by neutrophils. Many microorganisms that cause disease are able to avoid, neutralize, or survive the neutrophil response (Fig. 12.6). One strategy is to produce a polysaccharide capsule or a protein surface layer that prevents the neutrophil surface receptors from binding to the bacterial surface. This binding step is necessary to keep the microorganism from shooting away from the engulfing phagocyte, much as a fish slips away in water from grasping hands. Other microorganisms

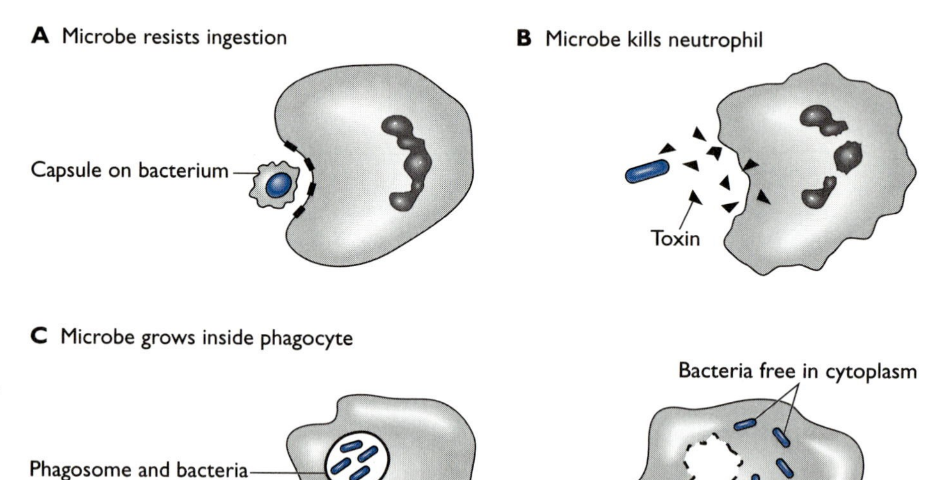

Figure 12.6 Mechanisms by which microorganisms avoid killing by neutrophils. (A) Some microbes resist ingestion by covering themselves with a capsule or protein layer that blocks binding of the neutrophil surface receptors. **(B)** Some microorganisms produce substances that kill or inactivate neutrophils. **(C)** Other microbes prevent fusion of the phagosome and lysosome and continue to live and replicate in the phagosome, ultimately rupturing it.

produce toxic proteins that kill or paralyze neutrophils. Still others prevent the phagosome-lysosome fusion and live happily in the phagosome or escape into the host cell cytoplasm (Fig. 12.6). Some bacteria can even survive the onslaught of the neutrophil's oxidative burst in the fused phagolysosome. How they survive killing by neutrophils is still not completely understood but it seems to involve cell wall components that neutralize the toxic chemicals produced by the neutrophil. Many bacteria also produce enzymes (**superoxide dismutase** and **catalase**) that together convert superoxide to water.

Once a neutrophil has generated its oxidative burst and released its toxic compounds, it dies. Thus, if a microbe can survive the first blast, it is left with a nice, nutrient-rich environment in which to grow. Microorganisms that are capable of growing in neutrophils and other phagocytic cells are among the most dangerous of all disease-causing microorganisms, because (as you will see in the next chapter) they undercut both the nonspecific and specific defenses. One example of such a bacterium is *Mycobacterium tuberculosis,* the microorganism that causes tuberculosis.

How neutrophils move to an infected site

Neutrophils usually move through the bloodstream with a rolling motion (Fig. 12.7). If the bacteria are infecting a mucosal surface or entering tissue, while the neutrophils are rolling merrily along through the blood vessels, how do the neutrophils find the invading bacteria? The answer to this question is the **complement system,** a set of proteins that is activated by invading microorganisms and attracts neutrophils to the site (Figs. 12.7 and 12.8). While the neutrophil moves to the site of infection, complement components and cytokines cooperate to activate the neutrophil's oxidative burst to give it greater killing power.

Complement activation. To see how this works, let us for the moment leave the neutrophils rolling along in the bloodstream, as yet unaware that

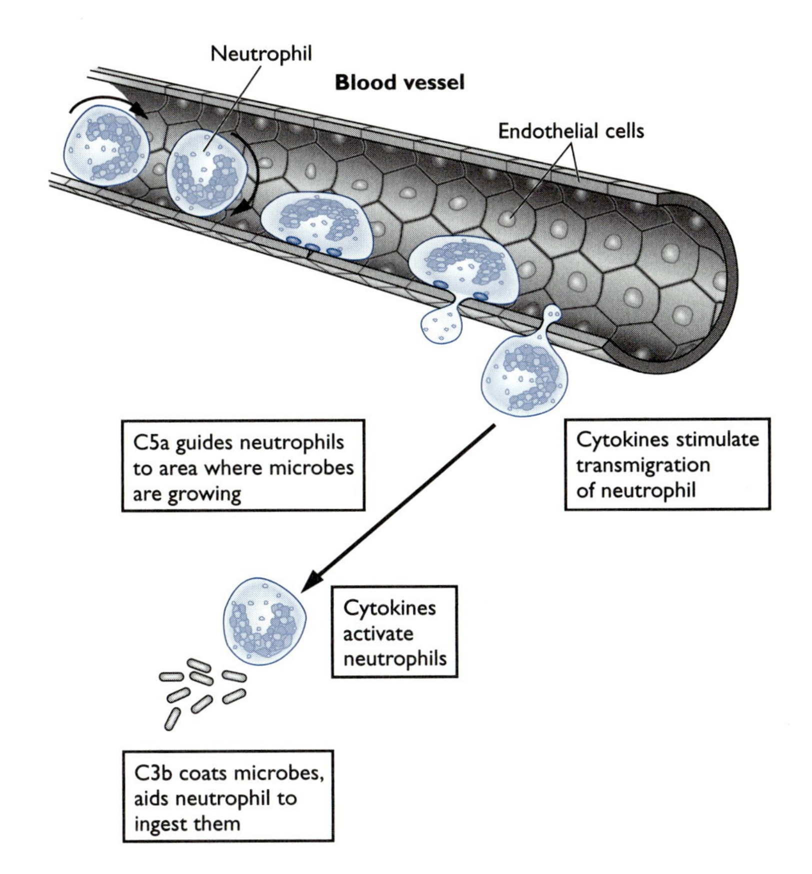

Figure 12.7 Movement of neutrophils from bone marrow through the bloodstream to the site of infection. Neutrophils arise in bone marrow and then enter into the bloodstream. As they move along through the blood vessels, they can respond to C5a by binding to receptors on cells lining the vessel, slowing and finally stopping. Neutrophils leave the bloodstream and follow a gradient of complement component C5a to the site where the invading microbes are located. Cytokines help to activate the neutrophils.

trouble is brewing, and move to the site where an infection is beginning to occur. The first internal defense the invading microorganisms encounter is complement. Complement is a set of proteins found in tissue and blood and is activated by molecules such as lipopolysaccharide (LPS) or lipoteichoic acid (LTA). That is, complement recognizes "signatures" widely found in the cell walls of gram-negative (LPS) and gram-positive (LTA) bacteria. Complement also responds to antibodies bound to invading microorganisms (Chapter 13). The interaction between **complement protein C3** and LPS or LTA triggers a series of proteolytic cleavages of C3 and other complement proteins (Fig. 12.8). The complement cascade involves a complex set of reactions, and it is easy to get lost in the details. Accordingly, we will focus only on the two most important cleavage events, the ones that cleave protein C3 into C3a and C3b and protein C5 into C5a and C5b.

The role of **C5a** is to attract neutrophils to the infection site by diffusing away from the infection site. Because many C5 molecules are cleaved to form C5a and C5b, a cloud of C5a radiates out from the site of complement activation. At some point, this cloud of diffusing C5a will encounter a blood vessel and diffuse into the bloodstream, where the neutrophils are located. Neutrophils respond to C5a by stopping their rolling motion and sticking to the blood vessel wall, where the concentration of C5a is highest (Fig. 12.7). Blood vessel walls are composed of cells (endothelial cells) that are only loosely bound together and can be pushed apart. The neutrophils squeeze

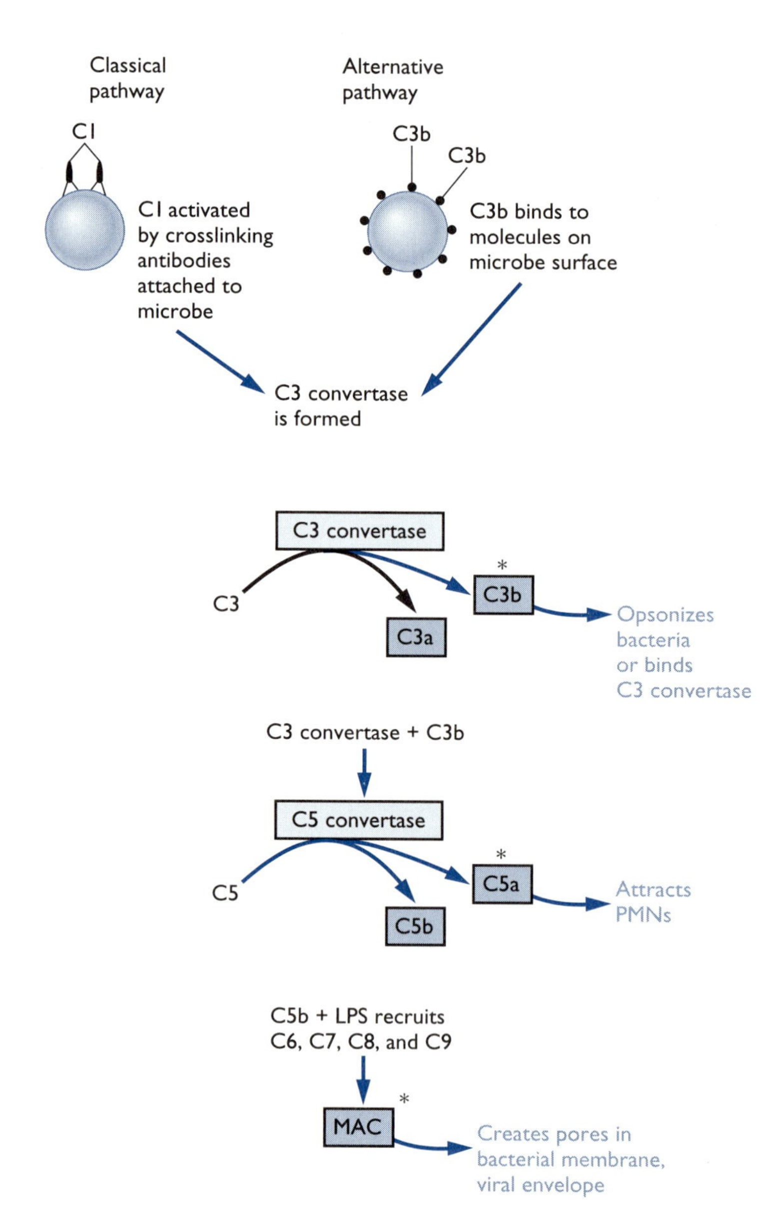

Figure 12.8 Main steps in activation of complement by the classical and alternative pathways. These two pathways differ only in the steps that initiate formation of C3 convertase, the enzyme that cleaves complement component C3 to C3a and C3b. Important activated products are C3b (which opsonizes bacteria), C5a (which attracts phagocytes to the area), and C5b–C9 [membrane attack complex (MAC), which inactivates enveloped viruses and kills gram-negative bacteria].

between endothelial cells and enter the tissue (transmigration), moving toward higher and higher concentrations of C5a (Fig. 12.7). C5a and cytokines stimulate the neutrophils to become activated so that they are more able to perform phagolysosome fusion, undergo the oxidative burst, and kill the microbes.

Meanwhile, back at the site of infection, complement cleavage product **C3b** is accumulating. C3b is a sticky molecule. One portion of C3b binds avidly to bacterial surfaces. Another portion of C3b interacts with receptors on the neutrophil surface, helping the neutrophil to ingest the bacteria it coats. This process is called **opsonization.** C3b does not bind to human tissues, because they are coated with a sugar called sialic acid. This effectively allows the C3b to distinguish between self and nonself, thus preventing the not-very-selective neutrophils from attacking human cells.

Collateral damage. This protection is only partial, however, and some collateral damage to human tissue does occur. Neutrophils are sloppy feeders and release some of their toxic contents into surrounding tissue. They also do some damage to blood vessel walls as they exit into tissue. The redness and swelling around an infected wound come mainly from leakage of blood fluids into tissue, as neutrophils exit the bloodstream. The pain most likely results from damage to nerve endings resulting from collateral damage that occurs when the neutrophil attacks the microorganisms and releases superoxide and other toxic substances into the surrounding area. As already mentioned, some bacteria have discovered how to use capsules to prevent themselves from being phagocytosed. Their capsules discourage the accumulation of C3b on their surfaces. Some bacteria even use capsules made of sialic acid (the coating of human cells) to further disguise themselves and reduce the effectiveness of C3b.

Even if neutrophils do not succeed in killing the invading microorganisms, they continue to arrive and attempt to contain the infection. The result is a spreading zone of damaged human cells, with bacteria growing in the interior. Macroscopically, this zone of destruction is evident as **pus** (mostly DNA and proteins released from dead human cells) and dead tissue (**necrosis**). In the worst case scenario, bacteria enter the bloodstream, and the infection, which was at first localized, spreads all over the body. Body-wide spread of a microorganism is called a **systemic infection.**

Direct killing of bacteria by complement components. Another action of complement, which we will mention only in passing, is the formation of the **membrane attack complex** (**MAC**), which consists of complement components **C5b, C6, C7, C8,** and **C9**. The MAC makes holes in the membranes of gram-negative bacteria, probably by forming pores. Bacteria that are killed by the MAC are called "serum sensitive," because the killing effect of serum was discovered first. Later, it was established that complement was the active ingredient in serum, but the old term has stuck. The MAC forms on LPS molecules and needs to be close enough to the cell surface for the pores to form. Some bacteria have altered their LPS structure so that the MAC does not form or, if it forms, does not damage the bacterial membrane. Such bacteria are called "serum resistant."

Cytokines

Before explaining why systemic infections can be lethal if not brought under control by antibiotics or the specific defenses, it is first necessary to describe another very important player in the nonspecific defense reaction, the cytokines. **Cytokines** are proteins produced by many cell types in the body, especially cells lining the blood vessels (endothelial cells) and **macrophages** (phagocytic cells associated with the specific defense system, which also play a killing role similar to that of neutrophils). Cytokine-producing cells have a protein receptor on their surfaces called **CD14,** which binds compounds specific to bacteria, such as LPS, peptidoglycan, and LTA, possibly with the help of other proteins. This binding event stimulates the cell to sound the alarm by releasing cytokines. Thus, as complement is being activated, endothelial cells and macrophages in tissue are spewing forth cytokines, which help to stimulate the neutrophils to greater killing power. Cytokines have even more diverse and complex roles in the specific defenses that will be explored in Chapter 13.

Septic Shock: When Neutrophils Run Wild

The neutrophil-complement defense is designed to control localized infections. As long as it succeeds in stopping an infection at an early point, a little collateral damage to tissue in a limited region is an acceptable trade-off. However, if the inflammatory response is triggered all over the body, as it is when microorganisms circulate in the bloodstream during a systemic infection, a serious and potentially lethal situation can result. This condition is called **septic shock.**

How septic shock develops

During septic shock, bacteria in the bloodstream trigger neutrophils to exit the blood vessels and become activated to greater killing activity all over the body. The consequences are illustrated in Figure 12.9. (This figure is quite complex, because a number of things happen simultaneously, each contributing to the development of septic shock.) Everywhere in the body, neutrophils attach to blood vessel walls and exit the bloodstream, causing leakage of blood fluids into surrounding areas. Neutrophils, activated too early by complement and cytokines, may even attack blood vessel walls, causing further leakage of fluids from the blood vessels. Neutrophil activi-

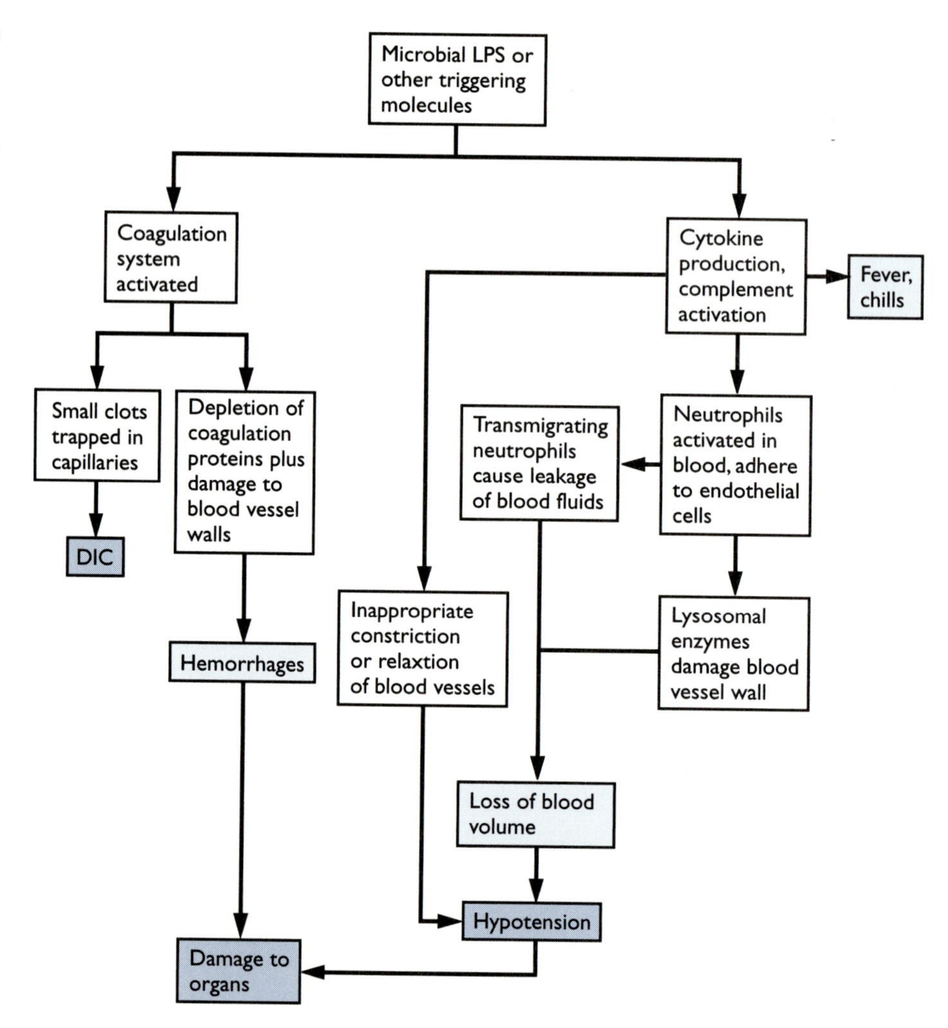

Figure 12.9 The events that occur in septic shock. The main symptoms of septic shock are hypotension, disseminated intravascular coagulation (DIC), internal hemorrhage, and organ failure.

ties also affect the contractile properties of blood vessel walls, reducing the ability to keep blood flow normal.

The massive loss of fluids, coupled with a reduced ability to control blood vessel contractions, results in a drop in blood pressure. A sudden large drop in blood pressure is one of the hallmarks of septic shock. Reduced flow of blood starves vital organs for oxygen and nutrients. Moreover, small hemorrhages occur because of damage to blood vessels. These damage vital organs, such as the lungs, kidney, and brain, with the net result that organs begin to fail. Often, patients with uncontrolled bloodstream infections experience the loss of function of one vital organ after another until death brings the whole process to a halt. Even in cases where the infection is brought under control by antibiotics, the longer the patient remains in the "shock" stage, the more likely that permanent damage to important organs, such as the heart, lungs, and kidney, can occur.

A final point about septic shock is worth making. Physicians tend to classify patients as either cured or not, as if all cured patients were equal. A person cured of a prolonged episode of septic shock can suffer later health problems associated with irreversible organ damage incurred during the shock episode. Such patients may even have reduced life expectancies. Early and effective intervention is critical.

Controlling septic shock

As recently as a few years ago, immunologists were confident that new compounds that blocked the action of cytokines would prove to be the miracle drugs to control septic shock and save countless lives. One after the other, these compounds have failed the efficacy test in clinical trials, bankrupting a few biotech companies in the process. It is still not clear why these compounds do not work, but one explanation is that when the steps leading to septic shock have gone beyond a certain point, it is very difficult to stop the process. More recently, clinicians have gone back to a simpler and intuitively more obvious solution: prevention or early detection of infection. Because the bacteria that cause septic shock are becoming more resistant to antibiotics, it is more and more important to identify as early as possible the causative agent and its susceptibility to antibiotics. Unfortunately, for economic reasons, hospitals have been scaling back or shutting down clinical microbiology lab services because of cost pressures exerted by health maintenance organizations. These organizations also discourage physicians from ordering expensive bacteriological tests. This trend may soon be reversed, as hospital administrators begin to realize the costs of serious infections, especially those leading to septic shock, for both hospitals and patients. Also, although physicians had long suspected that septic shock had serious long-term effects, supporting evidence for this belief is now beginning to appear.

Although neutrophils can obviously do great damage, it is important to recall that they perform a critical protective function. This fact is underscored by what happens when they are absent (Box 12–1).

Natural Killer Cells

The natural killer cell is important in the defense against viral infections and infections caused by intracellular bacterial pathogens. Little is known about this part of the defense system. The role of natural killer cells in controlling infections has been somewhat controversial, because not many of them are seen in infected areas. Possibly they are dying too quickly to accumulate, and more of them enter the area than is apparent from inspecting tissue specimens. In the laboratory, natural killer cells recognize cancer cells and virus-infected cells and kill them. They probably recognize proteins on the infected cell surface that appear only when a cell is under stress.

Recently, natural killer cells have been shown to play a role in the control of bacterial infections in which bacteria grow inside human cells. Natural killer cells do not kill by engulfing infected human cells, which are as big as they are. They attach to and kill cells by releasing the contents of granules that contain molecules toxic to human cells. This is the same killing strategy used by cytotoxic T cells of the specific defense system (Chapter 13), except that the cytotoxic T cells target infected cells more specifically. Like neutrophils, natural killer cells are born in the bone marrow and circulate in blood. They are also found in the lymph nodes and spleen, where the cells of the specific defense system are located, and may interact with them in some way.

What About Eukaryotic Microbes?

Little is known about nonspecific defenses against eukaryotic pathogens, although scientists believe that such defenses exist. The neutrophils are probably involved, because neutropenic patients are more prone to fungal infections than immunocompetent people. There is also a cell type called the **eosinophil,** which contains granules that are released during infections caused by worms. These granules contain compounds that are clearly toxic for some eukaryotic pathogens, but not much is known about their action. Research into nonspecific defenses against eukaryotic pathogens is a rapidly expanding area of interest in microbiology.

box 12–1 *Importance of Neutrophils: The Neutropenic Patient*

Some of the greatest successes of modern medicine inadvertently created new opportunities for microorganisms to infect humans. One example of this phenomenon is cancer chemotherapy. Current cancer chemotherapeutic agents are not highly specific. They kill all rapidly growing cells. So, in addition to hitting the tumor, these agents also take out rapidly dividing normal cells, such as neutrophils. For a period after a chemotherapeutic treatment, the neutrophil concentration in a patient's blood drops to nearly zero, a condition called neutropenia. Also, damage to another rapidly dividing population, the intestinal mucosal cells, reduces their effectiveness as a barrier to infection. The result is that cancer patients become susceptible to infections caused by bacteria or fungi that normally could not survive the defenses of a healthy person. These infections often are caused by members of the patient's own microbiota or by usually innocuous microorganisms in soil or water. Unfortunately, many of these same microorganisms are also resistant to antibiotics. Scientists are working on ways to make cancer chemotherapeutic agents "smarter"—able to target tumor cells specifically rather than killing all rapidly dividing cells. Another approach is to shorten the period of low neutrophil concentrations by stimulating bone marrow to produce neutrophils more quickly after a chemotherapy treatment. The plight of neutropenic patients illustrates how effective neutrophils normally are in keeping healthy people free of infection.

chapter 12 at a glance

Nonspecific defenses of the human body

Site	Defenses	Function
Skin	Dry, acidic conditions <37°C	Limit bacterial growth
	Sloughing cells	Remove bacteria
	Resident microbiota	Compete for colonization sites
Hair follicles, sweat glands	Lysozyme, toxic lipids	Kill bacteria
Beneath skin surface	Skin-associated lymphoid tissue (SALT)	Specific defense against bacteria
Mucosal surface	Mucin layer	Physical barrier, traps bacteria
Mucin layer	Lysozyme	Digests bacterial peptidoglycan
	Secretory immunoglobulin A (sIgA)	Prevents bacterial attachment to mucosal cells, helps to trap bacteria in mucin
	Lactoferrin	Binds iron, prevents bacterial growth
Mucous membrane	Sloughing cells	Remove adherent bacteria
	Tight junctions	Prevent bacteria from invading between mucosal cells
Beneath mucosal membrane	Mucosa-associated lymphoid tissue (MALT)	Produces sIgA Specific defense against bacteria
Blood, tissue	Neutrophils	Engulf, kill bacteria
	Complement components	Opsonize bacteria (C3b), guide neutrophils (C5a)
	Natural killer cells	Kill infected human cells, tumor cells

Examples of bacterial strategies for countering or avoiding the nonspecific defense system

Defense Strategy	Microbial Response
Neutrophil follows C5a trail	Bacteria produce a protease that degrades C5a
Neutrophil engulfs bacteria	Antiphagocytic capsule on bacteria prevents uptake by neutrophils Sialic acid capsules mimic human tissue
Phagosome fuses with lysosome to form phagolysosome	Ingested bacteria prevent phagolysosome fusion
Phagolysosome releases barrage of toxic chemicals	Cell wall of bacterium detoxifies or protects from toxic chemicals Bacteria produce superoxide dismutase, catalase
Complement membrane attack complex (MAC) kills gram-negative bacteria by making holes in their membranes	Bacteria change lipopolysaccharide, become serum resistant (resistant to formation of effective MAC)

Questions for Review and Reflection

1. In what sense are skin, mucosal surfaces, mucin, neutrophils, and complement nonspecific?

2. Why does it make sense that the nonspecific defenses should be connected to the specific defenses (for example, skin is connected to SALT)? In the set of responses to infection that occur, what are the general roles of the nonspecific and specific defenses relative to each other? [Hint: Consider the order in which microbes encounter these defenses.]

3. Most people view mucus as gross and unpleasant. Why might mucus production be elevated during an infection and what purpose does it serve?

4. What are the roles of complement in coordinating the neutrophil defense?

5. It is easy to see why bacteria that kill neutrophils might be dangerous invaders, but capsule producers that do not kill neutrophils are at least as dangerous. Why is this the case?

6. Why are neutrophils activated to maximum killing capacity only after they leave the bloodstream in the response to a localized infection?

7. Why would the human body have evolved a defense system that can kill it?

8. In what sense are natural killer cells nonspecific? What is their role?

9. Fungi can cause septic shock. What does this suggest about components of the fungal cell wall?

key terms

complement cascade activation by proteolytic cleavage of a group of proteins found in blood and tissue that help organize the neutrophil response; main roles are to attract neutrophils to the infected site (C5a) and aid neutrophils in ingesting microorganisms (C3b); complement components C5a and C6–C9 form a membrane attack complex that kills gram-negative bacteria

cytokines and chemokines proteins produced by many human cell types (especially endothelial cells, macrophages, and cells of the specific defense system), which organize the activities of the cells of the nonspecific and specific defense systems and activate neutrophils to greater killing activity during an infection

mucosal (epithelial) membranes layers of cells (often only one cell thick) that line the internal parts of the body

mucus (mucin) a sticky mixture of proteins and polysaccharides that covers the mucosal membranes and acts as a repository for proteins that are toxic for microorganisms

natural killer cells cells that attack and kill infected human cells by binding to them and pelting them with the same sort of toxic compounds found in neutrophils

neutrophils (PMNs) cells of the nonspecific defense system that are normally found in the bloodstream but migrate to the infected site during an infection; main role is to ingest and kill microorganisms

nonspecific defenses defenses of the human body that are always available and target invading microorganisms in general rather than specific ways

opsonization a process that involves the coating of microbes by proteins that help neutrophils to ingest them; the same proteins bind to receptors on the neutrophil as well as to the microbial surface; in this chapter, C3b was the opsonizing protein; in the next chapter, antibodies will also be seen to opsonize microorganisms

SALT, MALT, GALT specific defense systems that interact closely with the nonspecific defenses of skin (SALT) or mucosal surfaces (MALT, GALT); stands for skin- (mucosal-, gastrointestinal-) associated lymphoid tissue

septic shock a condition caused by a systemic infection, in which microbes trigger the activation of neutrophils all over the body; symptoms are the result of neutrophils exiting the blood vessels and damaging them

systemic infection microorganisms enter the bloodstream and move throughout the body

13 Fortress Human Body: The Specific Defenses

Even those who cannot afford designer clothes have the benefit of a designer immune system to back up the nonspecific defense system.

The nonspecific defenses, such as mucin, the washing action of fluids, and the inflammatory response, are quite effective in preventing or controlling many infections. Yet some microbes have developed strategies for bypassing the nonspecific defenses. The next line of defense is the **immune system,** a group of cells that respond in a specific fashion to each specific microbe. This group includes the cytotoxic T cells, T helper cells, B cells, and activated macrophages (Fig. 13.1). Also part of the immune system are cytokines and other proteins that regulate and activate the cells of the immune system.

Role of the Immune System in Combating Infectious Microbes

Need for specific defense system

The ability to respond to a specific microbe allows the immune system to tailor its defense strategy to the particular features of the invading microbe and to focus its forces on that microbe and its specific infection strategy. For example, one effective response to a viral infection is the enlistment of immune system cells, called cytotoxic T cells. **Cytotoxic T cells** recognize human cells infected with the particular virus and kill such cells selectively. If the invading microbe is not a virus but an encapsulated bacterium, a different type of response is needed: the production of protein complexes called antibodies that bind to the capsular polysaccharide and opsonize the bacteria, so that neutrophils can engulf them.

Timing of the specific defense response

A price must be paid for this specific tailored response. After the first exposure to an infectious microbe, days and even weeks can pass before the immune response reaches full strength (Fig. 13.2). This is a price paid only once, because after the immune response has been stimulated the first time, the immune system is primed to respond very quickly—within a day—to a second encounter with the same microbe. This is what is meant when a person is said to be immune to a specific type of infection. The quick response is possible because the first encounter generated "**memory cells**" that persist for years and allow the immune system to respond rapidly to any reappearance of the microorganism.

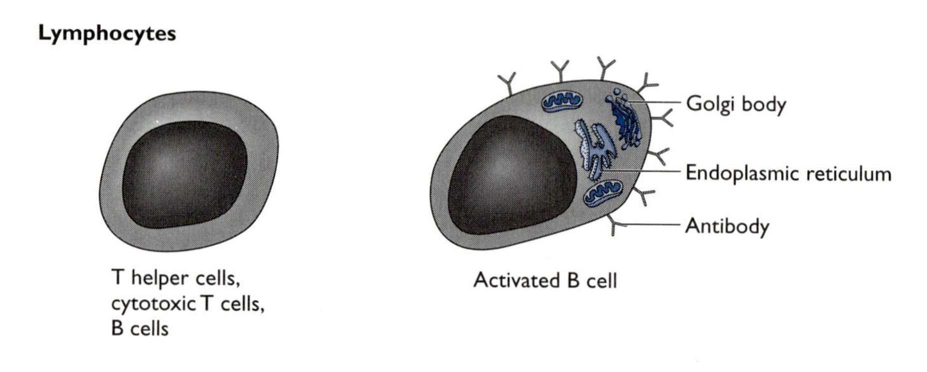

Figure 13.1 Cells important in responding to a specific microbe. T helper cells and cytotoxic T cells are not morphologically distinguishable. B cells are similar in appearance to T cells until they become activated B cells, the form that produces antibodies. Macrophages are easily distinguished from the lymphocytes. Macrophages differ from neutrophils in that they are larger and have a more compact nucleus.

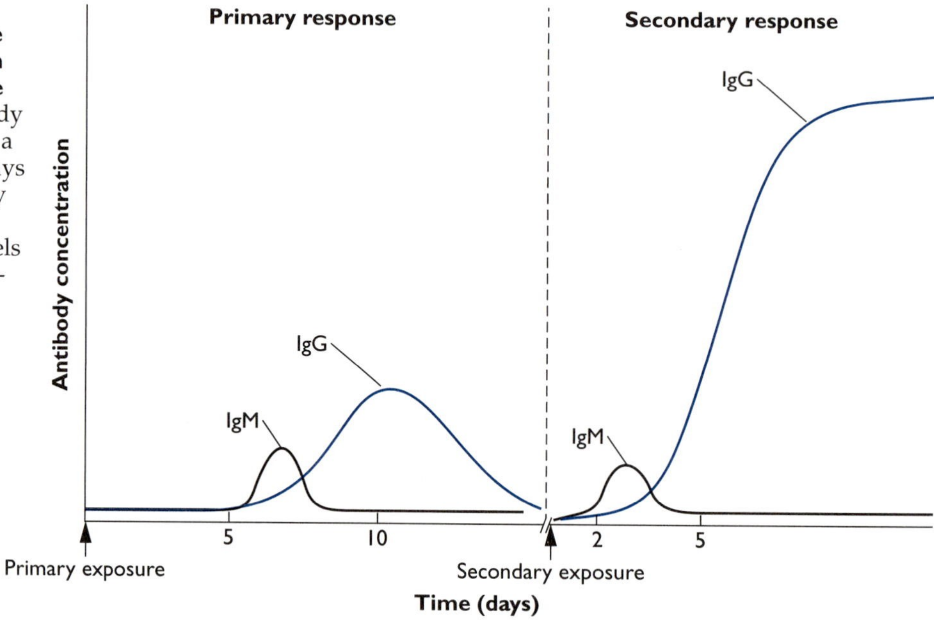

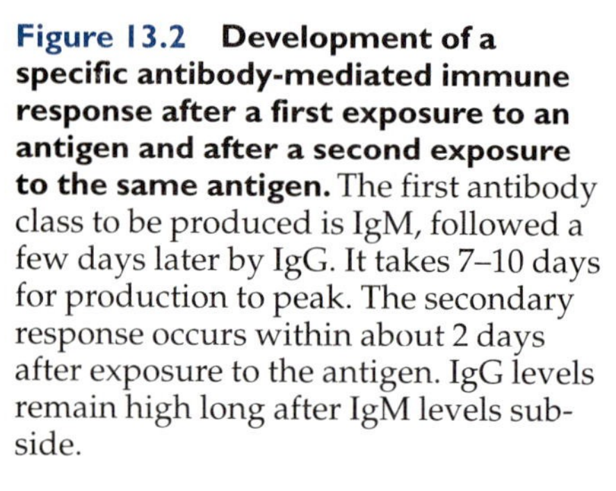
Figure 13.2 Development of a specific antibody-mediated immune response after a first exposure to an antigen and after a second exposure to the same antigen. The first antibody class to be produced is IgM, followed a few days later by IgG. It takes 7–10 days for production to peak. The secondary response occurs within about 2 days after exposure to the antigen. IgG levels remain high long after IgM levels subside.

Vaccination is a strategy for eliciting these specific defenses without actually having to endure the disease. Instead, a person is exposed to the vaccine, which is made of parts of a microbe, killed microbes, or microbes that have been crippled so that they do not cause the disease. These stand-ins for the real microbe trick the body into responding as if an infection had actually occurred. Vaccination will be described in more detail in Chapter 14. As in the case of the nonspecific defenses, enterprising microorganisms have developed ways to evade the specific defenses.

Antibodies

Characteristics of antibodies

Structure of antibodies. Antibodies are protein complexes produced by **B lymphocytes** (B cells). Antibodies recognize specific regions of molecules called **epitopes.** The molecule containing the epitope is called an **antigen.** Although an antigen may be any foreign molecule, from animal proteins to components of microbial cells, "antigen" will be used here to mean an infectious microbe or some part of it. The structure of an antibody monomer is illustrated in Figure 13.3. The antibody monomer consists of two heavy chains and two light chains. (The words "heavy" and "light" refer to the size of the protein, with the heavy chains being the larger.) The heavy and light chains are held together by strategically placed disulfide bonds and noncovalent interactions. Although the antibody structures in Figure 13.3 are drawn in the two-dimensional Y shape for convenience, the actual structure of an antibody is much more compact and folded in upon itself.

Figure 13.3 The structures of IgG, IgM, and IgA. The antigen binding sites (composed of one heavy and one light chain) recognize a specific epitope. The Fc region has a phagocyte receptor binding site and a complement binding site. In the body, IgG occurs as a monomer, IgM as a pentamer, and IgA as a dimer. The monomers of IgA are linked by J chains, as are the monomers of IgM.

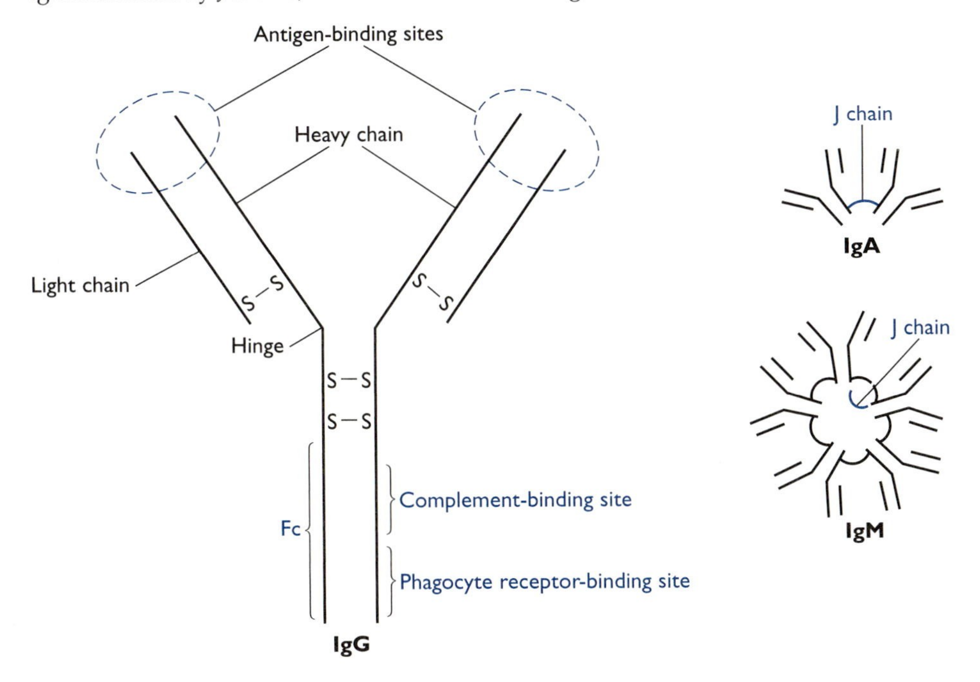

Each of the two **antigen-binding sites** of an antibody monomer recognizes and binds to the same epitope on the antigen (Fig. 13.3). Antigens can be quite large, but epitopes are small. For example, an epitope on a protein antigen will usually be about 4–16 amino acids in length. Microbes contain many epitopes. Not all of these are recognized by the immune system, which responds preferentially to a subset of the possible epitopes. Such epitopes are called "immunogenic." It is still not clear why some epitopes elicit a robust immune response, whereas others do not. Understanding how the immune system decides to recognize some epitopes but not others is important in vaccine design. For example, it is now possible to produce epitope-sized peptides synthetically. Peptides are not only much cheaper to produce than proteins, but they also make it possible to direct the immune response toward a specific region of an antigen. Scientists need this capability to design a vaccine that will stimulate an immune response that is optimum and protective.

Roles of antibodies

Some roles of antibodies are illustrated in Figure 13.4.

Opsonization. Just as activated complement component C3b binds to the surface of a bacterium and helps neutrophils ingest and kill the bacterium,

Figure 13.4 Roles of antibodies. For simplicity, antibodies are shown as monomers. **(A)** Antibodies may opsonize bacteria, thus facilitating ingestion by phagocytes such as neutrophils or macrophages. **(B)** Neutralization occurs when antibodies bind to toxins or viruses, thus preventing them from binding to host cells. **(C)** Antibodies can activate complement by binding to the surface of a microbe. **(D)** sIgA in mucus attaches to mucin components through its Fc region, thus leaving the antigen binding sites free to trap microbes and prevent them from reaching the mucosal surface.

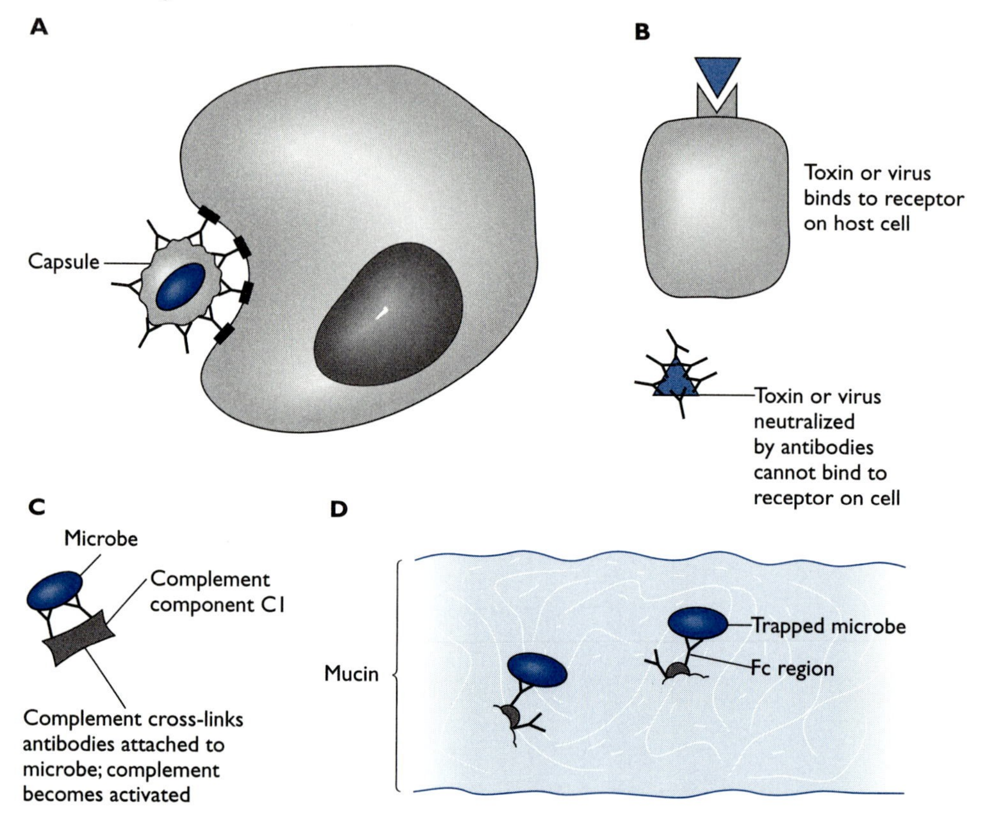

antibodies can also bind to microbial surfaces and help phagocytic cells ingest them. This is possible because the **Fc region** of the antibody (Fig. 13.3) binds to receptors on the surface of phagocytic cells. Why would the body need two types of opsonizing proteins, C3b and antibodies? A good example is provided by the encapsulated bacteria mentioned in Chapter 12. Neutrophils cannot ingest and kill encapsulated bacteria, because the capsules of these bacteria discourage the deposition of C3b on the surface. If the body responds to this emergency by producing antibodies that bind to the capsular surface, the neutrophils can then bind to the Fc portion of the antibodies and eliminate the invader.

A variation on opsonization occurs when antibodies bind to viral proteins displayed on the surface of an infected human cell or to the surface of a microbe too large to be ingested. Such antibodies allow phagocytes to bind to the surfaces of these target cells by attaching themselves to the antibody. Although the phagocytes cannot ingest the cells to which they are attached, they can bombard the surface of the target cells with the toxic compounds produced in their phagolysosomes. This type of attack is called **antibody-dependent cell-mediated cytotoxicity** (**ADCC**).

Neutralization. A second role antibodies play is to neutralize toxins or viruses. Toxins are proteins or other molecules that bind to and harm human cells. An example is the protein toxin that is responsible for the disease tetanus. The bacteria that cause tetanus grow in deep wounds, where they are protected from the immune system by dead tissue surrounding the wounded area. As the bacteria divide, they release a protein toxin that leaks into the bloodstream and circulates throughout the body. The toxin attacks neurons and causes lockjaw, a painful, spastic paralysis that is often fatal. Similarly, the bacterial disease diphtheria is caused by a bacterial toxin that enters the bloodstream and damages heart tissue. In this case, the bacteria grow in the throat instead of in the wound.

Toxins, which cause the damage in diseases such as tetanus and diphtheria, must be detoxified or **neutralized.** If they cannot bind to the receptors they recognize on the surfaces of human cells, toxins cannot harm the cells. Antibodies bind to the toxin and prevent this binding event. Antibodies are bulky molecules and, because of their bulk, physically block the interaction between toxin and human cell. Antibodies neutralize viruses in exactly the same way. Viruses must bind to receptors on a host cell before they can invade. Antibodies that bind tightly to viral surface proteins prevent the virus from attaching to its target cell.

Activation of complement. A third role of antibodies is the activation of complement. The complement cascade can be activated either by interacting with molecules on a microbe's surface or by interacting with an antibody bound to the microbe's surface (see Chapter 12, Fig. 12.8).

Trapping microbes in mucus. One type of antibody, **secretory immunoglobulin A** (**sIgA**), is secreted into mucin. There, the Fc portion of the antibody allows it to stick to mucin components, leaving its antigen-binding sites exposed. The antigen-binding sites bind incoming microbes and trap them in the mucin, preventing their access to the mucosal surface. An important immune protection against polio, a disease caused by a virus that enters the body by invading through the intestinal mucosa, is sIgA that binds to proteins on the viral surface.

Types of antibodies

Different antibodies specialize in the specific roles just described. They also specialize with respect to the time at which they are produced during an immune response and their location in the body.

IgG. The most abundant antibody type found in blood and lymph is **immunoglobulin G (IgG)**, an antibody monomer (Fig. 13.3). IgG is the only antibody type that crosses the placenta. Maternal IgG circulating in an infant's bloodstream is an important protective mechanism during the first year of life. There are four subtypes of human IgG (IgG1–IgG4), each of which differs in amino acid sequence and disulfide cross-linking of heavy chains, but they are otherwise very similar. These subtypes have somewhat different functions. IgG1 (the most abundant of the IgG subtypes) and IgG3 are the most effective at opsonizing microbes, probably because their Fc portions bind most tightly to receptors on phagocytes (Table 13.1). IgG2 and IgG4 opsonize less effectively than IgG1 and IgG3. IgG2 seems to specialize in binding to polysaccharide antigens. IgG4 may have a regulatory role. The different subtypes of IgG provide an illustration of how the body fine-tunes an immune response. Some division of labor among antibody types makes sense, given the number of roles played by antibodies.

IgM. Another type of antibody is **immunoglobulin M (IgM)**. IgM consists of five monomers that are connected. A peptide called the **J chain** is also connected to the five monomers (Fig. 13.3). IgM is the first antibody class to appear in response to a pathogenic microbe (Fig. 13.2), whereas IgG predominates in response to continued or subsequent exposure to the same microbe. The appearance of IgM early in an immune response is useful for diagnostic purposes because it rises and falls during the acute stage of an infection, whereas IgG levels may remain high after the infection has been resolved. IgM, however, is much less abundant than IgG and can thus be harder to detect. IgM is the antibody type that is best at activating complement.

IgA and sIgA. A third type of antibody, **immunoglobulin A (IgA)**, consists of two antibody monomers, joined by a J chain. It is found in low concentrations in serum and tissue. The main role of IgA in blood and tissue is to aid in the clearance of antigen-antibody complexes from the body. This is important because once antibodies have bound to a toxin, virus, or bacterial surface, these complexes need to be eliminated. If such complexes persist too long in circulation, there is a danger that the immune system will recognize them as foreign, precipitating an autoimmune response that attacks human cells.

Table 13.1 Protective Roles of Serum Antibodies IgG and IgM

Role	IgG1	IgG2	IgG3	IgG4	IgM
Neutralize toxins	+	+	+	+	+
Neutralize microbes (prevents binding to a target host cell)	+	+	+	+	+
Opsonization	+	–	+	–	–
Complement activation	+	+	+	–	+
Cross placenta	+	+	+	+	–

Another form of IgA, **secretory IgA (sIgA)**, is an important defense of mucosal surfaces. It is produced by the cells of mucosal linings in the body. sIgA is IgA with an additional peptide, called the **secretory piece,** which is added as the antibody is secreted into mucin (Fig. 13.5). As already mentioned, the main role of sIgA is to trap incoming microbes or microbial toxins in the mucin layer and prevent them from reaching the mucosal cells. sIgA is also secreted in breast milk. Thus, sIgA, like IgG, serves as an important protection for infants who have not yet developed their own set of immune responses.

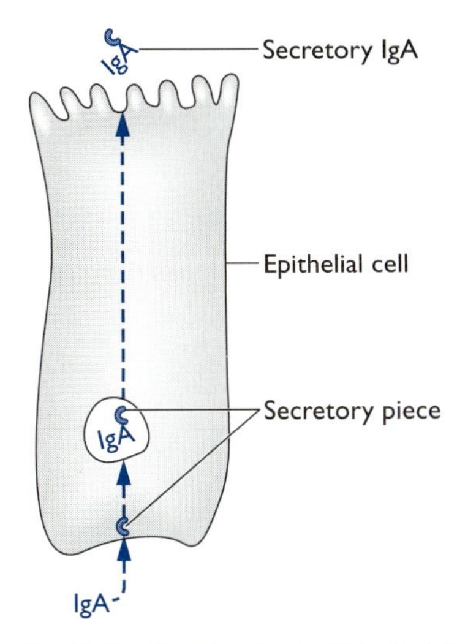

Figure 13.5 Synthesis of sIgA. IgA is produced in the area below the mucosal epithelial cells. It then binds to a receptor on the basal surface of the cell, and, as it is transported across the cell to the lumen, a small molecule (the secretory piece) is attached, thus making it into sIgA.

IgE. A fourth type of serum antibody is **immunoglobulin E (IgE)**. IgE is an antibody monomer and is thought to play a protective role against disease-causing worms. IgE is also involved in the allergic response. IgE binds to a human cell type called a mast cell. **Mast cells** are found throughout the body and are abundant near mucosal surfaces. If two IgE molecules bind closely enough together to be cross-linked by an antigen, the mast cell releases granules containing histamine and other vasoactive compounds (compounds that constrict or relax blood vessel walls), which in turn cause the allergic reaction. In the respiratory tract, this reaction takes the form of sneezing and excess mucus production. In excessive amounts, histamine and other mediators of allergy can produce asthma. In the intestinal tract, diarrhea and cramping are more typical symptoms. Some scientists speculate that the elimination of worms from the human population in developed countries may be linked to asthma and other allergic conditions. The IgE connection is one of the reasons for this suggestion.

Serological tests

Antibodies are the basis for an important type of clinical test, the **serological test.** The designation "serological" came from the fact that the first such tests detected antibodies in serum taken from a patient. Serum, the fluid that is left when blood clots, contains soluble components of blood, such as antibodies. Today, the term "serological test" is applied to any test that uses antibodies. Thus, serological tests may be used outside the clinical context for applications other than the diagnosis of disease. There are two main types of clinical serological tests: those that detect antibodies specific for a particular antigen and those that use antibodies to detect an antigen. Today, antigens in blood are also detected by polymerase chain reaction (PCR). Thus, for example, the most sensitive detection test for human immunodeficiency virus (HIV) in serum is PCR that amplifies a portion of the virus's genome. Here, we will consider some of the tests that involve antibodies.

Measuring antigen or antibodies in serum. Both types of serological tests are illustrated in Figure 13.6. A bewildering variety of these tests is now available; for simplicity, only one example of each of the two approaches is given. Note how the tests take advantage of the antibody structure. One type of test detects antibody in the serum by immobilizing a specific antigen to a plastic surface. If antibody specific to that antigen is present in that patient's serum, it will bind to the antigen. The bound antibody is detected by using a second antibody against the Fc portion of the bound antibody. The second antibody has a label and is available commercially. Differences in the Fc portion of IgG and IgM are used to design secondary antibodies that recognize one or the other. For detection of an antigen such as a virus in serum, a **"sandwich" technique** is used. Antibodies specific for that antigen are bound to

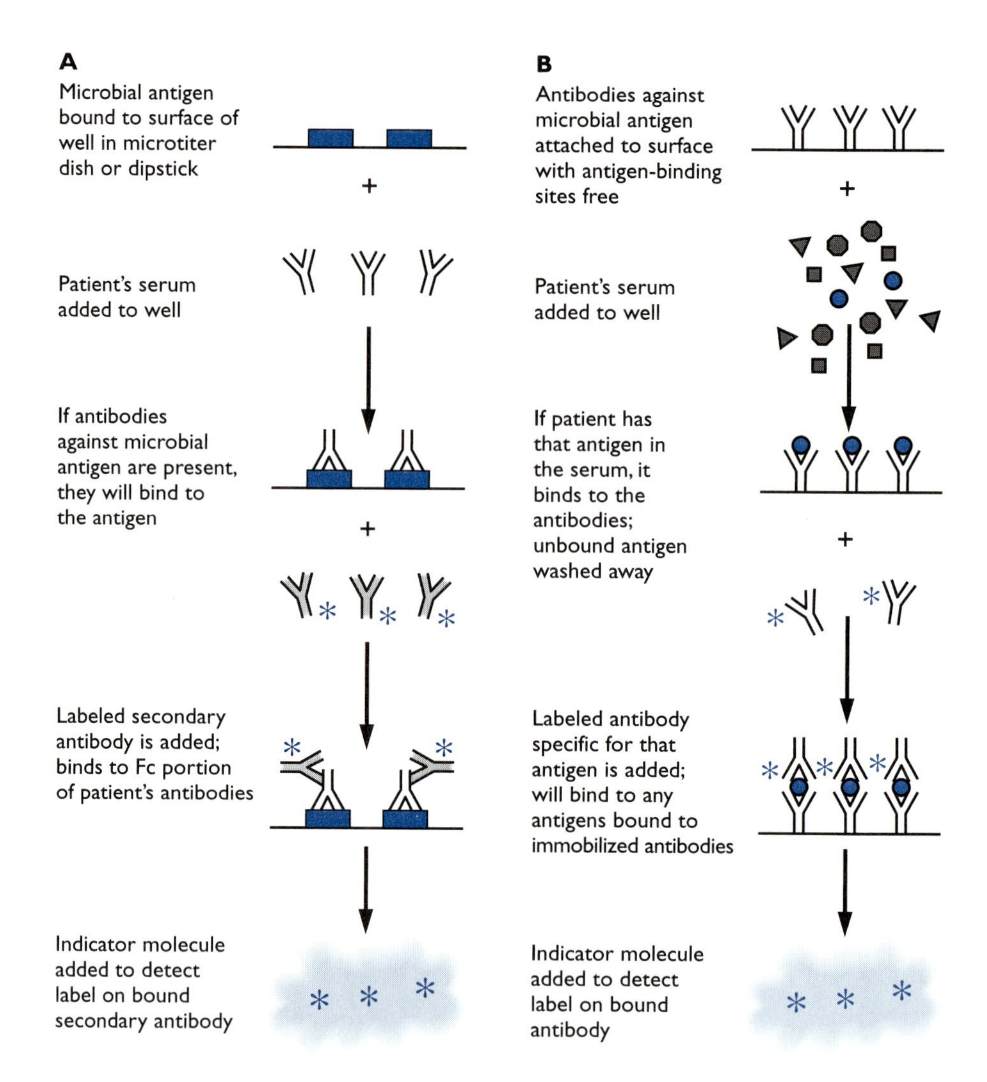

Figure 13.6 Types of serological tests. (A) A test that detects a specific antibody in a patient's serum. **(B)** A sandwich test that detects a particular microbial antigen in the serum.

plastic and "capture" the antigen. Then a second labeled antibody specific for that antigen is added and binds to another part of the same antigen.

Tests that detect antibodies in the patient's serum are used to determine whether a patient has mounted an immune response. These tests can be made quantitative by diluting serum and determining what dilution no longer contains detectable antibody. The inverse of this dilution is called a titer. The higher the titer, the higher the concentration of antibody. As previously stated, IgM is a good indicator of active infection. IgG can be, too, if at least two tests are done: one early in the infection and one later in the infection. If the IgG titer is significantly higher in the second serum sample than in the earlier one, this indicates that the concentration of IgG is rising—a sign that active infection is occurring. The test most commonly used in the initial testing for HIV infection detects antibodies against the virus in serum. The initial test uses a mixture of viral antigens bound to a paper matrix. A problem with this type of test is that serum antibodies can sometimes bind nonspecifically to the antigen mixture, producing a weak false-positive reaction. In such cases, a test that fractionates the antigens clarifies the diagnosis. This fractionation test is based on a technique called the Western blot (Fig. 13.7), which is widely used in research.

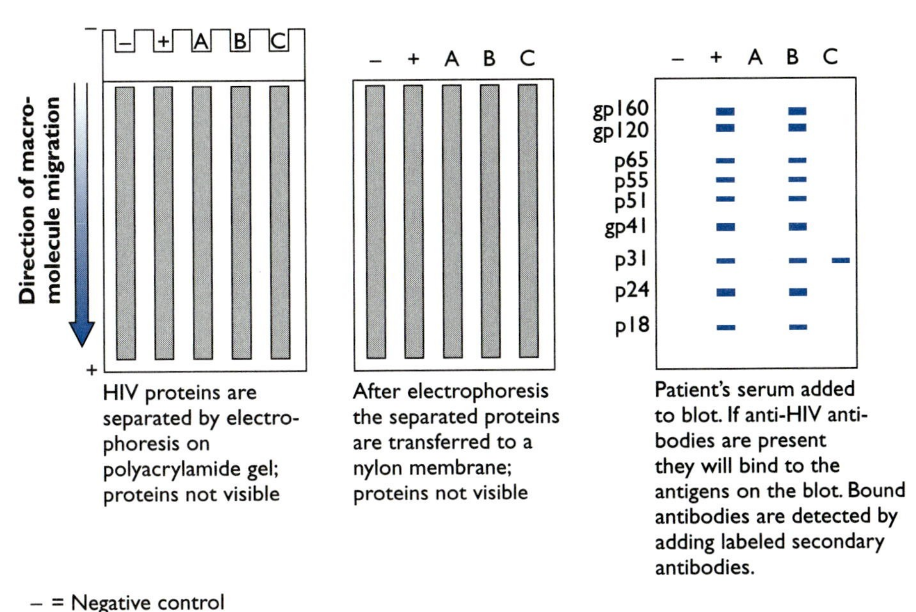

Figure 13.7 Western blot test for detecting antibodies against human immunodeficiency virus (HIV) in a patient's serum.

The Western blot. In a Western blot, the proteins (in the case of our HIV example, HIV proteins) are separated in a gel in an electric field according to size. The separated proteins are then transferred to a matrix often referred to as a membrane. The membrane containing the transferred proteins is called a **blot.** The blot is incubated with serum. Bound antibody is then detected by a labeled secondary antibody. Laboratory workers performing the Western blot test for HIV infection do not run a gel every time they want to do the test. Instead, they purchase strips of a blot from a commercial source and incubate the strips with the patient's serum. In research, scientists most often use the Western blot to detect a particular protein in complex mixtures of proteins. Instead of serum from a patient, they use antibody that has been generated against a particular antigen. For example, if a scientist wants to know whether a bacterium was producing enzyme A, he or she would separate proteins from the bacterium on a gel to make a blot, then incubate it with antibodies specific for enzyme A.

Cytotoxic T Cells and Activated Macrophages

Cytotoxic T cells and activated macrophages are components of what is called cell-mediated immunity or the cellular immune response. The primary role of **cytotoxic T cells** is to kill human cells infected with viruses or other intracellular pathogens. Cytotoxic T cells have receptors on their surfaces that bind specifically to a particular microbial antigen and kill only host cells displaying that antigen on their surface. Virus-infected cells, for example, often display viral proteins on their surfaces. The presence of these proteins is a signal to attack. The attack is reminiscent of the attack of antibody-bound phagocytes in ADCC, in that the attached cytotoxic cell secretes toxic substances that kill the target cell. The toxic substances produced by the cytotoxic T cells are somewhat different from those produced by the phagocytic cells. Thus, the cytotoxic T cell response and ADCC complement each other in an immune response to infection.

Like neutrophils, **macrophages** are phagocytic cells that engulf and kill microorganisms. Macrophages and neutrophils differ, however, in that macrophages are longer lived and more likely to be found in tissue than in the bloodstream. **Monocytes,** the precursors of macrophages, are found in the bloodstream. Another difference between macrophages and neutrophils is that macrophages participate directly in the specific defenses, first as the initial processors of foreign antigens and later as activated macrophages. As with different antibody types, different subtypes of macrophages probably perform these different functions. **Activated macrophages** have been stimulated by cytokines to higher levels of oxidative burst. Activated macrophages can kill microbes, such as the bacterium that causes tuberculosis, that are not killed by neutrophils and resting macrophages. The activated macrophage response is both specific and nonspecific. Activated macrophages do not differentiate between one antigen and another, but they are included in the specific defenses because their activation is part of the specific defense response.

Production of Antibodies, Cytotoxic T Cells, and Activated Macrophages

Processing the antigen: a critical decision point

Epitope display. How does the body choreograph the production of antibodies, cytotoxic T cells, and activated macrophages? The process begins with macrophages that process the antigen and display epitopes on their surfaces (**antigen-presenting cells, APCs**). Processing the antigen means that the macrophages break down a microbe into epitopes, bind some of those epitopes to a protein complex called the **major histocompatibility complex** (**MHC**), and finally display the MHC-epitope complex on the APC surface (Fig. 13.8). The reasons that the APC chooses some epitopes over others is not well understood, but this is obviously a critical step in the immune system's decision process. APCs all have the same function, breaking up and presenting antigens to initiate the immune response, but differ in location and shape and thus have different names. For example, the APCs of the brain are called **microglial cells,** and those of the liver are called **Kupffer cells.**

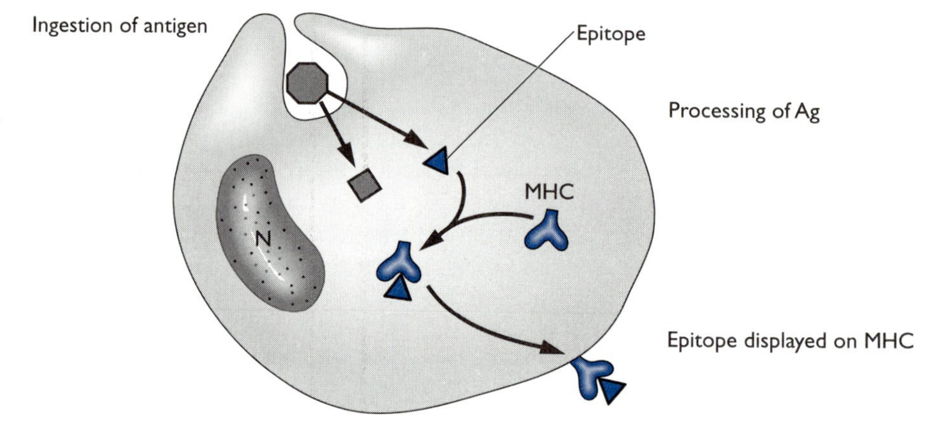

Figure 13.8 Antigen processing by an antigen-presenting cell (APC). The antigen is ingested and broken down into epitopes. Some of these epitopes bind to a major histocompatibility complex (MHC). The MHC-epitope complex is then displayed on the surface of the APC.

The MHC I/MHC II decision. There is an additional complexity to consider. The next step is for one of two types of T cells to recognize the MHC-epitope complex. If this is a cytotoxic T cell, the result will be activated cytotoxic T cells. If it is a T helper cell, the ultimate result will be antibody production. The APC directs which type of T cell will be most likely to interact with the MHC-epitope complex by using one of two different MHCs. **MHC I**-epitope complexes favor a cytotoxic T cell response, and **MHC II**-epitope complexes favor a T helper cell response.

Scientists are still trying to understand how the choice between MHC I and II is made, but some outlines of the process are beginning to appear. Recall that the cytotoxic T cell response is directed against microbes, such as viruses, that can invade the cell interior. When the APC takes up such an antigen, the antigen can escape from the phagosome. Such antigens elicit an MHC I response. Antibodies (the production of which is associated with T helper cells) work best on microbes or toxins that stay outside cells and are thus less likely to escape the APC phagosome by passing through its membrane. Such antigens elicit an MHC II response. This theory does not explain why viruses can elicit both a cytotoxic T cell and an antibody response, but it begins to make some sense of the MHC I/MHC II decision process.

Interaction between the APC and T cells

General features of the interaction. Depending on which MHC displays the epitope, the APC will next interact with the appropriate type of T cell (Fig. 13.9). The T cell recognizes not only the MHC type but also the shape of the MHC-epitope complex, which is the signature for that particular epitope. The protein complex on the T cell surface that performs this very specific recognition is called the **T cell receptor.** The body produces thousands of T cells with different T cell receptors, each of which recognizes a specific epitope. Interaction with the APC not only singles out one specific T cell, but then causes it to reproduce itself many times (**proliferation**). Later, some of these T cell clones will become the memory cells that allow later encounters with the same microbe to produce a more rapid immune response.

Note that in Figure 13.9, much more is going on than the specific interaction between T cell receptor and MHC-epitope on the APC. First, many other proteins make contact. These proteins are not specific for the epitope but instead ensure that the T cell is recognizing an APC, not another cell type. These interactions also ensure that only a T cell with just the right T cell receptor will bind tightly to the MHC-epitope and that the two will be able to interact stably. In effect, these other contacts are designed to prevent the immune system from interacting with any cells other than the appropriate ones, thus preventing an attack by the immune cells on the human body. For example, if a T cell accidentally recognized a protein complex on a tissue cell other than an APC, the result might be serious tissue damage.

Cytotoxic T cell activation. Tight binding of a T cell to an APC stimulates the APC to release cytokines (for example IL-1, TNFα). These cytokines stimulate the T cell to proliferate and become activated. Cytotoxic T cells (which have a CD8 protein on their surface) will be activated if the epitope displayed on the APC was complexed with MHC I. Cytotoxic T cells are good at recognizing infected cells, because virtually all cells of the body produce MHC I. An infected cell will display epitopes from the

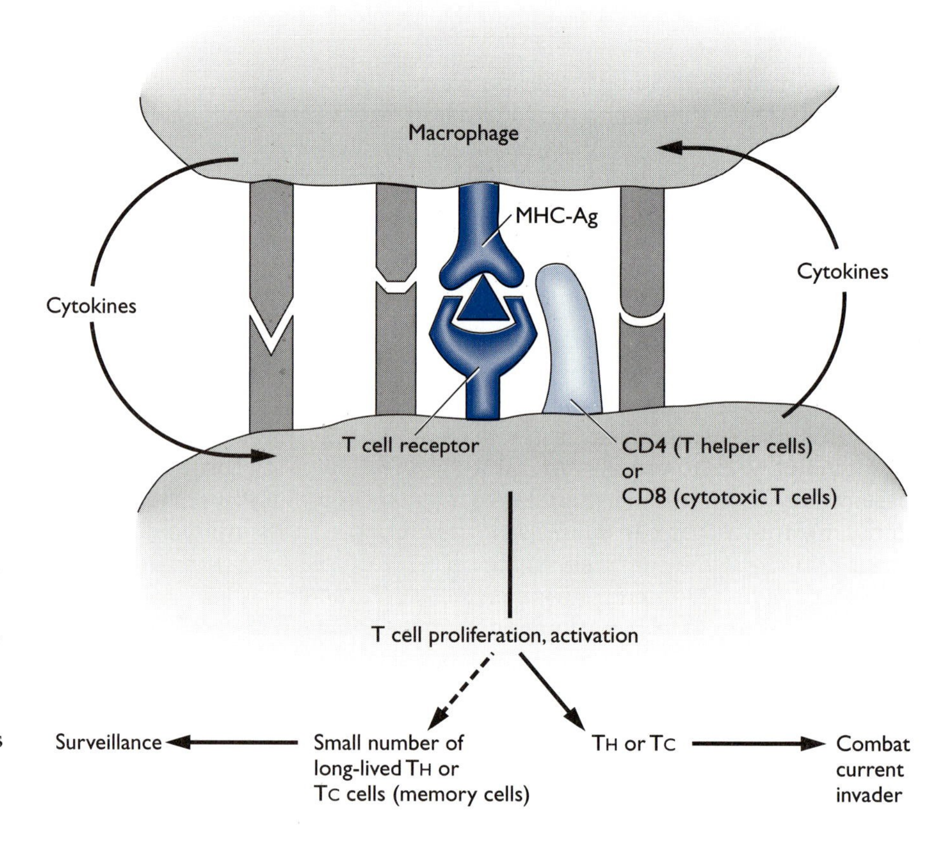

Figure 13.9 Interaction between antigen-presenting cells (APCs) and T cells. The T cell receptor will bind only to the major histocompatibility complex (MHC) presenting the epitope that it recognizes. Many other proteins on the two types of cells must also make contact. Cytotoxic T cells have CD8 proteins on the surface, whereas T helper cells have CD4 proteins.

invading microbe on its surface in a complex with MHC I. The cytotoxic T cell receptor will recognize this MHC I-epitope complex in much the same way it recognizes the MHC I-epitope complex on the surface of the APC. When a cytotoxic T cell binds to the surface of an infected cell, the cytotoxic T cell releases toxic compounds that kill the infected cell.

T helper cell activation: another decision point. If the APC presents an epitope in a complex with MHC II rather than MHC I, T helper cells are stimulated (Fig. 13.9). APCs stimulate T helper cells by a process similar to that used to stimulate cytotoxic T cells. That is, the T helper cell receptor recognizes the MHC II-epitope complex, and the binding between the APC and the T helper cell is aided by the CD4 surface protein on the T helper cell and by other protein-protein interactions between the APC and the T helper cell. Tight binding causes the APC to release cytokines, which stimulate the T helper cell to proliferate and become activated.

Two types of T helper cells, TH1 and TH2, descend from the same cell type, TH0 (Fig. 13.10). TH0 cells are directed to produce TH1 type or TH2 type T helper cells by the kinds of cytokines that are produced. The decision between TH1 and TH2 production is complex and is only now beginning to be understood. The pathway taken by the TH0 cell (that is, formation of TH1 or TH2) is very important. TH1 cells produce the cytokine, **interferon γ (IFN-γ)**, which activates macrophages and further activates cytotoxic T cells produced by the MHC I pathway. TH1 cells also stimulate B cells to produce opsonizing antibodies (predominantly IgG1 and IgG3).

By contrast, the TH2 cells activate **eosinophils,** a cell type associated with the response to worm infections. TH2 cells stimulate B cells to produce antibodies, but these are predominantly IgE. Both eosinophils and IgE are also associated with the allergic response. Thus, the TH1 type response is the most desirable in most microbial infections, because it stimulates both antibody production and cell-mediated immunity. The role of the TH2 type response is not understood, but presumably it has an important role or it would not have evolved. (The pathways shown in Fig. 13.10 are those believed to occur in humans. In mice, the response looks somewhat different. Thus, a diagram you may see elsewhere that seems to disagree with this one is probably the rodent pathway.)

Production of antibodies by B cells

Response to protein antigens. In order for a B cell to be stimulated to produce antibodies against protein antigens, it must be recognized by the appropriate T helper cell. B cells can act like APCs, in that they take up foreign antigens, process them, and present epitopes on their surfaces in complex with MHC II. B cells are one of only a few cell types that produce MHC II, so the use of MHC II aids T helper cells in finding the correct B cell. A T helper cell activated by an APC with a particular MHC II-epitope complex will recognize only a B cell with that same MHC II-epitope complex displayed on its surface (**cognate B cell**). Binding of a T helper cell to the B cell, like binding of the APC to the T cell, requires a set of multiprotein contacts (Fig. 13.11).

Tight binding between the B cell and the T helper cell stimulates the T helper cell to produce cytokines, which in turn stimulate the B cell to

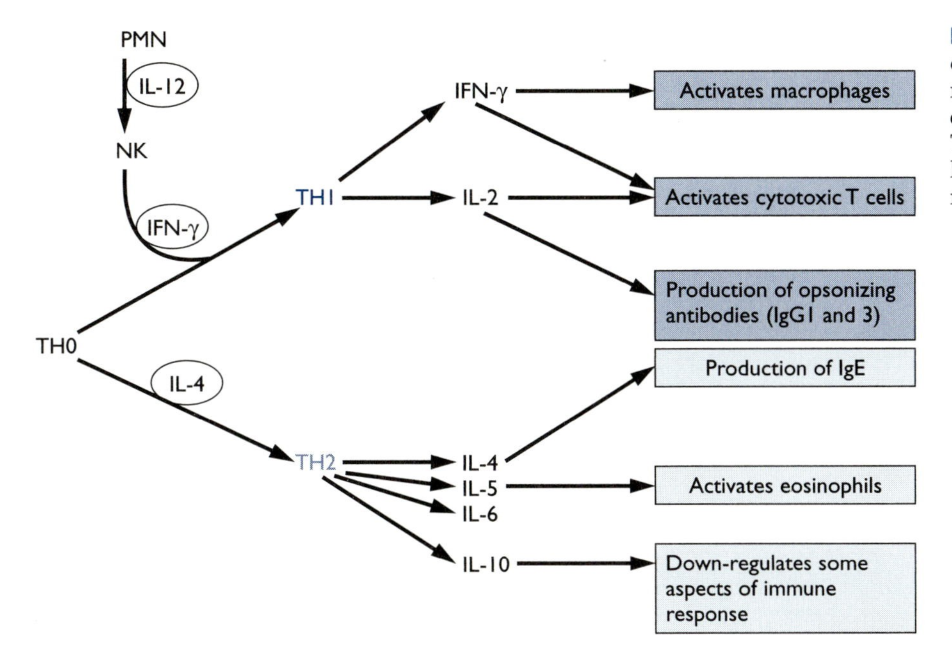

Figure 13.10 Development and roles of TH1 and TH2 cells. TH0 cells respond to various cytokines and differentiate into either TH1 or TH2 cells. The TH1 response ultimately leads to stimulation of antibody production and cell-mediated immunity.

proliferate and differentiate into the form that secretes antibodies. A small percentage of activated B cells become memory B cells. In subsequent encounters with the same antigen, the antigen will bind to antibodies on the surfaces of the memory B cells and stimulate the B cell directly to proliferate and secrete antibodies. This is a much more rapid response than the APC-initiated process. During an infection, both the APC-initiated response and direct B cell stimulation by antigen binding can occur, but the APC-initiated process is most important in the first encounter with a particular microbe, whereas direct stimulation of memory T helper cells or memory B cells by antigens plays a more important role in subsequent encounters.

Response to polysaccharide antigens. Bacteria that produce polysaccharide capsules pose a special problem for the immune system. The protective

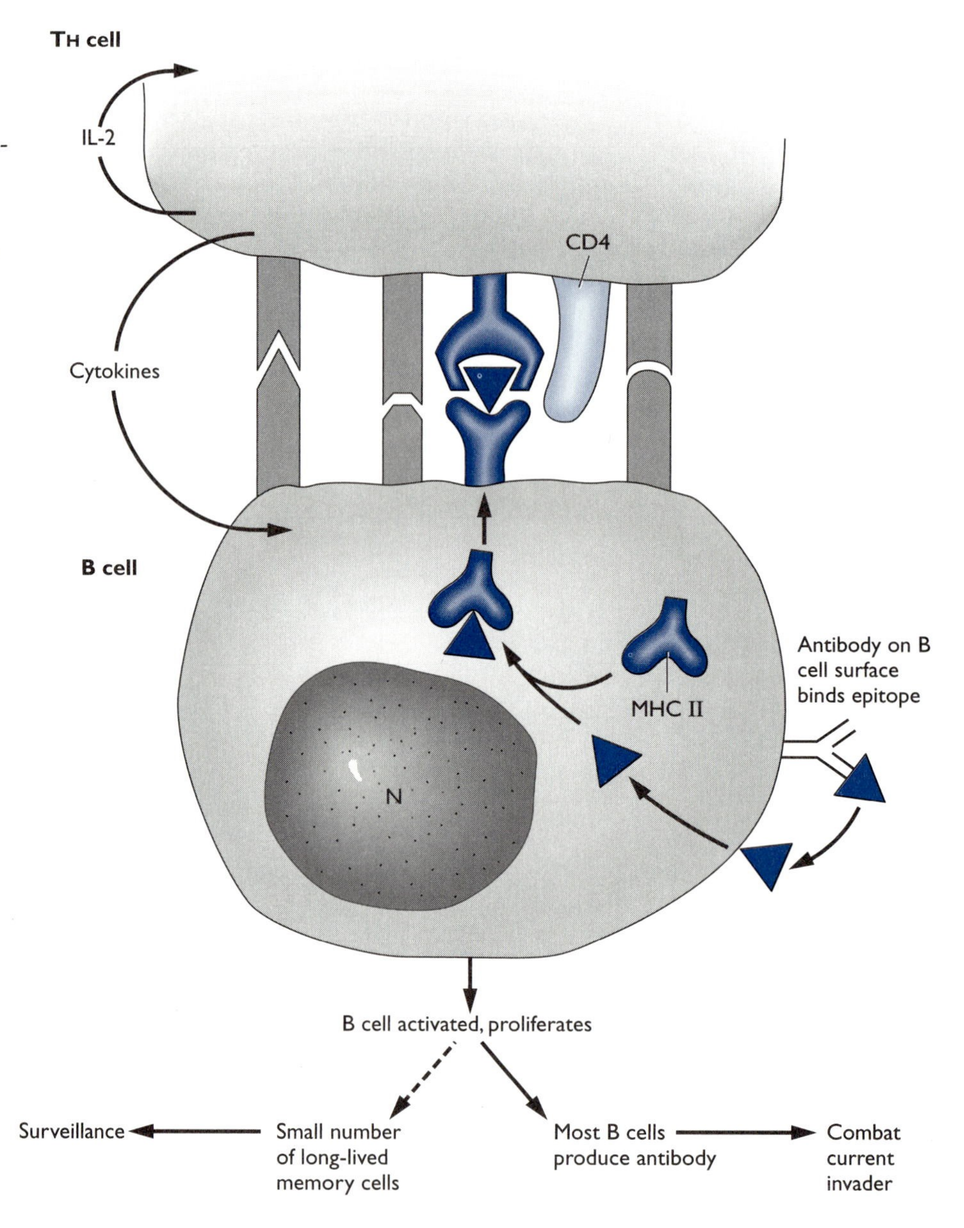

Figure 13.11 Stimulation of B cells to produce antibodies. The TH1 cell finds and activates a B cell presenting the same epitope on the major histocompatibility complex II (MHC II) complex (cognate B cell) as that on the APC that initially activated the TH1 cell. As in T cell activation, many surface proteins on the two cells must interact, but only a few are shown.

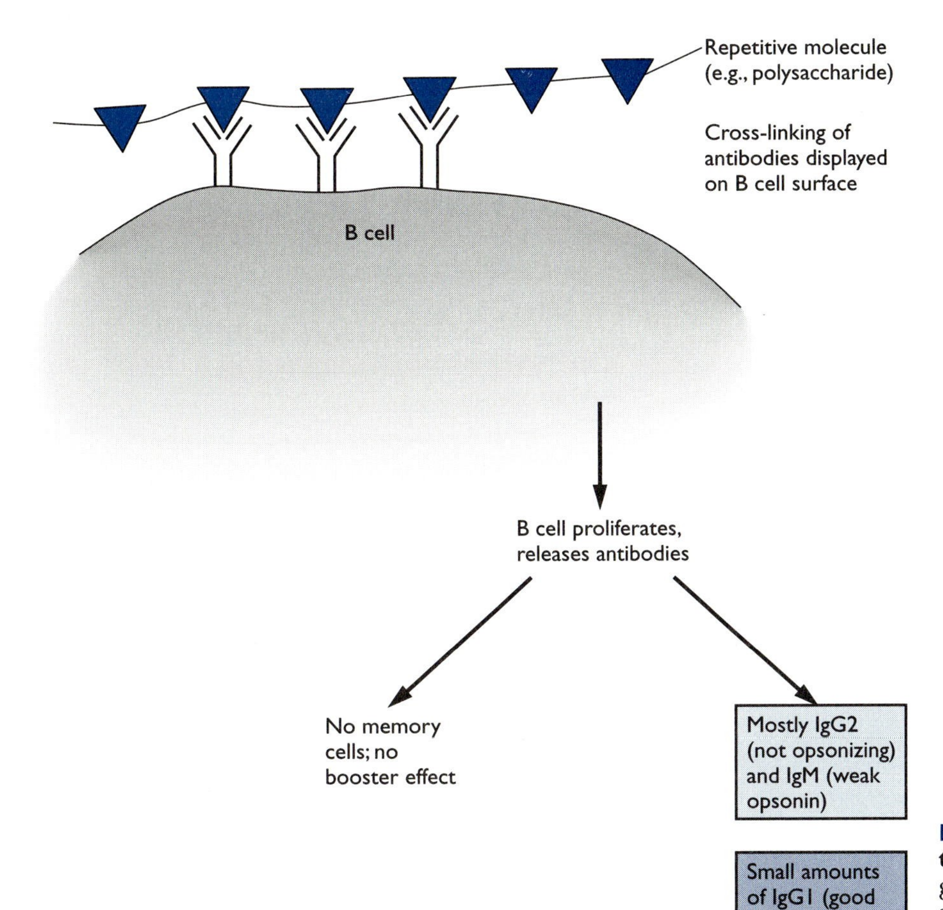

Figure 13.12 T-independent production of antibodies. T-independent antigens interact directly with B cells and stimulate only an antibody-mediated response.

response against encapsulated bacteria consists of antibodies that bind capsular polysaccharides and opsonize the bacteria. Polysaccharides are handled by a pathway different from that used to generate antibodies against protein antigens. Polysaccharides and B cells interact directly, with no involvement of T helper cells. This is called the **T-independent pathway** of antibody generation. Polysaccharides elicit an immune response in children and adults, but not in infants younger than 2 years.

T-independent antigens interact directly with B cells, so they provoke an antibody-mediated, but not a cell-mediated, response. Unactivated B cells display the antibody they produce on their surfaces. Cross-linking of these surface antibodies stimulates the B cell to increase production of the antibodies it produces and release them into the circulation (Fig. 13.12). Polysaccharides are characterized by repetitive epitopes, a trait that promotes cross-linking of the antibodies on the B cell surface.

Development of the Specific Defense Response

Fully effective specific and nonspecific defenses do not develop until human infants are 1–2 years old. During most of the first year, infants are protected at least partially by maternal antibodies transferred through the placenta during gestation or ingested after birth in breast milk. Circulating maternal antibodies can actually interfere with an immune response to some antigens, because high levels of antibody to a particular epitope will bind the antigen and prevent the infant from developing an antibody response to that epitope. The damping effect of circulating maternal IgG is the reason that some vaccines are not given to infants younger than 1 year. During the period of time when the protection conferred by maternal antibodies decreases and before the infant's own immune system develops, pathogenic microbes have a window of opportunity. This is one of the reasons that children younger than 2 years are especially vulnerable to infectious diseases. The nonspecific defenses also are not well developed in children younger than 2 years.

How Microorganisms Overcome or Avoid the Specific Defense System

We have mentioned a few examples of how microorganisms evade the immune system, but this subject has not received as much attention as microbial strategies for defeating or avoiding the nonspecific defense system. Because this is an important problem, we provide here an overview of known microbial strategies. Some microorganisms kill macrophages, thus reducing the number of APCs. Two examples are HIV, which infects monocytes (the precursors of macrophages), and *Mycobacterium tuberculosis,* the bacterium that causes tuberculosis. Some bacteria produce proteases that specifically cleave IgG or sIgA. The role of these proteases in disease is somewhat controversial. Cleaving sIgA would certainly make the antibody much less effective in trapping bacteria in mucus.

Another anti-antibody strategy is based on the fact that antibodies are effective in opsonizing bacteria or activating complement only if they bind in the correct manner, with their antigen-binding sites stuck to the bacterial surface. This leaves their Fc portions free to interact with phagocyte receptors or with complement. Some bacteria bind the Fc portions of antibodies, so that the antibodies are bound with the antigen-binding sites extending outward instead of attaching to the antigens on the bacteria. This strategy may have the added benefit of making the bacterium look like a human B cell displaying antibodies on its surface. A few microorganisms employ the rather expensive approach of continuously changing their surface antigens. As soon as antibodies develop against one version of a surface antigen, a few members of the population rearrange their DNA to express a different one. These members are selected because the antibody response is no longer effective.

Not all microbial defenses are directed against antibodies and their activities. Viruses can make themselves less evident to cytotoxic T cells by limiting the amount of viral antigens displayed on the cell surface. Some microbes elicit an inappropriate response, possibly by manipulating the TH1/TH2 pathway.

The crux of the problem is that bacteria and other microorganisms can evolve faster than humans. Thus, they can find ways to evade our defenses more easily than we can modify our defenses to thwart them. But the fact that most microorganisms are successfully put off by our defenses shows that, despite our comparatively slow evolutionary adaptation rate, these defense are quite effective.

Protective and Nonprotective Immune Responses

The human immune system is such an awesome evolutionary development that one can easily fall into the error of thinking of it as infallible. In fact, the immune system sometimes responds inappropriately. An example of this is provided by the response to tuberculosis. Tuberculosis is caused by a bacterium, *M. tuberculosis,* which can grow in and kill phagocytic cells. The only type of phagocytic cell that can kill *M. tuberculosis* is an activated macrophage. Antibodies are not only an inappropriate response but may even help the bacteria by facilitating their entry into neutrophils and non-activated macrophages. In many people infected with *M. tuberculosis,* however, the antibody response predominates. These are the people most likely to die of the disease. People whose immune systems instead mount an activated macrophage response are the ones most likely to survive.

A second example is provided by vaccines that actually make their target diseases worse. This happened in the case of a vaccine against respiratory syncytial virus, a cause of death in infants. The vaccine performed well in laboratory rodents, but in humans it apparently elicited the wrong kind of response, possibly because of a wrong turning at the TH1/TH2 decision point. Infants who were vaccinated got a more serious form of the infection. Learning how to engineer a protective response is a primary goal of modern vaccine research. Scientists are finding, however, that knowing how the body mounts an immune response is not enough. It is also essential to understand how the microbe in question interacts with the human body and is affected by various types of immune responses.

A third example of an inappropriate immune response is one in which the immune system begins to attack human tissues, a condition called **autoimmune disease.** For the immune system, autoimmune disease is the equivalent of septic shock for the nonspecific defenses: a response intended to be protective that turns destructive. Why would the immune system attack the body it is designed to protect? Some autoimmune diseases occur when a person is infected with a microbe that produces a protein very similar to proteins found on human cells. An example is the bacterium *Chlamydia trachomatis,* a cause of infertility and pelvic inflammatory disease. This bacterium produces copious amounts of a protein that is quite similar to a protein produced by stressed human cells. Cells being infected by *Chlamydia* are certainly stressed and may be attacked by the immune system reacting to the chlamydial protein. This response is thought to be partly responsible for the extensive damage to fallopian tubes that can occur during an infection caused by *C. trachomatis.* Another trigger of an autoimmune response is antigen-antibody complexes that remain too long in circulation. The complexes are recognized as foreign and attacked by the immune system. Some of the damage caused by syphilis may arise from this process.

The two examples just given are autoimmune responses thought to be caused by microbial infection. Most immunologists probably would not list microbial infections as the main cause of autoimmune disease. Certainly diseases such as lupus, arthritis, and inflammatory bowel disease, all thought to be caused by autoimmune responses, are not generally associated with microbial infections. Recently, however, some scientists have raised the question of whether many autoimmune diseases may in fact be precipitated by microbial infection. Their argument is that the immune system obviously evolved to combat microbial infections, so whenever the immune system misbehaves, a microbe must somehow be involved. It will be interesting to see how this debate plays out.

chapter 13 at a glance

The Immune System: An Overview

Component	Characteristics	Roles
Antibodies	Proteins that bind very specifically to certain antigens; Fc portion binds receptors on phagocytes, recognized by complement	Opsonization (mainly IgG*) Neutralize toxins, viruses (mainly IgG, sIgA) Antibody-dependent cell-mediated cytotoxicity (ADCC) (mainly IgG) Activate complement (IgM, IgG) Trap microbes, toxins in mucin (sIgA) Pass through placenta (IgG) Secreted in breast milk (sIgA)
Antigen-presenting cells (APC)†	Macrophages that degrade antigens, displaying them on major histocompatibility complex I (MHC I) (cytotoxic T cell response) or MHC II (T helper response)	Initiate the development of the immune response
Cytotoxic T cells	Cells that interact with APCs displaying MHC I	Kill infected human cells.
T helper cells	Cells that interact with APCs displaying MHC II	Induce B cells to produce antibodies; produce cytokines that activate cytotoxic T cells, activate macrophages to higher killing power.
B cells	Cells that interact with T helper cells or directly with polysaccharide antigens	Produce antibodies

*For simplicity, the different subtypes of IgG have been lumped together.
†Although B cells can act like macrophage APCs, they are not listed under APCs to emphasize the key role played by the macrophages in the development of the immune response.

Tests Based on Antibodies

Serological tests to detect antigen or antibody in serum
- Antigen detection tests
  - Antibodies bound to plastic trap antigen
  - Labeled antibodies bind trapped antigen
- Antibody detection tests
  - Antigen bound to plastic traps antibodies in serum
  - Labeled antibody binds to Fc portion of trapped antibodies
- Titer: the inverse of the highest dilution of serum that does not react positively on the test

Western blot: proteins separated by electrophoresis prior to binding with antibodies

chapter 13 at a glance *(continued)*

Examples of Microbial Strategies that Counter the Specific Defenses

Specific defense	Microbial counter-strategy
Macrophage processes antigens	Microbe kills macrophages
IgG opsonizes microbial cell	Microbe binds Fc portion to its surface, so IgG is wrong way around Ability to keep changing surface antigens (antigenic variation) so that antibodies do not recognize antigen
sIgA traps microbes in mucin	Microbial protease destroys sIgA
Cytotoxic T cells find and kill infected cells	Some viruses limit the display of their proteins on the human cell surface, making detection by T cells difficult
T helper cells dictate the type of response (TH1 or TH2)	Some microbes elicit an inappropriate TH response

Questions for Review and Reflection

1. What are the uses of the Fc portion of an antibody? Why is this region separated from the antigen-binding site?

2. What do antibodies do that complement does not do and vice versa?

3. Why are there so many kinds of antibodies?

4. Why does it make sense that sIgA does not activate complement?

5. Why is a single test for IgM useful for diagnosis of disease but two tests are required if IgG is used? What is an antibody titer and what does it mean?

6. Why is a Western blot stained with labeled antibodies rather than with a general protein stain?

7. Why do APCs have to be macrophages?

8. What is the function of T helper cells? Why don't APCs react directly with B cells?

9. Putting together what you learned in this chapter, describe a protective immune response and a nonprotective immune response to a virus that invades through the gastrointestinal tract.

10. What are the decision points in the immune response pathway? How do they resemble each other? How do they differ? In a nonprotective response, how might things have gone wrong?

key terms

adjuvant a compound, such as alum, that is added to a vaccine to stimulate the immune response to vaccine components

antibody monomer a protein complex with two antigen binding sites and an Fc portion that interacts with receptors on phagocytic cells or with complement components

antigen anything the immune system recognizes as foreign, including microorganisms

antigen-presenting cell (APC) a macrophage that processes antigens and presents them on its surface as part of a major histocompatibility complex (MHC)-epitope complex

autoimmune disease a disease that arises because the immune system begins to recognize human cells as foreign and mounts an attack on them

B cell a cell that produces antibodies after interaction with T helper cells; interacts directly with polysaccharide antigens

cytokines proteins produced by many cells of the immune system (as well as other cells) that activate or depress the cells of the immune system

cytotoxic T cell a cell that recognizes microbial antigens displayed on infected human cells and kills them

epitope a small portion of a protein antigen that is recognized by antibody antigen-binding sites and is displayed by the antigen-presenting cells on MHC

neutralization binding of antibodies to toxins or microbes, preventing them from binding to their target cell

opsonization binding of antibody or complement component C3b to the surface of a microbe and to receptors on the surface of a phagocyte; aids in phagocytosis

protective immune response response that successfully protects the body from microbial infection

secreted antibodies secreted immunoglobulin A (sIgA), a dimer

serum antibodies immunoglobulin G (IgG) a monomer; immunoglobulin M (IgM), a pentamer; immunoglobulin A (IgA), a dimer; and immunoglobulin E (IgE), a monomer

T helper cell cell that first interacts with the antigen-presenting cell (APC), then with B cells to stimulate them to produce antibodies

T-independent response response to antigens such as polysaccharides that are not processed through the APC-T cell pathway.

14 Vaccination

Jenner's cowpox vaccination procedure was ridiculed by some. The artist James Gillray even published a cartoon showing vaccinated people with small cow heads growing out of various parts of their bodies. Fortunately for the millions of people whose lives have been saved from deadly diseases such as smallpox and tetanus, and whose limbs were not twisted by polio, the public listened to Jenner and his advocates and not those who ridiculed vaccination.

Along with the discovery of microbes and the understanding of their roles in disease, the discovery of vaccination is surely one of the milestones of microbiology. Not only was vaccination an important new way of preventing disease, but its use led to the discovery of the immune system. It is worth spending a page or so taking a selective look at some of the events of the early days, not so much to recite historical occurrences but to give an idea of what it must have been like to be involved in such revolutionary discoveries.

The Origins of Vaccination

Edward Jenner (1749–1823), an English physician, was the first to practice what we now call vaccination. In the 1700s, smallpox was dreaded as a disfiguring and even fatal disease. Jenner noticed that milkmaids, who often had a much less serious disease called cowpox, seemed to contract smallpox less frequently than other people. This suggested to Jenner that cowpox might somehow prevent smallpox, but he had no idea how or why. Nonetheless, he began inoculating a series of neighbors and acquaintances with material taken from sores on milkmaids' hands. Today, this mode of operation would be entirely unacceptable, scientifically or ethically, but, at a time when smallpox was a very real and daily threat, many people willingly submitted themselves to Jenner's experimental treatment. Compared with other standard medical treatments of the time, such as removing large amounts of blood or drinking noxious and possibly dangerous brews, Jenner's skin scratches probably seemed like a very mild treatment.

Jenner died at about the time **Louis Pasteur** (1822–1895) was born. Pasteur, a French chemist, would one day give vaccination a scientific basis, although his initial interests lay elsewhere. Inadvertent results from an experiment that today would be considered unacceptably sloppy led him to make vaccination his ruling passion. Pasteur was trying to show that a certain bacterium caused fowl cholera, a disease often fatal to chickens. His original idea was to find out how the disease was transmitted and intervene at that level. He was repeating an earlier experiment, in which he had injected the supposed cause of fowl cholera into seven chickens. In that previous experiment, the birds all died—the expected outcome. But this time, the birds sickened, then recovered, possibly because the number of bacteria injected was too low. Pasteur then did what more scientists should do after a "failed" experiment: he went on vacation. When he returned, he was ready to repeat the experiment, but there was a problem. In his absence, no one had ordered new chickens.

Ever resourceful, he decided to reuse the chickens that had survived the previous experiment plus a couple of chickens not yet used in other experiments. This is the decision that would horrify modern-day scientists, because they know that the previously injected chickens had been changed irrevocably. Pasteur, of course, could not know this. He injected all the chickens with a fresh batch of bacteria. The two new chickens died, but the seven previously injected chickens lived. They did not even develop the usual symptoms of the disease.

Today, the reason is obvious: the surviving chickens had been vaccinated. But in Pasteur's day the immune system was unknown. What is amazing about this episode is that Pasteur did not shrug his shoulders and give up this line of investigation. Instead, with a brilliant and intuitive leap, he grasped the fact that successful survival of the first exposure to a disease might make the animals impervious to later exposures.

Pasteur went on to create vaccines against anthrax and rabies. At first, he was ridiculed by colleagues. He seems, however, to have had a knack for public relations. To silence sniping by detractors and competitors, he called in the press to witness a trial of his vaccine against **anthrax,** an often fatal disease of animals and, as we now know, of humans as well. The vaccine trial succeeded, making a big impression on the reporters, who spread the news of the discovery widely. For another aspect of past and present microbiology that is not often noted, see Box 14.1.

Types of Vaccines

Vaccines are nontoxic antigens injected or ingested to induce artificially the specific defenses against a particular microorganism. A good vaccine can provide life-long immunity to an infectious disease. Vaccines are not only much cheaper than diagnosis and treatment of infections already underway, but they prevent disease from occurring in the first place. In cases where there is no effective therapy for an infection, vaccines may be the only way to protect people from disease. Vaccines come in a variety of forms that range from whole microorganisms to proteins or other antigens isolated from these microorganisms.

Whole microbe vaccines

In the early days of vaccine development, all vaccines consisted of intact microorganisms treated in some way to make them less infectious. One method was to kill the microorganisms with heat or formaldehyde. These treatments left the antigens on the microbe capable of inducing an immune response but eliminated the ability of the microbe to cause disease. A problem with this strategy is that the microbes are dead. Living microorganisms, because they replicate in the body, usually elicit a more robust immune response and are more likely than killed organisms to cause production of memory cells. Living microorganisms provide longer stimulation of the

box 14–1 *The Courage of Microbiologists*

Life in the microbiology laboratory in Pasteur's day was fraught with dangers, such as handling infectious microbes for which no cure was known. Pasteur's assistants gamely held open the jaws of rabid dogs so that Pasteur could sample saliva, knowing full well that a slip leading to a bite could result in a painful and certain death. Pasteur and his assistants injected the bacterium that causes anthrax into a variety of animals. Anyone who has injected animals knows how easy it can be to stick oneself by accident. The fact that anthrax is currently a favorite with bioterrorists should give some idea of how dangerous anthrax is to humans. Such acts of courage are not limited to the days of Pasteur. In the ensuing years, microbiologists have entered areas where epidemics were underway and have worked in their laboratories with the causative microbes or their toxins.

Modern microbiologists have distinguished themselves, too. Long before there was any therapy for acquired immunodeficiency syndrome (AIDS), microbiologists—some not long out of college—performed experiments that required handling high concentrations of HIV in an effort to find a cure or a vaccine. During the outbreaks of drug-resistant tuberculosis in the 1990s, microbiologists quietly and courageously worked with bacteria that caused a nearly incurable form of tuberculosis. These teams performed their work for the same reason as earlier microbiologists: to find ways to protect humanity against new disease threats.

The experiences of these adventurous microbiologists illustrate two things that most people will find surprising. First, microbiologists often face risks and challenges that make the events of operating room dramas on TV seem tame by comparison. Second, microbiologists entering this mine-filled zone almost never become seriously ill. None of Pasteur's brave assistants is recorded as succumbing to anthrax or rabies. Incidents of laboratory-acquired HIV infections or multidrug-resistant tuberculosis have been rare, if indeed any cases have occurred. This illustrates the fact that knowledge of the threat faced and consistent application of common sense are very effective barriers against disease and death.

immune system. Despite this limitation, however, there have been a number of effective killed microbe vaccines.

A living microorganism can be made safer by using a less virulent relative that resembles it enough to trick the immune system into mounting a response against the virulent microorganism. The mild cowpox virus in the lesion material used by Jenner was just such a stand-in for the deadly smallpox virus. Another approach is to mutate the organism so that it is still able to replicate a few times in the human body but can no longer cause disease. Such mutated strains are called **attenuated strains.** Attenuated live viruses currently are used as vaccines against **measles, mumps, rubella,** and **polio.** The experience with polio vaccines illustrates the strengths and risks of live attenuated vaccines. The first vaccine against polio was a killed virus vaccine (**inactivated polio vaccine, IPV**). This vaccine had to be injected more than once. The live polio virus vaccine that supplanted it (**oral polio vaccine, OPV**) could be given orally as a single dose. But we now know that OPV can cause disease in immunocompromised people and others who have not been immunized, so it must be used with discretion.

One problem with whole microbe vaccines, especially bacterial ones, is that they may have side effects as a result of toxic substances, such as lipopolysaccharides (LPS), that come along with the protein antigens. An example is the old vaccine against **whooping cough,** caused by *Bordetella pertussis*, a gram-negative bacterium. This form of the vaccine, the P in the diphtheria-tetanus-pertussis (DTP) vaccine, has been routinely administered to infants for decades. It consisted of killed bacteria. This form of the vaccine had a number of side effects, including discomfort in the region of the injection site and, less frequently, convulsions. Although these side effects were usually of short duration, more serious side effects, such as neurological damage, occurred in rare instances. The solution to the side-effect problem was to create a subunit vaccine that contained only one or two protein antigens, not the whole bacterium.

Subunit vaccines

Subunit vaccines are made of a single antigen or a simple mixture of antigens. These vaccines provide an alternative to whole microbe vaccines in cases where the whole microbe is toxic or where such a complex mixture of antigens evokes an inappropriate immune response. Subunit vaccines have two drawbacks. One is their cost. It is much more expensive to extract and purify isolated antigens than to use intact microbes, a problem that may be remedied soon by recombinant vaccines. Another current drawback of subunit vaccines is that they must be administered more than once to mimic the repeated stimulation of the immune system by a living microorganism. Nevertheless, the safety and efficacy of subunit vaccines has made them increasingly popular.

Toxoids. The original subunit vaccines were bacterial protein toxins that had been treated to make them nontoxic but still immunogenic. Such proteins are called **toxoids.** The D and T vaccines against **diphtheria** and **tetanus,** respectively, are toxoids. The new pertussis vaccine, called **aP** for **acellular pertussis,** is a toxoid and, in some cases, also contains a surface adhesin protein. D and T are among the safest vaccines ever developed and have been very effective in preventing disease. Making the pertussis vaccine a subunit vaccine, one that seems to be as effective as the original killed bacteria vaccine, has made this vaccine much less likely to cause side effects.

Polysaccharide vaccines. The protective response against encapsulated bacteria is antibodies that bind capsular polysaccharides and help phagocytes to ingest the bacteria. Three polysaccharide vaccines currently in use are those against pneumonia caused by the bacterium *Streptococcus pneumoniae,* childhood meningitis caused by the bacterium *Haemophilus influenzae* type b, and meningitis caused by *Neisseria meningitidis.* One problem with polysaccharide vaccines is that polysaccharide antigens are not as immunogenic as protein antigens. A solution is to link a part of the polysaccharide covalently to a protein. This causes the immune system to process the vaccine as a protein, producing antibodies and memory cells to the carbohydrate portion as well as to the protein carrier. Such a vaccine is called a **conjugate vaccine.** Often the protein used is one of the toxoids D or T, so that an added bonus of vaccination is a booster for the D or T immunization.

Modern Controversies

Serious side effects from vaccines are rare. Otherwise, vaccines would never have been approved for administration to infants and children. Yet, because side effects exist, vaccines are sometimes blamed for a wide spectrum of medical problems in vaccinated children. Vaccines are administered during the first years of life, when most signs of congenital neurological disease begin to appear. Thus, parents who cannot accept a genetic explanation for their child's autism or paralysis or cerebral palsy sometimes seize on vaccines as likely explanations. And who can prove them wrong? Even if a vaccine only rarely produces side effects, who can deny with certainty that a child is not that one case in ten million?

To compound the problem, parents of young children today grew up in an era when vaccines had almost eliminated such previously dreaded diseases as diphtheria, whooping cough, and polio. Grandparents who remember children dying of diphtheria, or an infant nearly suffocating from the prolonged coughing fits of whooping cough, or a child paralyzed by polio have a hard time convincing their adult children that these diseases even existed. Out of sight, out of mind. These diseases are not on the evening news, and it is hard for parents to believe that the microorganisms are still around, waiting for the vaccine safety net to develop enough holes for children to fall through. Instead, many parents become overly concerned with the imagined risks of vaccines. When a parent alleges that his or her child has been damaged by a vaccine, however lacking in evidence such a claim might be, the event is likely to receive full—and often muddled—media coverage.

Some groups of people oppose vaccination entirely, trusting instead to what they call natural vaccination, contracting and surviving the disease itself. As long as their numbers remain small, the children of such parents are protected by the fact that most of their contacts are vaccinated, making it difficult for an epidemic to get started. But if enough people refuse vaccination for their children, epidemics of diseases like diphtheria, whooping cough, and measles may return. Measles and mumps, now vaccine-preventable diseases, can be very serious, especially for older children. Rubella was once a cause of birth defects in the children of women infected for the first time during pregnancy. No one wants to see these diseases return, yet they are sure to do so in a generation of insufficiently vaccinated children. One question that has not received much attention is whether it is wise for unvaccinated children, especially when they beome teenagers, to travel to countries where diseases such as measles, for example, are common.

Successful Vaccines

Smallpox vaccine

Table 14.1 lists successful vaccines now in use and shows their effects on disease incidence in the United States. Let us look beneath the surface of these various vaccine stories for a better understanding of what vaccines do and why they can have side effects.

The eradication of smallpox. The ultimate vaccine success story is that of the viral disease smallpox, which has been completely eliminated worldwide. This was the first time in history that an infectious disease had been eradicated by human intervention, and it serves as a reminder of how powerful a force for human health vaccines can be. More recently, world health agencies have been working together to eliminate another viral disease, polio.

After the eradication of smallpox, the vaccine was no longer administered. This created a new type of problem: the potential use of smallpox virus by bioterrorists. Although scientists describe smallpox as "eradicated," the virus is not really gone. Stocks of smallpox virus are stored in freezers in the United States, Russia, and probably in other locations. What a temptation for bioterrorists! Although limited stocks of the vaccine exist, not nearly enough is available to vaccinate everyone. In fact, the efficacy of these old vaccine preparations is in question.

Viral lifestyle makes development of some vaccines difficult. If scientists have succeeded in eradicating smallpox and nearly eradicating polio, why is there no vaccine to prevent acquired immunodeficiency syndrome (AIDS), another viral disease? The answer to this question illustrates the way in which the life cycle of a virus and the types of cells it infects affect the ease with which a vaccine can be developed. Smallpox virus infects human epithelial cells, but, after each round of replication, it is released from the cell

Table 14.1 Baseline 20th-century annual morbidity (cases of disease) before the vaccine became available and 1998 morbidity from nine diseases with vaccines recommended before 1990 for universal use in children

Disease	Baseline 20th-century annual morbidity	1998 morbidity	% decrease
Small pox	48,164*	0	100%
Diphtheria	175,885†	1	100%§
Pertussis	147,271¶	6,279	95.7%
Tetanus	1,314**	34	97.4%
Poliomyelitis (paralytic)	16,316††	0§§	100%
Measles	503,282¶¶	89	100%§
Mumps	152,209***	606	99.6%
Rubella	47,745†††	345	99.3%
Congenital rubella syndrome	823§§§	5	99.4%
Haemophilus influenzae type b	20,000¶¶¶	54****	99.7%

*Average annual number of cases during 1900–1904.
†Average annual number of reported cases during 1920–1922, 3 years before vaccine development.
§Rounded to nearest tenth.
¶Average annual number of reported cases during 1922–1925, 4 years before vaccine development.
**Estimated number of cases based on reported number of deaths during 1922–1926, assuming a case-fatality rate of 90%.
††Average annual number of reported cases during 1951–1954, 4 years before vaccine licensure.
§§Excludes one case of vaccine-associated polio reported in 1998.
¶¶Average annual number of reported cases during 1958–1962, 5 years before vaccine licensure.
***Number of reported cases in 1968, the first year reporting began and the first year after vaccine licensure.
†††Average annual number of reported cases during 1966–1968, 3 years before vaccine licensure.
§§§Estimated number of cases based on seroprevalence date in the population and on the risk that women infected during a childbearing year would have a fetus with congenital rubella syndrome (7).
¶¶¶Estimated number of cases from population-based surveillance studies before vaccine licensure in 1985 (8).
****Excludes 71 cases of *Haemophilus influenzae* disease of unknown serotype.

Source of table: 1999. Morbidity Mortality Weekly Reports. 48: 243–247.

to infect other cells. Because the virus spends a significant period of time outside the cell, antibodies that neutralize the virus have many chances to find and coat the viruses. The polio virus enters the body through the intestinal mucosa. Secretory immunoglobulin A (sIgA) easily binds and traps the virus in mucus, rendering it unable to reach the mucosa. By contrast, when human immunodeficiency virus (HIV) infects a cell, it can move directly from cell to cell, thus avoiding neutralizing antibodies after the initial infection event by hiding out in an area inaccessible to antibodies. Moreover, HIV strikes at T helper cells and monocytes, the precursors of macrophages. These two cell types are critical in the development of the antibody and cytotoxic T cell response. Thus, HIV destroys the cells that in a vaccinated person might be expected to combat the infection.

Polio vaccine

OPV, an attenuated viral vaccine, illustrates the advantages of a live oral vaccine. The earlier polio vaccine was IPV, which was given by injection. OPV elicits a better protective response than IPV, because it replicates for a period and thus stimulates the immune system more effectively. Also, the vaccine strain is shed in feces and will usually "infect" other members of the same family, thus spreading the benefits of vaccination to people who have not received the vaccine. Of course, if these other family members happen to be immunocompromised, then exposure to the vaccine may cause serious disease. OPV can cause disease even in healthy people, although this is uncommon. All cases of polio acquired in the United States today are the result of immunization with OPV. For this reason, IPV is now recommended for the entire immunization series.

In developing countries, OPV offers another advantage over IPV in its method of administration. In countries where sterile syringes are in short supply, the same needle is often used to vaccinate more than one person, a practice that can spread blood-borne pathogens such as HIV and hepatitis. Thus, vaccines that are administered orally or nasally, instead of by injection, are much needed.

Hepatitis virus vaccines and the measles-mumps-rubella vaccine

Despite continuing disappointments for scientists trying to develop vaccines against such difficult viral targets as HIV, some important antiviral vaccine advances have been made. For example, effective vaccines against hepatitis A and B viruses are now available. Both of these viruses attack the liver. **Hepatitis A virus** has been associated mainly with ingestion of contaminated foods in developing countries, but, as more foods cross international boundaries, the range of hepatitis A has been increasing. **Hepatitis B** is spread mainly by sexual transmission, especially among people with multiple partners, and by contaminated needles. Some parents have objected to the recommendation by pediatricians that all infants receive the hepatitis B vaccine (HBV). These parents understandably protest that "my child will never be touched by that kind of nastiness." In fact, children are not isolated objects but are affected by the health and actions of their parents and caregivers. What if a parent or caregiver is hepatitis B positive? Parents can transmit hepatitis B to their children through close contact. Moreover, what if the child's physician or dentist is a carrier of hepatitis B? Hepatitis B virus transmission from doctor to patient has been documented. The first hepatitis B vaccine consisted of purified formalin-treated surface antigens obtained from the serum of infected people. It has now been replaced by a subunit form of the vaccine. A subunit vaccine for hepatitis A is now available also.

Other successful antiviral vaccines include the measles, mumps, and rubella vaccine (MMR). Measles and mumps are common childhood diseases that usually are not very serious, although in some instances they can kill. Worldwide, measles is one of the leading killers of children, especially in developing countries. Rubella is more serious, because this type of measles can cause birth defects when pregnant women become infected. Because the components of MMR are attenuated live viruses, the vaccine only needs to be administered once. However, it is usually administered twice for reasons having to do with the demographics of vaccine coverage.

Influenza vaccine

Another useful antiviral vaccine is directed against the influenza virus. In most people, this vaccine does not provide complete immunity against disease. The reasons—and the reasons that the vaccine is recommended anyway—are instructive. Influenza is caused by **influenza virus,** one of the most rapidly mutating viruses, with surface proteins that undergo a constant series of changes. Sometimes large shifts in the amino acid sequence of envelope proteins occur. This variability means that as one strain of virus displaces another, the antibodies to the surface proteins of the old strain no longer neutralize the new virus as effectively. People immunized against last year's virus can become susceptible to this year's strain, although they may still be partially protected. Because of the genetic variability of influenza virus, people are advised to get flu shots every year. The vaccine changes with the identity of the strains currently making the rounds.

The fact that the strain of virus currently causing influenza can change from year to year places a strain of another sort on vaccine manufacturers and people who design vaccines. Because it takes a long time to produce enough new vaccine to meet the demand, manufacturers need about 8 months to produce a new version of the vaccine. Accordingly, every February, a committee meets to guess what strains will be making the rounds during influenza season in the following winter. This guess is based on what happened during the recently finished influenza season, and the guess can be wrong. If so, vaccinated people may be only partially immune and may suffer the disease, although usually in a less virulent form. When big shifts in envelope proteins occur, however, the old immune response is rendered completely ineffective, and the worst and most fatal outbreaks of influenza can occur, affecting vaccinated as well as unvaccinated people.

The long period of time needed to produce a new vaccine illustrates the importance of an area of research that is only now beginning to receive the attention it deserves: techniques for more rapid development of vaccine strains and for faster production of large amounts of vaccine. The technology used for vaccine production today has not changed much since the middle of the 20th century.

If the effectiveness of the influenza vaccine is such a crapshoot, why do physicians recommend it? For that matter, because the disease is usually more of a nuisance than a life-threatening condition in adults younger than 50 years, why should younger people bother to be vaccinated? One compelling reason is the **herd effect.** Elderly persons, the group most likely to die of influenza followed by pneumonia, have immune systems with declining potency. Vaccines often are not as effective in this age group as in younger adults and children. Because vaccinated nonelderly adults are less likely to contract influenza or, if they do, are likely to experience milder and shorter periods of infection, they form a protective barrier between influenza and more susceptible elderly individuals. The chances of exposure to the disease are effectively decreased for less immune-protected people. Employers favor flu shots for another reason: fewer days of work are lost. For this reason many large employers offer free influenza vaccinations to their employees.

Antibacterial vaccines

DTP vaccine. An example of a very effective vaccine against formerly common bacterial diseases is the trivalent vaccine DTP. The D and T are toxoids. Because they are proteins and thus do not escape the phagosome of an antigen-presenting cell, they elicit an antibody rather than a cell-mediated response. This is an effective response in the case of diphtheria and tetanus, because the symptoms of these diseases are caused by protein toxins produced by the bacteria growing in the throat (diphtheria) or in a wound (tetanus). The toxins, not the bacteria, enter the bloodstream and damage essential organs.

Vaccination with DT elicits an antibody response that neutralizes diphtheria and tetanus toxins. Thus, even if a person is repeatedly colonized by the bacteria that cause diphtheria or tetanus, no symptoms will develop because an immediate antibody response neutralizes circulating toxin. The DT vaccine produces long-term immunity to diphtheria and tetanus. However, due to the seriousness of tetanus a tetanus booster is usually administered to persons with the kinds of deep wounds associated with tetanus, to make sure that antibody levels are high enough to be protective.

Some new and more defined versions of the pertussis vaccine (P) have fewer side effects than the older form. These vaccines consist of a detoxified form of a toxin produced by *B. pertussis* (pertussis toxin) plus a surface adhesin of *B. pertussis.* These new subunit vaccines are called acellular pertussis vaccines (aPs) to distinguish them from the old whole-cell vaccine (P).

***Haemophilus influenzae* type b vaccine.** Another very successful antibacterial vaccine is the one that protects against childhood meningitis caused by the gram-negative bacterium *Haemophilus influenzae* type b (hence the vaccine name, **Hib**). *H. influenzae* type b was once the most common cause of meningitis in children and also caused **epiglottitis,** a rapidly progressing disease that can close the airway and lead to suffocation. Unlike the bacteria that cause diphtheria, tetanus, and whooping cough, *H. influenzae* does not produce a protein toxin. Instead, *H. influenzae* coats its surface with a polysaccharide capsule that discourages C3b binding and thus allows the bacteria to avoid ingestion by phagocytes.

Antibodies that bind to capsular polysaccharides allow phagocytes to ingest and kill the bacteria and thus prevent this destructive (and frequently fatal) disease from getting under way. Because *H. influenzae* causes meningitis primarily in children under the age of 5 years, it is necessary to immunize infants as early as possible. The problem is that infants do not mount a T-independent antibody response to polysaccharides. To solve this problem, the *H. influenzae* type b capsular polysaccharide antigens were attached covalently to a protein to produce a **conjugate vaccine.** In this form, the immune system processes the vaccine as if it were a protein, producing a robust antibody response that targets the capsular epitopes as well as those on the protein to which the capsular polysaccharides are attached.

Adjuvants

One problem with subunit vaccines is that because they are not replicating organisms, they do not continue to stimulate the immune system, a necessary prerequisite for a robust memory cell response. This defect is remedied in two ways. First, multiple doses of the vaccine are given. For the poor recipient who begins to feel like a pincushion, this method may not be appealing—but it works. The use of **adjuvants** is a second factor that helps to prolong the effects of subunit vaccines by prolonging stimulation of the immune response. Adjuvants appear to work by trapping the antigens in a chemical complex formed by the adjuvant, from which the antigens are released relatively slowly, thus prolonging exposure to the antigens.

The only adjuvant currently licensed for use in the United States is **alum,** an aluminum salt. Other adjuvants are under development. A particularly interesting vaccine delivery system consists of **microspheres** made of resorbable suture material or some other inert substance that gradually breaks down in the body. The vaccine antigens are encapsulated in the adjuvant material and are thus released slowly over a period of time. By administering a mixture of microspheres of differing sizes, it is hoped that the vaccine would be released over a long enough period that a memory response could be elicited with only one or two injections. This may solve problems such as the necessity of repeated DTaP administrations. Another strategy is to use molecular techniques to make a live, avirulent bacterial vaccine strain that produces the protein antigen (or portions of it) on its surface.

Timing of Vaccinations

To afford protection when it is most needed, a vaccine must be administered in a timely manner (Fig. 14.1). For example, because meningitis resulting from *H. influenzae* type b and hepatitis caused by hepatitis B virus can strike very young infants and maternal antibody levels are very low, vaccines for these diseases must be administered as early as possible. In the case of mumps, measles, and rubella, circulating maternal antibodies protect the infant for the first year of life. These antibodies decrease the strength of the immune response when vaccination is given too early. For this reason MMR vaccine is administered later than some other vaccines.

Note that although MMR does not need to be given more than once, it is recommended that a second dose be administered to children entering first grade or to children entering middle school or high school (Fig. 14.1). It is best to give MMR around age 2, because this allows the child's own immune system to take up where maternal antibodies leave off. Nonetheless, a significant number of children are not inoculated at this time. These children are somewhat protected from infection by the herd effect. As they grow older and especially after they enter school, however, the risk that they will be exposed rises significantly. The best plan is to administer the vaccine at the beginning of grade school, but, in areas where this is not feasible, administration at entry into high school is better than nothing.

Problematic Vaccines

The previous sections depicted vaccines as highly effective tools for preventing disease. Unfortunately, there have been less encouraging vaccine stories. As important as the successes, these stories illustrate the challenges vaccine scientists face now and in the future. Two examples will serve as illustrations: a vaccine against the main cause of bacterial pneumonia, the gram-positive bacterium *S. pneumoniae,* and a vaccine against tuberculosis, a disease caused by the gram-positive bacterium *Mycobacterium tuberculosis.*

Streptococcus pneumoniae vaccine

In the case of *S. pneumoniae,* we know why the vaccine is not as effective against the disease as one would wish. The vaccine is a polysaccharide vaccine intended to elicit antibodies against the polysaccharide capsule that protects *S. pneumoniae* from ingestion by phagocytic cells. In the case of *H. influenzae,* this problem was solved by covalently attaching the polysaccharide antigen to a protein. With *S. pneumoniae,* this is not so easy to do, because there are so many antigenic types of the capsular polysaccharide—nearly 100 at last count. A subset of these is associated with the most serious consequences of pneumonia, but the number of antigenic types is still too large to lend itself to a conjugate vaccine solution.

A recently approved conjugate vaccine targets the most common antigenic types of polysaccharide found in children with earache, also caused by *S. pneumoniae.* Childhood earache is rarely fatal, but the vaccine is important not only to prevent suffering but to reduce the pressure to treat earache with antibiotics. Demand from parents for antibiotic treatment of earache has been one of the strongest factors leading to increased resistance in *S. pneumoniae,* a bacterial species that is a major killer of the elderly. It remains to be seen whether this vaccine, which contains only six or seven antigenic types of capsular antigens, helps prevent adult cases of *S. pneumoniae,* which are much more likely to be fatal.

Age ▶ Vaccine ▼	Birth	1 mo	2 mos	4 mos	6 mos	12 mos	15 mos	18 mos	24 mos	4-6 yrs	11-12 yrs	14-16 yrs
Hepatitis B[2]	Hep B											
		Hep B			Hep B						Hep B	
Diphtheria, Tetanus, Pertussis[3]			DTaP	DTaP	DTaP		DTaP[3]			DTaP	Td	
H. influenzae type b[4]			Hib	Hib	Hib	Hib						
Polio[5]			IPV	IPV	IPV[5]					IPV[5]		
Measles, Mumps, Rubella[6]						MMR				MMR[6]	MMR[6]	
Varicella[7]						Var					Var[7]	
Hepatitis A[8]									Hep A[8] in selected areas			

Approved by the Advisory Committee on Immunization Practices (ACIP), the American Academy of Pediatrics (AAP), and the American Academy of Family Physicians (AAFP).

On October 22, 1999, the Advisory Committee on Immunization Practices (ACIP) recommended that Rotashield (RRV-TV), the only US-licensed rotavirus vaccine, no longer be used in the United States (MMWR Morb Mortal Wkly Rep. Nov 5, 1999;48(43):1007). Parents should be reassured that their children who received rotavirus vaccine before July are not at increased risk for intussusception now.

[1] This schedule indicates the recommended ages for routine administration of currently licensed childhood vaccines as of 11/1/99. Additional vaccines may be licensed and recommended during the year. Licensed combination vaccines may be used whenever any components of the combination are indicated and its other components are not contraindicated. Providers should consult the manufacturers' package inserts for detailed recommendations.

[2] **Infants born to HBsAg-negative mothers** should receive the 1st dose of hepatitis B (Hep B) vaccine by age 2 months. The 2nd dose should be at least 1 month after the 1st dose. The 3rd dose should be administered at least 4 months after the 1st dose and at least 2 months after the 2nd dose, but not before 6 months of age for infants.
Infants born to HBsAg-positive mothers should receive hepatitis B vaccine and 0.5 mL hepatitis B immune globulin (HBIG) within 12 hours of birth at separate sites. The 2nd dose is recommended at 1 month of age and the 3rd dose at 6 months of age.
Infants born to mothers whose HBsAg status is unknown should receive hepatitis B vaccine within 12 hours of birth. Maternal blood should be drawn at the time of delivery to determine the mother's HBsAg status; if the HBsAg test is positive, the infant should receive HBIG as soon as possible (no later than 1 week of age).
All children and adolescents (through 18 years of age) who have not been immunized against hepatitis B may begin the series during any visit. Special efforts should be made to immunize children who were born in or whose parents were born in areas of the world with moderate or high endemicity of hepatitis B virus infection.

[3] The 4th dose of DTaP (diphtheria and tetanus toxoids and acellular pertussis vaccine) may be administered as early as 12 months of age, provided 6 months have elapsed since the 3rd dose and the child is unlikely to return at age 15 to 18 months. Td (tetanus and diphtheria toxoids) is recommended at 11 to 12 years of age if at least 5 years have elapsed since the last dose of DTP, DTaP, or DT. Subsequent routine Td boosters are recommended every 10 years.

[4] Three *Haemophilus influenzae* type b (Hib) conjugate vaccines are licensed for infant use. If PRP-OMP (PedvaxHIB or ComVax [Merck]) is administered at 2 and 4 months of age, a dose at 6 months is not required. Because clinical studies in infants have demonstrated that using some combination products may induce a lower immune response to the Hib vaccine component, DTaP/Hib combination products should not be used for primary immunization in infants at 2, 4, or 6 months of age unless FDA-approved for these ages.

[5] To eliminate the risk of vaccine-associated paralytic polio (VAPP), an all-IPV schedule is now recommended for routine childhood polio vaccination in the United States. All children should receive four doses of IPV at 2 months, 4 months, 6 to 18 months, and 4 to 6 years. OPV (if available) may be used only for the following special circumstances:
1. Mass vaccination campaigns to control outbreaks of paralytic polio.
2. Unvaccinated children who will be traveling in <4 weeks to areas where polio is endemic or epidemic.
3. Children of parents who do not accept the recommended number of vaccine injections. These children may receive OPV only for the third or fourth dose or both; in this situation, health care professionals should administer OPV only after discussing the risk for VAPP with parents or caregivers.
4. During the transition to an all-IPV schedule, recommendations for the use of remaining OPV supplies in physicians' offices and clinics have been issued by the American Academy of Pediatrics (see *Pediatrics*, December 1999).

[6] The 2nd dose of measles, mumps, and rubella (MMR) vaccine is recommended routinely at 4 to 6 years of age but may be administered during any visit, provided at least 4 weeks have elapsed since receipt of the 1st dose and that both doses are administered beginning at or after 12 months of age. Those who have not previously received the second dose should complete the schedule by the 11- to 12-year-old visit.

[7] Varicella (Var) vaccine is recommended at any visit on or after the first birthday for susceptible children, ie, those who lack a reliable history of chickenpox (as judged by a health care professional) and who have not been immunized. Susceptible persons 13 years of age or older should receive 2 doses, given at least 4 weeks apart.

[8] Hepatitis A (Hep A) is shaded to indicate its recommended use in selected states and/or regions; consult your local public health authority. (Also see *MMWR Morb Mortal Wkly Rep.* Oct 01, 1999;48(RR-12); 1-37).

Figure 14.1 Timing of vaccinations. This figure illustrates the current complexity of the vaccination program for a child. Some of the most important research on vaccines today seeks to reduce this complexity by combining vaccinations and improving delivery of antigens. This figure is intended to be solely illustrative and may change as new information becomes available. Updated information about vaccination requirements can be obtained from healthcare providers.

The older anti-pneumonia vaccine, consisting of a mixture of 23 types of capsular polysaccharides (not the conjugated form), is currently being used to vaccinate elderly people, the group most likely to die of pneumococcal pneumonia. This brings us back once again to the influenza vaccine. Influenza predisposes people, especially the elderly, to develop bacterial pneumonia. Thus, influenza vaccination not only erects a firewall against influenza, but a barrier against pneumonia, which is usually a far more serious disease. Also, as *S. pneumoniae* becomes more and more resistant to antibiotics, the need for an effective vaccine intensifies. *S. pneumoniae* is not only a problem for the elderly. With the demise of *H. influenzae* type b as the main cause of childhood meningitis, *S. pneumoniae* has become a leading cause of meningitis in this age group, although fortunately it causes meningitis less often than *H. influenzae* once did.

Antituberculosis vaccine

Tuberculosis is another disease for which a vaccine is desperately needed. The anti-tuberculosis vaccine, BCG, takes its name from the discoverers of the vaccine. BCG stands for the **bacillus of Calmette and Guerin.** BCG is an attenuated strain of *Mycobacterium bovis,* which causes a tuberculosis-like disease in humans and cattle and has been widely used as an antituberculosis vaccine outside the United States. It has been shown to be fairly effective in preventing disseminated tuberculosis in infants, but its ability to prevent tuberculosis in children and adults is controversial. Field trials have yielded efficacy levels ranging from 0% protection to as high as 80% protection. Controversy over the efficacy of BCG and the fact that it makes people test positive on the tuberculosis skin test were the bases for deciding not to administer the vaccine widely in the United States. In the late 1990s, outbreaks of multidrug-resistant tuberculosis changed the risk picture completely. BCG is now administered to health care providers who care for patients with tuberculosis, especially in areas where drug-resistant strains of *M. tuberculosis* are becoming more common. A little protection is better than none at all.

The extreme variability in BCG field trial results is unusual. Intense investigation has ensued to explain this variability. There may be many reasons, including variations in vaccine lots, failure to store the vaccine properly, and genetic differences between different human populations. Whatever the explanation for the variability, a better vaccine than BCG is clearly needed. The growing drug-resistance of *M. tuberculosis* has made development of a vaccine all the more imperative. A good feature of BCG is its safety. It has been administered to hundreds of millions of infants, children, and adults with virtually no side effects. For this reason, scientists are studying its use as a delivery vehicle for recombinant vaccine against other diseases.

Passive Immunization

People who have not been immunized against a specific disease can still be protected by **passive immunization,** the injection of antibodies from another person or animal. Human antibodies injected into the bloodstream can remain in circulation for as long as several months. They do not elicit an immune response and, if their levels remain high for a long time, they can actually dampen an immune response to the antigen. Thus, the benefits of passive immunization are of limited duration. One example of the use of passive immunization is the treatment of unvaccinated people who have been exposed to diphtheria. The antibodies temporarily protect the person by neutralizing the diphtheria toxin until the bacteria can be cleared from the body. The old method of preventing hepatitis A, still used in some places, was injection of a fraction of pooled human serum (gamma globulin). Pooled serum is made by collecting serum from large numbers of healthy people, mixing these samples together and separating out the fraction that contains most of the antibodies. So many people have been exposed to hepatitis A worldwide that pooled serum carries enough antibodies to make it protective. People traveling to a country where hepatitis A is prevalent are often given such an injection, which then would be effective in neutralizing the virus for several weeks.

New Directions, New Causes for Hope

Manipulating the TH1/TH2 responses

To avoid ending this chapter on a depressing note, we include a brief look at some recent advances that may help to make vaccines more effective in the future. One new approach was mentioned in Chapter 13: using information about how the body recognizes antigens and mobilizes an immune response to design vaccines that elicit a specific kind of response. For example, interleukin-12 appears to stimulate a TH1-type response. If this proves to be correct, administration of interleukin-12, together with a vaccine, could make the vaccine more effective. Similarly, making soluble vaccines particulate by trapping them in or on carrier materials could help to stimulate a cell-mediated response.

Recombinant vaccines

The first use of cloning in vaccine production was to create subunit vaccines. For example, newer forms of the influenza and hepatitis B vaccines contain viral surface proteins that were produced by bacteria carrying the cloned genes that encode these proteins. Genetically engineered bacteria themselves are being developed as vaccines. Live oral vaccines are usually more desirable than injected ones (except for safety reasons, as in the case of OPV). Accordingly, much effort has gone into engineering strains of bacteria known to stimulate the gastrointestinal-associated lymphoid systems to serve as delivery vehicles for vaccines. One candidate bacterium is *Salmonella typhimurium. S. typhimurium* is a common cause of food-borne disease. It survives passage through the stomach and invades the Peyer's patches of the small intestine, which contain the cells of the gastrointestinal-associated lymphoid tissue. *S. typhimurium* has a well-developed genetic system, and scientists have learned to engineer it so that it displays proteins from other microbes on its surface. The problem has been to find an attenuated strain of *S. typhimurium* that does not cause disease yet survives long enough to stimulate the immune system.

Another bacterium under investigation as a vaccine vehicle is BCG, the attenuated strain of *M. bovis,* which has been used as a vaccine against tuberculosis. This vaccine is normally injected, but what makes it attractive as a vehicle is its record for safety and its ability to elicit a robust antibody response. In fact, extracts from *Mycobacterium* species have been used for years as an adjuvant (Freund's adjuvant) for producing antibodies in rabbits or mice. Efforts are underway to engineer BCG to display vaccine proteins, such as diphtheria or tetanus toxoids, on its surface.

Bacteria are not the only vehicles being tested for vaccine delivery. Vaccinia virus, once used as a vaccine against smallpox, is being engineered to display other vaccine proteins. The advantage of using this virus as a carrier is that the vaccinated person would become immune to smallpox in addition to whatever other antigens the virus is displaying. If bioterrorists are successful in gaining access to smallpox virus samples now stored in U.S. and Russian freezers, or if the virus escapes some other way, this could become an important advantage of the vaccine.

No live vaccine like those just described is close to being available for widespread use, but, if scientists are eventually successful in developing them, a great advance will have been made.

Plant delivery of vaccines

Another potentially useful biotechnology initiative is the creation of plants that produce protein antigens such as D and T. Two major problems with distributing vaccines in the developing world are cost and the need for injections. Vaccines that could be produced inexpensively and in an orally administered form could save millions of lives. The ultimate goal of expressing antigen proteins in plants is to develop a variety of banana, a fruit fed to infants in many countries, that produces vaccine proteins. Banana is now amenable to genetic engineering, but other problems remain to be solved. For example, as previously noted, no one has yet succeeded in delivering a subunit vaccine orally and eliciting a good immune response. A protein antigen without the aid of an adjuvant would have a poor chance of success. Nonetheless, this adventurous line of research is exciting. Often visionaries who have forged ahead with what everyone else was sure would not work have been the ones to open new frontiers. Also, by the time it becomes feasible to express bacterial or viral proteins in banana, new insights into ways to stimulate the gastrointestinal-associated lymphoid tissue may have been developed.

DNA vaccines

A new and adventurous approach to vaccine development is to inject DNA that encodes a vaccine protein (**naked DNA vaccines**) into human muscle cells. The DNA is adsorbed to gold particles and injected with an air gun into muscle tissue. Incredibly, this has been shown in animals to result in transient production of foreign proteins by the muscle cells. Display of the foreign antigen lasts only a month, but this is long enough to evoke a robust response. Because the display is localized, side effects associated with an immune attack on the body's own cells are minimized. Moreover, the DNA is injected into differentiated cells, not into the germ line cells.

If successful, the approach would have two important advantages. First, unlike live vaccines, DNA on gold particles can be stored dry and need not be refrigerated. Second, it is much less expensive to prepare pure DNA than to prepare a pure protein. As you might expect, there are already people anxious to rain on the parade of the naked DNA vaccines. DNA itself is immunogenic, and the suspicion is high that some diseases such as lupus are caused by an immune reaction to human DNA. Look forward to a major battle over side effects. As in the past, however, the public's tolerance of possible adverse side effects of a vaccine will be directly proportional to perceived risks of encountering the disease unprotected. People at high risk of developing AIDS, for example, will be much more tolerant of side effects than people who are less at risk for the disease.

More effective production and delivery of vaccines

Effective delivery of vaccines to the population at large continues to be a serious problem, even in developed countries. It would help to be able to administer several vaccines simultaneously in a single dose, preferably by the oral route or by inhalation, and thus eliminate the need for so many booster shots. One possible solution is the use of live viral or bacterial vaccine strains that present antigens of more than one pathogen. More immediately, however, researchers are trying to determine whether DTaP, Hib, and HBV could be given in a single dose. At present, they are administered during the same visit but are given by injection in different body sites. Orally administered forms of DTaP, Hib, and HBV are also being developed, and work on methods of administration for such vaccines by inhalation has begun.

A different type of delivery problem is illustrated by the influenza vaccine. If a new strain of influenza were to begin sweeping the world, pharmaceutical companies would need about 7–8 months to produce and deliver the new vaccine. Similarly, if smallpox were to break out again, it would take months to produce enough vaccine to supplement the limited stores that have been stockpiled. Research is now under way to identify methods to speed up the process of vaccine production and delivery to the people who need it.

The Larger Issues

Ultimately, the solutions to vaccine-related problems are not only scientific. They are economic and political. Perhaps the most hopeful development in recent years is that governments have now realized that vaccines are an excellent, low-cost health investment. Governments are exploring ways to protect pharmaceutical companies, which have been made understandably skittish by questionable lawsuits mounted by individuals who claim injuries to their children's health as a result of vaccine side effects. Governments also need to find ways to encourage pharmaceutical companies to spend more on vaccine research by making vaccines more profitable. The World Health Organization and the United Nations Children's Fund continue to remind the developed countries of their moral obligation to continue to develop vaccines needed by the developing world. Although everyone agrees with this philosophy, economic disincentives have made putting it into practice difficult. The developing world has a role to play, too. It is feasible to eradicate polio and measles just as smallpox was eradicated. The nations of the world have been cooperating to make the eradication of polio, now nearly complete, a reality. The eradication of measles is equally important, because measles infections kill nearly a million children each year, most of them in developing countries.

Delivering available vaccines as widely as possible in the developing world is only one of the global vaccine problems. Another is the lack of effective vaccines against such important diseases as AIDS, malaria, and tuberculosis. As AIDS ravages Africa and other countries too poor to afford anti-HIV medication (even the short course of AZT that reduces transmission from mother to infant during the birth process) the only hope is an effective vaccine. The increasing resistance of HIV to antiviral compounds and the increasing resistance of *M. tuberculosis* to antituberculosis antibiotics provide further compelling reasons for making vaccine development a high priority. If effective action is not taken, developed countries could one day find themselves as unprotected against diseases like HIV and tuberculosis as the developing countries are now.

chapter 14 at a glance

Types of Vaccines

Composition of vaccine	Type of vaccine	Example of vaccine type
Live attenuated microbe	Complex	Oral polio vaccine (OPV) MMR vaccine BCG antituberculosis vaccine
Killed microbe	Complex	Inactivated polio vaccine (IPV)
Toxoids, surface proteins	Subunit	DTaP vaccine New influenza vaccine Hepatitis B vaccine
Mixed polysaccharides	Subunit	*Streptococcus pneumoniae* vaccine (pneumonia)
Conjugate vaccine	Subunit	*Haemophilus influenzae* type b (meningitis) New *S. pneumoniae* vaccine (earache)

Examples of Vaccine Problems

Vaccine	Problem
S. pneumoniae mixed polysaccharide vaccine	No response in infants, reduced efficacy in elderly Too many serotypes cause disease, therefore conjugate vaccine is difficult to make
BCG vaccine	Variable protection rates in different populations Makes people positive on the TB skin test, reducing the test efficacy as a screening tool

Passive Immunization

Injection of antibodies from some other source

Used to treat exposed people who are not immune

Does not immunize the person

Acts rapidly, no waiting period

chapter 14 at a glance *(continued)*

New Initiatives

Development	Advantages	Potential problems
Cytokines to control response	Produce a more effective, directed immune response	TH1/TH2 decision point not well understood Cytokines can cause side effects
DNA vaccines	Cheaper vaccine Longer stimulation of immune system	Immunogenicity of DNA might cause autoimmune disease
Plant-delivered vaccines	Cheap, easy to administer Especially useful in developing countries	Long development time ahead Need for adjuvant
Recombinant vaccines	Vaccines that immunize against more than one disease Prolonged stimulation of immune system Most vehicles have an established safety record	Still in development
Improved delivery and production of vaccines	Faster response time to diseases like influenza Administration of multiple vaccines in a single dose Making oral versions of vaccines that are now injected	Still in development

Questions for Review and Reflection

1. Some groups refuse vaccination for religious reasons. In the bad old days of smallpox, such objections were swept aside in the name of the public good, but, in the more enlightened world, governments have been reluctant to forcibly vaccinate such people. Should they be allowed to refuse vaccination? What are the broader public health issues?

2. Rubella ("German measles") is primarily a childhood disease. Why then is it considered a greater problem for pregnant women?

3. In the United Kingdom in the 1980s, a media-induced scare about the alleged side effects of the P portion of the DTP vaccine caused many parents to demand that their children receive only the DT vaccine. Without looking back at the news accounts of the period, can you predict what happened next?

4. You are a scientist developing a vaccine against an important disease. Can you imagine circumstances in which your vaccine might actually make the disease worse?

5. In Africa in 1976, a frightening outbreak was caused by the Ebola virus. This virus caused a horrible death preceded by bleeding from the eyes and other orifices of the body. This outbreak centered around a vaccination clinic. Why is this not surprising, and how might it be prevented in the future?

6. How do the protective effects of vaccines against polio and other viruses differ from the protective effects of the successful antibacterial vaccines? Explain these differences.

7. Make an educated guess as to why the polio vaccine has been more effective than the influenza vaccine. [Hint: Consider why the influenza vaccine becomes ineffective.]

8. How does passive immunization differ from active immunization (vaccination)? Why doesn't passive immunization confer long-term immunity?

9. Scientists are talking about using passive immunization to treat infections caused by strains of bacteria so resistant to antibiotics that they are currently untreatable. How would passive immunization help, and why might it be preferable to vaccination for this particular application?

key terms

adjuvant a compound administered with an antigen to stimulate the immune response to that antigen

attenuated strain a strain of microorganism that has been mutated to decreased virulence but is still able to elicit an immune response

conjugate vaccine a vaccine in which a polysaccharide antigen is covalently linked to a protein, so that the immune system processes it like a protein and produces a memory cell response as well as a more robust immune response

DNA vaccine a vaccine consisting of DNA instead of proteins or polysaccharides; the DNA enters human cells and produces the protein antigen that actually stimulates the immune response

herd effect the phenomenon in which so many people in a population are immune to the disease that the causative microbe cannot infect enough susceptible people to cause an outbreak of the disease; protects the unvaccinated from exposure to the disease

immunogenic the ability of an antigen to elicit a strong immune response

passive immunization a technique in which antibodies from another source are injected into a person to provide transient protection against a disease to which the person is not immune; only works in cases where the protective immune response is an antibody response

subunit vaccine a single antigen or simple mixture of antigens used as a vaccine

toxoids protein toxins that have been treated to make them nontoxic but still immunogenic

vaccine a nontoxic antigen or mixture of antigens that elicits a protective immune response against a microbial disease

III microbes and humans out of balance: infectious diseases

The defenses of the human body, described in Part 2 of this book, are normally quite effective, but some microbes can get past human defense systems. Such microbes are able to cause disease because they have developed strategies for undermining or subverting the natural defenses of the human body. Humans, in turn, fight back with preventive measures, such as sanitation and vaccination, or with a direct attack that employs antimicrobial compounds.

The balance between microbes and humans is dynamic and always shifting. Scientists occasionally talk about winning the war with disease-causing microbes, as if a decisive final victory could be possible. So far, decades of intense effort on the part of some of the most brilliant scientists in the world have led to a reduced incidence of microbial infections but not complete elimination. In fact, as old diseases are vanquished, some new ones have appeared. The more we learn about microbes, the more we realize that the best we can hope for is a standoff—one that can only be maintained through constant vigilance.

15 Introduction to Infectious Diseases

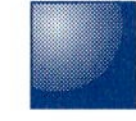

Despite all the emphasis on antibiotics as miracle cures for infectious disease, by far the greatest health miracle of the 20th century was devising new strategies for the prevention of infectious disease transmission.

Learning the ways in which infectious diseases could be controlled was one of the greatest medical triumphs of the last two centuries. Starting in the 1800s, with the realization that many diseases were caused by microorganisms, great strides have been made in identifying and preventing diseases that once took millions of lives. Vaccines were developed against some of the most feared diseases. The discovery of antimicrobial compounds made it possible, for the first time, to cure infectious diseases. Smallpox was eradicated, the first such global success. As late as 1970, the human victory over disease-causing microbes seemed unassailable. Then, the cracks in the façade of human invulnerability began to appear. Acquired immunodeficiency syndrome (AIDS), a new viral disease, emerged. A short time later, new bacterial diseases such as Legionnaire's and Lyme diseases were documented. Tuberculosis reappeared in developed countries. An epidemic of cholera, a disease previously seen mainly in Europe and Asia, swept through South America. Large and widespread outbreaks of food-borne disease began to occur with depressing regularity. Medication failed in some patients infected with bacteria that were resistant to many antibiotics. Humans were learning a new lesson. It doesn't pay to underestimate an adversary with a three-billion-year evolutionary head start. They're still here! An overview of the leading infectious disease killers, based on 1998 data from the World Health Organization, is shown in Figure 15.1.

Emerging and Reemerging Infectious Diseases

Complacency gives disease-causing microbes another chance

Scientists in Pasteur's time knew that effective prevention of disease required constant vigilance. By the 1950s, however, this principle did not seem so important. Physicians, elated by the success of antimicrobial compounds, began to think that infectious diseases were no longer a problem, except for those few viral diseases for which no effective drugs were available. Scientists stopped investigating life-threatening bacterial diseases, such as pneumonia and tuberculosis, and programs for the prevention of tuberculosis were dismantled. Pharmaceutical companies, faced in the

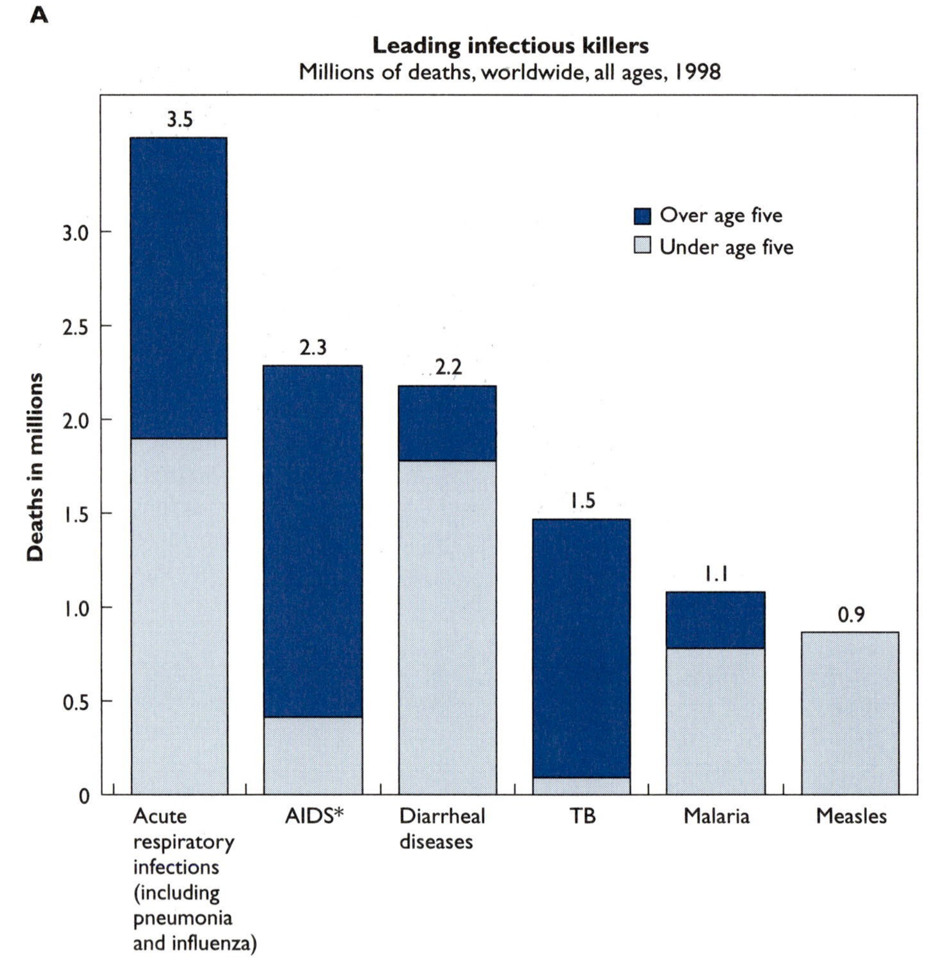

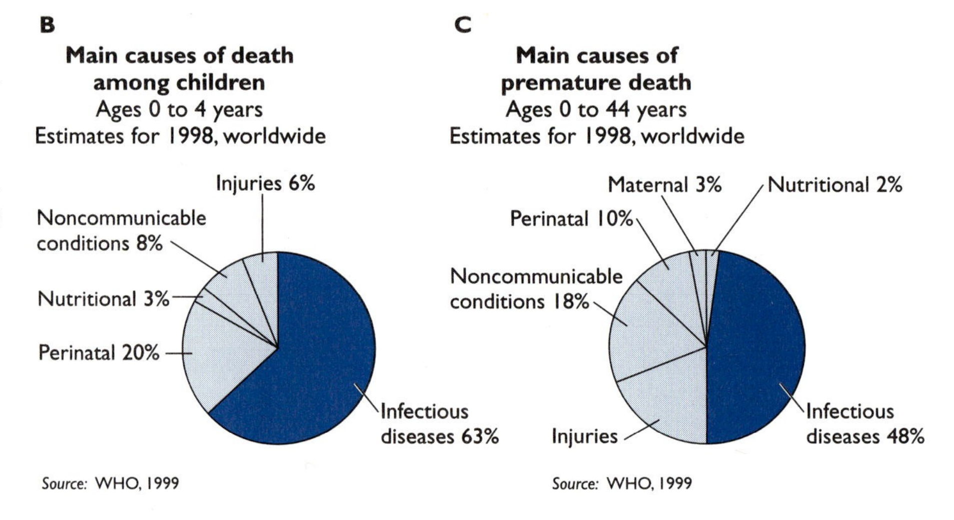

Figure 15.1 Role of infectious diseases as cause of death worldwide in 1998. (A) Microorganisms are responsible for millions of deaths each year. In 1998, the major killers of children younger than 5 years of age were acute respiratory infections and diarrheal diseases, whereas the major killers of those older than five were acute respiratory infections, acquired immunodeficiency syndrome, and tuberculosis. **(B)** In children between birth and 4 years, infectious diseases accounted for 63% of deaths. **(C)** Infectious diseases accounted for almost half (48%) of all deaths between birth and 44 years. (Data from the World Health Organization, 1999.)

1960s with a glut of antibiotics on the market, began to cut back or shut down completely their antibiotic research and discovery programs. This meant that by the 1990s, fewer and fewer antibiotics entered the market each year. Because it takes at least 10 to 20 years to bring a new antibiotic from the lab bench to the market, the impact of decisions made in the 1960s was not felt until much later. Complacency was not limited to the developed countries. Worldwide, the control of major infectious diseases continued to have a low priority compared with other investments, such as arms expenditures (Fig. 15.2).

Although the medical community forgot about bacteria and eukaryotic microbes, the microbes did not forget their human targets. Bacterial resistance to antibiotics rose steadily, until some strains of bacteria were treatable by only one antibiotic or were not treatable at all. Physicians were slow to admit that resistance to antibiotics was a problem, because they assumed there would always be new antibiotics to treat resistant strains. Hospital administrators were the first to become alarmed about resistance to antibiotics, noting the escalating costs and increased time patients spend in the hospital because of infections caused by resistant microbes.

Changes in disease patterns

The unexpected appearance of new diseases in the last decades of the 20th century was something of a shock to the medical community. In most cases,

Figure 15.2 Comparison of global military spending and spending on control and prevention of infectious diseases. Global military spending was far in excess of the amount spent on control and prevention of three of the leading infectious disease killers. By comparison, deaths resulting from those infectious diseases since 1945 far outstripped those resulting from wars. (Data from the World Health Organization, 1999.)

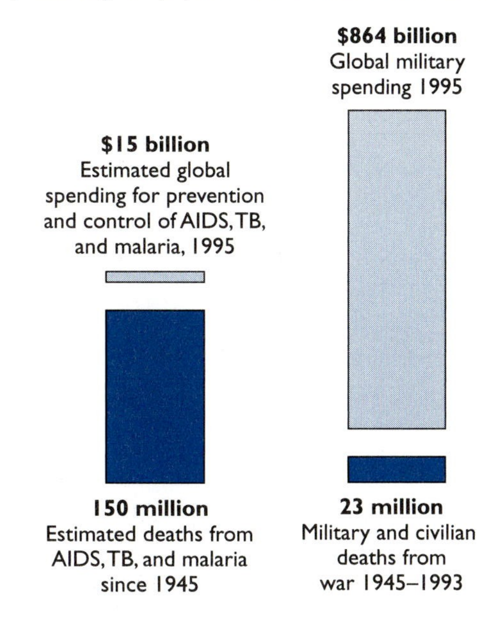

the microorganisms had been around for years, but the opportunities to infect humans had not been frequent enough to produce disease outbreaks. Changes in human practices that increased the frequency and conditions of contact with microbes created situations in which disease could occur. An example of this, the connection between air-conditioning and Legionnaire's disease, was introduced in Chapter 1. Medicine itself produced some new diseases. Cancer chemotherapy and organ transplants created new populations of people whose immune systems became severely compromised during attempts to stop tumor growth or prevent rejection of new organs. To make matters worse, highly susceptible people often were collected together in hospitals, thus increasing the likelihood of disease transmission. Microbes that had been thought to be incapable of causing human disease became a serious problem in these immunocompromised people.

Individuals infected with human immunodeficiency virus (HIV) swelled the ranks of those who were hypersusceptible to microbial infection. Crowding in homeless shelters and prisons increased the likelihood of disease spread. The increasing number of elderly people, whose immune systems had begun to deteriorate and who were often crowded together in nursing homes, provided still another disease opportunity for enterprising microbes. Crowded day care centers provided yet another susceptible population. The media began to publish alarmist reports of "killer bugs" and predicted an imminent return to the pre-antibiotic era. Physicians began to realize that new antibiotics were no longer appearing with the frequency they had a decade before. Alarm bells were going off everywhere.

The pharmaceutical industry sprang belatedly into action on the antibiotic discovery front, but with distinctly less enthusiasm than in the first half of the century. Antimicrobials had become much less economically attractive for these companies because it was more difficult to discover new antimicrobials. It had also become more expensive to conduct the clinical trials needed to establish safety and efficacy and to satisfy the requirements of regulatory agencies. Moreover, antibiotics usually cured patients after a short course of therapy. By contrast, other pharmaceuticals, such as antidepressants and blood pressure medications, were administered daily for long periods of time and were thus much more profitable. Anti-HIV medications were an exception to this rule, because they must be taken daily for life, but most antimicrobials were unprofitable compared with other categories of drugs. New antibiotics are now once again beginning to move toward the market, but it will take years to achieve the levels of effectiveness and abundance that characterized antibiotic availability in the 1950s and 1960s.

Tuberculosis, which had been virtually eradicated in developed countries, showed up again in big cities beginning in the 1980s. At the same time, new government policies led to the widespread discharge of patients from mental institutions, causing the ranks of the homeless to swell and contributing enormously to the spread of antibiotic-resistant strains of the bacterium that causes tuberculosis. Homeless shelters provided places where disease transmission could occur easily. The regimen of antituberculosis drugs was complicated, and many patients were unsupervised, leading to low compliance and the rise of drug-resistant strains. International travel boomed, and travelers became vectors of disease. Malaria cases appeared in developed countries, brought home by citizens traveling overseas. Travelers also brought penicillin-resistant strains of *Streptococcus pneumoniae,* the most common cause of bacterial pneumonia, from Europe to the Americas.

Fortunately, not all the news was bad

At the same time that health officials were waking up to the realization that infectious diseases had by no means been conquered, scientists discovering "new" infectious diseases were bringing renewed hopes for cures for diseases previously thought to be incurable. The first such breakthrough was the discovery that a gram-negative bacterium, *Helicobacter pylori,* causes most gastric and duodenal ulcers. This discovery led to an antibiotic cure for ulcers and has dramatically increased the quality of life for ulcer sufferers. In addition, *H. pylori* appears to be responsible for gastric cancer, one of the most deadly forms of malignancy. Thus, curing ulcer patients early may also reduce the incidence of gastric cancer. The *H. pylori* success story generated a veritable gold rush to find a microbial cause for other chronic incurable diseases. Atherosclerosis, rheumatoid arthritis, coronary artery disease, colon cancer, and inflammatory bowel disease are among the diseases now being reconsidered. How many actually prove to be microbial in origin remains to be determined, but if even one of these diseases is moved from the incurable to the curable list, a giant advance will have been made. Another sort of good news is eradication of a microbial disease. The first success was the eradication of smallpox, and polio is expected to be eradicated within the next year or so. For an interesting and novel attempt to eradicate a disease for which there is no vaccine but for which there is an effective antibiotic therapy, see Box 15.1.

What Do Microbes Want?

Many people, including a number of microbiologists, view disease-causing microbes as organisms that have evolved specifically to infect humans. For whatever reason, these organisms must be out to get us. This way of thinking has come to dominate studies of viral and bacterial diseases. But is it correct? Scientists who study protozoal pathogens present us with another possibility: that at least some human diseases occur because microbes adapted to one host find themselves in a different host to which they are not adapted. This is clearly the case with the protozoan *Toxoplasma gondii,* which can cause serious disease in humans.

T. gondii normally lives in cats, where it goes through a complex life cycle that causes the cat no discernable discomfort. When cattle or other farm animals ingest food contaminated by cat feces, the normally well-programmed life cycle of *T. gondii* goes awry. *T. gondii* is not excreted in feces, as it is in the cat, but migrates into the tissue of the farm animals. Ingestion of undercooked meat or inhalation of contaminated cat feces by humans can introduce the protozoan into the human body. Here, *T. gondii* is usually eliminated very rapidly, causing no problems unless the infected person is immunocompromised or is pregnant and has not been exposed previously to *T. gondii.* In the former case, *T. gondii* can wander at will through blood and tissue, causing a systemic infection that can be fatal. In pregnant women, *T. gondii* localizes to placental tissues and passes through the placenta to infect the fetus. There *T. gondii* cannot survive very long but can cause serious damage to the infected fetus. Clearly, this situation is as bad for *T. gondii* as for humans. To how many disease-causing microbes does this same model apply? Scientists disagree on the answer to this question.

It is remarkable that, after more than a century of study, controversy remains over *why* microorganisms cause disease. Scientists still have a lot to learn about the relationship between disease-causing microbes and humans.

box 15–1 *Eradication of Syphilis in the United States: a Bold New Initiative*

The National Institutes of Health (NIH) and the Centers for Disease Control and Prevention (CDC) have launched an ambitious campaign to eliminate syphilis from the United States. Other industrialized countries have already done this, but especially large problems are posed by the size and diversity of the U.S. population. Today, syphilis cases are at their lowest level ever, about 7,000 total cases as of 1999. This is no reason for complacency, however. Not only could syphilis reemerge, but the disease today takes its toll largely among the poor in the southeastern United States. Most of its victims are African Americans living in areas of poverty, where their access to health care, if it exists at all, is minimal. This situation is completely unacceptable, especially because the disease can be eliminated.

Supporters of the effort point out that, in addition to correcting one of the worst racial and geographical disease disparities in the United States, the NIH-CDC program will reduce transmission of human immunodeficiency virus (HIV), improve infant health, and save about $1 billion annually in health care costs. Syphilis is caused by a spirochete, *Treponema pallidum. T. pallidum* is spread among adults largely by sexual transmission. Pregnant women can transmit it to their infants, because *T. pallidum,* like *Listeria monocytogenes* and *Toxoplasma gondii* (Chapter 17), crosses the placenta and attacks the fetus. The results of such infections range from mental and physical abnormalities to fetal death. Syphilis can increase the spread of HIV, because the sores (chancres) that form near the area of initial transmission breach the skin or mucosa and make it easier for HIV to enter the blood.

The strategy for eliminating syphilis rests on a variety of approaches. These include increased community involvement and education, increased surveillance, increased resources for identification and treatment of people with syphilis, and a rapid and aggressive response to outbreaks. *T. pallidum* is one of the few bacteria that remain highly susceptible to penicillin, so a cheap and effective treatment exists. Also, there are several syphilis tests for screening high-risk populations.

The concentration of syphilis cases among poor African Americans living in the southern states is especially shameful in view of the history of syphilis treatment among the same group. Syphilis was widespread throughout the United States in the first half of the 20th century. In fact, as late as the 1960s, a syphilis test was required by most states before a marriage license could be issued. In the 1940s, the first effective antibiotic treatment for syphilis was discovered. A group of physicians became concerned that not enough was known about the progression of the disease, especially the later stages involving the nerves and muscles. These doctors made the argument to the agency that would later became the NIH that, in the interests of science, a group of men with syphilis should be left untreated and the progression of their disease observed. The men chosen were African Americans in Mississippi, most of whom did not know they had syphilis. They were given free health care for everything but syphilis, a ploy to explain the periodic physical examinations and tests they would undergo. It was not until 1965 that a medical professional objected to the study. Even so, this disgraceful situation did not become widely known to the public until 1972. A final injustice, small compared to the larger one, was that the study came to be known as the Tuskegee experiment, as if one of America's premier African American colleges had been responsible. In fact, the government employees running the study only used some of Tuskegee's facilities, and it is not clear whether university officials were aware of the study's focus.

Sources: Centers for Disease Control and Prevention. Syphilis Continues to Retreat: Nation Sets Sights on Elimination. CDC Reports All-Time Syphilis Lows and Concentration in 1% of U.S. Counties [press release, October 7, 1999]. Centers for Disease Control and Prevention, Atlanta, GA.

Jones, J.H. 1993. Bad Blood: The Tuskegee Syphilis Experiment. Simon and Schuster: New York.

Terminology

Many people find it hard to believe that they can be colonized with microbes capable of causing such serious diseases as pneumonia and meningitis without developing any symptoms. The fact that carriage without symptoms is far more common than disease is a testimony to the power of a strong defense system. The defenses of the human body usually keep potential disease-causing microorganisms under control. Breaches in these defenses alter this balance and leave a person open to a shift from colonization to disease. The terminology of infectious diseases reflects this understanding of the relationship between microbes and their human hosts. **Disease** (literally, dis-ease) is the term used to describe symptoms that result from colonization of the body by a pathogenic microorganism.

Infection is defined as colonization of the body by a microbe that is capable of causing disease, whether or not symptoms develop. A person who is infected but does not develop symptoms is called an **asymptomatic carrier** to emphasize the fact that, although no symptoms are evident, the person still carries and sheds microorganisms that may cause disease in others.

Asymptomatic carriers are often the most dangerous spreaders of a disease, because they do not stay at home but go about their usual business, bringing them into contact with people who may be more susceptible to the disease. The classic case of an asymptomatic carrier was that of Mary Mallon (d. 1938), known as Typhoid Mary. Mary was a cook who carried the bacterium that causes typhoid fever (*Salmonella typhi*) in her intestinal tract without developing the disease. She continued to shed the bacteria, however, and hundreds of people who ate food unintentionally contaminated by her hands died of typhoid fever. First identified as a carrier in 1907, she was eventually jailed to prevent her from working as a cook. To her death, however, Mary never believed that she was responsible for causing disease in others. How could someone who felt and appeared healthy possibly be carrying dangerous disease-causing bacteria? Today, we see a similar phenomenon in the case of asymptomatic HIV-infected people who continue to practice unsafe sex.

Microorganisms capable of causing disease are called **virulent** or **pathogenic** microorganisms. Again, it is important to emphasize that a virulent or pathogenic microorganism does not necessarily cause disease in the person it colonizes. The outcome of the encounter depends on the strength of the body's defense systems, on the number of microorganisms encountered and on the virulence of the infecting microbe. **Infectivity** describes the ability of a microbe to infect human hosts it encounters. The term **pathogenicity** describes the ability of a microbe to cause clinically apparent symptoms. The more likely the microbe is to cause symptoms in people it infects, the more pathogenic it is. **Virulence** describes the ability of a microbe to cause severe disease. A microbe can be pathogenic without being particularly virulent, if it causes a mild form of disease in most of the people it infects. The common cold is an example of a disease caused by an organism that is pathogenic but not virulent. HIV is an example of a microbe that is both pathogenic and virulent.

Measures of Infectivity and Virulence

How do scientists define infectivity and virulence in quantitative terms? A measure of infectivity is the **infectious dose 50 (ID_{50})**, which is defined as the number of microbes needed to infect 50% of animals exposed to them (Fig 15.3). The ID_{50} usually is measured using laboratory animals instead of human subjects, especially in the case of virulent microbes or microbes for which no effective therapy is available. Infectivity in human populations can be deduced from the attack rate of an infection. The **attack rate** is defined as the number of cases of clinically apparent disease divided by the number of susceptible people in the population. A measure of virulence is the **lethal dose 50 (LD_{50})**, which is defined as the number of microbes required to kill 50% of the animals exposed to them. A measure of virulence in human populations is the **case fatality rate** of a disease. The case fatality rate is defined as the number of deaths from a particular disease divided by the number of clinically apparent cases of that disease.

The practical importance of terms such as these is evident from discussions of bioterrorism. A popular focus for bioterrorists has been the

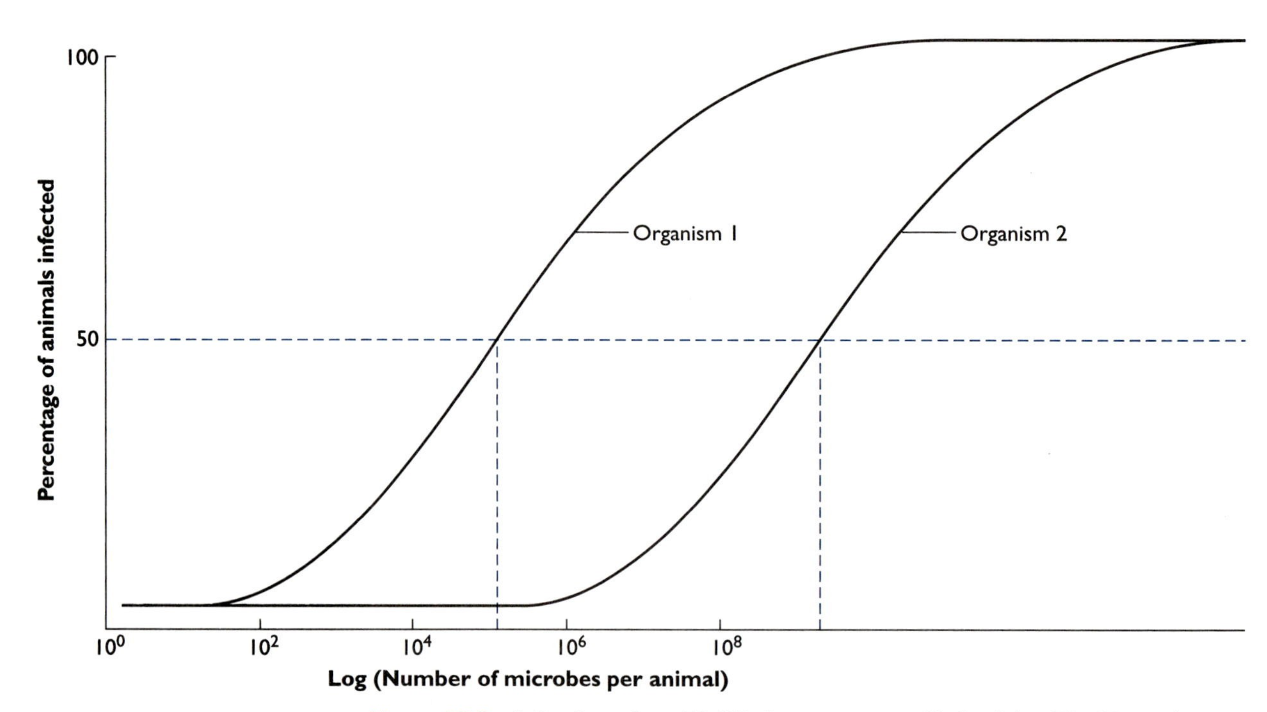

Figure 15.3 Infectious dose 50 (ID_{50}), a measure of infectivity. The ID_{50} is the number of microbes per animal required to cause infection in 50% of animals inoculated. The lower the number, the more infectious the microorganism. Organism 1 has an ID_{50} of approximately 10^5, so it is more infectious than Organism 2, which has an ID_{50} of approximately 10^9. Lethal $dose_{50}$ (LD_{50}) assays are interpreted similarly, except for the scoring of the number of animals that died at each dose.

bacterium that causes anthrax. *Bacillus anthracis* is a spore-forming gram-positive bacterium. The fact that it produces spores, which are quite hardy and survive for years, makes *B. anthracis* easy to store and disperse. *B. anthracis* is a very virulent bacterium that kills most of the people who develop symptomatic disease. Death results from septic shock. Anthrax clearly has a high case fatality rate. What no one knows, however, is the attack rate. That is, if a bioterrorist sprayed a room full of people with *B. anthracis* spores, how many people would develop symptomatic infection? Bioterrorists have tried spraying *B. anthracis* spores more than once, and so far no cases of anthrax have resulted. This could be the result of incompetence on the part of the bioterrorists, but it is more likely because *B. anthracis* has a low attack rate. By contrast, native Americans in the 16th century who came into contact with smallpox virus for the first time had a very high attack rate and a high case fatality rate. In Europe, where the disease had been around for centuries, the attack rate and case fatality rate were lower, probably because of some degree of immunity acquired from occasional contacts with individuals shedding low numbers of the virus.

Proving Cause and Effect in Microbial Infections

Today, the idea that organisms too small to be seen by the unaided eye could cause disease is taken for granted, but this was not always the case. During the 1800s, the hypothesis that microbes were the cause of diseases like tuberculosis or cholera was quite controversial.

Koch's postulates

In an attempt to devise a rational basis for proving that a specific microbe could cause a specific disease, the microbiologist **Robert Koch** (1843–1910) proposed a set of postulates for proving that a microorganism is responsible for a disease. **Koch's postulates,** first stated in 1882, are listed in Table 15.1. The first postulate states that the microorganism must be associated with lesions of the disease. The second postulate asserts that the microorganism can be isolated in pure culture from a person with the disease, preferably from the lesions associated with the disease. This has proved to be difficult in some diseases, because of a failure to cultivate the causative organism. Few people doubt that syphilis is caused by the bacterium *Treponema pallidum,* because antibiotic treatment that eradicates this bacterium from the body cures the disease. However, despite the fact that these spirochetes can be seen in lesions of the disease, the bacterium has never been grown in the laboratory. The third postulate states that the microorganism isolated from a person with the disease must be administered to another person or animal and shown to produce the disease in that person or animal. This is often the hardest postulate to satisfy, especially for diseases caused by microorganisms that are human specific and do not readily infect animals. Finally, Koch's fourth postulate states that the microorganism must be re-isolated in pure culture from the animal or human that was infected experimentally to satisfy the third postulate.

An amusing but little known historical fact is that the ink was hardly dry on Koch's postulates before the author himself complained about his third postulate. At that time, Koch had much evidence to suggest that the dreaded disease cholera was caused by a bacterium (later named *Vibrio cholerae*). Cholera was a significant disease at the time and was known to be spread through contaminated water. Identifying the cause would allow scientists and public health officials to pinpoint water sources that carried the cholera bacteria. Koch had been unable to infect any animals with *V. cholerae,* and his opponents did not hesitate to throw his third postulate in his face. Eventually he was vindicated by additional scientific studies.

Koch's postulates are still relevant today

Today, scientists are more prone to accept the idea that a particular microorganism causes a disease, especially if the appropriate antimicrobial therapy predicted by this understanding of the disease effects a cure. In fact, some scientists thought that Koch's postulates had been relegated to a dusty shelf of the library, when the controversy surfaced over whether *H. pylori* caused most gastric ulcers. Scientists and pharmaceutical companies

Table 15.1 Koch's Postulates

Postulate 1.	The microbe must be present in all people with the disease and should be associated with the lesions of the disease.
Postulate 2.	The microbe must be isolated in pure culture from a person who has the disease.
Postulate 3.	The isolated microbe, when administered to humans or animals must cause the disease.
Postulate 4.	The microbe must be isolated in pure culture from the human or animal infected to satisfy postulate 3.

with vested interests in the theory that ulcers were caused by stress found it hard to believe that bacteria might be the culprits. At issue were some of the most lucrative drugs sold by pharmaceutical companies, medications that had to be taken daily for life. The news that a patient who once spent thousands of dollars a year for anti-ulcer medication could now spend a few hundred dollars for a short course of antibiotics that would quickly cure the disease and prevent recurrences was bad news for some pharmaceutical companies. Understandably, if these companies were going to accept this new theory of ulcers, they wanted proof, with all the i's dotted and t's crossed. Koch's postulates assumed center stage once again, especially the third postulate. Eventually, animal models for gastric ulcers allowed scientists to satisfy Koch's third postulate. Moreover, the discovery of an antibiotic combination that eliminated the bacteria from the stomach and stopped ulcer recurrences provided even more convincing proof that ulcers were caused by *H. pylori.* But initial failure to satisfy the third postulate held up acceptance of *H. pylori* as the cause of ulcers for years.

What Makes a Microbe Pathogenic or Virulent?

Characteristics of microorganisms that allow them to cause disease are called **virulence factors** or **pathogenicity factors.** Defining these factors has been a major focus of research in recent years, and some new insights have been obtained. In retrospect, these answers seem intuitively obvious to those who view disease as a shift in the equilibrium between microorganisms and the defenses of the human body. The fact that they seemed revolutionary when first proposed gives a good indication of how much the view of infectious disease has changed over the past few decades. Some of the features of pathogenic microbes are listed in Table 15.2.

Virulence factors that allow microbes to cause an infection

A microbe must first get to the body site it targets. For example, a bacterium or virus that infects the intestinal tract must be able to survive passage through the stomach and must be able to attach to the intestinal lining to prevent being washed out of the site. Thus, most bacteria and protozoa that cause intestinal infections have flagella, which allow them to swim to the intestinal wall, and pili or other mechanisms of adhesion that help them adhere to the intestinal mucosal cells. Viruses that infect the intestine must be able to attach to and infect intestinal cells or to transit the intestinal wall and gain entry into the bloodstream. A microbe that can live inside a biting insect has a natural system for getting itself injected into the human blood stream.

As you learned in Chapters 12 and 13, the human body is protected from infection by a variety of defenses. A second characteristic shared by most

Table 15.2 Features of Pathogenic Microbes

Attach to host cells
Evade host defenses
Obtain iron and other essential nutrients
Produce symptoms

disease-causing microorganisms, is the ability to evade or subvert one or more of these defenses. Microbes have evolved a variety of mechanisms for doing this. Capsules protect microbes from phagocytosis. Some bacteria are able to live in and kill unactivated phagocytes. Other microbes make themselves invisible to the immune system by making their surfaces resemble host tissues. Still others attack the cells of the immune system directly.

A third characteristic of most disease-causing microbes is their ability to obtain essential nutrients from human tissues. Iron is a very important nutrient for virtually all microorganisms, but the concentration of available iron in human tissues and blood is low, because the body produces iron-binding proteins such as transferrin or lactoferrin that bind iron very tightly. Disease-causing microbes either have very efficient iron-uptake systems that allow them to scavenge what little free iron is available, or they are able to remove the iron from transferrin or lactoferrin. The human body is potentially a rich source of carbon and nitrogen. These elements, however, normally are not free but rather are found as part of complex molecules—polysaccharides or proteins. A number of microbes that infect tissue produce polysaccharidases or proteases, which damage tissue as they release carbon and nitrogen.

Nonprotein microbial toxins

Lipopolysaccharide (endotoxin) and lipoteichoic acid. In Chapter 12, the ability of bacterial surface molecules, such as lipopolysaccharide and lipoteichoic acid, to activate the inflammatory response was described. Lipopolysaccharide has been called endotoxin, because it is part of the bacterial cell (endo) and is only released during cell lysis. The term endotoxin was meant to differentiate it from exotoxins, protein toxins that are usually released from the cells without lysis. Yeasts and protozoa also elicit an inflammatory response, but the molecules that do this have not been well characterized.

For microbes that are unable to survive attack by neutrophils, molecules such as lipopolysaccharide are a liability. But for microbes that can survive the inflammatory response, this response is advantageous, because it liberates nutrients by killing tissue cells. In addition, if the region of dead tissue becomes large enough, it protects the bacteria from the immune system, because immune cells move through the blood or lymph to a site. If enough damage has occurred to cut off the blood supply, this will inhibit migration of immune cells to the site where the microbes are growing. Macrophages and neutrophils can migrate in tissue but do not do this as well in areas of extensive tissue damage. Similarly, antimicrobial compounds may not diffuse readily into a damaged site. This is why, in cases of gas gangrene, a bacterial disease characterized by extensive tissue damage, it is necessary to remove the dead tissue surgically (**debride** the wound) before antibiotics and the body's defense systems can bring the bacteria under control. In cases where debridement is not successful, amputation of the limb is necessary. Gas gangrene is an interesting example of the way in which bacteria cause damage. Here, tissue damage is not only the result of microbe-produced proteases and polysaccharidases (although these contribute to the damage), but also results from the physical pressure of the copious amounts of carbon dioxide produced by the bacterium *Clostridium perfringens* as it utilizes the carbon sources from the dead human cells.

Protein toxins (exotoxins)

A-B toxins. Protein toxins are individual proteins or protein complexes that attach to eukaryotic cells and damage them. These toxins are produced by bacteria. Fungi, algae, and some bacteria produce low molecular weight toxic substances that are not proteins (Chapter 9). Some protein toxins target cells of insects and other animals as well as cells of humans. Here we will focus on protein toxins that target human cells. A common type of protein toxin is the **A-B type toxin** (Fig. 15.4). These consist of a B portion, which binds to specific receptors on the surface of the human cell and facilitates the entry into that cell of the A portion, which actually does the damage. The B part, by itself, is nontoxic. The same is true of the A part by itself, because it must be introduced into the target cell by the B part. The B portion may consist of a single protein or a complex of several proteins. The A portion is usually an enzyme, consisting of one or two proteins.

Figure 15.4 Structure and mechanism of action of an A-B type toxin. The B portion of the toxin binds to the target host cell. The active A portion then enters the host cell, where it produces an effect. The B portion of many A-B type toxins consists of several subunits. The A portion often catalyzes an ADP-ribosylation reaction, in which ADP-ribose is transferred to a host cell protein, thus altering or inhibiting its activity.

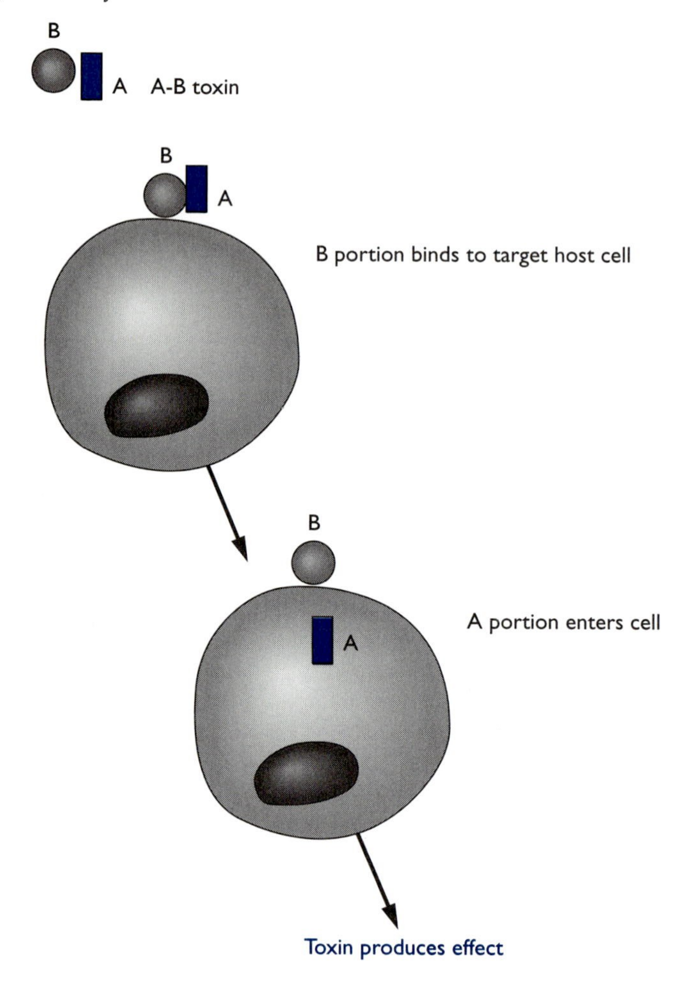

The A portion of most A-B type toxins catalyzes a reaction called **adenosine-diphosphate (ADP)-ribosylation,** in which it removes the ADP-ribosyl group from nicotinamide adenine dinucleotide (NAD) and attaches it to a human protein. ADP-ribosylation inactivates the protein. The results of this inactivation depend on what protein is ADP-ribosylated. In the case of *Escherichia coli* strains that produce a heat-labile enterotoxin called LT, a regulatory protein is inactivated, causing ion channels that control the flow of water across cell membranes to go out of control. The result is water loss from tissue, leading to dehydration and diarrhea. In the case of diphtheria toxin, the target is a step in which amino acids are added to a growing peptide chain. The result is cessation of protein synthesis and death of the cell. In the case of tetanus toxin, the proteins inactivated are nerve cell proteins that control the transmission of neuronal signals.

Most A-B toxins are secreted by the bacteria into the extracellular fluid. The toxin then finds and binds to the type of cell it targets. An interesting variation on this action that was discovered only in the past decade is a type of toxin delivery system called a type-III secretion system. Bacteria with type-III secretion systems bind to the target cell and inject the toxin directly into the cell interior (see Fig. 17.4 in Chapter 17). An example of a bacterium that uses this strategy is *Yersinia pestis,* the cause of plague.

Superantigens. Superantigens are another type of protein toxin. They attach to the T cell receptors of T helper cells and the major histocompatibility complexes (MHCs) of the antigen-presenting cells (APCs). Usually these cells associate only when the T cell recognizes an antigen that is being presented on the MHC of the APC. Superantigens force the cells into an unnatural association that they would not otherwise form, because they are not presenting or recognizing antigens at the time (Fig. 15.5). Whereas only a tiny fraction of a percent of T helper cells would normally be interacting with antigen-presenting cells during an immune response, superantigens can cause up to 20% of T helper cells to attach to antigen-presenting cells and become activated. This is why the toxins are called superantigens. The unnatural activation of so many T cells is toxic, because it results in the release of large amounts of cytokines by the immune cells. Such a release can cause a form of septic shock.

In the 1990s, lurid news articles described a new disease characterized by extensive tissue damage: "THE BACTERIUM THAT ATE MY FACE." The gram-positive bacterium *Streptococcus pyogenes* was responsible for the massive tissue destruction experienced by some people. *S. pyogenes* has caused a number of diseases over the past century, ranging from scarlet fever and rheumatic heart disease to wound infections. Scarlet fever and rheumatic heart disease have virtually disappeared, but *S. pyogenes* is still around. The modern strains that cause extensive tissue damage around a wound (called cellulitis) or deadly systemic infections, produce a superantigen. The cellulitis they cause was reminiscent of a disease called streptococcal gangrene, which was experienced in the early 1900s by soldiers with battle wounds. *S. pyogenes* is unusual, even among bacteria, for the variety of different conditions it has caused.

Roles of toxins in disease. Toxins can be the sole cause of disease symptoms or simply a part of the microorganism's array of virulence factors. Some examples of the roles of toxins in disease are shown in Figure 15.6. Food-borne disease caused by *Staphylococcus aureus* and botulism are examples of diseases in which the bacteria do not infect, but instead produce the toxin in food. The toxin is ingested with the food. Botulism is caused by a gram-positive, spore-forming bacterium *Clostridium botulinum*. *C. botulinum* is closely related to the bacterium that causes tetanus, *Clostridium tetani*.

Botulism toxin is a neurotoxin that causes a flaccid paralysis. Death is usually due to collapse of the respiratory system. The toxin of *S. aureus* is much less dangerous. It stimulates the nerve ends in the stomach that control peristalsis, causing projectile vomiting and severe abdominal pain. The disease lasts only 1 or 2 days and is not fatal. Note that the difference in lethality between botulism toxin and *S. aureus* toxin (called enterotoxin) is accounted for by the difference in the cells they target. Both target nerve cells, but their effects on the nervous system are quite different and thus have different impacts on the human body.

Diphtheria, caused by *Corynebacterium diphtheriae*, and tetanus are diseases in which the bacteria infect tissue but cause little local damage. Instead, disease symptoms are the result of a toxin released into the bloodstream. *C. diphtheriae* colonizes the throat, and *C. tetani* colonizes deep wounds. *C. diphtheriae* is spread from person to person. *C. tetani* is found in soil, especially soil contaminated by animal feces.

An example of a bacterium that infects tissues and causes damage by a number of mechanisms has already been mentioned: *S. pyogenes*, a common

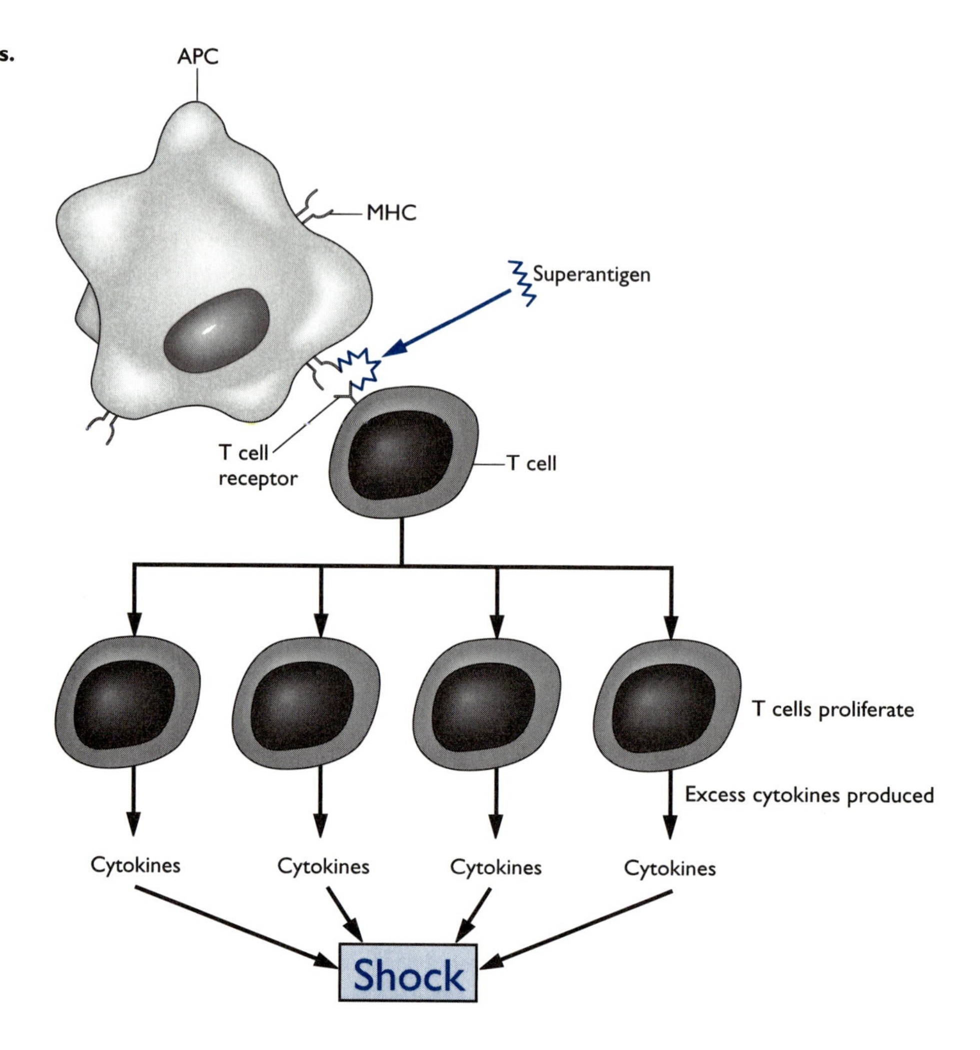

Figure 15.5 Action of superantigens. Superantigens cause binding of the major histocompatibility complex (MHC) of antigen-presenting cells (APCs) that are not presenting an antigen and the T cell receptor of T helper cells. Because the usual specificity of APC-T helper cell interaction is bypassed by the superantigen, a much higher than normal number of T cells is activated. This results in the production of large amounts of cytokines, which can cause shock and death.

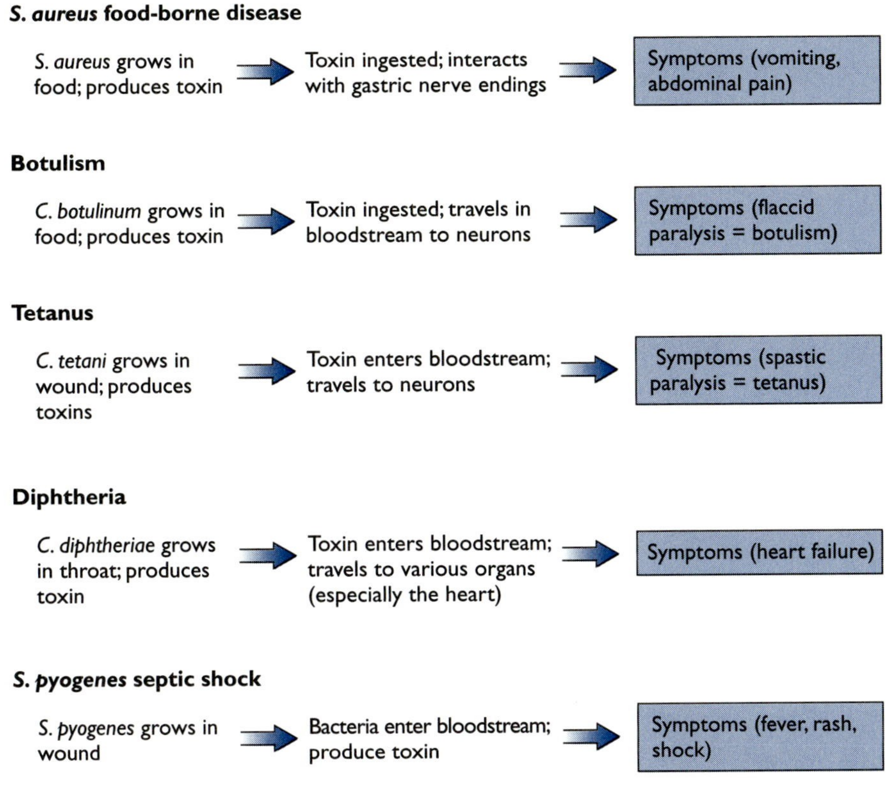

Figure 15.6 Roles of toxins in disease. *Staphylococcus aureus* and *Clostridium botulinum* produce toxins when they grow in food. The toxin is ingested and produces the effect. *Clostridium tetani* and *Corynebacterium diphtheriae* colonize the host and produce toxins once they are in the host. The toxins, but not the bacteria, enter the bloodstream and move to the target cells. *Streptococcus pyogenes* infects the host and enters the bloodstream, where it produces a toxin that is a superantigen. The actions of the superantigen result in shock.

cause of wound infections. It usually causes inflammation in the area of the wound, resulting in a red, painful area with lines radiating out from the wound site. These lines are caused by bacteria migrating in the tissue. Polymer degrading enzymes produced by the bacteria allow it to break down tissue and spread outward from the wound. In some cases, however, *S. pyogenes* enters the body through very tiny wounds and does so little damage in the wound area that the infected person does not realize at first that infection has occurred. The most dangerous infections caused by *S. pyogenes* are cellulitis and the ones where bacteria enter the bloodstream. *S. pyogenes* has lipoteichoic acid in its cell wall, which can trigger septic shock. The superantigen it produces also contributes to toxic shock. *S. pyogenes* is a human-specific pathogen that is carried in the mouth and throat, where it rarely causes disease. This location, however, allows it easy access to any wounds that might occur.

The mystery of toxin function. For those who believe that microbes are out to get us, toxins are easy to understand. On closer inspection, however, this view would need to embrace the idea that microbes are so vindictive that they do things to injure humans, even when these actions have no benefit for the microbes themselves. Consider botulism toxin or *S. aureus* enterotoxin. These toxins are produced in foods but do not help the bacteria to infect. Instead, the bacteria pass through the intestinal tract without causing infection, leaving the toxins behind to cause disease. Even diphtheria and tetanus toxin are difficult to understand as aids to microbial survival. Not all strains of *C. diphtheriae* produce diphtheria toxin. Yet the toxin nonproducers are just as successful as the toxin producers in colonizing the human throat. Similarly, tetanus toxin has no known function in aiding the bacteria to colonize wounds.

The story becomes more complex and more puzzling when you consider that many of the toxin genes are carried on bacteriophages. Thus only bacteria that are infected by a lysogenic phage produce the toxin. This means that toxin genes in the bacteria themselves did not necessarily evolve to ensure the survival of the bacteria but were acquired in chance encounters with bacteriophages. And why would a bacterial virus be carrying genes that encode a neurotoxin or a diphtheria toxin? One explanation is that the genes we call toxin genes and that are carried by bacterial viruses play some role in regulating the viral life cycle and just happen to be toxic to human cells. Perhaps some student reading this book will solve this long-standing mystery.

Where Do Pathogenic Microbes Come From?

Interest in the evolution of pathogens has been growing because of the need to understand how new diseases can continue to emerge and why new forms of "old" pathogens continue to appear. Understandably, most of the attention given to this topic has focused on recent evolution of pathogens. There are two ways evolution can occur in microbes: existing genes in the microbe's genome can mutate, or new genes can be acquired from other microbes (horizontal gene transfer).

Evolution by mutation

Viruses that infect eukaryotic cells seem to favor the first strategy, especially those viruses with RNA genomes. RNA is less stable chemically than DNA. Moreover, enzymes that reproduce RNA molecules do not have the kind of proofreading capability found in enzymes that reproduce DNA. The result is that RNA-copying enzymes tend to make more mistakes. Together, the chemical instability of RNA and the lack of proofreading functions in many of the viral enzymes that reproduce RNA give viruses with RNA genomes the potential to mutate very rapidly. Not surprisingly, two RNA viruses, HIV (the cause of AIDS) and influenza virus (the cause of influenza) are among the most rapidly mutating viruses known. Mutants of HIV have been seen to appear in the same patient over the course of the infection—a very short time span for evolution. Mutations in HIV and influenza virus help them to avoid the body's immune responses. HIV is also mutating to greater resistance against the new antiviral compounds. This type of evolution to resistance to antiviral compounds has been slower in the case of influenza virus but may well occur in the future. In one case, viruses become more virulent by horizontal gene transfer: influenza virus has a genome that comes in segments (see Chapter 16). The most virulent strains are ones that have acquired a new segment from another influenza virus. Such acquisitions can make them virtually invisible to the immune system that normally controls them.

Evolution by horizontal gene transfer

Bacteria also evolve by mutation, and this type of evolution has created strains resistant to an important class of antibiotics, the fluoroquinolones. The bacterium that causes tuberculosis became resistant to antituberculosis drugs by mutating the genes that are the targets for these drugs. Aside from these examples, however, bacteria seem to prefer to take advantage of horizontal gene transfer to acquire new traits. The reason is obvious. Changing by random mutation ensures that many unsuccessful mutants will die in the process, whereas acquiring a gene that has already developed in some other microbe is a lot less hazardous and more likely to produce immediate results. Antibiotic-resistance genes are clearly being transferred horizontally among bacteria. Pathogenic bacteria can acquire resistance genes from each other or from bacteria that do not cause disease but are present in the same site. The human intestinal tract, for example, contains a huge population of bacteria that do not usually cause disease in healthy people. Yet when these bacteria are exposed to antibiotics, they can become resistant and transfer these resistance genes to pathogenic bacteria that happen to pass through the intestinal tract.

Horizontal gene transfer can also increase the virulence of a bacterium. *Escherichia coli* provides a good example of this. Most strains of *E. coli* do not cause intestinal disease, but a subset of strains can do so. For example, a type of *E. coli* called *E. coli* O157:H7, which has been in the news because of its ability to cause bloody diarrhea and kidney failure in children, seems to have arisen when a strain of *E. coli* that could cause a mild form of diarrhea acquired toxin genes from *Shigella,* a bacterium that causes bloody diarrhea and kidney failure. The *Shigella* toxin genes entered *E. coli* on a bacteriophage. *Shigella* has been largely eliminated from the food and water supplies of developed countries, but *E. coli* is still very much in evidence. The study of pathogen evolution is still in its infancy, with most work focused on viruses and bacteria, but a better understanding of how pathogens arise may help to predict and prevent the emergence and spread of new pathogens.

Evolution of virulence

An interesting question that arises in connection with pathogen evolution is: which direction does pathogen evolution take—toward increased or reduced virulence? Most people take the pessimistic line and assume that pathogens evolve to greater virulence. From the microbe's point of view, however, this is not a very smart strategy. Severe diseases that kill rapidly not only kill the supplier of all sorts of goodies but also reduce the likelihood that the causative organism will be transmitted before the infected person dies. Pathogens that cause severe diarrheal disease are swept rapidly out of the intestine into a presumably less attractive and certainly less nutrient-rich environment. One could argue that the most successful pathogens are those that colonize the body without causing noticeable symptoms, establishing an asymptomatic carrier state that can persist for months or even years. If this view is correct, microbes that are initially virulent should mutate in the direction of lesser virulence, to a better fit with their human host. One case in which this may have occurred is syphilis. Historians of science suspect that the rapidly fatal disease called leprosy in the Middle Ages might actually have been syphilis. If so, the syphilis of today is a much milder disease, taking years rather than months to finish off the person it infects. Unfortunately, scientists have so far been unsuccessful in testing this hypothesis. If syphilis has indeed mellowed with age, the process took rather a long time from a human perspective: centuries. Of course, the change in the virulence of syphilis, if indeed it has occurred, could reflect better nutrition and less exposure to the pathogen rather than a change in the ability of the bacterium to cause disease.

Another confounding factor is the as yet inexplicable shift that can occur in the strains of microbes prevalent at any given moment in history. In the late 1800s, scarlet fever (a disease caused by the bacterium *S. pyogenes*) was a disease with a high fatality rate. It was characterized by a diffuse red rash, which gave the disease its name. Some parents lost child after child to scarlet fever. By the 1930s and 1940s, *S. pyogenes* had changed its spots somewhat. The rash had the same appearance, but scarlet fever had become a mild childhood disease. However, many young adults, especially military recruits living in crowded conditions, were contracting a form of *S. pyogenes* infection that damaged the heart valves (rheumatic fever). The disease had different symptoms and affected a different age range than those seen in scarlet fever. Some infected people died immediately, but many felt the effects of damaged heart tissue only much later in life, when they became susceptible to infections of the heart valves (endocarditis, an often fatal disease). Inexplicably, in the mid-20th century rheumatic fever virtually disappeared. This type of *S. pyogenes* has resurfaced during the past 10 years as a more severe form of scarlet fever in children. In adults, *S. pyogenes* is back in yet another form: what the tabloids dubbed "the flesh-eating bacterium," a severe form of tissue destruction called streptococcal gangrene. Another version of *S. pyogenes* that caused a rapidly fatal disease in adults also appeared recently, then disappeared. Scientists believe that this shifting pattern of *S. pyogenes* diseases has been caused by one strain of *S. pyogenes* supplanting another. Diseases can submerge as well as emerge, but we still don't understand why.

A better understanding of the co-evolution of humans and their attendant microbes is particularly important if we are to answer the question: how many disease-causing microbes are out there? Is there any limit to the number of microbes that are capable of causing human disease? Microbiologists are fond of stating that only a tiny minority of the microbes in nature cause disease, but is this really true? Protozoa are similar in many ways to the phagocytes that are one of the human body's main defenses against bacterial infection. This suggests that virulence factors first began to evolve not after the arrival of animals and humans on the evolutionary scene, but much earlier, when protozoa first appeared. If so, most types of microbes have experienced the selective pressures directing evolution toward a form that would, if opportunity presented itself, make them capable of causing infection. The ultimate virulence factor may prove to be the ability to grow at 37°C.

Diagnosis and Treatment of Infectious Diseases

Diagnosing microbial infections

Some **symptoms** (what the patient reports) and **signs** (what the physician observes) in infectious disease direct the physician to suspect a patient has a microbial infection. These symptoms and signs are listed in Table 15.3. All are indications of the inflammatory response being mounted by the body against the microbial invader. Before trying to decide what microbe is causing the infection, it is helpful to establish that the patient actually does have an infection. One test that helps to identify a bacterial infection is the concentration of neutrophils in the patient's blood. This usually is higher than normal in patients experiencing a bacterial infection, but, because the concentration of neutrophils varies widely among different people in different settings, a high neutrophil count is an indication rather than positive proof of infection.

The site of infection can be a valuable clue to what type of microorganism is causing the infection. For example, in people outside hospitals, painful urination and cloudy urine are almost always caused by *E. coli.* Similarly, a limited number of protozoa, bacteria, and viruses cause most cases of diarrhea. Because the results of diagnostic tests can take days, these early diagnostic indicators can be valuable in helping the physician to decide on immediate treatment. Another valuable diagnostic indicator is the person's location and occupation. Fever in a pet-shop worker who has frequent contact with exotic birds suggests the possibility of psittacosis, a bacterial disease rare in the rest of the population. Similarly, a health care worker who cares for tuberculosis patients or a person who has been living in homeless shelters in a big city are far more likely to contract tuberculosis than most other members of the population.

Despite these indicators, however, the symptoms of a microbial infection are often too nonspecific to serve as good diagnostic tools. There are exceptions to this, of course. Someone whose urine turns a dark brown is almost certainly infected with *Plasmodium,* a protozoal parasite that causes malaria. Similarly, genital warts caused by papilloma virus have a very distinctive appearance. Yet, such clear diagnostic indicators are the exceptions rather than the rule. Moreover, in immunocompromised people, symptoms can be even less clear, because of impairment of the body's defenses, which are responsible for many of the symptoms.

At one time, a good guess about the identity of the microbe responsible for the disease was all a physician needed to prescribe appropriate antimicrobial therapy. As the incidence of resistant strains has risen, making an educated guess about therapy is becoming more difficult. In critically ill patients, such as those manifesting signs of septic shock, the identity of the causative agent becomes less important than which antimicrobials will be

Table 15.3 Signs and Symptoms Commonly Associated with Infectious Diseases

Skin or mucosal lesions
Inflammation (red, swollen, painful area)
Pus or purulent discharge
Painful urination, vaginal itching and burning
Swollen lymph nodes, enlarged spleen
Fever
Generalized aches and pains
Changes in mentation (confusion, lethargy)
Unintended weight loss
Loss of appetite (anorexia)
Vomiting
Diarrhea, dysentery (bloody diarrhea)

effective. Here, antimicrobial susceptibility tests are critical. Such tests are most highly developed for bacteria, which cause the majority of serious infectious diseases. Similar tests can be used for yeast infections. Tests for susceptibility of viruses and protozoal pathogens are more complicated and are not performed routinely. This may change in the future. As HIV becomes more resistant to antiviral drugs, tests to detect mutations that confer resistance may be performed more frequently. Here, we will describe a few of the tests for antibiotic susceptibility of bacteria.

Measuring susceptibility of bacteria to antibiotics

Two types of tests are used to determine whether bacteria are susceptible or resistant to antimicrobial compounds. The first is the **disk diffusion test** (Fig. 15.7 and color plate 8). Disks containing a specified concentration of an antibiotic are placed on an agar plate that has been spread with the bacterium whose susceptibility is being tested. During the incubation period, the antibiotic diffuses outward from the disk, and, when the bacterial growth becomes visible, there may be zones of no growth around the disk (zones of inhibition). The size of the zone of inhibition indicates whether the microorganism is susceptible or resistant to the antibiotic. Establishing the cutoff between susceptible and resistant strains is rather complicated and involves consideration of not only the amount of antibiotic needed to inhibit the growth of the bacteria but also what concentrations of antibiotic are achievable in the site where the infection is localized.

A second type of test is the **microbroth dilution test.** The **minimal inhibitory concentration** (**MIC**) is determined from this type of test (Fig. 15.8A). A calibrated number of bacteria is inoculated into tubes containing increasing concentrations of the antibiotic. After incubation, the bacteria grow in some tubes but not in others. The antibiotic concentration in the first tube where no growth is observed is called the MIC. The microbroth dilution test is more quantitative than the disk diffusion test, but it is more expensive to run. The **E test,** a more quantitative variant of the disk diffusion test, is sometimes a more practical choice. In this test, plastic strips containing a gradient of an antimicrobial agent are placed on a plate spread with bacteria. The strips are imprinted with a scale showing the concentration of antimicrobial. The MIC is the value on the strip where the

Figure 15.7 The disk diffusion antimicrobial susceptibility test. Bacteria are spread on the surface of an agar plate to form a lawn. Disks containing known amounts of antibiotic are placed on the lawn. The plate is incubated, and the diameters of areas showing no growth are measured. The bacteria can be classified as susceptible or resistant to a specific antibiotic by looking up the diameters of no-growth zones in published tables.

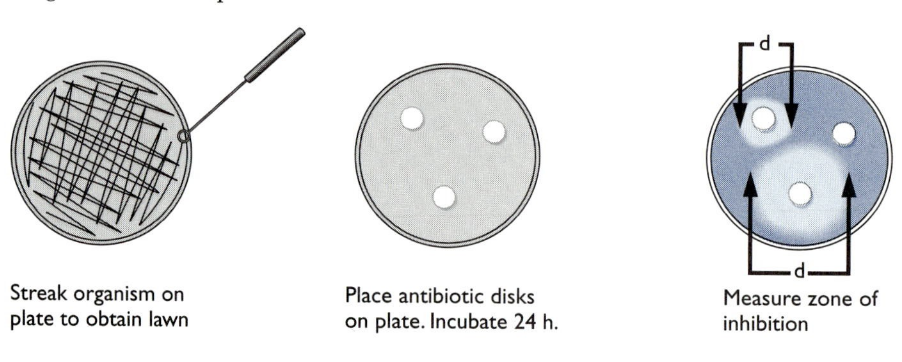

elliptical zone of inhibition intersects the strip (Fig. 15.8B). Physicians prefer a test that gives an MIC value in cases where bacteria may be poised on the line between resistance and susceptibility, because increasing the dose of the antibiotic to the appropriate level might effect a cure where the usual dose level would not.

Watch the Skies!

To end this chapter by taking a much broader view of infectious diseases, consider the questions faced by astrobiologists about whether any microorganisms found on distant planets might be able to cause disease here on

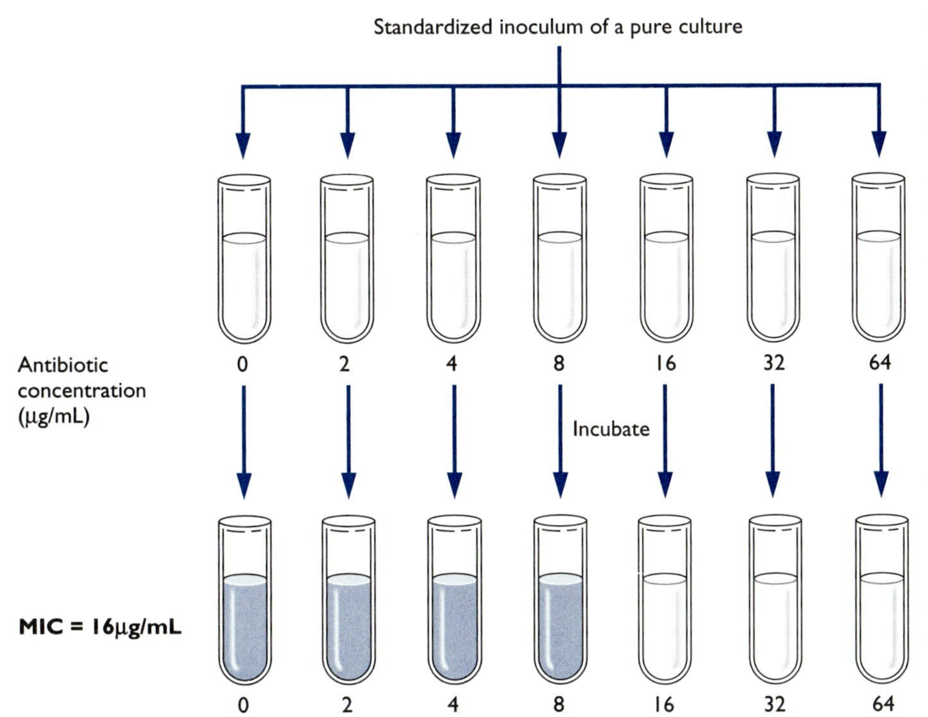

B

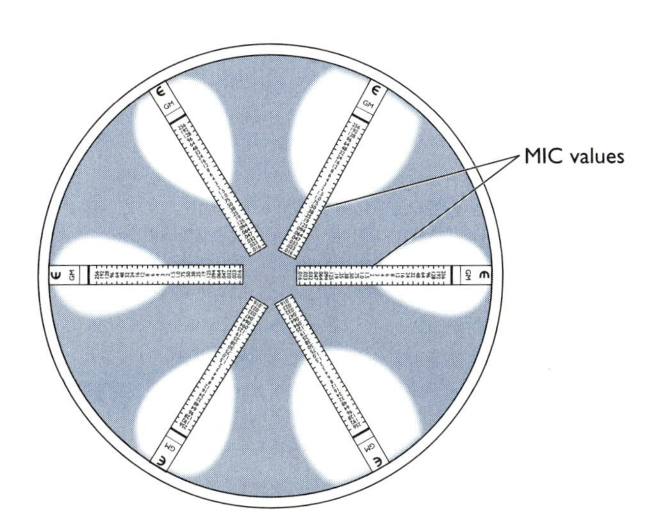

Figure 15.8 Antimicrobial susceptibility assays that give MIC values. **(A)** In the microbroth dilution minimal inhibitory concentration (MIC) assay, the same calibrated number of bacteria is inoculated into each of a series of tubes containing increasing concentrations of an antibiotic. The tubes are incubated for 24 hours and then checked for growth of the bacteria (as indicated by a change from clear to turbid). The concentration of antibiotic in the first tube where there is no detectable growth is the MIC. In this figure the MIC is 16 µg/mL. **(B)** In the E test, bacteria are spread on the surface of an agar plate to form a lawn as in the disk diffusion test. Strips containing calibrated amounts of antibiotic are placed on the lawn, and the plate is incubated. The MIC value is read from the strip at the point where the ellipse of no growth crosses the strip. In contrast to the disk diffusion assay (Fig. 15.7), the E test is more quantitative and provides the MIC directly.

earth. To prevent this, all sorts of containment features are being built into the space missions that will bring samples from other planets back to earth. Yet one could argue that a planet with an ambient temperature much hotter or colder than the temperature of the human body is unlikely to breed microorganisms that could colonize the human body. Specifically, how likely is it that viruses capable of infecting human cells would arise where the highest forms of life are primitive prokaryotes? Skeptics might accept this argument but would be quick to point to the bacteria, fungi, and algae that, although unable to infect the human body, can produce toxins that can harm it. This is the most likely extraterrestrial killer microbe scenario. Fortunately, it is the one that should be easiest to control by keeping the number of any extraterrestrial microbes as low as possible. Even earth-adapted microorganisms have not managed to create huge human die-offs or to suck all the oxygen out of the atmosphere. It seems unlikely that a microbe adapted to another planet would be able to do so.

Forget the Skies! Watch the Ground!

Just when you thought the world was safe, geologists have raised a new possibility that may send you back to your survival shelter. As minerals precipitate, forming calcites and other solid deposits, microorganisms become entombed within the precipitates. Scientists have long known that microorganisms can survive for long periods of time under adverse conditions. If the bacterium that causes tuberculosis can still remain infectious after decades entombed in calcified lesions in one person's lungs, how long could other entombed microorganisms last? Is it possible that as erosion wears down rocks or as dinosaurs and early humans are unearthed, that long dormant microbial pathogens could be released again into a world where the human immune system has no previous experience of them? This scenario sounds bizarre and is highly unlikely—but not impossible. Microbiologists have used the polymerase chain reaction to amplify the genomes not only of the 1918 influenza virus (from preserved lung specimens) but also of the bacterium *Yersinia pestis,* the cause of bubonic plague, taken from the teeth of plague victims buried in the Middle Ages. Where there is DNA or RNA, there is the possibility of live organisms or at least the transfer of their nucleic acids to modern organisms. We wouldn't want to alarm you unduly, but just to be on the safe side: don't become too intimate with your pet rock.

chapter 15 at a glance

Examples of changes that have allowed new disease patterns to emerge

Large hospitals with wards containing very sick patients

Cancer chemotherapy

Immunosuppressive therapy received by transplant patients

Human immunodeficiency virus (HIV) and its immunosuppressive effect

Increased number of large nursing homes

Increased number of elderly people

Examples of factors that encourage the re-emergence of old diseases

Homeless shelters

Crowded conditions in big cities

Increased international travel

Discontinuation of preventive programs

Proving disease causation and measuring infectivity and virulence

Koch's postulates
- Microbe must be isolated from an infected human or animal
- Inoculating an animal or human with the isolated microbe produces disease
- Microbe can be reisolated from experimentally infected animal or human

Measures of infectivity and virulence
- Experimental studies: infectious dose (ID_{50}), lethal dose (LD_{50})
- Surveys of disease in the human population: attack rate, case fatality rate

Examples of virulence factors

Adhesins such as pili

Ability to evade nonspecific and/or specific defense systems.

Strategies to obtain iron and other important nutrients

Toxins
- Lipopolysaccharide (LPS), lipoteichoic acid (LTA): septic shock, inflammation
- A-B toxins
  - Two protein components
  - Binding (B) portion (nontoxic): necessary for attachment to target cell
  - Active (A) portion (toxic): enters cell and damages it, usually by ADP-ribosylating some important cell component

(continued)

chapter 15 at a glance *(continued)*

Superantigens
- Force association of antigen presenting cell (APC) and T helper cell
- Cytokine release by APC and T helper cell is responsible for septic shock

Roles of toxins in disease
- Bacteria do not infect, toxin causes symptoms
  - Botulism (*Clostridium botulinum*)
  - *Staphylococcus aureus* food-borne disease: enterotoxin stimulates nerve endings in stomach
- Bacteria infect, but toxin solely responsible for symptoms
  - Diphtheria (*Corynebacterium diphtheriae*)
    - Bacteria colonize throat
    - Toxin enters the bloodstream, ADP-ribosylates a protein involved in protein synthesis
  - Tetanus (*Clostridium tetani*)
    - Bacteria colonize a deep wound
    - Tetanus toxin enters bloodstream, attacks neurons.
- Symptoms are the combined result of infection and toxin production
  - *Streptococcus pyogenes:* LTA plus superantigen cause septic shock

Evolution of microbial pathogens

Rapid mutation rate
- HIV
- Influenza virus

Horizontal gene transfer
- *E. coli* O157:H7 acquired toxin genes from another bacterium

Rise and fall in prevalence of different disease-causing strains of the same species.
- *Streptococcus pyogenes* causes different diseases today than a century ago

Diagnosis and treatment of disease

Signs and symptoms of infectious diseases often nonspecific (Table 15.2)

Patient history, occupational exposure

Isolation and identification of the causative microorganism

Susceptibility to antibiotics
- Disk diffusion test (semiquantitative)
- Microbroth dilution minimal inhibitory concentration (MIC) test (quantitative)
- E test (quantitative)

key terms

A-B toxins exotoxins with a nontoxic portion (B) that binds to a host cell and a toxic portion (A) that acts on the host cell to produce an effect

attack rate number of cases of clinically apparent disease divided by number of susceptible people in the population; measure of infectivity

case fatality rate number of deaths from a disease divided by the number of cases of clinically apparent disease; measure of virulence

colonization ability of a microbe to remain at a particular site and replicate there

disease specific symptoms arising from colonization of the body by a microbe capable of causing disease

disk diffusion assay method to determine susceptibility of bacteria to antibiotics; disks containing antibiotic are placed on lawn of bacteria and incubated; clear zones indicate susceptibility

exotoxins protein toxins produced by bacteria; usually secreted into the extracellular environment

ID_{50} infectious dose 50; number of microbes required to cause infection in 50% of animals exposed to them; measure of infectivity

infection colonization of the body by a microbe capable of causing disease

LD_{50} lethal dose 50, number of microbes required to kill 50% of animals exposed to them; measure of virulence

microbroth dilution minimal inhibitory concentration (MIC) assay quantitative test using two-fold serial dilutions of an antibiotic to determine the MIC of that agent for a particular microbe

pathogenicity ability of a microorganism to cause clinically apparent symptoms

superantigens toxins that force association of antigen presenting cells (APCs) and T helper cells causing release of cytokines

virulence ability of a microbe to cause severe disease

Questions for Review and Reflection

1. Can you think of some changes in human habits and conditions not mentioned in this chapter that might also contribute to the emergence or reemergence of infectious diseases?

2. Why do scientists usually use the 50% level in determining infections $dose_{50}$ (ID_{50}) or lethal $dose_{50}$ (LD_{50})? [Hint: look at the curves in Figure 15.1.]

3. Some diseases have an LD_{50} but not an ID_{50}. What type of disease would have this characteristic?

4. Which is worse, a disease with a low ID_{50} and high LD_{50} or a disease with a high ID_{50} and low LD_{50}? Give examples of such infectious diseases.

5. Koch's postulate 3 seems to be the most problematic of his postulates. What variation of this postulate is available today?

6. Koch's postulates were obviously not formulated with microbiota shift diseases in mind. How would you change Koch's postulates to apply to a microbiota shift disease such as periodontal disease?

7. Can you think of some ways bacterial protein toxins might be beneficial to you?

8. Why do physicians often prefer the MIC test to the disk diffusion test? What additional information does it provide?

9. Suppose you set up an MIC test and had no growth in any tube. How would you interpret this result?

16 The Lung, a Vital but Vulnerable Organ

When the main risk factors for acquiring an infection are breathing and handshaking, prevention becomes very difficult.

According to the United States Centers for Disease Control and Prevention (CDC), bacterial pneumonia is one of the leading causes of hospital admissions in this country. The only other conditions that bring more people to the hospital each year are pregnancy and heart surgery. For the elderly, pneumonia caused by bacteria is a common cause of death. Viruses also pave the way for other microbes to cause lung disease in people infected with influenza virus or human immunodeficiency virus (HIV). Respiratory tract infections do not strike only the elderly and immunocompromised. Almost everyone has a viral respiratory disease such as influenza or the common cold at some time during his or her life. Influenza is especially dangerous, because it can increase a person's risk of contracting bacterial pneumonia.

Respiratory Infections Get a Second Wind

Importance of respiratory tract infections

Tuberculosis, a dreaded respiratory disease that struck old and young alike, was for a time considered a disease of the past. In an ominous development, tuberculosis, for many years uncommon in developed countries, re-emerged as a major problem. Worldwide, tuberculosis is one of the top causes of death. The most troubling aspect of the re-emergence of tuberculosis is that drug-resistant strains of *Mycobacterium tuberculosis* began to appear in developed countries, because unsupervised patients had used antituberculosis therapy carelessly. Soon the developed countries were exporting multidrug-resistant *M. tuberculosis* to the rest of the world. Fortunately, as a result of a belated but effective response to re-emergent tuberculosis in the United States and other developed countries, tuberculosis seems once again to be under control.

The same thing cannot be said for influenza and pneumonia, which continue to occur at high rates. Controlling influenza may assume major importance in the near future. Every few decades since good records have been kept, outbreaks of especially virulent influenza have occurred. If this cycle continues, a new outbreak of "killer influenza" is imminent. In late 1997 and early 1998, this fear of a coming epidemic caused public health officials in

Source: Culver Pictures

Hong Kong to order the slaughter of thousands of chickens after several people succumbed to fatal influenza infections caused by a new strain of influenza virus. The chicken connection will be explained later in this chapter.

Social and economic impact of respiratory tract infections

An account of respiratory infections that focuses only on the death toll does not give a complete picture of their effects. Respiratory diseases remain a major cause of days lost at work or school. "Walking pneumonia," a bacterial lung disease that is rarely fatal, causes young adults to miss weeks of high school or college classes. Influenza is such a common cause of lost days at work and lost productivity that many employers offer free influenza vaccinations to employees. And, of course, the common cold (whose name says it all) continues to disrupt lives and complicate schedules. There is no question that respiratory diseases pose major public health and economic problems in both developed and developing countries.

Breathing and shaking hands: major risk factors

Respiratory diseases offer especially challenging prevention problems. With food-borne or water-borne diseases, people can protect themselves by cooking foods properly or by boiling water. With sexually transmitted diseases, people can control their risk by minimizing the number of sexual partners or by using condoms. But with respiratory tract infections, where a major risk factor is breathing, exposure is not so easy to control. A famous photograph shown here, taken during the 1918 influenza epidemic that killed millions worldwide, shows a group of Seattle policemen with their mouths and noses covered by gauze masks in a futile effort to prevent exposure to the virus. Such masks exclude large particles, but respiratory diseases are most often spread by tiny liquid droplets (**aerosols**) that readily pass through most masks. The kind of masks that filter out these particles are very expensive and uncomfortable to wear. Influenza virus can also be spread by contact with the hands of an infected person. In today's world, where shaking hands is an enshrined greeting ritual, this type of transmission is almost as difficult to prevent as airborne transmission. In this case, however, a person can gain some protection from frequent hand washing and not touching his or her face.

One way that respiratory infections can be prevented is to avoid contact with infected people, but this is increasingly difficult to do, especially in modern urban areas, where people are frequently herded into crowded subways, buses, elevators, airplanes, and—to wind down from all that crowding stress—bars and restaurants. Public transportation is the respiratory pathogen's best friend. In the case of some bacterial and fungal pneumonias, however, the threat may come not from without but from within. The microbes that cause pneumonia can colonize the throat and persist there for long periods. During that time, they are poised and ready for any breaches in the defenses that prevent oral bacteria from reaching the lungs.

The Physiology and Microbial Ecology of Respiratory Tract Infections

Why lung infections are so dangerous

Respiratory diseases, such as the common cold, influenza, and walking pneumonia, are usually not fatal, because the microbes target the upper respiratory tract. More serious are infections that occur in the lung, including severe cases of influenza and cases of pneumonia caused by bacteria or fungi. Lung infections can impair gas exchange, the lung's main function, but the most serious consequence occurs when bacteria or yeasts leave the lung and enter the bloodstream. Septic shock is the leading cause of death in pneumonia patients.

The lung is an organ that filters blood. Vessels circulate blood through the lung so that hemoglobin can release carbon dioxide and pick up oxygen. Blood vessels are separated from the air sacs of the lung by two thin membranes (Fig. 16.1). This arrangement makes it possible for gas exchange to occur, but the delicate membranes are very vulnerable to damage. Microbes that enter the lungs and elicit an inflammatory response not only impair gas exchange but may disrupt the membranes enough to allow blood to enter the lungs and microbes to enter the bloodstream.

Defenses of the lung and how microorganisms overcome them

Figure 16.1 depicts the defenses of the lung. There are two levels of defense: defenses designed to prevent microbes from entering the lung and defenses designed to eliminate microbes that get past the first line of defense and enter the lung.

Oral microbiota. An important protection of the lung is the resident microbiota of the oral cavity and nasopharynx. A pathogen that can colonize the mouth or nasopharynx has a better chance of causing infection than

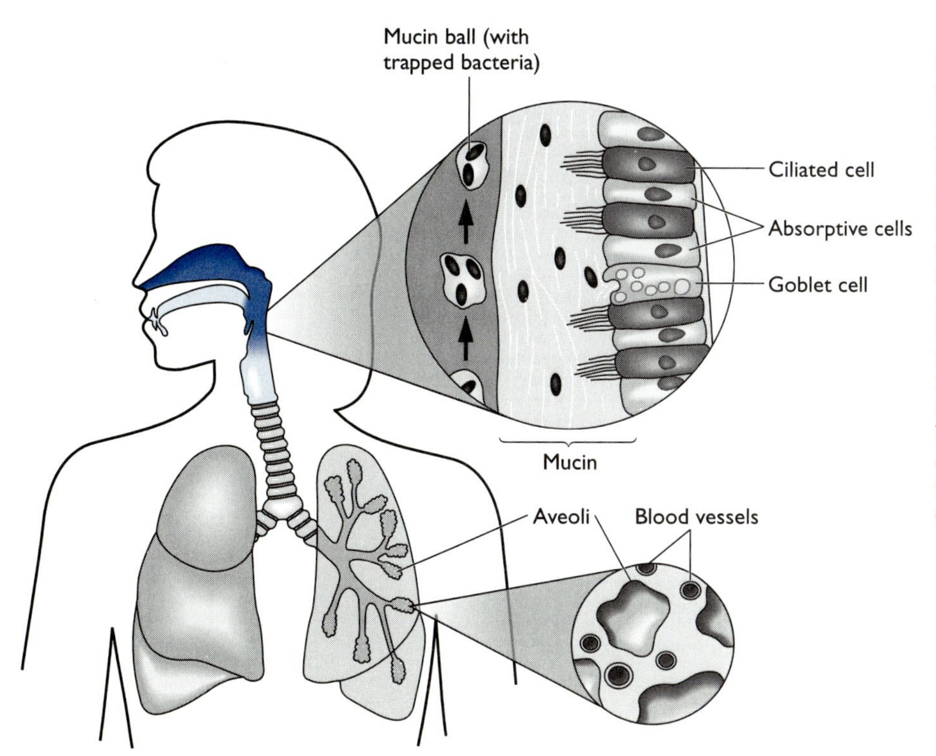

Figure 16.1 Defenses of the upper and lower respiratory tract. The nasopharynx is protected by a thick mucin layer that contains secretory immunoglobulin A (sIgA), lysozyme, and lactoferrin. Ciliated cells, interspersed along the membrane, move mucin blobs containing trapped microbes up and out of the airways. The resident microbiota of the oral cavity provides additional protection. The major defense against microbes that reach the lungs is the alveolar macrophage. Insets show the action of ciliated cells and the close proximity of blood vessels to alveoli. Inflammation can allow bacteria to breach the barriers that separate lung and blood cells and enter the bloodstream.

one that cannot, because high numbers of the pathogen are present and are constantly being inhaled. This increases the chances that the pathogen will eventually succeed in bypassing the defenses of the upper airway especially if a transient breach of host defenses occurs.

The importance of the **oral microbiota** as a defense against lung infections is evident from the fact that suppression of the resident microbiota by antibiotics or cancer chemotherapy is often linked to an increased incidence of pneumonia. Patients in hospitals for prolonged periods frequently experience a change in their oral microbiota, whether or not they are taking antibiotics. The reason for this is not known. This change in microbiota can be a problem, because the new bacteria that move in tend to be ones from the hospital environment. These are more resistant to antibiotics than the bacteria they replace. Finally, poor oral hygiene, lowered salivary flow in comatose patients, and the use of respirators can affect adversely the composition, density, and access of the microbiota to the lungs, predisposing the patient to a lung infection. For this reason, hospital caregivers pay special attention to the oral hygiene of those unable to do this for themselves.

The presence of an abundant microbiota in the mouth and throat may be protective, but it complicates the diagnosis of lung infections. **Sputum specimens** (material coughed up from an infected lung) can be contaminated by oral microbes. Thus, it is necessary to differentiate between organisms that are contaminants and the actual pathogen. Because sputum specimens can be contaminated so easily with saliva, microbiologists in the clinical laboratory check incoming sputum specimens to make sure that neutrophils are present, as would be expected if the material came from an infected area. Also, specimens can be taken with a bronchoscope or needle (transtracheal aspiration) to bypass areas heavily colonized by the microbiota and collect samples directly from the lung.

The ciliated cell defense. The air is full of microbe-laden particles that must be removed from inhaled air before it reaches the lung. This is the job of the **ciliated cell defense.** The mucosal membranes of the upper airway, like those of the intestinal tract, are protected by a thick **mucin** layer. The respiratory tract is curved, and the turbulence produced by air moving through this curved tube increases the likelihood that microbe-laden particles in the air will contact and stick to mucus at some point. A specialized cell type, the ciliated cell, has long protruberances that wave constantly in one direction. Their action propels blobs of mucin out of the airway. The **coughing and sneezing reflex** helps to remove large particles or excess mucus.

The combination of mucin and ciliated cells is normally quite effective in keeping microbes out of the lung. Only the smallest particles, such as those found in liquid aerosols, succeed in reaching the lung. This is the reason liquid aerosols are more frequently associated with the spread of lung infections than are dust or other airborne debris. Also, the small particles in liquid aerosols tend to remain suspended in the air for longer periods than larger particles and thus have a better chance of being inhaled.

Microorganisms evade the defenses of the upper airway, either by being carried on small particle aerosols or by taking advantage of natural breaches of the ciliated cell defense. Any practice or condition that impairs the natural defenses of the lung increases a person's susceptibility to infection. Influenza is often a precursor of secondary bacterial pneumonia, because the virus temporarily destroys ciliated cells. Smoking depresses the effectiveness of the ciliated cell defense. This is one of the reasons why people with long histories of smoking have a higher risk of bacterial and fungal infections than nonsmokers. Alcoholics who vomit after they have passed out and then inhale the vomitus may introduce not only oral bacteria but enough stomach acid to impair the ciliated cell defense. Invasive assisted-breathing systems, with tubes positioned in the airways of hospitalized patients, make it easier for microbes to bypass the defenses of the upper airway and cause pneumonia.

Defenses of the lung. Microbes that succeed in reaching the lung encounter the next respiratory defense, the lung macrophages (**alveolar macrophages**). During an infection, other cell types, such as neutrophils, natural killer cells, and cytotoxic T cells, will be recruited to the area to help macrophages clear the invader, but macrophages are on the scene from the beginning. This makes sense, because the critical hours needed to bring neutrophils to the site in numbers large enough to be effective would give the infecting microbes an unacceptable head start. Also, some of the lung macrophages are antigen-presenting cells that are on hand to initiate the specific defense response. If a person has been vaccinated or has experienced the disease before, antibodies will also arrive early on the scene and help the macrophages to ingest and kill the invading microbes.

Most microorganisms that cause serious lung disease evade the alveolar macrophage defense, either by having an **antiphagocytic capsule** or by being able to survive in and kill macrophages. Such microbes, in effect, have bypassed the nonspecific defenses. Because the macrophages and the neutrophils that rush to the site are unable to kill the microbes, the microbes continue to multiply in the site. The phagocytes, trying unsuccessfully to eliminate the invaders, damage lung tissues. This not only provides more nutrients for the microorganisms but also increases the risk that some members of the rapidly growing microbial population will enter the bloodstream. The protective defense against encapsulated microbes consists of antibodies that opsonize them and enable the macrophages and the neutrophils that rush to the site in the early stages of infection to ingest and kill them. The protective defense against microbes capable of growing inside macrophages is the **activated macrophage defense.** These defenses need to be deployed as rapidly as possible to prevent inflammatory damage in the lung. This is the reason that immunization is so critically important in preventing serious respiratory tract infections.

The missing defense: effective vaccines against the most serious causes of lung infections. In Chapter 14, the two examples of not-so-good vaccines were the vaccine against *Streptococcus pneumoniae* and the vaccine against *M. tuberculosis.* Certainly one of the greatest tragedies produced by the complacency of the period from the 1960s to the 1980s, described in Chapter 15, was the loss of decades of scientific research that might have produced more effective vaccines against these two serious diseases. Today, scientists are scrambling to make up for those lost years. Some hopeful signs exist, especially in the case of new *S. pneumoniae* vaccines, but this is no consolation to the families of people who have succumbed in the meantime. Also, as with antibiotics, it takes a long time to complete the clinical trials that test the efficacy and safety of new vaccines. A vaccine developed today will not be available for several years, at the very earliest.

Influenza, Still a Serious Health Threat

Importance of influenza

Influenza is estimated to cause about 20,000 deaths per year in the United States alone. Generally, death is caused not by the influenza virus itself, but by the subsequent bacterial pneumonia that develops in some people after a bout of influenza. In most years, the vast majority of fatalities occur in elderly people, but there have been influenza epidemics, such as that of 1918, when millions of people of all ages died. Subsequently, there have been two or three other exceptionally bad flu years. The last outbreak of "killer flu" occurred in the late 1970s. Public health officials are concerned about when the next epidemic will strike and are planning for coping with it when—not if—it comes. One important advance has been the development of a fairly effective vaccine against influenza, a vaccine that public health officials hope will take the punch out of the next wave of killer flu. As explained in Chapter 14, however, the Achilles heel of this vaccine is the continued and unpredictable evolution of new strains of the virus that will arise due to its high rate of mutation. This, together with the months-long lag time between detection of a new strain of influenza virus and delivery of a vaccine to the public, makes the vaccine less of a protection against a future flu epidemic than it could be.

Influenza is caused by an enveloped virus, the influenza virus. There are three types of influenza virus: A, B, and C. They differ mainly in the amino acid sequences of their envelope proteins. Type A is most likely to cause the large, multination outbreaks called **pandemics.** Types B and C cause milder disease and usually do not produce epidemics. The structure of influenza virus and its replication cycle were introduced in Chapter 10. To recap briefly, the envelope of influenza virus contains two types of proteins, neuraminidase and hemagglutinin. Hemagglutinin allows the virus to enter human cells by attaching to their sialic acid residues. Neuraminidase is important for viral exit from the cell. Both hemagglutinin and neuraminidase are the target of anti-influenza vaccines. The influenza genome has an interesting and significant property: it is not a single segment but is composed of 7–8 segments of single-stranded RNA. Each of these segments has its own protein covering. This segmented form of the genome makes possible large genetic changes in the virus, a phenomenon called antigenic shift.

Influenza viruses attack the ciliated epithelial cells of the respiratory tract. As already mentioned, these cells are an important defense against infection of the lung, and impairment of this defense increases the likelihood that bacteria can reach the lung and cause a secondary infection.

Antigenic drift and antigenic shift

Influenza virus is constantly changing. In fact, it is the fastest mutating virus known, with another RNA virus, HIV, as a close second. Viruses with RNA genomes generally mutate faster than viruses with DNA genomes, because the enzymes that reproduce RNA are less accurate than the ones that reproduce DNA. Small changes in the viral genome sequence can lead to changes in the envelope proteins, making them different enough to be less well recognized by the memory cells of the immune system (**antigenic drift**). People are advised to get flu vaccine each year, not just to bolster the immune response, but because the composition of the flu vaccine is changed from time to time to reflect differences in the surface proteins of the viruses currently circulating. Small changes resulting from mutations do not eliminate completely the protection conferred by an old vaccination or by infection with previous years' influenza strains. Thus, more people may get the flu when a new variant of influenza virus arrives that is only slightly different from the previous strain, but the disease should not be more severe, because of the partial protection from the earlier exposure. "Killer flu" arises from big genetic changes that make the virus virtually invisible to the immune system, so that no protection remains (**antigenic shift**).

Big changes in envelope proteins occur because of two features of the influenza virus: its **segmented genome** and the ability of influenza A viruses to infect many kinds of mammals and birds. Suppose, for example, that a pig is infected simultaneously with human influenza A virus and a porcine influenza virus. If, during the co-infection, the two types of virus replicate in the same cell, the predominantly human virus can pick up a genome segment from the avian virus (Fig. 16.2). This may result in a new influenza A virus with new envelope proteins that will not be recognized by the immune systems of most people in the population. The "new" virus can cause the most serious form of influenza. Scientists using modern molecular techniques have amplified and cloned a DNA copy of influenza genome segments from tissue specimens taken during the 1918 pandemic. By comparing the nucleotide sequence of the cloned DNA copies of these segments with sequences of human and animal influenza viruses, they concluded that the nucleic acid sequences of some of the 1918 strain genome segments matched pig influenza genome sequences more closely than human influenza genome sequences. This suggests that the 1918 virus contained a genome segment of pig origin. Mixing of viral genome segments occurs in other animals, too, such as the ill-fated Hong Kong chickens.

Bacterial Pneumonia

Importance of bacterial pneumonia

S. pneumoniae, also called the **pneumococcus**, is the most common cause of bacterial pneumonia and is frequently the cause of death (as a result of **secondary infection**) in fatal influenza cases. *S. pneumoniae* has an impressive record. According to the CDC, each year in the United States alone this one species of bacterium causes 3,000 cases of meningitis, 50,000 cases of bloodstream infection, 500,000 cases of bacterial pneumonia, and 7 million cases of ear infection. Not bad for a tiny bacterium, which, when viewed under a microscope, looks about as threatening as a beach ball. *S. pneumoniae* is a gram-positive coccus often seen in pairs (**diplococcus**) (Fig. 16.3). Its distinctive diplococcal appearance is useful for rapid diagnosis of the disease. Two other bacteria that cause pneumonia are *Klebsiella pneumoniae* and *Haemophilus influenzae* (so named because it was originally misidentified as the cause of influenza). Both of these bacteria are gram-negative rods. Gram-orientation notwithstanding, all three of these bacteria cause the same type of lung damage and symptoms, although *Klebsiella* pneumonia is somewhat more destructive than the other two.

Why, then, would a physician care which one of these bacteria was causing the disease? Until recently, *S. pneumoniae* was a lot less likely to be resistant to antibiotics than *K. pneumoniae* and *H. influenzae,* and good old

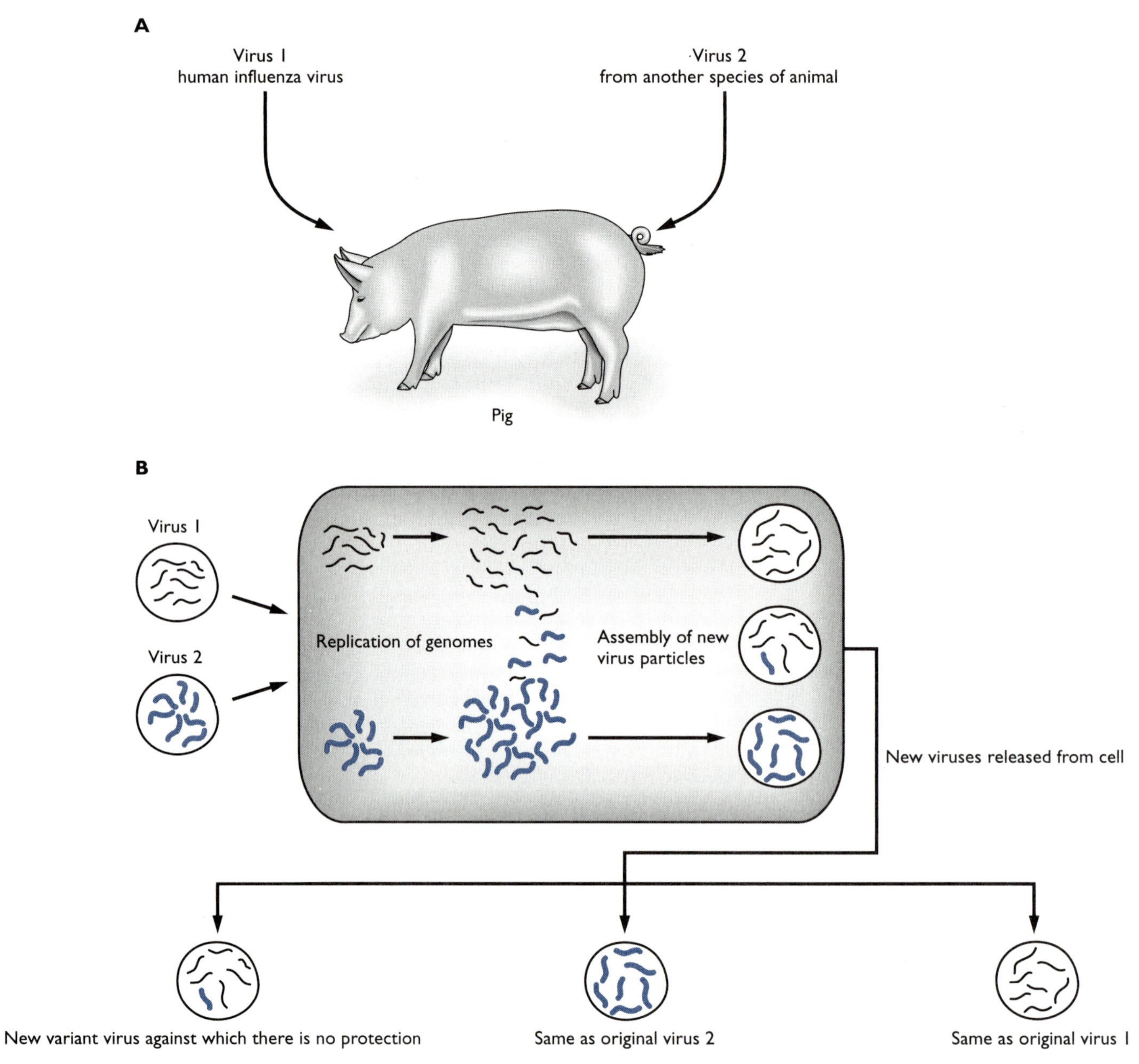

Figure 16.2 Antigenic shift in influenza virus. (A) Influenza viruses from different species simultaneously infect a single animal. **(B)** The two types of viruses replicate in a single cell. When the new viral particles are assembled, segments of the genome from different viruses occasionally are incorporated into the same virus particle. If a human influenza virus receives a genome segment from a swine or avian virus, this "new" virus may have new envelope proteins that allow it to bypass the immune system of most of the human population.

penicillin was usually effective. By contrast, more advanced drugs were needed to treat the gram-negative species. In recent years, the appearance of penicillin-resistant *S. pneumoniae* has changed this picture somewhat, but at least in certain regions of the United States, penicillin still works for many cases of *S. pneumoniae* infection. Because penicillin is inexpensive and has a good safety record, it is important to use this antibiotic in cases where it is still effective. Also, the use of penicillin contributes little to resistance to the more advanced β-lactam antibiotics.

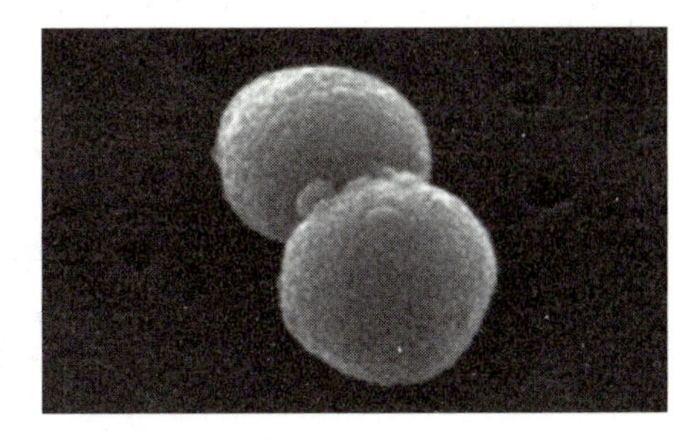

Figure 16.3 Scanning electron micrograph of *Streptococcus pneumoniae*. The organism usually occurs in pairs (diplococci). Courtesy of Janice Carr, Centers for Disease Control and Prevention, Atlanta, GA.

Virulence factors of bacteria that cause pneumonia

S. pneumoniae has a number of traits that help it to cause disease. It readily colonizes the human mouth and throat and is carried by many people, a trait that, after a case of influenza or some other defense-impairing event, gives it a chance to reach the lung. One of the reasons that *K. pneumoniae* causes pneumonia much less often than *S. pneumoniae* may well be because these gram-negative bacteria do not easily colonize the mouths and upper airways of otherwise healthy people. They seem to do best in hospitalized patients, whose oral microbiota tends to shift, for unknown reasons, to become more gram-negative.

Once in the lung, *S. pneumoniae, H. influenzae,* and *K. pneumoniae* would be killed rapidly if the alveolar macrophages could engulf them. All three of these bacterial species protect themselves by producing polysaccharide capsules, which prevent binding of complement component C3b to their surfaces and thus rob the phagocytic cells of the sticky coating necessary for phagocytosis. A nonimmune person who contracts a case of pneumococcal pneumonia has a 10%–15% chance of developing a bloodstream infection and septic shock. In some patients, *S. pneumoniae* goes straight to the brain and causes meningitis. An effective host response is to develop antibodies that bind to the capsular polysaccharides. The problems with the capsular polysaccharide vaccine were described in Chapter 14. Still, a partially effective vaccine is better than none.

A hopeful development on the *S. pneumoniae* vaccine front is the discovery of **PspA (pneumococcal surface protein A)**, a protein that appears to be exposed on the surface of the capsule. As a protein, it is a more effective antigen than capsular polysaccharides. Also, it may be less variable antigenically than the capsular polysaccharides. Vaccines based on PspA are being tested in animals and may soon enter clinical trials.

Tuberculosis: An Old Enemy Returns

How a once-controlled disease breaks out

In Chapter 1, the advent of a new type of bacterial pneumonia, Legionnaire's disease, was described. Just as "new" pathogens can emerge because of changing human practices, old ones can reappear for the same reason. **Tuberculosis**, a bacterial lung infection, is a good example of this phenomenon. Before 1950, people were acutely aware of tuberculosis, because it was a major cause of suffering and death in people of all ages. During the 1950s and 1960s a highly effective campaign to eradicate tuberculosis was carried out in all of the developed countries. In the United States, a visit to the shopping center, the precursor of today's malls, might give a shopper the opportunity to enter a booth or mobile van complete with an X-ray machine and the offer of a free tuberculosis test. This kind of aggressive marketing of diagnostic testing caused tuberculosis to become an uncommon disease in the United States and the rest of the developed world. But tuberculosis had not gone away, it was merely waiting in exile to stage a comeback.

Today, nearly one-third of the people in the world are infected with *Mycobacterium tuberculosis.* Many of these people do not have active disease but can later develop active disease if they become immunocompromised. In the developing world, tuberculosis is still an extremely common disease, especially where crowding and poor nutrition create the conditions under which it can thrive. Thus, the developed countries are small, relatively tuberculosis-free islands in a sea of people with the disease. When you live on an island, it is prudent to watch out for tidal waves.

In the modern world, travel over long distances is commonplace. Tourists move freely into and out of developed countries. Refugees or people looking for work enter the United States constantly and in large numbers. And with them they bring *M. tuberculosis.* Most of them are not even aware they have the disease. This is not an argument against admitting tourists and immigrants to the United States, but a fact of life that must be faced. Even if we were to build a huge wall around the country and let no one in, importation of tuberculosis would still be a problem as long as U.S. citizens were allowed to travel to foreign countries and return. Actually, the best justification for building a wall around the United States and other developed countries would not be to prevent the entry of immigrants but to prevent the export of **multidrug-resistant *M. tuberculosis.***

Just as benign neglect of tuberculosis prevention measures produced an upswing in cases, benign neglect also gave rise to drug-resistant strains of *M. tuberculosis,* a new development on the tuberculosis scene. Because the tuberculosis-control infrastructure was no longer in place, patients with the disease did not receive supervision to ensure that they followed the complicated, months-long drug regimen. Patient noncompliance created an environment conducive to the development of drug resistance.

Tuberculosis is a very contagious disease. On average, each person with an active case of tuberculosis infects at least 10 other people. In some cases, a single person has infected more than 200 people. Infected health care workers, teachers, and staff dealing with institutionalized populations are especially likely to infect large numbers of people. The risk of acquiring the disease from a person with an active infection increases with proximity to that person and the number of contacts. This principle is well illustrated by a rural New York state case involving a school bus driver who had a case of active tuberculosis for several months before it was diagnosed. Thirty-two percent of the 238 children who rode the bus were infected. Of those who spent the shortest amount of time on the bus (less than 10 minutes), 22% were infected compared to a 57% infection rate in those who spent the longest time on the bus (more than 40 minutes).

Characteristics of Mycobacterium tuberculosis

M. tuberculosis is a rod-shaped bacterium with a gram-positive type cell wall. It does not stain with Gram stain reagents, however, because its cell wall has a high lipid content, including some special lipids called **mycolic acids** (Fig.16.4). Like most respiratory pathogens, *M. tuberculosis* enters the lungs on small liquid droplets that help it to bypass the ciliated cell defense. In the lung, *M. tuberculosis* encounters the alveolar macrophages and is taken up but not killed by them. Instead, it multiplies in the macrophage that ingests it and kills the macrophage. The continued multiplication of bacteria in the region attracts more and more phagocytic cells as well as other cells of the immune system.

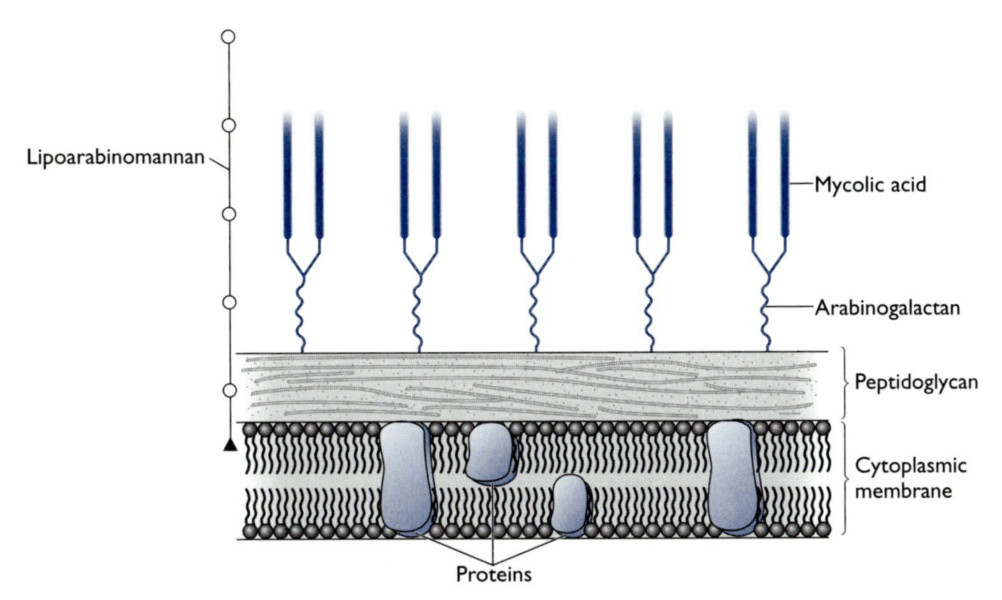

Figure 16.4 Cell wall of *Mycobacterium tuberculosis.* The thick cell wall of *M. tuberculosis* has an unusual structure and contains several unique sugars, including lipoarabinomannan and arabinogalactan. The high lipid content, especially mycolic acid, protects the bacteria from killing by phagocytes and elicits a strong inflammatory response. Although *M. tuberculosis* has a gram-positive cell wall, the high lipid content of the cell wall prevents it from staining with Gram reagents.

Although the macrophages do not succeed in controlling the infection, they continue to try to do so and, in the process, inflict considerable damage to tissue with their release of toxic substances. Gradually an area of dead tissue develops and is surrounded by a layer of macrophages and other cells of the immune system (Fig.16.5). This is called a **tubercle.** A tubercle that has been walled off and calcified is called a **granuloma.** This term explains the name immunologists use to describe the immune response to diseases like tuberculosis: a **granulomatous response.** At this point, the interior of the tubercle has a cheese-like consistency, a feature that helps to trap the bacteria. Activated macrophages are able to kill *M. tuberculosis.* If the activated macrophage response is brought into play early enough, the infection can be controlled, and the tubercles become **calcified.** Large lesions of this sort are visible on a chest X ray. People who develop tubercles that resolve early may not even know they have the disease. In the more serious form of the disease, where the immune response does not control bacterial growth, the interior of the tubercle **liquefies,** and the bacteria break out of the tubercle. At this point, *M. tuberculosis* can enter the bloodstream and spread to other organs of the body. If untreated, this form of the disease has a mortality rate of 50% or higher among adults. The mortality rate is even higher in infants and immunocompromised patients.

Most people infected with *M. tuberculosis* do not develop the active, symptomatic form of the disease. But this does not mean that they are out of danger. *M. tuberculosis* has the amazing ability to survive for years inside tubercles and can reactivate decades later if the infected person becomes immunosuppressed. This is called **reactivation tuberculosis.** Within the

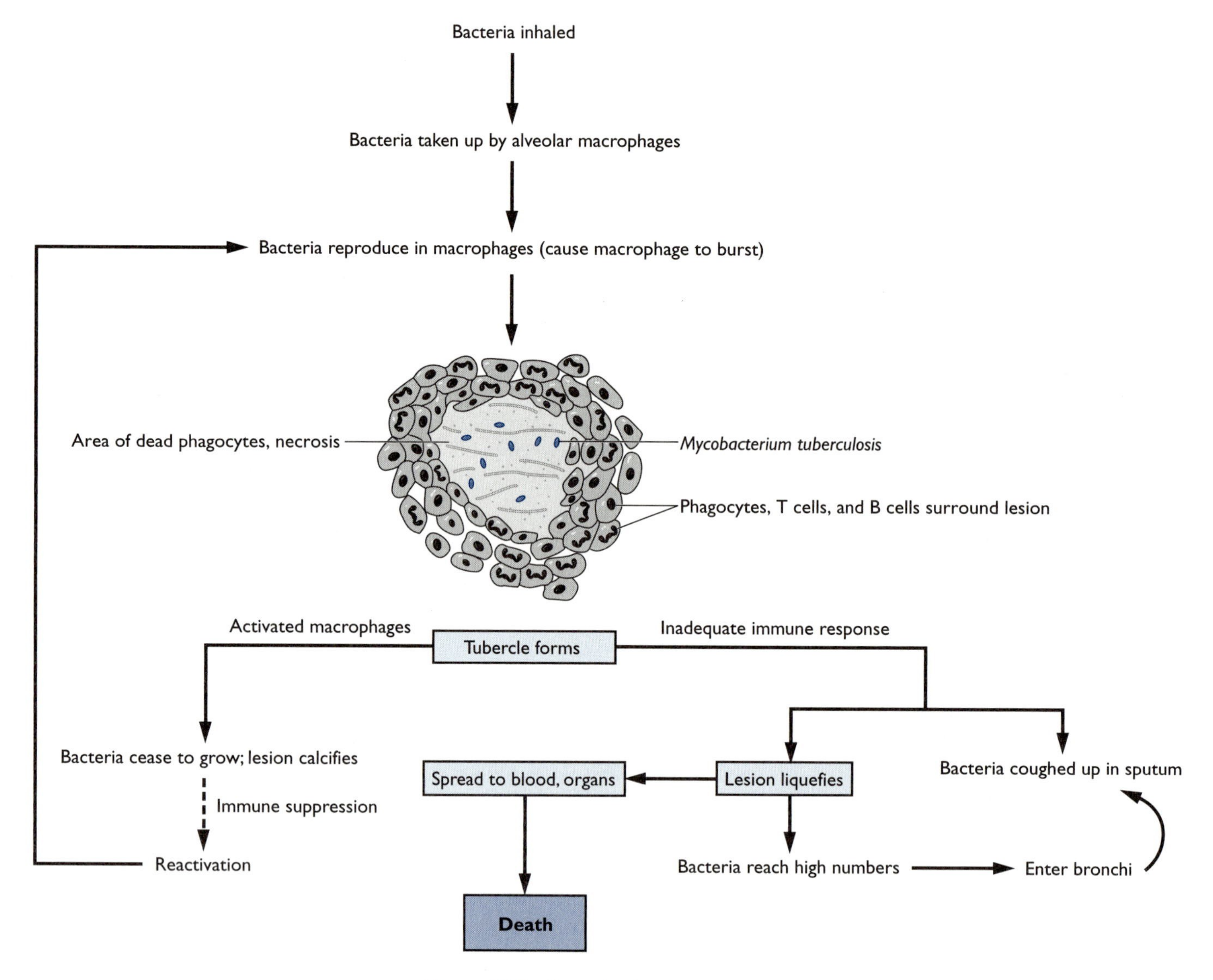

Figure 16.5 Steps in the development of tuberculosis. Bacteria inhaled in aerosols are taken up by alveolar macrophages, in which they multiply and eventually cause the macrophages to rupture. A cheese-like tubercle lesion forms. If the immune response is adequate, the lesion may calcify. If not, the contents of the tubercle may liquefy, in which case bacteria can escape from the lesion and spread to other organs or enter the bronchi, from which they are coughed up in sputum.

first 5 years after the initial infection is resolved, the risk of developing reactivation tuberculosis is about 5%.

M. tuberculosis systemic infections can involve bones and joints as well as organs. The bacteria produce painful malformations of bones and joints. Spinal tuberculosis, which produces abnormal spinal curvature in children, is a form of the disease that fortunately is very rare. The deformities of individuals called "hunchbacks" in earlier times may have been caused by this disease.

Tuberculosis tests

A simple test for tuberculosis is very useful as a screen for infected people. In this test, a minute amount of an extract of *M. tuberculosis* proteins called **purified protein derivative** (**PPD**) is injected into the skin. A person who has or has had tuberculosis will develop a red, raised lesion within 24–48 hours. If the lesion has a diameter greater than 5 mm, the person is considered tuberculosis positive. Even people whose disease resolved early and those who had the disease decades ago test positive on this skin test.

The skin test has limitations. People who have been vaccinated with the antituberculosis vaccine, **BCG**, will have a positive reaction on the tuberculosis skin test. As was already mentioned in Chapter 14, people vaccinated with BCG can still get tuberculosis. The skin test can be used to detect active disease in BCG-vaccinated people, however, because the lesions associated with infection are larger than those associated with vaccination. A second limitation of the skin test is that reactivity does not develop until 3–4 weeks after the initial infection has occurred. Immunocompromised persons, such as those with acquired immunodeficiency syndrome (AIDS), can fail to react to the test even if they have tuberculosis, a condition called **anergy.** Finally, a person who has had tuberculosis and has been cured of the disease will remain skin-test positive for many years. Thus, a positive test does not prove conclusively that a person has an active infection, but it helps to identify people who should be tested further for the disease.

Another rapid test for tuberculosis is to examine a sputum specimen that has been stained by a procedure called **acid-fast staining.** The cell wall of *M. tuberculosis* is so tough that procedures used to stain other bacteria, fail to stain it. Boiling the bacteria in basic fuchsin solution, a procedure that destroys the structural integrity of most other bacteria, is necessary to visualize *M. tuberculosis.* This is called the acid-fast stain. The main limitation of this test is that a person who is capable of spreading tuberculosis may have only a small number of bacteria in the smear examined under the microscope. It may be necessary to look at many microscope fields to find one or a few of the bacteria. One of many remarkable traits of *M. tuberculosis* is the ability of a single cell to cause extensive lung damage. This is because of the bacteria's ability to generate a powerful granulomatous response.

The gold standard for diagnosing a bacterial infection is cultivation of the bacterium in the laboratory. Cultivation is also a necessary prerequisite for determining whether the strain is drug resistant. Cultivation of *M. tuberculosis* is seldom done, however, except in large hospital laboratories or state public health laboratories. *M. tuberculosis* grows very slowly under laboratory conditions, and it can take weeks for colonies to appear, then more weeks to determine the drug-resistance pattern. This is why so much reliance has been placed on the skin test and the acid-fast staining test, despite their limitations. They give the first clue that a person may have tuberculosis. New molecular methods, currently under development, for detecting *M. tuberculosis* may provide a more accurate and rapid diagnosis in the future.

Tuberculosis therapy

Complexity and length of therapy. Standard therapy for people with active tuberculosis consists of at least three antibiotics. The four frontline antituberculosis drugs are **isoniazid, ethambutol, rifampin,** and **pyrazinamide.** Except for rifampin, these are not familiar antibiotics. Some of the more recently developed broad-spectrum antibiotics seem to have beneficial effects and may become part of future tuberculosis therapy. But historically and for the present, antituberculosis drugs are mostly tuberculosis-specific. This raises problems for companies manufacturing the drugs, because narrow-spectrum antibiotics are not nearly as profitable as those that can be used to treat many types of infections. Why is it necessary to use more than one antibiotic? The short answer, which should come as no surprise at this point, is that *M. tuberculosis* is hard to kill. Isoniazid kills the more rapidly growing bacteria, whereas the other three antibiotics eliminate the slowly growing or dormant bacteria in the tubercles.

A person being treated for tuberculosis takes three or four antibiotics almost daily for at least 6 months. Immunocompromised people, such as AIDS patients, must take the drugs even longer. Some of these drugs have side effects, including nausea. The fact that there are three or four separate pills to take makes it easy for a patient who is suffering side effects to take the pills selectively. It is in such patients that resistant strains of *M. tuberculosis* arise. *M. tuberculosis* mutates readily to resistance to individual antituberculosis drugs. An added benefit of taking more than one drug is that it virtually eliminates the likelihood of resistant strains arising. If patients take only one or two of the drugs at a time, the likelihood of resistance rises dramatically. This is the reason patients taking tuberculosis medications must be supervised, especially if they have problems such as alcohol, drug abuse, or a mental disorder that would make them more likely to be noncompliant.

The W strain. A particularly large and nasty outbreak of drug-resistant tuberculosis occurred in New York City between 1990 and 1993 and eventually spread to 11 hospitals. Most cases were caused by a strain of *M. tuberculosis* that was resistant to isoniazid, rifampin, ethambutol, and streptomycin. The strain, dubbed the **W strain**, was acquired in hospitals in virtually all of the cases. A total of 367 people were infected and 83% of them died. Death was unusually rapid, occurring 1–4 months after onset of the disease. Most of the infected people were HIV-positive, a fact that made them even more susceptible to severe infection. Even in healthy people, infection with multidrug-resistant *M. tuberculosis* had a high fatality rate. Some of the people who died were health care workers. So far, tuberculosis is the only disease healthy health care workers can get from AIDS patients except for HIV itself.

A sobering thought is that because the people who develop active infection are usually the tip of the tuberculosis iceberg, there are undoubtedly hundreds, perhaps thousands, of people walking around New York City with the W strain lying dormant in their lungs. These people could reactivate at any moment, producing new foci of infection. Another aspect of the W strain outbreak deserves comment: its cost. The cost of treating the 367 patients infected with the W strain is estimated to have exceeded $25 million dollars. This does not count the extra costs hospitals have incurred to install ventilated isolation rooms and special devices designed to protect health care workers.

Paucity of new antituberculosis antibiotics. The apathy toward tuberculosis that characterized the 1960s and 1970s was naturally accompanied by little activity in antituberculosis drug development. Not only were no new antituberculosis drugs developed, but virtually no work was done to find out how the existing drugs worked. Knowing a drug's mechanism helps scientist to develop analogs more easily. Scientists still do not know the mechanism of action of pyrazinamide and ethambutol. The mechanism of isoniazid action was discovered only within the past few years. Isoniazid is first activated by an enzyme called **catalase.** Normally catalase converts peroxide to water, but this enzyme also acts on isoniazid. The activated form of isoniazid inhibits the synthesis of mycolic acid, thus damaging the bacterial cell wall. One way the bacteria can become resistant to isoniazid is to eliminate production of catalase so that the drug is not activated.

When the upsurge in tuberculosis cases began in 1985, scientists scrambled to find other effective drugs. This search became more frantic as drug-resistant tuberculosis cases increased in numbers. Many of the newer antibiotics, such as the **fluoroquinolones,** had not even been tested for their efficacy against *M. tuberculosis.* The majority of antibiotics have no effect against *M. tuberculosis,* and no one knows why this is so. Luckily, more recent studies show that the fluoroquinolones may have some activity against *M. tuberculosis.*

Pneumocystis carinii: A "New" Fungal Respiratory Pathogen

Bacteria are not the only microorganisms ready to take advantage of new windows of opportunity created by changes in human activities. Fungi are equally avid fans of human vulnerability. Patients with AIDS provide the setting for the increased incidence of ***Pneumocystis carinii,*** a disease caused by a yeast. Previously thought to be a rare cause of disease, primarily affecting neutropenic patients, *P. carinii* is now much more common. Before the availability of effective anti-HIV therapy and prophylactic treatment for *P. carinii,* up to 80% of all AIDS patients developed *Pneumocystis* pneumonia, a disease that was a major cause of death in these patients. Today, *P. carinii* is less common in AIDS patients but is still a major killer among those who contract it.

Almost all aspects of *P. carinii* and the disease it causes have been matters of controversy and confusion. Initially, based on morphological characteristics, *P. carinii* was considered to be a protozoal parasite. Taxonomic studies employing ribosomal RNA (rRNA) sequence analysis finally settled the taxonomic status of *P. carinii:* it is most closely related to the yeast *Saccharomyces cerevisiae* (baker's yeast). However, the use of the terms trophozoite and cyst (terminology appropriate to protozoal parasites) had become so entrenched that these terms are still used instead of the more appropriate terms yeast and spore. Both forms are seen in the lungs of infected people and are diagnostic for the disease. Another perplexing question about *P. carinii* is where it is normally found. *P. carinii* has been found often enough in the upper respiratory tract of humans to cause some people to raise the question of whether it is a component of the resident microbiota. Inapparent infections with *P. carinii* are common, as evidenced by the fact that most children have antibodies to *P. carinii* antigens by the time they are 4 years old. Is *P. carinii* a human-specific pathogen? *P. carinii* appears not to be human specific, because an organism that resembles it closely enough to be designated *P. carinii* has been found in rats. So far, however, no evidence has been found to indicate that rat strains can infect humans or that animals are a common source of *P. carinii* infections.

In the lung, the trophozoite (replicating) form of *P. carinii* associates tightly with lung cells (**pneumocytes**, hence the name *Pneumocystis*). Microscopic examination of infected lung tissues suggests that this association is deleterious to the pneumocyte, raising questions about whether

P. carinii produces toxic substances that directly damage host tissues. *P. carinii* elicits an intense inflammatory response that may be responsible for lung damage. The result is a **foamy exudate**, and the lungs take on a "**honeycomb**" appearance. Currently, an antimicrobial combination, trimethoprim-sulfamethoxazole, is the treatment of choice. These antibiotics interfere with steps in the pathway that produces folic acid.

Examples of Less Serious but Common Respiratory Infections

As the title of this chapter suggests, the lung is indeed a vulnerable organ, and there are all sorts of dangerous microorganisms lurking in nature waiting to take advantage of breaches in lung defenses. But not all respiratory infections are equally dangerous. Fortunately, the most common respiratory infections are rarely fatal. An extremely common respiratory infection is the common cold, a disease caused by viruses. Another example of a fairly common, and usually nonfatal, respiratory infection is the bacterial disease "**walking pneumonia.**" This form of pneumonia is serious enough to put infected people out of commission for weeks and, on occasion, to hospitalize them. Deaths, though uncommon in immunocompetent people, do occur. We will end with these two infections, which also have the advantage of introducing some new microbial strategies for causing infection.

The common cold

No discussion of respiratory tract infections would be complete without a brief description of the common cold. Although influenza virus and the viruses responsible for the common cold produce many of the same symptoms, they do it in somewhat different ways. Many viruses cause colds and sore throats, but **rhinoviruses** are responsible for at least half of all colds and are also a common cause of sore throats. The structure and replication cycle of rhinoviruses are shown in Figure 16.6. Rhinoviruses are **naked icosahedral** viruses with a nonsegmented single-strand RNA genome. The **capsid** consists of four different proteins (VP1–VP4). VP1 mediates adherence to host cells and is the protein against which a protective antibody response is directed.

Two features of rhinovirus infection are worth noting. First, rhinoviruses provide an interesting example of how infectious microorganisms subvert host molecules to their own purposes. Rhinoviruses attach to a host cell protein called **intercellular adhesion molecule-1 (ICAM-1)**, which, as its name suggests, is involved in adhesion of host cells to one another. In particular, ICAM-1 mediates the adhesion of neutrophils to endothelial cells, the first step in the process by which neutrophils exit blood vessels. Normally, the number of ICAM-1 molecules expressed on the surface of a human cell is relatively low, but during inflammation, the number of ICAM-1 molecules on cell surfaces increases. Replication of viruses in respiratory epithelial cells triggers an inflammatory response, which results in an increase in rhinovirus receptors.

A second important feature of rhinoviruses is their variability. There are at least 100 different antigenic types of rhinovirus. Thus, a person who becomes resistant to one type can still be infected by the other 99+ types, not to mention all the other viruses that can cause colds. This is the reason no vaccine exists for the common cold. Two new antiviral drugs that block uncoating of the rhinovirus (WIN5284 and WIN51711) are now being

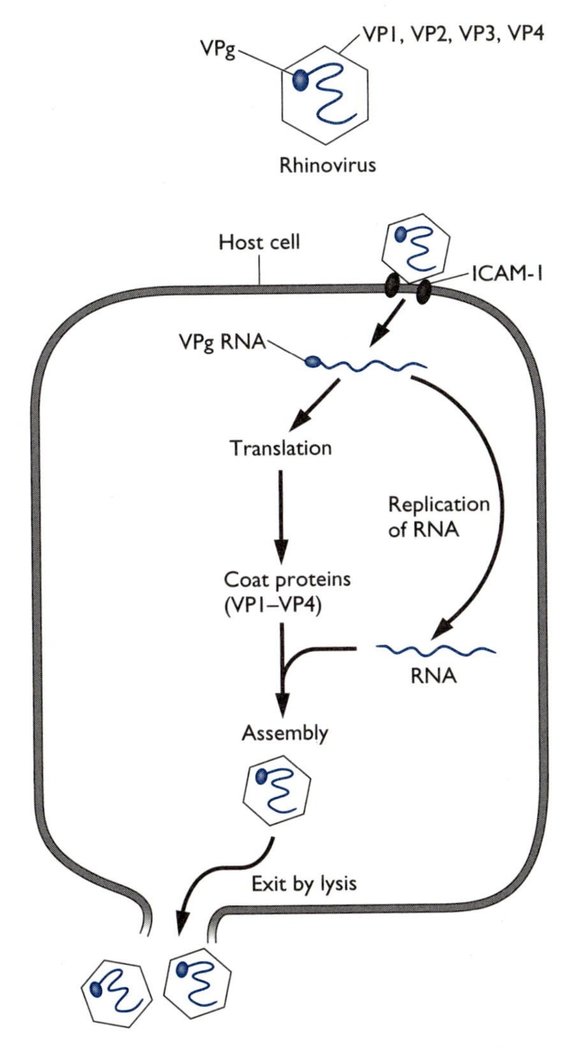

Figure 16.6 Structure and replication cycle of rhinovirus. Rhinoviruses do not have an envelope, so a protein in the capsid of the virus binds to intercellular adhesion molecule-1 (ICAM-1), a host protein. The nucleoprotein core is released directly into the cytoplasm, where the viral ribonucleic acid (RNA) is translated and replicated to form new viral genomes. The newly assembled virions are released by lysing the host cells.

tested for efficacy, but at best they will help with only some colds. Recently, a breakthrough of sorts has occurred, as scientists realized that they might be able to ameliorate cold symptoms by blocking the ICAM-1 molecules to which rhinoviruses attach. Compounds that do this are now being sought and tested.

In the absence of any effective treatment, folk remedies continue to be popular. These range from vitamin C and zinc tablets to herbal teas. None of these remedies have so far proven effective in clinical trials. The problem is that because colds are self-limiting and symptoms vary in severity, individual perceptions of a state of health can vary. One preventive method that does seem to work is to avoid touching your face during cold season, especially after contact with other people. Although colds might seem to be most easily spread by respiratory droplets, studies have shown that hand-to-face transmission is even more efficient. Frequent hand washing and avoiding hand-to-face contacts appear to be the most effective preventive strategy yet.

Mycoplasma pneumoniae: one cause of "walking pneumonia"

A particularly common cause of respiratory infection in college students is ***Mycoplasma pneumoniae.*** The disease it causes is called "walking pneumonia," because it is debilitating but usually not lethal. People with "walking pneumonia," however, may be so ill that they have to remain in bed for weeks. For a busy college student, this disease can spell the loss of an entire semester. *M. pneumoniae* is clearly a bacterium, but it differs from other bacteria in a couple of ways. First, *M. pneumoniae* is one of the few bacteria that lacks a peptidoglycan cell wall. The fact that this bacterium can survive in nature is probably explained by its close association with human cells. The mildest cases of disease occur when infection is limited to the upper airway. *M. pneumoniae* can also invade the lung and cause infection there, but this form of the disease is less common. Scientists are still not sure how *M. pneumoniae* evades the human defense systems. Possibly its close association with human cells is part of the answer.

A second unusual feature of *M. pneumoniae* is that, like human cells but unlike most bacteria, *M. pneumoniae* has cholesterol in its cytoplasmic membrane. Its human-cell-like cytoplasmic membrane may aid *M. pneumoniae* in associating closely with human cells and foiling the attempts of the human defense systems to detect it. *M. pneumoniae* has one of the smallest genomes so far found in a bacterium. It contains only about 500 genes, whereas most bacteria have 3,000–6,000 genes. *M. pneumoniae* is treatable with antibiotics such as erythromycin and tetracycline. The earlier treatment is given, the better it works. Usually antibiotics are given only in the more severe cases of disease.

Chlamydia pneumoniae

Another tiny bacterium that causes human respiratory disease is ***Chlamydia pneumoniae.*** *C. pneumoniae* has a two-stage life cycle (Fig. 16.7). It replicates inside human cells, in a form called **reticulate bodies.** Then it differentiates into smaller, denser-looking forms that do not replicate (**elementary bodies**). Elementary bodies are tough survival forms that can persist outside human cells until they find a new cell to infect. They attach to the new cell target and are taken into it in a vacuole, where they return to the reticulate body form and begin dividing again. *C. pneumoniae,* like *M. pneumoniae,* has no peptidoglycan layer but otherwise resembles a very small gram-negative bacterium.

C. pneumoniae was identified fairly recently as a common cause of pneumonia. The pneumonia can be severe, but the mortality is low. *C. pneumoniae* has been in the news recently because of a possible connection with atherosclerosis. So far, the results of experimental studies have been mixed, but the thought that even some cases of heart disease might be curable or preventable by antibiotics is enough to cause considerable excitement. Another species of *Chlamydia, C. trachomatis,* is a common cause of bacterial sexually transmitted disease (see Chapter 18).

What Can Be Done?

Respiratory diseases continue to be a major cause of morbidity and mortality. Vaccines may be the answer in some cases, such as *S. pneumoniae.*

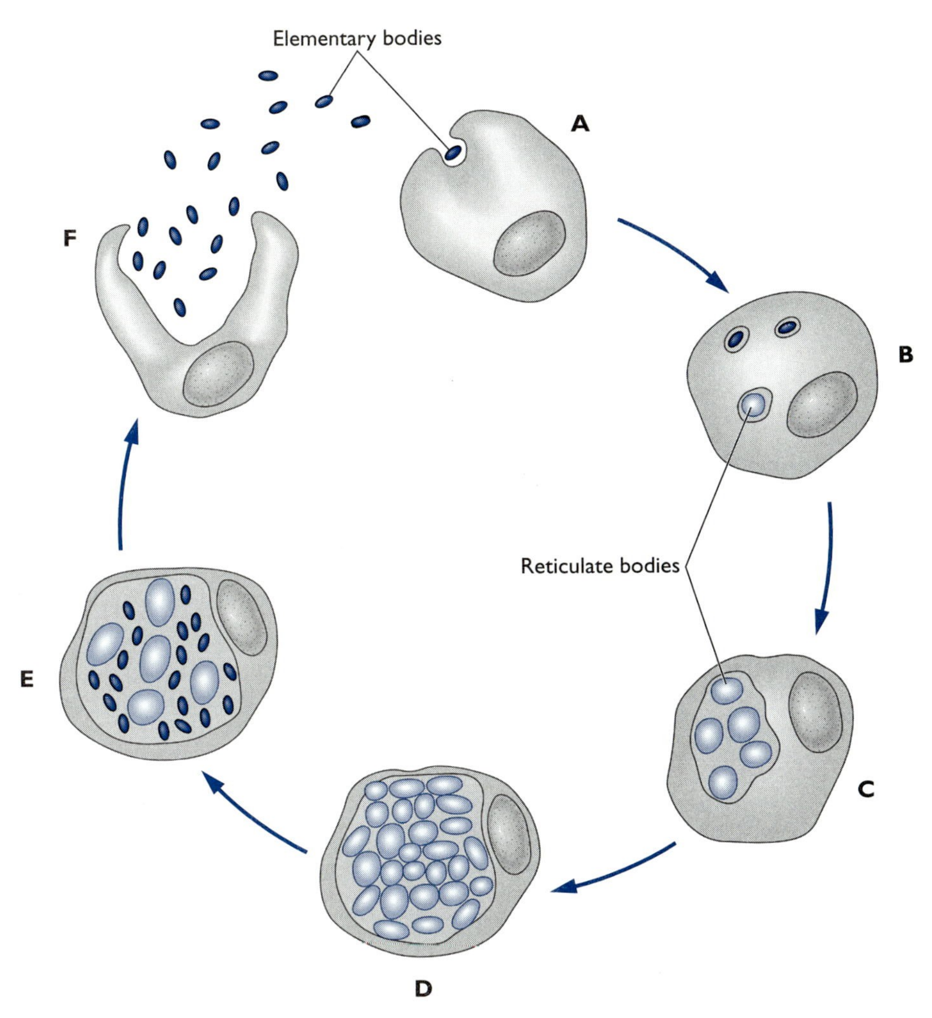

Figure 16.7 The life cycle of *Chlamydia pneumoniae*. **(A)** The survival form, the elementary body, is taken up by host cells. In the host cell, the elementary body changes to the reticulate body **(B)**, which is the actively dividing form **(C)**. After reaching saturation level **(D)**, the reticulate bodies convert to elementary bodies **(E)** which are released into the environment by the host cell **(F)**.

Antimicrobials will suffice for the control of others, unless resistance to antibiotics becomes too pervasive. Passive immunization (Chapter 14) may help in the case of encapsulated bacteria and some viruses. New therapies that reduce the toll of AIDS and influenza can also help prevent superinfections by other microbes, just as the new triple-drug therapy for HIV infections has reduced the incidence of *P. carinii* infections. The most important step to take, however, is to prioritize diseases by their actual importance, rather than allow special-interest groups, whether representing scientists, patients, or health care providers, to distort public health efforts to control disease. As is the case with all infectious diseases, education of the public and preventive measures, such as hand washing, will likely prove to be our best protection against respiratory infections.

chapter 16 at a glance

Major Defenses of the Respiratory Tract

Oral microbiota

Ciliated cells

Mucin

Coughing and sneezing reflex

Alveolar macrophages

Activated macrophages

Comparison of Major Organisms that Cause Respiratory Tract Infections

Species	Reservoir	Characteristics	Disease	Susceptible populations	Virulence factors
Rhinoviruses	Humans	Nonenveloped Nonsegmented ssRNA genome	Common cold	All people	Binds to intercellular adhesion molecule-1 (ICAM-1) Elicits inflammatory response
Influenza virus	Humans Birds Mammals	Enveloped Segmented (7–8) ssRNA genome	Influenza (A: pandemic; A, B: epidemic)	All people Most serious in elderly (influenza plus bacterial pneumonia)	Hemagglutinin (attachment) Neuraminidase (budding)
Streptococcus pneumoniae (pneumococcus)	Humans	G+ diplococcus	Pneumonia Meningitis	All people, especially <1 y old, >60 y old	Polysaccharide capsule (anti-phagocytic) Inflammatory cell wall components
Mycoplasma pneumoniae	Humans	Bacterium Lacks peptidoglycan Sterols in membrane	Pneumonia (walking pneumonia)	Mainly 5–20 years	Attachment to host cell causes toxicity Reason for toxicity unknown
Chlamydia pneumoniae	Humans	Bacterium Lacks peptidoglycan Elementary body Reticulate body	Pneumonia Possibly atherosclerosis	All people	Intracellular parasite Elicits an inflammatory response
Mycobacterium tuberculosis	Humans	Atypical G+ rod Acid fast High lipid content of cell wall Mycolic acids	Tuberculosis	All people, especially very young and very old	Grows in macrophages Elicits granulomatous response
Pneumocystis carinii	Humans (?)	Yeast, but forms called trophozoite and cyst	Pneumonia	AIDS patients Immuno-compromised people Malnourished children	Attaches to and kills pneumocytes

key terms

alveolar macrophages macrophages that occupy the lung and provide an immediate defense against any microbes that pass the defenses of the upper portion of the respiratory tract

antigenic drift and shift mechanisms by which influenza virus avoids host immune response; mutations result in antigenic drift; reassortment of segments of genome result in antigenic shift

bacterial pneumonia infection of the lung caused by bacteria, often following an influenza infection; may lead to bloodstream infection or meningitis; *Streptococcus pneumoniae* is the most common cause

ciliated cell defense cells lining the upper respiratory tract that propel blobs of mucin (with trapped, attached microbes) up and out of the airway

common cold upper respiratory tract infection caused by large array of viruses; approximately half of cases caused by rhinoviruses; symptoms result from host inflammatory response

influenza common respiratory disease that damages defenses and predisposes patient to secondary bacterial pneumonia; widespread outbreaks are called epidemics or pandemics

***Mycoplasma* pneumonia** "walking pneumonia" caused by small bacterium that adheres tightly to host cells; debilitating but usually not fatal

***Pneumocystis* pneumonia** pneumonia commonly caused by *Pneumoncystis carinii;* usually occurs in immunocompromised patients; damages pneumocytes

reactivation tuberculosis immune suppression allows bacteria to break out of tubercle, cause active disease

tuberculosis bacterial lung infection caused by *Mycobacterium tuberculosis,* which multiplies in macrophages and creates an area of dead tissue (tubercle) that may be walled off in a granuloma; may remain dormant or may spread to other tissues

tuberculosis skin test a diagnostic test in which purified protein derivative (PPD) from *M. tuberculosis* is injected into the skin; postive response is a red, raised lesion; false-positive response sometimes occurs in immunized people

Questions for Review and Reflection

1. Why are bacteria that can grow inside macrophages so dangerous?

2. A fairly effective vaccine against influenza is available and antiinfluenza drugs are available. Why then do health officials still worry about an outbreak of "killer flu"?

3. Can you think of a way to treat multidrug-resistant bacterial pneumonia in an immunocompromised patient?

4. Explain how the mouth's resident microbiota, which is normally protective, can actually cause lung infections under some conditions?

5. What are the limitations of the skin test for tuberculosis? If the test has so many limitations, why do physicians rely so heavily on it to diagnose tuberculosis?

6. Some people who are exposed to *Mycobacterium tuberculosis* do not develop symptoms, even though they become tuberculosis-positive on the skin test. What are some arguments for and against treating such people?

7. How does *Pneumocystis* pneumonia differ from that caused by bacteria?

8. Why are physicians urged to refuse patients' demands for antibiotics to treat colds and flu?

9. Why is there as yet no vaccine against the common cold? If there were, what would its limitations be?

10. How do *Mycoplasma pneumoniae* and *Chlamydia pneumoniae* differ from most other bacteria?

17 Gastrointestinal Tract Infections

The fact that some bacteria can live happily in the human stomach and others readily transit this formidable barrier should generate a new appreciation for the ability of microbes to adapt to just about any condition.

Human triathlons are grueling, but they pale in comparison with the gauntlet faced by microorganisms entering the human stomach and intestinal tract. First, microorganisms spend 2–3 hours in a vat of hydrochloric acid. Then they are thrown into the small intestine, the microbial equivalent of Grade III whitewater rapids. Finally, the survivors enter the big city of the microbial world, the human colon, a setting that teems with competitive and aggressive bacteria not eager to welcome new residents. These are the challenges faced by the microorganisms that cause intestinal infections. Oddly enough, the first two legs of the triathlon seem to be less daunting than the third, because most intestinal pathogens target the small intestine. At least one bacterium, *Helicobacter pylori* does not even get past the first leg of the triathlon, opting instead to remain in the stomach and upper part of the small intestine (duodenum). Although *H. pylori* can reside in the stomach for years, possibly a lifetime, it is not as tough as this evidence might at first suggest. However, this bacterium has developed a strategy for taking advantage of a niche made possible by the anatomy of the stomach.

Gastric Ulcers: *Helicobacter pylori*

The Helicobacter pylori revolution

Gastric ulcer disease, long considered to be the result of stress or unhealthy eating habits, proved in the end to be caused by a bacterium, ***H. pylori.*** **Ulcers** are inflamed regions of the stomach lining that can cause substantial pain and even death if bleeding develops. Moreover, the disease, if untreated, is chronic. That is, a person who develops ulcers will continue to have ulcers throughout life. The realization that ulcers are caused by bacteria and could thus be cured with antibiotics was a revolution in gastroenterology and in the lives of ulcer patients.

That discovery also created a stir among clinical microbiologists. Most microbes that cause human disease face conditions that seem relatively mild: a wet, near-neutral environment, 37°C temperature, lots of nutrients, and—for viruses—plenty of different cell types to infect. Because human tissues provide such hospitable settings for microorganisms, the defense systems of the human body must be powerful and vigilant. The stomach, however, offers an extreme low-pH environment, as well as hydrolytic enzymes that digest organic material. Such conditions would seem to make additional protection unnecessary. Yet *H. pylori* has developed a strategy for not only surviving the initial onslaught of acid and enzymes but for persisting in the stomach for decades.

Strategy for survival in the stomach

Surviving to reach the stomach mucosa. How does *H. pylori* do so well in the human stomach? Its ability to survive seemed surprising at first, even to scientists familiar with the organism. In the laboratory, *H. pylori* dies rapidly when exposed to pH conditions similar to those found in the stomach. The bacterium solves this problem by getting out of the acid interior of the stomach as rapidly as possible and into a more habitable area: the mucin layer that covers the gastric mucosa. Cells of the stomach lining are just as susceptible to acid as most bacteria, but this lining can tolerate the acidic contents of the stomach, because its cells secrete a thick, slightly alkaline mucin layer that covers and protects the vulnerable gastric mucosal cells. Protons do not diffuse very readily through this gel matrix, so the pH of the mucin layer can be maintained near 7, the pH range favored by *H. pylori.*

Food entering the human stomach is a rich source of sugars and amino acids that can be used by *H. pylori.* Moreover, *H. pylori* is a **microaerophile,** preferring low concentrations of oxygen. Near the surface of the stomach lining, oxygen concentrations are likely to be low because of oxygen utilization by human cells and slow diffusion of gases into the thick mucin layer.

H. pylori still had to solve the problem of how to reach the mucin layer. *H. pylori* is a curved gram-negative rod with a tuft of flagella at one pole. It uses these flagella to move rapidly to the mucin layer. Because the distances involved are large for a bacterium, *H. pylori* takes some time to reach the mucosa. For protection from stomach acid on this long voyage, *H. pylori* secretes an enzyme, **urease,** which cleaves **urea** into ammonia and CO_2. The CO_2 diffuses away, but the ammonia remains near the bacteria, forming a cloud of ammonia around the bacterium that neutralizes stomach acid in its immediate vicinity (Fig. 17.1). In the absence of acid, *H. pylori* is quite susceptible to being killed by ammonia, so the urease response clearly is designed to function in the stomach interior. *H. pylori* can regulate this potentially toxic response, so that when it reaches the neutral environment of the stomach mucus blanket, urease production is shut off.

Colonization of the mucin layer

After reaching the mucin layer, the bacteria burrow into the mucin. Some travel all the way to the surface of the stomach lining and attach to mucosal cells. Here *H. pylori* exhibits another adaptation to its human host. The lipopolysaccharide (LPS) of *H. pylori* does not elicit the type of intense inflammatory response elicited by the LPS of *Escherichia coli.* Presumably, this is because part of the sugar portion of LPS mimics an oligosaccharide, called **Lewis antigen,** found on human cells. The ability of *H. pylori* to mimic human cells may explain why most colonized people do not develop ulcers.

People who do mount an inflammatory response to *H. pylori* are the ones who develop ulcers. The inflammatory response is not successful in eliminating the bacteria, because most of the bacteria are in the mucin layer, beyond

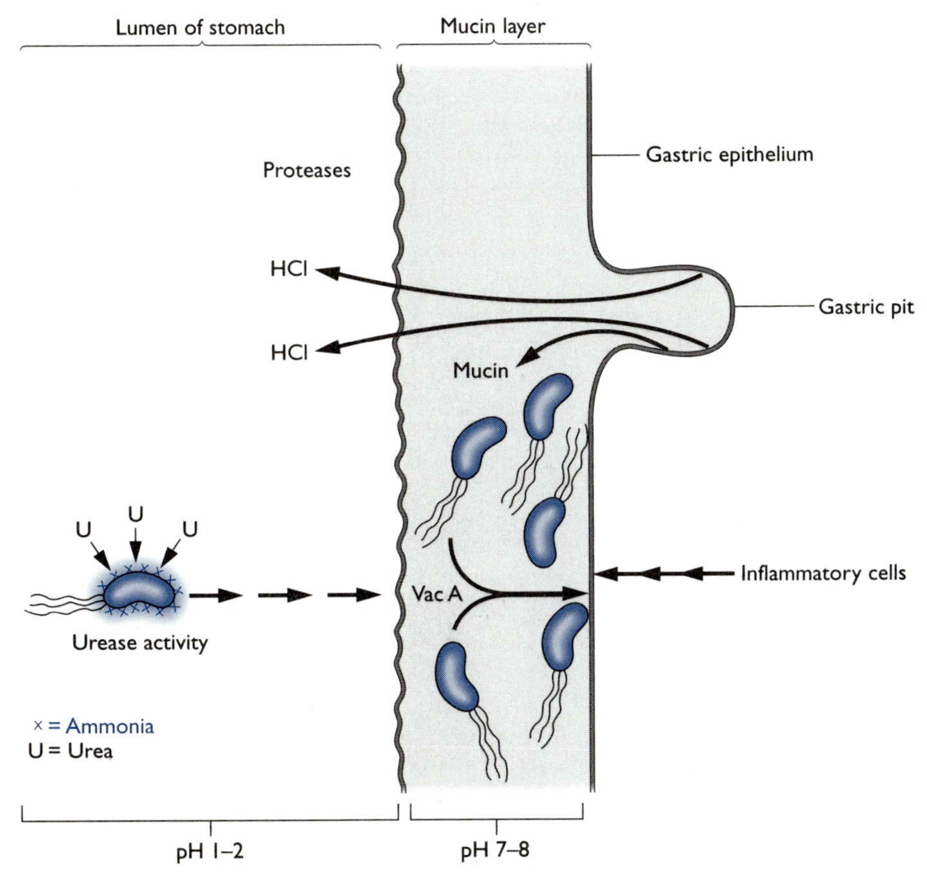

Figure 17.1 Progression of *Helicobacter pylori* infection of the stomach. The bacteria use urease to produce a cloud of ammonia that protects them from stomach acid until they can reach the neutral pH environment of the mucin layer. They grow in the mucin layer, sometimes eliciting an inflammatory response on the part of the host. Vac A toxin, produced by *H. pylori* growing in the mucin layer, may contribute to the inflammatory response by damaging mucosal cells.

the reach of the neutrophils. If the inflammatory response continues, especially if it grows stronger, the neutrophils damage the gastric mucosa sufficiently to produce an ulcer. If *H. pylori* LPS is relatively noninflammatory, why do some people mount an inflammatory response? This is an unanswered question. Two possible answers are, first, that people differ genetically in their response to *H. pylori* LPS and, second, that some strains of *H. pylori* are better able to evoke an inflammatory response than others. Both answers could be correct. Some *H. pylori* strains produce a protein toxin that kills gastric cells. The role of this toxin in ulcers remains controversial, but it may make the inflammatory response to the bacteria more severe. This toxin is called **vacuolating toxin A (VacA)**, because when it is applied to tissue culture cells, the cells develop large **vacuoles** before dying. The numbers of bacteria may also be important. An LPS of *H. pylori* may be less toxic than that of *E. coli* when both are administered at the same concentration, but may become toxic at higher concentrations. Strains that cause ulcers simply may be able to multiply more successfully than those that do not.

Ubiquity of Helicobacter pylori infection

About 40% of people in the United States are colonized with *H. pylori* by the age of 50. A person whose stomach is colonized with *H. pylori* has a 10% chance of developing ulcers at some time in his or her life. The fact that only a fraction of those colonized with *H. pylori* actually develop the disease made it difficult initially for scientists to accept the idea that *H. pylori* could be a significant cause of ulcers. In developing countries, colonization rates are higher, and colonization is seen even in young children. Yet the incidence of ulcers is lower in these countries than in developed countries. This has led scientists to speculate that early colonization may allow the body to come to terms with *H. pylori,* thus becoming less likely to mount an inflammatory response.

Finding a cure for ulcers

When scientists and physicians finally began to accept the idea that *H. pylori* causes ulcers, the hunt was on for an antibiotic cure. Initially, the results were disappointing. Antibiotics that killed *H. pylori* in the laboratory did not work when administered to human patients. How could this happen? First, not much was known about the delivery of antibiotics to the mucin layer of the stomach. For years scientists had been sure that no bacteria could grow in the stomach, so almost no pharmacological experiments had been conducted to determine antibiotic behavior there. An antibiotic that was destroyed by acid or stayed in the stomach contents might not get to the site where the bacteria were growing. The story of the rocky road to finding the right therapy is instructive. Studies of *H. pylori* grown under laboratory conditions showed that a number of antibiotics were able to prevent its growth. Yet these antibiotics did not cure ulcer patients. In retrospect this is not surprising. *H. pylori* growing in a test tube under optimal conditions is not necessarily the same bacterium as *H. pylori* growing in the stomach mucin layer. Accordingly, it took some time to develop an effective antibiotic therapy. The therapy consisted of at least two antibiotics to eliminate the bacteria, plus a compound that helped speed healing of the ulcer. A number of antibiotic combinations are currently in use, and all are quite effective.

Detecting Helicobacter pylori infection

Ulcer-like symptoms may not always be the result of bacterial infection. Sometimes ulcers are caused by overuse of aspirin or by alcohol abuse. To determine whether a patient is likely to benefit from antibiotic therapy, a physician first needs to ascertain whether an infection is actually underway. A rapid method for determining whether a person's ulcers are caused by *H. pylori* is a breath test that detects ammonia produced by the bacterial urease. Normally, little or no ammonia is in stomach contents or expired in the breath, so the presence of ammonia is a good indicator that the bacteria are present and active. The **ammonia breath test** can give false negatives, however. A better, but more time-consuming, diagnostic test looks for antibodies against *H. pylori* antigens in the patient's blood. This test can be problematic, too, because many people who have such antibodies are asymptomatic.

An added bonus accompanied the discovery that *H. pylori* causes ulcers. Ulcers are strongly associated with the eventual development of **gastric cancer,** a particularly nasty and potentially fatal form of cancer. How inflammation proceeds to unbridled growth of gastric cells remains a mystery, but eliminating the ulcers and the cause of the ulcers effectively moots the problem. No ulcers, no cancer. For those who belong to the "prevention is superior to intervention" school of medicine, the verdict is clear: early treatment of ulcers to short circuit possible gastric cancer is definitely a good idea.

An even better way to prevent both ulcers and gastric cancer would be to avoid acquiring *H. pylori* in the first place. No convincing evidence indicates that *H. pylori* is spread in food or water, but neither is there clear evidence that it is spread from person to person. The route of transmission remains uncertain, despite numerous studies. An alternative approach to preventing ulcers could be to colonize children early, just as children are colonized in developing countries, where gastric ulcers are rare. Although the logic of this approach has some appeal, carrying out such experiments would be considered highly unethical. If scientists discover a way to make *H. pylori* nontoxic but still capable of colonizing mucin, then colonizing people with the nontoxic variant could provide a competitive barrier to more dangerous strains.

Food-Borne and Water-Borne Diseases

Changes in the food industry affect food-borne and water-borne infections

Whether or not *H. pylori* is spread by food and water, many other microbial pathogens are spread by these routes. During the past several years, large outbreaks of diseases caused by ingestion of contaminated food and water have led to serious questions about the safety of the food and water supplies. Statistics on **food-borne disease** are not adequate to reveal whether the total number of cases of food-borne disease has increased or decreased over the past 50 years. Until recently, government agencies focused on large outbreaks and did a good job in tracing this type of disease. However, the suspicion is strong that the number of incidents of disease involving only one or a few people may be larger than previously suspected. Only recently have government agencies begun collecting information on this larger background of sporadic cases.

Centralization of the food supply and large outbreaks of food-borne infections. One thing is clear: the size of outbreaks of food-borne infections has increased dramatically in recent years. In 1950, the typical outbreak occurred on a local scale, after a church social or company picnic where contaminated food was served. These smaller outbreaks continue to occur, but, since 1980, a number of multistate outbreaks have involved thousands of people who consumed processed foods that theoretically should have been safer than foods prepared at home. Between 1950 and 2000, food production, processing, and distribution underwent massive centralization. For example, chickens raised in sheds containing thousands of birds now go to centralized packing plants, where hundreds of thousands of pounds of meat are processed every day and meat is packaged to be sent to supermarkets all over the country.

Ironically, although the procedures for safe handling and packaging food have improved dramatically, the centralization of the food industry insures that when safety lapses do occur, their effects will be felt by many people. Moreover, the news media are much more likely to report a multistate outbreak involving a fast food chain than an outbreak following a church social in Lickskillet, Kentucky. Thus the impact of outbreaks on public perception is larger today than ever before.

The water link. Although media attention has focused on outbreaks traced to meat, especially hamburger and chicken, experts admit that vegetables are potentially as dangerous as meat. Most meat-associated contamination problems can be solved by proper cooking, but this is hardly a solution for lettuce or fruit, usually consumed raw. Outbreaks associated with contaminated raspberries and other produce items have made the news a number of times in recent years.

These outbreaks are often assumed to be associated with importation of food from countries with poor standards of hygiene, but this is not necessarily the case. Certainly, some instances have occurred in which imported food was contaminated, such as the case of contaminated raspberries from Guatemala in 1996. The contamination of the raspberries probably occurred because the fruit was washed with local water and packed in ice made from the same water, which, as in many less developed countries, was contaminated with human and animal feces.

Even in the United States, irrigation water sprayed on vegetables or used to deliver fungicides is not of tap-water quality and may be contaminated with animal fecal bacteria entering the groundwater from manure runoff. This type of contamination may explain the outbreak of disease in Japan associated with radish sprouts. Radish seeds were exported from the United States to Japan, where their sprouts were used in school lunches. The seeds were contaminated with *E. coli* O157:H7, a particularly pathogenic strain of *E. coli*. Thousands of children developed bloody diarrhea. Some died. Debate continues over whether contamination occurred in the United States before the seeds were shipped or in Japan where the seeds were sprouted, but the fact that contamination could have occurred in either developed country is cause for concern.

Water-borne infectious diseases. Outbreaks of **water-borne infections** have also become larger. Big cities in the United States are struggling to maintain old and failing sewage systems, many of which were constructed in the 19th century. Sewage treatment facilities are now run more efficiently and effectively than ever before, but, as with a centralized food supply system, only a few mistakes in treatment at a centralized water supply can result in large outbreaks. One example of this phenomenon is the outbreak of diarrhea caused by the protozoal parasite, ***Cryptosporidium parvum,*** that occurred in 1993 in Milwaukee, Wisconsin. In this region that one beer maker advertises as "the land of sky-blue waters," a temporary glitch in the sewage treatment process led to an outbreak of diarrhea involving many thousands of people. Not all cases of water-borne disease involve tap water or water sprayed on crops. For an unusual water-borne disease outbreak associated with a new style of sports event, see Box 17.1.

Microbes that cause food-borne and water-borne infections

Many different microbes can cause food-borne or water-borne disease. In this chapter, we will look at some examples that best illustrate the complex interaction between such microorganisms and the human body, starting with the bacterial food-borne pathogens. In Chapter 15, some toxin-associated food-borne diseases were described. In this chapter, the focus will be on

box

17–1 *A Human Triathlon Acquires a Fourth Phase*

In the 1980s and 1990s, athletic competitions that combined several types of sports—bicycling, swimming, running—became very popular. Such sports competitions have an ancient history, but large events drawing crowds of participants and spectators are a late 20th-century development. Triathlons are designed to test the physical strength, stamina, and versatility of the participants. A triathlon held in Springfield, Illinois, in the summer of 1997 added a new and unplanned competition: a test of the participants' antimicrobial defense responses. A week or so after the completion of the Springfield triathlon, many participants developed a fever and flu-like symptoms, the unplanned fourth leg of this test of strength and endurance. The causative microbe turned out to be the bacterium ***Leptospira interrogans***. This spirochete has long been known as a cause of disease in developing countries where field workers tend crops that have been contaminated with the urine of rodents or other wild animals.

The filtration system of the mammalian kidney normally prevents the excretion of bacteria and other larger microbes in urine. But *L. interrogans* has a corkscrew motility and is so thin that it can pass through the filtration system of the kidney and be excreted in urine. Apparently, the Springfield lake used for the swimming part of the triathlon had been contaminated by the urine of infected rodents and other wild animals. Swimmers ingested the contaminated water unintentionally or had enough contact with it to develop this unusual—for the developed world—disease. *L. interrogans* may be much more common in natural water supplies than most people realize. But cases of disease experienced by isolated swimmers could easily be misdiagnosed as the flu.

Recreational waters, as the U.S. Centers for Disease Control and Prevention designates water sources from park drinking fountains to lakes, are potential reservoirs for infectious microbes. Couch potatoes take heart. If breathing outdoor air is hazardous to your health (Chapter 16) and swimming is, too, perhaps the strategy of sitting and snacking in front of a TV screen makes some sense as the newest human defense against infectious disease. Just make sure those snacks are well cooked!

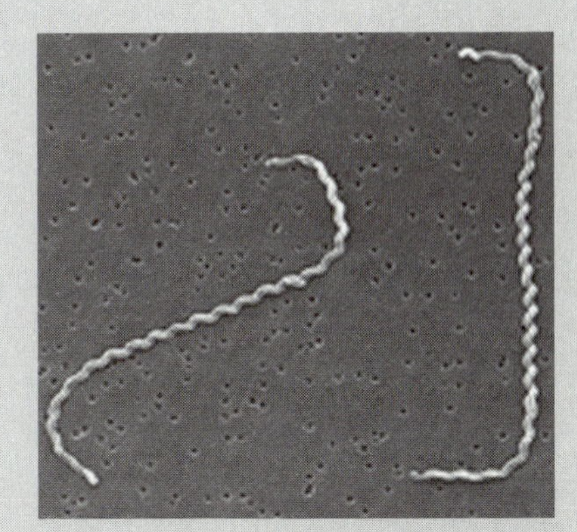

Scanning electron micrograph showing the structure of *Leptospira interrogans*. Courtesy of Rob Weyant, Centers for Disease Control and Prevention, National Center for Infectious Diseases, Atlanta, GA.

food-borne diseases in which microbes set up an infection (Fig. 17.2). At present, the most common causes of serious food-borne outbreaks are all bacteria: ***Salmonella, Campylobacter*** and ***E. coli.*** Another bacterium, ***Listeria monocytogenes,*** is a much less common cause of disease but is a cause for concern because of the high case fatality rate with which it is associated. Viruses, such as rotaviruses and Norwalk virus, probably cause more cases of diarrheal disease than bacteria, but these episodes are usually self-limiting and rarely fatal—at least in developed countries. In developing countries, rotaviruses are a major cause of infant deaths.

One type of food-borne virus that can cause quite serious disease in adults is **hepatitis A,** which can lead to serious liver damage, and sometimes to liver failure. Protozoa are also significant causes of food- and water-borne disease. Once thought to be uncommon in the developed world, two protozoa, ***Giardia intestinalis*** (arguably the cutest protozoan of all; see Fig. 17.9) and *C. parvum,* are now acknowledged to have caused major outbreaks of intestinal disease in developed countries. As an indication of how ubiquitous these protozoa are even in the United States, consider the fact that one occupational hazard faced by police divers in New York City is infection with these and other protozoa due to unintentional ingestion of water during dives.

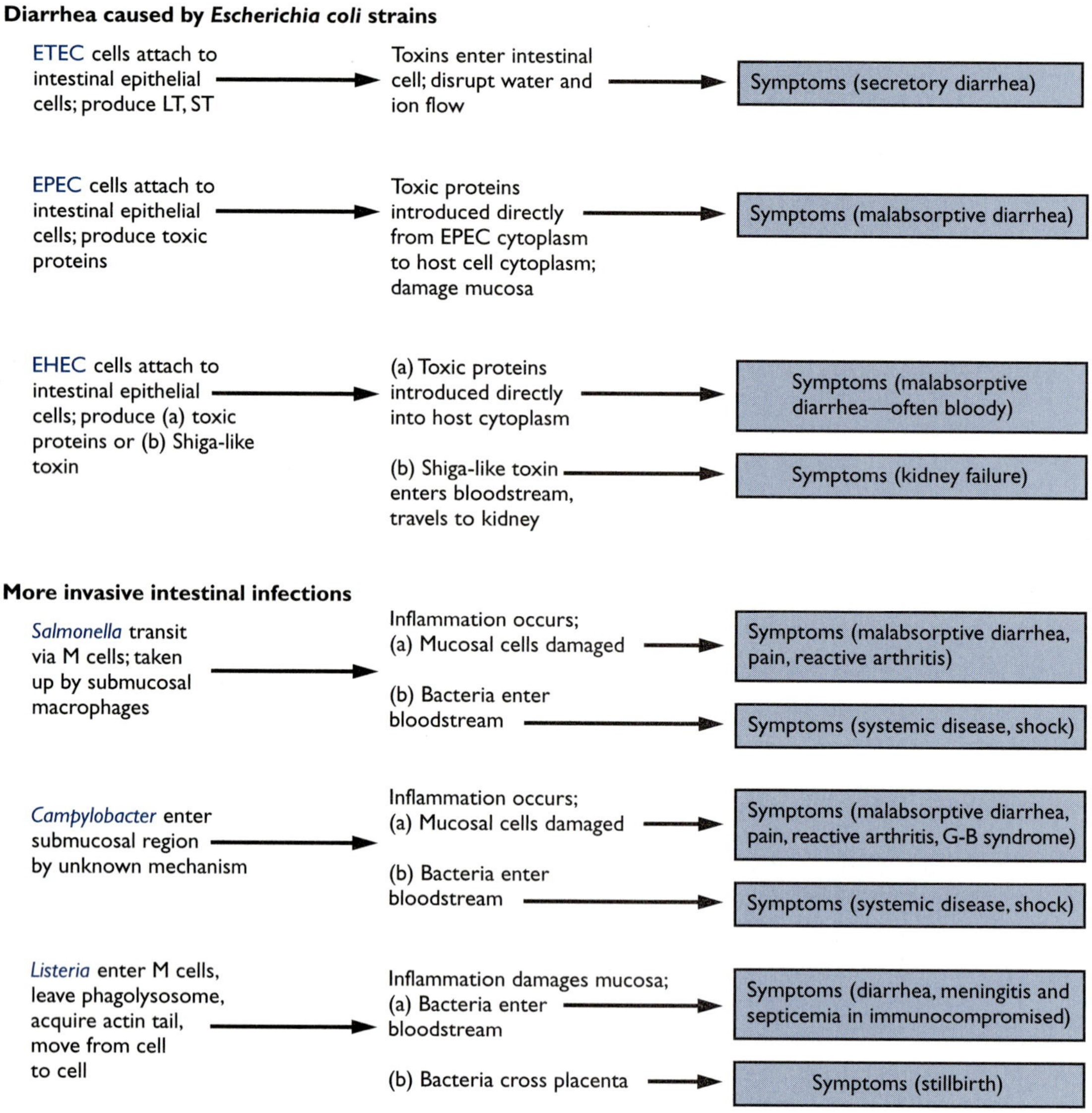

Diarrhea caused by *Escherichia coli* strains

ETEC cells attach to intestinal epithelial cells; produce LT, ST → Toxins enter intestinal cell; disrupt water and ion flow → Symptoms (secretory diarrhea)

EPEC cells attach to intestinal epithelial cells; produce toxic proteins → Toxic proteins introduced directly from EPEC cytoplasm to host cell cytoplasm; damage mucosa → Symptoms (malabsorptive diarrhea)

EHEC cells attach to intestinal epithelial cells; produce (a) toxic proteins or (b) Shiga-like toxin → (a) Toxic proteins introduced directly into host cytoplasm → Symptoms (malabsorptive diarrhea—often bloody)

(b) Shiga-like toxin enters bloodstream, travels to kidney → Symptoms (kidney failure)

More invasive intestinal infections

Salmonella transit via M cells; taken up by submucosal macrophages → Inflammation occurs;
(a) Mucosal cells damaged → Symptoms (malabsorptive diarrhea, pain, reactive arthritis)
(b) Bacteria enter bloodstream → Symptoms (systemic disease, shock)

Campylobacter enter submucosal region by unknown mechanism → Inflammation occurs;
(a) Mucosal cells damaged → Symptoms (malabsorptive diarrhea, pain, reactive arthritis, G-B syndrome)
(b) Bacteria enter bloodstream → Symptoms (systemic disease, shock)

Listeria enter M cells, leave phagolysosome, acquire actin tail, move from cell to cell → Inflammation damages mucosa;
(a) Bacteria enter bloodstream → Symptoms (diarrhea, meningitis and septicemia in immunocompromised)
(b) Bacteria cross placenta → Symptoms (stillbirth)

Figure 17.2 Mechanisms by which food-borne bacteria cause disease. The *E. coli* strains attach to host epithelial cells and produce toxic compounds. The toxins enter the host cells by different mechanisms. Enterotoxigenic *E. coli* (ETEC) causes secretory diarrhea with no inflammation, whereas enteropathogenic *E. coli* (EPEC) and enterhemorrhagic *E. coli* (EHEC) cause malabsorptive diarrhea because they damage epithelial cells. EHEC also produces a Shiga-like toxin, which can enter the bloodstream and cause kidney failure. *Salmonella, Campylobacter,* and *Listeria* invade the submucosal regions causing gastroenteritis. *Listeria* enters the bloodstream and can cause meningitis and septicemia in immunocompromised patients. *Listeria* also is able to cross the placenta and cause stillbirths.

Symptoms of food-borne and water-borne infections. A common symptom of infections of the small intestine is **diarrhea.** The cells of the small intestine normally control the flow of water, because they have **ion pumps** that maintain the flow of ions, and thus of water, across the membranes of mucosal cells. Retaining water in tissues is very important, because the human body is basically just organized sea water. Loss of water from tissue (**dehydration**) can kill. This is especially true for infants, who dehydrate more rapidly and more extensively than adults. The most common therapy for severe diarrheal disease is **rehydration therapy,** an effort to counter dehydration. Fluids are administered either by mouth or as intravenous solutions, with a goal of returning water to tissue. In addition to holding water, the small and large intestines actively absorb water. The small intestine accounts for at least 80% of the water absorbed each day. The colon is less active but makes a significant contribution. If the action of microorganisms on mucosa disrupts the ion pumps that prevent water loss or disrupts the absorption of water by intestinal cells, unusual loss of water results, causing diarrhea.

There are two types of diarrhea: secretory and malabsorptive. **Secretory diarrhea** is characterized by loss of water from tissues as a result of disruption of the activities of the mucosal ion pumps and usually does not involve damage to the mucosal cells or inflammation. **Malabsorptive diarrhea** occurs when mucosal cells are damaged, either directly by the infectious agent or secondarily by the inflammatory response to infection. Such damage prevents water absorption. In the small intestine, some types of pathogens can also disrupt the absorption of nutrients, leading to vitamin deficiency and malabsorption of fats (**steatorrhea**). If damage to mucosal cells during malabsorptive diarrhea becomes severe enough, bloody diarrhea (**dysentery**) develops.

The acid tolerance response

As already described, *H. pylori* has developed a strategy for evading the toxic effects of exposure to stomach acid. So have some other bacteria, but their protective response is much less well understood than that of *H. pylori.* If bacteria such as *Salmonella* and *E. coli* are exposed briefly to a slightly acidic pH ($pH < 5$), they become able to withstand much lower pH environments (pH 1–3) for a couple of hours—long enough to get through the stomach. This phenomenon, called the **acid tolerance response,** appears to consist of a furious pumping of protons out of the cytoplasm to keep the cytoplasm from acidifying to the point where the bacteria die.

How is this response induced? There are two possible scenarios. First, many foods, including sausages, sauerkraut, and pickles, have a slightly acidic pH. The human taste for slightly acidic foods may help prepare bacteria to transit the stomach successfully. A second possibility, which may explain the fact that nonacidic foods such as chicken or salad can be conduits of food-borne disease, is that when food enters the stomach, the pH rises somewhat above pH 1–2. Thus, bacteria entering the stomach in large meals experience transiently the sort of mildly acidic conditions that trigger the acid tolerance response, before they experience the lower pH values associated with undiluted hydrochloric acid. An important additional factor to keep in mind is numbers. Suppose that stomach acid kills 99.99% of microorganisms that enter the stomach. Put differently, only 0.01% survive. If 10^5 microorganisms enter the stomach, however, that still leaves 10 microbes that survive.

The many faces of Escherichia coli

E. coli is a gram-negative rod-shaped bacterium encountered previously in Chapter 11 as a normal denizen of the human colon. Most strains of *E. coli* are harmless and may even be protective, in that they prevent colonization by more virulent strains. The minority of *E. coli* strains that cause disease provide excellent examples of the ways in which slight variations in genetic makeup can allow bacteria to cause different symptoms. *E. coli* is also a particularly good example of the way in which a bad bacterium can turn worse by acquiring new traits through gene transfer from other bacteria.

Enterotoxigenic *E. coli* strains. Enterotoxigenic *E. coli* (ETEC) strains cause secretory diarrhea. ETEC is the most common cause of what is commonly called "**travelers' diarrhea,**" most often experienced by people who travel from developed to developing countries. For these mostly healthy adults, ETEC is, at worst, an unpleasant nuisance. For infants in the developing world, however, ETEC is a major threat to survival and a leading cause of infant deaths worldwide. ETEC strains differ from ordinary strains of *E. coli,* because, in addition to producing pili that allow the bacteria to attach to the ileal mucosa, they produce one or more protein toxins that act on intestinal cells, causing the cells to lose water. One of these toxins is called **heat-labile toxin** (**LT**), because it is inactivated more quickly by heat than the other toxin, **heat-stable toxin** (**ST**). Scientists now know that the reason for this difference is that LT is composed of proteins, whereas ST is a short peptide.

LT is an **A-B type toxin** (see Chapter 15), which is excreted by bacteria attached to cells of the small intestine. The B part of LT must first bind to intestinal cells so that the A part of the toxin can enter the cell and disrupt cellular functions (Fig.17.3). The **B part** consists of five protein subunits that interact to form a pore-like structure in the intestinal cell membrane, facilitating the movement of the **A part,** which has enzymatic activity, into the cell. Inside the cell, the A part of LT binds to a human cell enzyme called **adenylate cyclase** and locks it into the "on" position. The adenylate cyclase produces **cyclic adenosine monophosphate** (**cAMP**) from ATP. cAMP is an important regulatory molecule. Its level controls many cellular functions, including the ion pumps that keep water flowing in the right direction. As cAMP builds up, the ion pumps malfunction, and water and ions no longer held by the cells flood into the small intestine. ST has much the same effect, except that it consists of a single peptide that somehow causes **cyclic guanosine monophosphate** (**cGMP**) levels to rise in intestinal cells. cGMP is another important regulatory molecule, and, as cGMP levels rise, the ion pumps cease to function. Either way, diarrhea is the ultimate result. LT and ST illustrate an important point: two toxins that look dissimilar and act at different levels can nonetheless have the same net effect at the organ or tissue level.

ETEC strains appear to have gotten their LT toxin genes from the bacterium that causes the epidemic disease **cholera** (the gram-negative bacterium, ***Vibrio cholerae***). *V. cholerae* is closely related to *E. coli.* It produces a toxin (**cholera toxin**) with an amino acid sequence similar to that of LT and it functions in the same way as LT. The origin of ST remains a mystery.

Enteropathogenic *E. coli* strains. EPEC strains cause a malabsorptive diarrhea. EPEC is another major diarrheal killer of infants in developing countries and a contributor to travelers' diarrhea. EPEC strains bind intestinal cells via pili and somehow damage the surfaces of those cells. The mechanism of damage is still under investigation, but recent work has revealed a fascinating feature of these bacteria: once they bind to the surface of a eukaryotic cell, they can form a channel connecting their cytoplasm with that

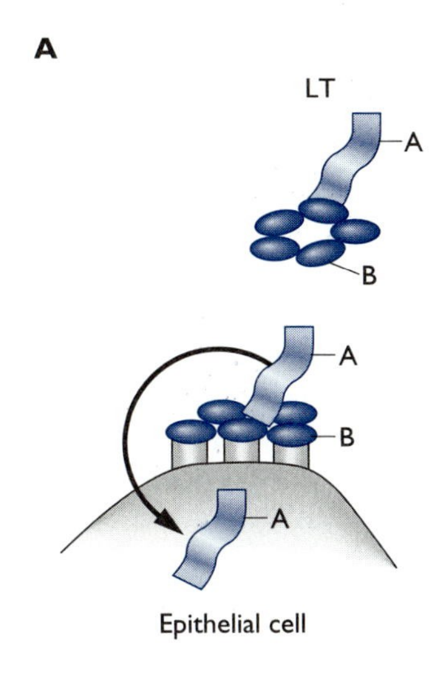

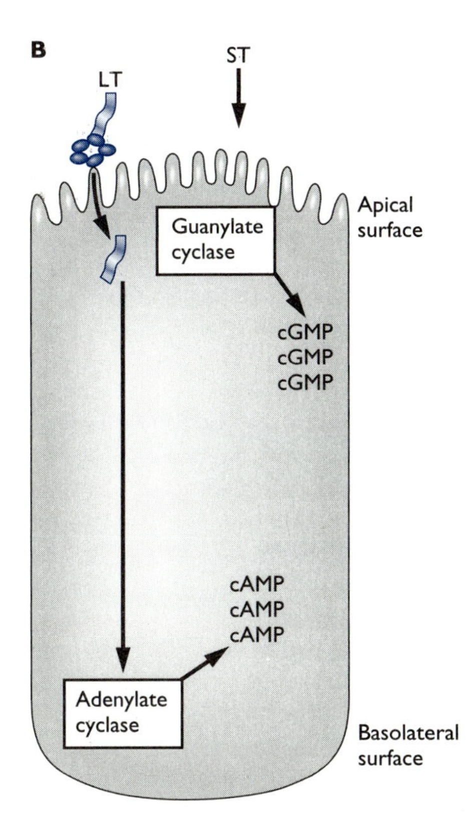

Figure 17.3 Structure of heat-labile toxin (LT) of *Escherichia coli* and mechanism of LT and heat-stable (ST) toxins. **(A)** LT is composed of five B subunits and one A subunit. **(B)** The B subunits of LT bind to the surface of an intestinal epithelial cell, and the A subunit enters the cell. The A subunit alters the activity of adenyl cyclase, causing excess production of cAMP. ST causes guanylate cyclase to produce excess amounts of cGMP. Elevated levels of cAMP and cGMP affect the ion pumps that maintain the balance of water and ions, resulting in secretory diarrhea.

of the human cell. This is called a **type III secretion system.** Through this channel they inject toxic proteins into the human cell (Fig. 17.4). This strategy protects the toxins from the host's antibody response. Antibodies against an extracellular toxin such as LT should be protective if they have access to the toxin, because they can bind the toxin during its extracellular phase and prevent it from attaching to human cells. If the toxin is never exposed to the extracellular fluid, however, antibodies cannot reach and neutralize it.

Enterohemorrhagic *E. coli* strains. Developed countries were proud of the low incidence of childhood deaths resulting from diarrhea until a new form of *E. coli*, **enterohemorrhagic *E. coli*** (**EHEC**), appeared. This is the infamous *E. coli* O157:H7 or, as it is popularly known in the media, "killer *E. coli*" (Fig. 17.5). EHEC causes a malabsorptive diarrhea that can become bloody

Figure 17.4 Delivery of EPEC toxic protein to human intestinal cells by a type III secretion system. The EPEC cell uses pili to bind to intestinal epithelial cells. Once the cells are bound, EPEC makes tighter contact with the surface of the epithelial cell and forms a channel that makes a direct connection from the EPEC cytoplasm to the intestinal cell cytoplasm. Toxic proteins produced by EPEC can then be injected directly into the human cell without ever being exposed to antibodies that might be present in the external environment.

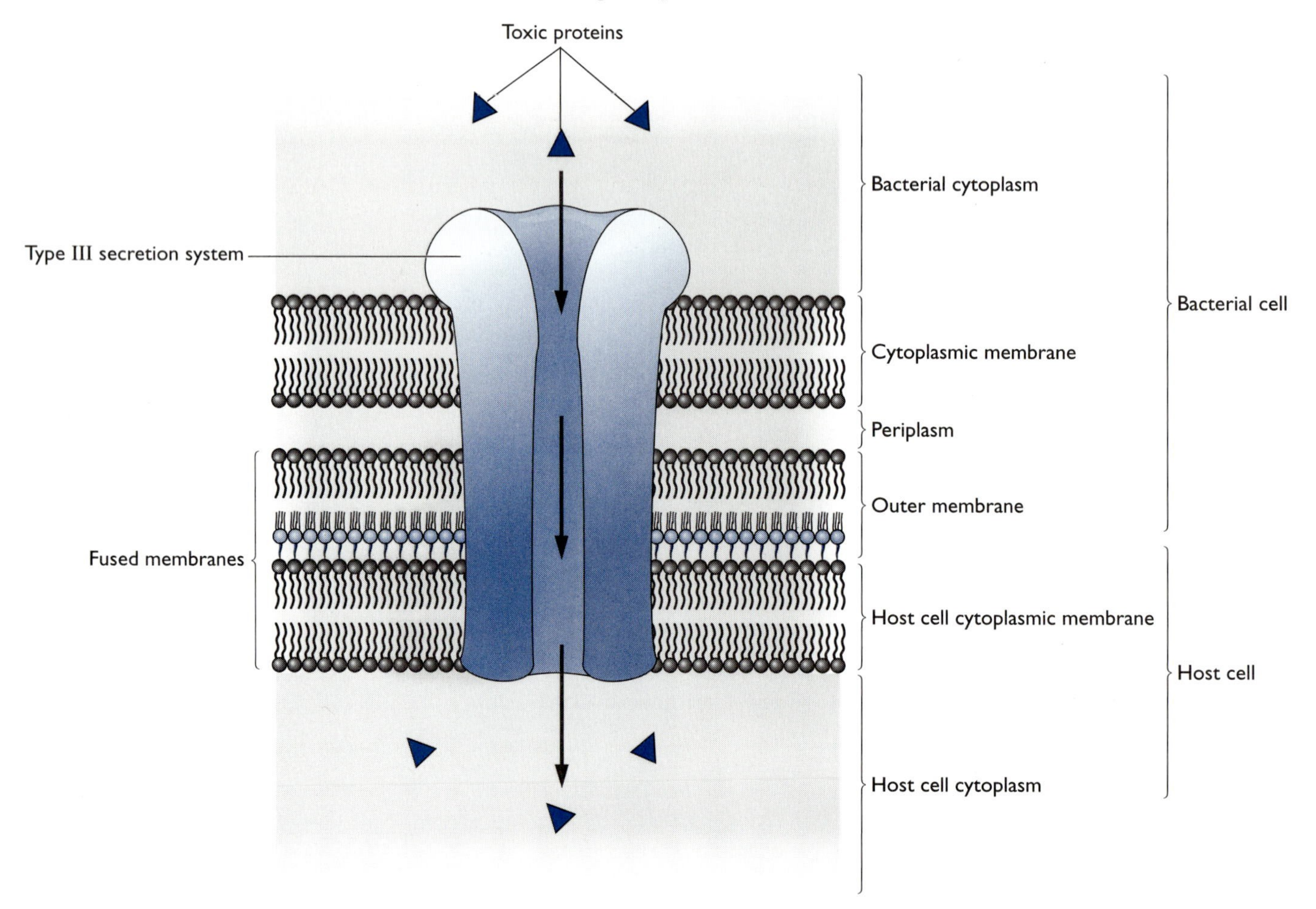

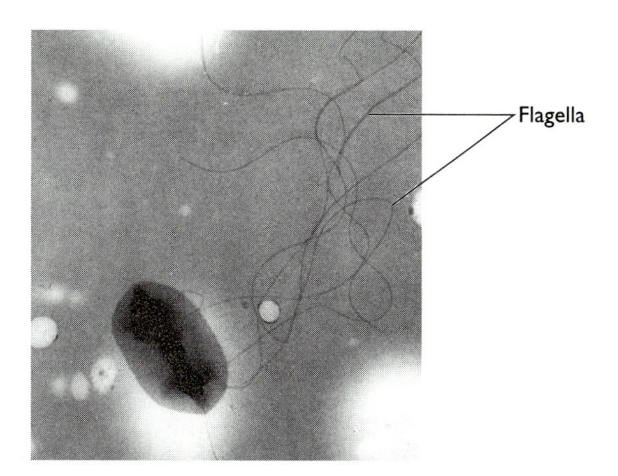

Figure 17.5 ***Escherichia coli* O157:H7.** This transmission electron micrograph of *E. coli* shows the flagella (H antigen). The O antigen of lipopolysaccharides is too small to be seen at this magnification. (Courtesy of Peggy S. Hayes, Centers for Disease Control and Prevention, Atlanta, GA.)

(dysentery). EHEC strains are more deadly than ETEC or EPEC strains because of the type of toxin they produce (the gene for which was probably acquired from *Shigella*, a notorious cause of dysentery). The toxin, called Shiga-like toxin, enters the bloodstream and damages the kidney. *E. coli* O157:H7 and other strains with similar traits have caused a number of deaths in children.

Although Shiga-like toxin is harmful to many different cell types, it seems to exert its most dramatic effects on **endothelial cells** (blood vessel cells). Damage to blood vessels in the kidney results in hemorrhages and limited blood flow to the area, causing kidney damage. Shiga-like toxin, circulating in the bloodstream in the vicinity of the intestine and depositing in the kidney, is thought to be the cause of kidney failure. For some reason, EHEC strains have their most serious effects on children, although some fatal cases have also occurred among elderly people.

The genes encoding Shiga-like toxin are carried on a bacteriophage. If the bacteriophage infects an EPEC strain and integrates into the bacterial chromosome (**lysogeny**), the strain then carries genes encoding the toxin. As discussed in Chapter 15, a number of bacterial toxin genes have proved to be carried on bacteriophages. Why bacteriophages would carry genes that have no obvious benefit to them remains unclear. Perhaps the toxins serve another function scientists have not yet discovered. Carriage of toxin genes on a bacteriophage may explain an initially mysterious phenomenon. In some children, treatment of EHEC diarrhea with antibiotics actually seemed to make the disease worse. One possible explanation is that antibiotics stress the bacteria, inducing the bacteriophage to jump ship by going into its lytic mode. During lytic growth, many copies of the phage genome and proteins are made, resulting in an upshift in toxin production.

What does the designation O157:H7 mean? Today, different strains of *E. coli* would be typed by pulsed field gel electrophoresis or a similar technique. In the days before these techniques were available, scientists turned to serological tests to distinguish benign *E. coli* strains from pathogenic strains. The O in O157 refers to the **O antigen** (carbohydrate portion) of LPS. The O antigens of different strains vary sufficiently to be useful for typing strains of *E. coli.* These different types of O antigens are recognized by different antibodies. The **H** is a **flagellar antigen,** and different H antigens are also recognizable by the antibodies that distinguish between them.

Treating bacterial diarrhea. Antibiotics usually are not used to treat diarrhea caused by *E. coli* strains. For one thing, the disease is self-limiting, and antibiotics do not shorten its duration significantly unless taken very early in the infection. For another, antibiotics may actually increase the amount of toxin produced by EHEC strains. The best therapy is rehydration. That is, the patient is given oral or intravenous fluids to replace fluids lost as a result of diarrhea. In EHEC infections, dialysis and other supportive therapy to compensate for poor kidney function are the only treatments currently available to patients with the severe form of the infection. Although the use of antibiotics to treat diarrhea is generally not recommended, people who visit their physicians before traveling to high-risk locations will generally receive antibiotics to take if and when they begin to experience diarrhea. The argument for this therapy is that limiting the diarrhea to a day or two reduces the amount of time the patient is out of commision and lessens the likelihood of severe symptoms.

Salmonella and Campylobacter

Salmonella and *Campylobacter* are two very different gram-negative bacteria, but they will be considered together, because they are currently two of the most common causes of food-borne illness in developed countries. Also, they have in common an ability to cause systemic disease as well as gastroenteritis. The systemic form can be fatal. *Salmonella* has been well studied, but little is known about *Campylobacter.* Only within the last decade have scientists begun to realize that *Campylobacter* is a much more common source of food-borne disease than previously thought. *Salmonella* is a gram-negative motile rod that is very closely related genetically and metabolically to *E. coli. Campylobacter* is a motile gram-negative comma-shaped rod that looks like and is closely related genetically and metabolically to *H. pylori. Salmonella,* like *E. coli,* can grow either aerobically or anaerobically, whereas *Campylobacter,* like *H. pylori,* prefers a very low oxygen environment (microaerophile). One of the reasons scientists missed *Campylobacter* for so long was that they were accustomed to incubating plates inoculated with stool specimens under normal atmospheric conditions. Only when they started incubating plates in a low-oxygen atmosphere with an increased CO_2 atmosphere did *Campylobacter* leap to the fore as a major food-borne pathogen.

Diseases caused by *Salmonella* and *Campylobacter.* Several species of *Campylobacter* cause disease (*C. coli* and *C. jejuni,* for example). We will refer to *Campylobacter,* the genus, rather than naming individual species, because what applies to one appears to apply to all or most of them. The nomenclature for *Salmonella* is quite complicated and is still changing. *Salmonella enterica* serotype *enteritidis* is the most common cause of **gastroenteritis.** *S. enterica* contains many other serotypes, such as *typhimurium.* Another species, ***S. typhi,*** causes **typhoid fever,** a much more serious disease than gastroenteritis. We will use the term *Salmonella* here to mean the types of *Salmonella* that cause gastroenteritis.

The disease caused by *Campylobacter* is similar to that caused by *Salmonella*: diarrhea, which can be accompanied by severe abdominal pain. *Campylobacter* infection is somewhat unusual in that one possible complication of the disease is **Guillain-Barré (G-B) syndrome,** a relatively rare neurological disease characterized by progressive weakening of the muscles. Although other microbes can trigger G-B syndrome, *Campylobacter* is now thought to be responsible for nearly half of all G-B cases in the United States. Most strains of *Salmonella* and *Campylobacter* are virulent for humans, although asymptomatic carriage of both organisms is commonly seen when outbreaks occur. *Salmonella* and *Campylobacter* are not virulent for most of the animals that carry them. Farm animals, especially chickens, may be the natural hosts of both of these genera of bacteria.

Virulence traits of *Salmonella* and *Campylobacter*. Once the bacteria reach the small intestine, they swim to the mucosa and invade underlying tissue. The way in which *Campylobacter* invades is not known, but *Salmonella* has a well-established affinity for **M cells** and enters the body through the **Peyer's patches** (Chapter 13; Fig. 17.6). After transiting the M cells, *Salmonella* encounters macrophages, which ingest the bacteria and try to kill them. *Salmonella* is not readily killed by ordinary macrophages and grows inside them. Only an activated macrophage, which has greater killing power than an unactivated one, can kill *Salmonella.*

The pain associated with the diarrhea caused by these two pathogens is probably the result of the inflammatory response elicited by bacteria growing in the tissue underlying the M cells. The inflammatory response also damages mucosal cells, producing a malabsorptive diarrhea. A fairly common complication of this type of infection is a **reactive arthritis,** elicited by

Figure 17.6 Steps in a *Salmonella* spp. infection and consequences for the human host. *Salmonella* moves through the mucin layer to the M cells of Peyer's patches. The bacteria pass through the M cell and are taken up by macrophages, in which they multiply and then lyse the macrophage. This causes an inflammatory response by the host. In a small percentage of cases systemic disease occurs.

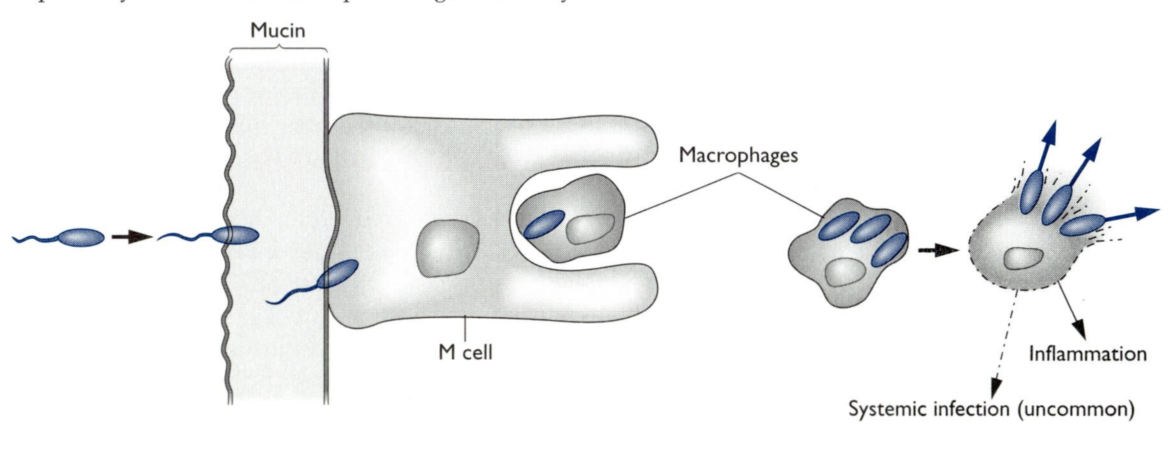

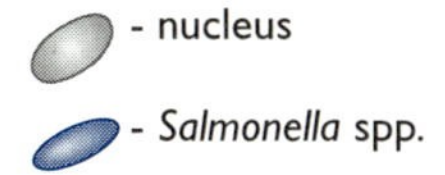

circulating bacterial antigens rather than by infection of the joints. The arthritis subsides spontaneously, and no long-term damage to the joints occurs. In the majority of infected people, the inflammatory response is successful eventually in limiting the spread of the bacteria, but in some people (the very young, the elderly, and the immunocompromised), systemic disease can develop and lead to septic shock and death. In cases where systemic disease develops, antibiotic therapy is essential.

***Salmonella* is not all bad.** *S. typhimurium* targets and stimulates the cells of the gut-associated lymphoid tissue and is easily manipulated genetically. These factors have made *S. typhimurium* a favored base for live vaccines. A considerable amount of work is being done to identify mutants of *S. typhimurium* that are able to invade and replicate a few times in Peyer's patches but do not cause symptomatic disease. Genes from other pathogens will be cloned into the vaccine strains so that *S. typhimurium* expresses the appropriate antigens and stimulates an sIgA response against them.

Listeria monocytogenes, an insidious and dangerous bacterial pathogen

Listeria monocytogenes is a gram-positive, motile rod that is potentially more dangerous than *Salmonella* and *Campylobacter. L. monocytogenes* can grow over a wide range of temperatures, including those in household refrigerators. In fact, when refrigerated, *L. monocytogenes* may increase in numbers dramatically. *L. monocytogenes,* like *Salmonella,* enters the body through M cells and invades the underlying tissue (Fig.17.7). In contrast to *Salmonella,* which transits the mucosal cells to reach the macrophages, *L. monocytogenes* breaks out of the phagolysosome of mucosal cells and moves from one cell to another, by polymerizing human actin. The bacterial cell acquires an actin "comet tail" as it moves through the cytoplasm.

Listeria infections usually are brought under control rapidly, and most people experience only mild flu-like symptoms. In the immunocompromised individual, however, *L. monocytogenes* can persist in the bloodstream and cause septic shock. In this case, lipoteichoic acid (LTA), instead of LPS, is the trigger, but the outcome is the same. Another serious condition can develop in women who are pregnant. The placenta is normally a very effective barrier against pathogens and keeps the fetus free of infection. *L. monocytogenes* is one of the few pathogens that can cross the placental barrier. Virtually nothing is known about how the bacteria do this, but one of the first signs of an outbreak of *L. monocytogenes* food-borne disease is stillbirths or babies infected in the uterus. The disease is made all the more tragic by the fact that this sign of an outbreak usually occurs too late to do anything to avert it.

Listeria infections are much less common than infections caused by *Salmonella, Campylobacter,* and *E. coli,* but cases may be more frequent than we think. Most of the data collected by the Centers for Disease Control and Prevention (CDC) and others has focused on outbreaks involving relatively large numbers of cases. These agencies are only beginning to monitor individual cases of disease. Whatever the outcome, pregnant women are advised to avoid soft cheeses and to eat only food that has been heated to near boiling and to avoid deli foods like cold cuts. At least one outbreak resulted from consumption of coleslaw made with contaminated cabbage, so produce as well as animal products can be a source of infection.

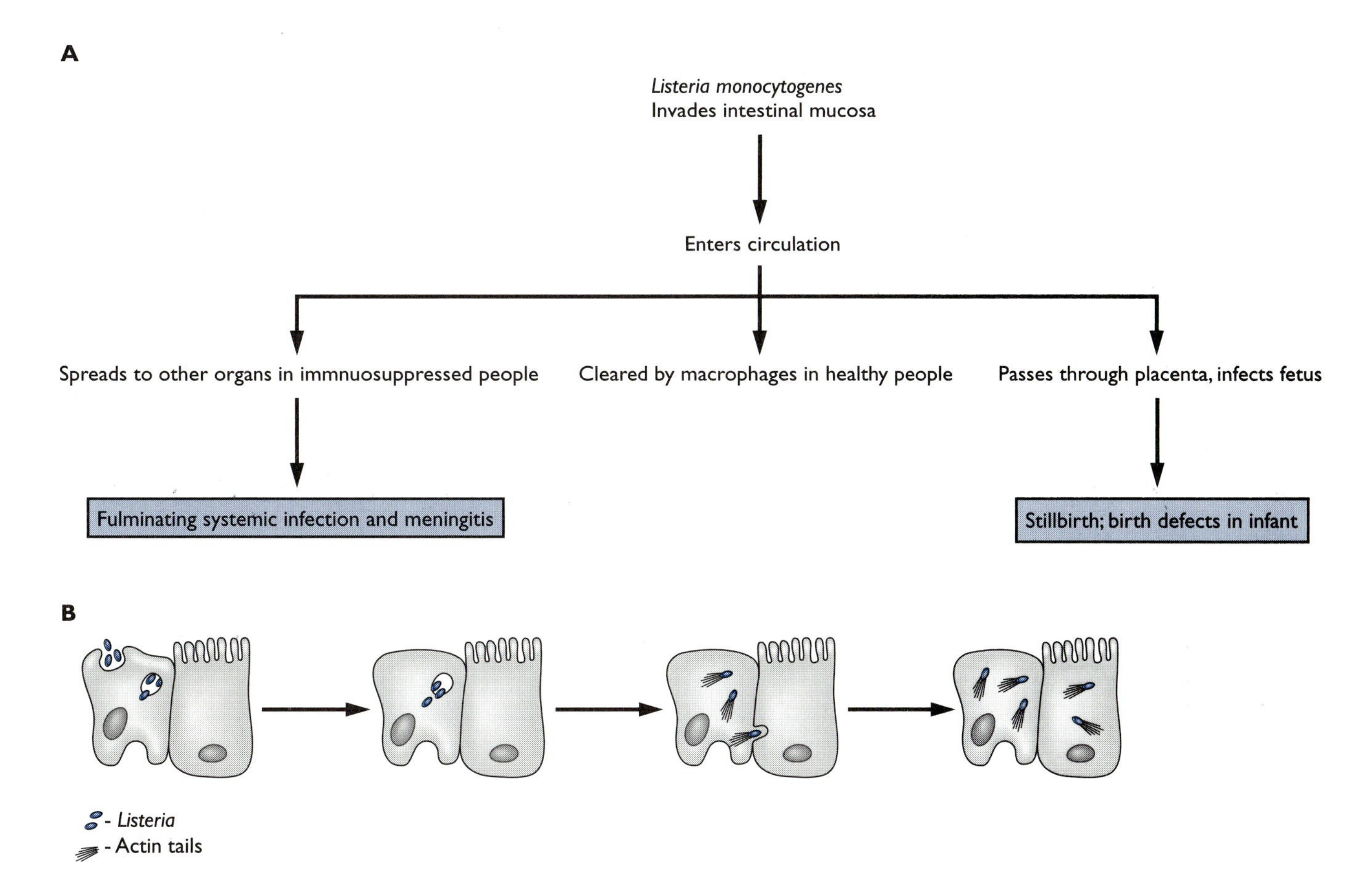

Figure 17.7 Steps in a *Listeria monocytogenes* infection and consequences for the human host. (A) Overview of infection. Bacteria that invade the mucosa usually are cleared rapidly in healthy adults but can cause systemic infections in immunocompromised patients; in women who have not developed an immune response, they can cross the placenta to cause stillbirth or birth defects. **(B)** *L. monocytogenes* enters the body through M cells. It breaks out of the phagolysosome and polymerizes human actin to form a comet tail, and then penetrates adjoining intestinal epithelial cells, eventually reaching the bloodstream.

When Small Is Not Beautiful: Viruses that Cause Food-Borne Disease

Viruses are a common cause of food-borne and water-borne gastroenteritis. This condition is sometimes called **acute infectious nonbacterial gastroenteritis (AING,** not to be confused with angst, although a person with AING can experience justifiable angst). In adults, AING is an unpleasant but self-limiting condition characterized by fever, vomiting, malaise, and diarrhea. AING is much more serious in infants, who are more likely than adults to experience severe dehydration. In the United States alone, more than 200,000 infants and young children are hospitalized each year for dehydration brought on by viral diarrhea. Deaths are uncommon among well-nourished infants living in areas where medical attention is available, but diarrhea is a major cause of death in infants worldwide. Currently, two of the most common causes of viral gastroenteritis in the United States are Norwalk (and Norwalk-like) virus and rotavirus. It should be kept in mind, however, that a common diagnosis in cases of AING is still "unknown etiology." Clearly some infectious agents remain to be identified.

Norwalk virus. Viruses are sometimes named for the place where the first-described cases of the disease occurred. **Norwalk virus** was named for an outbreak of AING that occurred in Norwalk, Ohio, in 1972. This is one of the smallest viruses known (25–30 nm in diameter) with a round shape and no envelope. Norwalk virus has an RNA genome. Because it has not yet been cultivated in tissue culture cells, little is known about its replication strategy. Norwalk virus was identified initially by electron microscopy of stool specimens from people with the disease. A number of Norwalk-like viruses, that is, viruses having the same size and shape as the Norwalk agent, have been identified subsequently. The capsid of the Norwalk virus consists of a single protein subunit arranged in a crystalline array over the viral surface. Immunological assays for detecting viral particles in stool specimens are now available.

Because the disease is self-limiting, no treatment is indicated, except for rehydration therapy. Nonetheless, tracing the outbreak to its origin, such as a food handler who is shedding the virus, is important in preventing further outbreaks. Food handlers are the source of most outbreaks of Norwalk virus disease. Almost any type of food can be involved, and the food handler may not have symptomatic disease. Shellfish and public water supplies have also been sources of outbreaks. One reason food handlers are so often implicated in spreading this disease and many others is that they are all too often poorly paid and have no medical insurance or paid sick leave. Thus, a person with unpleasant but not disabling symptomatic disease may come to work anyway.

Rotavirus. Rotavirus is a major cause of AING in infants and young children, both in the United States and worldwide. Adults, too, especially the elderly, can suffer from rotavirus gastroenteritis. Cases in otherwise healthy adults tend to be mild or inapparent. The virus was named for the characteristic wheel-shaped morphology of its capsid proteins when viewed in cross-section (Fig. 17.8). Rotaviruses are naked viruses with a segmented double-stranded RNA (dsRNA) genome. The rotavirus replication cycle starts with adsorption of the virus to the cell surface and entry of the nucleoprotein core. The virus carries its own RNA-dependent RNA polymerase, which produces messenger RNA for protein synthesis and copies of both strands to provide genomes for new virus particles. Viruses exiting the cell kill the cell. Rotaviruses infect the tips of small intestinal villi. Infected cells die and are sloughed prematurely, causing the villi to shrink. Damage to the mucosal surface, combined with a smaller area of absorption, causes diarrhea from decreased water absorption.

Rotaviruses grow poorly in tissue culture cells. Thus, electron microscopy or serological methods must be used to diagnose outbreaks. There are three types of rotavirus, types A, B, and C. Type A is seen in the United States, where outbreaks have occurred most frequently in daycare centers and nursing homes. In daycare centers, virus transmission occurs both by contact and by ingestion of contaminated food or water. Only a small number of viral particles are needed for transmission, so the virus spreads readily between children, especially those in whom sanitary toilet habits are not yet well learned. Rotavirus outbreaks, like influenza virus

A

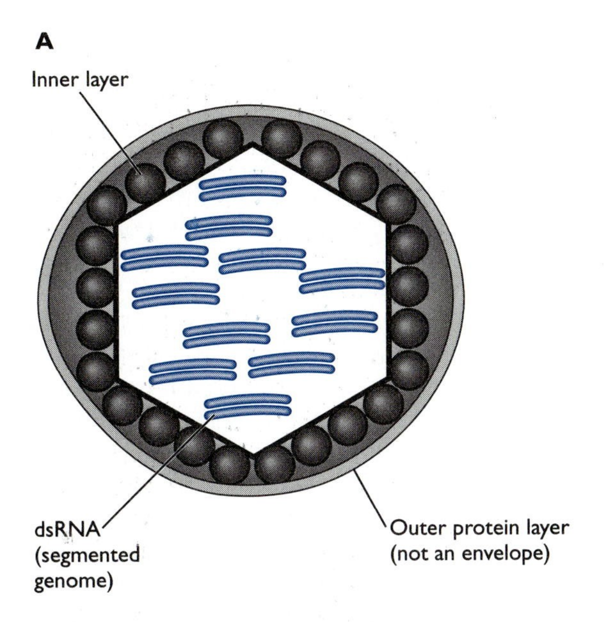

B

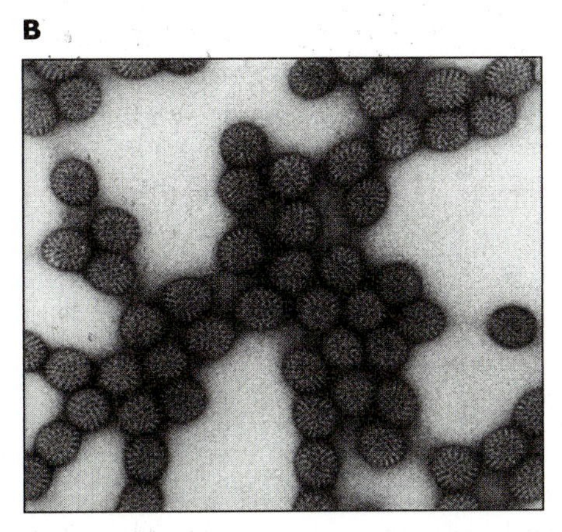

Figure 17.8 Structure of rotavirus. (A) Rotavirus has a segmented, double-stranded RNA genome. It is a nonenveloped virus with a protein outer layer surrounding its capsid. **(B)** Transmission electron micrograph of a rotavirus, showing the protein outer layer. (Courtesy of the Centers for Disease Control and Prevention, Atlanta, GA.)

outbreaks, have a seasonal pattern and are more common in the late fall and winter than in the summer months.

Although a rotavirus vaccine is now available, it has been problematic from the start. At first, the vaccine appeared to work poorly, but later studies showed that increasing the dosage appreciably increased efficacy. With higher dosages, however, side effects began to appear. A small number of vaccinated infants developed potentially dangerous intestinal blockages, and routine use of the vaccine was terminated. This vaccine might still be beneficial in countries where rotavirus diarrhea kills many infants each year. The story of this vaccine illustrates the difficult dilemma that arises when the ill health of a few in a developed country is used to justify the withdrawal of a preventive measure that could protect millions in developing countries.

Hepatitis A virus. Hepatitis, a disease that can lead to liver failure, is a far more dangerous food-borne disease, at least for adults, than AING. Hepatitis viruses are the most common causes of viral inflammation of the liver (hepatitis). These viruses include hepatitis A, B, C, delta, and E. The symptoms of hepatitis are fever, nausea, vomiting, and jaundice. Despite the similarity in names and symptoms, the different hepatitis viruses belong to different virus families and are spread by different routes. **Hepatitis A,** a naked virus with a single-stranded RNA genome, is spread by ingestion of contaminated food or water and causes a disease called infectious hepatitis. **Hepatitis B** is an enveloped virus with a partly double-stranded DNA genome, and is transmitted mainly by blood or blood products, shared needles, or sexual contact. **Hepatitis C** is an enveloped virus with a single-stranded RNA genome that is currently transmitted mainly via blood and blood products. Even without going on to delta and type E, it is evident that this is a very heterogeneous group of viruses with only one common feature: the ability to damage the liver. We will focus here on hepatitis A, because it is spread via food and water.

Hepatitis A virus is highly invasive. Unlike Norwalk virus and rotavirus, which remain in the mucosal cells, hepatitis A virus penetrates the intestinal mucosa and reaches the liver, where it replicates in liver cells. About 2–4 weeks after initial infection, symptoms begin to appear and are linked to damage done to the liver by virus replication. The severity of disease is not as great as that caused by hepatitis B virus. **Jaundice** results if the liver is damaged to such an extent that bile pigments, normally removed from the blood by the liver, are instead deposited in tissues and eyes. During the incubation period, before the appearance of overt symptoms, viruses enter the biliary system and are shed in the feces. The disease is extremely infectious, because the high numbers of viruses shed in feces can readily contaminate water or hands and thereby contaminate food. Many cases are mild, and the disease is usually self-limiting. A person who has had infectious hepatitis will continue to shed the virus for a few weeks after the symptoms subside.

Hepatitis A can be spread by food handlers and also is found in contaminated shellfish beds. The disease is widespread in the United States and is even more common in developing countries. About 20%–50% of the adult population of the United States has antibodies against hepatitis A virus, so there may be many more asymptomatic cases than were suspected previously. No therapy is available for infectious hepatitis, but a new vaccine is recommended for people who travel frequently to countries with a high incidence of hepatitis A and who may not have become naturally resistant to the disease.

Protozoa that Cause Diarrhea

Many protozoa cause intestinal disease, but as examples we have chosen a few protozoa with disease strategies that illustrate some of the ways protozoa cause infections. These protozoa also happen to cause major health problems in the United States. Two, *Giardia* and *Cryptosporidium,* have been implicated in a number of outbreaks associated with consumption of contaminated water. These microbes also can be acquired from food that has been washed with contaminated water. Thus, infections resulting from these pathogens can be a particular problem in produce. ***Cyclospora,*** the protozoal parasite made infamous by contaminated raspberries, is similar in many ways to *Cryptosporidium* and is only mentioned here in passing. The other protozoal parasite covered here, *Toxoplasma gondii,* is not a common cause of symptomatic disease. However, it shares with *Listeria* the ability to cross the placenta. Symptomatic infections can have disastrous outcomes and are thus taken more seriously than some more common diseases that have few serious consequences.

Giardia intestinalis: the cause of giardiasis

Symptoms of the disease. Giardiasis is a diarrheal disease that can range in severity from a mild diarrhea with flatulence, anorexia, and crampy abdominal pain to a copious, sometimes explosive, fatty (greasy, foul-smelling) diarrhea that can be quite debilitating. In poorly nourished infants and children, *Giardia* diarrhea can cause malnutrition. It can be bad news for healthy adults, too, because excessive dehydration and weight loss can be fatal. The causative agent of giardiasis is the protozoal parasite *Giardia intestinalis* (Fig. 17.9). Chapter 9 featured a drawing of *G. intestinalis.* It is reproduced here because the life cycle of this protist has considerable

Figure 17.9 Structure of different stages of *Giardia intestinalis* and its effect on the host intestine. (A) *G. intestinalis* occurs in a flagellated form, the trophozoite, and a cyst form, the survival form in the environment. The flagellated form attaches to host intestinal cells by means of a sucking disk. **(B)** *G. intestinalis* attaches to the cells lining the mucosal surface. It damages individual cells and causes the villi to be blunted. This damage decreases absorption of fluid and nutrients. *G. intestinalis* does not cause systemic disease.

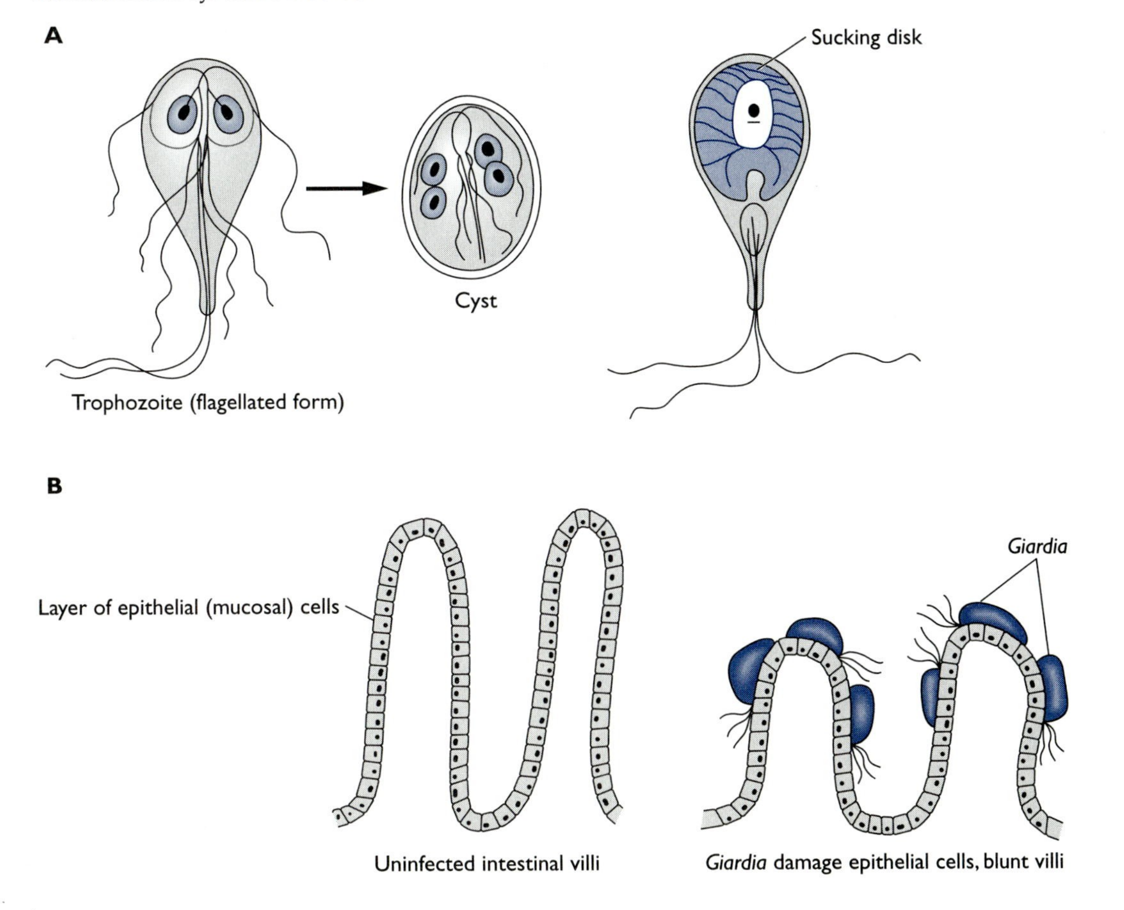

importance for its ecology. *G. intestinalis* has two forms, the **trophozoite** (replicative) form, which colonizes the upper part of the small intestine, and the **cyst** (survival) form, which is metabolically inert and does not replicate but can survive for long periods in the environment. Cysts play a comparable role to bacterial and fungal spores in the spread of disease. The trophozoite is pear-shaped, with two nuclei that look like eyes, leading to its deceptively appealing appearance. It also has flagella. Motility presumably is important for reaching the mucosal surface of the upper part of the small intestine. The severity of the disease depends on the number of cysts ingested and on the immune status of the host. Some research suggests that people who have been exposed to *Giardia* develop partial immunity to reinfection.

On its ventral surface, *G. intestinalis* has a **sucking disk,** which is used to anchor the parasite to the mucosal surface. The sucking disk damages the mucosal cells, resulting in a decrease in their absorptive capacity. In people carrying a heavy load of *G. intestinalis,* decreased absorption of water leads to the copious diarrhea associated with the disease, and decreased absorption of fats gives the diarrhea its characteristic "fatty" appearance. Decreased absorption of protein, vitamins, and other nutrients causes malnutrition. Substances not absorbed in the small intestine may be degraded in the colon by the colonic microbiota. The degradative activities of the colonic microbiota probably account for the fact that the smell of stools from patients with giardiasis is unusually foul. Unpleasant as this may be, it is a useful diagnostic indicator. Although most of the organisms remain on the surface of the intestine, some may work their way into the secretory tubules of the intestine and may even reach the gall bladder. However, giardiasis is not nearly as invasive as other protozoal diseases of the intestinal tract, such as toxoplasmosis.

Acquisition of giardiasis. *Giardia* is found worldwide and infects wild animals as well as humans. In the United States, *Giardia* cysts are found in many lakes and streams where wild animals (especially beavers) shed the cysts into the water. A good way to acquire *Giardia* in the United States is to drink the water from clear mountain streams during a hiking trip in the Rocky Mountains—or become a New York City police diver. In the United States, giardiasis is a disease that infects people in any socioecomic group. About 5% of the residents of Aspen, Colorado, were found to be shedding *Giardia* cysts. Many of the victims of giardiasis in the United States are recreational hikers and bikers.

Giardia can also be acquired from tap water where sanitation procedures are inadequate. Cysts of *Giardia* are somewhat resistant to chlorine, and outbreaks of giardiasis have occurred in cities or towns where lapses in chlorination procedures provided windows of opportunity. So far, such outbreaks have been relatively rare in the United States. Many cities, however, have aging sewer systems, badly in need of repair, and concerns have grown about the possibility that outbreaks of diseases such as giardiasis could occur more commonly in the future. Giardiasis can also be spread by contact. Outbreaks of giardiasis have occurred in daycare centers, and sexual transmission through oral-anal contacts between male homosexuals has been reported. The possibility that *Giardia* could contaminate produce is a cause for concern, especially in view of the fact that the same feces-contaminated water that contains *Giardia* may also contain hepatitis A virus, *Salmonella,* and other pathogens.

Diagnosis and treatment. The traditional approach to diagnosis of giardiasis is visualization of the trophozoite or cyst form in iodine-stained smears from fecal specimens. One problem with this approach is that the number of trophozoites and cysts shed in feces varies, and these diagnostic forms may not be seen in every stool specimen. At least three stool specimens must be examined from each patient. In addition, specimens must be examined while they are fresh, because the fragile trophozoites are rapidly destroyed outside the body. In recent years more effective tests have been developed that require only a single stool specimen. An enzyme-linked immunosorbent assay (ELISA)-based test is used widely.

Giardiasis is usually a self-limiting disease but can be quite debilitating. Symptoms can last for long periods of time (weeks to months) if untreated. For this reason, treatment of severe cases is desirable. The two most commonly used therapeutic agents are quinacrine (Atabrine, Sanofi Winthrop Pharmaceuticals) and metronidazole (Flagyl, SCS Pharmaceuticals). Quinacrine intercalates in DNA and presumably acts by inhibiting DNA synthesis. Metronidazole causes DNA strand breakage. Metronidazole is one of the few antimicrobial compounds that works against both protozoa and some bacteria.

Cryptosporidium parvum: a common water-borne pathogen

C. parvum, is, as the name suggests, a tiny protozoal parasite. C. *parvum* may be small, but it has a very complex life cycle, much more complex than the big, lumbering *G. intestinalis* (Fig. 17.10). **Oocysts,** the survival form of *C. parvum,* analogous to the cysts of *G. intestinalis,* are ingested in water or food. In the intestine, the oocyst develops into the **sporozoite** form, which enters the epithelial cells of the small intestinal mucosa. In the mucosal cells, the parasite goes through asexual and sexual replication cycles before releasing oocysts. In an otherwise healthy adult, the parasite only succeeds in completing a few rounds of replication before it is eliminated by host defenses. In immunocompromised people, the parasite can continue to replicate, giving rise to a profuse, cholera-like diarrhea that can be life threatening.

Like *G. intestinalis, C. parvum* has an animal reservoir, although in this case the reservoir is domestic animals (cows, pigs) instead of wild animals. Human carriers also shed the organism into water supplies used for drinking and recreation and have been important contributors to recent outbreaks. *C. parvum* oocysts frequently are isolated from water supplies, especially recreational water, such as drinking fountains and lakes used for swimming and boating. *C. parvum* oocysts are even more resistant to killing by chlorine than *G. intestinalis* and are thus not easy to eliminate from drinking water.

C. parvum may be a more common cause of diarrhea in children than previously thought, but, unless the child has some underlying condition that reduces the immune response, the illness is usually mild and self-limiting. The fact that the parasite was not recognized earlier is probably because most cases are so mild. Only when severe, life-threatening cases appeared in patients with acquired immunodeficiency syndrome and in other immunosuppressed patient populations was the role of *C. parvum* as a human diarrheal pathogen brought to the attention of the medical community.

C. parvum presents the same problems for diagnosis as *Giardia.* Traditionally, diagnosis was made by visualizing the oocysts in stained slides of feces. The oocysts are small and easily missed, so examination of feces does

A

***Cryptosporidium* spp. life cycle**

Oocyst ingested with contaminated water

Sporozoite released in intestine

Enters intestinal epithelial cells

Goes through sexual and asexual replication

Oocysts produced; shed in feces

Mild diarrhea in immunocompetent people

Severe, life-threatening diarrhea in immuno-compromised patients

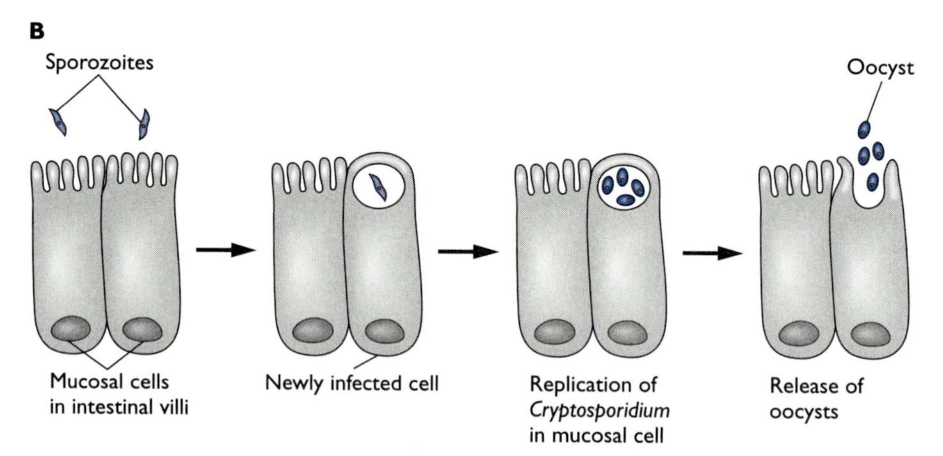

Figure 17.10 Steps in a *Cryptosporidium parvum* infection and its replication cycle. (A) The infective form (oocyst) is ingested with water and develops into a sporozoite, which enters the epithelial cell. This causes mild diarrhea in healthy people and severe diarrhea in immunocompromised patients. **(B)** During an infection, the sporozoites enter intestinal epithelial cells and replicate either sexually or asexually. The end result of either is the oocyst form, which is released into the lumen of the intestine. For simplicity, only two stages of the complex life cycle are shown here. Replication damages the epithelial cells, resulting in diarrhea.

not reliably reveal the parasite. A serological test is now used routinely. No established therapy is available for *C. parvum*. Oral and intravenous rehydration therapy can be used to help the patient to survive long enough for the infection to be cleared, although the parasite is not readily cleared by immunocompromised people.

The Listeria of the protozoal world: Toxoplasma gondii

Toxoplasmosis is an invasive disease caused by *T. gondii*. *T. gondii* resembles *C. parvum*, in that it has a complex life cycle that includes sexual as well as asexual forms (Fig. 17.11). *T. gondii* undergoes its complete life cycle only in the cat. Cats with a *T. gondii* infection, which is almost always asymptomatic, shed cysts of *T. gondii* in their feces. These cysts can be ingested involuntarily by farm animals or by people who clean litter boxes. In humans and in animals other than cats, *T. gondii* does not complete its life cycle to the cyst stage, but ends up as trophozoites in the tissue of the animal. Trophozoites occur in two forms: tachyzoites, which grow quickly, and bradyzoites, which grow slowly. Ingestion of undercooked meat carrying the tachyzoites is the most common route of acquisition of toxoplasmosis by humans. In the average healthy adult, toxoplasmosis is asymptomatic or is experienced as a mild disease resembling flu.

Toxoplasmosis is an extremely common infection in humans. At least 40% of most human populations (and as high as 90% of some populations) have antibodies to *T. gondii*, indicating previous infection. Groups with the highest incidence of antibodies to *T. gondii* are those with the greatest preference for raw or undercooked meats. Toxoplasmosis is a serious disease for only two groups of people: immunocompromised individuals and women who are pregnant and have not been previously exposed to the organism.

Figure 17.11 Progression of a *Toxoplasma gondii* infection and consequences for the human host. A person acquires the infective form (tachyzoite) from undercooked meat or cysts from cat feces. Tachyzoites are taken up in membrane vesicles by host cells, where they replicate and then lyse the cell. Newly formed tachyzoites infect other host cells. The degree of damage to the host depends on the host's immune status. In healthy people, symptoms are rare, but in immunocompromised patients serious systemic infections can occur. In women not previously exposed, *T. gondii* can cross the placenta and infect the fetus, producing birth defects.

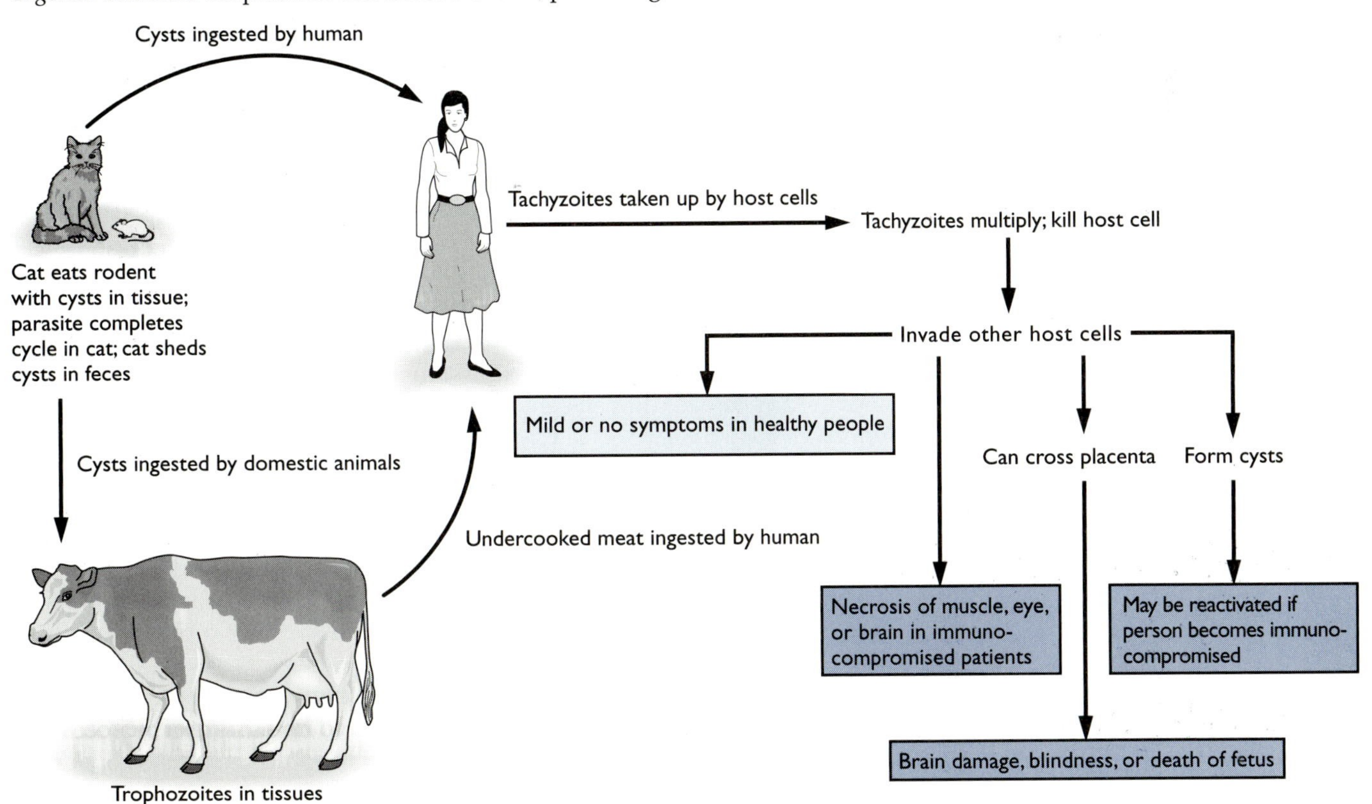

Progression in humans. In humans, ingested *T. gondii* tachyzoites first penetrate the gastrointestinal mucosa and enter the circulation. If cysts are ingested instead of tachyzoites, the cysts first become tachyzoites, then invade. *T. gondii* can invade many types of cells, including monocytes, which then carry the parasite throughout the body. Initial invasion occurs by a process that does not disrupt the host cell membrane but ends with the parasite encased in a membrane inside the host cell. In this membrane vesicle, the tachyzoite multiplies and eventually kills the host cell, releasing tachyzoites to infect other host cells. If the disease process continues, areas of necrosis develop around sites where the parasite has lodged. Areas of necrosis in muscle, eye, or brain give rise to the muscle pains, blindness, and neurological symptoms seen in the severe form of the disease. In host tissues, tachyzoites can form cysts. These cysts may persist in the tissue for many years, and reactivation of the infection can occur at a later date if the host's immune system becomes compromised, for example, as a result of Hodgkin's disease, leukemia, or other malignancy.

Effects of infection. In normal children (past the infant stage) and adults, the immune system rapidly brings the infection under control, so that the disease is either asymptomatic or very mild. By contrast, people with a defective immune response can develop a much heavier parasite load, and many more parasites end up in the tissues. *T. gondii* infections are also a problem in women who are pregnant, because *T. gondii* is one of the few microbes that can cross the human placenta. A fetus has little protection against *T. gondii* and usually develops an overwhelming infection, which can result in death. Survivors often are blind or brain damaged.

At one time, it was thought that only infections contracted during the first trimester of pregnancy could cause fetal death or damage. This idea arose from the fact that women who contracted toxoplasmosis in the second or third trimesters often gave birth to apparently healthy infants. Such children, however, although seemingly healthy at birth, exhibit abnormalities later in life, including mental retardation and learning difficulties. Thus, the current view is that toxoplasmosis at any point during pregnancy, even late in the third trimester, is a serious matter. Infections of newborns can also be serious, and recovery from disease contracted at this stage is seldom complete.

Treatment options. Treatment of toxoplasmosis in pregnancy is not always successful. Pharmaceutical agents that kill tachyzoites of *T. gondii* are available, including inhibitors of folate synthesis, such as pyrimethamine and sulfa drugs. The problem is that unless the infection is diagnosed very early, before transplacental transmission takes place, serious irreversible damage may already have been done by the time therapy begins. Thus, prevention is the primary focus. Cooking meat thoroughly and avoiding ingestion of raw milk are effective ways of avoiding infection by *T. gondii*. Cats that remain indoors and are fed on dry food are unlikely to become infected. Litter pans should be emptied daily, flushing contents down the toilet, and the pans disinfected with boiling water. Cats that stay outdoors for prolonged periods and hunt wild rodents for food are most likely to be infected. Pregnant women who do not have antibodies indicating a previous exposure to *T. gondii* should avoid contact with such cats.

Only a fraction of pregnant women are actually at risk for fetal infections, because most will already have had a mild or asymptomatic case of toxoplasmosis and will thus mount a protective response in time to prevent the systemic stage of the infection that precedes the in utero infection. However, a serological test for previous toxoplasmosis is a prudent element in the prenatal workup. A woman with prior exposure to toxoplasmosis has very little chance of contracting an in utero infection. If a rise in anti-*T. gondii* titer is detected during pregnancy (indicating that a new infection has occurred), therapy can be attempted.

The relative rarity of in utero toxoplasmosis indicates that the statistical chance of all the various risk factors (no prior exposure, ingestion of contaminated meat, contact with a cat who happens to be infected) occurring in a single case is small. However, rare is a relative term. To the 3000 or so women who have in utero infections each year in the United States, the relative rarity of the infection is not much of a consolation.

Summing Up

Don't let this chapter turn you into a food and water hypochondriac. Given the number of potential food- and water-borne pathogens (of which this chapter has described only a small portion), it is truly amazing that we are not sick every single day of our lives. The food industry and water treatment plants have done a very good job of protecting us from disease, but they cannot provide a safety net entirely without holes. The average food and water consumer has some responsibilities, too. First, know the enemy and act accordingly. If you are concerned about bacterial contamination of chicken or turkey, use basic common sense. Don't cut up salad greens on that cutting board that is smeared with raw chicken juices. Cook meat thoroughly. Unfortunately, vegetarians are not immune to food-borne disease. As mentioned earlier, water contaminated by animal or human bacteria can and does contaminate fruits and vegetables.

Second, don't fall for media hype and hysteria, such as that exemplified by the cover of *Time* magazine (Aug. 3, 1998) that depicted a stream of bacteria emerging from a faucet. This type of reporting is counterproductive and inaccurate. Sooner or later, every food-processing or water-processing company will have cases of food-borne or water-borne disease. The food industry and the water-processing industry are trying hard to meet the challenge, but it is going to take time. Fortunately, the federal government finally is taking a new interest in food-borne and water-borne pathogens. The CDC has always been on top of the issue, but now the United States Food and Drug Administration (FDA) and Department of Agriculture (USDA) have started to get into the picture.

Hazard Analysis and Critical Control Point Program. A new program now widely accepted by the food industry, the **hazard analysis and critical control point (HACCP) program,** has great promise for improved food safety. Previously, only the end product of a food-processing stream was assessed for safety. As you learned in Chapter 4, such tests could take days or even weeks. Meanwhile, trucks carrying the contaminated meat or vegetables could be motoring along to supermarkets. End product monitoring has made necessary the large recall operations that frightened consumers and eroded the public's confidence in the food supply. The HACCP program does not rely on end-product analysis but instead identifies steps in the food-processing stream where contamination is most likely to occur and monitors product safety at those points. Not only does this approach raise red flags indicating potential contamination events early and warn companies of potential weak points in their production chains but also ensures that what goes out in the trucks has been precertified for safety. Any company that has experienced a big recall of its product, not to mention the potential lawsuits, understands why this protection system, especially when backed up by the gravitas of the FDA, is a good thing for the food industry.

Antibiotic resistance. Because antimicrobial compounds are not recommended for most food-borne infections, it may come as a surprise to find that antibiotic-resistant bacteria have become an important issue in food safety. In earlier times, understandably, the focus was on food-borne and water-borne pathogens that caused intestinal disease. Recently, the FDA and the European Union have begun to realize that there is another level of food-borne disease: antibiotic resistant bacteria in the food supply (Box 17.2).

box 17–2 *Antibiotic Resistant Bacteria Become a Food Safety Issue*

Mention bacteria and food safety and most people think of food-borne pathogens such as *Salmonella* and *Campylobacter*. These bacteria certainly continue to be food safety problems. In particular, a strain of *Salmonella typhimurium*, strain DT104, has caused deaths. DT104 is resistant to five antibiotics. The lethal form of *Salmonella* infection is the systemic form, in which the bacteria enter the bloodstream. Such infections must be treated with antibiotics to prevent death from septic shock.

A different type of problem was suggested by scientists interested in possible consequences of feeding antibiotics to animals as growth promoters. The suggestion was that antibiotic-resistant intestinal bacteria, such as *Enterococcus* species, which do not cause immediate intestinal disease, could move through the food supply and into the human intestine. *Enterococcus* is a serious cause of postsurgical infections. Such infections usually involve a bacterium from the human microbiota, either that of the patient or that of a health care worker. A person colonized with a strain of *Enterococcus* that is resistant to many antibiotics would have a higher risk of later developing a difficult-to-treat post-surgical infection than someone colonized with antibiotic-susceptible strains.

Unfortunately, Europeans have launched—unintentionally—a large, multinational experiment that may show whether such concerns are real. European citizens and politicians, diverted by concerns over the hypothetical health hazards of genetically-engineered food plants, failed to notice a significant action taken by the European Union (EU). With little debate or public scrutiny, an antibiotic called avoparcin was approved for use as a growth promoter. Avoparcin has never been approved for agricultural use in the United States for a variety of reasons, one of the best of which is that it cross-selects for resistance to vancomycin. Vancomycin-resistant enterococci (VRE) are almost impossible to treat successfully, as U.S. hospitals have learned from VRE outbreaks. European hospitals have experienced few problems with VRE because of restricted use of vancomycin.

Avoparcin had been in use in Europe for only 2 years when scientists began to isolate VRE from foods in grocery stores and then from the intestines of adults in urban areas. In 1998, the EU abruptly banned the use of avoparcin along with several other antibiotics used in agriculture that cross-selected for resistance to front-line human-use antibiotics. The incidence of VRE in foods and human intestinal contents has dropped, but VRE is still around. It remains to be seen whether any colonized people develop postsurgical VRE infections. All concerned agree, however, that this experiment should not have been performed in the first place.

chapter 17 at a glance

Common Gastrointestinal Infections

[For information about food-borne diseases caused by toxins, not infection, see Chapter 15.]

Ulcers—lesions in stomach or duodenum

Diarrhea—loss of fluid or reduced uptake by mucosal cells

Dysentery—bloody diarrhea, indicating pronounced inflammation

Gastroenteritis—inflammation of the lining of the stomach or small intestine

Comparison of Gastrointestinal Pathogens

Species	Reservoir	Characteristics	Disease	Susceptible populations	Virulence factors
Helicobacter pylori	Humans	G– curved rod	Gastric and duodenal ulcers Gastric cancer	All people (?)	Urease Vac A toxin Elicits inflammatory response
Escherichia coli	Humans Other animals (?)	G– rod Facultative	Diarrhea [Enterotoxigenic *E. coli* (ETEC), Enteropathogenic *E. coli* (EPEC)] Dysentery Kidney damage [Entero-hemorrhagic *E. coli* (EHEC)]	All people	Heat-labile toxin (LT) (cholera-like toxin; ETEC) Heat-stable toxin (ST) (peptide, causes increase in host cell cGMP; ETEC) Alters host cell cytoskeleton (malabsorptive diarrhea; EPEC) Shiga-like toxin; EHEC
Salmonella	Many animals Humans	G– rod Motile Facultative Closely related to *E. coli*	Gastroenteritis (systemic disease in some people)	All people	Invasion of Peyer's patches Ability to survive in phagocytes Elicits inflammatory response
Campylobacter	Humans Many animals	G– curved rod	Diarrhea Severe abdominal pain Guillian-Barré (G-B) syndrome	All people	Invasion of mucosa by unknown mechanism
Listeria monocytogenes	Soil Humans Other animals	G+ rod Highly motile Polymerizes actin for intracellular movement Grows at refriger-ator temperatures	Listeriosis	Fetus Neonates Immuno-compromised people	Invasion of Peyer's patches Invasion of mucosal cells Crosses placenta

(continued)

chapter 17 at a glance (continued)

Comparison of the main types of microorganisms *(continued)*

Species	Reservoir	Characteristics	Disease	Susceptible populations	Virulence factors
Norwalk virus	Humans Water Shellfish	Small Nonenveloped virus ssRNA genome	Gastroenteritis (single-source outbreaks)	All people	Unknown
Rotavirus A, B, C	Humans Water	Nonenveloped virus Segmented dsRNA genome Wheel-like appearance in cross-section	Gastroenteritis	Primarily infants Young children Elderly	Destroy cells at villus tip Cause shortened villi
Hepatitis A virus	Humans Water Shellfish	Nonenveloped virus ssRNA genome	Infectious hepatitis	All people	Invades liver Kills liver cells
Giardia intestinalis	Humans Many animals	Protozoan Trophozoite and cyst stages	Giardiasis (steatorrhea)	All people	Sucking disk damages intestinal cells
Cryptosporidium spp.	Humans Domestic animals	Protozoan Complex life cycle	Cholera-like diarrhea	All people, especially immunocompromised patients	Growth on mucosa may damage mucosal cells
Toxoplasma gondii	Cats Many animals	Protozoan Complex life cycle	Toxoplasmosis	Fetus Immunocompromised people	Ability to cross placenta

key terms

acid tolerance response after exposure to slightly acidic conditions, some bacteria become able to withstand much lower pH environments for prolonged periods; permits them to transit the stomach without being killed

AING acute infectious nonbacterial gastroenteritis, a self-limited condition characterized by fever, vomiting, malaise, and diarrhea; caused by viruses

dysentery type of diarrhea in which stools contain blood and mucus; associated with *E. coli* O157:H7 and related EHEC strains

gastroenteritis inflammation of the lining of the stomach or small intestine, especially caused by bacterial infection

giardiasis diarrheal disease caused by *Giardia intestinalis;* ranges from mild to severe

M cells cells of the gut-associated lymphoid tissue located in the Peyer's patches of the intestine; used by some intestinal pathogens such as *Salmonella* and *Listeria* as an entryway into submucosal regions

malabsorptive diarrhea damage to mucosal cells that impairs water uptake

secretory diarrhea water and ion loss but no intestinal cell damage; caused by ETEC

steatorrhea malabsorption of fats leading to an odorous, fatty diarrhea

toxoplasmosis invasive disease caused by *Toxoplasma gondii;* may cross placenta and infect fetus

ulcers inflamed regions of the stomach or duodenal lining that can cause serious pain and bleeding; caused by *H. pylori*

Questions for Review and Reflection

1. As scientists worked to find a cure for ulcers, they first tested the effects of antibiotics on bacteria grown in the laboratory. How did they do this? What results made them think the antibiotics would work? [Hint: review the last sections of Chapter 15.]

2. There is controversy about the infectious dose$_{50}$ (ID$_{50}$) for *E. coli, Salmonella,* and *Listeria* food-borne disease. Generally, however, from what you know about these diseases, you can make a rough estimate of the ID$_{50}$ and lethal dose$_{50}$ (LD$_{50}$), the attack rate, and the case fatality rate of these diseases. What would be your estimate and why?

3. If new surveillance data suggest that *Listeria* infections are much more common than data from large outbreaks suggest, how would this affect your answer to the previous question?

4. The Centers for Disease Control and Prevention has launched a program called Foodnet, which is attempting to determine the incidence of such key food-borne diseases as salmonellosis in the United States. Why is this a more difficult undertaking than determining the incidence of bacterial pneumonia or influenza?

5. If you were hired to set up a hazard analysis and critical control point (HACCP) program for a hamburger production corporation, how would you proceed? What information would you need to decide on the critical checkpoints? Keep in mind that checking at every point is not economically feasible and that microbiological tests can take several days to yield results.

6. If hepatitis A and other hepatitis viruses are so common and no epidemics of liver failure are evident, why is there any need to administer a vaccine?

7. In this and earlier chapters, a major theme has been prevention. Suppose you set out on a campaign to protect people from food-borne and water-borne diseases. Because you are only one scientist, you will need to rank order your disease targets and decide at what points to intervene (for example, public education, food industry monitoring). How would your program look?

8. *Giardia* is cleared much less rapidly from the intestine than *Salmonella, Campylobacter,* and the diarrhea-causing viruses. Generate a hypothesis about why this is the case.

9. In an outbreak of food-borne disease, many people who consumed the contaminated product do not develop symptoms. On the other end of the spectrum, some people die. Apply the principles learned in previous chapters to explain this phenomenon.

10. Food-borne or water-borne diseases are so called because food or water is the usual mode of spread. Is it possible that the same pathogens that cause these diseases might be spread by other routes, and, if so, what are they?

18 Urogenital Tract Infections

The pathogens described in Chapters 16 and 17 were mostly equal opportunity infectors of humans, with the exception of age biases for the very young and the very old. The pathogens described in this chapter are almost all sexist in the sense that their effects on women are more serious than their effects on men. This bias arises from differences in the anatomy of the urogenital tracts of men and women and from the fact that women give birth, thus placing the baby at risk for infection.

The urinary and genital tracts provide challenges to infecting microbes. These challenges differ from those posed by the gastrointestinal tract or the lung. Thus, it should come as no surprise that most of the microbes that cause urogenital tract infections are different from those that cause infections in these other sites. The one exception among the microbes described in this chapter is *Escherichia coli,* a major cause of both intestinal tract infections and **urinary tract infections** (**UTIs**). However, the strains that cause UTIs are different from those that cause intestinal tract infections. It is interesting to note that the bladder resembles the small intestine in the sense that fast flow of contents is an important defense. Thus, the type of microbes able to colonize the small intestine might also be able to infect the bladder.

Urinary Tract Infections

The purpose of the **urinary tract** is to filter blood and remove metabolic wastes that could otherwise have toxic effects. The kidneys carry out this filtration process. Because potentially toxic substances must be eliminated without undue loss of water or essential minerals, this process is quite complicated. The result is a highly concentrated mixture of nitrogenous wastes and salts called **urine.** Urine is then conveyed through tubes called **ureters** to the **bladder,** where it is stored before disposal by urination (Fig. 18.1). To perform its function, the urinary tract must open occasionally to expel urine from the body, thus providing an opportunity for enterprising microbes to gain access to the **urethra** and bladder. Urine is a very rich source of nutrients for microbes that can accommodate a slightly acidic environment (pH 5–6).

Anatomy and defenses of the urinary tract

The male-female divide. To avoid microbial (especially bacterial) invasion of the urinary tract, the human body has evolved several defense mechanisms. The opening of the urinary tract has a sphincter action, which prevents the entry of microorganisms into the urethra. An important difference between the opening of the male and female urinary tracts is that the

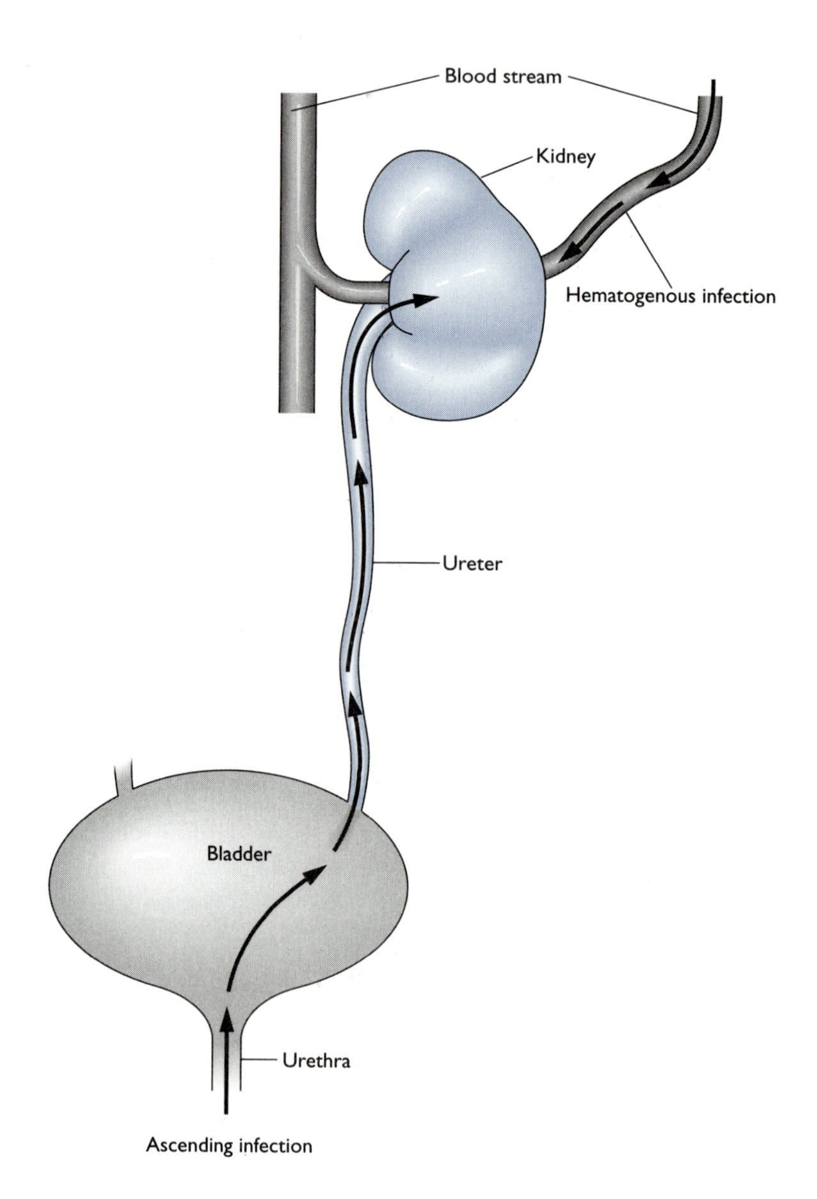

Figure 18.1 Anatomy of the urinary tract. Most infections of the urinary tract are ascending. The bacteria move from the urethra to the bladder, where they proliferate. In some cases they move up the ureters to the kidney. Less commonly, the kidney may be infected from the bloodstream (hematogenous infection).

opening of the male urinary tract is in the relatively dry area at the end of the penis, far removed from the anus. By contrast, the opening of the female urinary tract is located near the opening of the vaginal tract, a warm moist area easily colonized by bacteria. The female urethra is also near the anus, which serves as a rich reservoir of microbial pathogens. These anatomical differences between men and women are probably the main reason why the incidence of bladder infections in women is much higher than that in men. Not all females are equally likely to develop UTIs, however. Young girls, pregnant women, and women in the 20–40-year-old age group seem to be unusually susceptible to UTIs. In fact, 20% of women aged 20–40 have at least one UTI per year.

The reason for this age distribution is unknown. One guess is that young girls who have slight anomalies in their urethras that allow stagnant puddles of urine to form may be more prone to infection. Later in life this condition might not make so much difference. The debate continues over whether sexual activity predisposes women in the 20–40-year-old age group to UTIs, because of increased risk of introducing bacteria into the urethra. This may contribute, but so do other factors. Women who use diaphragms for contraception have a higher incidence of UTIs, apparently because nonoxynol-9, the active ingredient in the spermicides normally used with diaphragms, inhibits the growth of lactobacilli. **Lactobacilli** are an important protection of the vaginal tract. They produce **lactic acid,** which makes the vagina acidic, and **peroxide,** which kills or inhibits the growth of many microbes. Probably the main causative factor of UTIs is colonization of the vaginal tract by a microbe that is capable of causing such infections.

The urethral **sphincter** is very effective in preventing microbes from entering the urethra. Also, when the urethra opens during urination, the flow of urine is outward, so that the washing action of urine helps to clear any microbial invaders that might have gotten close to or past the sphincter. Because the washing action of urine is such an effective antimicrobial defense, microbes that manage to reach the bladder must penetrate the mucus layer that covers the mucosal membrane lining the bladder and adhere to mucosal cells. Urine has high concentrations of a glycoprotein, **Tamm-Horsfall glycoprotein,** which contains carbohydrate residues similar to those found on bladder mucosal cells. Because most microbial adhesins bind to sugar residues, Tamm-Horsfall glycoprotein, like mucin, may bind microbes and prevent them from attaching to the bladder mucosa. These combined defenses usually keep the bladder sterile between transient episodes of low level contamination.

Consequences of UTIs: from bladder to worse. Infections of the urethra (**urethritis**) or bladder (**cystitis**) may be unpleasant but seldom cause irreversible damage to tissue in these areas. The most serious consequence of a bladder infection is that the infection may ascend to the kidneys, causing a kidney infection (**pyelonephritis**). This type of spread is called an **ascending infection.** The kidneys are normally sterile. Because of their concentrating function, the fluid in the kidney has high osmolarity and a high concentration of urea. Activities of phagocytes and immune cells are inhibited in this high salt environment, making the kidney one of the areas of the body that is underprotected by the immune system. The body takes special precautions to prevent microorganisms from reaching the kidney, and the defenses of the lower urinary tract are normally successful in achieving this aim. The kidney filters blood. If inflammation allows bacteria to cross the endothelial wall, microorganisms growing in the kidney can gain access to the bloodstream. Moreover, inflammation of the kidney can interfere with liquid excretion and give rise to edema and hypertension. Infections can flow in the other direction. In people with a bloodstream infection, bacteria from the bloodstream can infect the kidney and ultimately the bladder. This is called a **hematogenous infection.**

Microbes that cause urinary tract infections

Identity of causative microbes. The most common causes of UTIs are bacteria. Infections acquired by healthy people who are not hospitalized are called community-acquired infections. Community-acquired UTIs are almost always ascending infections. More than 80% of community-acquired infections in people with no abnormalities of the urinary tract are caused by strains of *E. coli* called **uropathogenic *E. coli*** (**UPEC**). For many years, gram-negative bacteria, such as *E. coli* and its relatives, caused virtually all

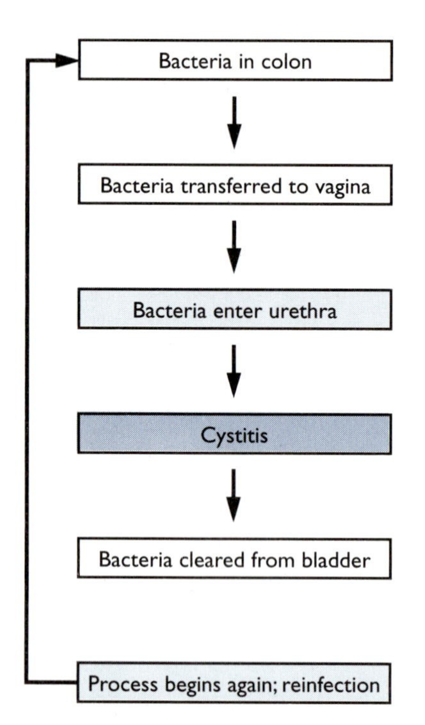

Figure 18.2 Source of microbes that cause community-acquired urinary tract infections. Most cases occur in girls or women. Members of the colonic microbiota (for example, certain strains of *Escherichia coli*) move the short distance from the anus to the vagina, where they become established. Next they proceed from the vaginal area to the urethra and then to the urinary bladder. A bladder infection is called cystitis. The bacteria may be cleared from the bladder by antibiotic therapy, but, if they remain in the colon or vagina, may reinfect the urethra and bladder.

cases of community-acquired UTIs. Today, a gram-positive species, ***Staphylococcus saprophyticus,*** is the second leading cause of UTIs in sexually active young women. With the exception of *S. saprophyticus,* which is found widely in the environment, the bacteria that cause community-acquired UTIs are all normally harbored in the colon (Fig. 18.2). The female vaginal tract is near the anus, and colonization of the vagina, especially near the vaginal opening, is probably an important step in the infection process. This initial colonization allows the bacteria to take advantage of any breaches in the defenses of the urinary tract.

The spectrum of **hospital-acquired** UTIs is different from that of **community-acquired** infections. *E. coli* and related gram-negative intestinal bacteria are still major causes of UTIs, but soil bacteria, such as *Pseudomonas aeruginosa,* which are almost never seen as causes of community-acquired infections, are prominent causes of hospital-acquired infections. Gram-positive bacteria, especially *Enterococcus* species, and the yeast *Candida albicans* also cause a significant number of hospital-acquired UTIs. (The reasons that the spectrum of hospital-acquired infections might be different from that of community-acquired infections will be explained in Chapter 19.)

Diagnosis of UTIs. Most physicians treat patients with UTIs on the basis of probability: if the infection is community-acquired, it is almost certainly *E. coli.* But in women with recurrent infections and in whom the initial antibiotic treatment fails, identification of the causative bacterium and determination of its antibiotic susceptibility profile become important. Because the urethra empties in the vulvar region, urine passes through a bacteria-rich location as it is collected for testing. A woman asked to provide a urine sample is thus instructed to clean the opening of the urethra with an alcohol pad and then let some urine flow before taking the sample. This **clean catch** procedure helps to reduce contamination of the urine specimen by vaginal bacteria.

The clinical laboratory first tests the urine specimen with a dipstick that has two parts. One detects **leukocyte esterase,** an enzyme found in neutrophils. The other detects **nitrate reductase,** an enzyme commonly found in *E. coli* and its relatives. This test, which takes only minutes, tells the clinical microbiologist whether the specimen should be processed further. If both tests are positive, the clinical microbiologist can be relatively confident that inflammation is being caused by the types of bacteria that are typically found in urine from infected patients. The next steps are to streak the urine onto agar plates to obtain a pure culture of the bacteria and then to determine the antibiotic susceptibility of the isolate (see Chapter 15).

How urinary pathogens cause disease

The virulence factors of uropathogenic strains of *E. coli* have been studied in detail. As we have seen in the case of the small intestine, where the flow of fluids was also an important host defense, bacterial attachment to bladder mucosal cells is critical for colonization of the urinary tract. Uropathogenic *E. coli* that cause cystitis produce so-called **type 1 pili,** which have a protein called **FimH** on the tips. FimH recognizes **mannose receptors** on the surface of a variety of epithelial cells. These include intestinal, vaginal, urethral, and bladder mucosal cells. Thus, the type I pili allow the bacteria to colonize all of the pertinent tissues necessary to cause a UTI. A subset of *E. coli* strains that cause kidney infections produces an additional type of pilus, called **P** or **Pap** (for **pyelonephritis-associated pili**). These pili recognize an α-D-Gal

(1,4)-β-D-Gal (globobiose; Fig. 18.3) moiety of a glycolipid found on the surface of bladder and kidney cells. Pyelonephritis is a more serious disease than cystitis or urethritis and can be life threatening. Therefore, *E. coli* strains that produce Pap pili are associated with more serious illness.

Bacteria closely associated with the bladder mucosa shed lipopolysaccharides (LPS) when they lyse, which elicits an inflammatory response and attracts numerous neutrophils to the area. The inflammatory response is probably responsible for most of the symptoms of urethritis and cystitis. These include fever, malaise, and a burning sensation during urination. Uropathogenic *E. coli* also produce a capsule that is thought to prevent phagocytosis of the bacteria by neutrophils, thus prolonging the inflammatory response as neutrophils try unsuccessfully to eliminate the bacteria.

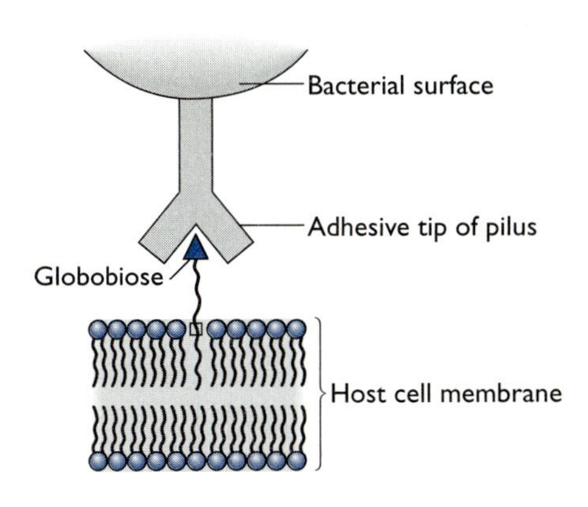

Figure 18.3 Binding of P pili to globobiose. The P pili recognize a disaccharide, globobiose, which is anchored by a lipid tail to the phopholipid membrane of the host cell.

Diseases of the Genital Tract

Anatomy and defenses of the male and female genital tracts

Most male genital infections occur in the urethra or on the surface of the penis. Infections of other parts of the male genital tract do occur, but they are uncommon except in elderly males. An example of an ascending infection is **epididymitis,** an infection of the epididymis, part of the sperm-collecting apparatus. The long male urethra and the flushing action of urine help to prevent microbes from reaching the upper parts of the male genital tract (Fig. 18.4).

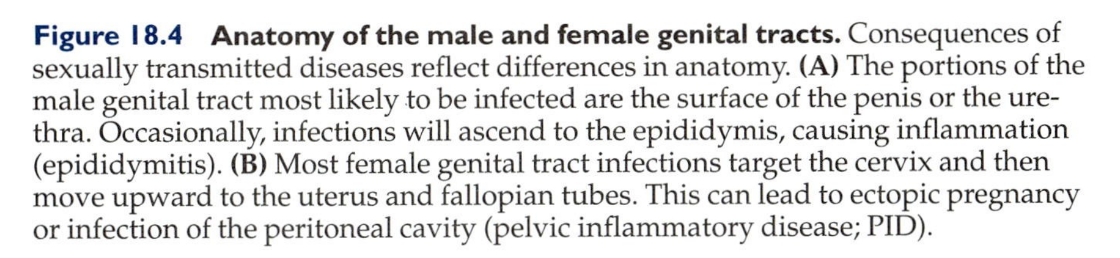

Figure 18.4 Anatomy of the male and female genital tracts. Consequences of sexually transmitted diseases reflect differences in anatomy. **(A)** The portions of the male genital tract most likely to be infected are the surface of the penis or the urethra. Occasionally, infections will ascend to the epididymis, causing inflammation (epididymitis). **(B)** Most female genital tract infections target the cervix and then move upward to the uterus and fallopian tubes. This can lead to ectopic pregnancy or infection of the peritoneal cavity (pelvic inflammatory disease; PID).

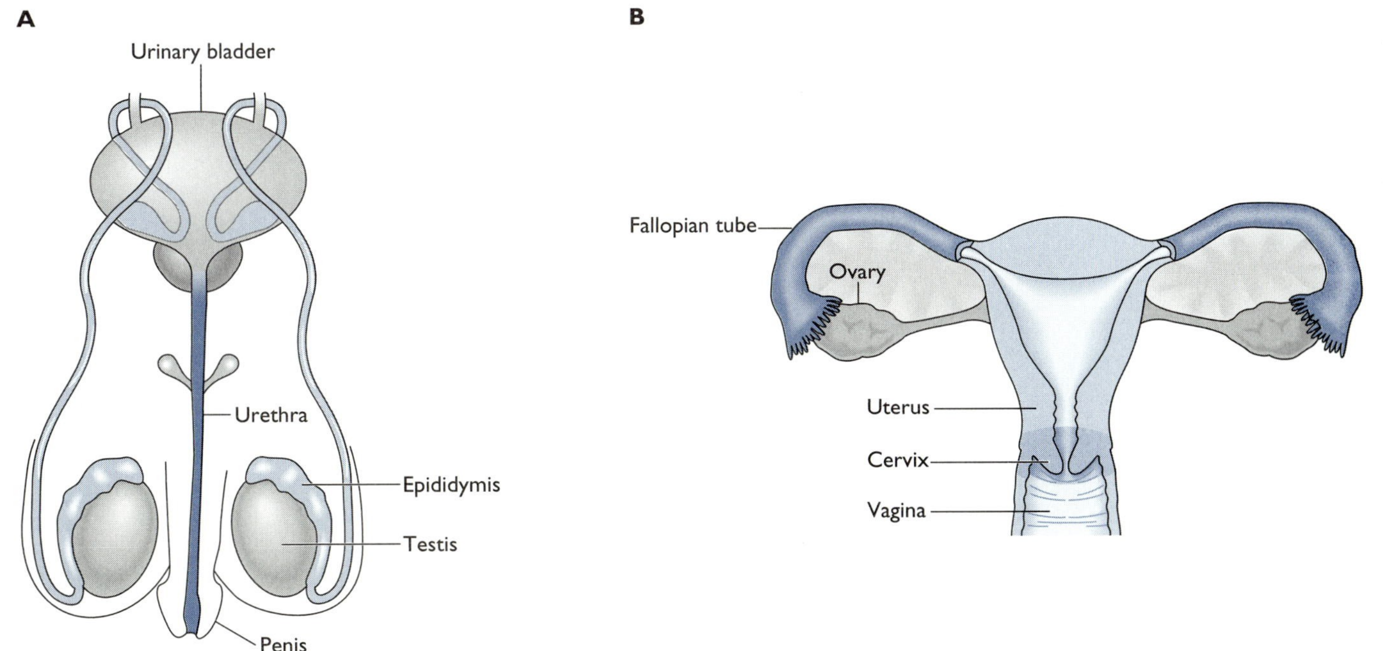

Defenses of the female upper genital tract. The female upper genital tract also is well protected from microbes unless the cervix becomes infected. If cervical infection occurs, then the female upper genital tract very commonly becomes infected as well. To see why this might be the case, consider the anatomy of the female genital tract, illustrated in Figure 18.4. The fallopian tubes are joined to the uterus, which connects to the vagina by a short opening in the cervix. The vagina is open to the exterior environment. In order for sperm to have access to eggs produced by the ovaries, there must be an opening in the cervix. Such an opening, however, provides a potential conduit by which bacteria infecting the cervix can enter the female upper genital tract. The cervical opening is normally filled with a mucus plug that is effective in preventing the entry of microorganisms into the uterus. For the uterus to fulfill its function, however, this plug cannot always be impermeable. In the middle of the menstrual cycle, the cervical mucus changes in viscosity sufficiently to allow sperm to enter the uterus. Bacteria enter the uterus on sperm. Also, the cervical plug is compromised during menstruation, but, in this case, the flow outward of menstrual blood (which also contains substances that inhibit the growth of microorganisms) helps to keep vaginal contents from ascending into the cervix.

Despite the apparent vulnerability of the female upper genital tract, infection of this area is not a frequent occurrence. Undoubtedly bacteria from the vaginal tract are introduced fairly frequently into the uterus, but the mucus lining the uterine mucosa and the inflammatory response clear out most of these invaders before they can become established.

Breaches in these defenses. A relatively small number of microbes can overcome these defenses. Once such a microbe has infected the cervix, the chance of ascending infection becomes quite high. Thus, the key to protecting the upper genital tract is either not to acquire such microbes in the first place or to treat cervical infections as quickly as possible. Unfortunately, because the cervix has few nerve endings and because women normally have some vaginal discharge, infections of the cervix often go undetected. The infected woman does not seek medical attention, because she does not know she is infected.

The vaginal tract normally harbors a dense population of lactobacilli, bacteria that ferment glycogen from vaginal mucosal cells to produce lactic acid. They are responsible for the acidic pH of the vagina (pH 5). Also, as noted previously, some lactobacilli also produce hydrogen peroxide, a bactericidal substance that may be even more effective than the acidic pH of the vagina as a protective barrier against colonization by pathogens. The vaginal mucosa is made of **squamous cells,** layers of flattened cells that are different from the tightly packed columnar cells lining the airway, intestine, and bladder. Squamous cells are believed to be suited to the vagina, because they are able to better resist the wearing effect of intercourse-associated friction than are the columnar-cell membranes. The cervical mucosa, however, is composed of **columnar epithelial cells.** Many pathogens seem to prefer columnar epithelial cells, a fact that may explain why most genital tract infections in women target the cervix instead of the vaginal wall.

Consequences of genital tract infections

Increased risk of HIV infection. One consequence of genital infections, such as the viral disease herpes or the bacterial diseases gonorrhea and chlamydial disease, is the increased risk of acquiring a **human**

immunodeficiency virus (**HIV**) infection. On average, a person must be exposed to HIV about eight times for infection to occur. Herpes viruses that produce lesions in the skin of the penis or the interior surface of the vagina can worsen these odds dramatically, by making it easier for HIV to enter the body. Bacterial infections of the cervix that evoke a potent inflammatory response not only cause lesions in the cervical tissue but also bring to the site cells of the immune system that are the targets of HIV attack (monocytes and T helper cells). A statistical analysis of heterosexual transmission of HIV estimated that the incidence of HIV infections in the heterosexual population could be decreased by as much as 90% if herpes and bacterial infections could be prevented or eliminated from the female genital tract.

Pelvic inflammatory disease and infertility. Microbes, especially bacteria, that infect the cervix can take advantage of breaches in the protective barrier that guards the upper genital tract and can cause infection in the uterus or fallopian tubes. Note that there is a gap between the ovaries and the ends of the fallopian tubes (Fig. 18.4). Thus, bacteria that ascend the fallopian tubes can enter the peritoneal cavity and cause infection there. Infection of the fallopian tubes and peritoneal cavity (**pelvic inflammatory disease, PID**) can cause pain and fever and may progress to a serious, life-threatening condition if bacteria enter the bloodstream.

Another consequence of inflammation of the fallopian tubes is the possibility of scarring, so that a fertilized egg can no longer pass through the tubes. Under normal conditions, the egg is fertilized in the fallopian tubes, then moves to the uterus, where it develops (Fig. 18.5). If movement through the fallopian tubes is blocked, the egg can begin to divide in the fallopian tube or in the peritoneal cavity, a condition called **ectopic pregnancy.**

Figure 18.5 How infection can cause ectopic pregnancy. As shown on the left side of the figure, eggs normally are fertilized in the fallopian tube **(A)**, then move down the tube **(B)** and become implanted in the uterus **(C)**. However, as shown on the right side of the figure, infection can cause blockage of the fallopian tube so that the egg develops in the tube **(D)** or in the peritoneal cavity **(E)**. This is called ectopic pregnancy, a condition that can cause shock and death.

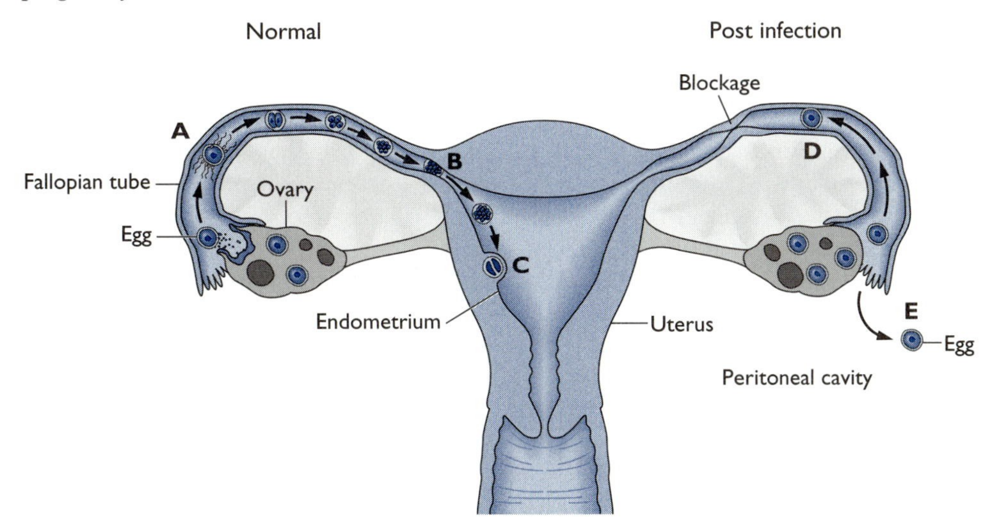

In an ectopic pregnancy, the fetus is aborted naturally, and, if quick action is not taken, extensive bleeding can cause a woman to go into shock and die. The ectopic pregnancy rate has been rising in recent years, an indication of how common undiagnosed and untreated genital infections have become. Even if ectopic pregnancy does not occur, a woman with fallopian tube scarring may become infertile. According to *The Hidden Epidemic,* a 1998 report from the National Academy of Sciences' Institute of Medicine, widespread undiagnosed and untreated genital tract infections in young women today could well turn into an epidemic of infertility in the future.

Cancer. A few viruses are capable of causing cancer, probably because they integrate their genomes into human DNA, a process that can cause a human cell to lose its growth rate control. Two such viruses are sexually transmitted: **human papilloma virus** (**HPV**) and **hepatitis B virus** (**HBV**). HPV causes cervical cancer, an especially dangerous type of cancer. This is what the PAP test is designed to detect. HBV causes liver cancer and usually is spread by sexual transmission or shared needles.

Fetal and neonatal infections. Infants born to women with genital tract infections are at risk. The fetus is sterile, microbiologically speaking, and the baby first encounters the microbial world when passing through the vaginal tract. The eyes and lungs of a newborn are especially vulnerable to infection. In the case of herpes-infected women, the risk of transmission to the infant is quite low. However, such transmissions, when they do occur, are potentially serious. The risk of mother-to-infant transmission of HIV, either during or after birth, is much higher. Recent studies have shown that administration of the anti-HIV drug **azido-deoxythymidine** (**AZT**) to the mother through the period before and after birth substantially reduces the chance of transmission. Bacteria can also cause fetal or neonatal infections. The bacteria responsible for gonorrhea can cause eye infections in newborns, and the bacterium that causes chlamydial disease also can cause eye infections and can infect the lungs of newborns.

Genital herpes virus infections

Perceptions of genital herpes

In the 1970s and 1980s, the public, unaware of the HIV epidemic in the making, was shocked by its first modern encounter with an incurable, sexually transmitted disease: genital herpes. For many who contracted herpes, the dim prospect of having herpes for life was compounded by the public stigma that immediately attached itself to the disease. Many people found the emotional consequences of herpes to be worse than the physical effects of the disease itself. Infected people formed support groups and sought psychiatric counseling to cope with mental distress. Today, people are not as concerned about genital herpes. Even before the antiviral compounds acyclovir and famcyclovir became available to treat the lesions caused by herpes viruses, it had become clear that in the vast majority of infected people, recurrences waned in seriousness over time and, in some cases, ceased entirely. Many couples in which one member is infected with herpes have found that infection of the other partner is by no means inevitable and can be prevented with a little forethought. Moreover, the advent of HIV swept the herpes hysteria aside. In the face of a far greater threat, a more commonsense attitude prevailed. The herpes experience shows, however, as did the initial reaction to acquired immunodeficiency syndrome (AIDS), that the medical reality of a disease can be so distorted by social reactions that the actual suffering of affected individuals is compounded unnecessarily.

Spread and development of symptoms

Herpes simplex virus type 2 (HSV-2), the most common cause of **genital herpes,** is spread primarily by sexual contact. **Herpes simplex virus type 1 (HSV-1),** the main cause of oral lesions (**cold sores**), is spread by non-intimate contact and is normally contracted early in life. These distinctions are not absolute. HSV-1 can also cause genital lesions, and HSV-2 can cause oral lesions. HSV-2 causes about 80% of genital herpes cases. The initial episode of genital herpes occurs about 12 days after exposure. Painful vesicles that progress to pustules and crusting appear on the glans penis, the vulva and cervix, or the anus, depending on the site of inoculation. Local lymph nodes may be tender and enlarged. The ulcers heal spontaneously after 1–3 weeks. An infected person is contagious while the lesions are present; infectivity decreases as the lesions disappear but is still a danger. At least one or a few recurrences are common, especially after periods of hormonal change or stress. Recurrences are usually less severe than the initial infection. Eventually, an infected person will become asymptomatic for long periods, but a recurrence is always possible under stressful conditions or if immunosuppression occurs later in life.

The symptoms caused by the virus and its ability to cause recurrences are explained by its strategy for replication and survival in the host (Fig. 18.6). Both HSV-1 and HSV-2 replicate initially in epithelial cells. Viral replication lyses the infected cells as the viruses bud from them. Killing of epithelial cells in the early stages of the infection elicits an inflammatory response that produces the lesions that characterize the disease. An important component of the host response is the cytotoxic T cell response. The cytotoxic T cell response would completely eliminate the viruses from the body if it were not for the fact that the viruses have an escape hatch: they can also enter the sensory neurons that underlie the epithelial cells and move from neuron to neuron to the nearest **ganglion** (trigeminal ganglion in oral infections, sacral ganglion in genital infections). Spread through neurons occurs by cell-to-cell transmission instead of by lytic destruction of infected cells. During this neural phase, herpes viruses are invisible to the immune system. Herpes virus, like a bank robber, goes underground after a destructive phase.

The latent stage

Once the viruses enter ganglial cells, they cease to replicate actively and enter a **latent state.** The mechanism of latency is not understood, nor is there much information about how the viruses are released from the latent state. During the latent state, the viral genome is detectable in the cell and some expression of viral proteins occurs, but active viral replication ceases. HSV does not display its antigens on the surfaces of neurons. Thus, cytotoxic T cells cannot recognize and kill infected neurons.

Viral replication can be reactivated by a variety of stimuli (stress, trauma, sunlight, menstruation, immunosuppression) and leads to a recurrence of symptoms, as the viruses move down the neurons to epithelial cells and begin once again to replicate there. The lesions subside spontaneously, an indication that host defenses are eventually effective in suppressing the epithelial cell stage of the infection. The recurrent nature of the disease, however, demonstrates that elimination of the latent state is much more difficult.

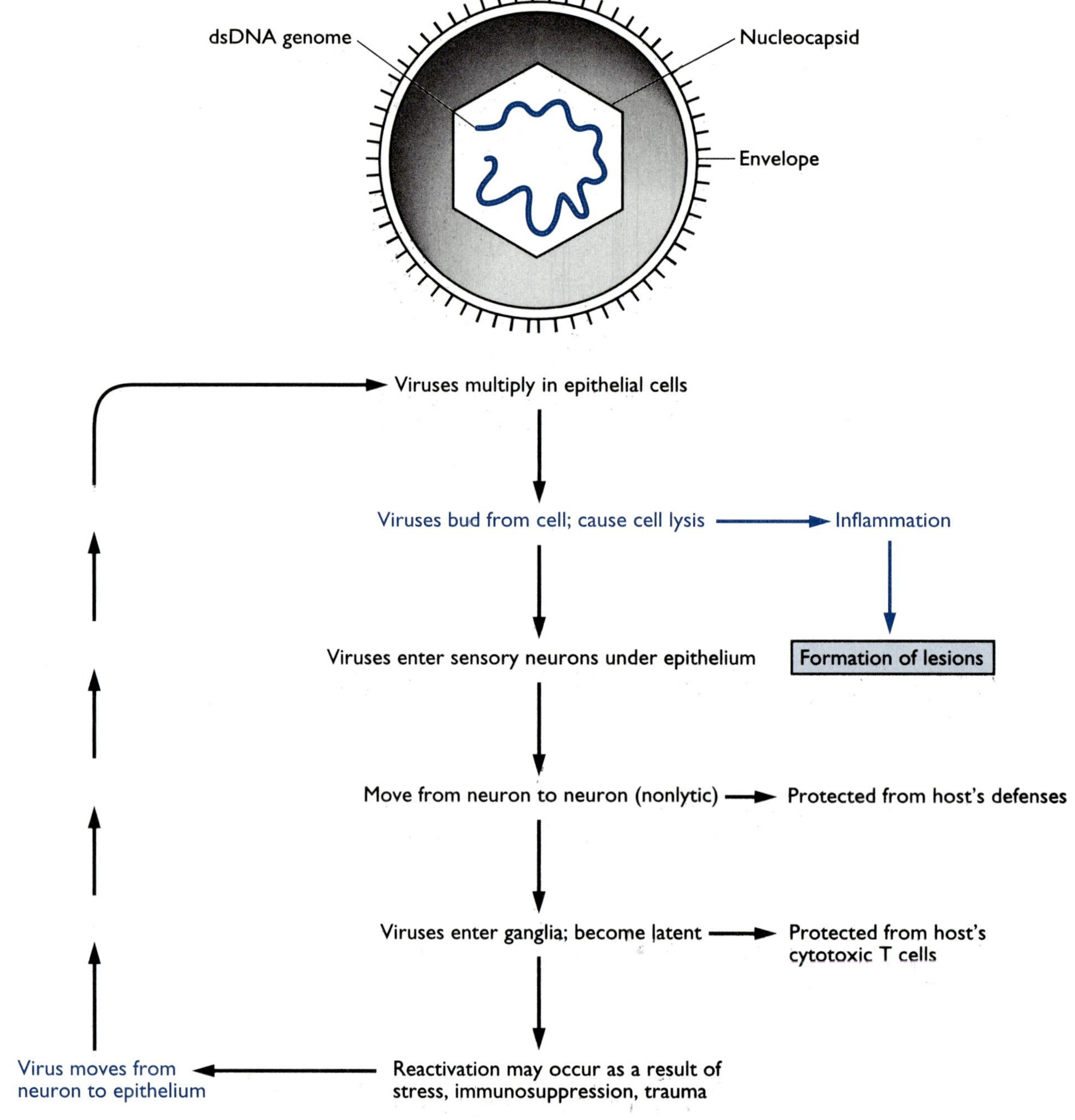

Figure 18.6 Structure of herpes simplex virus (HSV) and progression of infection. The viral envelope fuses with the host cell membrane, and the viral nucleoprotein core is released into the cytoplasm and enters the nucleus. The viral DNA is replicated in the nucleus, and viral proteins are synthesized in the cytoplasm. New viral particles are assembled in the nucleus, bud out of the nucleus, and lyse the host cell. Viruses enter sensory neurons and travel to ganglia, where they remain latent. Under certain conditions, the virus may leave the neuron and travel in the reverse direction to epithelial cells, in which the infection is reactivated.

HSV and its replication strategy

The structure of HSV is shown in Figure 18.6. HSV is an enveloped virus with an **icosahedral capsid** and a linear double-stranded DNA (dsDNA) genome. The virus binds to a host cell receptor, which is thought to be heparin sulfate proteoglycan, a polysaccharide-protein complex that is part of the extracellular matrix. Attachment is followed by fusion of viral and host cell membranes, and the viral nucleoprotein core is released into the host cell cytoplasm. At this point, a protein called **VHS** is released into the cytoplasm and shuts off protein synthesis by cleaving host cell mRNAs that are being translated by ribosomes associated with the endoplasmic reticulum. The **nucleoprotein** core enters the nucleus via pores in the nuclear membrane, and the viral genome is circularized to form a covalently closed circle. Expression of viral genes is regulated in a cascadelike manner, a fact that explains how the viruses coordinate production of new proteins and genomes during active replication.

After replication of the viral genome and production of viral proteins, the viral capsids are assembled, and the virus leaves the cell either by budding and breaking away from the cell (in epithelial cells) or by transfer to an adjacent cell (in neurons). In both cases, from the cytoplasmic membrane of the host proteins the virus acquires an envelope that contains viral envelope proteins.

Anti-HSV therapy

HSV encodes not only its own DNA polymerase but also two enzymes that process nucleotides, **thymidine kinase** (which adds a phosphate group to thymidine) and **ribonucleotide reductase** (which changes the nucleotide sugar from ribose to deoxyribose). These viral enzymes are needed to provide enough nucleotides and DNA polymerization capability for explosive viral replication in epithelial cells. The viral thymidine kinase and DNA polymerase provide useful targets for the antiherpes drug acyclovir. **Acyclovir** is a nucleoside analog. The mechanism of action of acylovir is shown in Figure 18.7. Acyclovir mimics a thymidine molecule. HSV thymidine kinase adds a phosphate to acyclovir to produce the nucleotide. The reason acyclovir does not injure human cells, which have their own thymidine kinase, is that human thymidine kinase recognizes acyclovir poorly or not at all. Human cell enzymes, however, do add two other phosphates to phosphorylated acyclovir. The viral DNA polymerase has a higher affinity than the human DNA polymerase for acyclovir triphosphate. Thus acyclovir triphosphate is incorporated preferentially into viral DNA. The end result is inhibition of viral DNA synthesis. A newer and more effective form of the acyclovir type of antiherpes drug is **famcyclovir.** A different type of antiviral drug, which also targets the viral DNA polymerase, is **phosphonoformate** (**Foscavir**). This drug acts directly to inhibit viral DNA polymerase without the need for activation by the viral thymidine kinase.

Acyclovir acts best against rapidly replicating viruses, but it does not kill all viruses, especially those lying latent in neurons. Thus, although acyclovir can reduce the duration of symptoms and the infective period, it does not effect a cure. In addition, acyclovir-resistant mutants of HSV have appeared. These generally arise in immunocompromised patients receiving prolonged antiviral therapy. Resistance is most often the result of a mutation that reduces production of viral thymidine kinase. This type of mutation is possible, because, despite the enhancement of HSV replication by its own thymidine kinase, viral thymidine kinase is not absolutely required for

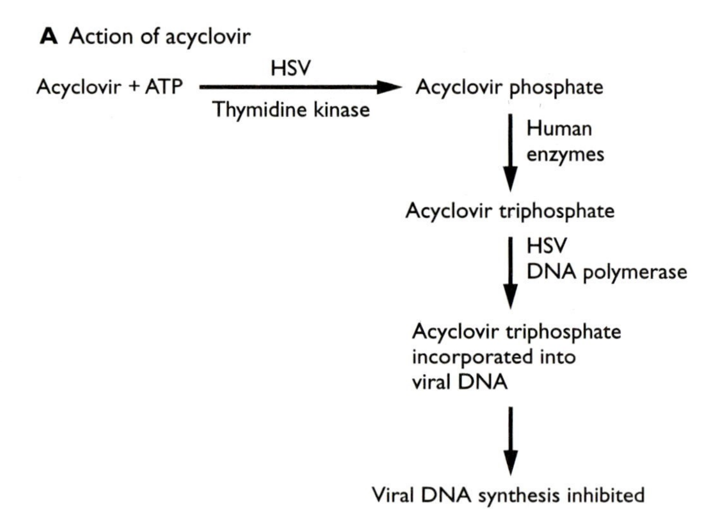

B Structure of acyclovir

$CH_2OCH_2CH_2OH$

Acyclovir

Figure 18.7 Mechanism of action of acyclovir and its structure. (A) Herpes simplex virus (HSV) thymidine kinase adds a phosphate to acyclovir. Human cell enzymes add two more phospates to form acyclovir triphosphate. HSV DNA polymerase incorporates acyclovir triphosphate into the viral DNA. This inhibits synthesis of viral DNA. **(B)** Structure of acyclovir.

viral replication. So far, the thymidine kinase-deficient mutants have proved to be less virulent than the wild type, but it is conceivable that mutations might arise that do not compromise virulence. Some other mutations seen in acyclovir-resistant mutants alter viral DNA polymerase so that it does not incorporate acyclovir triphosphate so readily.

Human immunodeficiency virus

HIV infection and the progression to AIDS

AIDS is caused by HIV. For many years, one of the major AIDS mysteries was why some people infected with HIV experienced long asymptomatic periods, lasting for years before the onset of severe manifestations of disease, whereas others experienced rapid disease progression, often resulting in death within 1 or 2 years. Moreover, some people repeatedly exposed to HIV never developed the disease at all. New insights into human genetic differences have begun to provide an explanation for these differences. To understand why some people are more susceptible to HIV than others, it is first necessary to understand how HIV causes AIDS.

The structure of HIV and an overview of the progression of the disease are shown in Figure 18.8. HIV enters the body through small tears or lesions in the vaginal, rectal, or oral mucosa. The virus quickly begins to replicate in T helper cells and monocytes, the precursors of macrophages. The replication

Structure of HIV

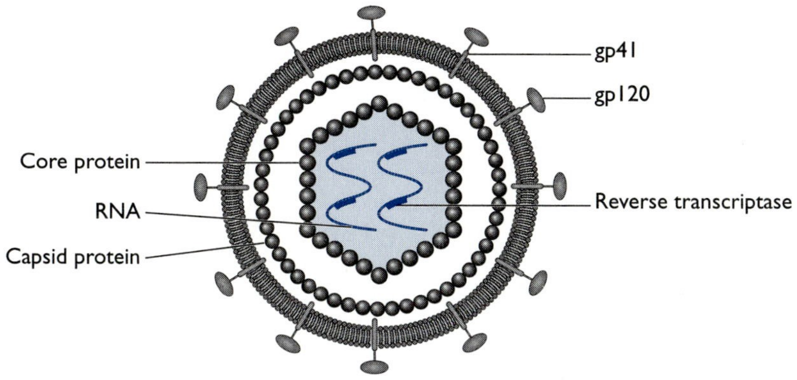

gp41
gp120
Core protein
RNA
Capsid protein
Reverse transcriptase

Progression of HIV infection

gp 120 of HIV attaches to CD4 receptor on host cells (T helper cells, monocytes)

↓

Asymptomatic phase until T helper cells decrease to <200/mm^3

↓

Increased susceptibility to opportunistic pathogens (especially *Pneumocystis carinii*)

↓

T helper cells decrease below 100/mm^3

↓

Other opportunistic infections, weight loss, persistent fever, diarrhea, Kaposi's sarcoma, dementia

↓

Death

Figure 18.8 Structure of human immunodeficiency virus (HIV) and progression of infection. The gp120 protein on the envelope of HIV attaches to receptors on host T helper cells and monocytes. Fusion of the viral envelope with the host cell membrane releases the nucleoprotein core into the host cell. Viral replication kills these cells. As the number of T helper cells diminishes, the patient becomes more immunocompromised and susceptible to opportunistic infections (acquired immunodeficiency syndrome; AIDS).

cycle of HIV is quite complex and is not presented here in detail. One aspect of its replication cycle is worth noting, however. HIV is an RNA virus. During its replication cycle, it makes a DNA copy of itself, which then serves as a template for production of new viral genomes and viral proteins. The enzyme that makes a DNA copy of the RNA viral genome is called **reverse transcriptase.** Before reverse transcriptase was discovered, the central dogma of molecular biology was that information flowed in a single direction, from DNA to proteins via mRNA. This rule no longer holds, as reverse transcriptase showed information can flow from RNA to DNA. On a more practical level, reverse transcriptase made it possible to use polymerase chain reaction (PCR) to amplify RNA genomes, by a process called RT-PCR (Box 18.1).

Viral invasion strategy

After entering the body, the virus is easily detectable in the bloodstream for a brief period. The level of virus in the blood then declines until it becomes undetectable. During this period, the virus moves from the bloodstream to the lymph nodes, where cells of the immune system are localized. HIV specifically attacks host cells carrying a cell surface protein called **CD4.** The cells with CD4 on their surfaces are T helper cells and monocytes. CD4 is a receptor for cytokines, proteins that regulate the activities of immune cells.

box 18–1 *Reverse Transcriptase-Polymerase Chain Reaction: a Technological Innovation with Widespread Applications*

A previous chapter described some of the problems scientists faced when they tried to cultivate viruses. Tissue culture, in addition to being expensive and difficult, carried a safety problem. Scientists working with large amounts of live viruses not only risked infecting themselves, but also risked infecting others by accidental releases. It is a tribute to scientists in the early days of human immunodeficiency virus (HIV) research that such events were extremely rare, but even today hazards exist. One solution would be to clone the viral genome or portions of it. Such clones would not cause a health hazard and would be much easier to manipulate than live viruses in scientific experiments. But cloning is a technique that, by its nature, is specific for double-stranded DNA. It is ironic that HIV itself provided the solution to the problem it was causing scientists. The discovery of reverse transcriptase allowed scientists working on RNA viruses like HIV and influenza virus to make and clone DNA replicas of these viral genomes.

Reverse transcriptase made PCR, also once a DNA-specific procedure, applicable to RNA molecules. In RT-PCR, the first step is to make a DNA copy of the RNA molecule using HIV reverse transcriptase, then amplify that copy by PCR. This technique made possible the most sensitive detection test currently available for detecting HIV in patients' blood and tissue. Moreover, RT-PCR has allowed researchers to obtain portions of the viral genome from stored tissue specimens containing RNA viral genomes, then amplify, clone, and sequence these genomes.

As often happens with scientific innovations, researchers working with other systems began to grasp the potential utility of RT-PCR. In Chapter 6, a technique for monitoring gene expressions by making gene fusions was described. Suitable reporter genes have not been found for many organisms. Where such technology is available, it may not be very sensitive. RT-PCR has now become a popular method for detecting messenger RNA (mRNA) transcripts in bacteria and eukaryotic cells. In fact, this technology has given rise to a powerful technique for genome sequencers. A scientist interested in neurons is not concerned with every single gene in the human genome, only in those expressed in neurons. The number of expressed genes is usually only a small percentage of the total genes in the genome. By using RT-PCR, scientists can selectively amplify the expressed genes by targeting mRNA instead of DNA. This technique is called **expressed gene sequence tagging (est)**.

The alert reader may wonder how scientists can amplify a segment of mRNA without knowing the sequence. This problem stopped scientists at first, but not for long. Recall that an important part of PCR is the primers. The solution was to use random pieces of DNA sequence 6–10 base pairs in length as primers. This makes the first step in RT-PCR somewhat nonspecific, but does not present a problem when the researcher is interested in all of the expressed genes.

Recently, scientists have learned that although CD4 is essential for viral attachment to its target cells, it is not sufficient. Other proteins called chemokine receptors must be present to facilitate the binding of the virus to the cell surface (**coreceptors**). **Chemokines** are another family of proteins, like cytokines, that help to regulate the immune system. Some people lack functional chemokine receptors on their T helper cells and monocytes. These are the people who seem to be either immune to HIV infection or, if infected, have very long asymptomatic phases. Efforts are now under way to develop drugs that block the interaction between HIV and the coreceptors necessary for binding of the virus to host cells.

During the asymptomatic period, the concentration of T helper cells in blood remains stable at about 200–500 per mL, close to the normal level of 500–1000 per mL. Entry into the symptomatic stage is signaled by a decrease in T helper cell concentration below 200 per mL. At this point, the infected person begins to exhibit increased susceptibility to a variety of opportunistic pathogens, especially the yeast *Pneumocystis carinii*. As the concentration of T helper cells decreases below 100 per mL, other opportunistic infections become common. This later stage of the disease is also characterized by weight loss, persistent fever, prolonged diarrhea, Kaposi's sarcoma (a cancer of the blood vessels), other cancers, and dementia. Death is caused by an overwhelming secondary microbial infection or cancer or a combination of both.

Once the symptomatic period begins, the patient usually dies within a year if left untreated. The mechanism that triggers the transition from the asymptomatic phase to the symptomatic phase of the disease is unclear, but one factor might be interaction of HIV with other common pathogens. Infection with other pathogens stimulates the host's inflammatory response, and the cytokines produced during inflammation enhance the replication of HIV.

Explaining AIDS symptoms

Many of the symptoms of AIDS can be understood by considering the cellular targets of HIV: T helper cells and monocytes. Some T helper cells stimulate B cells to produce antibodies. Others stimulate macrophages to become activated. T helper cells also stimulate other T cells to become cytotoxic T cells. Thus, loss of T helper cells eliminates the antibody response, the activated macrophage response, and the cytotoxic T cell response. Loss of T helper cells combined with loss of monocytes eliminates virtually all of the immune responses, leaving the person dependent only on the nonspecific defenses. The person is then highly susceptible to infection by a number of microbes, including many that do not cause infections in healthy people. Cytotoxic T cells are also important for eliminating incipient tumors, one explanation for the increased risk of cancer in people with HIV infections.

Viral replication in a T helper cell rapidly kills the cell, but the way in which the viruses kill the cell is unclear. Replication of the virus in monocytes is slower than in T cells, and infected monocytes continue to survive for a long period after the initial infection. Monocytes move to different parts of the body, where they differentiate into a number of different phagocytic cell types, for example, Kupffer cells of the liver and microglial cells of the brain. Thus, infected monocytes carry the virus to many different parts of the body, including the brain—probably the cause of the dementia seen in some AIDS patients.

Replication strategy of HIV

The replication cycle of HIV has been studied in great detail in the hope of finding ways to stop viral replication. The replication process has turned out to be quite complex and will only be outlined here. HIV is an enveloped single-stranded RNA (ssRNA) virus with a diploid RNA genome. The viral genome encodes genes for **envelope proteins (env)**, **capsid proteins (gag)**, and enzymes involved in viral multiplication (**pol**). The viral genome also encodes several **regulatory proteins (tat, rev, nef)**. Protein products of env, gag, and pol are cleaved by a viral protease to produce proteins with different functions (Fig. 18.9). Proteolytic cleavage of the env gene product, for example, produces two proteins, **gp120** and **gp41.** gp120 is exposed on the viral surface and mediates attachment of the virus to CD4 on T helper cells and monocytes. gp41 is embedded in the envelope and is linked to gp120 by a sulfhydryl bond. gp41 is thought to stabilize gp120 and is also the protein that mediates fusion of the viral and host cell membranes before viral entry into the cell and uncoating. The viral protease is a target of some of the new anti-HIV drugs.

The nucleoprotein core of the virus contains, in addition to the diploid genome, proteins that are important for viral replication. The viral enzyme, reverse transcriptase, makes a DNA copy of the viral RNA. The name **retrovirus,** which has been given to HIV and viruses with a similar replication strategy, comes from the reverse flow of information from RNA to DNA via reverse transcriptase. The reverse transcriptase is also a target for anti-HIV drugs. **Integrase** mediates integration of the DNA copy of the viral genome into the host cell genome. When the DNA copy of the virus has integrated into the host cell genome, it is transcribed at low levels by the cell's own **DNA polymerase, PolII.** After production of new viral genomes and viral proteins, the viral capsid either buds through the host cell membrane or mediates fusion of the infected cell with another cell.

HIV-1, HIV-2, and simian immunodeficiency virus (SIV)

Two types of human HIV, **HIV-1** and **HIV-2,** can cause AIDS. HIV-1 was isolated in 1983 in the United States and France. HIV-2 was first isolated in 1985 in West Africa. The HIV-1 and HIV-2 viruses differ primarily in that

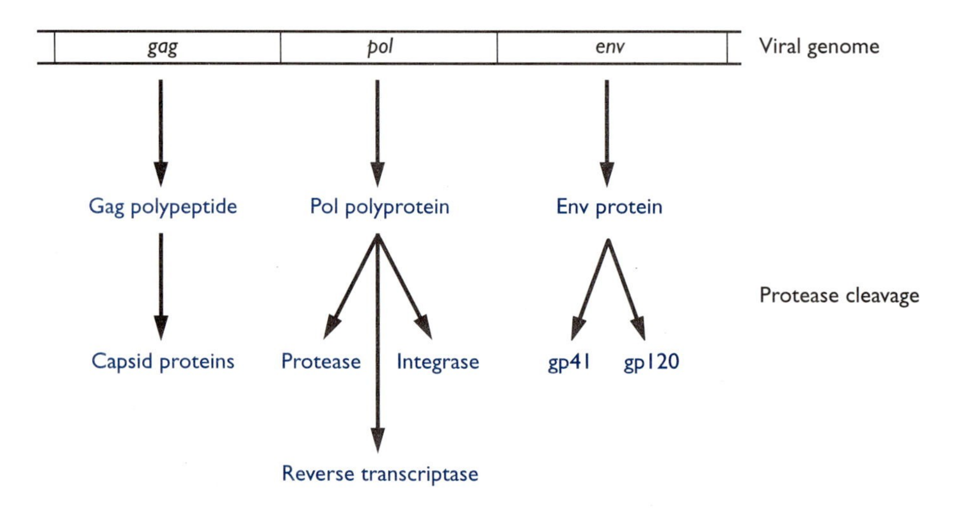

Figure 18.9 Cleavage of HIV polyprotein by viral protease. The RNA genome of HIV is translated into several large inactive polyproteins, each of which is cut by a viral protease into smaller active proteins.

HIV-2 has a gene, **vpx,** not seen in HIV-1. The function of this gene is not known. Otherwise, the two types of HIV have the same gene organization and share DNA sequence identity of about 45%. HIV-2 is closely related to a virus found in some African monkeys, **simian immunodeficiency virus (SIV;** 80% sequence identity). This similarity between HIV-2 and SIV has given rise to the speculation that HIV might have evolved from a monkey virus that entered the human population. Contaminated blood from infected monkeys may have entered through wounds on the hands of humans who were butchering monkeys for food. HIV-1 is related to a virus found in chimpanzees and probably entered humans the same way HIV-2 did. HIV-1 has one of the highest known viral mutation rates, and different isolates can differ appreciably from each other, even isolates from the same person. The mutation rate is high enough that the current form of the virus could have evolved within the past 20–100 years. The first known case of human HIV (diagnosed retrospectively from frozen serum samples) occurred in Zaire in 1959.

Spread of HIV

HIV is spread between humans by anal, oral, or vaginal intercourse, by shared needles, or by contaminated blood products. Pregnant women can spread the disease to their infants during birth or by breast feeding. There is no evidence for spread by aerosols, saliva, casual contact, or insects. Information about preventing the disease by use of latex condoms or clean needles (new or bleach-decontaminated) is currently being spread as widely as possible. The campaign encouraging condom use by homosexual men in the United States has had some success, leading to a decline in the incidence of HIV infection within this group. Unfortunately, this decline was offset by an increase in the number of heterosexually acquired cases, so that the incidence of HIV infection remains high. Teenagers have proved to be particularly resistant to campaigns promoting the use of condoms.

No vaccine against AIDS is available, and none is in the late stages of development, although researchers still believe that a vaccine is feasible. The fact that HIV targets immune cells makes it difficult to devise a vaccine. Antibodies that prevent the virus from binding host cells might appear at first to provide a protective response, but antibodies bound to the virus can facilitate the ingestion of viral particles by phagocytic cells such as monocytes, leading to infection of the cells. Also, HIV can mediate fusion of an infected cell with another cell and can thus spread without coming into contact with antibodies in blood. A cytotoxic T cell defense that targets infected cells might be effective early in the infection process, but, at later stages, cytotoxic T cells would have the same undesirable effect as the virus itself: killing T helper cells and monocytes. It has been suggested that a vaccine that does not completely block infection by HIV could still be effective simply by reducing the viral load to a level where the patient does not have symptoms and is unlikely to transmit the virus to others.

Scientists who study the population biology of infectious diseases warn that, even if an HIV vaccine is developed and demonstrates an effectiveness equal to that of vaccines against such diseases as polio or diphtheria, the result will not necessarily be a cure-all. Even the most effective vaccines do not protect everyone who is vaccinated. A 95% protection rate is considered very good but still leaves 5% of vaccinated people vulnerable. Also, immunity should be viewed in relative terms. Vaccines stimulate the immune response, but they do not render the body completely invulnerable. How many times could the immune system respond successfully in controlling exposure after exposure to the virus? Not 100% of the time.

In some diseases the effects of the development of a successful vaccine were augmented by changes in human habits, such as reduced crowding and improved nutrition. Even if a good HIV vaccine becomes available, the vaccine barrier may not hold if large numbers of people begin again to have frequent unprotected sex.

Treatment of HIV: have a HAART

Prior to 1995, the only drugs available for treatment of HIV were **nucleotide analogs,** such as AZT, **dideoxycytosine** (**ddC**), and **didanosine** (**ddI**). All these compounds are phosphorylated by host enzymes. HIV-1 reverse transcriptase incorporates them into viral DNA, causing premature termination of the DNA copy of the viral RNA. Termination occurs because reverse transcriptase cannot add another nucleotide to the nucleotide analogs. AZT and similar drugs helped to slow but did not stop the progression of AIDS. AZT has been successful, however, in reducing transmission of HIV from mother to infant.

A major breakthrough in HIV treatment was the discovery and approval of **protease inhibitors** (for example, **saquinavir, ritonavir,** and **indinavir**). These compounds block the action of the protease that cleaves the viral polyproteins to produce active viral proteins. A protease inhibitor is administered along with two reverse transcriptase inhibitors (**triple drug therapy**). This therapy has been dubbed **HAART** for **highly active antiretroviral therapy.** The triple drug therapy has been found to reduce concentrations of HIV-1 in blood to undetectable levels and slows the rate of progression of AIDS dramatically.

Triple drug therapy presents several problems. First, the cost is more than $10,000 per year. The expense has created two classes of AIDS patients: those who can afford the drug and those who cannot. Widespread protest has arisen over what some see as efforts by the pharmaceutical companies to profit at the expense of AIDS sufferers, their insurance companies, and the providers of last resort, federal and state governments. A second problem is that side effects have forced some patients to discontinue therapy. Third, and most disappointing, is the discovery that patients who discontinue use of the drugs usually experience an almost immediate rebound of viral levels in the blood. Hope had been high that the triple drug regimen would truly cure patients who could tolerate (and afford) it, by eliminating the virus from their systems. Finally, drug-resistant forms of the virus have already begun to appear. In spite of these problems, HAART can be considered successful in part, because it has reduced hospital stays, opportunistic infections, and the death rate from AIDS. HAART is not a cure, but, at least for those who can afford HAART, HIV has become more like a chronic disease.

The realization that therapy may be a lifelong proposition for most HIV-infected people has raised questions about the economic impact of AIDS therapy. Will the increasing cost of treating HIV-infected people displace money from infectious disease programs that protect children and young adults from other life-threatening infectious diseases? Already, funding for AIDS research has changed the funding structure sufficiently to produce underfunding in other important areas, such as research into cancer, heart disease, and antibiotic-resistant bacteria. These questions do not have easy answers. Nor do the much more troubling questions about what will happen in Africa and in Southeast Asian countries where AIDS is rampant but little money is available for therapy, prevention, and public education.

Viruses and Cancer: Hepatitis B (HBV) and Human Papilloma Viruses (HPV)

After all the bad news about HIV, a little good news is in order. The discovery that some common forms of human cancer are caused by viruses has led to preventive measures that may reduce the incidence of these cancers in the future. Evidence linking viral sexually transmitted diseases and cancer is strongest in the case of HBV and HPV. Both HBV and HPV are DNA viruses. Their features are compared in Table 18.1. HPV is a nonenveloped icosahedral virus with a circular dsDNA genome. HPVs replicate preferentially in epithelial cells and cause warts. Many of these warts are benign, but infections resulting from three types of genital HPV affect cervical epithelial cells and are associated with cervical cancer. These three types are **HPV 16, 18,** and **31.** Cervical cancer is a serious and common type of cancer in women. Fortunately, benign warts are raised and have a different appearance from the flat warts associated with cervical cancer. Thus, potentially cancerous warts can be identified fairly easily by visual examination and eliminated by cauterization. The mechanism by which certain types of HPV cause cancer is not understood in detail but is linked to integration of the viral genome into the host cell chromosome and interactions between proteins encoded by viral DNA and host proteins that control host cell proliferation. Unchecked proliferation leads to cancer.

Table 18.1 Comparison of Characteristics of Human Papilloma Virus and Hepatitis B Virus

Characteristic	Human papilloma virus (HPV)	Hepatitis B virus (HBV)
Genome	Circular, double-stranded DNA	Partially double-stranded DNA
Envelope	No	Yes
Capsid structure	Icosahedral	Icosahedral
Target	Epithelial cells	Liver cells
Disease	Warts Cervical cancer	Hepatitis Cirrhosis Liver cancer
Mechanism for causing cancer	Viral genome integrates in host DNA Viral proteins block control of host cell proliferation	Virus causes mutations in host genes that regulate cell growth Increases susceptibility to exogenous carcinogens
Method of transmission	Sexual intercourse	Sexual intercourse Shared needles Transfusion
Treatment	Cauterization	None; but can be prevented by vaccine
Vaccine	No	Yes

HBV is an enveloped icosahedral virus with a DNA genome. HBV infections are transmitted by sexual intercourse (especially anal intercourse), by sharing needles, by transmission from mother to infant during and after birth, and by blood transfusions or contaminated blood products (for example, those used to treat hemophiliacs). The viruses move from the point of entry to the liver, where they multiply. The result can be acute hepatitis, resembling that caused by hepatitis A virus, or the disease can be chronic and nearly asymptomatic. Long-term liver damage from viral replication can lead to **cirrhosis** and death. The other serious outcome of HBV infection is liver cancer. The precise nature of the connection between liver cancer and HBV infection has not been established. However, it appears that HBV causes mutations to accumulate in host genes that regulate cell growth, thus making liver cells more likely to develop into tumors because of exposure to exogenous carcinogens, such as the fungal toxin aflatoxin.

An effective vaccine against HBV infection is now available worldwide. The vaccine elicits a protective antibody response, in which antibodies coat the viral surface, rendering them unable to attach to and enter host cells. The HBV vaccine initially was given to infants and, in some cases, to members of certain high-risk populations, such as male homosexuals and people who use injectable drugs. Only within the last few years has HBV vaccination been recommended for a larger population at risk for HBV infection: sexually active teenagers.

Bacterial Diseases of the Genital Tract

Incidence of bacterial sexually transmitted diseases. An estimated 4 million new cases of *Chlamydia trachomatis,* 1.4 million cases of gonorrhea, and 45,000 cases of HIV infection are diagnosed each year in the United States. The distribution of gonorrhea and chlamydial disease is shown in Figure 18.10. Gonorrhea is concentrated in the southeastern part of the United States, while chlamydia infections are much more widely spread. The incidence of chlamydial infection is probably at least three times as great as the numbers reported to CDC. **Gonorrhea** is caused by *Neisseria gonorrhoeae,* a gram-negative diplococcus. *C. trachomatis* causes **nongonococcal urethritis** (**NGU**), a disease with symptoms indistinguishable from those of gonorrhea. *C. trachomatis* has a gram-negative-like cell envelope but lacks peptidoglycan. Whereas *N. gonorrhoeae* can grow outside human cells, *C. trachomatis* can only grow inside human cells. *C. trachomatis* is not the only cause of NGU but is responsible for 30%–50% of cases. Many cases of NGU cannot yet be tied to a specific infectious agent, leading scientists to suspect that other NGU agents remain to be discovered.

Symptoms of gonorrhea and NGU in adults. Gonorrhea and NGU contracted as a result of heterosexual intercourse appear as urethral infections in men and cervical infections in women. The progression of these diseases is different in men and women. In men, symptoms of urethral infection include painful urination and production of a thick, yellowish exudate. Infected men usually experience symptomatic disease, although 10%–15% of cases are asymptomatic. In women, inflammation of the cervix may cause lower abdominal pain and an increased discharge. As many as 85% of women, however, have no noticeable symptoms. Asymptomatic infections in women are especially frequent in the case of *C. trachomatis*

A Gonorrhea — reported cases per 100,000 population, United States, 1998

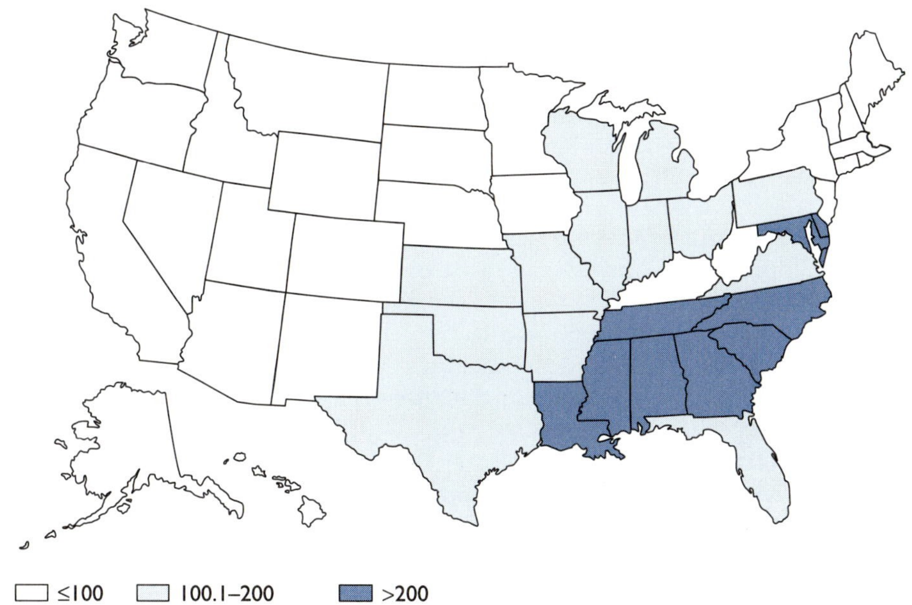

B Chlamydia — reported cases among women per 100,000 population, United States, 1998

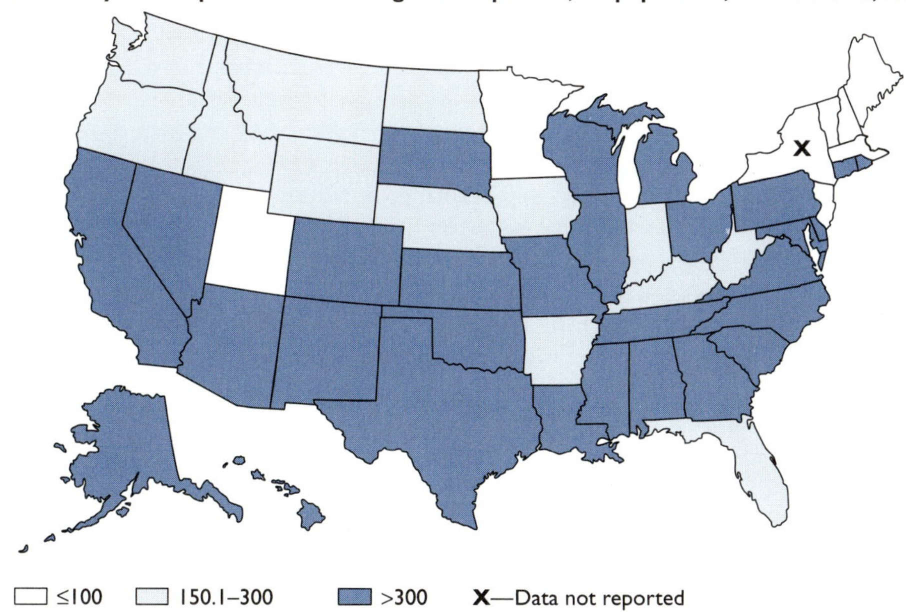

Figure 18.10 Maps of the United States showing the distribution of reported cases of gonorrhea (A) and chlamydia (B). The numbers of cases shown are probably an underestimate of the actual incidence of these diseases. Although gonorrhea has been brought under control, at least partially, in many areas of the country, chlamydial disease remains widespread. [From 1999. Centers for Disease Control and Prevention. CDC summary of notifiable diseases, United States, 1998. *MMWR* 47:29,35.]

infections. Disease acquired by anal intercourse leads to rectal inflammation and discharge.

Gonorrhea can also be acquired during oral sex. In such cases, symptomatic disease takes the form of a sore throat, which can be mistaken easily for a viral sore throat. Most people do not go to the doctor for a sore throat, and, even when they do, may be misdiagnosed. Yet quick and accurate diagnosis and treatment are important. Whether they have symptoms or not, people with gonorrhea can transmit the disease to others. Moreover, undiagnosed and untreated infections can have serious complications.

In men, infection of the urethra may spread to the epididymis, the tube connecting the testes to the urethra, causing painful swelling. This is a relatively rare event. In women, infection of the cervix can ascend into the uterus (**endometritis**) and fallopian tubes and eventually into the peritoneal cavity, causing PID. At least 1 million cases of PID per year are reported in the United States alone. PID is characterized by mild-to-severe abdominal pain and bleeding in the middle of the menstrual cycle, although some cases are asymptomatic. The infection may develop into sepsis. Even in mild or asymptomatic cases, the infection can cause scarring of the fallopian tubes, the leading cause of infertility and ectopic pregnancy. The chance of sufficient damage to cause infertility or ectopic pregnancy increases with each successive infection, from about 15% after the first episode of PID to about 70% after the third episode.

Once the cervix is damaged, the uterus is more easily colonized by other bacteria. Therefore, subsequent bouts of PID need not be caused by these sexually transmitted organisms. Women who develop endometritis and PID may not have had symptomatic cervicitis, and the symptoms of endometritis and PID may be the first signs of an infection. In approximately 2% of gonorrhea cases in both men and women, systemic infection can result in septic arthritis, which can lead to destruction of the affected joint if untreated. In a small fraction of gonorrhea cases, the organism can cause endocarditis or even meningitis.

Implications of bacterial genital infections for infants. Another serious complication of gonorrhea and NGU can occur in infants who acquire the disease during passage through the infected birth canal. The bacteria infect the infant's eyes. In the case of *N. gonorrhoeae* infections, blindness can result (**ophthalmia neonatorum**). Strains of *C. trachomatis* circulating in the United States today cause a self-limiting purulent conjunctivitis, which does not produce lasting damage to the eye. About 2%–6% of infants born in the United States develop *C. trachomatis* conjunctivitis, and this remains the most common cause of conjunctivitis in newborns. Some strains of *C. trachomatis* in other parts of the world, however, can cause blindness. *C. trachomatis* can also cause pneumonia in neonates, an even more serious disease. Pneumonia can develop when the infant inhales bacteria during the first breaths after birth or when the infant inhales organisms being shed from the eyes during a case of conjunctivitis, which may last for up to 3 weeks. Prophylactic antibacterial eye drops are given to newborns to prevent these complications.

N. gonorrhoeae* and *C. trachomatis *N. gonorrhoeae* (also called gonococcus) is a fragile organism that is nutritionally fastidious and does not survive well in the environment. This explains why gonorrhea is only spread by intimate contact and not on towels or by casual contact. *N. gonorrhoeae* is highly adapted to survive and grow in the cervix, urethra, throat, and rectum. Pili and other surface adhesins allow *N. gonorrhoeae* to adhere to columnar epithelial cells. Pili mediate the initial attachment, and other surface proteins mediate a closer association between the bacteria and the mucosal cells. The bacteria then enter the mucosal epithelial cells. The subsequent fate of the bacteria is not clear, but some seem to gain access to underlying tissue. The fact that gonococci adhere preferentially to columnar epithelium explains why infections of the genital tract localize to the cervix, which has a columnar epithelium, as opposed to the vagina, which has a squamous epithelium.

N. gonorrhoeae is a gram-negative bacterium but has an endotoxin that differs from the classical LPS found in most gram-negative bacteria. The *N. gonorrhoeae* LPS equivalent has a lipid A-like component and a core polysaccharide, but its O-antigen is a short oligosaccharide instead of the large, branched O-antigen structure of classic LPS. Accordingly, the LPS of *N. gonorrhoeae* is called **lipooligosaccharide (LOS)**. LOS has the same activities as LPS; that is, it activates the inflammatory response. LOS probably is responsible for most of the symptoms of gonorrhea and its complications. If a specimen taken from an infected site is viewed under the microscope, numerous neutrophils will be seen, many containing diplococci. It remains unclear whether *N. gonorrhoeae* survives inside the neutrophils or the ingested bacteria are on their way to destruction.

An sIgA response that prevents attachment of the bacteria to host mucosal cells should be protective. Yet people who have had gonorrhea do not become immune to reinfection, despite having mounted strong sIgA responses to the bacteria. The sIgA response does not stop reinfection, because *N. gonorrhoeae* has developed the ability to circumvent this response by making changes in proteins on its surface. This phenomenon is called **antigenic variation.** The bacteria do this in an interesting way. They have multiple copies of genes encoding surface proteins. These surface protein copies differ sufficiently from each other that antibodies raised against one version of the protein will not recognize other versions. Only one of these genes is expressed at any given time. From time to time, the bacteria replace the gene being expressed with a variant copy of the gene (Fig. 18.11). Bacteria with the new version of the surface protein will no

Figure 18.11 Mechanism of antigenic variation in *Neisseria gonorrhoeae*. The bacteria carry several copies of a gene that encodes the pilin subunit of pili or other surface proteins. Only one of these genes is expressed at any one time (*pilE*). The others are not expressed because they lack a promoter; that is, they are silent (*pilS*). Occasionally the *pilE* and *pilS* genes are exchanged by homologous recombination, and a new *pilE* is expressed. Because the amino acid sequences of the proteins encoded by the two genes differ enough that antibodies to one do not bind to the other, this substitution allows the bacteria to escape the host's immune response.

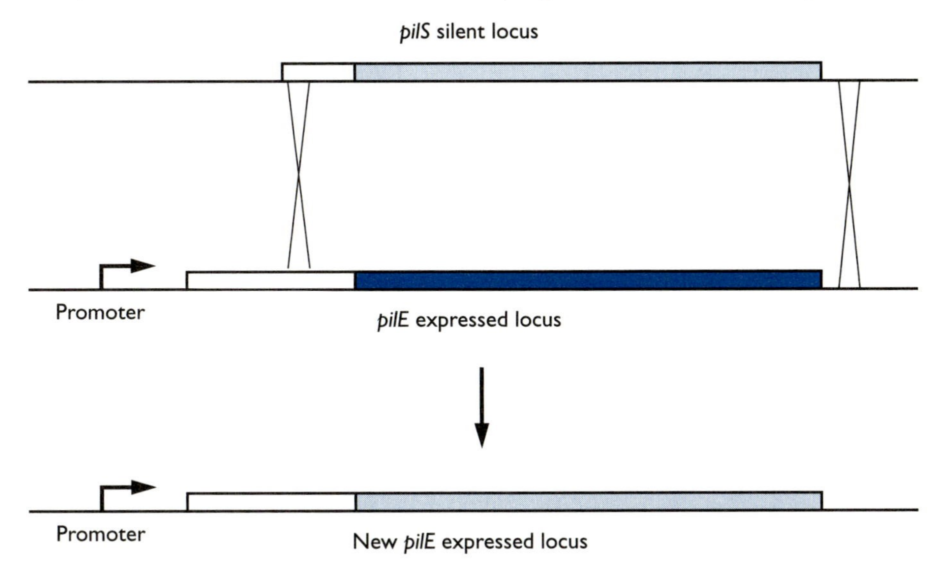

longer be affected by the sIgA that is present, and thus will be selected and soon become the major population.

The replication strategy of *C. trachomatis* is the same as that of *C. pneumoniae* (Chapter 16, Fig. 16.7). *C. trachomatis* has a gram-negative type outer membrane but no detectable peptidoglycan. Also, *C. trachomatis* is smaller than most bacteria and thus usually cannot be seen with an ordinary light microscope. The bacterium exists in two forms: the elementary body, which is metabolically inactive but tough enough to survive under conditions in which the other form, the reticulate body, would be killed. The elementary body is the form that moves from cell to cell and invades the cell. Inside a membrane-coated vacuole, the elementary bodies begin to turn into reticulate bodies, which divide to make many new cells. As the vacuole grows, elementary bodies begin to appear. Eventually the vacuole merges with the human cell membrane and releases the elementary bodies into the extracellular environment to hunt for the next cell.

In contrast to viral replication, which often kills the cell, chlamydial replication usually does not do much direct damage. Instead, the damage caused by the bacteria is inflicted by the pronounced inflammatory responses they elicit. Because inflammation is the culprit, it is not surprising that microbes as different as chlamydia and gonococci could cause such similar diseases.

Diagnosis and treatment. Diagnosis of gonorrhea is made by Gram staining pus taken from an infected site. Even in women without detectable symptoms, inflammation of the cervix can usually be seen when a physician examines the area with a speculum. If *N. gonorrhoeae* is present, numerous gram-negative diplococci are seen in the pus specimen. Nothing but neutrophils will be seen if the disease is caused by chlamydia. Seeing the gonococci, however, does not rule out the presence of chlamydia, because 10%–20% of men and 30%–40% of women who have gonorrhea are also infected with chlamydia. Until recently, the test for chlamydia was invasive and uncomfortable. To sample a male urethra, a small brush was inserted at least twice to get enough cells to detect chlamydia. This process, called urethral stripping, was as unpleasant as it sounds, and, understandably, men were not eager to be tested. In fact, few men who knew someone who had undergone the procedure were willing to undergo it themselves. For women, speculum examination and sampling of the cervix are not as painful, but many find the procedure embarrassing and unpleasant. A urine test for chlamydia has been introduced recently and should revolutionize testing for chlamydia. This new test uses PCR to amplify the bacterial DNA, which is then detected by DNA probes. At least one available test can detect both chlamydia and *N. gonorrhoeae* in the same sample.

Despite decades of intense efforts, no vaccine is available to prevent gonorrhea, because variation of surface antigens has made it difficult to develop an effective vaccine. Current efforts focus on highly conserved regions of surface proteins. Unfortunately, these regions tend to be buried in the final folded form of the protein and thus are not accessible to antibodies. *N. gonorrhoeae* was once susceptible to a number of antibiotics, including penicillin and tetracycline. Resistance to these antibiotics is now common in many areas of the country.

No vaccine is available for *C. trachomatis,* although tetracycline and erythromycin are effective treatments. So far, resistance has not emerged in this genus of bacteria, possibly because of its intracellular lifestyle. Because co-infection with *N. gonorrhoeae* and *C. trachomatis* is common, patients must be treated for both when mixed infections are suspected. Currently, ciprofloxacin is recommended for the treatment of gonorrhea and doxycycline for NGU. A relatively new drug, azithromycin, has been found to be effective against both gonorrhea and NGU. It has the advantage of requiring only one or two doses, a fact that makes patient compliance much better than with antibiotics that must be taken daily for a week or two. However, azithromycin is expensive and usually out of reach of the poorer public clinics. The argument has been made that azithromycin is actually more cost effective than doxycycline, because better patient compliance results in fewer repeat visits. The high cost, however, remains a barrier for many clinics and patients.

Yeast Infections

C. albicans is normally found in low numbers in the vaginal tract. Suppression of the resident microbiota and/or hormonal changes allow *C. albicans* to overgrow and produce an inflammatory response in the vulva and vaginal tract. This results in a profuse discharge, which may be curd-like in consistency. Infection is accompanied by burning and itching. The vulva and vaginal mucosa are inflamed, and whitish patches may be seen on the vaginal mucosa. This condition is called *C. albicans* vaginitis and is most often seen in women who have diabetes, have been taking antibiotics, are sexually active, or are in the third trimester of pregnancy. The influence of hormonal changes can be seen not only from the higher incidence in pregnant women but also from the fact that, in women who are not pregnant, the symptoms frequently worsen just before menses.

C. albicans vaginal infections can recur. It is not clear whether reinfection results from carriage of the virulent strain in the colon or from asymptomatic carriage by a sexual partner. *C. albicans* vaginitis is a noninvasive disease but can cause considerable discomfort to the infected woman. Infants passing through an infected birth canal can acquire *C. albicans* infections of the mouth (thrush) or eyes (conjunctivitis), but these infections are usually not serious. *C. albicans* infections of the vaginal tract, like candidiasis in other areas of the body, are most likely to be severe in immunocompromised people, such as those with AIDS. The best therapy for vaginal candidiasis is azole suppositories, which are sold across the counter and do not require a doctor's prescription.

Trichomoniasis

Another cause of vaginitis is the protozoan, **Trichomonas vaginalis** (Fig. 18.12). In contrast to other protozoa described in previous chapters, *T. vaginalis* has no cyst stage and exists only as a flagellated trophozoite (**trichomonad**). For this reason, it dies quickly outside the host and must be transmitted by intimate contact. Most cases of *T. vaginalis* vaginitis are acquired by sexual transmission. Growth of *T. vaginalis* in the vaginal tract causes an inflammatory response that is responsible for the common symptoms of vaginal discharge (which is runny and frothy), vaginal itching, and burning. The vaginal mucosa may have bright red, punctate lesions, and urination may be painful. In tissue culture cells, *T. vaginalis* adheres to the cell surface and causes a contact-dependent killing. Binding of *T. vaginalis* also causes the tissue culture cells to detach from the dish, another indication that the cells are dying. The way in which *T. vaginalis*

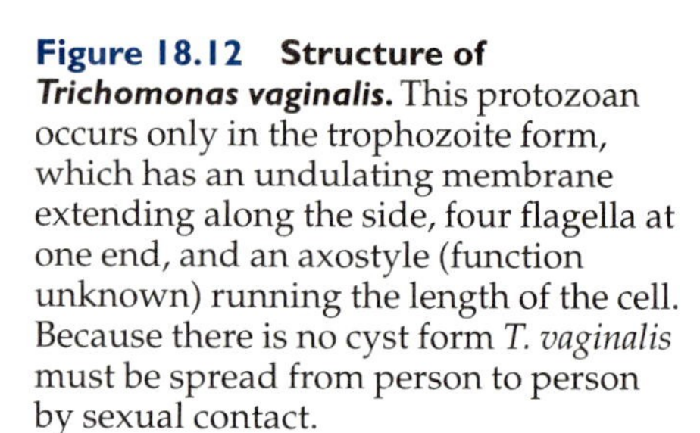

Figure 18.12 Structure of *Trichomonas vaginalis*. This protozoan occurs only in the trophozoite form, which has an undulating membrane extending along the side, four flagella at one end, and an axostyle (function unknown) running the length of the cell. Because there is no cyst form *T. vaginalis* must be spread from person to person by sexual contact.

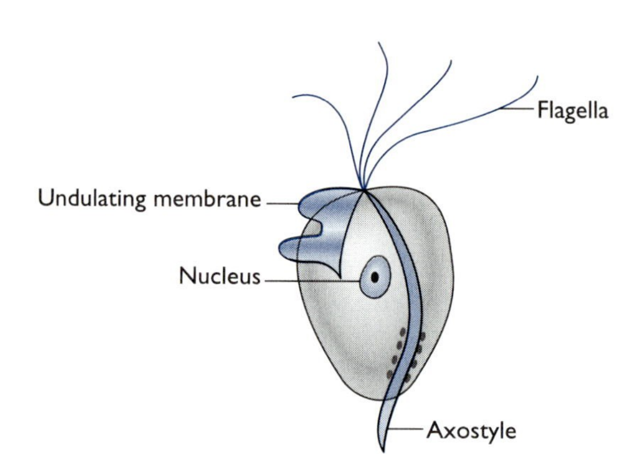

causes these effects is not known, but killing of vaginal cells could be responsible for eliciting the inflammatory response and producing the symptoms of the disease.

T. vaginalis colonizes the male urethra but usually does not cause inflammation. Symptomatic disease in men is rare, but asymptomatic carriers spread the disease. The male partner of any woman who has trichomoniasis should also be treated. Unlike herpes, gonorrhea, NGU, and syphilis, *T. vaginalis* vaginitis is a mild noninvasive disease. Its main effect is vaginal discomfort, and it has no known complications. Although a statistical association has been noted between the incidence of trichomoniasis and cervical cancer, no cause-and-effect relationship has been proven. Also, some scientists believe that, in pregnant women, trichomoniasis can lead to premature rupture of the placental membranes, but such an association remains to be demonstrated rigorously.

T. vaginalis infections in neonates are rare, but eye and lung infections have been reported. It is recommended that the parasite load in the vaginal tract be reduced before birth, for the safety of the infant and the comfort of the mother. Metronidazole is the drug of choice for treatment of trichomoniasis. Oral metronidazole is not recommended for pregnant women, especially during the first trimester, because of potential damage to the fetus. Vaginal suppositories are used instead.

Emerging into the Light

Diseases of the urogenital tract, especially those that focus on the genital tract, are difficult for many people to confront and discuss. These diseases have been associated in the public mind with sexual promiscuity, and people who acquire these infections are often stigmatized. Yet many cases of genital tract infections arise from transmission between partners who are usually sexually monogamous. One small lapse on the part of one or the other partner places both at risk. In some cases, such as viruses transmitted through the blood supply, completely abstinent individuals may acquire a disease that is identified in the public mind as sexually transmitted. A few patients have even contracted hepatitis B or HIV from their doctors or dentists. Another group of innocent victims includes infants born to infected mothers who may not realize that they are endangering their children. General public support for treating genital tract infections just like other infections, with no social stigma, could have important and beneficial health care ramifications.

chapter 18 at a glance

Structure of the urinary tract

Kidney

Ureters

Urinary bladder

Urethra

Defenses of the urogenital system

Urinary tract
- Urethral sphincter
- Tamm-Horsfall glycoprotein
- Rapid flow of urine
- Mucus layer

Female genital tract
- Vaginal microbiota
- Squamous cells of vagina
- Mucus layer
- Cervical mucus plug

Comparison of Urogenital Tract Pathogens

Species	Reservoir	Characteristics	Disease	Susceptible populations	Virulence factors	Treatment	Vaccine
Herpes simplex-2 Herpes simplex-1	Humans	Enveloped virus Linear dsDNA genome	Genital herpes (possible link to cervical cancer)	All people	Latent state in neural ganglia Regulated expression of genes	Acyclovir Phosphono-formate	No
HIV-1 HIV-2	Humans	Enveloped virus Diploid ssRNA genome dsDNA copy made integrates in host genome	AIDS	All people (except for a genetically immune subpopulation)	Targets T helper cells and monocytes (CD4 receptor) Coreceptors required	Azido-deoxthmi-dine (AZT) dideoxcystine (ddC) didanosine (ddl) Protease inhibitors	No
Human papillo-mavirus	Humans	Nonenveloped Circular dsDNA genome	Genital warts [human papil-loma virus (HPV) 16, 18, 31 linked to cervical cancer]	All people	Grows in epithelial cells Integrated copy may cause cancer	Cauterization	No
Hepatitis B virus	Humans	Enveloped Partially dsDNA genome	Hepatitis (possible link with liver cancer)	All people	Targets liver cells Causes liver damage	None	Yes
Escherichia coli; UPEC strains	Humans (resident microbiota)	Gram-negative Facultative bacteria	Urinary tract infections Cystitis	Community-acquired (young girls, 20–40 yr old women) Hospital-acquired	Type 1 pili (FimH) P pili (kidney infection) Lipopoly-saccharide	Ampicillin Trimethoprim-sulfamethoxazole	No

(continued)

chapter 18 at a glance *(continued)*

Comparison of Urogenital Tract Pathogens *(continued)*

Species	Reservoir	Characteristics	Disease	Susceptible populations	Virulence factors	Treatment	Vaccine
Neisseria gonorrhoeae	Humans (asymptomatic carriers)	Gram-negative diplococcus Nutritionally fastidious LOS	Gonorrhea Pelvic inflammatory disease (PID) Infant blindness	All people (gonorrhea, arthritis) Women (PID) Neonates (infant blindness)	Pili Antigenic variation Elicits inflammatory response	Ampicillin Tetracycline (some resistant strains) Azithromycin	No
Chlamydia trachomatis	Humans (asymptomatic carriers)	Obligate intracellular parasite Gram-negative-like cell envelope but no peptidoglycan Elementary body Reticulate body	Nongonococcal urethritis (NGU) PID Infant conjunctivitis and pneumonia	All people (NGU) Women (PID) Neonates (conjunctivitis, pneumonia)	Elicits inflammatory response	Doxycycline Erythromycin Azithromycin	No
Candida albicans	Humans (resident microbiota, men asymptomatic)	Budding yeast	Vaginitis (curd-like discharge) Thrush (infants, children)	Women (especially diabetics, pregnant women) After antibiotic use	Elicits inflammatory response	Azole	No
Trichomonas vaginalis	Humans (men asymptomatic)	Flagellated protozoan Trophozoite stage No cyst stage	Vaginitis (copious runny, frothy discharge)	Women	Kills host cells by contact dependent mechanism Elicits inflammatory response	Metronidazole	No

key terms

acquired immunodeficiency syndrome (AIDS) caused by the human immunodeficiency virus (HIV); may be sexually transmitted or acquired from contaminated blood (intravenous drug use or transfusions); virus attacks T cells and monocytes

cancer HBV causes liver cancer and HPV causes cervical cancer

cystitis infection of the urinary bladder; in community-acquired cases, source of microbes is usually colonic microbiota (*Escherichia coli* and related species); in hospital-acquired cases, causative agents are more varied

endometritis infection of the uterus; spread from cervical infection

genital herpes caused mainly by HSV-2; acquired by sexual transmission; initially cause lesions on penis, vulva, cervix or anus; replicate in epithelial cells then enter neurons and travel to ganglia where they become latent; may later be reactivated by variety of stimuli

gonorrhea infection of urethra in men and cervix in women (may lead to pelvic inflammatory disease); caused by *Neisseria gonorrhoeae*; may be asymptomatic, especially in women

nongonococcal urethritis (NGU) infection similar to gonorrhea; caused by *Chlamydia trachomatis* and other unidentified microbes; often asymptomatic

pelvic inflammatory disease (PID) infection of the fallopian tubes; may extend to the peritoneal cavity; commonly caused by *Neisseria gonorrhoeae* and *Chlamydia trachomatis*; may lead to ectopic pregnancy or infertility

pyelonephritis infection of the kidney; may develop from ascending urinary tract infection or may be acquired from hematogenous spread

urethritis infection of the urethra

urinary tract infection (UTI) infection of any portion of the urinary tract from urethra to kidney; usually caused by bacteria and occasionally by fungi or protozoa.

vaginitis inflammation of vaginal tract caused by *Candida albicans* or *Trichomonas vaginalis*

Questions for Review and Reflection

1. Is there any urogenital tract disease that could be identified as "equal opportunity" with regard to symptoms in men and women? What are the characteristics of the microbe that make the disease less sexist?

2. Why was the discovery that papilloma virus and hepatitis B virus cause cancer good news? What do these viruses have in common that might help explain why they cause cancer?

3. Azithromycin has the advantage that one or two doses are sufficient to cure chlamydia and/or gonorrhea. Why is this an advantage? Given that gonorrhea and chlamydia are likely to cause widespread infertility cases that eventually could engender large medical expenses, should the pharmaceutical company that makes azithromycin be compelled to offer this antibiotic at a lower cost so that clinics in poor areas could afford it? Or should the government get involved? Argue both sides of this question.

4. Suppose you wanted to develop a vaccine against *C. trachomatis*. How would you proceed?

5. Why has it been so difficult to develop a vaccine against HIV? How might the knowledge that some people are genetically immune to HIV help in the design of new drugs or prevention strategies?

6. Recently, trials of short-term AZT regimens to prevent mother-to-infant transfer of HIV came under attack from people concerned about medical ethics. These trials took place in Africa and Thailand. In these trials, one group of pregnant women was given a placebo and another was given AZT. Such a trial design would never be approved in the United States and Europe. Why would this trial design be unacceptable in developed countries? Why did some scientists in developing countries defend it?

7. Consider the pathogens other than HIV described in this chapter. Do your standards for the design of human trials change? Why or why not?

8. Scientists are excited about the hepatitis B vaccine and what they hope is the hepatitis C vaccine to come. Yet parents did not rush to have their children vaccinated. In fact, undercoverage is an emerging problem. How would you convince a parent that these vaccines are a good idea?

19 Nosocomial and Iatrogenic Infections

Modern medicine has created new ways to correct many health problems and in the process is prolonging human life and improving its quality. These same medical miracles have also provided new opportunities for microbial infections, some caused by microbes incapable of infecting healthy humans. Such infections, which can be fatal, continue to multiply in numbers and in varieties of microbes involved. The microbes can come from unexpected sources. To cope with microbial challenges hospital infection control specialists have become a new breed of detective.

Breaches in the body's defenses create openings that allow microorganisms that would not normally cause disease to infect. This is especially true in hospitals, where surgical wounds and other medical interventions create such breaches. Worse, microorganisms that flourish in the hospital environment on surfaces, in the air, or on the bodies of hospital care workers tend to be more resistant to antibiotics than bacteria outside the hospital. Infections acquired in hospitals have been given a special name, **nosocomial infections.** The name comes from the Greek words *noso,* for disease, and *komeion,* for taking care of. These infections, then, are acquired in a place where people with diseases are cared for. A related term, **iatrogenic** infections, is derived from the Greek word *iatro,* for physician. These are infections caused by physicians or other health care workers, whether in or out of hospitals. Some iatrogenic infections are unintentionally caused by lapses in aseptic techniques. Others are inescapable consequences of life-saving treatments such as cancer chemotherapy that temporarily render the patient immunocompromised. The problem of nosocomial and iatrogenic infections is serious. If such infections are not brought under control rapidly, the infected patient may die. Even if the infection is finally brought under control after a prolonged course of therapy, the patient may experience long-term debilities: scarring, stroke, or irreversible damage to the lungs or other vital organs (see Chapter 12 for an explanation of the consequences of bloodstream infections).

Features of Nosocomial and Iatrogenic Infections

Origins of the nosocomial infection problem

Before aseptic techniques were introduced, most patients whose body cavities were opened during surgery died of postsurgical infections. Such surgeries, accordingly, were procedures of last resort, and even very sick patients were reluctant to submit themselves to such risky therapy. Surgical wards were permeated with the smell of rotting flesh arising from patients with infected surgical wounds. A major advance came in 1867, when Joseph Lister (1827–1912) published results on his discovery that spraying carbolic acid (phenol) in the air and having surgeons and their assistants wash their hands in liquid carbolic acid dramatically reduced the rate of postsurgical infections. This was hard on the surgeons and their assistants, because phenol is a highly toxic and noxious substance, so much so that its use is strictly limited today by chemical safety regulations even in laboratory settings. During the late 19th and early 20th centuries, disinfectants safer than phenol were introduced, but the incidence of postsurgical infections remained unacceptably high. Only with the discovery of antibiotics did surgery become the relatively low-risk procedure it is today. Antibiotics were so successful, in fact, that hospital personnel became overconfident and began to skimp on such basic hygienic procedures as hand washing. Today, antibiotic-resistant bacteria are forcing hospital care workers to relearn such simple but effective procedures as hand washing and good aseptic technique.

Consequences of nosocomial infections extend beyond the infected patient

Postsurgical and other hospital-acquired infections are troubling to hospital administrators, because well-publicized cases not only generate lawsuits but undermine public confidence in hospitals. This has been especially true in the case of large hospitals with intensive care wards, where nosocomial infections are most likely to occur. Nosocomial infections also trouble hospital administrators and insurance companies, because patients who acquire infections in the hospital will stay longer and thus be more expensive to treat.

What are the average hospital patient's chances of incurring a nosocomial infection? Estimates place the number of cases of nosocomial infections each year in the 2 million range, about 3% of hospital patients. Infection rates are lower in small hospitals with mostly routine surgeries and higher in large hospitals, where more complicated and lengthier surgeries are performed on much sicker patients and where patients spend more time in intensive care units (ICUs).

Rates of infection also vary with the type of procedure and the area of the body involved. Surgeries involving normally sterile tissues not in contact with grossly contaminated areas, such as the colon, are the safest, with an infection rate of only about 2%. But in surgeries involving the abdominal cavity, where the contents of the colon may have contaminated surrounding tissue, the rate of postsurgical infections can approach 10%. Likewise, in surgeries that leave the patient open for hours, especially certain types of heart surgery in which the jagged surfaces of cut-open rib cages can breach the security net provided by the double-gloved hands of the surgeon, postsurgical infections can be as high as 10%. The surgeon is also a factor. Physicians who take chances by cutting short the surgical scrub or leaving the operating suite for prolonged periods during the surgery will have higher rates of postsurgical infections than surgeons who observe strict aseptic procedures.

How high do rates of infection rise before patients begin to avoid a hospital? Currently, hospitals do not voluntarily disclose rates of postsurgical infections, although consumer groups are suing to obtain this information. To be fair to the hospitals, the situation is difficult. Someone must care for seriously ill patients and those needing risky and complex surgeries. To penalize the hospitals that undertake these roles by scaring away patients would be counterproductive for both hospitals and patients. As is always the case when infectious diseases are involved, prevention is the best solution, and hospitals are now taking aggressive steps to decrease nosocomial infection rates.

Anticipating and preventing nosocomial infections

Although some settings are more likely than others to produce nosocomial infection, disease patterns can change unexpectedly. Another complicating factor is that the types of microbes responsible for these infections can change. Two changes that have occurred during the last two decades illustrate this problem. In the 1950s and 1960s, gram-positive bacteria (for example, *Streptococcus pyogenes* and *Staphylococcus aureus*) were common causes of infection. In the 1970s and 1980s gram-negative bacteria (for example, *Escherichia coli* and *Pseudomonas* species) emerged as the primary causes of nosocomial infections. Currently, gram-positive cocci have begun to predominate once again (for example, *S. aureus*, *Staphylococcus epidermidis*, and *Enterococcus* species), and species previously thought to be incapable of causing infections (for example, *S. epidermidis* and *Enterococcus* species) have become serious threats. The gram-negative bacteria are still around, but infection control specialists now have gram-positive bacteria and an increasing diversity of bacterial species to contend with as well.

A serious recent development has been the rise of **antibiotic-resistant bacteria.** This development further complicates diagnosis and treatment decisions. Having to go through two or more antibiotics to find the one that works means that the patient has the infection longer, increasing the chances that permanent damage or death can occur (Box 19.1).

An additional problem is that nosocomial infections can arise from unexpected sources, as two examples in this chapter illustrate. In one case, contaminated mouthwash was the culprit. In the other, bacteria growing in an anesthetic solution were responsible.

Responsibility does not rest solely with the hospital. The patient's family has a role to play in decreasing the incidence of nosocomial infections. The U.S. Centers for Disease Control and Prevention (CDC) have gone on record urging family members to question staff when they see clear examples of risky health care behavior, such as failure to wash hands, failure to change gloves between patient examinations, or careless handling of intravenous lines. Aggressive campaigns mounted by the CDC and hospitals against breaches of hygienic procedures among staff have proven partially effective. No one seems, however, to have considered the possibility that another effective method of dealing with the problem might be to provide positive rewards for improved behavior and to alleviate some of the time and economic pressures that encourage careless behavior by harassed hospital staff members.

This chapter explores some examples of nosocomial infections. These examples illustrate the variety of ways in which such infections can occur even in a well-run hospital and how far reaching their effects can be (Fig. 19.1). They also illustrate a continuing theme: breaching the defenses of the body creates opportunities for enterprising microbes.

box 19–1 Emerging Resistance in a Major Nosocomial Pathogen

Staphylococcus aureus, a gram-positive coccus normally found in the human nose and various other body sites, is a major cause of nosocomial infections. The name, *S. aureus* (*aureus* is the Latin word for gold), comes from the golden color of the colonies it forms. *S. aureus* is also a common cause of wound infections and food-borne disease in the community, but it is of greatest concern when it strikes already sick hospital patients. Some strains of *S. aureus* have become resistant to many antibiotics. Such strains have been called methicillin-resistant *S. aureus* strains (MRSA), although they are actually resistant to other antibiotics as well. Vancomycin has been the drug of choice for treating MRSA infections, but recently MRSA and some other *S. aureus* strains have begun to become less susceptible to vancomycin. These strains have been called VISA strains, for vancomycin intermediate-susceptibility *S. aureus*. Intermediate susceptibility means that the strain is not fully resistant but has a minimal inhibitory concentration value (see Chapter 15) approaching the resistance cutoff.

In January 2000, the U.S. Centers for Disease Control and Prevention (CDC) reported a case of a fatal nosocomial VISA infection. This was not the first such report, but it illustrates how dangerous these strains can be to very sick patients. A 63-year-old woman with a history of kidney problems and dialysis developed an *S. aureus* bloodstream infection 13 days after admission to the hospital for treatment connected with her kidney problem. Unfortunately, the *S. aureus* was a VISA strain. It was also resistant to β-lactam antibiotics (penicillin, oxacillin; Chapter 4), macrolides (clindamycin, erythromycin; Chapter 6), ciprofloxacin (a fluoroquinolone; Chapter 5), and rifampin (Chapter 6). The strain had intermediate susceptibility to chloramphenicol and vancomycin. The patient was treated with intravenous vancomycin but died 10 days after her disease was diagnosed.

Following the CDC guidelines, the hospital immediately undertook a survey to find the source of the VISA strain. None of the health care workers or family members surveyed was colonized with this or any other VISA strain. Thus, the strain probably came from the woman's own microbiota. She had been treated previously with vancomycin. If the hospital infection control team had identified a carrier other than the woman herself, steps would have been taken to avoid spreading the strain to other patients in the hospital.

The appearance of VISA strains causes concern inside hospitals and out. Such strains are dangerous for hospitalized patients, as illustrated by the case just described, but are also dangerous for otherwise healthy people in the community. In 1999, the CDC reported the deaths of four children who had died of MRSA between 1996 and 1999. The children had no known underlying condition that would have made them immunocompromised. When MRSA becomes VISA, the situation could become even worse and the number of treatment failures could increase. New antibiotics recently approved or under consideration for approval by the U.S. Food and Drug Administration should help in the fight against VISA. However, the loss of vancomycin, a frontline drug that has been successful in treating infections by *S. aureus* and other gram-positive cocci in the past, would be a major setback.

Sources: 2000. Morbidity and Mortality Weekly Rep. **48:**1165–1166.

Herold, B. C. et al. 1998. Community-acquired methicillin-resistant *Staphylococcus aureus* in children with no identified predisposing risk. JAMA **279:**593–598.

Infectious Disease Problems in Intensive Care Units (ICUs)

Nosocomial infections are most likely to occur in ICUs. Patients in ICUs usually have one or more underlying conditions that predispose them to disease. Moreover, ICU patients may be housed in relatively crowded conditions because of the need for close supervision of a number of patients. Nursing homes offer some of the same opportunities for infection: elderly people with less active immune systems, housed in crowded conditions. Thus it is not surprising that many of the same infectious disease problems that arise in ICUs also arise in nursing homes, especially large ones with poorer patients.

Indwelling urinary catheters

Because ICU patients are often comatose or incontinent, many are fitted with indwelling urinary catheters. The catheter, which is inserted into the urethra, keeps the bladder drained of urine, and the urine is collected in a bag that is changed regularly. Catheters bypass two major defenses of the urinary tract: the sphincter, which normally closes off the urethra to outside

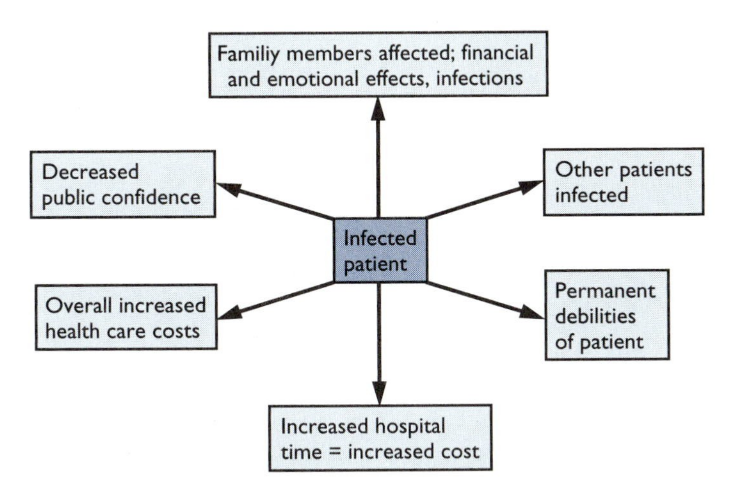

Figure 19.1 Consequences of nosocomial infections. The infected patient is most directly affected, but infection can be spread to family members and to other patients. Nosocomial infections can increase the time the patient must stay in the hospital, thus increasing costs. Patients also may never fully recover. Such infections can have significant impacts on family members (both financial and emotional) and can decrease public confidence in hospitals.

contamination, and the washing action of urine during normal urination. Catheters may also leave in the bladder small puddles of stagnant urine, which provide propitious places for bacteria to multiply. Given these factors, it is not surprising that patients with indwelling catheters in place for long periods of time frequently develop urinary tract infections. Bacteria form **biofilms** on the exposed surface of the catheter and grow along the catheter surface into the urethra. Catheter-associated urinary tract infections may not be symptomatic, and, even when they are, the patient may not be well enough to be aware of them. The danger is that even an asymptomatic bladder infection may turn into a bloodstream infection if the causative microbes ascend to the kidneys, where the inflammation they cause may allow them to enter the bloodstream. Nosocomial urinary tract infections can be caused by a variety of microbes. Among the most frequent causes are *E. coli* and related gram-negative bacteria, *Pseudomonas aeruginosa,* and *Enterococcus* species (Fig. 19.2). The yeast *Candida albicans* can also

Figure 19.2 Scanning electron micrograph of *Enterococcus* spp. Courtesy of the Centers for Disease Control and Prevention, Atlanta, GA.

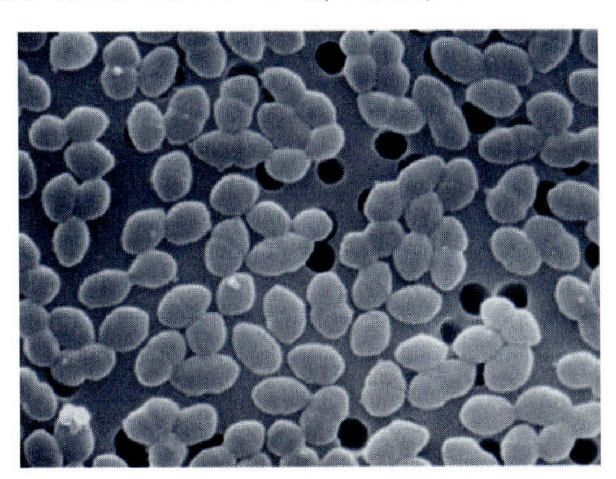

cause nosocomial urinary tract infections, a fact that indicates the degree to which some ICU patients are immunosuppressed.

Venous catheters

Plastic catheters inserted into veins are common means for introducing antimicrobials or other therapeutic agents into the bloodstreams of patients and for feeding comatose patients. These catheters pose nosocomial infection problems because they bypass the intact skin defense. Skin bacteria, such as *S. epidermidis,* form biofilms on the surfaces of these catheters and use the catheters as introduction points into the bloodstream. *S. epidermidis* usually does not cause infections in otherwise healthy patients, but in very sick patients or in patients with defective heart valves or heart valve implants this species can be lethal. Infectious disease problems associated with plastic implants are described later in this chapter.

Ventilator-associated pneumonia

Seriously ill patients may require devices that support breathing (ventilators). Oxygen is forced into the lung either intermittently or constantly to keep the patient's blood well oxygenated. Some oxygen delivery systems have short tubes that are simply inserted into the nose, but the more problematic ones have tubes that are inserted into the main stem of the bronchus (Fig. 19.3). Tubes inserted so deeply in the upper airway bypass an important defense of the lungs, the ciliated cells. Unless ventilators are carefully disinfected, the tubes themselves may introduce bacterial pathogens directly into the lungs. Although disinfecting ventilators might seem easy, the task is made more difficult by the partial resistance of bacteria, such as *P. aeruginosa* or *Staphylococcus* species, to commonly used disinfectants. Thus, if the disinfection protocol is not followed precisely or if the solutions are not at the right strength, some bacteria may slip through what is otherwise an effective procedure. Another source of infection is the air being pumped

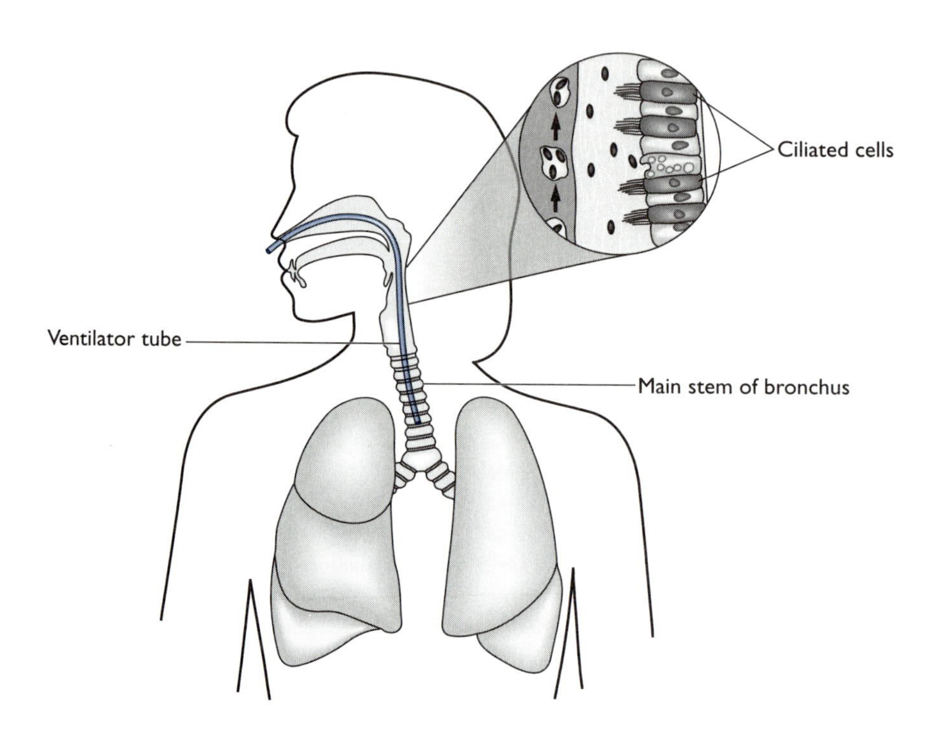

Figure 19.3 Ventilator tube in respiratory tract. Ventilator tubes extend into the main trunk of the bronchus. This bypasses the ciliated cells of the upper respiratory tract and allows microbes to be delivered directly from the environment to the lower portions of the respiratory tract.

into the lung. This air is humidified, so droplets of contaminated water can be introduced into the lung. Bacteria normally found in the mouth may even gain access to the lung via the ventilator. The length of time the ventilator is in place is also a factor. Breathing tubes are used in surgery to support respiration in anesthetized patients, but such use is brief and thus unlikely to cause infection problems. Ventilator tubes that are in place for weeks or longer are more likely to be associated with infections.

For an example of how insidious **ventilator-associated pneumonia** cases can be, consider an outbreak of ventilator-associated pneumonia reported to the CDC in 1998. The number of cases in one hospital had been unusually high, even for ventilator patients. When the bacterium responsible for causing the infections was cultivated and identified, it proved to be the same organism in all cases, *Burkholderia* (formerly *Pseudomonas*) *cepacia*, a relative of *P. aeruginosa*. *B. cepacia* is normally found in soil and water and is considered safe enough for use as a bioremediation agent to degrade pesticides and herbicides. *B. cepacia* had been identified as a cause of lung infections in cystic fibrosis patients but previously had not been one of the bacteria identified as responsible for ventilator-associated pneumonia.

A survey of materials and people who had contact with the infected patients revealed that the source of *B. cepacia* was the mouthwash being used for oral hygiene (Fig. 19.4). Because some ICU patients are comatose and cannot maintain oral hygiene, build-up of bacteria in the mouth becomes a problem. Obviously, bacterial build-up on the teeth and tongue in ventilator patients is dangerous, because these microorganisms could enter the lungs, where defenses are already impaired. Accordingly, as part of daily hygiene care, hospital staff members swabbed mouthwash on the patients' teeth and gums. The mouthwash being used in the hospital did not contain alcohol, because alcohol can dehydrate oral tissues. The absence of alcohol, however, meant that there was no barrier to growth of a bacterial contaminant in the mouthwash. Additional investigation revealed that the mouthwash had been contaminated by a water source used at the plant where it had been bottled. This example shows that ICUs not only may involve unexpected microbes, but that these microbes may come from unexpected outside sources.

Ventilator-associated pneumonia is much more likely to be caused by the more common nosocomial culprits, such as *P. aeruginosa* or *S. aureus*. The *B. cepacia* case shows how difficult it is for even the most diligent nurses and infectious disease specialists to foresee and prevent nosocomial infections.

Neonatal intensive care units

Perhaps the most heart-wrenching hospital-acquired infections are those that occur in another high-risk patient population, infants in **neonatal ICUs.** Many of these small bundles of mortality are already fighting not only the effects of prematurity but cocaine residues, acquired immunodeficiency syndrome, or organ abnormalities. Hand washing by hospital staff, unfortunately, is often as underused in this setting as in other areas of hospitals. The reason is easy to understand. Hospital staff members, worried about the more immediate problems of their small charges, can all too easily forget that infectious disease is also a potential threat. As in the case in adult ICUs, infections can come from unexpected sources (Box 19.2). A recent outbreak of *P. aeruginosa* infections thought to be transmitted on the fingernails of nurses, especially those with long nails, provides yet another example of an unexpected mode of transmission.

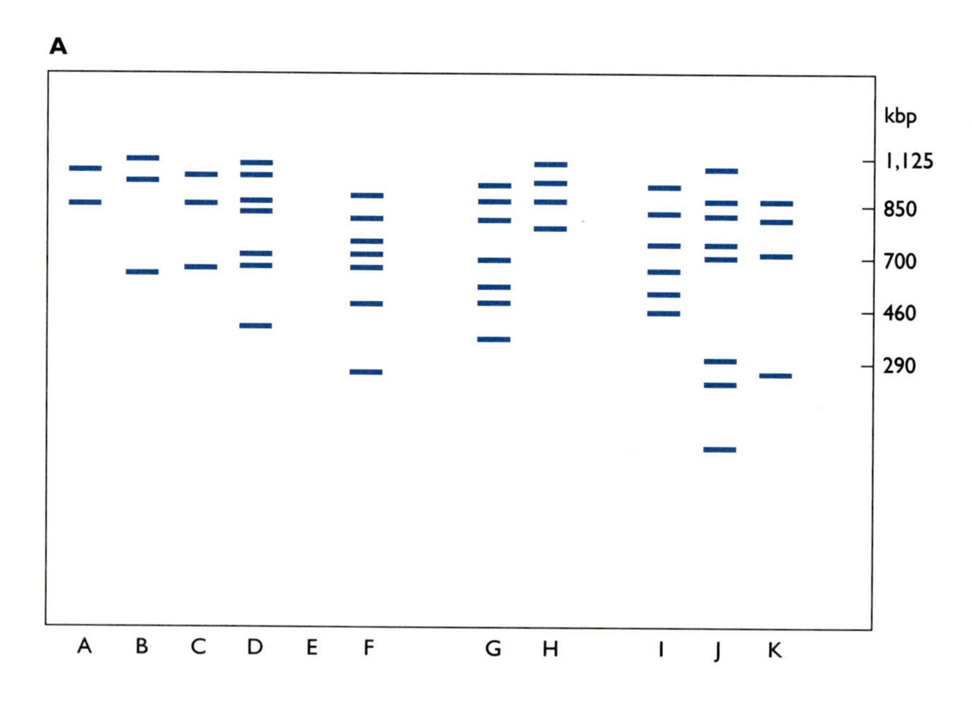

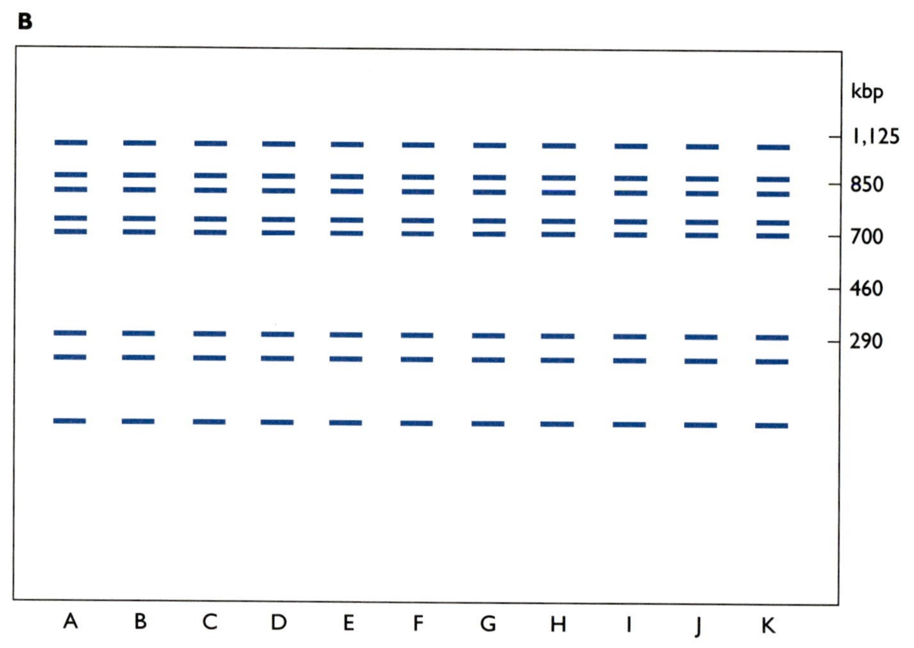

Figure 19.4 Results of pulsed field gel electrophoretic analysis of bacteria isolated from patients involved in the mouthwash outbreak. (See Chapter 3 for a review of PFGE.) **(A)** Comparison of migration of DNA fragments from different strains of the same species not associated with the outbreak except for the mouthwash isolates (lane J). Note that the pattern of migration is unique to each strain. **(B)** Comparison of migration of DNA fragments from different isolates obtained from sick patients (lanes B–K) and an isolate obtained from the mouthwash (lane A). Note that the strains have an identical pattern, a clear sign of a single-source outbreak.

Postsurgical Infections

Often the causative agents of postsurgical infections are microbes normally found on the human body, introduced either by the surgeon or by later contamination of wounds by the patient's or nurse's own microbiota. A major cause of such infections is *S. aureus,* which commonly colonizes the nose and can easily be spread from staff to patient or from patient to patient. Keeping patients clean is not as easy as it might seem. First, washing and bathing often must be done while the patient remains in bed. Bacteria on the sheets or on bed rails are not eliminated by this procedure. More seri-

box

19–2 *Breast Milk as a Source of Infection*

In one reported case, twins born prematurely were fed milk expressed from their mother's breasts, a standard procedure in such cases. The breast milk that should have been so beneficial proved to be deadly. The infants developed a fatal disease that began with intestinal necrosis and progressed to septic shock. An investigation of the sources of the infection revealed that the mother's milk had been contaminated with *S. epidermidis*, a common skin bacterium. *S. epidermidis* usually cannot infect healthy people and has not been known to cause intestinal infections that progress to septic shock. In this case, however, *S. epidermidis* appears to have done the unexpected once again.

The contamination problem apparently arose because the breast milk had not been refrigerated. Because *S. epiderimidis* is everywhere on skin, it would be virtually impossible to prevent it from entering expressed breast milk. Because of the lack of refrigeration, the bacteria had increased dramatically in numbers in the milk before being fed to the babies. Pasteurization of the milk (heating it to 60°C for 20 minutes) might have prevented this tragedy, but mother's milk is generally not pasteurized for a very good reason: pasteurization would impair the activity of the sIgA molecules that are an important part of the beneficial effect of mother's milk. Mother's milk remains the best option for neonates, but this unusual case underscores the fact that proper refrigeration and careful handling are absolutely essential for safety. Such cases are rare enough that hospital care workers would not have expected this possible hazard to premature infants.

Source: **Griffiths-Jones, A.** 1997. Caring for neonates. *Nursing Times* **93:**75.

ously, staff members caring for patients who have indwelling urinary catheters or who require bedpans may inadvertently create aerosols from bedpan use or from changing outflow bags containing contaminated urine. The most important protection against postsurgical infections, aside from a surgeon who takes bacterial infections seriously, is alert health care workers who notice and react aggressively to early signs of infection, such as fever.

Postsurgical infections from a contaminated anesthetic

Still another example of infections caused by contamination occurred between 1990 and 1993. Seven different hospitals experienced outbreaks of bloodstream infections. The CDC investigated these outbreaks of postsurgical infections, and the initial findings were confusing. Different bacteria were isolated from different patients, a finding that suggested at first that there was no single source of infection as there was in the case of the contaminated mouthwash. Nor was a single surgeon or a single nurse associated with all of the infected patients. All of the patients in all of the hospitals, however, had received the anesthetic propofol (Diprivan). Propofol is unusual among anesthetics, because it is lipid-based. If contaminated and left at room temperature, this type of anesthetic can support the growth of many bacteria, because it is such a rich medium. Propofol is injected, so if it becomes contaminated, the microbial contaminant is introduced directly into the bloodstream. This lead at first proved elusive, because microbiological testing of sealed bottles showed that the anesthetic was sterile when it reached the hospital. Microbiological analyses of syringes containing propofol, however, showed that the contamination event had occurred between the opening of the sealed bottle and the injection of the anesthetic into the patient.

Additional investigation of events between the opening of the bottles of propofol and injection into patients revealed not only the route of infection but exposed a serious problem that plagues modern hospitals. Because of time pressures imposed by high patient loads and inadequate numbers of support staff, physicians and other health care workers may take shortcuts in standard procedures designed to protect the patient from infection. The investigative team saw health care workers filling syringes without using aseptic technique, carrying these syringes in ungloved hands from one room to another, leaving them sitting for hours at a time at room temperature, reusing syringes, and other lapses in good aseptic practice. Vials of propofol, contaminated by one careless act during the filling of the syringe, sat at room temperature, allowing bacteria to grow, so that the next patients injected with propofol from that vial were given even more contaminated injections. Use of the syringe from the first contaminated vial to withdraw propofol from a second, newly opened vial, led to further contamination events.

To be fair to harried staff members, however, it is important to point out that such cases are unusual, a fact that makes the shortcuts seem innocuous. Also, staff members are often under pressure to carry out more procedures than can fit comfortably in the amount of time allotted. It is easy to understand, for example, why staff members might have filled a number of syringes in the morning for use during the surgeries planned for that day. Unfortunately, in the propofol-related infection cases, some of the infected patients died or sustained long-term physical damage because of such shortcuts.

The infected surgeon

A problem that has received very little attention in the media but is well known to hospital infection control specialists is that of surgeons who are infected with human immunodeficiency virus (HIV), hepatitis B virus, or hepatitis C virus. At present, the surgeon's infection status is kept secret, out of respect for individual privacy rights. HIV-infected surgeons usually are assigned other duties fairly quickly by hospitals, but the hepatitis problem persists. Some cases of hepatitis B or hepatitis C virus transmission from surgeon to patients have been reported. Although such cases have been rare, they are troubling to patients and hospital administrators. What are the limits to a surgeon's right to practice when he or she is infected? This is an especially sensitive issue if the surgeon performs heart surgery or other surgeries in which contamination of the patient by the surgeon's blood can occur.

Neutropenic Patients: Cancer Chemotherapy

Neutropenia is a condition in which a person has an abnormally low number of neutrophils in circulation. Before the development of cancer chemotherapy, only people with certain genetic defects or severe malnutrition experienced neutropenia. **Cancer chemotherapy,** which kills not only the rapidly dividing cells of the tumor but also other rapidly dividing cells, such as neutrophils, produces a temporary neutropenia. As with other infections, the length of neutropenia and the patient's condition are critical. Patients undergoing cancer chemotherapy for the first time are at a lower risk of infection than patients in whom cancer has returned and are experiencing chemotherapy on top of an already weakened immune system. Also, the longer the patient receives therapy, the greater the risk of infection.

People with abnormally low neutrophil counts have 5–10 times as many infections as they experience when their neutrophil counts return to normal. Neutropenic patients not only contract more infections but experience more serious infections because an important defense against invading bacteria has been lost. Almost any bacterium can infect neutropenic patients, including *S. epidermidis,* a common skin inhabitant, and *Enterococcus* species, bacteria normally found in the human colon. Thus, the patient is in danger from his or her own microbiota. Soil and water bacteria, such as *P. aeruginosa,* can also infect neutropenic patients. Not only are the bacteria just mentioned encountered frequently in daily life, but they are also notorious for resistance to many antibiotics, a fact that makes infections caused by them difficult to control.

The ideal solution to the chemotherapy-induced neutropenia problem would be "smart" cancer chemotherapeutic agents that target tumor cells for destruction but leave the neutrophils alone. Such drugs are currently under development but are probably years from the market. In the meantime, restoring neutrophils as soon as possible after chemotherapy treatment seems the most promising approach. **Granulocyte colony-stimulating factor (G-CSF)** is a protein so named because, under *in vitro* conditions, it stimulates neutrophils (also called granulocytes) to reproduce more actively and stimulates bone marrow to replace dead neutrophils as rapidly as possible. G-CSF preparations and related compounds are currently available for use in getting chemotherapy patients' neutrophil levels back to normal as quickly as possible.

Implant-Associated Infections

Plastic heart valves have been a boon to many patients who would have died without these prosthetic devices that keep damaged hearts working effectively. Unfortunately, contamination of a valve implant during surgery can occur, especially when proper aseptic procedures are not observed meticulously. The protection against contamination should be two-fold. First, the surgeon temporarily eliminates potentially dangerous skin microorganisms by performing the surgical scrub. Then gloves are donned to give an added measure of protection against possible contamination of the implant by the surgeon. The reason both measures are necessary is that heart surgery involves sawing open the rib cage, creating small bone slivers that can penetrate gloves. The metal wires used to rejoin the ribs also can penetrate the barrier of protection offered by gloves. Thus, the skin underneath must be ultra clean to protect against bacteria that might escape through glove breaches.

Contamination of a valve implant is difficult to treat for two reasons. First, the plastic valve discourages migration of neutrophils and macrophages that would normally mop up small bacterial spills. Second, bacteria—those highly adaptive creatures—have already evolved to form biofilms on plastic surfaces. Because bacteria in biofilms, especially mature biofilms, are less susceptible to antibiotics than free-living bacteria and because the phagocytic cell defense is impaired, contamination of a valve or other plastic implant can be disastrous for the patient, creating a nidus of infection (a condition called endocarditis) that can leak bacteria into the bloodstream and cause localized damage to the heart. Although recent research has shown that prompt antibiotic therapy for infection can sometimes save the implant, a well-developed colonization usually requires a second operation to remove the contaminated valve, resulting in additional damage to the heart and increased risk of a bloodstream infection and septic shock.

The microbial culprits in implant-mediated infections are some of the same ones responsible for infections of neutropenic patients: *S. epidermidis,*

S. aureus, and *Enterococcus* species. Staphylococci in particular seem to have taken to plastic with a vengeance. They adhere to the plastic, actually causing pits and crevices to appear. After the first layer of bacteria binds to the plastic, a biofilm quickly forms as other bacteria are bound to the underlying layer of bacteria by a polysaccharide matrix.

Burn Patients

In one sense, infections in burn patients are not truly nosocomial, because these infections would occur as readily outside a hospital as in. Because patients with serious and extensive burns are usually taken to hospitals, however, their infections tend to reflect whatever microbes are in the hospital environment. Patients who have survived the initial trauma of severe burns are at high risk for infection for two reasons. First, the barrier of intact skin has been disrupted. To make matters worse, the skin-associated lymphoid tissue (SALT), the arm of the immune system that underlies skin, has been severely impaired or destroyed and does not return until the burned tissue heals. A burned area is wet with plasma oozing from the tissue, and is an ideal culture medium for bacteria. Two bacteria that are the most common causes of burn infections are *S. aureus* and *P. aeruginosa*. *S. aureus*, a gram-positive coccus, is commonly found in or on the body (especially in the nose) and is a hardy microorganism that survives for long periods of time in the hospital environment. *S. aureus* is a virulent pathogen that causes disease even in healthy people. Diseases caused by *S. aureus* range from food-borne disease caused by toxin-producing strains (Chapter 15) to wound infections and systemic infections. By contrast, *P. aeruginosa*, a gram-negative rod normally found in water and soil, usually does not cause infection in healthy people. Nonetheless, in burn patients *P. aeruginosa* is every bit as dangerous as *S. aureus*.

Both *S. aureus* and *P. aeruginosa* tend to be resistant to a variety of antibiotics, making infections difficult to bring under control. Accordingly, burn patients must be watched very carefully for early signs of infection. Because a burned area is already red and swollen, the signs of inflammation are not easily seen, but the appearance of pus can suggest that an infection is under way. *P. aeruginosa* produces several pigments that fluoresce under ultraviolet light. A Woods lamp, which produces such light, can be used to illuminate the wound in the dark to detect the tell-tale fluorescence that suggests a *P. aeruginosa* infection. The major concern in cases of burn infections is not so much the superficial infection, although such infections can slow the healing process, but the possibility that the bacteria may enter the bloodstream and cause septic shock.

A second problem is that burn patients may have inhaled very hot air and smoke during exposure to fire. Thus the nonspecific defenses of the lung may be impaired. Bacterial pneumonia is another disease to which burn patients are especially susceptible. Once again, the biggest danger is that bacteria will enter the bloodstream and cause septic shock.

What is Being Done

Over the past century, many important medical advances have prolonged life and increased the quality of life for millions of patients. The question now is how to preserve these impressive advances without paying an unduly large infectious disease bill. The answer to this question is clear.

Health care workers need to return to the mindset of an earlier era that focused on prevention.

Hospital infection control specialists are now formulating plans for isolation of patients, prevention of infection by interrupting the transmission chain, and bolstering the defense systems of temporarily immunocompromised patients. The good news is that hospitals are now taking the nosocomial infection problem quite seriously and are moving in effective ways to reduce it. Unfortunately, these remedies will be expensive initially, but so are the consequences of hospital-acquired infections, both to the patient's health and the hospital's bottom line. If, as seems inevitable, lawyers decide to insert themselves more aggressively into the hospital-patient relationship and begin suing hospitals over hospital-acquired infections, the financial incentive to clean up the hospital's act will become even stronger.

chapter 19 at a glance

Extent of the nosocomial infection problem

Infections may be fatal

Damage to patient who survives may be permanent

Added expense

Changing spectrum of causative agents
- 1950s–1960s: predominantly gram-positive bacteria
- 1970s–1980s: predominantly gram-negative bacteria
- Today again predominantly gram-positive bacteria
- Microorganisms today more likely to be resistant to antimicrobial agents

Prevention
- Enforcement of proper hygiene
- Infection control specialists monitor episodes of infection
- Early and accurate diagnosis

Intensive care units: selected examples of problems

Indwelling urinary catheters bypass defenses of bladder, urethra

Venous catheters bypass defenses of skin
- Main causative agent: *Staphylococcus epidermidis*

Ventilator-associated pneumonia: mouthwash example shows unpredictability
- Mouthwash swabbed on teeth and gums for hygiene
- Some mouthwashes that contain no alcohol are contaminated at plant
- Intubation bypasses airway defenses
- Cause of pneumonia: *Burkholderia cepacia*, gram-negative soil bacterium

Neonatal intensive care
- Lack of conscientious hand washing still a problem
- Bacteria may be transmitted on long nails
- Infants have underlying conditions making them susceptible to infection
- Another example of an unexpected source of infection: mother's breast milk

chapter 19 at a glance

Postsurgical infections

Infection of the surgical wound during or after surgery
- Heart surgery or surgery near the bowel particularly hazardous
- Length of time required for the operation is important
- Example of contaminated anesthetic
  - Contaminated by mishandling
  - Anesthetic used was a good growth medium for bacteria
  - Bacteria injected into patients' bloodstreams with anesthesia

Transmission of *Staphyloccus aureus* and other bacterial pathogens from the patient's microbiota or from staff to patient

Neutropenic patients

Cancer chemotherapy kills neutrophils as well as cancer cells

Increased risk of infection by many different microorganisms

Possible solutions
- "Smart" cancer drugs, target tumor only
- Granulocyte colony-stimulating factor (G-CSF) and other proteins that stimulate production of new neutrophils

Implant-associated infections

Plastic surface somewhat protected from defense systems

Bacteria in blood attach to implant and form a biofilm

Inflammation damages heart tissue; bacteria can leak into bloodstream

Implant may have to be removed and replaced

Often caused by *Staphylococcus* species or *Enterococcus* species (both gram-positive cocci)

Burn patients

Burn obliterates skin barrier, skin defenses (skin-associated lymphoid tissue [SALT])

Area is wet, exposed; aids growth of bacteria

Most common infectious agents in wound infections
- *Pseudomonas aeruginosa*, gram-negative soil bacterium
- *S. aureus*, gram-positive human nasal microbiota

Inhalation of very hot air and smoke reduces airway defenses
- Pneumonia
- Bloodstream infection

key terms

biofilms community of microorganisms attached to a surface, such as plastic in a catheter or implant; held together by polysaccharide matrix; resistant to antimicrobial agents

iatrogenic infections infections caused by physicians or other health care workers; may occur in or out of hospitals

intensive care units wards for very sick patients many with conditions predisposing them to infections; areas usually crowded

neutropenia condition in which patient has abnormally low number of neutrophils; often occurs in patients receiving cancer chemotherapy

nosocomial infection an infection arising from hospital sources; source may be microbiota of patient or staff; contributing factors are poor aseptic techniques, surgery, neutropenia, time in intensive care, catheters, burns

urinary catheter plastic tube that keeps bladder drained of urine; bypasses urethral sphincter and washing action of urine, predisposing patient to urinary tract infections

venous catheter plastic tube inserted into veins to deliver therapeutic agents or nutrients to patients; bypasses skin, giving microbes direct access to bloodstream

ventilator-associated pneumonia pneumonia that occurs when tubes to deliver oxygen are inserted into main stem of bronchus; bypasses ciliated cell defense of upper respiratory tract

Questions for Review and Reflection

1. Nosocomial infections have a maddeningly unpredictable quality, but they do have some similar themes. Identify those themes for the examples given in this chapter. Then make an attempt to predict the next unexpected source of nosocomial infections.

2. Pulsed field gel electrophoresis (PFGE) was introduced in Chapter 3 in connection with a meningitis outbreak. Pulsed field gels from the mouthwash outbreak are provided in Figure 19.4. How do you interpret these data? What if there had been a slight change in the migration of a single band of one of the strains isolated in the outbreak? What would you have done then?

3. Explain, based on what you learned in Chapter 3, how the investigative team progressed from initial samples to PFGE. What would a PFGE analysis have looked like in the anesthetic outbreak described in this chapter? Would PFGE have been necessary in this case?

4. What would you do about a hospital care worker who is colonized with multidrug-resistant *Staphylococcus aureus* or *Staphylococcus epidermidis* or *Enterococcus* species? Keep in mind that they were colonized because their jobs put them at risk for such colonization.

5. Pursuant to question 4, what would you do about surgeons who are infected with hepatitis B, hepatitis C, or human immunodeficiency virus (HIV)? A dentist who has the same infection profile? Should all health care workers be tested for these infections?

6. Organ transplant patients were not covered in this chapter, but they too are at higher than normal risk for infection. Why?

7. Another topic not covered in this chapter is the hospital kitchen. At least one incident of transmission of multidrug-resistant *S. aureus* in hospital food has been reported. Hospital food is bad enough on the culinary and esthetic levels. Consider its potential to cause serious nosocomial infections. What can be done to prevent such infections?

8. Medical students have asked the authors why the term "nosocomial" infection has become so popular when "hospital-acquired" infection is so much more explicit. The authors gave an answer that some would consider to betray unseemly levity: If your doctor informs you that he is treating you for a nosocomial infection, you will call him your savior. If he tells you he is treating you for a hospital-acquired infection, you are more likely to call your lawyer. Do you agree with this or are there better reasons for using terms like nosocomial?

20 Arthropod-Borne Diseases and More Zoonoses

Intact skin is a powerful defense against infection, but some microbes bypass this defense by living in arthropods that inject them through the epidermis and into the dermis or even into the bloodstream. In most such infections, the arthropod transmits the infectious microbe from other animals to humans. Other microbes transmitted from animals to humans can infect without the intervention of insects, through a respiratory route or through cuts in skin. One such pathogen, the cause of anthrax, has attracted the attention of bioterrorists.

Arthropod-borne diseases have caused human suffering and death for many thousands of years. Some examples covered in this chapter are given in Table 20.1. Bubonic plague, in which fleas transmit the bacterial disease from rats to humans, is a classic example. Bubonic plague is still around but, for the most part, has been brought under control. Other arthropod-borne diseases seem to be on the rise. Lyme disease, caused by a tick-borne spirochete, has emerged as a significant human disease in certain parts of the United States and Europe. Malaria, a mosquito-borne disease, has reemerged in many parts of the world. Dengue fever, a viral insect-borne disease, is a major cause of deaths from infectious disease in Latin America, Africa, and the Indian subcontinent. **Zoonoses,** infections acquired from animals or their products, were introduced in Chapter 17, in which a number of food-borne infections of animal origin were described. This chapter introduces some other zoonoses that are not spread by food but by arthropods, inhalation, or inoculation through breaks in the skin (Table 20.1).

Although arthropod-borne diseases and zoonoses are of primary interest because they pose serious threats to human and animal health, they also provide fascinating examples of the ways in which microbes can adapt to two or more different hosts and move regularly and successfully from one to the other. This chapter explores these interactions and explains some of the special problems involved in controlling arthropod-borne diseases and zoonoses.

Table 20.1 Examples of some common arthropod-borne and zoonotic infections

Arthropod-borne diseases
Bubonic plague (*Yersinia pestis*, a rod-shaped gram-negative bacterium)
Lyme disease (*Borrelia burgdorferi*, a spirochete)
Ehrlichiosis (*Ehrlichia* species, bacteria that grow only inside host cells)
Malaria (*Plasmodium* species, protozoa with a complex life cycle)
Dengue fever (Dengue virus)
Other zoonotic infections
Anthrax (*Bacillus anthracis*, a gram-positive spore-forming bacterium)
Brucellosis (*Brucella* species, small gram-negative coccobacillus)
Rabies (Rabies virus)
Bartonellosis (may be arthropod-borne) (*Bartonella henselae*, a gram-negative bacterium)

Microbes Transmitted by Arthropods

Advantages for the microbe

No viruses, bacteria, or protozoa are known to be capable of invading intact skin without assistance. Previous chapters described various underlying conditions, including wounds, burns, and venous catheters, that allow these microbes to bypass the skin defense. From the microbe's perspective, however, these conditions leave too much to chance. An almost guaranteed means of penetrating intact skin is for the microbe to be carried by an arthropod that takes a blood meal and, in the process, injects the microbe into the human dermis or even into the bloodstream.

Need for complex adaptation

Not many microbes have managed to use arthropods as injection vehicles. Insects such as flies or roaches may carry microbes as passive passengers on their feet. To make use of a biting arthropod, however, a microbe must be able first to survive and proliferate in the arthropod's gut, where the blood meal goes, and then enter the salivary glands or otherwise position itself for injection into the human body. Most of the known arthropod-transmitted microbes have this ability to adapt to and reproduce in the arthropod that transmits them. As expected from the close association between microbes and the arthropod, the microbe usually is carried by only one or a few related species of arthropods. Although microbes carried by arthropods generally prefer one or a few closely related species of arthropods as hosts, the same arthropod can carry more than one type of microbe, as will be seen from examples in this chapter.

Many microbes, such as human immunodeficiency virus (HIV), hepatitis viruses, or *Streptococcus pneumoniae,* that are found in the human bloodstream at various times during the disease process are not spread by arthropods. This fact probably is explained by their inability to survive and reproduce in the arthropod, an alien environment compared with the human body. Why would proliferation be necessary? The answer lies in the infective dose 50 (ID_{50}). Arthropods that ingest blood take in very small quantities containing low numbers of microbes. If the ingested microbe dies quickly in the arthropod, as is the likely fate of viruses that cannot infect arthropod cells, or if the microbe is present in numbers too small to infect when re-injected into the next human victim, no infection will result.

Insects and plant diseases

Although the examples of insect-borne diseases presented in this chapter are all infections of humans or other animals, insects are also important vectors for spreading plant diseases. Plants have a defense much like intact human skin, in the form of rigid cell walls not readily penetrated by microbes. Insects can help microbes bypass this barrier in two ways. The insect can directly inject microbes carried in its gut or salivary gland, or, when the insect feeds on the plant, a wound may be created that is readily invaded by microbes. This second route can be a vehicle for entry of microbes into the human body but is a far more important mode of transmission in plants. The rot on fruit, for example, often is the result of bacteria colonizing wounds made by insects. Birds and wind damage can have the same effect.

Examples of Arthropod-Borne Diseases

Bubonic plague

Bubonic plague continues to elicit fear and capture the imagination. Although modern outbreaks have occurred in Southeast Asia during the Vietnam War and in the Middle East, the Black Death of the Middle Ages continues to stand as a potent symbol of infectious disease catastrophes. In the Middle Ages, the plague had a particular horror, because people had no idea how it was acquired and so could take no effective measures to stop its spread. Theories about plague transmission ranged from divine vengeance to direct transmission from one person to another. We now know that this latter theory was getting closer to the truth but missed a critical piece of the story: plague also could be transmitted from rats to humans by means of rat fleas. This explains the failure of an attempt in 1383 by city officials in Marseilles to prevent the entry of plague into their city by quarantining sailors on incoming ships for 40 days. The rats unfortunately traveled freely from ship to shore by running down the ropes anchoring the ships to the docks. Bubonic plague has spawned many interesting stories, but one popular legend has been proven false by killjoy historians: the children's poem "Ring Around the Rosy" is not about bubonic plague but arose much later.

Yersinia pestis*, the bacterium that causes plague.** The bacterium that causes bubonic plague is ***Yersinia pestis, a small gram-negative rod. In the laboratory, *Y. pestis* is difficult to grow, because it requires many nutrients and is not very hardy. In fleas and mammals, however, *Y. pestis* is in its element and is a tenacious and destructive microorganism. The normal habitat of *Y. pestis* is wild animals, especially wild rodents. Rats are not the only hosts. Prairie dogs are a major reservoir in the western United States. *Y. pestis* has even been found in such unexpected animals as domestic cats and wild antelopes. It can kill its mammalian hosts. Epidemics of the disease occur in prairie dogs from time to time, and many infected animals die. The persistence of *Y. pestis* in rodent populations, however, indicates the possibility of an asymptomatic carrier state of some sort. The routes of transmission of *Y. pestis* are summarized in Figure 20.1.

Transmission of *Y. pestis*. The bacteria are spread from rodent to human by fleas. In contrast with other examples given in this section, *Y. pestis* does not live in happy equilibrium with its flea host but can kill the flea. After a flea ingests *Y. pestis* in a blood meal from an infected animal, the

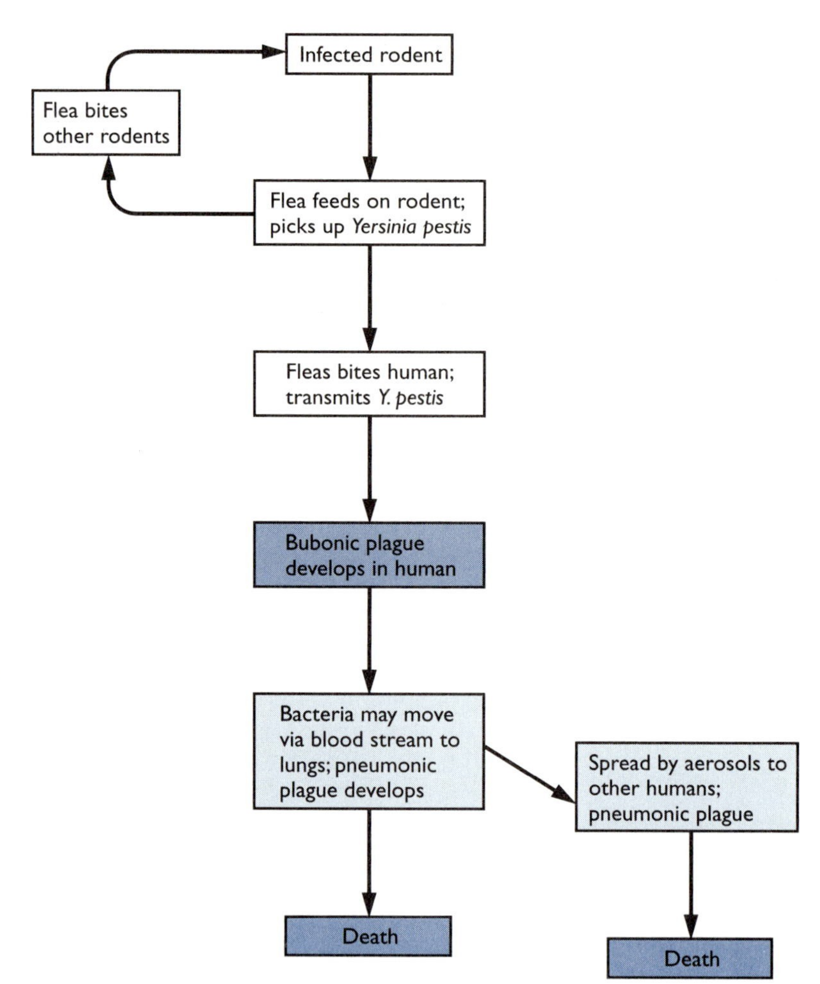

Figure 20.1 Routes of transmission of *Yersinia pestis*. Wild rodents serve as the primary reservoir of *Y. pestis*. The rat flea is the vector that can transmit the bacteria from one rodent to another or to other animals, including humans. In the human the bacteria travel in macrophages to the lymph nodes, where they cause the formation of buboes (bubonic plague). If the bacteria enter the bloodstream and travel to the lungs, pneumonic plague results. Pneumonic plague can be transmitted from human to human.

bacteria grow in the flea's gut. In the insect gut, the bacteria cause the ingested blood to clot, blocking the insect's ability to ingest another blood meal. Although this can kill the flea, it is also a strategy for transmission of the bacteria to a new host. When the flea tries to feed on a new host, it attempts to free its intestinal tract of the clot by regurgitating it. In this way, the insect is forced to spew bacteria-containing gut contents into the bloodstream of the animal or human on which it is feeding. This is an unusual mechanism of transmission; most microbes that are arthropod-transmitted either enter the salivary gland or take advantage of some naturally occurring arthropod activity that spreads the microbe, such as fecal contamination of the wound.

Pathogenicity. Once in the mammalian body, *Y. pestis* avoids being killed by the host defenses by invading and growing inside macrophages. Activated macrophages can kill the bacteria, but unactivated macrophages are the major site of bacterial growth in the body. Because many macrophages are located in lymph nodes, where they serve as antigen-presenting cells, the lymph nodes are major sites of infection (Fig. 20.2). The consequent inflammatory response causes the lymph nodes to swell, producing visible lumps, called **buboes,** under the armpit, on the neck, or in

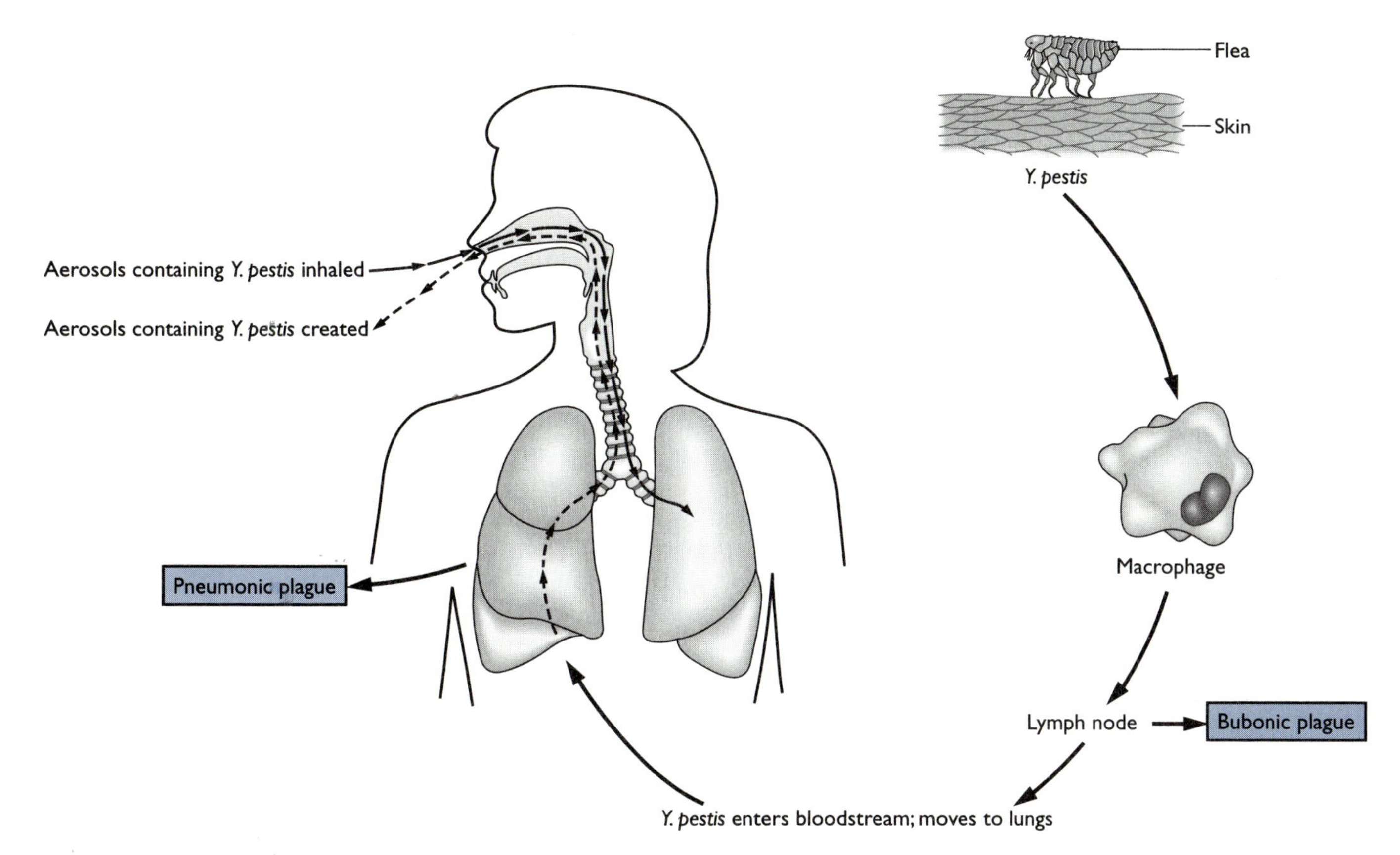

Figure 20.2 Routes of entry and movement of *Yersinia pestis* through the human body. The flea injects *Y. pestis* through the skin, where it is engulfed by macrophages and transported to the lymph nodes. In the lymph nodes the bacteria proliferate. They may enter the bloodstream and move to the lungs. The bacteria also may be inhaled in aerosols.

the groin. The lungs also contain many macrophages. *Y. pestis* moves from the bloodstream to the lungs, where the bacteria reproduce actively in lung macrophages. This phase of plague is called **pneumonic plague.** From the lungs, bacteria can be spread directly from human to human by aerosols. Death is caused by septic shock. During septic shock, the accumulation of small clots in the capillaries can cause localized necroses by blocking the circulation. This is probably the source of the blackish color seen near the lips and extremities of plague victims, the symptom that gave plague the name "Black Death."

Prevention and treatment. Prevention of transmission is, as with so many other infectious diseases, the best strategy for controlling plague. Rodent control is important in areas where plague becomes a problem. In the western United States, cases of plague are unusual (about 10–15 cases a year), and the main victims are hikers or hunters, who are most likely to come into contact with wild rodents. The most efficient preventive strategy in this case is to warn potential victims to avoid contact with these rodents. *Y. pestis* remains susceptible to many antibiotics, although resistant strains have begun to appear. If antibiotic therapy is administered early in the disease, survival rates are high and the patient usually has no long-term deleterious effects. Because plague is an uncommon disease in most parts of the

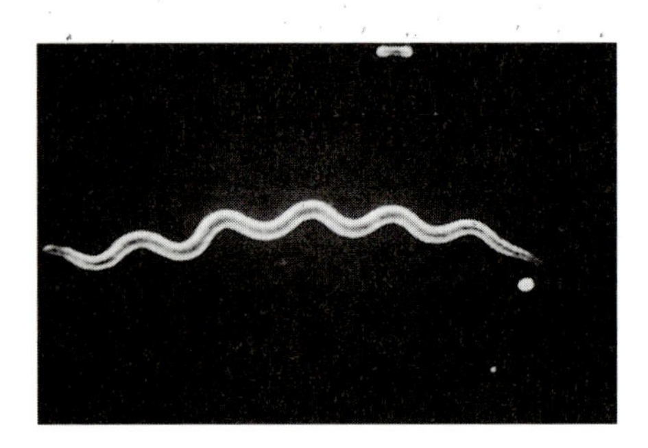

Figure 20.3 Microscopic view of *Borrelia burgdorferi* spirochetes. Courtesy Dr. Edward P. Ewing, Centers for Disease Control and Prevention, Atlanta, GA.

world today, however, physicians may not diagnose the disease in time. Education of physicians in areas where *Y. pestis* is endemic in the rodent population has helped to cut the number of cases in which plague is identified only posthumously. Public education is also important during the occasional small outbreaks in which, if reporting of the facts is not careful and accurate, public hysteria is sure to follow.

Lyme disease

A disease arising from a complex cycle involving tick, deer, mouse, and humans, **Lyme disease** has been much in the news in the United States. Although it is currently the most common cause of arthropod-borne disease in the United States, it is not nearly so widespread or likely to be encountered as the news accounts imply. Lyme disease, so-called because it was first identified in Lyme, Connecticut, is a systemic disease caused by the **spirochete *Borrelia burgdorferi*** (Fig. 20.3). Untreated Lyme disease may cause neurological problems and painful arthritis in some people, but the infection is rarely fatal. *B. burgdorferi* lives mainly in white-footed mice. It may live in other mammals, but the mouse is the main reservoir from which human infections are acquired. Deer are important elements in the transmission of *B. burgdorferi* but not because they become infected with the bacteria. Instead, the deer provide an important host for the adult form of the ticks that spread *B. burgdorferi.* The transmission of this disease is an example of the way in which differing ecologies of the tick, the bacteria, and the mammalian hosts can overlap to maintain the bacteria in the environment and spread it. In this case, the main mammalian players are deer and mice. Humans seem to have gotten accidentally in the way of this fundamental cycle of *B. burgdorferi* maintenance and transmission.

The tick component. The three stages of the tick that carries *B. burgdorferi, Ixodes dammini,* are depicted in Figure 20.4. Each stage takes a blood meal, which plays a key role in the tick's development. Adults mate and take a blood meal. Then, the female lays eggs and dies. The eggs develop into larvae, which in turn take a blood meal that stimulates them to molt into the nymph stage. This progression from adult to nymph takes a year. Nymphs that survive the winter take a blood meal before becoming adults to complete the tick life cycle. A critical feature of the feeding habits of these three different forms is that the adults prefer deer, an animal that is not infected with *B. burgdorferi.* If the larvae and nymphs had the same preference, *B. burgdorferi* would not have established itself. But the larvae and nymphs are happy to feed on white-footed mice and other mammals infected with *B. burgdorferi. B. burgdorferi* is rarely, if ever, transmitted from adult female to larvae transovarially, so it must be maintained by being reacquired by each new generation of ticks. If a larva becomes infected, it develops into an infected nymph. The infected nymph can then transmit the bacteria to mice or to humans when it takes a blood meal prior to becoming an adult.

Deer ticks are pool feeders. That is, instead of injecting the bacteria into the mammal on which they are feeding, deer ticks create a small wound in the dermal layer of skin, into which blood flows. If this blood contains *B. burgdorferi,* the tick acquires the bacteria. Bacteria taken up by larvae enter the midgut of the tick. Later, they move to the salivary gland, so that when the tick feeds on a new host, bacteria are released in the tick's saliva and enter the lower layers of the host's skin.

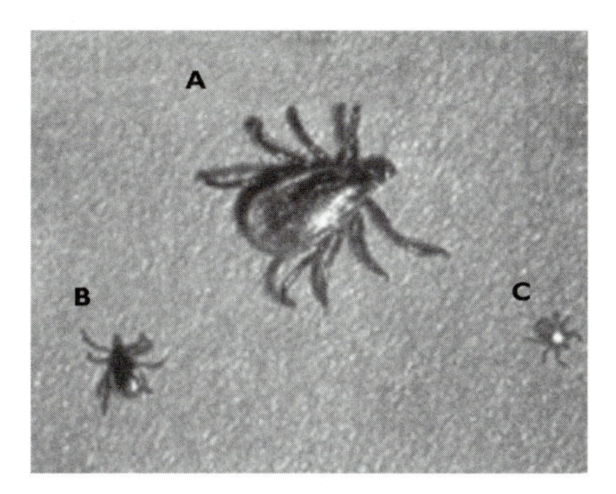

Figure 20.4 The three stages of the deer tick *Ixodes dammini.* **(A)** The adult stage feeds on white-tailed deer. **(B)** The nymph stage, which develops from the larval form, spends the winter on the forest floor. **(C)** The larval stage feeds on white-footed mice. Courtesy of the Centers for Disease Control and Prevention, Atlanta, GA.

In going from mammal to tick, the bacteria change their surface proteins. A surface protein called **OspA** (**outer surface protein A**) becomes prominent, and others that were prominent in the mammalian body (**OspC**) disappear. When the tick takes another blood meal, the switch back from OspA to OspC begins. The switch between OspA and OspC is undoubtedly only one of many changes in the tick-to-mammal transition, but it illustrates how the bacteria are reacting to their imminent entry into a new host.

In humans, spirochetes that have been inoculated into the dermis multiply transiently in the skin, moving out from the site of inoculation. In most infected people, the body's inflammatory response produces a characteristic **ring-shaped lesion** around the site of infection. Then the bacteria enter the bloodstream and spread to tissues all over the body. One result of infection is a temporary joint inflammation that moves from one joint to another. More serious effects include neurological damage that can result in paralysis and vision loss.

The mammalian component. Lyme disease is a classic case of an **emerging infectious disease.** The bacteria, engaged in a mouse-to-deer cycle of tick feeding, have been around a long time, probably for thousands of years or more. What happened to cause this disease to enter the human population? First, restrictions on deer hunting in the northeastern United States caused an explosion in the deer population. Second, the reforestation of some of these same areas resulted in an increase in oak trees, with a concomitant increase in acorn production. The white-footed mouse feeds on acorns, so an increase in acorns not only expanded the mouse population but attracted the mice into new areas. Acorns also were attractive to deer. At the same time these changes occurred, humans began to build more and more houses in areas where deer and mice roamed. Deer and mice came into yards, shedding their ticks, which proceeded to feed on the next warm-blooded creature to come within their range: humans. Lyme disease is not transmitted directly from human to human. The main reason for the emergence of this infectious disease is that humans unintentionally have come between ticks and their natural hosts. Another, less discussed factor in the rise of Lyme disease to medical prominence is that it has tended to be a disease of the upper and upper-middle classes, affecting people skilled at mobilizing media attention and health care resources.

Prevention and treatment. Diseases like Lyme disease can be prevented from entering the human population by controlling the wild animal populations, controlling the tick population, and vaccinating humans. Controlling the deer and tick populations are certainly viable options. However, the same well-to-do people who, because of the location of their houses, are at highest risk for contracting Lyme disease also have a long record of opposition to hunting animals (the most obvious way to control the deer population) and to the use of insecticides (the most obvious way to control the tick population). Thus, attention has tended to focus on vaccination and treatment with antibiotics.

The U.S. Food and Drug Administration approved a vaccine against *B. burgdorferi* in 1998. This vaccine has a significant problem: it targets OspA, which is produced by bacteria in the tick, not by bacteria in the human body. Scientists who rushed to develop a vaccine did not realize this at first, because OspA was produced in bacteria growing in laboratory medium. The OspA vaccine is not worthless, because some killing or neutralization of the bacteria may occur in the tick when the blood meal enters the tick gut. New forms of the vaccine that target OspC are currently being developed. Until a better vaccine is available, the only option that remains is treatment. Treatment is most effective when the disease is diagnosed early. However, some infected people do not exhibit the classic bulls-eye rash or the migrating arthritis-like pains that are symptomatic of early disease. Serological tests are improving but are not yet very reliable.

Fellow travelers. The same *Ixodes* ticks that carry *B. burgdorferi* also carry a very different type of bacterium: ***Ehrlichia*** species. *Ehrlichia* species cause an acute febrile disease (ehrlichiosis) with symptoms that resemble a severe case of the flu: fever, headache, and joint and muscle pains. In contrast with Lyme disease, which is almost never fatal, disease caused by *Ehrlichia* species is fatal in about 2% of infected people who are not treated in a timely manner. *Ehrlichia* species differ from *B. burgdorferi* in a number of ways. *Ehrlichia* are obligate intracellular pathogens that grow only inside neutrophils. Their natural reservoirs include a variety of animals, such as dogs and deer. Treatment, too, is different. Yet both *Ehrlichia* and *B. burgdorferi* are spread by the same type of tick. In some cases, ticks carry both pathogens and can transmit a double dose of infection.

Malaria: an insect-borne protozoal disease

Lyme disease understandably is a concern to those who have it or are at risk, but it is not a life-threatening disease in most victims. By contrast, **malaria** is a leading killer of humans worldwide. Malaria is caused by a protozoan, *Plasmodium.* Four *Plasmodium* species infect humans, but this chapter will focus on *Plasmodium vivax,* the species that causes most cases of malaria, and *Plasmodium falciparum,* the most deadly species.

Plasmodium species are spread by the *Anopheles* mosquito. *Plasmodium* species have a very complex life cycle, both in humans and in the insect. The human part of this life cycle is illustrated in Figure 20.5. A similarly complex cycle occurs in the insect. In contrast with plague and Lyme disease, in which the form of the microbe introduced into the human body by

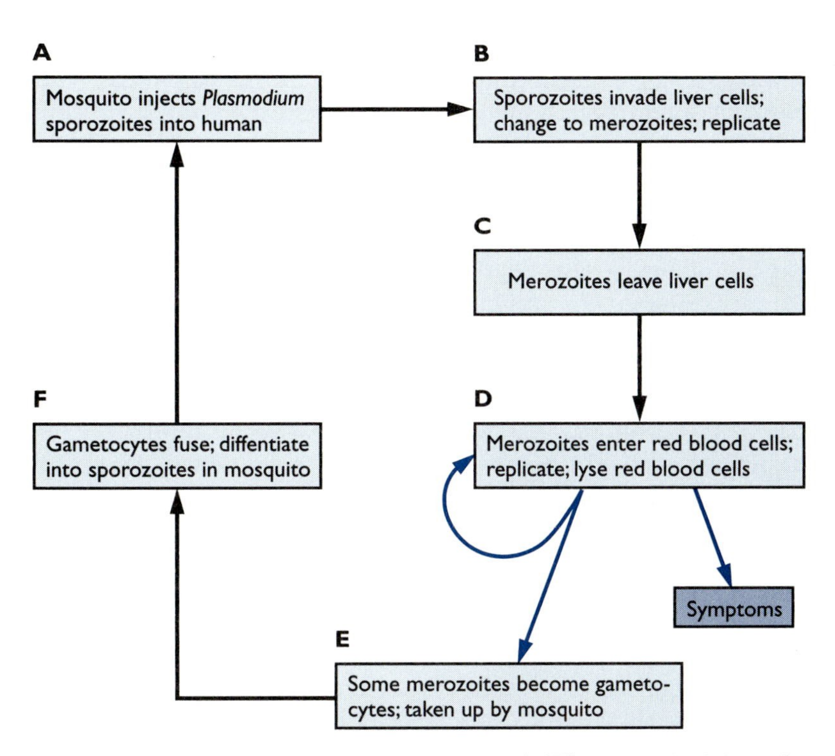

Figure 20.5 ***Plasmodium* life cycle in the human.** **(A)**The mosquito injects the sporozoite form of the parasite into the human bloodstream. **(B)** Sporozoites invade liver cells, where they change into the merozoite form and replicate. **(C)** Merozoites leave the liver cells, enter red blood cells, replicate, and lyse the red blood cells. **(D)** Most merozoites enter new red blood cells; repeat cycle. **(E)** Some merozoites differentiate into macro- and microgametes, which are taken up by mosquitoes. **(F)** In the mosquito, the gametocytes fuse and differentiate into sporozoites, which are ready to begin the cycle all over again.

one arthropod looks the same as the form later ingested by another arthropod, the injected form of *Plasmodium* (the sporozoite) is different in appearance from that later ingested by an insect (Fig. 20.6). This apparent difference between bacterial and protozoal pathogens may not be as great as it seems at first. Bacteria, too, undergo internal changes when moving from insect to mammal and vice versa, but these changes are not so apparent visually as those in the shapes of the various stages of *Plasmodium.*

Although the steps in the life cycles of *P. vivax* and *P. falciparum* are the same at the level depicted in Figure 20.5, these species differ in two ways that make a big difference in symptoms and therapy. In both species, when

Figure 20.6 **Forms of *Plasmodium* found in the human body.** **(A)** The sporozoite is the form injected by the mosquito. **(B)** The merozoite is the form that lyses red blood cells.

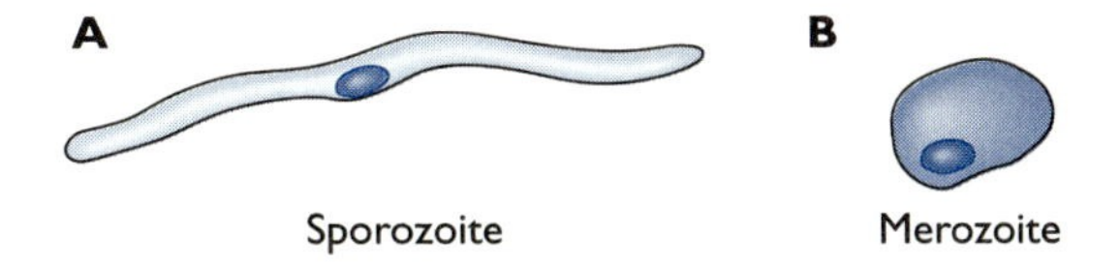

the **sporozoite** form enters the human body, it first invades liver cells, where it makes the transition to the **merozoite** form. The merozoite form does damage by replicating inside first liver and then red blood cells and lysing them in the process. Later, the sexual forms (**macro-** and **microgametes**) appear. *P. vivax* continues its replication cycle in the liver, even after some merozoites have left the liver and invaded red blood cells. In contrast, *P. falciparum* merozoites cycle only once in the liver, then invade the red blood cells, and never return to the liver.

Another difference between *P. vivax* and *P. falciparum* is in their preferences for red blood cells. *P. vivax* preferentially invades older red blood cells, a minority of the population. Growth of the merozoites in these cells causes them to lyse. Release of red blood cell components into the blood elicits a cytokine response that produces a high fever and chills. Because the *P. vivax* merozoites then must wait for another generation of older red blood cells to develop, invasion and lysis occur periodically. This is the reason for the periodic episodes of fever and chills that are characteristic of this type of malaria. Episodes of fever and chills can recur for years. *P. falciparum* invades all stages of red blood cells. Thus, the lysis of red blood cells is more extensive, and the fever and chills are more constant. Extensive lysis of red blood cells is the reason *P. falciparum* malaria is the form most likely to be fatal. If the lysis of red blood cells is massive enough, high levels of hemoglobin will be excreted in the urine, which turns dark brown. This is known as **blackwater fever** and can lead to kidney failure. Dying blood cells trigger widespread release of cytokines and other proteins that affect the tone and integrity of the blood vessel walls. A situation much like the septic shock caused by bacteria can occur and result in death. The brain is the organ most affected in severe cases of *P. falciparum* malaria.

Prevention and treatment. Because humans are the main reservoir for *Plasmodium* species, control of the insect vector is critical in preventing transmission. Insecticide spraying campaigns and efforts to eliminate stagnant areas of water where mosquitoes breed have been important public health tools for preventing the spread of malaria. A massive DDT spraying campaign in the 1950s and 1960s resulted in an impressive drop in the incidence of the disease. But a combination of emerging insecticide resistance in the mosquito population and a backlash in developed countries against DDT have reversed these gains. Developing countries in the malaria zone (the tropics and subtropics) have experienced a resurgence of malaria. Alternative prevention strategies, including insecticide-impregnated mosquito nets, have been tried without much success. Progress toward a vaccine has been slow and frustrating, and few promising candidates are in sight.

Successful treatment of *P. vivax* malaria must achieve three goals: stop the symptoms by killing the merozoites in red blood cells, kill the protozoa in liver cells to prevent recurrences of *P. vivax* malaria, and kill the gametocytes that can be taken up by a mosquito, thus rendering the patient noninfectious. Chloroquine, a commonly used antimalarial, kills the blood forms of the plasmodia but not the liver form. Another drug, primaquine is best at killing the liver form but is not so effective on the other forms of the parasite. In addition, primaquine has serious side effects in some patients. To cure *P. vivax*, it is necessary to use a combination of drugs. In the case of *P. falciparum*, only one drug is needed, because there is no continuing liver stage. Another complication of the treatment of malaria is that some

Plasmodium strains have become resistant to chloroquine, the most popular and least toxic therapy. Resistance to chloroquine is now widespread in some areas. One unusual type of control of malaria results from a human genetic alteration (see Box 20.1)

Dengue fever: an insect-borne viral disease

Dengue fever is a viral disease transmitted by *Aedes aegypti* mosquitoes (Fig. 20.7). The mosquito is widespread in all parts of the tropics and semitropics. Most transmission is human to human via mosquitoes, but a wild animal reservoir (for example, monkeys) may play a role in maintaining the virus in nature. The disease is characterized by a sudden onset of fever, headache, and severe pains in the joints, back, and limbs (Fig. 20.8). Because of the excessive pain experienced in muscles and joints, the disease has been called **breakbone fever**. A rash may also appear. In many infected people, the symptoms last only a few days and then resolve spontaneously, but dengue fever can also kill. The most serious form is **dengue hemorrhagic fever** or **dengue shock syndrome.** This can occur if a person is infected with a second, different strain of the virus. In this form of the disease, viral replication in monocytes and macrophages, its target cells, can lead to release of

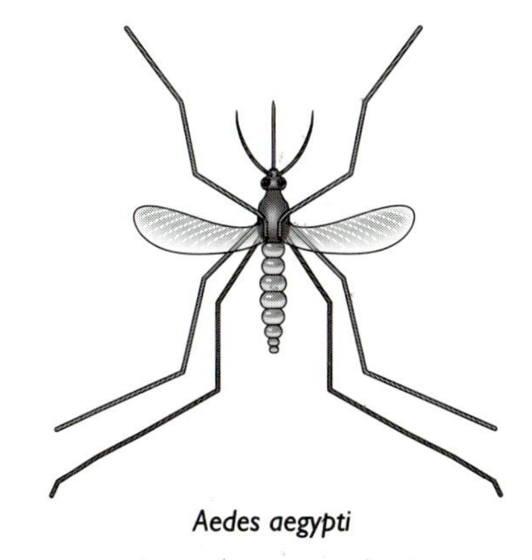

Figure 20.7 Schematic drawing of *Aedes aegypti*. *A. aegypti* is the vector of dengue fever. Adapted from a drawing by James M. Steward, Centers for Disease Control and Prevention, Atlanta, GA.

box 20–1 *Sickle Cell Disease and Malaria: an Explanation for the Existence of Some Genetic Diseases?*

Sickle cell disease is a genetic disorder in which a change in the amino acid sequence of the protein hemoglobin leads to a defect in red blood cell shape that can allow the sickle-shaped red blood cells to clog the capillaries, causing pain and death. A person with sickle cell disease has two copies of the defective gene. A person with only one copy of this gene (sickle cell trait) has no health problems under normal conditions. Some evidence suggests that people with sickle cell trait are more resistant to malaria than are people with two wild-type copies of the gene. This could arise from the fact that hemoglobin is involved in recognition and invasion of red blood cells by *Plasmodium* merozoites. An epidemiological observation that bolsters this connection is that sickle cell trait is found much more commonly in people of African descent than those of European descent. Europe has experienced malarial outbreaks but never to the extent of the constant and intense exposure to malaria in Africa.

Recently, two other examples of the same phenomenon have been proposed, both of which are quite controversial. One is a proposed connection between partial genetic immunity to smallpox and genetic immunity to human immunodeficiency virus (HIV) infection. It is now well established that a subset of the human population is genetically immune or partially immune to HIV infection, because the cell targets of HIV lack a chemokine coreceptor or have a mutant form of that coreceptor. The chemokine coreceptor stabilizes the interaction between HIV and its primary receptor, CD4. Recent studies of a close relative of the smallpox virus have suggested that the virus might use some of the same cell-surface receptors as HIV virus. This has led to the speculation that repeated epidemics of smallpox in Europe might have selected for a genetic trait that conferred partial resistance to smallpox virus, a trait that is now conferring partial resistance to HIV infection. The fact that resistance or partial resistance to HIV seems to be highest in people of European descent squares with the fact that, although Africans certainly suffered from smallpox, their exposure was not as extensive or as long in duration as that of Europeans. The other proposed example is a link between the cystic fibrosis gene (homozygous) and resistance to the bacterial disease typhoid fever.

Even if both the HIV–smallpox hypothesis and the cystic fibrosis–typhoid fever hypothesis are proven wrong, it is intriguing to consider the possibility that infectious diseases might have selected for genetic alterations that in one copy provided protection from disease but in two copies gave rise to genetic disease. This idea certainly makes more sense than the prevailing notion that genetic diseases arose through random mutations with no selective pressures to fix them in the population.

Source: Travis, J. 2000. Survivors' benefit? *Science News* 157:63.

cytokines and other proteins that make the blood vessels leaky and produce many of the same symptoms as septic shock caused by bacteria. Dengue fever has been epidemic in many parts of the world, especially in tropical and subtropical areas. Today, dengue fever has reemerged in Latin America, where it is now a common cause of disease. Reemergence is a direct result of the reappearance of the mosquito *Aedes aegypti*, the insect vector that carries the virus.

Dengue fever is caused by dengue virus, an enveloped RNA virus with a simple replication cycle. The replication cycle is somewhat unusual, in that it occurs in the cytoplasm, in close association with the endoplasmic reticulum. The virus does not bud from the cell but lyses the cell during its exit phase. The virus has a single-stranded RNA genome that initially acts

Figure 20.8 Progression of dengue virus disease. The *Aedes aegypti* mosquito injects the virus into a human. The person develops symptoms, which resolve spontaneously. If the person is infected with a second strain of the virus and that second strain differs from the first, the virus may replicate in monocytes and macrophages and cause dengue hemorrhagic fever.

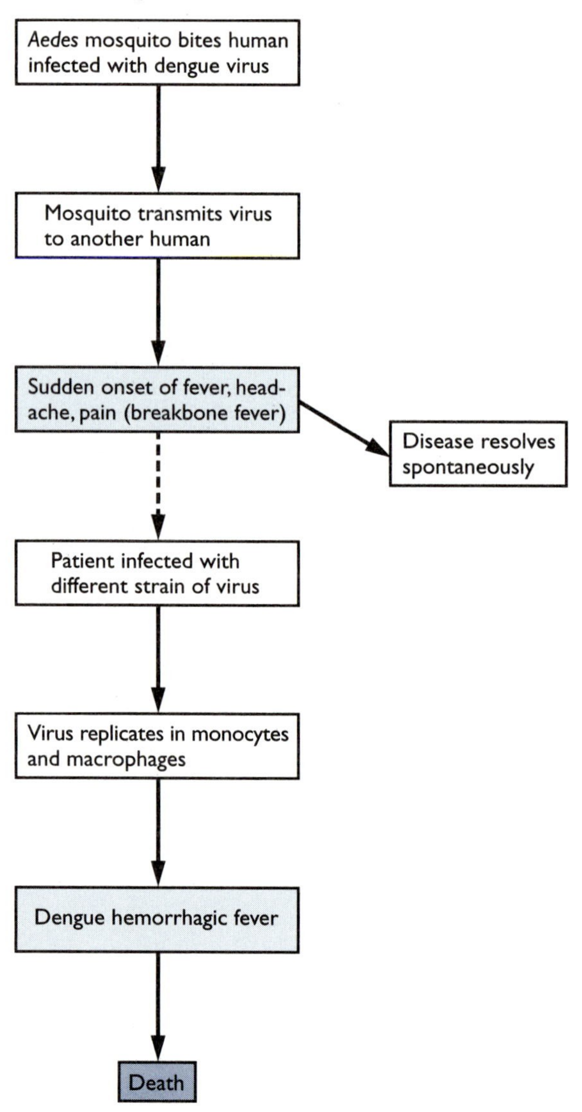

as a messenger RNA (mRNA) template for the synthesis of viral proteins. The RNA genome also serves as a template for synthesis of a complementary copy of itself. This complementary copy then serves as a template from which another copy is made. This second copy is the viral genome.

Because the cell targets of dengue virus are monocytes and macrophages, antibodies that recognize the virus can be harmful and not protective to the person that produces them. Antibodies bound to the virus facilitate entry into these cells, because the cells contain receptors for the Fc portion of antibody and use the antibodies to bind and internalize viral particles. The antibodies remain bound to the envelope proteins, but, once the nucleoprotein core is released, these antibodies no longer have any effect on viral replication. This feature of dengue virus pathogenicity probably explains the observation that second infections are more serious than primary ones. This feature also makes it difficult to develop a vaccine, although vaccine development efforts are currently underway.

No antiviral therapy is available for dengue fever. The only option is prevention, which means insect control and preventing contact with insects. Preventing contact with insects is a virtual impossibility for poor people who must work in agricultural areas where standing water provides an ideal breeding ground for mosquitoes. For people who develop the most serious form of the disease, only supportive therapy is available. This can be effective for those who can afford it, but for those beyond the reach of health care systems, supportive therapy is not an option. Insect control remains the best available approach.

More Zoonotic Infections

Although food-borne zoonotic infections still challenge public health controls, zoonoses spread by inhalation, contact, or animal bite have been brought under control in most countries. The temptation is strong to ignore these "old" diseases as solved problems, but it is instructive to consider a few of them in order to understand how a zoonotic infection can be controlled successfully. These may serve as guides for ways to solve newer or reemerging zoonotic infection problems. Finally, at least one of them, anthrax, has attracted the attention of bioterrorists.

Old enemies: anthrax, brucellosis, and rabies

Pasteur's first two successful vaccines were against anthrax and rabies. Development of a vaccine against anthrax was important, because many farmers were losing their animals to this disease. Moreover, as later became evident, the bacterium that causes anthrax could cause human disease. The motivation for developing a vaccine against the rabies virus came from the horrible and uniformly fatal disease this virus caused in humans. In his youth, Pasteur had been shocked by the painful deaths of fellow villagers attacked by a rabid wolf. This experience inspired his passionate commitment later in life to developing a rabies vaccine. Brucellosis, originally recognized as a disease that caused cattle, sheep, and swine to abort spontaneously, was later found to cause a serious and sometimes fatal human disease. Eliminating brucellosis from livestock in developed countries was a major victory for farmers and for people exposed to the disease.

Anthrax. Anthrax is caused by *Bacillus anthracis*, a gram-positive spore-forming bacterium. In animals, *B. anthracis* causes a disease that can be fatal.

In humans, it can cause either a skin disease, characterized by a single painful lesion with a black center, or a serious and rapidly fatal lung disease. In the past, those most likely to develop anthrax were people working with meat or hides from infected animals. More recently, anthrax has been the focus of bioterrorists, a fact that has caused interest in the disease to rise. Later in this chapter, the virulence factors of *B. anthracis* will be described in connection with possible bioterrorist use of anthrax. A vaccine against *B. anthracis*, made of whole inactivated bacteria, is now available. This vaccine is very effective in preventing anthrax in animals and was instrumental in eliminating the disease from farm animals. It has not been thoroughly tested for efficacy and safety in humans, to whom boosters must be given every 6 months.

Brucellosis. Brucellosis is a bacterial disease caused by *Brucella* species. *Brucella* species are tiny, gram-negative **coccobacilli** (short rods that sometimes look like cocci). Different *Brucella* species cause infections in cattle, pigs, and goats. One species of *Brucella* infects dogs, especially those bred in kennels, but is an uncommon cause of human disease. In animals, brucellosis is a chronic disease that persists for long periods, sometimes for life. *Brucella* does not kill the animal but instead infects the male and female reproductive tracts, causing abortion in females and sterility in males. Obviously, this type of infection is of great concern to farmers and to animal breeders. In the United States, cattle are monitored routinely for brucellosis, and infected animals are killed. Pig breeders also keep an eye on their animals and eliminate infected animals that cannot be cured. Because large numbers of bacteria are shed in the urine and milk of infected animals, the disease is transmitted readily to other animals. Farmers must take quick and effective action to protect uninfected animals.

In humans, *Brucella* species cause a completely different disease. The bacteria do not localize in the human genital tract and rarely cause genital complications. Instead, the infection is systemic, with bacteria found in many parts of the body. The bacteria spread throughout the body in neutrophils, and they invade and grow inside macrophages. Thus, an infection can occur wherever there are macrophages (for example, in the lung, spleen, and lymph nodes). Brucellosis can resemble the flu, but symptoms, including a relapsing fever and fatigue, can persist for long periods and seriously impair a person's quality of life. If diagnosed correctly, brucellosis can be treated with antibiotics. Because the disease is not common in the United States, the few cases that do occur are often misdiagnosed.

Veterinarians or farmers, who can acquire the bacteria through small cuts in their hands, are most likely to acquire brucellosis. Another group at risk is slaughterhouse workers, who inhale aerosols containing the bacteria and can develop lung infections, which then become systemic. Although brucellosis is now uncommon in livestock in developed countries, scattered cases still occur. The bacteria are still around but are under control. For this reason, active monitoring and control programs remain important. Brucellosis became an issue in the western United States in the 1990s, when it was discovered that a herd of buffalo carried the disease. Most of the animals in the herd were killed because of concerns about the spread of the disease to cattle and to humans.

Rabies. The discovery of effective control and treatment for rabies infections, once one of the most feared of all microbial diseases, was a triumph for

Pasteur and his colleagues and is still important today. In the past, rabid dogs were a major cause for concern because of their close proximity to humans. Routine canine vaccination has made dogs an uncommon vehicle for spread of the disease to humans. Today, wild animals, such as bats and raccoons, are the main reservoirs for the rabies virus. Rabies is transmitted most often from animal to human by bites, but people exposed to caves in which large numbers of infected bats live can acquire the disease by inhalation. As always, an understanding of risks involved in various human activities is the key to prevention. People who have frequent contact with bats, however indirect, should either receive the rabies vaccine or be monitored closely for signs of infection. Once the symptoms of the disease develop, the patient almost always dies.

Rabies virus is a large, enveloped RNA virus with a bullet-shaped appearance (Fig. 20.9). In mammals, the virus is highly specific for nerve tissue. In a person who has been bitten by a rabid animal, the disease does not manifest itself immediately. Instead, months may pass before the symptoms appear. The reason for this seems to be that the virus must replicate to achieve sufficient numbers for its physiological effects to become apparent. The long incubation time of the disease (on average 20–90 days) makes possible a rather unusual treatment: immunization of the infected person *after* the initial infection has occurred. Today, the treatment of suspected rabies victims includes a combination of (a) making sure the person was actually exposed to the virus by examining the allegedly rabid animal, (b) passive immunization with antibodies that neutralize the virus, and (c) administration of the antirabies vaccine. Passive immunization covers the patient in the period during which the vaccine is eliciting an immune response.

If neutralizing antibodies are so effective in preventing infection of nerve tissue and the subsequent development of symptoms, why do people

Figure 20.9 Electron micrograph of rabies virus. The bullet-shaped virus has a ribonucleoprotein core surrounded by an envelope studded with glycoprotein spikes. Copyright K.G. Murti/Visuals Unlimited.

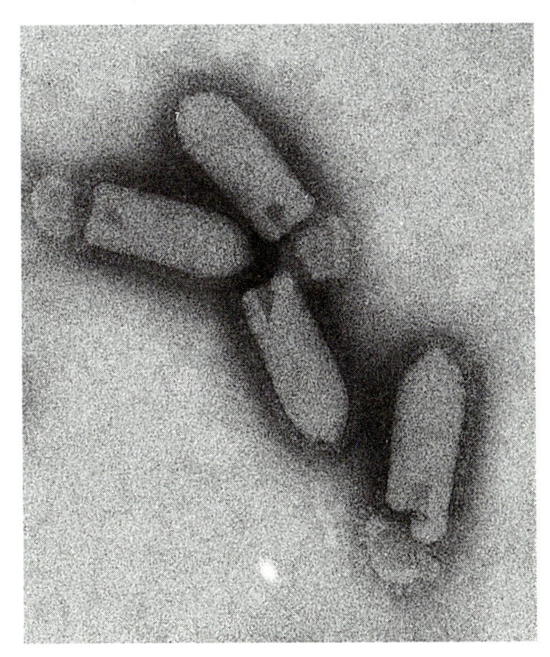

naturally infected with rabies virus die so frequently–especially if their immune systems have enough time to respond to the infection? One answer is that the nervous system, for understandable reasons, is off limits to the immune system most of the time. A virus that replicates specifically in nerve tissue will be hidden from the immune system.

Is administering the passive vaccination and vaccine at the same time likely to be counterproductive because they will cancel each other out? This is certainly a legitimate cause for concern. The problem is solved by administering the passive vaccination in a separate body site from the vaccine. This solution seems to have been successful, because development of rabies is extremely rare in people treated according to current protocols.

Prevention of rabies in animals is an obvious way to prevent human rabies cases. In developed countries, dogs—formerly one of the most common sources of human rabies—are now vaccinated routinely. An oral vaccine is distributed in various baits to wild animals, such as foxes, wolves, and raccoons. Bats remain an unsolved problem in this chain of prevention. Nonetheless, this campaign to eradicate rabies virus from animal populations has been quite successful in developed countries.

New problems: anthrax again, Bartonella, and porcine retroviruses

Anthrax and bioterrorists. Anthrax, a problem already on the way to being solved in Pasteur's time, has returned as a cause of concern because of threats by bioterrorists to use *B. anthracis* as a weapon. During the Cold War, scientists in the Soviet Union and possibly other countries began experimenting with this bacterium as an agent of germ warfare, so *B. anthracis* has long been on the list of possible destructive agents.

B. anthracis has two features that make it attractive to germ warriors. It is a spore former, and **spores** remain active for years. This feature makes it easier to store and disseminate the bacteria. In addition, the most serious form of anthrax is deadly. Inhalation of *B. anthracis* spores can lead to a serious lung infection that can become systemic. This form of the disease is nearly always fatal if untreated. Systemic anthrax may be difficult to diagnose because initial symptoms resemble those of flu. Failure to diagnose this rapidly fatal disease quickly enough makes treatment ineffective. Bioterrorists have hinted that antibiotic-resistant strains of *B. anthracis* have been developed. Whether this is true remains to be seen, but the threat of anthrax is certainly not one to be taken lightly.

B. anthracis has two types of virulence factors that explain its ability to cause disease. It produces an antiphagocytic capsule, which is very unusual, in that it is composed of proteins instead of polysaccharides. It also produces two toxins that are important for virulence. One is **edema toxin** that produces swelling (edema) when injected into the skin. This toxin may be responsible for the black-centered lesion seen in **cutaneous anthrax. Lethal toxin** is thought to contribute to the rapidly developing shock that is the cause of death in people with **systemic anthrax.** The lethal toxin consists of two protein subunits, a B protein that binds to human cells and a toxic A protein, which the B protein injects into the cell. The B protein subunit has been called **protective antigen,** because animals that have antibodies against this protein are protected from the disease. The antibodies bind the B part of the toxin and prevent it from binding to target cells. The human cell receptor for B subunit has not been identified but is clearly a receptor found on many types of human cells, because lethal toxin attacks many different cell types.

The mechanism of action of the A protein remains unknown but appears to be a proteolytic enzyme with an as yet unidentified human cell target.

The finding that B protein is an effective vaccine in animals has raised hopes that it could serve as an effective vaccine in humans. The problem with this vaccine strategy, however, is that it would be too late to administer vaccine *after* a bioterrorist attack. The most effective countermeasure in people who have been subjected to a bioterrorist anthrax attack could be passive immunization, the injection of preformed antibodies into the bloodstream to bind up toxin being produced by the bacteria. Keeping large stores of antibodies ready for an attack presents a challenge, unless such an attack is small or localized.

The form of the vaccine used in animals has side effects in humans, and its efficacy in humans has not been tested carefully. This fact surfaced in a debate that ensued from attempts by the U.S. military in 1999 to vaccinate troops who might be exposed to anthrax bioweapons. Some soldiers objected to forced vaccination with a vaccine for which efficacy and resulting side effects had not been clearly established.

Bartonellosis. A new zoonotic infection, first noted in patients with acquired immunodeficiency syndrome (AIDS) but possibly more widespread, is caused by the gram-negative bacterium, *Bartonella henselae.* Before the advent of AIDS, this bacterium was considered so obscure that most older medical microbiology texts gave it little more than a paragraph's attention. The most evident symptom of the disease, aside from fever, is a type of lesion that occurs around blood vessels in the skin, resulting in large, lump-like lesions that may occur in multiples. The lesions in interior regions of the body, such as the spleen or spinal column, are much more likely to cause serious problems.

Bartonellosis was first recognized as a distinct disease in patients with advanced HIV infections but is now being reported in individuals who are fully immunocompetent. At first it was thought that the bacteria were transmitted directly from cats to humans, and so the disease was named cat-scratch disease. However, it now appears that the disease requires an insect vector, the cat flea. The primary reservoir of *B. henselae* is cats, especially young cats infested with fleas. The mechanism of transmission is still not clear, but the bacteria are probably introduced by traumatic inoculation with flea feces, as would occur from a cat scratch. In some areas, many cats have *Bartonella* in their bloodstreams but appear to be completely healthy, so the bacteria may not have an adverse effect on infected cats. An obvious way to avoid this disease is to avoid cats, but the popularity of these pets makes this solution impractical. Finding that a vector is involved makes avoiding the disease more straightforward: keep cats free of fleas. Some health workers have urged AIDS patients to get rid of their cats, but people who are ill become more attached than ever to their companion animals. Moreover, cats and other pets have beneficial effects on the outlook and well being of people who are seriously ill.

Another approach to controlling bartonellosis is rapid and accurate diagnosis, followed by effective treatment of the disease. Unfortunately, *B. henselae* is very difficult to cultivate in the laboratory. The disease it causes is difficult to diagnose in patients who do not manifest skin lesions, because good detection tests are not yet on the market. Work is now underway to make diagnosis easier. The recognition that cat-scratch disease may not be transmitted directly from cats to humans but instead requires a vector alerts us to the fact that other zoonotic diseases may also have an arthropod vector that has not yet been recognized.

Xenotransplantation in trouble. Human organs are often in short supply for people needing transplants. Many people die each year waiting for a suitable human organ to become available. The suggestion that animal organs (**xenotranplantation**) might be used instead has caused considerable excitement. Although such a prospect was unthinkable in earlier days because of organ rejection problems, new advances in preventing graft rejection have made it possible, in theory, to transplant organs from animals into humans. Initially, the preferred animal donors were baboons or other primates, but objections were raised because of the possibility that primate viruses might be able to infect humans. As mentioned in an earlier chapter, evidence suggests that HIV might originally have been a monkey virus (SIV) that entered the human population. No one really wanted experiments on organ transplant patients to serve as tests of the hypothesis that viruses from monkeys could cause yet another serious epidemic disease in humans. To make matters worse, the immunosuppressive drugs given to transplant patients make them especially vulnerable to infection, so if any virus could jump from monkeys to humans, these are the types of people in whom such a jump could most easily occur.

A second candidate animal was the pig. In fact, pig neural cells had already been injected into the brains of patients with Parkinson's disease without apparent negative effects. Interest in transplanting whole organs from pigs to humans caused great excitement until a 1997 report of the isolation from pig cells of two retroviruses that could infect human tissue culture cells. These viruses were called **PERV**s, for **porcine endogenous retroviruses**. Retroviruses are viruses that integrate into the chromosome of host cells as part of their replication strategy. In a person infected with a retrovirus, the virus could thus conceivably become part of the human germ line. Moreover, the fact that HIV is a retrovirus and that some human retroviruses are known to cause cancer caused additional concerns about the safety of organ transplants. Patients desperate for a transplant might well be willing to take such risks, but health care workers were concerned that these patients could serve as reservoirs for spread of serious disease into the larger human community.

At present, xenotransplantation is being hotly debated and, in most countries, no transplants are being permitted until there is some resolution of the safety issue. The debate over xenotransplantation illustrates how an old problem like zoonotic infections can take on a new face in connection with evolving medical technologies.

chapter 20 at a glance

Features of arthropod-borne diseases and zoonoses

Diseases involving a wild animal reservoir are difficult to control.

Diseases involving a domestic animal reservoir are preventable by vaccination or culling.

Arthropod-borne diseases can be controlled by controlling the arthropod vector.

Diagnosis takes into account history and occupation (for example, travel, veterinary practice).

Comparison of microbes and diseases in this chapter

Microbe (disease)	Characteristics	Arthropod vector	Mammalian host(s)	Mode(s) of transmission	Control measures
Yersinia pestis (plague)	Gram-negative bacterium Grows in macrophages Buboes, septic shock	Rat flea	Rodents, other wild animals, humans	Rat to human via rat flea Human to human by aerosols	Avoid contact with infected animals Rodent control, antibiotics
Borrelia burgdorferi (Lyme disease)	Spirochete Elicits inflammatory response Invasive Joint and neurological effects	*Ixodes* tick	White-footed mice (main host) Other animals Humans	Mouse to human via *Ixodes* tick	Avoid contact with ticks Remove ticks Vaccine Antibiotics
Plasmodium vivax, falciparum (malaria)	Protozoan, complex life cycle Sporozoite injected Merozoites infect RBCs (old ones, *P. vivax*; all stages, *P. falciparum*); initial cycle in liver cells (continuous cycle, *P. vivax*; one cycle, *P. falciparum*) Gametocytes acquired by insect	*Anopheles* mosquito	Humans main reservoir	Human to human via *Anopheles* mosquito	Insect control Antimalarial compounds Need to eliminate blood forms and gametocytes (both *P. vivax* and *P. falciparum*) and liver form (*P. vivax*).
Dengue virus (dengue fever)	Enveloped RNA virus Infects monocytes and macrophages	*Aedes* mosquito	Humans main reservoir, some wild animals such as monkeys	Mostly human to human via *Aedes* mosquito	Insect control Supportive therapy No effective antivirals available
Bacillus anthracis (anthrax)	Gram-positive bacterium, spore-former Antiphagocytic capsule Toxins (edema, lethal)	None	Domestic animals	Animal to human by aerosols	Vaccinate domestic livestock animals Human vaccine available but not very effective

(continued)

chapter 20 at a glance *(continued)*

Comparison of microbes and diseases in this chapter *(continued)*

Microbe (disease)	Characteristics	Arthropod vector	Mammalian host(s)	Mode(s) of transmission	Control measures
Brucella species (brucellosis)	Gram-negative coccobacillus Grows inside macrophages Localized to reproductive tract in animals Systemic in humans	None	Domestic animals	Animal to human by aerosols or through cuts	Eliminate sick animals Antibiotics to treat human disease
Rabies virus (rabies)	RNA virus, bullet shape Invades neural tissue	None	Wild animals, bats Most dogs vaccinated	Animal to human by bite or inhalation (bat guano)	Vaccinate dogs Oral vaccination of wild animals (with bait) in areas where disease is common
Bartonella henselae (bartonellosis)	Gram-negative bacterium Causes lesions around blood vessels AIDS patients most affected	Fleas (?)	Cats	Animal to human by scratch or bite contaminated with flea feces	Avoid or treat flea-infested cats Treat human disease with antibiotics

key terms

arthropod-borne disease disease spread from human to human or animal to human by an arthropod vector (insects, ticks); bypasses intact skin defense

anthrax potentially fatal disease in domestic animals, caused by *Bacillus anthracis*, a spore-forming bacterium; can cause skin lesions or fatal lung disease in humans; used by bioterrorists

bartonellosis disease carried by cats; transmitted to humans by cat scratches contaminated with flea feces; most common in immunocompromised patients

brucellosis disease that targets reproductive systems of cattle, pigs, and goats; caused by *Brucella* species; causes systemic disease in humans

bubonic plague disease caused by *Yersinia pestis*; transmitted from rodent to human by fleas and from human to human by aerosols

dengue fever viral disease transmitted from human to human by *Aedes aegypti* mosquitoes; dengue hemorrhagic fever is the most serious form

ehrlichiosis acute febrile disease caused by *Ehrlichia* species; transmitted from a variety of animals (for example, dogs) to humans by ticks

emerging infectious disease newly recognized infectious disease; often the result of changes in human activities

Lyme disease systemic disease caused by the spirochete *Borrelia burgdorferi*; transmitted from white-footed mice to humans by *Ixodes* ticks; deer not infected but allow adult tick to take necessary blood meal

malaria systemic disease caused by *Plasmodium* species; transmitted from human to human by *Anopheles* mosquitoes; complex life cycle of parasite includes sporozoites, merozoites, gametocytes (sexual forms)

pneumonic plague a form of plague that can be transmitted from human to human by aerosols

rabies fatal viral neurological disease transmitted to humans primarily by bites of infected animals (or by inhalation of aerosols from bat guano); vaccine available

xenotransplantation transplantation of organs from animals to humans

zoonosis infection acquired from an animal or its products; may be transmitted by an arthropod vector or acquired directly by inhalation or through cuts

Questions for Review and Reflection

1. What are some common features of the arthropod-transmitted diseases discussed in this chapter? Are all arthropod-borne diseases also zoonoses?

2. Did the microbes covered in this chapter have any common features? If so, what are they?

3. How can you explain the fact that the most serious kind of malaria is the easiest to cure? What is the general principle behind your answer?

4. In the 1990s, an outbreak of an acute respiratory disease occurred on an Indian reservation in the southwestern United States. The disease was caused by hanta virus and was spread from deer mice to humans. There is no evidence for human-to-human transmission. Moreover, it was noted that the mice relied on piñon nuts, the abundance of which varies from year to year. No effective vaccine or antiviral compounds have been developed for hanta virus. Yet very few cases of hanta virus disease have been reported in recent years. Make an educated guess as to the way(s) in which this was accomplished. What would you do as a public health official to prevent the spread of hanta virus disease?

5. To answer the previous question, what examples from this chapter did you draw upon for guidance? Explain your answer.

6. A person who has been bitten by a rabid animal will be given rabies vaccine. Usually vaccination must be given *before* the infection occurs. Explain the reason for this apparent anomaly in the case of rabies treatment. Explain what relevance this example might have for treatment of human immunodeficiency virus infections.

7. If you were asked to help solve the *Bartonella* problem, what would you propose? What emotional as well as health and scientific factors would you take into account?

8. In this chapter, the destruction of a herd of *Brucella*-infected buffalo was cited. Argue the pros and cons of this decision. What were the possible negative consequences of not destroying the buffaloes? If you were on a mission to save the buffaloes but also had some consideration for the farmers, how would you have handled this situation?

IV microbes in the environment and in industry

Interesting as human-microbe interactions are, they cover only a subset of microbial activities. Microbes have similarly close, and often beneficial, interactions with arthropods, plants and nonhuman animals. Some of the interactions between microbes and arthropods or animals that are responsible for human disease have been covered in Chapter 20. In the next 5 chapters, we will explore some interactions of microbes with each other, of microbes with arthropods, and of microbes with plants that are either beneficial or neutral for humans. Chapter 21 covers what have been called "microbial communities" and raises the question—for which there is still not a good answer—of whether microbes form communities in the same sense that plants and animals do. Chapter 21 also describes some exciting innovations that are making it possible to investigate microbial communities in new ways. In Chapter 22, the bright world of photosynthesis is described. We'll be honest. Photosynthesis is one of those topics that most microbiology students find as dull as dirt. Yet without photosynthesis, this book would not be here and you would not be here to read it. We try to present photosynthesis as the interesting and important topic it should be. In fact, we introduce some of the most fascinating bacteria on earth, the cyanobacteria, which played a starring role in Chapter 2 and are now revealed to have a more sinister side. In Chapter 23, we explore a topic that is unknown even to many microbiologists, the evolving story of the intimate associations of microbes with arthropods. Chapter 24 looks at some associations between bacteria and plants, including one in which bacteria use sex to force a plant to feed them. Finally, in Chapter 25, we return to the attempts of scientists to harness the diversity of the microbial world, the world of biotechnology.

21 Analyzing Microbial Communities

Are microbes social creatures like us, or do they simply look out for themselves, moving into sites where they can predominate and make a good living? Is there always interspecies warfare, or can different species cooperate for the greater good? The older view portrayed microbes as ruthless individuals. Newer information raises doubts about this assumption, but the question is still open.

In the early days of microbiology, scientists were intrigued by the complex mixtures of microbes in environmental samples they viewed through microscopes. They soon realized, however, that working with such mixtures was very difficult, because cultivating them in laboratory media caused the composition of the population to change dramatically. Because these changes were difficult to control, it seemed more reasonable to isolate individual microbes in pure culture and work with pure cultures instead of the assemblage of microbes found in the original site. This decision would lead in the next century to extraordinary progress in understanding microbial physiology and genetics.

What is a Microbial Community?

Only in the latter part of the 20th century did scientists become frustrated with the limitations of the pure culture approach. This frustration had two sources. First, scientists were beginning to suspect that microbes growing as pure cultures in laboratory media did not behave as they did in their natural habitats. Not only were nutrients more limited and more irregularly available in most natural settings than in laboratory media, but interactions with other microbes in the natural site were not possible in a pure culture. A second and more important source of frustration was that a comparison of microscopic observations of the original specimen with what scientists managed to cultivate from that specimen made it clear that only a tiny fraction of the original microbes were being cultivated. Even the most persistent and talented cultivators of microbes were missing most of the original microbial population in their efforts to isolate pure cultures. Some of the reasons for this were discussed in Chapter 3. What could be done to remedy this situation? For a description of a unique summer course that introduces students to analysis of microbial communities and microbial diversity, see Box 21–1.

The rRNA gene revolution

The first step in finding a solution was to apply the new molecular methods, specifically the molecular yardstick approach being used to identify pure culture isolates (Chapter 2). A breakthrough came when scientists realized that just as it was possible to amplify a ribosomal RNA (rRNA) gene from a pure culture, it would be possible to amplify rRNA genes from a mixture of microbes (Fig. 21.1). Of course, such an amplification, in contrast to an amplification of rRNA genes from a pure culture, produced DNA fragments that contained many different rRNA gene sequences, all of the same size. The solution was to separate the different rRNA genes by cloning them into a cloning vector (Chapter 5). When the mixture of vectors, each containing a single rRNA gene, was used to transform *Escherichia coli,* each resulting transformant now had a single cloning vector containing a single cloned rRNA gene. By sequencing a number of rRNA genes and comparing them with genes in the databases, a picture of the composition of the microbial population began to emerge. In effect, scientists ended up with a list of microbes that might be present in the population.

Because this type of analysis was not dependent on the ability to grow the microorganisms on laboratory media, it was possible to identify tentatively not only the microbes that were present but also the microbes that were not being cultivated. Sometimes the sequences of rRNA genes were close to those of cultivated organisms, but just as often the sequences were very distant from known organisms, suggesting that whole phylogenetic groups of microbes had been missed completely in previous cultivation attempts. In many environmental settings, less than 1% of the organisms for which rRNA gene sequences were obtained were recognizable as cultivated strains or had identical sequences to microbes isolated from that environment by classic cultivation methods. This was true for bacteria and archaea and even more true

box

21–1 *The Microbial Diversity Summer Course at the Marine Biological Laboratory at Woods Hole, Massachussetts*

No discussion of the analysis of microbial diversity would be complete without mentioning a unique 7-week summer course given each year in June and July at the Marine Biological Laboratory in Woods Hole, Massachusetts. Founded by marine microbiologist Holger Jannasch, whose preferred area of study was the microbes found in hydrothermal vents located deep on the ocean's floor, the course has introduced thousands of young scientists to the wonders of microbial diversity. The Woods Hole area provides a variety of interesting study sites, including the Great Sippiwisset Salt Marsh. During the first 3 weeks of the course, students are introduced to a variety of cultivation techniques and to molecular approaches to population analysis, such as polymerase chain reaction/sequencing of ribosomal RNA (rRNA) gene sequences. A unique feature of the course is the next 4-week period, during which students carry out their own individual research projects on material taken from the area. Samples include not only material taken from soil or salt marsh sands but material taken from the bodies of marine animals. Most of these samples have not been characterized previously in detail, so the projects are truly research projects. The course usually is limited to graduate students and postdoctoral trainees, but undergraduates have been admitted from time to time. The students come from all over the world, giving the course a human cultural diversity as well as a microbial culture diversity.

For more information about the course, consult the Marine Biological Laboratory Web site: www.courses.mbl.edu, and go to the Microbial Diversity Course page. The Marine Biological Laboratory also offers a number of other interesting courses, descriptions of which also can be accessed from the Web site.

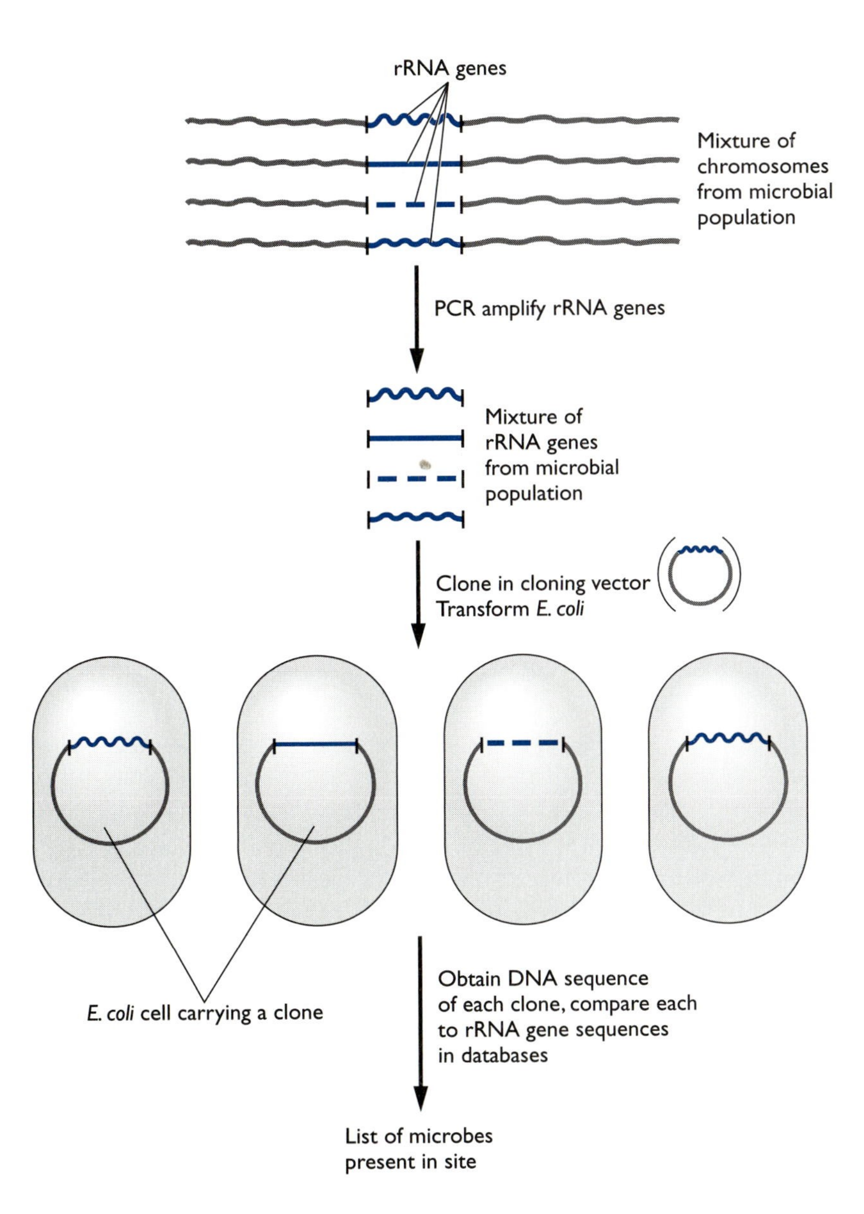

Figure 21.1 Steps in the process of polymerase chain reaction (PCR)/sequencing of ribosomal RNA (rRNA) genes from a mixed population of microbes. The sample is first treated to lyse the microbes in the sample, so that their DNA is released. PCR is used to amplify the rRNA genes from the DNA in the mixed population. Primers can be chosen to amplify most or all microbes (the so-called "universal" primers) or to amplify selected populations, such as bacteria or archaea or eukayrotes. After amplification, the mixture of rRNA gene sequences is cloned into a cloning vector (Chapter 5) and transformed into *Escherichia coli* to separate the individual rRNA genes in the sample. The plasmids are isolated, as described in Chapter 5, and the rRNA gene from each one is sequenced. The sequences are then submitted to a computer database to determine the nearest known microbial relative.

for the eukaryotic microbes. Except for eukaryotic human pathogens and a few model organisms, such as *Dictyostelium* and *Tetrahymena,* few successful attempts had been made to isolate and identify environmental eukaryotic microbes. That such microbes were abundant in nature, however, was evident from microscopic examination of environmental samples.

Limitations of the rRNA gene approach to characterizing microbial communities

Just as cultivation has its drawbacks for analyzing microbial communities, the rRNA gene sequence approach has limitations of its own. These are listed in Table 21.1, along with corresponding drawbacks of the cultivation approach. Both approaches have their strengths and weakness. The big advantage of cultivation is that, if successful, the organism is available for study in the laboratory, whereas with the polymerase chain reaction (PCR)/sequencing approach, the scientist is left with only a possible name. The great advantage of PCR/sequencing over cultivation is that it provides clues as to what microbes are not being cultivated. Combining the two approaches is clearly the winning strategy.

Table 21.1 Comparison of drawbacks of the polymerase chain reaction (PCR)/sequencing approach to characterizing microbial communities with those of the cultivation approach

PCR/sequencing of ribosomal RNA (rRNA) genes	Cultivation of microbes from sample
Not all microbes are lysed	Not all adherent microbes are released from particles so that they are in the liquid plated on the cultivation medium
Not all rRNA genes are amplified equally, and some are missed	Not all microbes are cultivated on the media chosen
Chimeras can be produced	
Amplification of contaminating extraneous DNA	Medium that does not support growth of microbes in the sample may support growth of contaminants from the laboratory environment
Inhibiting substances in the sample poison the PCR reaction	Inhibiting substances inadvertently included in the medium kill microbes that might grow on it otherwise
Interpretation of sequence data; how to define a species	Interpretation of traits of cultivated microbe, including its rRNA gene sequence; how to define a species
Phylogenetic identity of a microbe does not necessarily give information about its metabolism; metabolism may vary within the same phylogenetic group identified by the rRNA gene sequence analysis	Metabolic tests alone may place genetically unrelated microbes in the same species

The first problem with the rRNA gene sequence approach is that only those microbes that were lysed successfully to release their DNA would be represented, because PCR will amplify only genes in free DNA. Any DNA still inside an intact microbe will not be amplified. Techniques used to disrupt such easily lysed bacteria as *E. coli* proved to be unsuccessful with many other microbes. This problem has now been solved by using a machine called a **bead beater.** Beads of glass or silica are placed into a small tube containing the sample to be analyzed. The mixture is then shaken violently for a few minutes. The collisions of the beads create enormous shear forces that tear apart virtually any microbe. Oddly enough, although this type of shearing might be expected to shear DNA as well as microbes, the fragments of DNA released are still large enough to contain intact rRNA genes. A corresponding drawback for cultivation procedures is that microbes tightly adherent to surfaces of particles may not be released into the liquid part of the sample that is plated on the agar medium.

A second drawback of the rRNA gene sequencing approach is that, in a mixture of genes from different microbes, some rRNA genes are amplified by PCR more efficiently than others. The reason for this is not known. Also, partial PCR products from different genes sometimes can hybridize with each other, producing **chimeras** (Fig. 21.2). These usually are easy to spot if the whole cloned fragment is sequenced, because the two portions of the chimera will align with different sequences in the databases. The corresponding drawback to the cultivation approach is that the medium used usually does not support the growth of all microbes equally, and some will not grow at all in it. Another artifact of the PCR approach arises from the fact that PCR will amplify exogenous DNA that is introduced inadvertently by the investigator from laboratory reagents or the laboratory environment. Of course, the cultivation approach has this same problem, because microbial contaminants from the environment can be introduced into the medium if

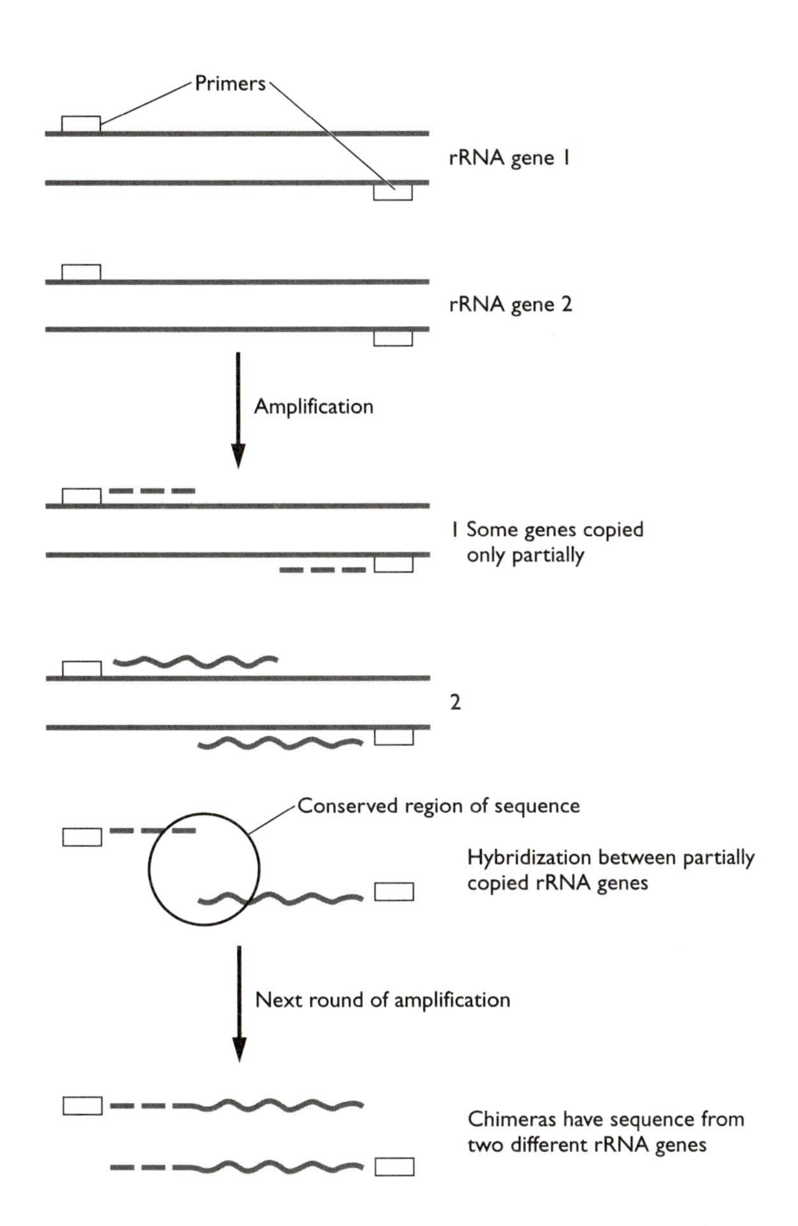

Figure 21.2 How chimeras can be produced in a polymerase chain reaction (PCR). Incomplete DNA copies of the region to be amplified are produced. Because ribosomal RNA (rRNA) genes contain highly conserved regions as well as variable ones, the sequences in the conserved regions of two different genes can hybridize with each other, creating a double-stranded region that will serve as a primer for DNA polymerase to recognize. The molecules produced in the next round of amplification will contain sequences from each of the two different original sequences.

the investigator's aseptic technique is not good enough. Even if the medium does not support the growth of most of the microbes in the sample of interest, it may support the growth of contaminants and yield isolates that were not in the original sample.

A third drawback of the PCR/sequencing approach is that some samples, especially those containing soil or sediment, contain substances that inhibit the PCR reaction. In such cases, no amplified DNA fragments are obtained from the PCR reaction. The culprit in soil and sediment samples appears to be a set of complex polymers, called **humic substances,** which give soils and sediments their dark color. This problem can sometimes be solved by diluting the sample to reduce the concentration of the interfering substance to below inhibitory levels, still leaving enough DNA to amplify. Alternatively, kits are now available for purifying the DNA in a sample, so that it is separated from other polymers in the mixture. The corresponding drawback in cultivation methods is that the medium used may contain substances such as heavy metals that inhibit the growth of some microbes (Chapter 3).

One problem that is more of a quandary than a drawback is how to analyze the results of rRNA gene sequence data. How different in sequence must the rRNA gene from one organism be from that of another to be considered as representing a new species or a new genus? Currently, a 5% difference is considered sufficient to indicate a separate species, but controversy remains about this figure. This problem is compounded by the fact that many prokaryotes have multiple rRNA genes. These genes usually are virtually identical to each other in sequence, but in some cases have been found to differ by 1%–4%. The corresponding problem for cultivation is identifying the isolated microbes and deciding how to define a species on the basis of phenotypic traits. In the early days of identification based on phenotypic traits, scientists put *Salmonella typhi* (the cause of typhoid fever) and *Salmonella typhimurium* (a cause of diarrhea) into different species, based on their different traits and the different diseases they caused. We now know that these species are at least 97% identical at the DNA sequence level. Today, scientists have become "lumpers" and will usually place two bacteria in the same species if they share at least 70% identity at their chromosomal DNA sequences.

A final drawback of the rRNA gene sequence approach to analyzing microbial communities is that it gives only an indication of what species of microbes are found in the site. It is usually not possible to deduce the metabolic capabilities of the microbe from this sequence information, because there are many examples in which a microbe with one type of metabolism, such as the ability to obtain energy from sunlight or the ability to reduce sulfate (SO_4) to hydrogen sulfide (H_2S), will have very similar rRNA gene sequences to microbes without this trait. A mystery remains as to why metabolically similar microbes do not always cluster phylogenetically. The corresponding drawback in cultivation-based methods, if only phenotypic traits are used to define species, is that dissimilar microbes that are genetically quite distant from each other have sometimes been placed in the same species—another indication that metabolism does not always follow phylogeny.

Given the potential problems with the rRNA gene sequencing approach, scientists generally agree that this is only a first step to analyzing a microbial community. Its great strength is that it gives clues as to what type of microbes may have been missed in cultivation attempts and suggests new media to use for further cultivation efforts. By narrowing down the possible conditions for media composition and atmosphere, scientists may be better able to cultivate what previously could not be cultivated because certain types of media and conditions were not tried previously. Confronted with such a vast array of possible media and conditions (Chapter 3), scientists rarely try more than a small subset of these when they attempt to cultivate microbes from an environment. The results of rRNA gene sequence analysis are interpreted most safely when they are considered as suggesting hypotheses about which microbes are in a site instead of accepted as proof of community composition. Combined with other types of analysis, this approach becomes a powerful tool for analyzing microbial communities.

Microscopy

Phase-contrast and light microscopy. The first microscopes simply magnified whatever organisms were present in a specimen. Modern microscopes have many more capabilities, such as improved resolution and the ability to view specimens under different wavelengths of light. Once again, microscopes are at the center of microbiological analyses of environmental samples. To understand why such capabilities can provide important information about the microbes in a sample, consider the different ways such microbes can be viewed and what information this can give about the microbes present in a sample. A phase-contrast microscope illuminates a specimen with ordinary light. Light is diffracted differently by a microbe than by its surrounding liquid, and the phase-contrast microscope collects only certain diffracted wavelengths (Fig. 21.3). The result is that the microscope not only magnifies the microbes but also makes their edges and features clearer.

Phase-contrast microscopy is used to view microbes in a drop of liquid covered by a cover slip (**wet mount**). This type of analysis gives information not only about the shape of a microbe but also whether it is motile and whether that motility is likely to derive from flagella or gliding. An organism

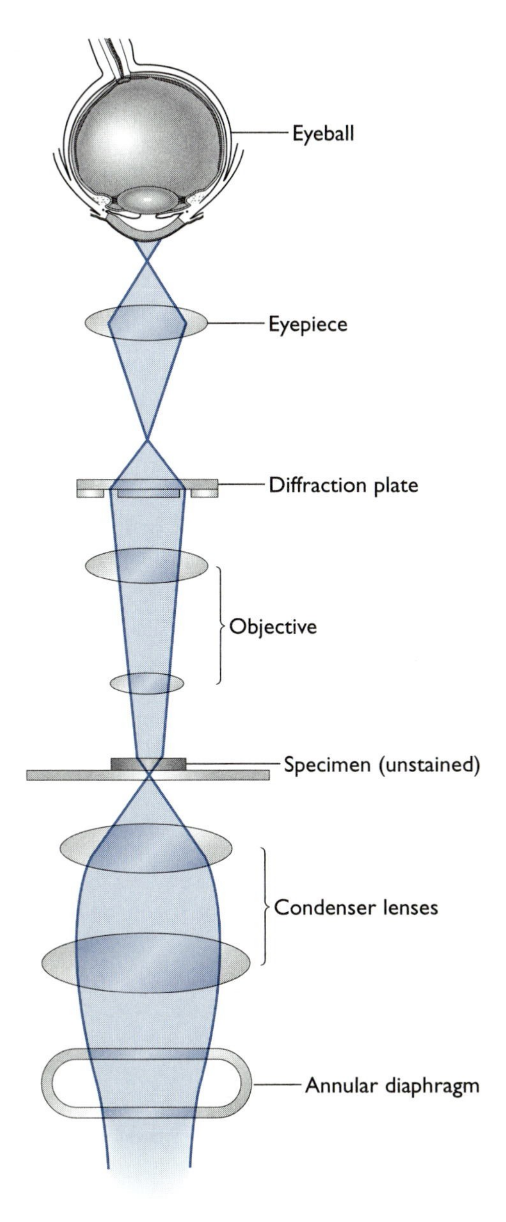

Figure 21.3 Diagram of a phase-contrast microscope. Phase-contrast microscopy has an advantage over light microscopy, which usually requires that bacteria be stained in order to be seen. Phase-contrast microscopy takes advantage of the slight differences in cell density and refractive index and converts them into easily detected variations in light intensity. This variation in light intensity is amplified by a diffraction plate located above the objective lens. The result is that the bacteria appear as dark images against a light background.

that swims rapidly through the medium, especially with a wiggling or tumbling motion, is probably motile because of flagella. An organism that moves more slowly over surfaces is exhibiting a gliding motility. Assessing motility is a bit tricky for beginners, because **Brownian motion,** which results from the flow of liquid as the microscope slide plus cover slip are compressed by the objective and which creates zones of flow within the microscope field, may make it appear that microbes are moving when they are simply being buffeted by artificial currents. The state of the microbes being examined also is critical. A common mistake made by students studying marine microbes for the first time is to use distilled water instead of salt water to make a wet mount. If you want to see bacteria lysing before your eyes, this is definitely the way to proceed. A more subtle effect involves the oxygen concentration in the liquid trapped between a cover slip and the slide surface. Over the viewing time, the liquid in a wet mount can become anoxic. Bacteria that require oxygen will become sluggish. A sign that this has happened is that the organisms remain very actively motile around an air bubble or at the wet mount edges, which are exposed to air, but are not motile in other parts of the wet mount.

Other information can be obtained from phase-contrast microscopy. **Sulfur granules** and **gas vacuoles** reflect light to give a microbe a "Christmas tree light" appearance (Color Plates 12, 13). The presence of sulfur granules inside a bacterium indicates that its energy metabolism is based on sulfur compounds. Spores also refract light in a very distinctive way, so that spore-producing bacteria are easy to identify. "Visual junk" in the background sometimes obscures the presence of microbes. An example of this situation occurs when iron-oxidizing bacteria or microbes attached to an insect's intestinal wall are being sought. In the former case, crystalline forms of the metal to which the bacteria are attached may create a visual pattern against which bacteria disappear. In the latter case, cells of the intestinal wall have the same effect.

The simplest form of microscopy is to illuminate the slide directly with light (**light microscopy**). The problem with such direct illumination is that bacteria are not visible, because light passes through them. Thus, the bacteria must be stained in some way. In clinical microbiology, the Gram stain is most commonly used. Unfortunately, the Gram stain is very unreliable when applied to environmental microbes. Many environmental bacteria with gram-positive type cell walls, for example, stain gram-negative. Thus, the Gram stain is not used nearly so commonly to visualize environmental microbes as it is to visualize clinical isolates.

Fluorescence microscopy. A fluorescence microscope is used to view organisms containing molecules that can be activated by one wavelength of light to emit light of a different wavelength (Fig. 21.4). This type of microscope displays those microbes that contain fluorescent compounds as bright forms against a dark background. In the case of samples that contain a lot of debris, microbes can be visualized more clearly and can be distinguished from the background by staining them with dyes that intercalate with DNA and fluoresce only under this condition. One such stain is 4',6-diamidino-2-phenylindole (DAPI), which interacts with DNA and, in the intercalated state, gives off a blue-white fluorescence. If the background does not fluoresce at the same wavelength, the bacteria will appear as ghostly bluish forms against a dark background.

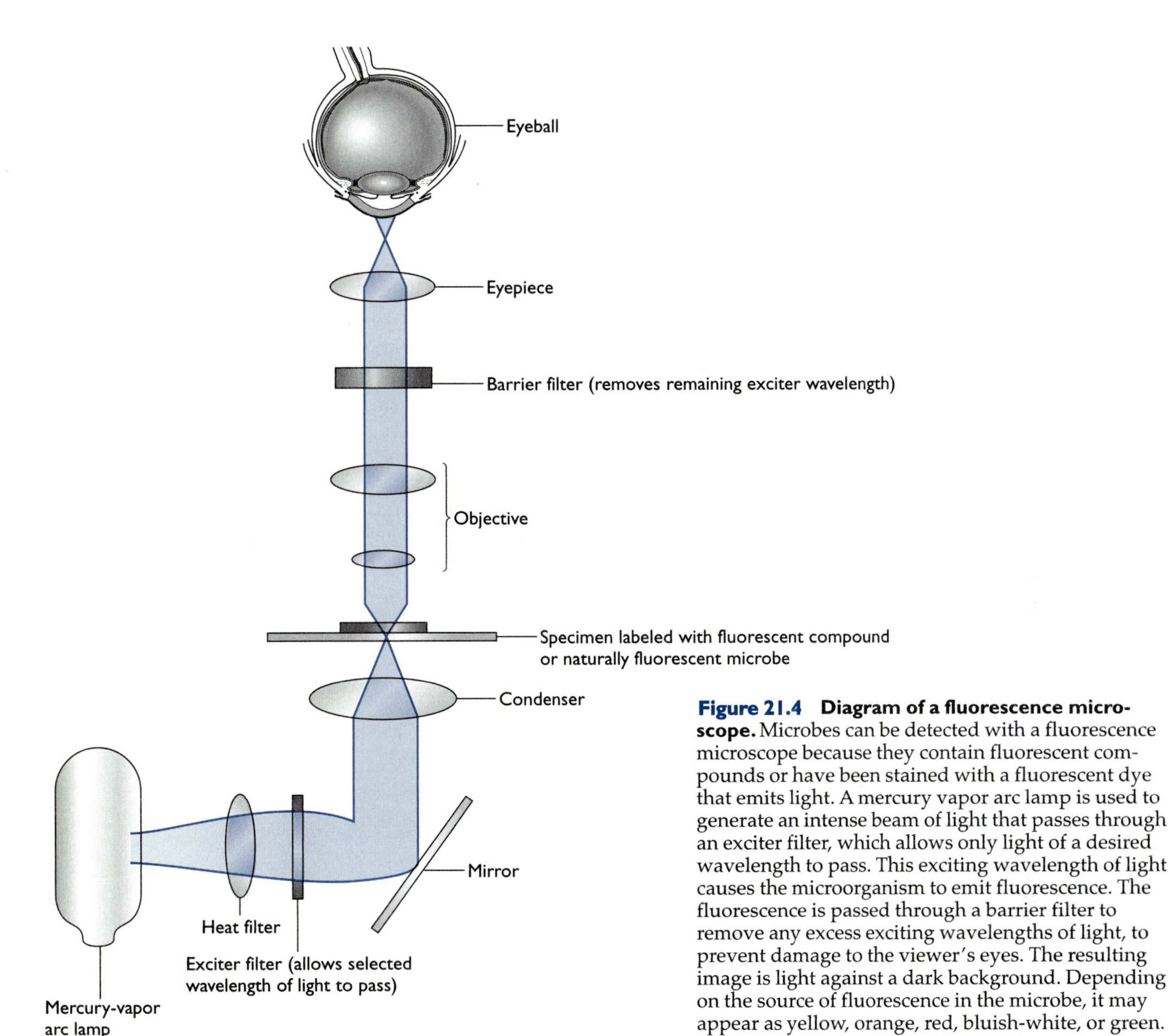

Figure 21.4 Diagram of a fluorescence microscope. Microbes can be detected with a fluorescence microscope because they contain fluorescent compounds or have been stained with a fluorescent dye that emits light. A mercury vapor arc lamp is used to generate an intense beam of light that passes through an exciter filter, which allows only light of a desired wavelength to pass. This exciting wavelength of light causes the microorganism to emit fluorescence. The fluorescence is passed through a barrier filter to remove any excess exciting wavelengths of light, to prevent damage to the viewer's eyes. The resulting image is light against a dark background. Depending on the source of fluorescence in the microbe, it may appear as yellow, orange, red, bluish-white, or green.

Photosynthetic bacteria contain pigments that fluoresce red when green light is used to illuminate the microscope field. Nonphotosynthetic organisms do not fluoresce under this condition and are thus not seen. This trait can be used to identify the photosynthetic members of a bacterial population (Color Plate 10B). Archaea that produce methane (CH_4) have a distinctive set of enzyme cofactors that cause these organisms to fluoresce blue when illuminated with the appropriate wavelength of light. This differs from DAPI staining, in that no stain is needed to obtain this type of fluorescence.

A microscopic examination is at once exciting and frustrating. Many microbiologists have looked into the teeming complexity of a microbial population in the initial microscopic analysis, then labored hard in their cultivation techniques, only to find that the microbes they have cultivated bear no discernible resemblance to those apparent in the initial examination. Useful as classical microscopy is as an analytical tool, it falls short in revealing what species might have been present in the original sample.

A scientist with a sign on the lab door that says "gone fishing" may actually be hard at work in the laboratory. A new technique called **fluorescent in situ hybridization (FISH)** allows scientists to assess the identity of microbes seen under a microscope. Specifically, it allows a scientist to assess the accuracy of PCR/sequence data of the actual population. Suppose you have identified five new rRNA sequences from a community analysis of a site and you want to check how common these sequences are in the actual population. Simply take the sequences you have obtained, use them to design 30–50 bp DNA hybridization probes (fragments of single-stranded rRNA genes that hybridize to rRNA and to the rRNA genes of an organism). These fragments are labeled with a fluorescent dye (**rDNA probes**). Steps in this type of analysis are provided in Figure 21.5. Microbes in a sample are treated with methanol to make large holes in their membranes, which allow the labeled

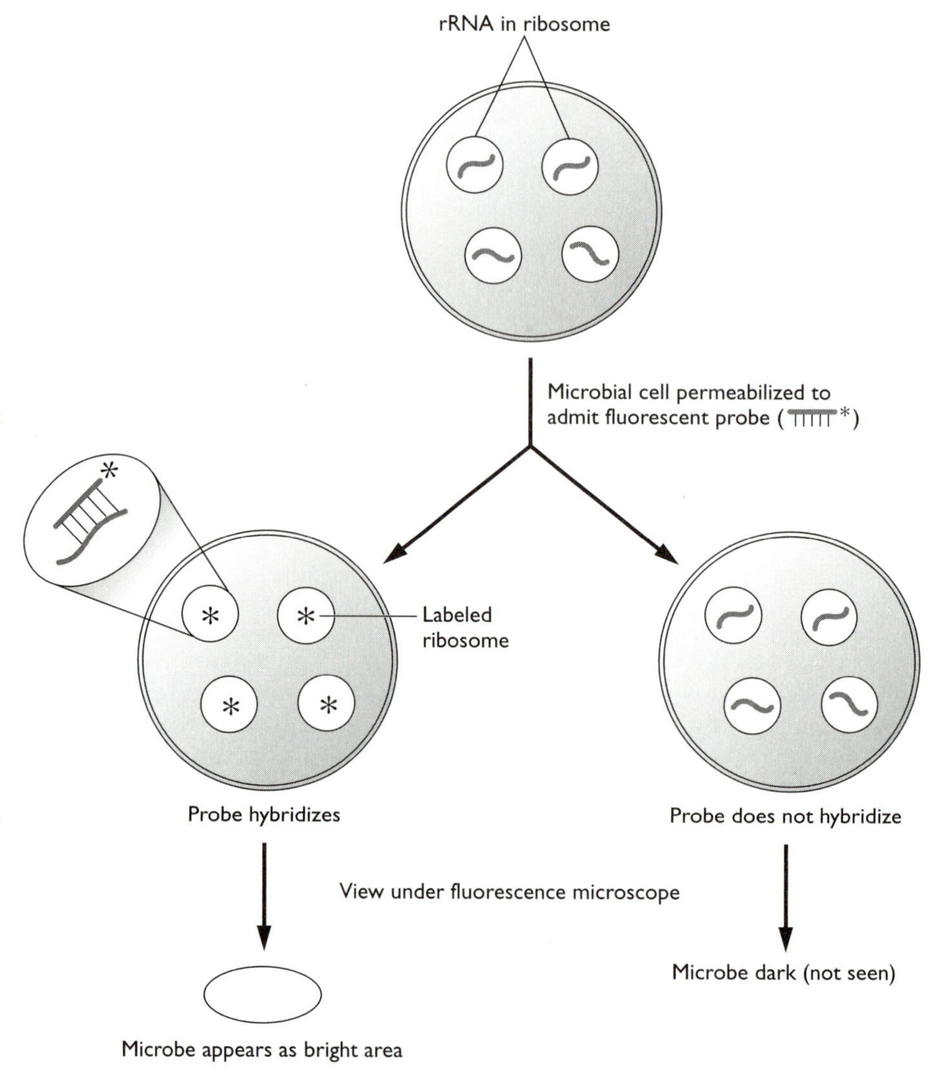

Figure 21.5 Fluorescence in situ hybridization (FISH). The goal of FISH is to use fluorescently labeled probes that detect microbes with different ribosomal RNA (rRNA) sequences in a sample containing a population of bacteria. The first step is to attach microbes in the sample to the slide. This is usually done by coating the slide with poly-lysine or some other coating to which microbes adhere tightly. Microbes cn the slide are then treated with a methanol solution that makes holes in their membrane (permeabilizes them so that they can receive the probe). A fluorescently labeled single-stranded DNA probe with a specific rRNA gene sequence is added to the slide. The labeled probe diffuses into the microbial cell through the holes produced in its membrane by the methanol treatment. If the rRNA in the ribosomes of the microbe hybridizes to the probe, the probe will bind tightly enough that it is not washed off. A wash step removes unbound probe. Microbes with rRNA that has reacted with the probe will become fluorescent and will appear as bright spots when viewed under a fluorescence microscope. The rRNA in ribosomes is the target for the probe instead of the rRNA gene, because there are many more copies of rRNA than of the rRNA gene, thus making the technique much more sensitive. If the probes are labeled with different compounds that fluoresce at different wavelengths, more than one type of microbe can be probed for.

probes to enter the cells but keep the morphology of the cell intact. Enough copies of rRNA are in the ribosomes of rapidly growing cells to react with the labeled rDNA probe and give a visible signal when the labeled probe binds to its target. This type of analysis helps to settle questions of abundance of a particular microbial type in a sample. If, for some reason, a particular rDNA sequence was amplified preferentially, FISH makes it clear that this sequence is not representative of the population. Similarly, if none of the amplified sequences stain a specific microbial component of the community, it becomes clear that the rRNA gene sequence of this member of the community was not detected by the PCR/sequence analysis.

The possible identity of the "missing" members of the microbial population can be obtained by using probes that identify certain phylogenetic groups of microbes. An example of such a probe is one that hybridizes to rRNA genes from most gram-positive bacteria. Sets of such probes are now widely available. FISH detects only the most numerically predominant members of a population, because minor components are far outnumbered and thus may not show up in most fields examined by the microscopist. Other problems that may arise in connection with a FISH analysis are background fluorescence in the sample (for example, as a result of chlorophyll), organisms that fluoresce naturally at the wavelength used to visualize the bound hybridization probe, and failure to make all of the microbes in a sample permeable. And, as with PCR/sequencing of rRNA genes, the identity of a microbe determined by FISH does not necessarily give information about the metabolic activities of that microbe.

Life in a Gradient

Metabolic types of microbes

A microbe is classified metabolically in terms of its carbon source, electron donors, and electron acceptors. **Autotrophs** are bacteria that use carbon dioxide (CO_2) as a carbon source. They vary in their electron donors and acceptors, but all have in common that they can grow with carbon dioxide as a sole carbon source. Autotrophs are found among the bacteria and the archaea. An example of an autotroph is the **methanogenic archaea,** which use hydrogen to reduce carbon dioxide to methane to obtain energy and also use carbon dioxide as a carbon source. A microbe that derives its carbon and energy from oxidation of more complex organic nutrients, such as sugars or amino acids, is termed a **heterotroph.** Heterotrophs that use complex organic compounds are all bacteria, but some archaea can use simpler organic compounds such as acetate.

Lithotrophs are microbes that use inorganic compounds as electron sources for generation of energy. Lithotrophs are found among both bacteria and archaea. Examples of lithotrophs covered in this chapter are sulfur-oxiding bacteria. Microbes that obtain energy from light are called **phototrophs.** These are covered in detail in Chapter 22. Most of the phototrophs are bacteria or eukaryotes. Some archaea can use light energy, but they do it in a different manner, using a photosynthetic machinery not based on chlorophyll or chlorophyll-like molecules such as those used by bacteria, algae, and green plants. Sometimes these terms are combined. For example a lithoautotroph is a microbe that uses carbon dioxide as a carbon source and an inorganic compound as an electron donor. A photoautotroph is a microbe that uses carbon dioxide as a carbon source and light as its source of energy.

A carbon cycle based on the one-carbon compounds, methane and carbon dioxide

Overview of the cycle. Metabolic analysis has been around for years but now assumes new importance when coupled with the rRNA gene sequence analysis. It provides a guide to what types of microbes might be expected to predominate in a particular site. Chemical analysis of soils or of sediments and the water overlying them reveals metabolic end products that provide clues as to what types of activities are being carried out by resident microbes. An example of the sort of profile that might be seen in a sediment and overlying water layer from a freshwater lake or swamp is shown in Figure 21.6. The consumption of polysaccharides in the sediment by oxygen utilizers causes oxygen to be depleted. In the anoxic sediment, methane is detected, indicating that acetate or hydrogen and carbon dioxide are being consumed by the methanogens and used to produce methane. Methane, which diffuses upward into the water overlying the sediment, disappears in the vicinity of the oxic/anoxic interface, where the oxygen level is too low to favor spontaneous oxidation of methane to carbon dioxide. This indicates the presence of bacteria capable of oxidizing methane to carbon dioxide.

Methanogenesis. Steps in the reduction of carbon dioxide to methane are shown in Figure 21.7. Methanogenic archaea attach a carboxyl group to the first of a special set of cofactors. As the process proceeds, the carboxyl group is transferred to other cofactors and reduced in a stepwise manner from the most oxidized form of carbon, carbon dioxide, to the most reduced form, methane. In the process of these tightly coupled reduction steps, protons are pumped outside the cell to provide adenosine triphosphate (ATP) for cellular activities. This explains how methanogenic archaea gain energy, but how do they make cellular components? The answer is shown in the expanded pathway illustrated in Figure 21.8. Some carbon dioxide can be taken to acetyl-S-CoA, and hence through pathways that produce amino acids, sugars, and other essential cell components (see Chapter 8).

Methane oxidation. Bubbles containing methane and carbon dioxide rise from the sediment and move into the zone above the sediment, where

Figure 21.6 Chemical profiles of a sediment, the water overlying it, and the types of microbes that probably are present. Note that the methanogens are located in the anoxic zone, because they are poisoned by oxygen, and the methane oxidizers are located at the oxic/anoxic interface, because they require low oxygen levels but cannot compete with higher oxygen levels that would spontaneously oxidize methane.

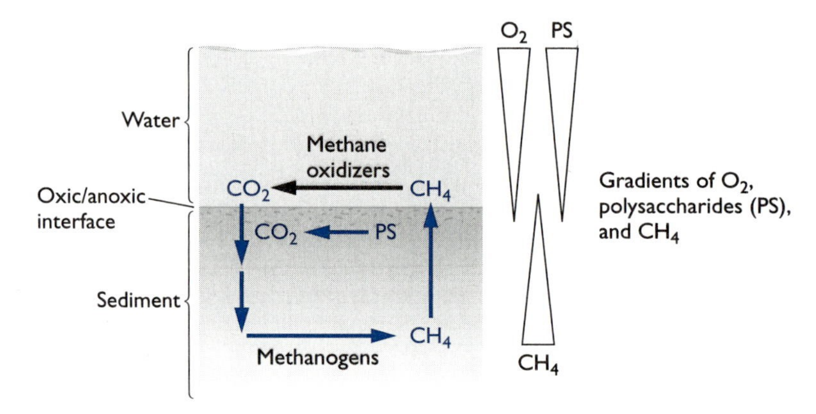

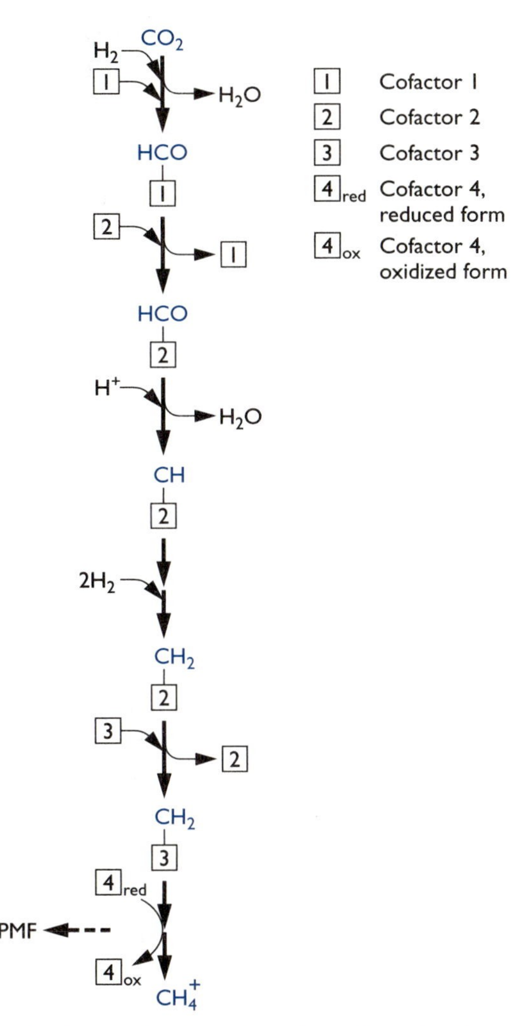

Figure 21.7 Steps in the production of methane from hydrogen and carbon dioxide. Carbon dioxide is first fixed to a cofactor as a formyl group (first reduction), then reduced in steps as it is passed to other cofactors that help the enzymes carrying out the reductive reactions to do their job. Although methanogenesis is a reductive reaction, it is nonetheless coupled to an electron transport system, as illustrated here, and thus generates proton motive force (PMF).

molecular oxygen is present. In this zone, some of the methane disappears. What happens to the rising methane in bubbles from the swamp? In the region above the sediment is a group of bacteria, the **methane oxidizers,** that use methane as a source of energy and carbon. This fascinating group of bacteria oxidize methane to formaldehyde, which is an intermediate not only in the generation of energy but also in the synthesis of organic components of the bacterial cell (Fig. 21.9). **Formaldehyde** is a very toxic compound for all living things. It interacts with proteins in such a way as to inactivate them. Yet the methane oxidizers have been able to use formaldehyde as a reactive intermediate in their energy and carbon metabolism without pickling themselves in the process. The strategy that these gram-negative bacteria use to protect themselves is to perform the methane-to-formaldehyde reaction in the periplasm, then further oxidize the formaldehyde to formate and carbon dioxide, which are nontoxic, in the

Figure 21.8 Steps in the pathway used by methanogens to make acetyl-S-CoA, which serves as an entryway into synthesis of more complex organic molecules important for cell function. For simplicity, some steps are omitted, as indicated by multiple arrows.

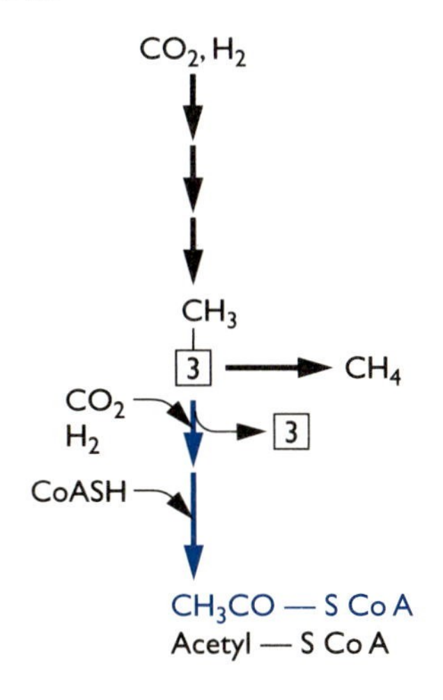

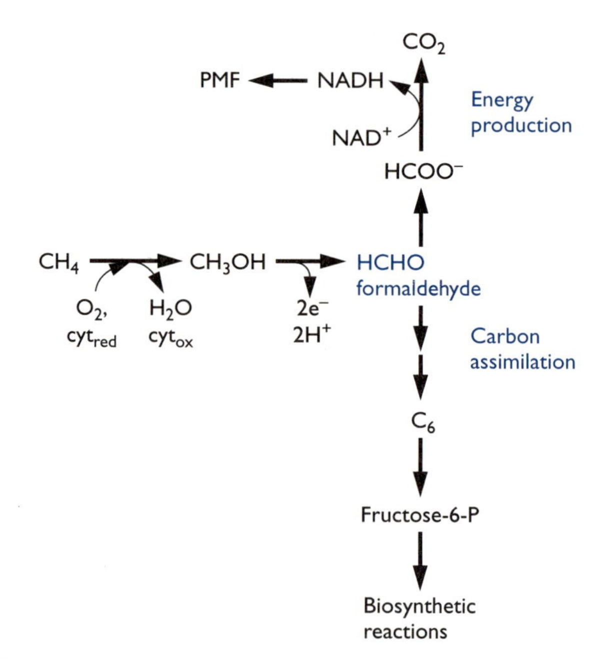

Figure 21.9 Steps in the pathway methane oxidizers take to use methane as a source of energy and carbon. Note that formaldehyde is the key intermediate in the pathway and is formed in the periplasm of these gram-negative bacteria to protect them from its toxic properties. cyt_{red}, reduced cytochrome; cyt_{ox}, oxidized cytochrome.

membrane or use proteins in the cytoplasm associated with the membrane (Fig. 21.9).

Importance of methane oxidizers. Methane oxidizers are not just curiosities; they have taken on considerable political and economic importance. Methane is a greenhouse gas, and, according to the Kyoto Accords (an agreement on emissions limitations approved by many nations), countries must limit their methane emissions. Methane is produced by the burning of fossil fuels, but most of it is of microbial origin. Rice paddies, landfills, and the rumens of cattle are major sources of methane gas. How can these methane emissions be controlled? One strategy is to enlist the methane-oxidizing bacteria, which convert methane to carbon dioxide, itself a greenhouse gas but less active than methane. Efforts to decrease the production of methane (by draining rice paddies more often) or increasing oxidation of methane (by adding a layer of soil to landfills where the methane-oxidizers can scrub some of the methane) are currently under study. Methane, a highly flammable gas, also is being channeled for use in cooking and heating in developing countries, where cow manure is abundant and can serve as a source of methane. Methane can be fun, too (Box 21–2).

The sulfur cycle

Sulfate-reducing bacteria. In freshwater environments, methanogens tend to dominate as consumers of hydrogen produced by the carbohydrate oxidizers. In marine environments, however, the high concentration of sulfate allows a different type of bacteria, the **sulfate-reducing bacteria,** to act as hydrogen consumers. In anoxic sediments, these bacteria use sulfate as a

box 21–2 *How to have fun with methane: the Volta exercise*

Allesandro Volta (1745–1827) first discovered in 1778 that methane was a flammable gas. This was the same Volta who also played an important role in early discoveries about electricity and after whom the unit "volt" was named. Today, scientists young and old sometimes honor Volta in a more light-hearted way for his discovery of the flammability of methane. To participate in such revels, you need a large plastic funnel, a stopper with a hole in it that fits snugly in the funnel's outlet, some matches, a freshwater pond or swamp with a nice thick dark ooze on the bottom of it, and clothes you don't mind parting with. When the night is pitch dark, wade into the water and let your feet sink into the ooze. Try not to fall over (but in the end, making a fool of yourself is part of the fun). Put the stopper into the funnel, invert the funnel, and fill it with water (see Figure). Put your finger or thumb firmly over the hole in the stopper. Kick the ooze to release bubbles, which, if you are lucky, should contain some methane. The bubbles will be trapped in the funnel and will gradually displace the water under the funnel to form a gas pocket (see figure below). Push down on the funnel to compress the gas so that it will rush out of the opening when you release your finger. Have a compatriot standing by with a lighter. Remove your finger from the hole, releasing the gas, to which your assistant puts the flame, and you may be rewarded with a dramatic column of flame (Figure). Be sure to keep your head out of the way, unless scorched eyebrows and hair are your preferred look.

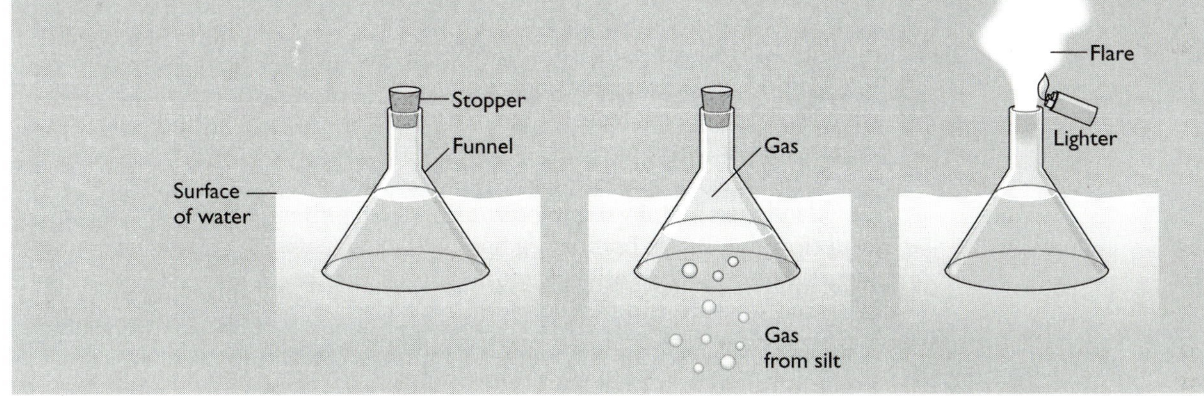

terminal electron acceptor, in a process called **dissimilatory sulfate reduction.** This term comes from the fact that sulfate is used as a terminal electron acceptor instead of being incorporated into cysteine and other sulfur-containing compounds, a process labeled **assimilatory sulfate reduction** (Fig. 21.10). Note that this assimilation process requires two ATP molecules to activate the sulfate to **phosphoadenosine phosphosulfate** (**PAPS**), so that it can be reduced to sulfite (SO_3^{2-}), which will be reduced by nicotinamide adenine dinucleotide phosphate (NADPH) to the sulfide that will be incorporated into cysteine. Assimilation is a reaction that consumes energy in the form of both ATP and NADPH. Dissimilatory sulfate reduction, by contrast, produces energy, although initially it, too, requires ATP to activate the sulfate.

Bacteria that carry out dissimilatory sulfate reduction and use sulfate as an electron acceptor must first use ATP to activate sulfate to form an energy-rich compound called **adenosine phosphosulfate** (**APS**), the same reaction as the first in the assimilatory pathway. The energy in APS is used to power the reduction of sulfate to sulfite. After this reduction step, which consumes two electrons, sulfite can then accept eight more electrons. This

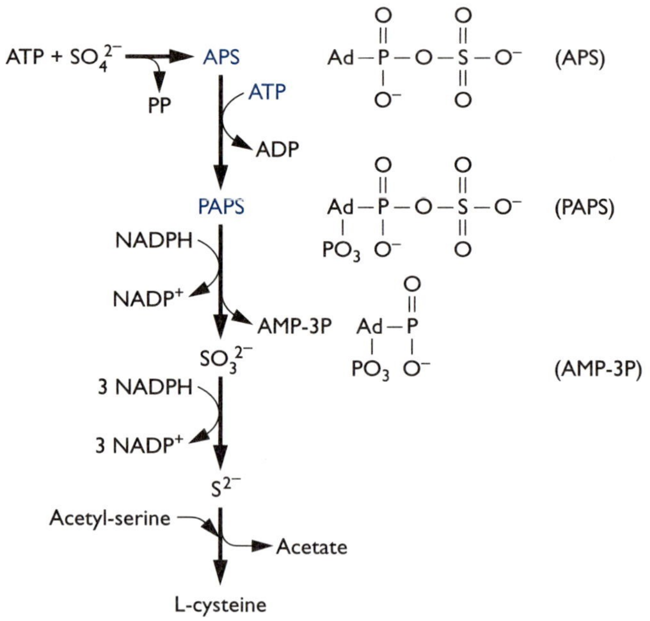

$ATP + SO_4^{2-} \rightarrow$ APS (PP released)

APS + ATP $\rightarrow$ PAPS + ADP

PAPS + NADPH $\rightarrow$ SO_3^{2-} + $NADP^+$ + AMP-3P

SO_3^{2-} + 3 NADPH $\rightarrow$ S^{2-} + 3 $NADP^+$

S^{2-} + Acetyl-serine $\rightarrow$ L-cysteine + Acetate

(APS) Ad–P(=O)(O⁻)–O–S(=O)(=O)–O⁻

(PAPS) Ad(PO$_3$)–P(=O)(O⁻)–O–S(=O)(=O)–O⁻

(AMP-3P) Ad(PO$_3$)–P(=O)–O⁻

Figure 21.10 Assimilatory sulfate reduction. Sulfate is reduced in order to provide sulfide that can be incorporated into the amino acid cysteine. Many bacteria carry out this reaction. Note that this pathway consumes energy but does not produce it.

process pumps protons out of the cell and creates a **proton motive force (PMF)**, which can be used to make ATP. The sources of the protons and electrons that are used to generate this PMF and reduce sulfate to sulfide are organic compounds such as lactate (Fig. 21.11).

The sulfate reducers use hydrogen instead of NADPH as a source of electrons and protons. Hydrogen diffuses into the periplasm, where a periplasmic enzyme, **hydrogenase,** converts hydrogen to two protons and two electrons. The electrons are then transferred to a cytochrome, which passes them ultimately to sulfite to produce sulfide. Sulfate reducers that use this pathway obtain their protons and electrons from hydrogen generated from fermentation of organic compounds such as lactate (Fig. 21.11). In addition to providing hydrogen, which is used as a source of protons and electrons for the electron transport system, the fermentation of lactate to acetate also produces two molecules of ATP from ADP.

Sulfide-oxidizing bacteria. As was the case with the methane cycle, sulfide is produced in anoxic sediments, because growth of sulfate reducers is inhibited by oxygen. Even if this were not the case, sulfate reducers would have trouble competing with bacteria that utilize oxygen or nitrate as a final electron acceptor, because these latter reactions generate energy more efficiently. The sulfate reducers can compete successfully, however, with the methanogens, because sulfate reduction is energetically more efficient than reduction of carbon dioxide to methane. This, together with the high concentration of sulfate in sea water, explains why sulfate reducers generally dominate over methanogens in marine sediments, whereas methanogens are more likely to dominate in freshwater environments, where the concentration of sulfate is low.

Most sulfide oxidizers require oxygen to oxidize sulfide to sulfur or sulfate, because they use sulfide or sulfur as the electron donor and oxygen as the terminal electron acceptor. Some sulfide oxidizers are photosynthetic. Although the sulfide- and sulfur-oxidizing prokaryotes require oxygen, they cannot compete with the spontaneous oxidation of sulfur compounds where oxygen levels are too high. Thus these prokaryotes are found at the oxic/anoxic interface that forms as a result of oxygen diffusion into the water from the surface and oxygen consumption by bacteria in the process of carbohydrate utilization in the sediment (Fig. 21.12; Color Plate 11). A gradient of sulfide also forms as a result of reduction of sulfide in the anoxic sediments, creating a combination of gradients in which the favorable zone for growth of sulfide oxiders can be fairly narrow. In stratified lakes and ponds, bodies of water that remain unstirred long enough to allow such stable gradients to form, collection of water samples at various depths may reveal a pink-colored layer with turbidity that indicates the presence of bacteria. These bacteria are pink because they contain pigments involved in photosynthesis (Chapter 22). When viewed under a microscope, they may be seen to contain bright yellow bodies consisting of elemental sulfur that has precipitated inside the cell (Color Plates 12, 13). Sulfide-oxidizing bacteria use a variety of strategies for oxidation, but, in all these strategies, protons are removed from the sulfur compound and the remaining electrons are passed to cytochromes, which use oxygen as a terminal electron acceptor. An example of a strategy for sulfide oxidation is shown in Figure 21.13. Note that the oxidation of sulfide occurs in the outer membrane. This is probably necessitated by the fact that sulfides are toxic for most living things and would be toxic to the sulfide oxidizers as well if high enough concentrations built up in the periplasm or cytoplasm.

The dark side of sulfide oxidation. Sulfide oxidation has a dark side in some settings. In mine drainage from sites containing high sulfide ores, sulfide oxiders, such as *Thiobacillus thiooxidans,* oxidize sulfur compounds to form sulfite (sulfuric acid). Their activities can lower the pH of the drainage water to pH 1, a condition at which these bacteria grow very well. The sulfuric acid they produce can decimate wildlife in the area of the mine drainage site. If enough iron is present, the iron combines with the sulfuric acid to form iron sulfide (pyrite) which precipitates and stains the water a yellow-iron color. In pipes where sulfur compounds are part of the fluid stream, sulfide

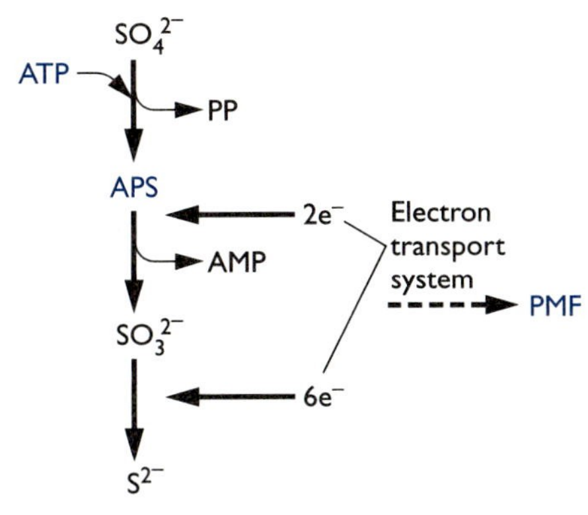

Figure 21.11 Dissimilatory sulfate reduction, the pathway used by sulfate-reducing bacteria as a way to create proton motive force (PMF) and produce adenosine triphosphate (ATP). Hydrogen derived from oxidation of organic compounds such as lactate diffuses into the periplasm, where hydrogenase converts it into protons (which remain outside the membrane, generating the PMF), and the electrons are transferred to cytochromes and eventually used to reduce sulfite to sulfide. Note that, as in the case of assimilatory sulfate reduction, sulfate first must be activated to adenosine phosphosulfate (APS) to make it capable of being reduced to sulfite, a form of sulfur that can be reduced by the cytochromes disposing of their electrons.

Figure 21.12 Sulfur cycle in marine environments, where sulfate is abundant in seawater. Sulfate is reduced to sulfide in the anoxic sediment layers. It diffuses upward into the water overlying the sediment, where it is oxidized by the sulfate oxidizing bacteria. This cycle can also occur within a sediment if the oxic/anoxic layer lies within that sediment (Chapter 22).

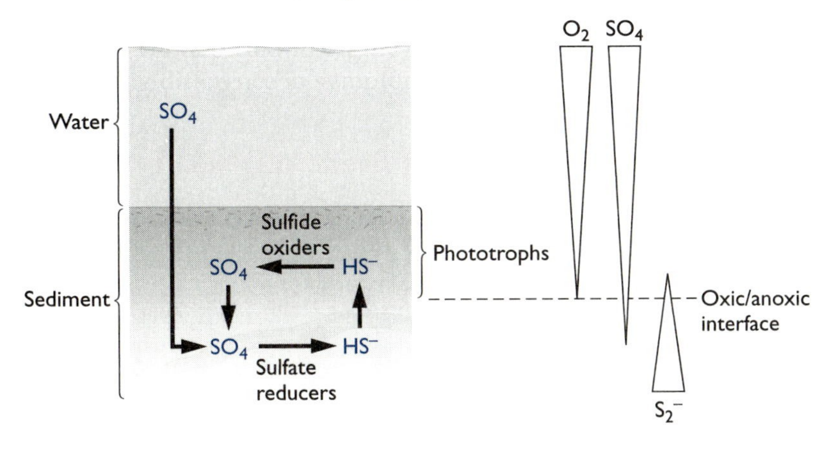

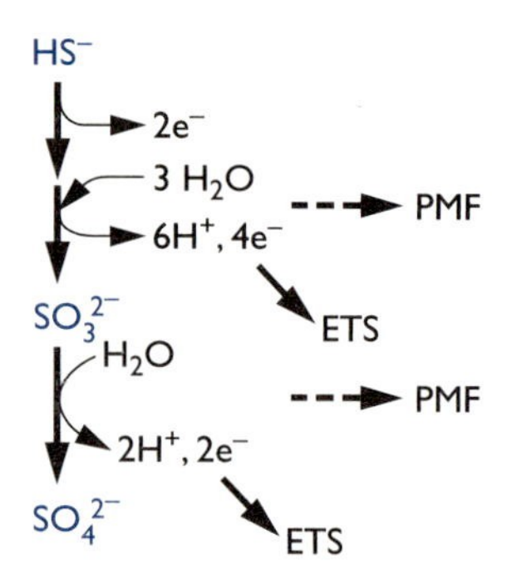

Figure 21.13 Pathway for sulfide oxidation. Note that this pathway uses oxygen as the terminal electron acceptor, so sulfide-oxidizing bacteria need oxygen. Yet, like the methane oxidizers, they cannot compete with the spontaneous oxidation of sulfide that occurs in highly oxygenated water or air, so they locate themselves near the oxic/anoxic boundary. PMF, extrusion of protons; ETS, electron transport system uses O_2 to dispose of electrons.

oxiders can produce enough sulfuric acid to undermine the integrity of the pipe material and produce distortions of the tubing or leaks in it. Microbially induced corrosion of metal pipes is a continuing problem for the pipe industry, especially when the pipes carry toxic byproducts from an industrial process or radioactive material in effluent from a nuclear power plant.

Global Chemical Cycles and Their Significance

In Chapter 1, great emphasis was placed on the role of microbes as recyclers. Here, the outlines of the recycling activities of microbes begin to emerge more clearly as a balance between production of end products by one group of microbes and utilization of those end products by others. Such cycles are an important balance for maintaining spaceship Earth. In the next chapter, another part of the global carbon cycle will be introduced: the fixation of carbon dioxide by photosynthetic microbes. An area of continuing controversy in recent years has been global warming. There is little doubt that the atmospheric concentration of carbon dioxide and other greenhouse gases, such as methane, has increased. The only thing controversial about this observation is how high carbon dioxide concentrations need to be before some fairly drastic global changes ensue, such as changes in ocean currents that cause the oceans to go anoxic. This has happened before in Earth's history, with disastrous consequences for oxygen-requiring marine animals. Schemes to reduce atmospheric carbon dioxide and methane are numerous. Some suggested solutions are heavy on engineering technology, such as pumping carbon dioxide into coal or oil fields, but the most successful solutions are likely to be those that recruit microbes to restore the balance they maintained in earlier times.

Often students coming for the first time to the variety of metabolic pathways found in prokaryotes are put off by what looks like a dry recital of pathways. In fact, such pathways are the life-blood of industry, which exploits them either to produce important products using microbes or to prevent microbial processes such as the corrosion of pipes and acid mine drainage. Such pathways are also becoming increasingly important for people interested in remediating a variety of problems, ranging from buildup of carbon dioxide in the atmosphere to eliminating toxic compounds from a toxic waste dump. Engineering solutions to such problems often prove to be expensive and sometimes risky. For example, a suggestion to clean up Cape Cod, Massachusetts, ground water contaminated by pollutants from a local military base by piping water out of the ground and treating it to remove the pollutants was criticized by residents who feared that such massive pumping would bring salt water into previously freshwater environments. Recruiting microbes to aid in the effort is likely to be far cheaper and more environmentally friendly. The problem with this approach is that there is still too little information about microbial metabolic pathways to design such solutions. For example, it has been observed by people interested in biodegrading jet fuel and other organic pollutants that bacteria degrading these compounds often make something toxic into something even more toxic. Why do they do this? Are scientists looking at the wrong microbes or is there something about the metabolic pathaway that needs to be understood in order to encourage the complete breakdown of toxic compounds into nontoxic products? This question is a challenge for scientists of the future.

chapter 21 at a glance

Comparison of methods for analysis of microbial populations

Method	What it reveals	Limitations
Polymerase chain reaction (PCR)/sequencing of rRNA genes	Hypotheses about the species and genus of microbes found in the population	Incomplete lysis of members of the population PCR inhibitors in the sample Selective amplification of some rRNA gene sequences Interpretation of sequence data
Fluorescent in situ hybridization (FISH)	Abundance of a particular type of microbe in the sample Physical association of different microbes	Naturally fluorescent compounds in the sample. Incomplete permeabilization of some microbes Specificity of hybridization probe under conditions used
Metabolic (end-product analysis)	Metabolic types of microbes in the site Location of different metabolic types	No information about species of microbes responsible for activity

Metabolic types of microbes

Term	Examples	Characteristics
Autotrophs	Methanogenic archaea, photosynthetic bacteria, and eukaryotes	Use carbon dioxide as sole carbon source Variety of electron donors and acceptors
Methane/methanol-oxidizing bacteria		Oxidize C1/C2 to carbon dioxide to generate energy and biomass Formaldehyde is an intermediate
Heterotroph	Polymer-utilizing bacteria Bacteria and archaea that use simple carbon compounds such as acetate	Oxidize carbon compounds more complex than carbon dioxide Variety of electron donors and acceptors
Lithotroph	Sulfide-oxidizing bacteria (e.g., *Thiobacillus* species)	Use oxidation of inorganic compounds to obtain energy Most use oxygen as an electron acceptor
Sulfate-reducers (heterotrophs)	Sulfate-reducing bacteria (e.g., *Desulfovibrio* species)	Use inorganic compounds as electron acceptors Sulfate-reducers are obligate anaerobes, but some reducers of inorganic compounds (e.g., nitrate) can also use oxygen as an alternate electron acceptor (Chapter 24)

key terms

assimilatory sulfate reduction process carried out by many bacteria; used to produce the sulfur-containing amino acid cysteine, not for energy production

autotrophs microbes that obtain cell carbon from carbon dioxide

dissimilatory sulfate reduction process used to produce energy via an electron transport system that uses sulfate or sulfite as an electron acceptor; bacteria that use this pathway are obligate anaerobes

fluorescence microscopy detects microbes that fluoresce under certain wavelengths of light; useful for identifying photosynthetic bacteria (in which pigments fluoresce when illuminated by green light) or methanogens (some of which have cofactors that fluoresce blue-white); also used to detect fluorescence from the DNA of microbes stained with 4',6-diamidino-2-phenylindole (DAPI) or other stains that become fluorescent when they intercalate with DNA

fluorescent in situ hybridization (FISH) microbes on a slide are permeabilized to allow fluorescently labeled hybridization probes to enter the cell and hybridize with ribosomal (rRNA) sequences in the cell; used to identify phylogenetic types of microbes in a sample; gives quantitative information as well as information on the physical association of microbes

heterotrophs microbes that obtain their carbon from complex organic molecules

lithotrophs microbes that obtain energy by oxidizing inorganic compounds such as sulfide and sulfate

methanogens archaea that obtain energy and carbon by reducing carbon dioxide; some can utilize other carbon sources such as acetate, but the ones that utilize hydrogen and carbon dioxide are featured here

methane-oxidizing bacteria bacteria that obtain energy and carbon by oxidizing methane; some can also utilize methanol, but, for simplicity, the methane-oxidizers are featured here; unique in that formaldehyde is an intermediate in the oxidation pathway

phase-contrast microscopy reveals shape, motility, and intracellular granules, including sulfur granules and gas vacuoles

phototrophs microbes that obtain energy from light; can be autotrophs or heterotrophs (Chapter 22)

phylogenetic group cluster of organisms with related ribosomal RNA (rRNA) gene sequences

Questions for Review and Reflection

1. Why is the cloning step necessary in the polymerase chain reaction (PCR)/sequence analysis of a microbial population?

2. If your PCR/sequence analysis revealed only bacterial ribosomal RNA (rRNA) sequences but you suspected that archaea were also present because methane is detected in the sample, how might you detect the methanogenic archaea? (Hint: Consider the design of your PCR primers.)

3. Could genes other than rRNA genes be used to obtain information about the metabolic activities of some of the microbes in the sample? Give two examples and explain how PCR/sequencing could provide such information. What are some limitations of this approach?

4. How does phase-contrast microscopy differ from fluorescence microscopy? What information does each of these types of microscopy give? How does fluorescent in situ hybridization (FISH) differ from 4',6-diamidino-2-phenylindole (DAPI) staining and fluorescence microscopy to detect naturally fluorescent microbes?

5. Several limitations of the PCR/sequencing approach to characterizing a microbial community were listed in this chapter. Which ones are offset by FISH analysis? Does FISH analysis have some of the same limitations of PCR/sequencing? If so, what are they?

6. Does FISH analysis have limitations that are not a problem with PCR/sequencing? What are they and how might they be overcome?

7. Compare and contrast methane production (carbon dioxide reduction) and sulfate reduction. What types of microbes carry out these processes? Do these processes have anything in common?

8. Compare and contrast methane oxidization and sulfide oxidation. What are the characteristics of the microbes that carry out these reactions, and why do they prefer the oxic/anoxic interface to the more oxygenated upper layers of soil or water overlying a sediment?

9. How could an analysis of chemical end products in a sediment and the water that overlies it be used to supplement a PCR/sequencing analysis of the same samples? What is learned from each type of analysis in this case? Would a FISH analysis give any additional information? If so, what type of information?

10. Suppose you succeed in isolating several kinds of microbes from a microbial community. How could you determine how abundant your isolates are in the original sample? Would information about end products produced by your isolates be useful in answering this question?

11. The term "microbial community" is often used to describe microbial populations such as those described in this chapter. In what sense are these microbes a "community"? Do microbes have to be in close physical association with each other to form this type of community? In other words, what is your answer to the question posed in the short provocative statement that starts the chapter?

22 Phototrophic Microbes: Food Source for Planet Earth

For most of us, the sun evokes thoughts of vacation fun, but the sun has a far more primitive and important significance. Sunlight is the source of energy used by phototrophic microbes and plant chloroplasts (once free-living microbes themselves) to produce the biomass that is the basis of the food chains which ultimately support the higher forms of life.

Phototrophs are microbes that use photosynthesis to generate energy. Dense masses of **phototrophs** can be quite colorful because of the presence of chlorophyll and other pigments. The green scum in a pond or green streamers anchored to rocks on a stream bottom are most likely a combination of algae and cyanobacteria. A pink scum on sand in a salt marsh harbors the so-called purple phototrophic bacterium. A particularly lovely manifestation of phototrophs and their interactions can be seen by cutting into the sand of the Great Sippiwisset salt marsh near Woods Hole, Massachusetts, in the area between the shoreline and where plant life begins. A series of bands of different colors can be seen in the sand (Color Plate 9). These and similar assemblages of microbes are called microbial mats, because the microbes are stuck together by a polysaccharide slime that gives the layers a mat-like consistency. If you grasp a portion of the mat, it can be pulled up as a unit, much like a segment of carpet.

Microbial Mats Formed by Phototrophs: Another Example of Life in a Gradient

Gradients of oxygen, methane, sulfate, and sulfide were discussed in Chapter 21. An example of a sulfur cycle that includes phototrophs is a good way to start this chapter, which focuses on the lifestyles of the small and phototrophic. A closer look at the microbial mat taken from the salt marsh shows a number of colored layers atop a black layer (Fig. 22.1). The green band (and sometimes a peach band) at the top of the microbial mat is composed of cyanobacteria and diatoms. These microbes carry out a form of photosynthesis known as **oxygenic photosynthesis,** which splits water into hydrogen (H_2) and oxygen (O_2). The oxygen is released, and the hydrogen is used by the microbes as a source of electrons. If a piece of the mat is taken back to the laboratory and submerged in an aquarium under a strong

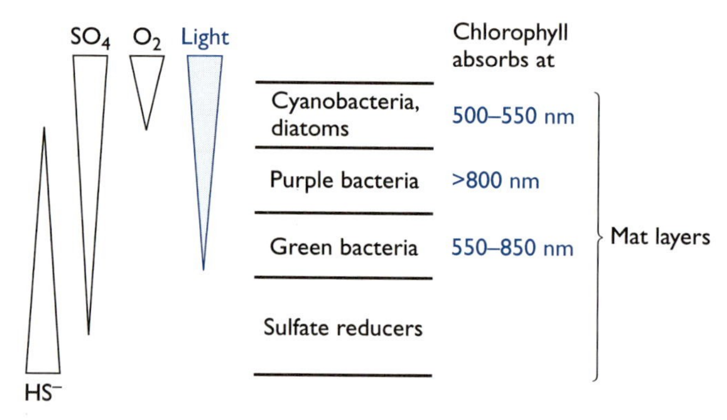

Figure 22.1 Illustration of the factors that cause microbial mat layers to form. The mats form in gradients of sulfide (HS^-), sulfate (SO_4), oxygen, and light. The top layers contain the phototrophs. The oxygenic phototrophs, which do not require anoxic conditions for photosynthesis (and in fact produce oxygen themselves) occupy the top layers. They absorb some wavelengths of light (500–550 nm), leaving other wavelengths for the lower layers. The purple and green phototrophs, which require anoxic conditions for photosynthesis, are in the next layers. In this example of a microbial mat, the purple and green phototrophs use sulfide as an electron source and thus position themselves in the sulfide gradient. They also absorb different wavelengths of light, a fact that also contributes to their position in the mat. The lowest layer is occupied by the sulfate-reducing bacteria, which produce sulfide used by phototrophs in the higher layers. The bottom layer is black, because sulfide complexes with iron to form a grey-black color.

light source, bubbles of oxygen will begin to appear within a few hours. These bubbles come from oxygen produced by the microbes of the top layers. Oxygenic phototrophs use a form of **chlorophyll** identical to that in green plants, and oxygenic phototrophs were almost certainly the microbes plants recruited as the organelles now known as chloroplasts.

Next in the stack of colored bands is a pink band, which consists of bacteria that carry out a different type of photosynthesis, one that does not produce oxygen (**anoxygenic photosynthesis**). These have been called purple bacteria because of the pink-purple color of their pigments. Purple bacteria in this layer are using sulfide as an electron source, a process that will be described shortly. Sometimes another layer of purple bacteria may be seen, but for simplicity these bacteria will be treated as a single band. Beneath the pink layer lies a second green band, composed of another type of anoxygenic phototroph, the green bacteria. These bacteria also oxidize sulfide. The microbes of the pink and lower green band contain **bacteriochlorophyll**. Although the bacteriochlorophyll performs the same function in photosynthesis as chlorophyll, converting light energy into chemical energy, it has somewhat different properties. The lowest band is a gray-black band. This is formed not by phototrophs but by sulfate reducing bacteria, which produce the sulfide used by anoxygenic phototrophs of the upper layers. Some of the sulfide reacts with iron to form a gray-black precipitate, which gives the layer its color.

Why do the mat layers form in such an orderly fashion? The bands reflect the effects of chemical gradients that form in this part of the marsh: a gradient of sulfate (from the seawater that floods this area twice daily), a gradient of sulfide (produced by the sulfate-reducing bacteria), a gradient of oxygen (from the atmosphere and the oxygenic phototrophs), and a gradient of sunlight. This is shown schematically in Figure 22.1. The chlorophyll of the oxygenic phototrophs absorbs light in the wavelength region of 500–550 nm. The bacteriochlorophyll of the purple bacteria absorbs light wavelengths in the range above 800 nm, light that filters through the microbes of the top layer. The bacteriochlorophyll of the green sulfur bacteria in the next layer, which are also anoxygenic phototrophs, absorbs light in the 550–850 nm wavelength range. The purple sulfur and green sulfur bacteria are also seeking the optimum place in the sulfide and oxygen gradient to carry out sulfide and sulfur oxidation. The oxygenic phototrophs in the top layer produce oxygen during the day but are inactive at night. In the dark, the anoxygenic phototrophs become oxygen-respiring heterotrophs and consume oxygen, making the lower layers anoxic. This anoxic environment, which is maintained to a lesser extent during the day because of limited oxygen diffusion, is an ideal place for the sulfate reducers (black layer) to reduce sulfate from sea water and obtain energy from the proton motive force (PMF; Chapter 21).

In freshwater environments, where sulfate and sulfide levels are low, another type of anoxygenic phototroph predominates. Here, purple and green bacteria are found. At one time these were called purple or green "nonsulfur" phototrophs to distinguish them from the purple or green sulfur phototrophs that oxidize sulfide or sulfur. More recent evidence has suggested that many of the "nonsulfur" phototrophs could actually oxidize sulfur compounds but at lower concentrations than the sulfur phototrophs. For simplicity, we will simply use the terms purple and green bacteria or purple and green phototrophs to distinguish the two. Purple bacteria are found in ponds and lakes. Green bacteria are also widely distributed and have been found in alkaline hot springs.

Cultivating Phototrophs

Before looking into the physiology of the phototrophic bacteria, it is helpful to know something about their characteristics. Because scientists first encounter the microbes they study in their attempts to cultivate them, this is a good place for us to start, too. How would you set about isolating a phototroph? An obvious difference between cultivation of phototrophs and cultivation of other types of microbes is that the phototrophs require a light source. Other differences set the phototrophs apart. Some phototrophs grow well on the surface of an agar plate. Others, however, grow best within an agar matrix, made of agar that has been treated to eliminate soluble contaminants (washed agar). Accordingly, scientists often use a technique called the "agar shake" to fish out such microbes. An environmental sample is inoculated into a medium containing molten agar and then mixed. The agar solidifies to form a solid tube of agar, in which the microbes are suspended. The agar tube is then placed near a light source. The medium contains very few nutrients, so that heterotrophs do not overgrow the slow-growing phototrophs.

If phototrophs that require sulfur compounds are likely to be present, a drop of sulfide is placed at the bottom of the tube after mixing the agar and the sample to allow a sulfide gradient to form. After a week or so, colonies begin to appear in the agar. One striking feature of these colonies is their colors: green, red, yellow, pink, and brown (Color Plate 14). This feature of photosynthetic microbes is already informative. These microbes are producing pigments. Some of these pigments are chlorophylls, components of the photosynthetic machinery itself. Other pigments absorb ultraviolet light that could be harmful to the microbes. That is, the pigments constitute a sunscreen for light-requiring organisms and protect their proteins and DNA from the harmful effects of short wavelength light, which in sunlight accompanies the longer wavelength light needed for photosynthesis.

To isolate cyanobacteria, the oxygenic phototrophs, a somewhat different strategy can be employed, one that takes advantage of the gliding motility some of these bacteria exhibit. Although these bacteria can be cultivated on agar plates, an amusing way to tease them out of an environmental sample is to take advantage of their tendency to move away from the dark toward the light. This scheme is shown in Figure 22.2. An environmental sample is placed on a section of a petri dish, and that part of the dish is covered by foil. The dish is placed next to a light source. After a day or two, the cyanobacteria will migrate out of the dark zone and into the light. Cyanobacteria that form multicellular filaments large enough to be discernible to the unaided eye can be "captured" with tweezers for further cultivation.

Anoxygenic Photosynthesis

Anoxygenic photosynthesis is thought to be the most primitive form of photosynthesis, because the process is simpler than the photosynthesis used by oxygenic phototrophs. In addition, anoxygenic phototrophs require anoxic conditions to perform photosynthesis. Thus, they could have emerged as phototrophs in the period before cyanobacteria began to introduce molecular oxygen into Earth's atmosphere. One caveat to this simplistic evolutionary argument is that these same phototrophs often can use oxygen in the dark to grow heterotrophically. Because their photosynthetic systems are simpler, however, we will start with them, whether they were evolutionarily first or not.

Figure 22.2 Simple scheme devised by scientist Kurt Hanselmann for luring motile cyanobacteria out of microbial mats and other samples. Multicellular cyanobacteria have a gliding motility that allows them to move over solid surfaces, such as the floor of an empty petri dish. Covering the sample containing the cyanobacteria with foil makes it dark, and it quickly becomes anoxic. Cyanobacteria do not like either the dark or anoxic conditions, so they migrate into the lighted area of the petri dish, where they may be trapped for further study.

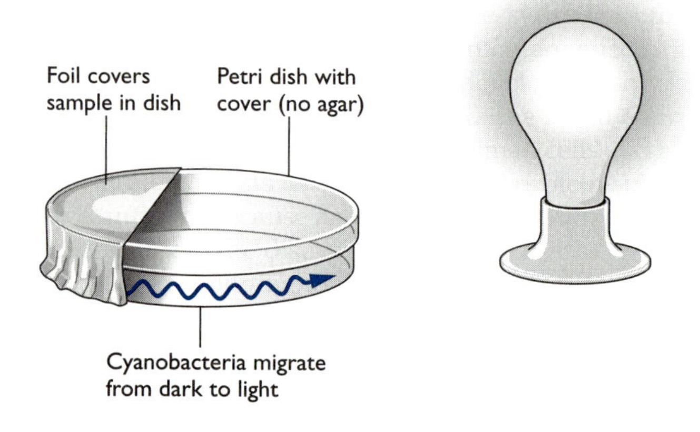

Purple photosynthetic bacteria

An example of a purple nonsulfur photosynthetic bacterium is *Rhodobacter sphaeroides. R. sphaeroides* normally is found in lakes and ponds with low sulfide content. It has been studied intensively by scientists because of its metabolic flexibility. In high light, anaerobic conditions, *R. sphaeroides* grows photosynthetically. Under these conditions, *R. sphaeroides* can use hydrogen as a source of electrons, and carbon dioxide (CO_2) or organic compounds as carbon sources. In the dark, if oxygen is present, *R. sphaeroides* grows heterotrophically on organic compounds, with oxygen as an electron acceptor. Thus, *R. sphaeroides* can grow either photoautotrophically or heterotrophically. *R. sphaerodes* and a related species, *Rhodobacter capsulatus*, have been important models for studying how such versatile bacteria can switch from one metabolic strategy to another.

The type of anoxygenic photosynthesis practiced by *R. sphaeroides* is illustrated in Figure 22.3. This example introduces the organization of photosynthetic systems. It is easiest to understand this process by thinking of photosynthetic systems as having two components: an electron transport system similar to those described in Chapter 8 and a **photosystem** (also called a reaction center). The electron transport system pumps protons and contributes to the PMF, which can be used to generate ATP. The photosystem converts light energy into chemical energy, in the form of an energized electron, which can be fed into the electron transport system. In the case of the purple phototrophs, the electron transport system is termed the **bc complex.**

The photosynthetic process starts in the photosystem. Purple and green bacteria have only one photosystem, whereas the oxygenic phototrophs have two. Purple and green bacteria also differ from the oxygenic phototrophs in that they have bacteriochlorophyll, whereas the oxygenic phototrophs have the same sort of chlorophyll as green plants. The functions of these chlorophylls are the same. Using a photon of light, two molecules of a bacteriochlorophyll molecule (called P_{870}, because it absorbs photons of wavelength 870 nm) activate an electron to a more electronegative state. That is, the E_0' value of the electron changes from about +0.4 volts to about –0.8 volts. Recall

Figure 22.3 Type of photosynthesis used by purple bacteria. In the photosystem, light energy is converted to chemical energy in the form of an activated electron. The activated bacteriochlorophyll, P^*_{870}, transfers its electron to ubiquinones (UQ), which eventually interact with the bc complex to pump protons and generate proton motive force. The electrons are then returned to P_{870} by a mobile cytochrome c (cyt c). This type of photosynthesis is cyclic, in the sense that the same electron is cycled through the photosystem and the bc complex. Nicotinamide adenine dinucleotide phosphate (NADPH) is generated in a separate reaction, which requires energy (not shown). For simplicity, only key steps in the pathway are shown.

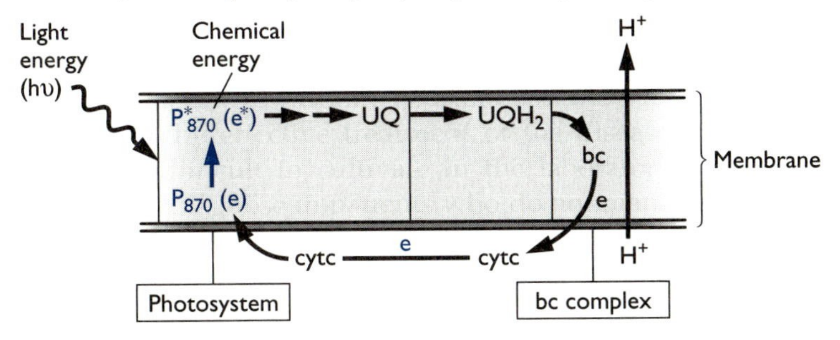

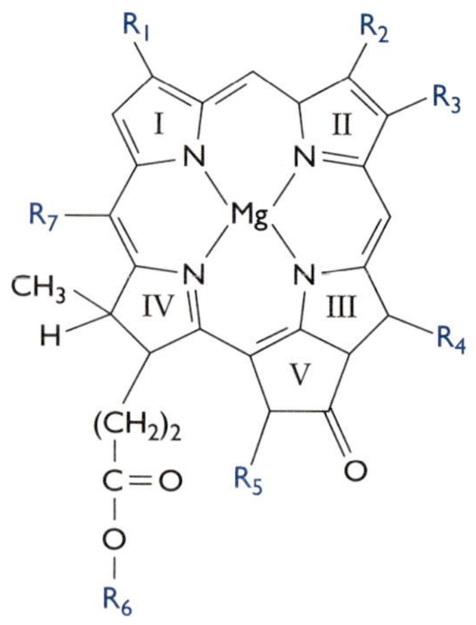

Figure 22.4 The structure of chlorophyll. Chlorophyll and bacteriochlorophyll have very similar structures, as might be expected from the fact that they have similar metabolic functions. The shared aspects of structure are shown here. The differences between different bacteriochlorophylls and between these molecules and chlorophyll lie in the R groups, outlined in blue in this figure.

from Chapter 8 that the more negative the E_0' value, the more able is the electron carried by a compound to use that electron to reduce other compounds. The structure of bacteriochlorophyll is shown in Figure 22.4. After the bacteriochlorophyll molecules produce an activated electron, the electron is transferred through a series of steps to quinones called **ubiquinones.** These quinones migrate from the light harvesting center, which contains the chlorophyll molecules and their scaffolding proteins, to the bc complex, which pumps protons out of the cell, The electrons, now with a much more positive E_0' value, are passed to a cytochrome c, which conveys the electron back to the photosystem. Thus, in contrast to the electron transport systems described in Chapter 8, where electrons flow linearly from nicotinamide adenine dinucleotide phosphate (NAD(P)H) to a terminal electron acceptor such as oxygen, the electron flow in the purple bacteria is cyclic. The electron donor is activated P^*_{870}, and the electron acceptor is unactivated P_{870}.

Bacteria that use this type of photosynthesis have a problem. Their photosynthetic systems provide energy in the form of PMF, which can generate ATP, but they need some way of making NAD(P)H, which is required for many cellular biosynthetic processes. They make NAD(P)H by using an enzyme, NAD(P)H-ubiquinone oxidoreductase, to transfer electrons and protons from the ubiquinone to $NADP^+$ to produce NADPH. Thus, although the electron flow is mostly cyclic, some electrons are removed from this cyclic flow in the interests of making NAD(P)H. The electrons removed from the cyclic flow must be replaced. Such replacement electrons are derived from inorganic compounds, such as sulfides or from organic compounds.

Because the E_0' value of reduced ubiquinone is more positive than that needed to reduce $NADP^+$ to NADPH, this reaction requires energy, which is probably provided by the PMF. Thus, some of the energy generated by photosynthesis is used to make NADPH separately from the process of photosynthesis. As will be evident shortly, the green bacteria and the oxygenic phototrophs have managed to make synthesis of NADPH a more immediate byproduct of photosynthesis.

Green bacteria

As was the case with the purple bacteria, green bacteria use bacteriochlorophyll molecules to convert light energy into chemical energy (in the form of a negative E_0' electron), which is then used to reduce a series of compounds, ultimately reducing **ferredoxin,** a protein with an E_0' negative enough to reduce $NAD(P)^+$ to NAD(P)H without requiring additional energy input (Fig. 22.5). Alternatively, the reduced ferredoxin can pass its electrons to a quinone-cytochrome electron transport chain in a bc complex similar to that used by the purple bacteria, which pumps protons and ultimately returns the positive E_0' electron to the unactivated bacteriochlorophyll molecule. Because the electron flow can be either cyclic (for PMF generation) or noncyclic (for NADPH production), electrons must be added to the system from some other source to compensate for electrons channeled to NADH production. These electrons come from oxidation of sulfur compounds (as shown in Fig. 22.5) or from oxidation of organic compounds.

Oxygenic Photosynthesis

The photosynthetic strategy used by the oxygenic phototrophs, exemplified by cyanobacteria, is, in effect, a combination of the strategies used by the

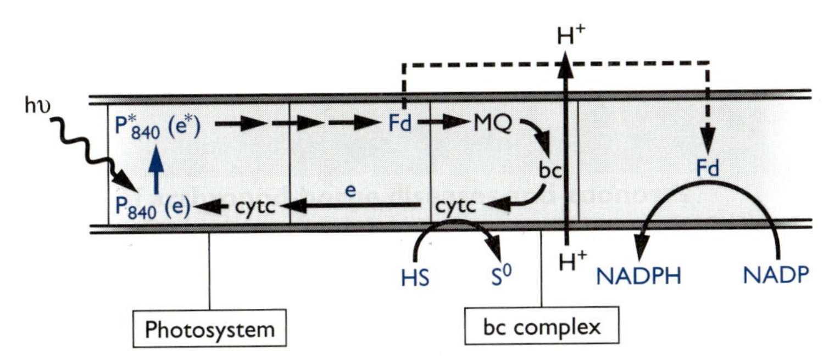

Figure 22.5 Type of photosynthesis used by green bacteria. The main differences between this and the type of photosynthesis used by purple bacteria include the involvement of ferredoxin (Fd), which can either participate in a cyclic reaction to produce proton motive force (PMF) or produce nicotinamide adenine dinucleotide phosphate (NADPH), and the use of sulfide as an electron source to compensate for those electrons siphoned off for NADPH production. Also, the type of quinone used, menaquinone (MQ), is somewhat different in structure to that used by the purple bacteria (ubiquinone), but plays the same role. This type of photosynthesis can be cyclic (PMF production) or noncyclic (NADPH production). For simplicity, only key steps in the process are shown.

purple and green oxygenic phototrophs, except that water is the source of electrons (Fig. 22.6). Human engineers have tried for years to split water into hydrogen and oxygen to produce energy but are nowhere near doing this as efficiently as the oxygenic phototrophs. The process by which light energy is used to remove hydrogen from water to serve as a source of electrons that will be activated to a more negative E_0' value is not well understood. This part of the photosynthetic process is carried out by **photosystem (or reaction center) II,** in which the chlorophyll molecule absorbs light of wavelength 680 nm. This chlorophyll (P_{680}) can produce an electron with an E_0' value low enough to reduce quinones, but not low enough to reduce $NADP^+$ to NADPH. Photosystem (or reaction center) II is used mainly to

Figure 22.6 The type of photosynthesis used by oxygenic phototrophs. In this case, there are two photosystems, I and II. Photosystem II derives electrons from water and produces molecular oxygen in the process. This part of the process is noncyclic. As with the other types of photosystems, quinones (PQ) convey electrons between the photosystem and the bf complex (equivalent to the bc complexes of Figures 22.3 and 22.5), which pumps protons to generate proton motive force (PMF). Plasticyanin (Pc) conveys electrons to photosystem I, where the electron is activated a second time to a state where it can ultimately reduce ferredoxin (Fd), which can either transfer electrons to the bf complex for more PMF generation or be used to reduce NADP to NADPH. For simplicity, only key steps in the process are shown.

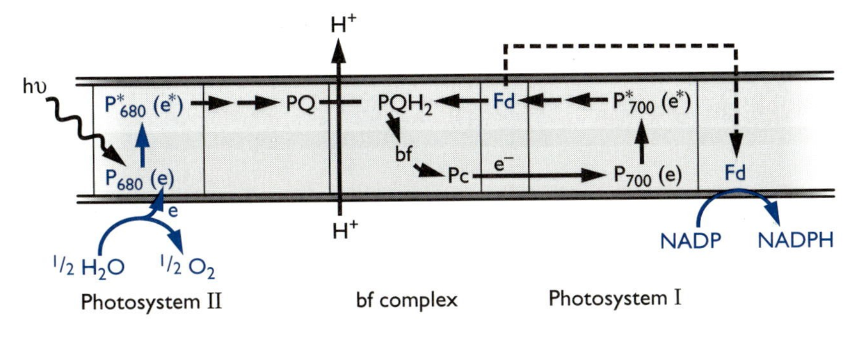

generate energy, as quinones transfer electrons to cytochromes and pump protons to produce the PMF.

Instead of returning the electron to the P_{680} of photosystem II, the oxygenic phototrophs take advantage of the fact that the electron still has a more negative E_0' value than it did at the start of the process. An electron transfer protein called **plastocyanin** (Pc) transfers the electron to the chlorophyll molecule of a second photosystem, **photosystem I.** This chlorophyll (P_{700}) absorbs light of a slightly longer wavelength than P_{680}. The electron that is transferred to P_{700} is activated to an E_0' value of –1.0 volts, negative enough to reduce ferridoxin, which can in turn reduce $NADP^+$ to NADPH. The electron left over from this reaction then can be recycled and returned via plasticyanin to unactivated P_{700}, generating more PMF in the process. Thus, the oxygenic phototrophs, like the green bacteria, produce NAD(P)H as well as PMF directly from photosynthesis instead of having to use PMF energy to power the production of NAD(P)H.

Cellular Location of the Photosynthetic Apparatus

The components of photosynthetic systems are not in the cytoplasmic membrane but instead are localized to **intracytoplasmic membranes** that can be extensions of the cytoplasmic membrane or separate membrane-bound vesicles (Fig. 22.7). Thus, the PMF forms across the membranes of these vesicles instead of across the cytoplasmic membrane as in the electron transport systems described in Chapter 8. The use of membrane vesicles presumably increases the area of membrane available for photosynthetic production of energy and NAD(P)H far beyond what would be possible with the cytoplasmic membrane alone. Light penetrates membranes very readily, so that access to intracellular compartments is not a problem. The use of vesicles also leaves the cytoplasmic membrane free for its other activ-

Figure 22.7 The photosynthetic machinery is located in intracytoplasmic vesicles, not in the cytoplasmic membrane. In the case of the purple bacteria **(A)**, these vesicles are usually invaginations of the cytoplasmic membrane, which have a very different protein composition than the cytoplasmic membrane itself. Here, the invaginations are shown as vesicles, but they may also appear as stacks of thin, finger-like vesicles. In green bacteria and in cyanobacteria, the vesicles are separate from the cytoplasmic membrane **(B)** but closely associated with it. The proton motive force from photosynthesis is generated across the membranes of these vesicles.

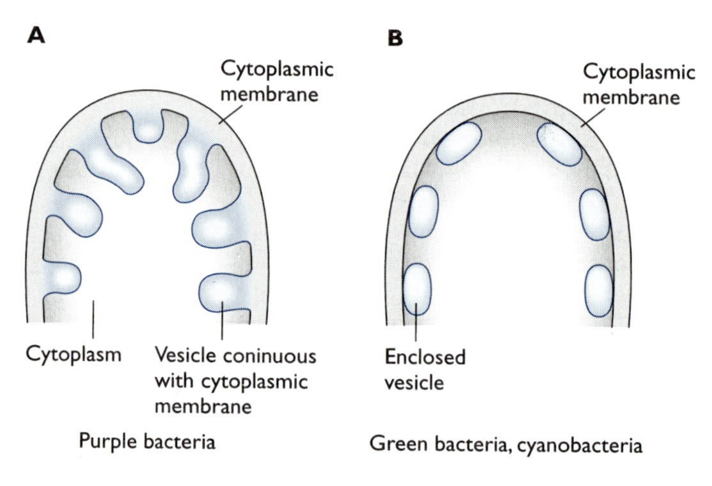

ities, such as transport of nutrients, secretion of proteins, and sensing systems that respond to environmental signals (Chapter 4).

Fixation of Carbon Dioxide by the Calvin Cycle

Bacteria capable of growing with carbon dioxide as their sole carbon source incorporate the carbon dioxide into more complex organic compounds using the **Calvin cycle.** This pathway is not found exclusively in photosynthetic microbes but is introduced here because it is used by many of the phototrophs as a way of generating organic carbon compounds. The Calvin cycle uses the enzyme **rubisco** (for ribulose bisphosphate carboxylase) to combine carbon dioxide with the C-5 compound ribulose bisphosphate, generating two molecules of the C-3 compound, 3-phosphoglycerate (Fig. 22.8). This initial reaction consumes no energy or NADPH, but the subsequent reactions, in which 3-phosphoglycerate is converted to glyceraldehyde 3-phosphate, consume both ATP and NADPH. This pathway, which consumes large amounts of energy and NADPH, accounts for the need for autotrophs that produce large amounts of NADPH and ATP. The fact that photosynthesis is very efficient at providing both types of compounds explains why so many phototrophs are capable of autotrophic growth.

Of the six molecules of glyceraldehyde 3-phosphate created in each turn of the Calvin cycle, only one is channeled into the synthesis of cell components. The remaining five are converted by a complex series of reactions back to the C-5 compound that is combined by rubisco with carbon dioxide (Fig. 22.8). Thus, five C-3 compounds are converted to three C-5 compounds so that the ribulose-bisphosphate is available once more to fix more carbon dioxide.

Figure 22.8 The Calvin cycle, a widely used method for incorporating carbon dioxide into cell biomass. Carbon dioxide is combined with a five-carbon compound (C-5) to form six three-carbon compounds (glyceric acid-phosphate) by the enzyme rubisco. Adenosine triphosphate (ATP) and nicotinamide adenine dinucleotide phosphate (NADPH) are then consumed in the steps that convert glyceric acid-phosphate to glyceraldehyde 3-phosphate, which is an intermediate in glycolysis and can be used as a substrate for biosynthetic reactions. One molecule of this C-3 compound is used for biosynthesis. The remaining five molecules return to the Calvin cycle to be converted into the C-5 compound that is a substrate for rubisco.

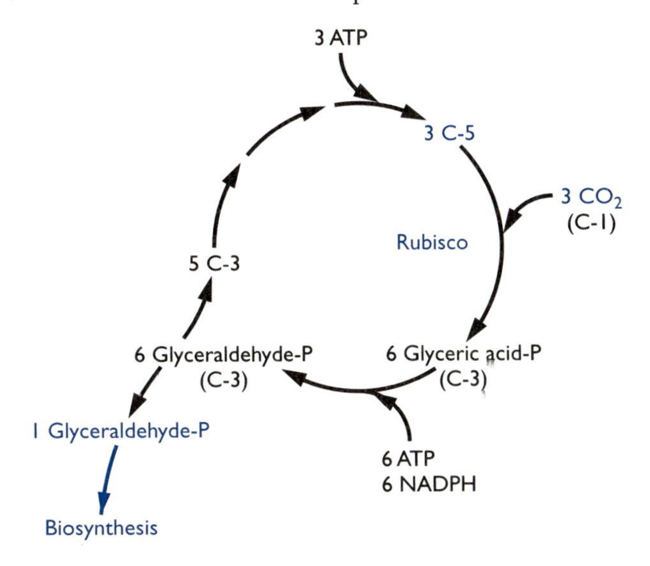

Nitrogen Fixation

The incorporation of atmospheric nitrogen (N_2) into nitrogenous compounds needed by microbes, a process called **nitrogen fixation,** as in the Calvin cycle, is not limited to phototrophs. This seems a good place to introduce it, however, because this pathway is widespread among the cyanobacteria. Cyanobacteria are the powerhouses of the photosynthetic world. They create biomass from sunlight (energy source), water (electron source), atmospheric carbon dioxide (carbon source), and atmospheric nitrogen (nitrogen source). It is no wonder that cyanobacteria, along with diatoms and algae, which have similar capabilities, are the basis of so many food chains. Green plants can carry out oxygenic photosynthesis and carbon dioxide fixation but have lost the ability to fix nitrogen. How this happened in evolution is unclear. If cyanobacteria were the origin of chloroplasts, then plants should have obtained nitrogen fixation capability as well as oxygenic photosynthesis and carbon dioxide fixation. In Chapter 24, you will learn how some plants have returned to the bacterial world to regain nitrogen fixation capability by luring nitrogen-fixing bacteria into their roots.

Nitrogenase

The oxygenic phototrophs can obtain nitrogen from atmospheric nitrogen gas. The reduction of nitrogen gas to ammonia, which can be used to make amino acids and other nitrogenous compounds, is catalyzed by **nitrogenase,** an enzyme complex. The reactions catalyzed by nitrogenase are illustrated in Figure 22.9. Fixation of atmospheric nitrogen, like fixation of atmospheric carbon dioxide via the Calvin cycle, is a very expensive process, which consumes many molecules of ATP and NADPH. Only microbes that can produce ATP and NADPH very efficiently could afford to use both a Calvin cycle and nitrogenase. This capability, however, allows the oxygenic phototrophs to grow on sunlight, water, and atmospheric nitrogen, resources that are abundant in the open oceans and other places where more complex nitrogen-containing compounds may be present only in very low amounts.

Separating nitrogen fixation from oxygenic photosynthesis

One problem that has to be solved by the nitrogen-fixing oxygenic phototrophs is that nitrogenase is very sensitive to oxygen and requires an

Figure 22.9 Reduction of nitrogen gas (N_2) to ammonia (NH_3) by the enzyme complex nitrogenase. Nitrogenase has two components, I and II, each of which is composed of two protein subunits. Nitrogen gas is reduced to ammonia by reduced component I (I_{red}) of nitrogenase, which is oxidized (I_{ox}) in the process. Oxidized component I must be reduced to reduced component I in order to allow the nitrogen fixation process to continue. This is accomplished by nitrogenase component II, which restores I_{red}. II_{ox} must itself be regenerated. This is accomplished by ferridoxin (Fd). The recycling of component I consumes ATP.

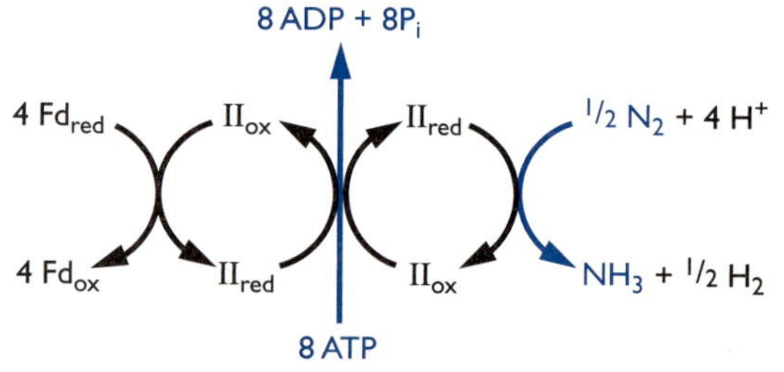

anoxic environment to function effectively. The oxygen-producing cyanobacteria have at least two strategies for solving this problem. One strategy, adopted by the filamentous (multicelluar) cyanobacterium *Anabaena* (Fig. 22.10; Color Plate 10), is to use different cells for different functions. *Anabaena* produces three types of cells: vegetative cells (which divide on a regular basis), heterocysts (a terminally differentiated cell), and akonetes (cyanobacterial spores). The vegetative cells are actively photosynthetic but do not produce nitrogenase. The heterocysts are not actively photosynthetic but instead are dedicated to nitrogen fixation. The nitrogenous compounds provided by nitrogenase in the heterocysts diffuse into adjacent vegetative cells to satisfy the nitrogen requirements of these actively dividing cells. Heterocyst-producing cyanobacteria can "count," in the sense that when the number of vegetative cells rises to the point that the flow of nitrogen into these cells begins to become too low, a vegetative cell in that region of the chain of cells differentiates into a heterocyst, so that the ratio of heterocysts to vegetative cells is maintained and the heterocysts are evenly spaced in the chain of vegetative cells. The way in which cyanobacteria accomplish this complex task is still a matter of intensive research.

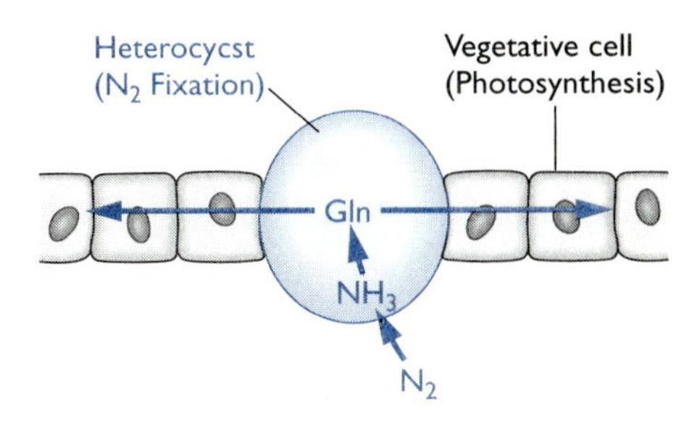

Figure 22.10 How filamentous cyanobacteria solve the problem of keeping nitrogenase separate from oxygen. The filamentous cyanobacteria solve this problem by physically separating photosynthesis and nitrogen fixation. These bacteria have two cell types, the vegetative cell, which divides and carries out photosynthesis and carbon dioxide fixation, and the heterocyst, a terminally differentiated cell that carries out nitrogen fixation but not photosynthesis. In the heterocyst, nitrogen is first reduced to ammonia by nitrogenase, then ammonia is incorporated into glutamate and glutamine (Gln) as described in Chapter 8. Glutamine is transferred to the vegetative cells, which use it as a nitrogen source for biomass synthesis.

Many cyanobacteria, such as the single-celled cyanobacteria that are so numerous in the open ocean, do not produce heterocysts. These bacteria solve the problem of separating nitrogen fixation from photosynthesis by a temporal strategy, which has created an exciting new area of research into the circadian rhythms of bacteria. An example of a cyanobacterium that uses this strategy is the single-cell cyanobacterium *Synechococcus*. *Synechococcus* grows photosynthetically during the day, generating ATP and NADPH. In the dark, the photosynthetic apparatus ceases to function and nitrogenase becomes active.

Such a metabolic light-dark pattern is the bacterial equivalent of the **circadian rhythms** of animals. Circadian rhythms are physiological cycles with a 24-hour periodicity. The "clock" that sets the circadian rhythms is endogenous, responding to external signals to keep it set, but with an innate periodicity of its own. Once thought to be restricted to eukaryotes, circadian rhythms are now known to be found in at least this one denizen of the bacterial world and may thus have had a very ancient origin. Although the genes involved in the cyanobacterial clock are different from those found in insects and mammals, the mechanism seems to be the same. This mechanism is not understood in detail and is thus shown very abstractly in Figure 22.11. The periodicity is established by a balance between proteins that activate certain genes (positive factors) and proteins that turn off expression of these genes (negative factors). When the input signal (e.g., sunlight) is present, the positive factors predominate. With cessation of the signal, the negative factors predominate. Whether positive or negative factors predominate affects the output, which can be expression of genes such as those that encode the photosynthetic apparatus or those that encode nitrogenase. It will be interesting to see whether future research reveals additional examples of bacterial (and archaeal) circadian rhythms.

The Dark Side of Oxygenic Phototrophs

Earlier in this book (Chapter 2) and in this chapter, the cyanobacteria were presented as humans' best friends in the microbial world. Cyanobacteria put the first molecular oxygen in Earth's atmosphere, making possible the evolution of oxygen-consuming organisms, and they became the chloroplasts that made photosynthetic eukaryotes (including plants) possible.

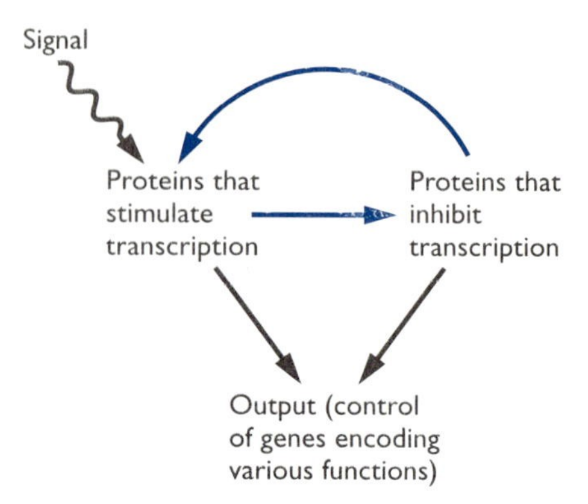

Figure 22.11 The single-celled cyanobacteria, such as *Synechococcus* species, use a different strategy (circadian rhythms) to solve the problem of separating oxygen-generating photosynthesis from nitrogenase. They separate photosynthesis and nitrogen fixation temporally. These processes occur as circadian rhythms. Researchers are currently investigating how circadian rhythms are maintained. This simple figure shows the current understanding of how the process may work. An input signal, such as sunlight, stimulates the production of proteins that have a positive effect on the expression of certain genes (output), such as those involved in light-associated processes like photosynthesis. They also trigger the production of negative regulators that tend to shut them off. As long as the signal continues to signal the positive components, the output controlled by these components stays on. But with the cessation of the input signal (e.g., darkness), the negative factors predominate and other genes are expressed. The circuit outlined in blue is the one thought to maintain the periodicity by alternating a buildup of positive factors and negative ones. This simple model for the clock that maintains circadian rhythms is already controversial and will probably be replaced by a more sophisticated alternative.

Moreover, they now help feed the creatures of planet Earth by being at the base of most food chains. A lot of us would love to have a resumé like that. Nonetheless, it is important to note that these essential and usually benign bacteria have a dark side. For reasons that are unknown, many, under certain conditions, produce substances that are toxic for humans and animals. Every year, in the United States and Europe, dogs and livestock die because they consume water from ponds or puddles containing cyanobacteria. These deaths are not common, and human deaths due to cyanobacteria toxins are even more rare, but the close proximity of cyanobacteria to human and animal water supplies has made public health officials wary of this potentially deadly source of water pollution.

To understand why this as-yet-hypothetical problem might concern public health officials, consider the potency and targets of the major cyanobacterial toxins. The cyanobacterial toxins are small molecules, but they can pack a big punch. For example, the cyanobacterial toxin, saxitoxin, is one of the most potent toxins known, approaching botulism toxin in its potency (Fig. 22.12). Saxitoxin is a neurotoxin, which impairs neurological function in humans and can kill in high enough doses. Do you want cyanobacteria that produce this toxin growing in a reservoir near you? Other cyanotoxins impair liver function and are also among the more potent toxins known.

Scientists have established that cyanotoxins are produced by many species of cyanobacteria but only at certain times. The specific conditions that trigger toxin production have not been identified. One intriguing question is what these molecules, which are so toxic for humans, do for the cyanobacteria. Are they some byproduct of their photosynthetic lifestyle that allows the bacteria to have a competitive edge or are they accidental

Figure 22.12 Cyanobacterial toxins. (A) Comparison of the relative toxicities of cyanobacterial toxins saxitoxin and microcystin with potent bacterial toxins such as botulinum, tetanus, and diphtheria toxins and cobra venom. **(B)** The structure of saxitoxin, which most closely resembles an alkaloid. Microcystin (not shown) is a cyclic peptide. Cyanobacterial toxins can affect the heart, liver, or nervous system.

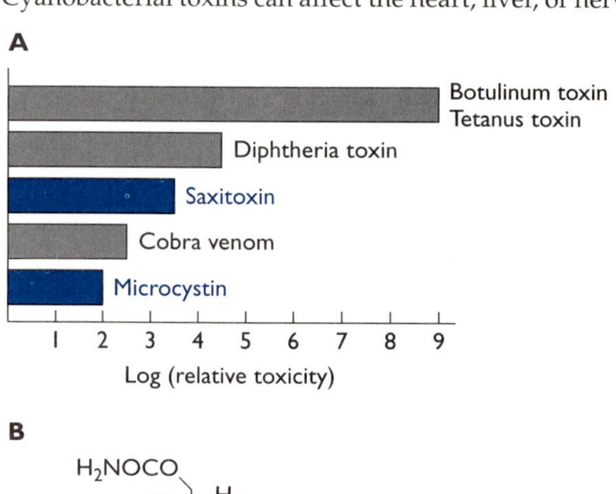

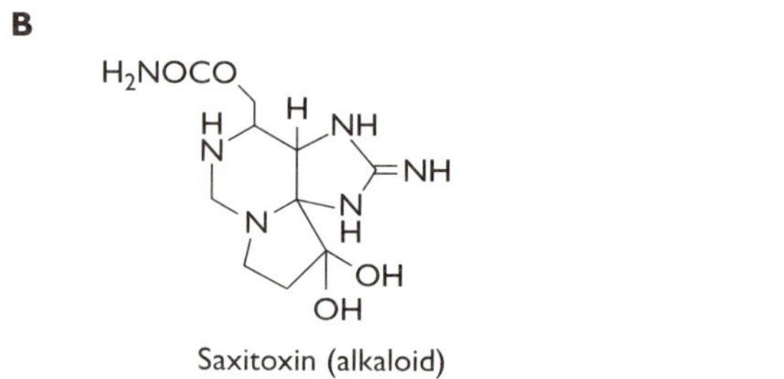

products of cyanobacterial blooms in which the bacteria are releasing incompletely degraded cell components?

As in the case of metabolic pathways such as sulfide oxidation, learning the steps in the process of photosynthesis often strikes students as an exercise created by professorial sadists to drive them crazy with boredom. In fact, properly understood, photosynthesis is the basis for most, perhaps all, life on Earth. Anyone who cares about dogs or whales should be interested in the ultimate source of their food supply, even though this food supply passes through many organisms before it reaches such large animals, which graze high up on the food chain. The fact that this life source is also potentially toxic for these same animals—and humans—should also be a matter of interest. Cyanobacterial toxins have been treated as a curiosity by many scientists, but in fact it is possible that long-term exposure to low levels of these compounds may have health effects that have been missed completely. Scientists know that the fungal toxin aflatoxin makes people more susceptible to liver damage cause by hepatitis viruses. Are there any such synergistic effects involving cyanotoxins? Shouldn't someone be looking into this, especially in view of the increasing number of cyanobacterial blooms associated with nitrogen runoff from agriculture and human sewage? Perhaps you will be the one to open up this area of public health research. Whether or not cyanotoxins prove to be a significant source of human or animal morbidity and mortality, it is important to realize that the absence of evidence is not the same as evidence of absence. Just because few scientists seem interested in pursuing this line of research does not mean that there is not an important public health problem waiting to be discovered and remedied.

chapter 22 at a glance

Comparison of phototrophs

Type of phototroph	Characteristics	Electron donor	Source of carbon
Anoxygenic phototrophs	Oxygen not produced Single photosystem, can be cyclic or noncyclic	Sulfur compounds, hydrogen (H_2)	Carbon dioxide (CO_2), organic compounds
Purple bacteria	Pink-purple colonies Single cyclic photosystem with bacteriochlorophyll Make nicotinamide adenine dinucleotide phosphate (NADPH) separately from photosynthesis	H_2, organic compounds Sulfur compounds (sulfur and some "nonsulfur" phototrophs)	CO_2, organic compounds
Green bacteria	Single-celled, photosystem that can be cyclic (proton motive force [PMF] generation) or noncyclic compounds (NADPH production)	Sulfur compounds, H_2	CO_2, organic compounds
Oxygenic phototrophs (e.g., cyanobacteria)	Can be single-celled or multicellular (filamentous) Two photosystems, one noncyclic (photosystem II) and one cyclic (photosystem I) Chlorophyll-based system like green plants, which generates oxygen, Fix nitrogen (nitrogenase)	Water	CO_2

Components of photosystems and their functions

Component	Characteristics	Role
Photosystem	Chlorophylls and the proteins that interact with them	Convert light energy into chemical energy Start the transfer of electrons to other compounds
Chlorophyll, bacteriochlorophyll	Complex organic compound, associated with proteins but is the active molecule	Convert light energy to chemical energy in the form of an energized electron, which is passed to other compounds
Quinones	Organic molecules that carry electrons and protons	Carry electrons from chlorophyll to other components of the system
Cytochromes	Proteins that carry electrons	Carry electrons to terminal electron acceptor
bc or bf complex	Proteins that pump protons and transfer electrons	Generation of proton motive force (PMF)
Ferridoxin	Protein capable of reducing NAD(P) to NAD(P)H^+	Production of NAD(P)H^+
Plasticyanin	Protein involved in electron transfer in cyanobacteria	Transfer electrons between photosystem II and photosystem I
Intracellular membranes	Invaginations of cytoplasmic membrane or self-contained intracytoplasmic vesicles surrounded by a membrane	Site of photosynthesis

key terms

anoxygenic photosynthesis photosynthesis that transforms light energy into chemical energy, using electron donors such as inorganic or organic compounds

bacteriochlorophyll the type of chlorophyll used by purple and green phototrophs to convert light energy into the chemical energy stored in an activated electron

Calvin cycle process used by microbes and plants to incorporate carbon dioxide into biomass; consumes large quantities of ATP and nicotinamide adenine dinucleotide phosphate (NADPH); rubisco is the enzyme that serves as a marker for this metabolic capability

chlorophyll used by oxygenic phototrophs and green plants to convert light energy into chemical energy; similar in structure to bacteriochlorophyll, with the same function

circadian rhythms biological cycle with a periodicity of 24 hours; seen in some unicellular cyanobacteria as well as in eukaryotes; allow the cyanobacteria to separate nitrogen fixation and photosynthesis temporally

green bacteria bacteria that form colonies with a green color; use a type of photosynthesis that produces NADPH as well as ATP through proton motive force; electron flow is mostly cyclic, with some electrons diverted to make NADPH; electrons removed during NADPH production are replaced by oxidation of inorganic or organic compounds

nitrogenase oxygen-sensitive protein complex that catalyzes the reduction of atmospheric nitrogen to ammonia; used by oxygenic phototrophs and other microbes to acquire nitrogen for the complex nitrogenous compounds needed to form biomass

oxygenic photosynthesis form of photosynthesis that splits water to obtain electrons and produces molecular oxygen; has two photosystems, one noncyclic (photosystem II) and one cyclic (photosystem I), which produces NADPH

purple bacteria bacteria with colonies that have a pink-purple color and which use a cyclic type of photosynthesis that does not concomitantly produce NADPH

Questions for Review and Reflection

1. Photosynthesis can be quite confusing when first encountered in a biology course. To make things easier, focus on the similarities. How are all the phototrophs the same? In what respects (e.g., electron source, number of photosystems) do they differ?

2. How is photosynthesis similar to the electron transport systems described in Chapter 8? How are they different? What do photosynthetic systems do that simple electron transport systems cannot?

3. What does it mean to say that photosystems convert light energy into chemical energy? Explain in terms of E_0'. (Oh, no! Energetics again!)

4. Look beyond the photosynthetic strategies and ask how purple, green, and oxygenic phototrophs make a living by acquiring carbon, nitrogen, and energy. These are highly successful life forms. They must have something going for them. What is it?

5. What does it mean that both oxygenic photosynthesis and circadian rhythms have their origins in the bacterial world? Or that the first terminally differentiated cells occurred in a bacterium?

6. Photosynthesis is an amazing process. If we could duplicate it in the laboratory, we could create an entirely new sector of the biotechnology industry. Make an educated guess as to why this goal has proved so elusive.

7. Cyanobacteria have both the Calvin cycle and nitrogenase, two very energy-expensive systems. Why does it make sense that cyanobacteria would be able to support such systems? What advantages do they confer?

8. In Chapter 1, the explanation for the "dead zone" started with blooms of photosynthetic bacteria. Although cyanobacteria can use atmospheric nitrogen as a source of nitrogen, they can also use nitrate and ammonia as nitrogen sources. Why would the availability of nitrate or ammonia cause a cyanobacterial bloom in a region where such bacteria had been rather sparse before?

9. Your roommate is taking an introductory microbiology course and has developed a very bad attitude toward learning about photosynthetic metabolism. You are concerned that this bad attitude will translate into an equally bad grade on the next exam. How would you convince your roommate that learning about photosynthesis can be a good thing, even a fun thing?

23 Invertebrate–Microbe Interactions

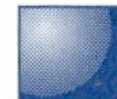

We have grown so accustomed to thinking of bacteria inside tissues as disease that it comes as a surprise to realize that for many invertebrates, as well as some plants, bacteria in their tissues are essential for life. Such close, mutually beneficial associations found so commonly in invertebrates make one wonder whether similar examples may be found in the mammalian world.

The term "**symbiosis**" was coined to mean a close association between two organisms. In this chapter, one of the organisms will be a bacterium and one will be a eukaryotic cell of an invertebrate. Symbiotic relationships can have a deleterious effect for one member of the pair. Such associations are called **parasitic.** Disease-causing microbes interacting with the human body provide an example of a parasitic relationship. If one member of the symbiotic relationship benefits, but the other member is neither harmed nor benefited, the relationship is called a **commensal** relationship. If both partners benefit, the relationship is called a **mutualistic** relationship. Symbiotic organisms can also compete with each other for nutrients, a relationship that may be harmful to both members.

It is important to realize that these terms are often applied somewhat arbitrarily in that they reflect only the current state of knowledge about the interactions of the symbiotic pair. For example, bacteria that normally live in the human colon (Chapter 11) are called commensals, because it is clear that they benefit from nutrients supplied by their human host. Yet, as more is learned about the ways in which bacterial activities in the colon benefit the human partner, this description of the relationship may have to be upgraded to one of mutualism. Despite this nomenclatural uncertainty, terms such as commensalism and mutualism are useful for denoting the general nature of the interaction in a way that emphasizes its positive side. This chapter will describe several interactions between microbes and invertebrates, most of which are beneficial to both partners. To show how complex such interactions might become, however, an example involving three partners (a worm, a bacterium, and a caterpillar larva) will show that in some cases the interaction between two of the partners may be parasitic, whereas that between two others of the trio is mutualistic.

Two other terms describing symbiotic relationships indicate the location of the partners relative to each other. In some of the examples given in this chapter, the microbe resides inside the cells of the invertebrate. Such microbes are called **endosymbionts.** In other examples, the microbe is in close association with its invertebrate host but lives outside the host's cells. Such microbes are called **exosymbionts.**

Oases at the Bottom of the Ocean

Hydrothermal vents: a surprising explosion of life in a desert

Explorations of the ocean's bottom, away from the land masses, at first revealed a site where life seemed to be very sparse. A few widely spaced brittle starfish and a very occasional fish were all that was to be seen. This picture made sense, in view of what was known about the fate of biomass produced near the ocean's surface. This biomass was consumed by animals in the surface waters and excreted as fecal pellets. The fecal pellets, in turn, were ingested by deeper-living creatures, partially used, and then excreted. After many rounds of ingestion and excretion, not much was left to reach the bottom of the ocean. Imagine the surprise of scientists, then, when Alvin, the manned submersible vehicle from the Woods Hole Oceanographic Institute (Woods Hole, MA), first encountered hydrothermal vents, which bloomed with life on the ocean bottom. Hydrothermal vents occur at places where spreading has thinned the Earth's crust sufficiently for cracks to appear, allowing super-heated water from basalt and magma (~350°C) to well up through the crust. Water emerging from such a vent spews sulfides and other inorganic compounds from the Earth's interior into the cold overlying waters of the ocean. Sulfides and other minerals precipitate as the hot water from the vent mixes with the cold water (4°C) of the ocean, forming a cloud of what looks like smoke. Precipitating minerals from the vent waters eventually form cones, called chimneys, out of which the gray-black sulfide "smoke" spews (Fig. 23.1). Such chimneys are called "black smokers" by the scientists. Other vents emit warm rather than hot water, and still other vents, called cold seeps, leak mineral-laden cold water. Each of these has its associated fauna, but we will focus here on the more dramatic hydrothermal vents.

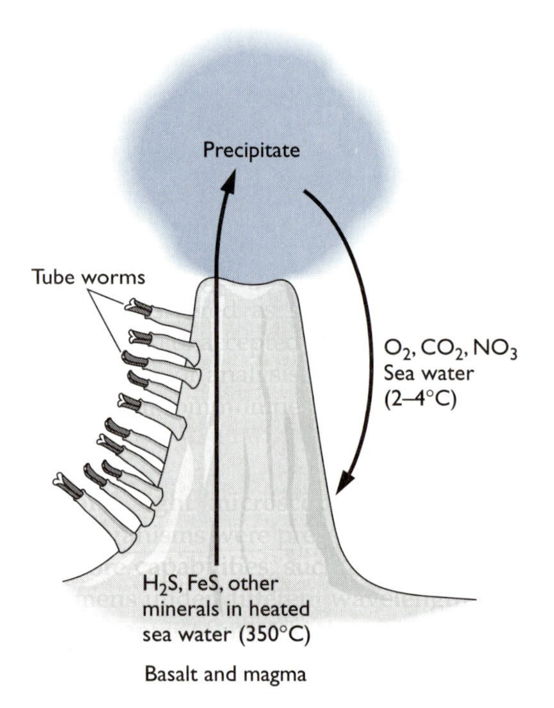

Figure 23.1 Structure of a hydrothermal vent black smoker. Tube worms, shown here only in one location, cover the surface of a black smoker. Super-heated water spews through the vent chimney, bringing sulfides and other compounds from the underlying rock layers into the cold waters of the overlying ocean. Some of the minerals precipitate, forming the "smoke" and building up the smoker chimney. The tube worms harvest sulfides, oxygen, carbon dioxide, and other compounds from the surrounding mixture of upwelling sulfide-rich water and ambient sea water.

A black smoker might not seem like a very promising site for life. Not only are sulfides toxic for most living things, but the huge extremes of temperature encountered by any animal growing on or near the surface of such a smoker would place major stresses on living tissues. Yet, the black smokers are covered by a profusion of animals, at a density of biomass that rivals that found in a rain forest. The diversity of the life, however, is rather limited. The surface of the smoker is covered with hundreds of meter-long worms called tube worms (Fig. 23.1). The best-studied species of tube worm is *Riftia pachyptilla.* The name "tube worm" comes from the fact that the animal is encased within a semirigid tube, which helps protect its vulnerable body from predators. Also found near black smokers are small shrimp and crabs.

Metabolic interactions: Riftia and its bacterial endosymbiont

Tube worms are obviously very tough eukaryotes, because they are living in an environment where the gradient of temperature ranges from a high of 70–80°C to a low of ~10°C farther from the vent surface, not to mention the high pressures experienced at the sea bottom. How these animals manage to cope with such high pressure, high temperatures, and steep temperature gradients remains to be discovered. A closer look at the anatomy of a tube worm at first raised another puzzling question. The worm has no mouth or gut. Instead, it exposes its gills to the sulfide-rich waters around the vent, taking up sulfide, oxygen, and carbon dioxide into blood vessels that circulate the blood to the lower body of the worm.

In the worm's blood is a form of hemoglobin that binds not only oxygen but also sulfide and carbon dioxide. The hemoglobin carries these elements to a mass of cells called the trophosome. Inside the cells of the trophosome are bacterial endosymbionts (the *Riftia* endosymbiont) which obtain energy by oxidizing sulfide to sulfate, a process described in Chapter 21 (Fig. 23.2). Adenosine triphosphate (ATP) and nicotinamide adenine dinucleotide phosphate (NADPH) resulting from sulfide oxidation are used to fix carbon dioxide via the Calvin cycle of the endosymbiont. The source of nitrogen for *Riftia* is less certain, but the symbiont is thought to be providing nitrogenous compounds for the worm as well as organic compounds such as sugars. The carbon and nitrogen compounds produced by the endosymbiont are used to feed the worm.

The *Riftia* symbiont has not yet been cultivated in the laboratory. How, then, did scientists figure out which reactions were being catalyzed by this gram-negative bacterium? Pathways like the one for carbon dioxide fixation and sulfide oxidation have enzymes that are specific to that particular pathway. Rubisco, for example, is associated only with carbon dioxide fixation and not with any other known pathway. Scientists obtained crude preparations of the endosymbiont from tissues of the trophosome and tested them for enzymes such as rubisco and adenosine diphosphate sulfurylase (an enzyme involved in sulfide oxidation). To determine whether these enzymes were of bacterial origin, scientists determined the DNA sequences of the genes encoding them and found that they were closely related to known bacterial genes. Sulfide oxidation has never been found in eukaryotes, so its presence is considered a strong indication that the organism responsible is a bacterium. However, because so little is known about eukaryotic microbes and their metabolic capabilities, it is wise to err on the side of caution when making such sweeping assumptions. What does the endosymbiont gain from its interaction with the tube worm? First, the tube

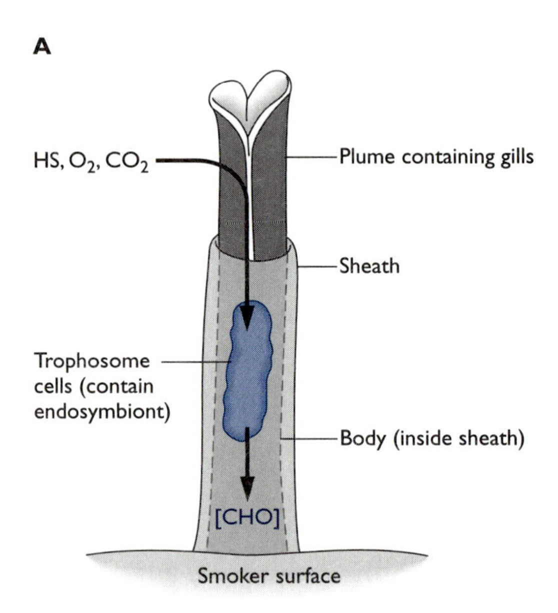

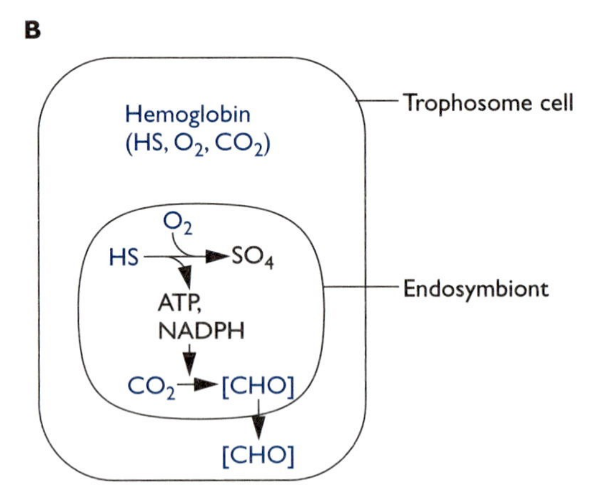

Figure 23.2 Anatomy of the tube worm *Riftia* (A) showing the sheath, the gills and the flow of nutrients in blood from the gills to the trophosome cells, and **(B)** nutritional interactions between the trophosome cells and the bacterial endosymbiont. The endosymbiont oxidizes HS^- to SO_4^{3-} to produce adenosine triphospate (ATP) for fixation of carbon dioxide into organic carbon compounds [CHO], which are provided to the cells of the worm.

worm's hemoglobin provides it with the substrates it needs to obtain energy and carbon. Second, the interior of the tube worm's cells provides a protective environment that shields the bacterium from protozoa and other predators.

Just as the origin of the tube worms and other vent animals is a mystery, so is the origin of the endosymbionts. Do the endosymbionts have a free-living form that "infects" the worms at some stage? Or are the endosymbionts inherited by juveniles and persist because they divide along with tube worm cells? So far, no one has seen any tube worm juveniles, although they undoubtedly exist. Growing tube worms in the laboratory might answer such questions, but this will be difficult, because tube worms die quickly when brought to the low pressure environment of the ocean's surface.

Chemosynthesis at the vents: a mode of primary production independent of sunlight?

Until the discovery of the vent animals, the dogma that photosynthesis was the original source of all complex carbon and nitrogen compounds was firmly in place. The photosynthetic organisms have been called the "primary producers" to emphasize their central role at the base of all food chains. The discovery of the vent animals raised the question of whether another source of primary biomass production exists, one based on chemosynthesis instead of photosynthesis. That is, the vents act like the sun, in that they provide a source of energy for fixation of carbon dioxide and nitrogen, just as photosynthesis does in the upper layers of the oceans and on the surface of land masses. This concept is somewhat controversial, because it is not clear whether the vents are entirely self-sufficient. For example, sulfide oxidation requires oxygen, and virtually all of the molecular oxygen comes from photosynthesis by phytoplankton or plants. Is there a form of sulfide oxidation that does not require molecular oxygen? Scientists are seeking an answer to this question. If the answer is yes, the vents could be considered primary producers of biomass independent of photosynthesis.

Let Your Little Light Shine: *Vibrio fischeri* and the Squid *Euprymna*

An unusual type of mutualistic reaction

In the case of *Riftia* and other mutualistic interactions covered in this chapter, the interaction between microbe and invertebrate tends to be a nutritional one. That is, the bacteria provide food for the invertebrate host, and the host provides food for the bacteria. An example of a very different type of interaction, in which the host provides the bacteria with nutrients but the bacteria provide the host with protection against predators, is that of the tiny squid *Euprymna* and its bacterial exosymbionts. Far from the ocean's bottom, in the shallow warm ocean waters off the coast of Hawaii, lives the inch-long squid, *Euprymna scolopes* (Fig. 23.3). *E. scolopes* is a night feeder and swims above the bottom of the shallow shore waters in search of food. To predators on the bottom, which might want to feast on a succulent *E. scolopes,* the squid should be a clear target because of the shadow it casts against the twilight or starlit sky. However, the squid uses a novel strategy to hide itself from predators.

The strategy is a simple one. The squid has a special organ, called a light organ, which acts as a kind of low-wattage flashlight. The light organ is organized similarly to an eye, in that it has ink sacs that reflect light produced in the organ so that the light shines out of the organ in a downward direction. In this way, the squid produces a low level of light, similar to that from the twilight or starlit sky, so that it disappears against this dimly lighted background. The source of the light is the light-producing (**luminescent**) bacterium *Vibrio fischeri.* Luminescent microbes are widespread

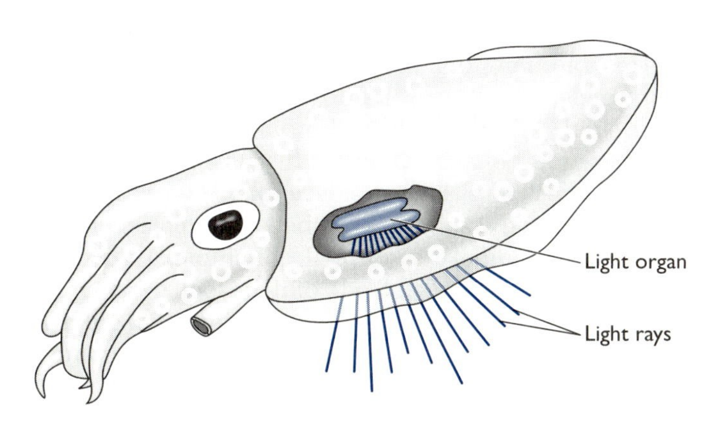

Figure 23.3 Anatomy of the squid *Euprymna*, showing the internal location of the light organ (in cut-away section) and the emergence of light in the downward direction. *Vibrio fischeri* in the light organ are responsible for the bioluminescence.

in nature and use the same strategy for luminescence seen in fireflies and fish with luminescent organs. An enzyme complex, luciferase, carries out a reaction that is opposite to photosynthesis. Instead of converting light energy into chemical energy, the luminescent organisms convert chemical energy into light.

Luciferase: the enzyme complex that produces light

The luciferase reaction is shown in Figure 23.4. ATP and NADPH are consumed in a series of reactions that produce light. Molecular oxygen is also consumed in this reaction. The importance of oxygen is graphically demonstrated by using a culture of *V. fischeri,* which has been grown in broth culture without shaking and has thus depleted the medium of oxygen. This culture is turbid, but in the dark it exhibits no luminescence. If the culture is placed in a tube, with some air at one end, and the tube is inverted in a dark room to let the air pass through the culture, the tube suddenly starts to glow. The glow is transient, of course, because it ends when the bacteria run out of oxygen.

Expression of the genes encoding the enzymes needed for the luminescence reaction is controlled by a quorum-sensing system (Chapter 6). As the concentration of bacteria in the light organ increases, the autoinducer they produce builds up to a level that allows expression of the genes encoding the luciferase complex to be expressed. Thus, the luciferase complex is only produced when the density of bacteria is sufficiently high for a useful level of light to be emitted. This prevents energy from being wasted in the energy-expensive luciferase reaction when that reaction is not productive.

Interactions between the squid and the bacteria

The interaction between the squid and the bacteria is more complex than this description of luminescence in the laboratory makes it sound. First, the light organ contains a pure culture of *V. fischeri,* something very uncommon in nature, especially for an animal exposed to the thousands of bacterial species found in sea water. How does *E. scolopes* select *V. fischeri*? The first layers of specifity seem to be mediated by a mucus gel the squid secretes when it wants its light organ to be colonized. Other species of bacteria beside *V. fischeri* can bind to this mucus gel, but only *V. fischeri* and a few other species can migrate along the gel to the light organ. Inside the light organ, the squid exerts still another selective pressure: it produces a myeloperoxidase, similar to that found in human phagocytic cells, that kills most bacteria—but not

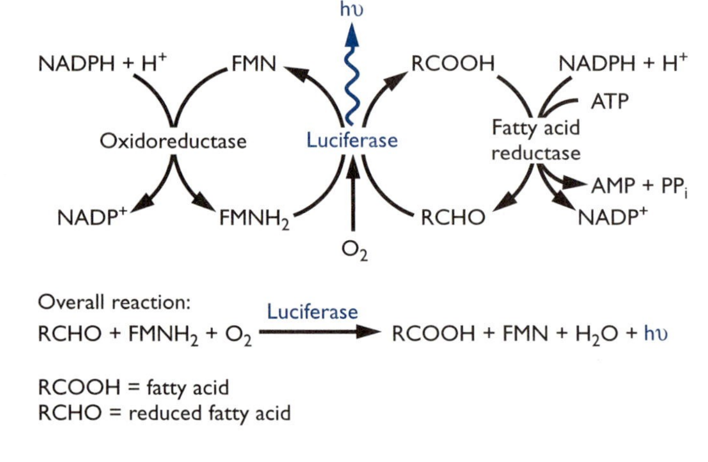

Figure 23.4 Luciferase reaction responsible for biolumescence. An oxidoreductase oxidizes NADPH as it reduces flavin mononucleotide (FMN) to $FMNH_2$. At the same time, NADPH + H^+ and ATP reduce a fatty acid, which together with $FMNH_2$ and oxygen interact with luciferase to produce light (hv). These reactions are reminiscent of part of an electron transport system, except that instead of serving as the electron acceptor, oxygen interacts with the enzyme luciferase to generate light.

V. fischeri. Thus, incoming bacteria must master a series of hurdles that seems tailor-made to select for *V. fischeri*. In this way, the squid ensures that only luminescent bacteria of a specific kind will colonize its light organ.

Once in the light organ, *V. fischeri* benefits by receiving from the squid amino acids and possibly other organic carbon and energy sources. The squid is presumably supplying molecular oxygen too, because without oxygen, light would not be produced. The details of the *V. fischeri–E. scolopes* mutualistic interaction are still being worked out, but it is clear that the interactions are complex and successful in that they result in colonization of the light organ with the type of bacterium that can protect the squid from predators. What the bacteria gain from this is rather transient, because *E. scolopes* expels the bacteria every morning and reacquires them at night. *V. fischeri* has the advantage that it can live as a free-living bacterium in the environment, so ejection from the light organ is presumably not a problem for the bacteria. Whether the environment of the light organ is more congenial from the bacterium's point of view than the surface of a fish or seawater is unclear. *Euprymna* continues to be protected, and the bacteria continue to come back for another round of colonization.

Bacterium–Aphid Symbiosis: A Tale of Natural Genetic Engineering

Characteristics of the bacterium–insect association

Mutualistic interactions in which a bacterium feeds an invertebrate do not occur only at the bottom of the sea but are probably quite common in other environments. An example of a terrestrial mutualistic association is provided by the aphid, an insect that is the bane of gardeners. Aphids feed on the sap of plants. Plant sap is an abundant and easily accessible food source for an insect that can bore through plant tissue to get at it, but sap is basically just sugar-water. It is deficient in amino acids and vitamins that the insect needs to grow and reproduce. Aphids solve this problem by having specialized cells, called bacteriocytes, that contain a bacterial endosymbiont, *Buchnera aphidicola* that produces amino acids for the insect (Fig. 23.5).

Overcoming feedback inhibition

B. aphidicola is a gram-negative bacterium that can synthesize a number of amino acids and vitamins. In Chapter 6, the concept of feedback inhibition was introduced. Feedback inhibition is a mechanism bacteria use to prevent synthesis of excess amino acids and other compounds that require energy to produce. When the end product of the pathway begins to build up, it inhibits the activity of one or more enzymes in the biosynthetic pathway,

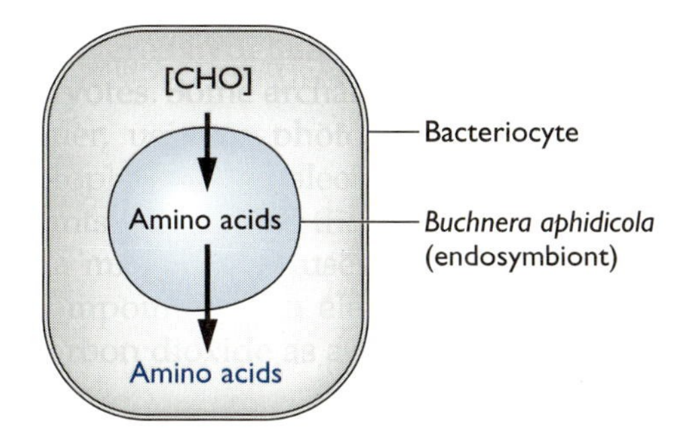

Figure 23.5 Nutritional interactions between the bacterial endosymbiont of aphids, *Buchnera*, and the aphid bacteriocytes. The aphid supplies sugars and other organic compounds [CHO] to the bacteria, which in turn produce amino acids to supplement the nitrogen-poor plant sap diet of the aphid.

thus shutting down its own synthesis. Yet, *B. aphidicola* apparently produces more amino acids than it needs, and these excess amino acids are used by the insect as supplements to its plant sap diet. In theory, feedback inhibition should prevent this kind of interaction, but, in the case of *B. aphidicola,* does not. *B. aphidicola,* like the *Riftia* endosymbiont, has not been grown outside of insect cells, but scientists were able to solve the mystery of how feedback inhibition has been short-circuited by examining DNA sequence information from the bacterium's chromosome and plasmids.

In the case of the pathway for synthesizing tryptophan, for example, the genes encoding pathway enzymes are similar to those of *Escherichia coli,* a bacterium that controls tryptophan utilization by feedback inhibition. The genes of *Buchnera* are similar enough to those of *E. coli* to expect that the proteins they encode would be subject to feedback inhibition. To short-circuit this inhibition, *Buchnera* has acquired a multicopy plasmid that carries genes encoding the proteins in the tryptophan biosynthetic pathway that are subject to inhibition by excess tryptophan (Fig. 23.6). Because there are multiple copies of these genes as a result of the multiple copies of the plasmid that carries them, much more inhibitable enzyme is produced and thus much more tryptophan is needed to shut down tryptophan synthesis. All the insect has to do is to use up the excess amino acids produced by the bacteria fast enough to keep the level of tryptophan below the new higher threshold for shutting down the pathway.

If a human scientist had manipulated a bacterium genetically to overcome feedback inhibition of amino acid biosynthetic pathways, it would be called genetic engineering. Clearly, bacteria were practicing genetic engineering of their own millions of years before humans were even a gleam in nature's eye.

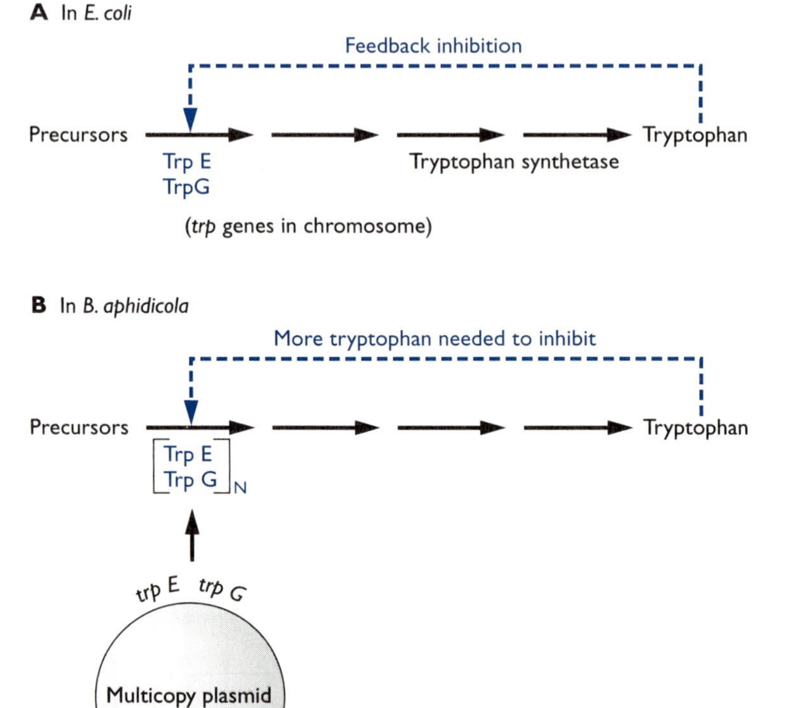

Figure 23.6 How *Buchnera* has reduced the effect of feedback inhibition so that it can produce enough amino acids to feed the insect as well as itself. The example of tryptophan biosynthesis is used here. Two genes that encode feedback-inhibited enzymes in the tryptophan biosynthetic pathway, *trpE* and *trpG,* have been moved to a multicopy plasmid and thus produce more of the enzymes than would be produced if these genes were still in the chromosome. Because more TrpE and TrpG proteins are being produced, it takes higher concentrations of tryptophan to shut down the biosynthetic pathway, thus allowing more tryptophan to be produced by the bacteria. Removal of tryptophan by insect cells also helps to keep tryptophan concentrations below the feedback inhibition level.

DNA sequence information also suggests how *Buchnera* benefits from the association. *Buchnera* has genes that suggest it uses sugars or other simple organic compounds as a carbon source. The aphid presumably provides the compounds needed by the bacteria as carbon and energy sources. Also, as in the case of the tube worm endosymbionts, the insect cells provide a protected environment for the bacteria.

A Genital Tract Infection of Insects with Many Consequences

The sexual games played by the bacterium Wolbachia *insure its propagation: fertility incompatibility*

Wolbachia is another example of a bacterial endosymbiont of insects, but the consequences of carriage of this bacterium vary from one insect to another and are not necessarily beneficial for the insect. In general, these effects operate to insure the spread of *Wolbachia,* which is transmitted transovarially from an adult female to her progeny. Although some evidence suggests a horizontal spread of *Wolbachia* from one insect to another, transovarial spread is clearly the dominant mode of transmission. To ensure its transmission within the insect species it has infected, *Wolbachia* cannot exert a harmful effect on its insect host and may even be beneficial in some insects. This uncertainty about the effects of *Wolbachia* on different insect hosts makes it difficult to categorize the interaction as mutualistic, parasitic, or commensal.

The interaction between *Wolbachia* and the fruit fly *Drosophila melanogaster* has been studied most. The interaction between *Wolbachia* and *Drosophila* seems to be commensal, because *Wolbachia* clearly is using the fruit fly as a convenient way to propagate itself, with no apparent adverse effects for the fruit fly. However, as more is learned about this interaction, a different picture could emerge. *Wolbachia* infects the genital tract cells of the fruit fly. Carriage of *Wolbachia* by insect cells alters the fertility pattern of the insect. This alteration is illustrated in Figure 23.7. Normally, an uninfected male and an uninfected female mate to produce numerous progeny. Once *Wolbachia* has entered the insect population, however, the fertility picture begins to change. An infected male and an uninfected female produce few or no progeny. This appears to be the result of subtle effects of *Wolbachia* on the sperm of the infected male, effects that make the fertilized egg unable to develop normally. By contrast, an infected female can mate productively with either an uninfected male or an infected male to produce numerous progeny. The changes *Wolbachia* makes in the chemistry of the female reproductive cells somehow "rescue" the defect in sperm from infected males but do not affect the reproductive capacity of sperm from uninfected males. The mechanism of the rescue of the sperm defect in infected males is currently under investigation. The effect of *Wolbachia* on *Drosophila* is called **fertility incompatibility,** because it insures that once *Wolbachia* has entered the insect populations, infected females are much more likely to produce progeny in crosses involving any male than uninfected females, which can mate productively only with uninfected males.

An intriguing feature of the *Wolbachia–Drosophila* interaction is its implications for taxonomy. Entomologists use breeding patterns as one criterion for defining a species. If only *Wolbachia*-infected females can breed productively, once *Wolbachia* has become widespread in a population, does this mean that *Wolbachia* has created two species of *D. melanogaster,* one comprised of uninfected females and uninfected males and one comprised of infected females and either infected or uninfected males?

A

Uninfected female × uninfected male = Progeny.
Uninfected female × infected male = Few progeny.
Infected female × uninfected male = Progeny.
Infected female × infected male = Progeny.

B Model

Bacteria in testes alter sperm.
Sperm fertilizes egg, but embryo development is impaired in uninfected female.
Bacteria in egg of infected female "rescue" sperm defect.

Figure 23.7 Fertility incompatibility caused by *Wolbachia* in sex cells of the fruit fly *Drosophila melanogaster*. (A) Once *Wolbachia* has entered the insect population, it skews reproductive patterns, so that infected females can produce progeny in crosses with either infected or uninfected males, whereas uninfected females can only breed productively with uninfected males. This gives infected females a reproductive advantage that causes infected insects to predominate in the insect population. **(B)** Current model for how fertility incompatibility works. Eggs from infected but not uninfected females can rescue the defect in sperm from infected males. The mechanism of this rescue is under investigation. In other insects, interactions between *Wolbachia* and its arthropod host can produce parthenogenetic arthropods (wasps) or transform males into reproductively proficient females (wood louse). All these interactions serve to increase the number of infected females in the population. Ovarian transmission of the bacteria is the dominant mode of spread.

Another intriguing feature of *Wolbachia* infection arises from the fact that *D. melanogaster* is a model insect that has been used widely in research on neurobiology and development. It has been estimated that as many as one-third of experimental colonies of *D. melanogaster* may be infected with *Wolbachia*. What effects, if any, does this infection have on the results of experiments being conducted with the infected insects? If the effects of *Wolbachia* infections are limited to the genital tract, consequences for studies of neurobiological or developmental phenomena may be affected minimally, if at all. If the consequences of infection are more global, effects on the outcome of such studies could be significant.

Effects on other arthropods

Fertility incompatibility is only one of the effects of a *Wolbachia* infection. In female wasps, the chemical changes wrought by a *Wolbachia* infection make the females parthenogenetic. That is, they are able to produce daughters without mating. *Wolbachia* is located in the cytoplasm of the female reproductive cells and is readily transmitted from mother to daughter. Infection of the cytoplasm of male reproductive cells, by contrast, is a reproductive dead end for the bacteria. Thus, bypassing males entirely helps to insure the persistence and spread of *Wolbachia* in the insect population.

Yet another effect is seen in insects such as the wood louse. In this case, infection by *Wolbachia* can cause a male to turn into a reproductively proficient female, probably by affecting hormone production in the male genital cells. In other insect species, *Wolbachia* infection may actually kill the males.

It Takes Three to Tango: The Case of the Glowing Caterpillar Larvae

A three-way interaction involving an invertebrate hunter that uses bacteria to kill its invertebrate prey

As a final example of bacterium–invertebrate interactions, consider an interaction that involves two invertebrates: one that is the hunter and one that is

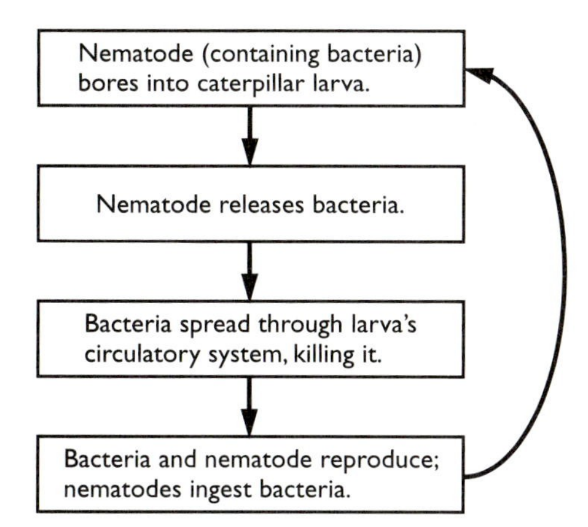

Figure 23.8 Interactions between *Photorhabdus* (bacterial exosymbiont), *Heterorhabditis* (nematode host), and their prey (caterpillar larvae). Bacteria growing in the caterpillar larvae luminesce, probably because of the abundance of energy sources and oxygen in the larval tissue, but it is not clear whether luminescence plays a role in the symbiotic interactions involved in infection of the caterpillar larvae.

the prey. The bacteria, of course, are on the side of the hunter (Fig. 23.8). If you are adventurous enough to venture into the woods at night and look very closely at the insect fauna, you might see a puzzling site: faintly glowing caterpillar larvae. These caterpillar larvae are not making a new fashion statement. Their insides are being liquefied by bacteria that, like *V. fischeri,* are capable of luminescence. The bacterium, *Photorhabdus luminescens,* is unable to penetrate the tough covering of a caterpillar larva on its own. It is conveyed into the caterpillar larva by its invertebrate partner, the nematode *Heterorhabditis. Photorhabdus* is an exosymbiont of *Heterorhabditis,* which lives inside the nematode's gut. In the nematode gut, the bacteria produce a protein that has a high content of cysteine and lysine, two relatively uncommon, but essential, amino acids. This protein probably helps the nematode acquire these amino acids.

Heterorhabditis is a parasite of caterpillar larvae. It uses the larvae as a place to lay eggs, which then develop into adult nematodes. *Photorhabdus* also plays a role in this reproductive cycle. The nematode has mouth parts that allow it to bore through the tough wall protecting the caterpillar larva. Inside the larva, *Heterorhabditis* releases its bacterial symbiont, which spreads through the caterpillar larva's circulatory system, killing the larva and liquefying larval tissue. The nematode uses this larval "soup" as a convenient place in which to lay eggs, which develop into thousands of new nematodes. The newborn nematodes acquire the bacterial exosymbiont and exit the now defunct caterpillar larva.

The faint glow emanating from the dying caterpillar larva is caused by *Photorhabdus.* As is evident from Figure 23.4, the energy-expensive process of luminescence requires substantial inputs of carbon and energy sources. The luminescence is an indication of how rich the larval interior is in nutrients—so rich that the bacteria are able to produce light. Is the luminescence simply a sign of the exuberance of bacteria confronted with a rich nutritional buffet or does it serve some other function? Scientists have speculated that the dim light emanating from the dying larvae might attract other larvae to the site, but there is little evidence for or against this interesting hypothesis.

How Widespread Are Invertebrate–Microbe Symbioses?

No systematic survey has yet been made of invertebrates to see how commonly the types of symbioses described here occur in nature. Yet the number and variety of known examples suggest that such associations might be very common indeed. In earlier chapters, other examples were given. Termites and wood-eating cockroaches rely on their intestinal protozoal and bacterial exosymbionts to convert wood chips into acetate, which can be used as a source of carbon and energy by the insect. In the case of arthropods that transmit human disease, such as rat fleas (plague), ticks (Lyme disease), and mosquitoes (malaria and dengue fever), the microbe has a life cycle in the insect as well as in the humans or animals it infects. This indicates a very complex and specific interaction between the arthropod and the microbe it carries. In the case of the fleas that carry *Yersinia pestis,* the cause of plague, this association is deleterious to the insect. In other cases of arthropod-transmitted disease, however, the arthropod is either unaffected or may even benefit. Beneficial effects of microbes carried by disease-transmitting arthropods have now begun to attract the attention of scientists interested in interrupting the transmission of the microbe to humans. The goal of this research is to find inexpensive, nontoxic interventions to replace currently used insecticides, which are known to be harmful to humans and animals. Antibiotic treatment may be too expensive and too likely to increase levels of resistance in the target bacteria, but introducing nonpathogenic microbial competitors to drive out the pathogenic microbes may be possible in some cases.

The number of invertebrate–microbe associations found to date raises questions about whether similar associations are more common in vertebrates than previously known. The existence of beneficial exosymbionts in humans and animals, the microbiota of the body (Chapter 11), is well documented. Are there cases of beneficial endosymbionts as well? Certainly, parasitic endosymbionts, such as *Chlamydia* species, carry out their life cycles inside human cells (Chapter 16), but little work has been done to look beyond cases of endosymbionts associated with disease. It is not easy to find such endosymbionts. The discoverers of the *Riftia* endosymbiont had to examine thousands of thin-sections with an electron microscope to find the endosymbionts, and at first they were not sure whether the coccoid cells they were seeing were bacteria or eukaryotic cells. Today, by using polymerase chain reaction amplification of bacterial 16S ribosomal RNA genes and fluorescent in situ hybridization (Chapter 21), it has become easier to screen for endosymbionts. So far, however, such surveys of cells from healthy humans or other mammals have not been undertaken.

chapter 23 at a glance

Comparison of symbiotic associations*

Symbiotic association	Location of the bacterial symbiont	Provided by the bacterial symbiont	Provided by the host
Riftia–symbiont (mutualistic)	Endosymbiont	Oxidizes sulfide to produce energy Fixes carbon dioxide Reduces nitrate to ammonia	Subtrates for bacterial activities Protected environment
Euprymna–Vibrio fischeri (mutualistic)	Exosymbiont	Produces light (protects squid from predators)	Amino acids, oxygen Protected environment
Wolbachia–Drosophila (parasitic or mutualistic?)	Endosymbiont	Fertility incompatibility Promotes persistence of bacteria in the insect population	Protected environment Nutrients
Aphid–*Buchnera* (mutualistic)	Endosymbiont	Provides amino acids for aphid	Provides sugars, other organic compounds Protected environment
Photorhabdus–Heterorhabitis (mutualistic)	Exosymbiont	Provides cysteine, lysine for nematode Kills and liquefies larva to aid nematode reproduction	Introduces bacteria into larva, which is a rich source of nutrients

*In all the examples covered in this chapter, the microbial symbionts were bacteria, but as seen in earlier chapters, protozoa can also play a role as symbionts (e.g., in the termite hindgut).

Examples of specialized adaptations of host invertebrate to accommodate bacterial symbionts

- *Riftia* (tube worm): cells of the trophosome harbor bacteria; hemoglobin conveys sulfide, oxygen, and carbon dioxide to cells containing endosymbionts
- *Euprymna* (squid): light organ designed specifically to hold a pure culture of *V. fischeri* and stimulate luminescence of the bacteria
- Aphid: bacteriocytes hold *Buchnera*; furnish organic compounds and harvest amino acids produced by the bacteria
- *Heterorhabditis* (nematode): intestinal tract provides congenial environment for bacterial exosymbiont *Photorhabdus.*
- *Drosophila* (fruit fly): changes in insect sex cells seem to be directed by *Wolbachia*, but the specific localization of the bacteria to these cells suggests that some features of the insect cells favor *Wolbachia* invasion and intracellular growth in ways other cell types do not

key terms

bioluminescence energy-consuming pathway that converts chemical energy into light; found in bacteria and eukaryotes; catalyzed by the luciferase enzyme complex

commensal association symbiotic association that is beneficial to one partner but neither beneficial nor deleterious to the other(s)

endosymbiont microbe that lives inside the cells of its symbiotic partner

exosymbiont microbe that lives outside the cells of its symbiotic partner

fertility incompatibility result of an infection of *Drosophila* by the bacterium *Wolbachia*, which makes uninfected females reproductively incompatible with infected males

mutualistic association symbiotic association that is beneficial to both (all) partners

parasitic association symbiotic association that is beneficial to one partner at the expense of the other(s)

nutritional interactions one symbiotic partner (an invertebrate in the examples used in this chapter) provides carbon sources to its endosymbiont or exosymbiont; these sources are converted by the endosymbiont or exosymbiont into nutrients required by the invertebrate

symbiotic association close association of two (or more) organisms, which may be beneficial to one or both (all)

Questions for Review and Reflection

1. What is the role of the *Riftia* hemoglobin in the association between *Riftia* and its endosymbiont? Make an educated guess as to why is it important that this hemoglobin binds sulfide and oxygen in such a way as to keep them physically separate until they are passed to the endosymbiont?

2. In an earlier chapter (Chapter 9), the microbes of the termite hindgut and their association with the termite were described. Which of the symbiotic associations described here most closely resembles this symbiotic association? In each of the examples provided in this chapter, the relationship was tentatively described as mutualistic, commensal, or parasitic. Describe a new scientific finding that would change this classification scheme.

3. How does the *Euprymna–Vibrio* symbiosis differ from all the other interactions described in this chapter?

4. In Chapter 11 (not the bankruptcy kind but the one in this book), the microbial populations of the human body were described. Which of the associations described here is most like the association between the microbes of the human colon and the human body?

5. Coral reef animals have an algal endosymbiont that is essential for survival. Considering what you learned about algae and their activities (Chapters 10 and 22) and extrapolating from the symbioses covered in this chapter, how would you guess what the algae are contributing to the survival of coral reef animals? What might be happening when corals bleach and die?

6. The described effects of *Wolbachia* carriage varied considerably from one insect to another. How could such a range of outcomes occur? What do they have in common?

7. *Photorhabdus luminescens*, like *Vibrio fischeri*, is capable of luminescence. How do the roles of luminescence appear to differ in the life cycles of these two bacteria and their invertebrate hosts?

24 Plant–Microbe Interactions

First student: ***What do you know about nitrogen-fixing bacteria in plants?***

Second student: ***Nitrogen-fixing bacteria in plants?! I didn't even know nitrogen was broken!***

Microbes, especially bacteria and fungi, play a number of important roles in fostering the growth and health of land plants. These roles range from controlling the balance of different nitrogenous compounds in soil (**the nitrogen cycle**) to forming symbiotic associations with plant roots that provide nitrogenous compounds directly to the plant. Not all interactions between microbes and plants are beneficial. Microbes are a significant cause of plant disease. In this chapter, the focus will be on interactions between bacteria in plants, because these interactions have been well studied and offer examples of the widest range of possible interactions. In terms of disease, however, it is important to note that plant viruses and fungi are at least as important as bacteria as pathogens. Fungi also form beneficial associations with trees and other plants, triggering plants to increase root surfaces and thus increase the capacity to absorb nutrients from the soil. Interactions between fungi and plants have been more difficult to study in depth than those involving bacteria and viruses because of the greater ease with which bacteria and viruses can be genetically manipulated.

The Nitrogen Cycle in Soil

Nitrification and denitrification

Nitrogen in soil exists in a variety of forms, from the most reduced, nitrogen gas (N_2) and ammonia (NH_3), to the most oxidized forms, nitrite (NO_2^-) and nitrate (NO_3^-). The balance between these different forms of nitrogen is maintained by bacteria that oxidize ammonia to nitrite and nitrate (**nitrifiers**), reduce nitrogen gas to ammonia (**nitrogen-fixing bacteria**), or reduce nitrate and nitrite to nitrogen gas (**denitrifiers**; Fig. 24.1). Nitrogen compounds enter the soil in a number of ways. Farmers apply nitrate or ammonia as fertilizer. Urea from animal wastes or applied as chemical fertilizer is cleaved by bacterial enzymes called ureases to produce ammonia and carbon dioxide. Plants that contain nitrogen-fixing bacteria in their roots store nitrogenous compounds from which ammonia is released when these plants are plowed under and decompose.

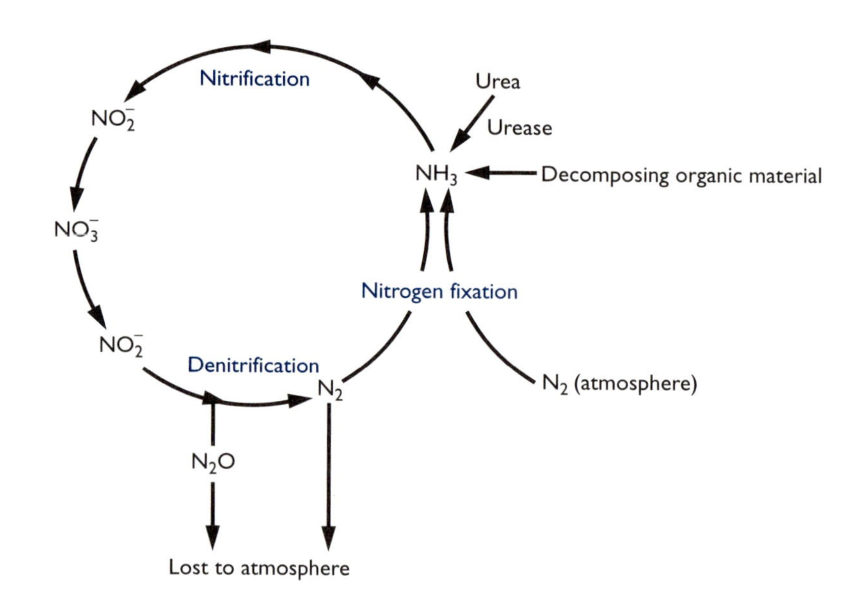

Figure 24.1 Overview of nitrogen cycle in soil. Nitrification (oxidation of ammonia to nitrite and nitrate) requires oxic conditions. Denitrification (reduction of nitrate to nitrogen gas) and nitrogen fixation (reduction of nitrogen gas to ammonia) require anoxic conditions. Nitrogen fixation can be carried out by free-living bacteria or bacterial endosymbionts in plant roots. Other sources of ammonia are urea in animal wastes, which is cleaved to ammonia and carbon dioxide by bacteria and rotting organic material, from which ammonia is released.

Plants can use either ammonia or nitrate as a source of nitrogen. Why, then, does it matter which form predominates in soil? An adverse consequence of denitrification in soil is that two of the products of denitrification, nitric oxide (N_2O) and nitrogen are gases, which can be lost from the soil, depleting it of nitrogen (Fig. 24.1). Also, nitric oxide is a greenhouse gas that also engages in reactions with oxygen that deplete the ozone layer. Under some circumstances, this aspect of nitrate reduction could play a beneficial role. In Chapter 1, the consequences of nitrate leaching from farmlands into rivers, creating "dead zones" resulting from algal blooms far downstream, was described. A scheme for preventing nitrate from reaching nearby water supplies takes into account the fact that the balance of ammonia to nitrate is determined by the oxygen content of the soil (Fig. 24.2). Oxidation of ammonia requires oxygen to serve as a terminal electron acceptor. Thus, nitrification is most likely to occur in well-aerated soils. Reduction of nitrate to nitrogen gas occurs under anaerobic conditions, which are most likely to occur in waterlogged soil. Creating zones of waterlogged soil between the

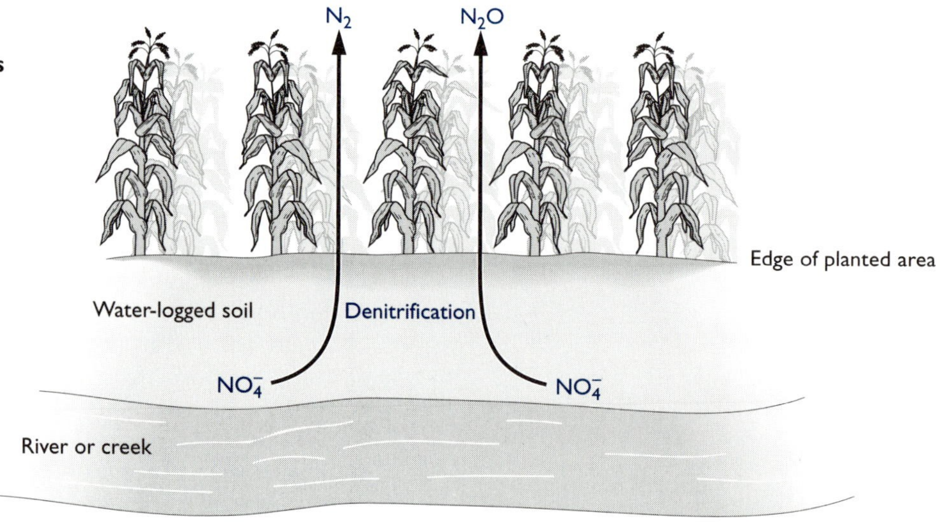

Figure 24.2 Scheme for using denitrification to remove nitrate from farm runoff that may cause problems such as the dead zone in the Gulf of Mexico. Waterlogged soil between the edge of the field and the creek or river provide the anoxic conditions that encourage evolution of nitric oxide (N_2O) and nitrogen (N_2), which are lost to the atmosphere.

cultivated field and the water source allows at least some of the nitrate to be reduced and lost as gaseous nitrogen. Of course, this has the drawback of increasing the amount of greenhouse gases if nitrous oxide is the primary form of gas. Also, as already pointed out, such a solution is wasteful, in that it depletes the land of nitrogen, thus necessitating the application of more fertilizer. Although this solution to the nitrate runoff problem is clearly not perfect, it stands as a good example of how scientists are trying to manipulate the activities of bacteria in the soil to solve problems created by human practices.

More questions, few answers

Scientists are applying molecular techniques such as those described in Chapter 21 to characterize bacterial populations, including the nitrifiers and denitrifiers, in different soils. A surprising finding came from recent studies that compared soils under cultivation and soils once cultivated that have been returned to wilderness. The microbial populations in soils that were once cultivated do not return immediately to the composition found in soils that had always been wilderness. Although some change does occur in the microbial populations of land that is returned to wilderness, even after 7 years, the populations of the soils that were once under cultivation remain different from their always-wild counterparts.

Given the role of microbial populations in soil fertility and the close associations between microbes and plants, this finding raises some interesting questions. Is it possible that the reason that some wild plant species do not return after agricultural soil has been removed from cultivation is that the right microbial population has not been reestablished? Taking the soil microbial communities into account brings reforestation efforts into a different perspective. Another interesting question, which has not yet been addressed, is what effects wildfires or slash and burn agriculture have on microbial populations of soil and how rapidly these populations return to normal after a fire has burned itself out.

Ammonia oxidation

The steps in ammonia oxidation are shown in Figure 24.3. As with sulfide oxidation (Chapter 21), this process is centered around an electron transport system that pumps protons out of the cell and passes electrons to oxygen, the terminal electron acceptor. Ammonia is first oxidized to hydroxylamine (NH_2OH) in a reaction that consumes oxygen. Hydroxylamine diffuses across the cytoplasmic membrane to the periplasmic space, where it is further oxidized to nitrous acid (HNO_2). This reaction results in the release of four protons to the outer surface of the cytoplasmic membrane, generating a proton motive force (PMF). Electrons from this reaction are channeled to an electron transport system, consisting of quinones and cytochromes, and finally are used by a cytochrome oxidase to reduce a molecule of oxygen to water. The cytochrome oxidase pumps two more protons in the process. A similar set of reactions oxidizes nitrite to nitrate, releasing protons in the process and channeling electrons through an electron transport system that uses oxygen as a terminal electron acceptor (Fig. 24.3).

Nitrate reduction

Escherichia coli and many other bacteria can use nitrate as an electron acceptor if no oxygen is present. Nitrate is reduced to nitrite by these organisms. Some soil bacteria, such as *Paracoccus denitrificans*, can reduce nitrate all the

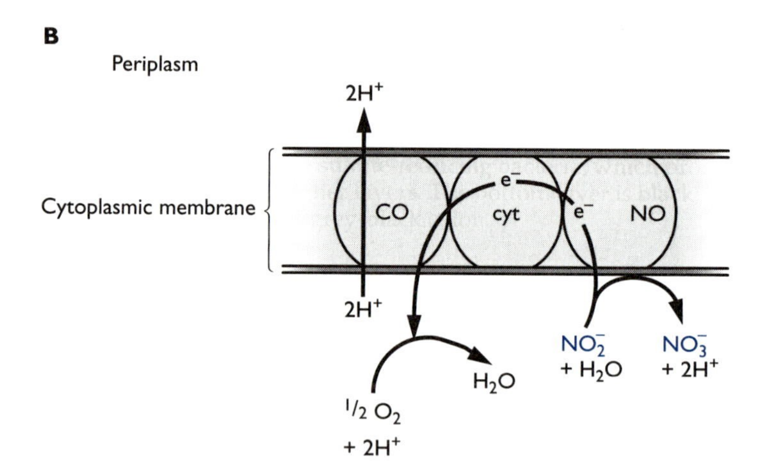

Figure 24.3 Oxidation of ammonia and nitrite. (A) Oxidation of ammonia to nitrite, using an electron transport system with oxygen as the terminal electron acceptor. Oxygen is also consumed in the initial oxidation step. **(B)** Oxidation of nitrite to nitrate. Note that the end result of these pathways is net transport of protons out of the cell. CO = cytochrome oxidase; ETS = electron transport system; HAO = hydroxylamine (NH_2OH) oxygenase; AMO = ammonia (NH_3) monoxygenase; NO = nitrite (NO^-_2) oxidase; cyt = cytochrome.

way to nitrogen gas. In all these reactions, the nitrogen compounds are used as electron acceptors for an electron transfer system. The reason nitrate reduction occurs only under anaerobic conditions is that oxygen, when it is present, is used instead of nitrate as the preferred electron acceptor, because more energy is gained by using oxygen as the electron acceptor.

Nitrogen fixation

Many soil bacteria can reduce nitrogen gas to ammonia and thus contribute to the fertility of the soil. The reactions involved in nitrogen fixation were described in Chapter 22 for photosynthetic bacteria. Nitrogen-fixing nonphotosynthetic bacteria use the same strategy for reducing nitrogen to ammonia. Recall that **nitrogenase**, the enzyme complex that carries out nitrogen fixation, consumes large amounts of energy and is very sensitive to oxygen. Nitrogen-fixing soil bacteria often are found in close association with the roots of plants, in a region called the rhizosphere. It is likely that the activities of these bacteria contribute in a variety of ways to the nutrition of the plants with which they associate, including the production of ammonia. Up to one-fifth of the organic material a plant makes from photosynthesis and carbon fixation is released through the roots into the soil, presumably for the purpose of attracting beneficial bacteria and providing them with sources of carbon and energy to support such activities as nitrogen fixation. In addition to providing nitrogenous compounds, these bacteria may also compete with disease-causing bacteria, thus playing a protective role, much as members of the microbiota of the human body do. The concentration of bacteria in the rhizosphere is considerably higher than that in soil away from plant roots. Among these bacteria will be oxygen consumers. Thus, the rhizosphere may be relatively anoxic, even when surrounding soils are well aerated. Otherwise, nitrogen fixation would not occur, because of the sensitivity of the nitrogenase complex to oxygen.

The most efficient use of nitrogen-fixing bacteria is seen in the case of plants that recruit bacterial endosymbionts to their root cells. In this case, nitrogen gas is reduced to ammonia and incorporated into more complex nitrogenous compounds within the cells of the plant itself. An example of this type of mutualistic symbiosis is provided in the next section.

Rhizobium: a Bacterium that Makes Plants Self-Fertilizing

Crop rotation to renew the fertility of agricultural soils

The rotation of crops is an old agricultural practice designed to restore the fertility of the soil. A nitrogen-depleting crop like corn was succeeded by a crop of clover or alfalfa that added nitrogen to the soil. Clover and alfalfa could restore soil fertility because they have bacterial endosymbionts that reduce atmospheric nitrogen gas to ammonia, thus increasing the amount of fixed nitrogen in the soil when the crop is plowed under.

Interactions between a plant and its bacterial endosymbiont

The formation of a symbiotic association in which nitrogen-fixing bacteria enter the root cells of certain plants involves a complex set of steps in which the bacteria and the plant communicate using chemical signals. In the process, the bacteria and the plant undergo rather dramatic physical and metabolic alterations. A visible sign of this interaction is the formation of **nodules** on the plant root. These nodules provide an environment conducive to the nitrogen-fixing activities of the bacteria. The interiors of the nodules have a red color, caused by a type of hemoglobin that is important in the nitrogen-fixation process. The complexity of the nodule-forming process is fascinating to scientists but is bad news for those who had hoped to use genetic engineering to make plants like corn capable of acquiring bacterial endosymbionts to become self-fertilizing.

As an example of such a symbiotic relationship, we will use alfalfa and its bacterial endosymbiont, *Rhizobium meliloti*. Other plants that incorporate nitrogen-fixing bacteria in their roots are associated with different bacterial species. The plant-bacterium interaction is clearly a very specific one. *R. meliloti* is a gram-negative rod-shaped bacterium that can grow as a free-living bacterium in soil or enter a symbiotic association with the root cells of alfalfa. In the soil, *R. meliloti* is motile and has an aerobic or microaerophilic metabolism. Under these conditions, *R. meliloti* does not fix nitrogen. The steps by which free-living *R. meliloti* are recruited to become endosymbionts of alfalfa are shown in Figure 24.4. The plant releases flavonoids, compounds with a cyclic structure that are recognized as signals by *R. meliloti*. When the bacteria come close enough to a flavonoid-emitting plant root to experience high levels of flavonoid, the flavonoid alters gene expression of certain bacterial genes, an act that initiates the interaction between the bacterium and the plant root.

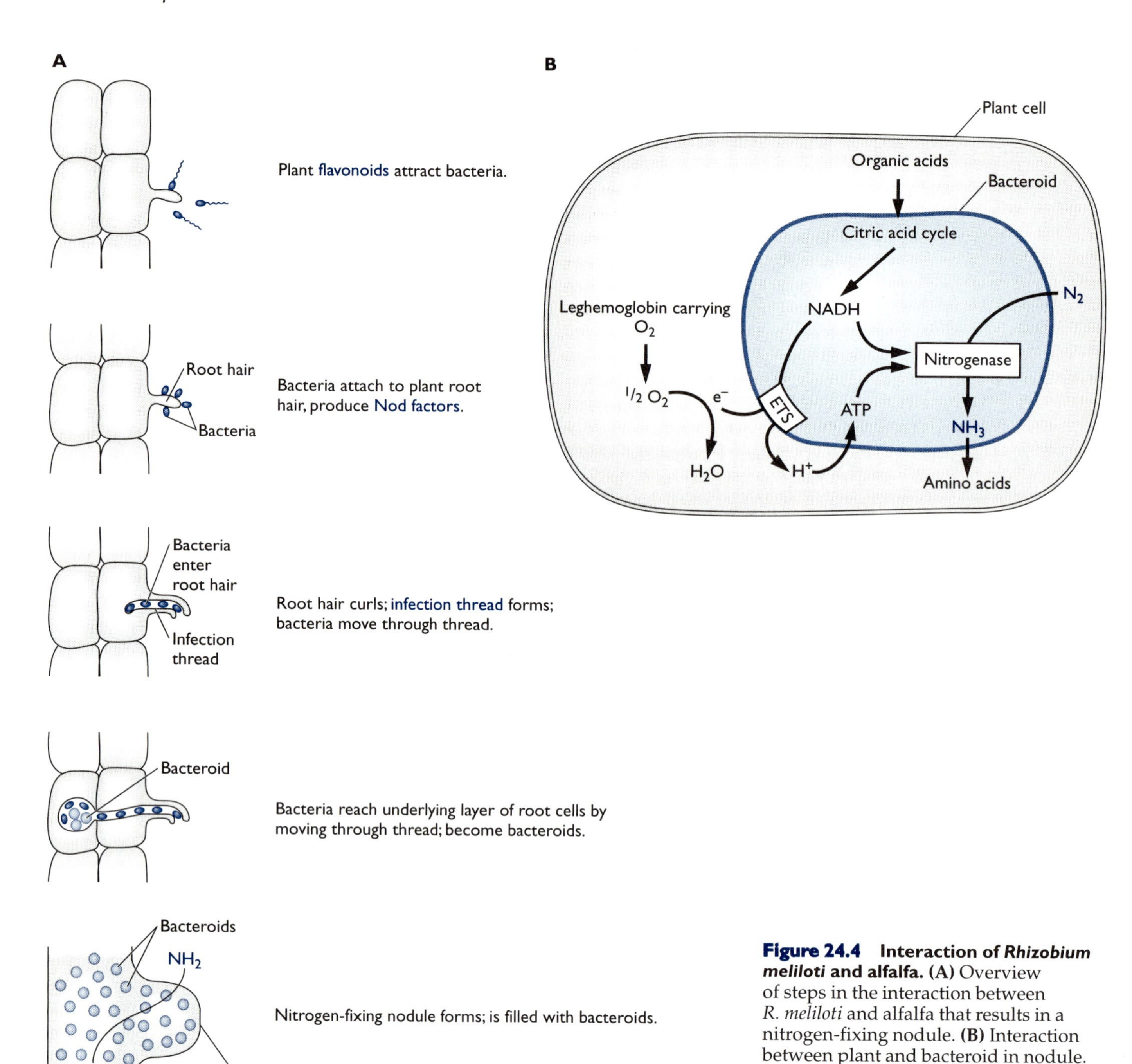

Figure 24.4 Interaction of *Rhizobium meliloti* and alfalfa. (A) Overview of steps in the interaction between *R. meliloti* and alfalfa that results in a nitrogen-fixing nodule. **(B)** Interaction between plant and bacteroid in nodule. Leghemoglobin = globin (from plant cell) and hemin (from bacteroid).

The first bacterial response to the flavonoid signal is production of oligosaccharides called Nod factors (for **nodulation factors**). The structure of a Nod factor is shown in Figure 24.5. Nod factors derive their name from the fact that when they interact with proteins on the surface of the root cells, they cause the plant to initiate a series of changes that ultimately results in a nodule containing the bacteria. In response to Nod factor, a root hair curls and a structure called the infection thread forms. *R. meliloti*, which have become nonmotile and have attached themselves to the root hair, begin to enter the curled root hair and follow the elongating **infection thread** into

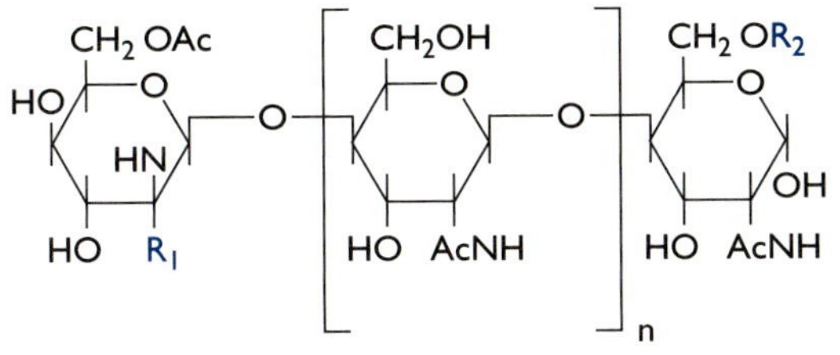

Figure 24.5 Basic structure of a Nod factor. Different rhizobia produce oligosaccharides with different types of substitutions, a feature that explains in part the specificity of certain *Rhizobium* species for certain plants. R_1 = lipid; R2 = SO_4^{2-}, Ac, or H.

underlying root cells (Fig. 24.4). The root cells begin to divide and alter physiologically to form the nodule that will house the bacteria. Meanwhile, the bacteria have changed in shape, from rods to a larger, more spherical form, and become capable of fixing nitrogen. This form of the bacterium, which appears to have become terminally differentiated so as to be incapable of living outside the root nodule, is called a **bacteroid**.

In the nodule, the plant feeds the bacteroids malate and other citric acid cycle intermediates. The bacteroids use these compounds, together with an electron transport system, to make nicotinamide adenine dinucleotide phosphate (NADH) and ATP to support the reduction of nitrogen gas to ammonia. Ammonia is exported to the plant cells, which use it to make glutamate and glutamine, that are in turn passed to other parts of the plants for use in making nitrogen-containing cell components. To obtain the maximum amount of ATP, the bacteroids use an electron transport system that employs oxygen as an electron acceptor. Yet oxygen poisons nitrogenase. This problem is solved by **leghemoglobin,** an unusual hemoglobin responsible for the red color of the nodule interior. The globin (protein) portion of this hemoglobin is provided by the plant, and the hemin portion is provided by the bacteria. The leghemoglobin binds oxygen and carries it to the electron transport system, so that the oxygen can be reduced by the electron transport system but is not free to interact with nitrogenase, which is in the cytoplasm of the bacteroid.

As already mentioned, the interaction between *Rhizobium* species and plants is usually quite specific. That is, a species of *Rhizobium* will form nodules only on one or a few closely related plant species. This specificity is probably mediated at several levels, such as recognition of certain flavonoids by the bacteria, production of specialized Nod factors by the bacteria, attachment of the bacteria to root hairs, and ability to initiate invasion of the plant root. The details of how this specificity is determined are still under study. In this chapter, we have used a *Rhizobium* species as an example. Rhizobia are not the only nodulating bacteria. *Bradyrhizobium* species, a slower-growing type of nodulating bacteria are equally important in forming symbiotic associations with plants. Other nodulating bacteria may remain to be found, but the rhizobia (now called the synrhizobia) and the bradyrhizobia are the best understood.

Sex for Food: the *Agrobacterium*–Plant Interaction

An interaction between bacteria and plants that is nearly as complex as that between *R. meliloti* and alfalfa is an interaction initiated by the gram-negative bacterium, *Agrobacterium tumefaciens*. In this case, however, the relationship appears to be parasitic, with the bacteria benefiting from the interaction and the plant not benefiting. *A. tumefaciens* is the cause of crown gall tumors on plants. Crown gall tumors are considered a plant disease by

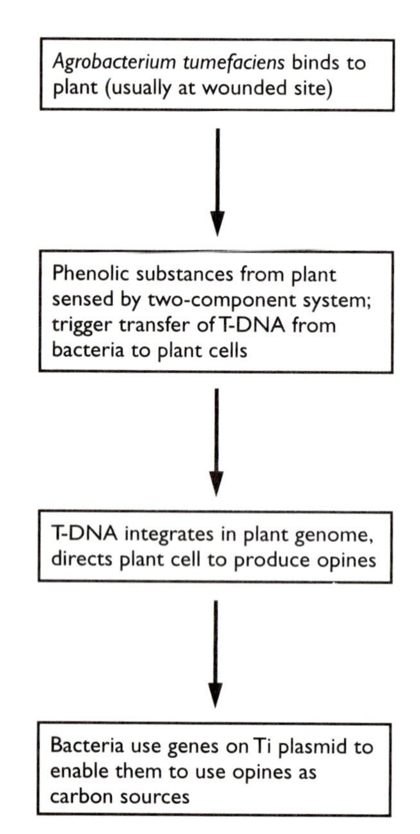

Figure 24.6 Steps in the production of a crown gall tumor by *Agrobacterium tumefaciens.*

producers of ornamental plants, because the galls cause unsightly protuberances on ornamental plants. The plant does not seem to be terribly inconvenienced, however, and is not stunted or killed even when the crown gall tumor is a large one. Because no clear benefit has been discovered for the plant in having a crown gall tumor, the relationship between *A. tumefaciens* and the plants it affects will be described as a commensal one.

The interaction between *A. tumefaciens* and the plant is unusual, in that the interaction between the bacteria and the plant is genetic rather than metabolic, as was the case with the *Rhizobium*–plant interaction. The steps in crown gall tumor formation are shown in Figure 24.6. The bacteria first attach themselves to the plant, usually in an area of wounded tissue. The preference for wounded tissue is because wounded tissue exudes phenolic compounds such as flavonoids, which trigger *A. tumefaciens* to undertake a series of steps that ultimately results in introduction of bacterial DNA into the plant and creation of the crown gall. Many of the key genes involved in the interaction between *A. tumefaciens* and the plant are carried on a large plasmid, called the Ti plasmid (Fig. 24.7A). Phenolic compounds produced by wounded plant tissue activate a two-component regulatory system encoded on the Ti plasmid. This two-component system in turn activates expression of two different types of genes.

One set of genes mediates the excision of a segment of the Ti plasmid called the T-DNA, and a second set of genes mediates the transfer of T-DNA by conjugation to plant cells. In the plant cell, the T-DNA enters the nucleus and integrates into the plant genome. The T-DNA contains genes encoding plant hormones, genes that have promoters recognized by plant cells. In the plant, expression of these genes and production of the hormones they encode causes the plant cell to divide, leading to formation of the tumor. The T-DNA also induces the plant cells in the tumor to produce opines, unusual derivatives of amino acids and sugars that can be used by *A. tumefaciens* as sources of carbon and energy (Fig. 24.7B).

Meanwhile, back in the bacterium, the Ti plasmid encodes enzymes that allow the bacteria to catabolize opines so that they can be used as sources of carbon and energy. Why does the T-DNA elicit opine production, when production of sugars by the plant would have the same beneficial effect for the bacteria? Opines are unusual compounds, and few bacteria besides *A. tumefaciens* carrying a Ti plasmid can use them. Thus, any other bacteria that might be in the area are not serious competitors for these compounds. In effect, the Ti plasmid allows the bacterium to use bacterial sex (conjugation) to "farm" the plant cells, creating a tumor that provides nutrients for the bacteria. Hence the title of this section, "sex for food." So far, the strategy used by *A. tumefaciens* seems to be unique to this genus, but further exploration of plant-associated bacteria may yield other examples of similar strategies.

In the two interactions between bacteria and plants just described, the plants either benefit from the association or are not harmed significantly. Plants in other types of association with bacteria may not fare so well (Box 24–1).

Infections of Plants

Diseases of plants can be caused by viruses, bacteria, fungi, or nematodes. In this section, we will focus on bacterial infections of plants, because recent research has shed new light not only on the mechanisms by which bacteria cause infections but also on the plant's defense system. Plant pathogens,

A

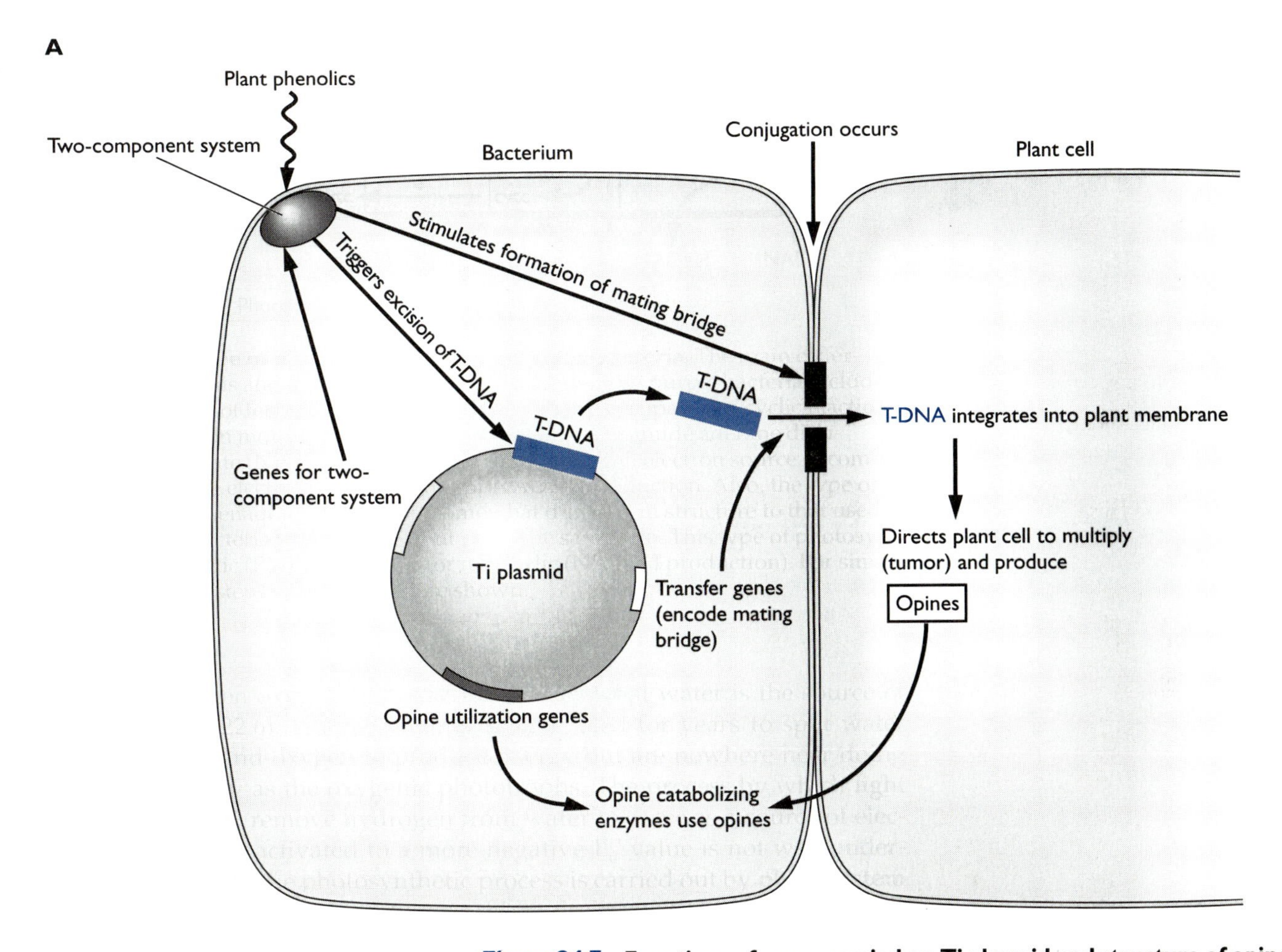

B

$$
\begin{array}{l}
\qquad\; NH \\
\qquad\;\, \| \\
NH_2-C-NH-(CH_2)_3-CH-COOH \\
\qquad\qquad\qquad\qquad\quad\;\; | \\
\qquad\qquad\qquad\qquad\quad\; NH \\
\qquad\qquad\qquad\qquad\quad\;\; | \\
\qquad\qquad\qquad\quad CH_3-CH-COOH
\end{array}
$$

Figure 24.7 Functions of genes carried on Ti plasmid and structure of opine. **(A)** Role of genes carried on *Agrobacterium tumefaciens* Ti plasmid in crown gall tumor formation. The two-component regulatory system encoded on the plasmid senses plant phenolic compounds from wounded plant tissue and initiates the formation of the mating bridge that will transfer the T-DNA by conjugation into the plant cytoplasm and ultimately into the plant nucleus. T-DNA encodes plant hormones that cause the plant cells to divide, producing the tumor. Cells of the tumor produce opines that can be used as carbon sources by the bacteria because of opine-catabolizing proteins encoded by the Ti-plasmid. **(B)** Structure of an opine, a condensation product of a sugar and an amino acid. Few bacteria can utilize these unusual carbon sources, limiting the competition between *A. tumefaciens* and other bacteria for opines.

whether viral, bacterial, or fungal, are usually specific for certain species of plants. Scientists have been interested in understanding this specificity, because such an understanding might suggest new ways to prevent plant diseases that take a major toll on economically important crops. The observed specificity could arise from positive factors that allow a bacterial pathogen to target a particular plant species, such as the ability to adhere to that plant's tissue or to produce toxic compounds that harm only that plant's tissues. Or the specificity could arise from the ability of most plant hosts to resist infection by a particular bacterium, restricting its range to those few plant species that do not mount the appropriate defense reaction. The latter hypothesis seems to be the correct one.

box 24–1 Ice Nucleation, an Unusual Mode of Microbial Damage to Plants

Fruit growers in areas where frost is uncommon are particularly vulnerable to loss of crops when frosts do arrive. A major factor in fruit damage associated with frost is the gram-negative bacterium, *Pseudomonas syringae*, which can cause plant diseases in some plants but lives as a commensal on others. *P. syringae* does, however, have an unusual property that allows it to damage plants even when it does not infect plant cells. It has a protein in its outer membrane that causes ice crystals to form at temperatures slightly above freezing. Because the difference between loss and retention of a fruit crop is often a matter of a few degrees, this trait makes the bacteria a serious threat to fruit when temperatures hover around freezing. Presumably the bacteria benefit from damage caused by ice crystals, which undermine the integrity of the plant tissues, causing nutrients to be released.

One of the earliest products of biotechnology was the isolation of a mutant of *P. syringae* that lacked the ice-nucleation protein. The strain was called "ice-minus," because it did not trigger ice formation. This mutant could still colonize the plant surface as well as wild type and thus could be applied to fruit crops early in the season to prevent later colonization by the wild-type ice-nucleating strain. The ice-minus mutant has been used successfully to prevent frost damage.

Although *P. syringae* has been a problem for farmers, it has its beneficial uses, too. Ski resorts use ice-nucleating strains of *P. syringae* to aid in production of artificial snow.

The defense system of plants: the hypersensitive response

The nonspecific and specific defense systems of humans and other mammals were described in Chapters 12 and 13. Plants once were thought not to have such defense systems, but recent research has brought this assumption into question. Plants have a defense response, the hypersensitive response, which has features analogous to the human immune response. Like the immune response, the response to disease-causing microbes is specific for one or a few closely related microbes. This response was first observed in the case of fungal infections but is now much better understood in the case of bacteria. The hypersensitive response occurs when plant cells recognize a microbial pathogen as a dangerous invader (Fig. 24.8).

Bacterial plant pathogens use a secretion system very similar to the type III secretion systems found in animal pathogens (Chapter 17). Type III secretion systems inject toxic proteins directly into eukaryotic cells. In the case of plant pathogens that have such systems, such as *Erwinia* species, the nature of the toxic compounds is still not well understood, but it is clear that some plant cell targets of these pathogens recognize one or more of these injected proteins specifically. The plant proteins involved in recognition of the bacterial proteins are called **R proteins** (for resistance proteins). When an R protein binds an injected bacterial protein, it triggers a signal cascade that causes the death of plant cells in the vicinity where the invading microbes are located. This response creates a dry, dead area on the plant that is not at all conducive to the growth of bacteria. Recent findings that toxic forms of oxygen are involved in this response have suggested similarities to the oxidative burst of human neutrophils, which is an important part of the human nonspecific defense system. The R protein response is analogous to the specific response mediated by the T cell receptors on human T cells. The T cell receptor interacts specifically with an antigen displayed on a macrophage and activates T cells to become cytotoxic cells or stimulates other cells of the immune system. These responses can also contribute to localized damage that is designed to kill or wall off the invading microbe.

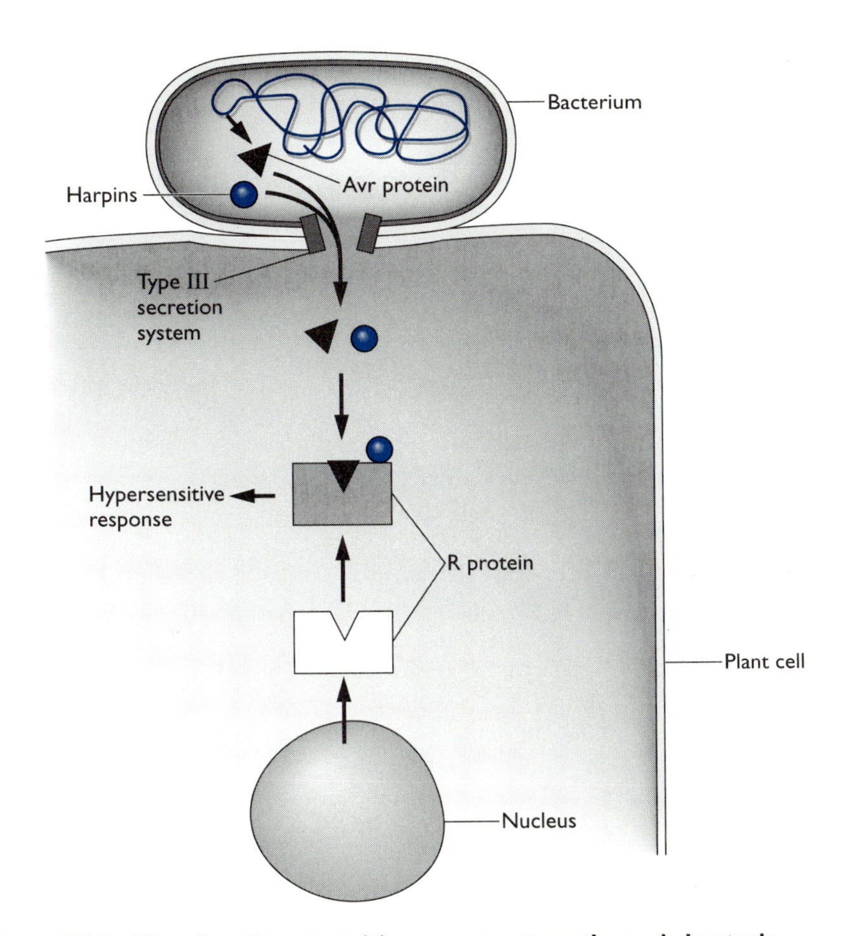

Figure 24.8 The plant hypersensitive response to pathogenic bacteria. A similar response is thought to occur against fungi and some viruses. Bacterial pathogens use a type III secretion system (encoded by *hrp* genes) to inject toxic proteins into the plant cell cytoplasm as part of their initial invasion infection strategy. If the plant contains a resistance (R) protein that recognizes one of the proteins injected by the bacteria, the interaction with the plant R protein and the bacterial protein elicits a localized cell death of plant cells and release of toxic oxygen derivatives that limit the movement of the bacteria in the plant. Proteins that are recognized by plant R factors are called avirulence (Avr) proteins, despite the fact that they may function as virulence proteins in plants that do not have the appropriate R factor. Harpins, proteins produced by the bacteria, stimulate the R factor response to avirulence proteins in a way that is still not understood. Nor is their function clear in the infection process in plants not protected by a hypersensitive response to the bacterial pathogen.

Avirulence genes of microbes: resistance genes of plants

The bacterial genes that encode the proteins that evoke the hypersensitive response in the plant have been called **avirulence genes,** because a bacterium that expresses them and injects them into the plant triggers the hypersensitive response that effectively contains the infection. Thus, even if the bacterium has toxic proteins that enable it to infect plant tissues, it is unable to proceed in the infection process because it is contained at the infection site. Therefore, production of compounds, even ones toxic for the plant, that are recognized by a plant R protein, effectively renders the bacterium incapable of infecting that plant host. If, however, the plant R factors do not recognize avirulence proteins, the bacterial invasion can continue and spread throughout the plant to cause serious systemic infections. The genes that encode the proteins that form the type III secretion system are called *hrp* genes (for hypersensitive response protein genes). Some *hrp* genes not involved in formation of the type III secretion system encode proteins called **harpins.** The specific role of harpins is still not understood, but their presence stimulates the interaction of plant R proteins with avirulence (Avr) proteins and facilitates the triggering of the hypersensitive response. What their role is in bacterial colonization and infection of the plant in cases where plant R proteins do not recognize the bacterial Avr proteins is under investigation. Harpins must have some more constructive role in facilitating bacterial colonization and infection besides helping the plant to prevent the infection.

The terminology, Avr, is confusing to people accustomed to terms such as "virulence genes," which denote genes encoding proteins that promote infection and successful evasion of the mammalian defense systems. This confusion arises because the term Avr is used to describe the plant's side of the infection experience. If the plant recognizes the Avr proteins and this allows the plant to limit the spread of invading bacteria, that bacterium is, from the perspective of the plant, rendered "avirulent." To make matters more confusing, Avr proteins can actually act as virulence proteins, which promote the progression of disease, in plants that do not mount a hypersensitive response to that particular bacterial pathogen.

Scientists are interested in applying the hypersensitive response to protect plants against disease-causing microbes by introducing R proteins that interact with and neutralize microbial pathogens that cause important plant diseases. One drawback of this strategy has been that bacteria (and presumably other microbes) can mutate their *avr* genes so that the plant's R proteins no longer recognize the Avr proteins they produce. Nonetheless, the R protein strategy for protecting plants against economically important pathogens is a promising approach to limiting plant disease without the use of antibacterial or antifungal chemicals.

chapter 24 at a glance

Comparison of plant–microbe symbioses covered in this chapter

Symbiotic partners	Type of symbiosis	Factors provided by plant	Factors provided by bacterium
Nitrogen cycle bacteria in soil in close association with plant roots (rhizosphere)	Mutualistic (except production of gases such as nitric oxide (N_2O) that can enter atmosphere and deplete soil of nitrogen) Exosymbionts	Organic compounds that attract microbes to rhizosphere near plant roots	Convert nitrogenous compounds to forms usable by plants
Rhizobium–plant (nodulation of roots)	Mutualistic (endosymbiont)	Plant flavonoids induce bacteria to produce Nod factors Provide lectins that bind Nod factors to start nodulation Form infection thread to guide bacteria Form nodule where bacteroids fix nitrogen Provide organic compounds to bacteroids Provide globin portion of leghemoglobin that prevents oxygen from damaging nitrogenase	Produce Nod factors that initiate nodulation in plant root Bacteroids fix nitrogen to provide nitrogenous compounds for plant Provide hemin portion of leghemoglobin
Agrobacterium–plant (crown gall tumor)	Commensal (?) No clear damage to plant except for unsightly tumor Exosymbiont	Plant responds to hormones encoded by transferred bacterial DNA (T-DNA) by forming a tumor that produces opines that can be used as carbon sources by the bacteria	Ti plasmid responds to plant phenolic compounds by making mating bridge through which T-DNA is transferred by conjugation to plant Provides genes that encode enzymes for catabolism of opines made by plant
Bacterial plant pathogens–plant	Parasitic Bacteria damage plant tissue, invade plant cells and circulation (unless prevented by hypersensitive response by plant)	Plant provides a host for plant pathogen to infect, unless it protects itself with a hypersensitive response Plant R proteins recognize injected bacterial proteins and activate hypersensitive response that limits spread of bacteria in plant	Bacteria provide Hrp proteins that form type III secretion system that injects toxic proteins into plant cells Some of these proteins may be recognized by R proteins (avirulence factors) Harpins aid in stimulation of hypersensitive response

Metabolic pathways

Nitrate reduction: occurs under anoxic conditions; some soil bacteria can reduce nitrate to nitrite, then to nitric oxide (N_2O) and nitrogen (N_2); nitrate and other intermediates serve as final electron acceptors

Ammonia oxidation: oxygen-consuming reaction that involves an electron transport system that uses oxygen as the final electron acceptor

Nitrogenase: enzyme complex that consumes nicotinamide adenine dinucleotide phosphate and adenosine triphosphate to reduce nitrogen to ammonia (NH_3) (see Chapter 22 for details)

key terms

avirulence genes bacterial genes encoding proteins that are recognized by plant R proteins, triggering the plant-protective hypersensitive response

bacteroids forms of *Rhizobium* and *Bradyrhizobium* species that are endosymbionts of plant root cells and that have differentiated to reduce nitrogen gas to ammonia in plant root cells

denitrification the reduction of nitrate to nitrous oxide and nitrogen gas by soil bacteria

flavonoids plant-produced compounds that serve as signals to *Rhizobium* to start the nodulation process or to *Agrobacterium tumefaciens* to initiate the transfer of T-DNA from the bacteria to the plant.

***hrp* genes** bacterial genes encoding the proteins that form the type III secretion system, which injects proteins into the plant cell cytoplasm; also encode harpins that aid in eliciting plant hypersensitive response

hypersensitive response protective reaction by the plant, wherein plant R proteins recognize bacterial avirulence proteins and harpins, triggering local tissue damage that prevents the bacteria from spreading

nitrification oxidation of ammonia to nitrite and nitrate by soil bacteria

nitrogen fixation reduction of nitrogen gas to ammonia by an enyzyme complex, nitrogenase

Nod factors bacterial oligosaccharides that trigger nodulation in plants that attract nitrogen-fixing bacteria to enter their roots

nodulation infection of root hairs of some plants by bacteria capable of nitrogen fixation, thus rendering the plant self-fertilizing

opines unusual carbon sources produced by crown gall tumor cells and used as a food source by *Agrobacterium tumefaciens*.

rhizosphere space around the roots of plants where concentrations of microbes are highest as a result of the presence of organic compounds secreted by plant roots

T-DNA segment of a Ti plasmid that is transferred to the plant by conjugation and contains genes that encode plant hormones, directing the plant cells to multiply and produce opines

Ti plasmid plasmid that encodes most of the genes needed for *Agrobacterium tumefaciens* to elicit formation of a crown gall tumor by a plant; also contains genes that allow *A. tumefaciens* to catabolize opines produced by plant tumor cells

Questions for Review and Reflection

1. What do nitrification and denitrification have in common? In particular, how do the pathways of ammonia oxidization and nitrate reduction resemble each other?

2. Why might plants provide carbon and energy sources for rhizosphere bacteria that are not endosymbionts? Is this altruism or self-interest?

3. How are bacteroids different from free-living *Rhizobium* species, and how do these differences reflect the different roles and lifestyles of these two forms of the same bacterium?

4. Do you agree with our classification of the relationship between *Rhizobium meliloti* and alfalfa as a mutualistic one? Why? Or could this interaction be better described as a parasitic one in which the plant enslaves the hapless bacteria?

5. If you were going to use genetic engineering to make a *Rhizobium* species nodulate a corn plant (not a natural host), what would you have to be able to do? Why has this very desirable goal so far eluded scientists?

6. Assume you are a scientist interested in understanding why the interaction between a species of *Rhizobium* and a particular plant is so specific. How might this specificity be determined?

7. A scientific report, which proved to be unsupported by the data, claimed that crown gall tumors made plants more resistant to some insects. If future research should prove the existence of such a phenomenon, how would this change your view of the nature of the interaction between *Agrobacterium tumefaciens* and plants?

8. How does the hypersensitive response of plants work to prevent disease? Why can a plant look diseased when this response is working successfully? In what sense does the hypersensitive response resemble the mammalian immune response, and how do plant pathogenic bacteria resemble mammalian bacterial pathogens?

9. What are avirulence genes and why are they given this name? What other bacterial genes are necessary to trigger the hypersensitive response in a plant? What do these genes normally do for the bacteria that make them, and, if they prevent the bacteria from infecting plants, why have they not been lost as a result of selective pressure?

10. In what senses are nodulation by *Rhizobium meliloti* and crown gall tumor formation by *Agrobacterium tumefaciens* infections of plants?

25 Biotechnology

Genetic engineering has given scientists a way to make specific and controlled changes in bacteria, plants, and animals, in contrast to traditional approaches, which relied on random mutations or uncontrolled mixing of genes. Genetic engineering also gives scientists a new edge in the race to provide new pharmaceuticals and improve agricultural yields.

The term biotechnology can often conjure up images of cloned sheep and genetically engineered agricultural products. In fact, today's biotechnology industry is heavily focused on such technologies as microbial fermentation to produce amino acids, antibiotics, and other pharmaceutical proteins, as well as the bioremediation of toxic waste sites with microbes. It is still common for such processes to employ microbes isolated from nature instead of microbes that have been genetically engineered. Nonetheless, this chapter will focus on the use of genetic engineering to produce new products and processes, because the biotechnology industry is increasingly dominated by genetic engineering approaches, including areas that formerly relied exclusively on naturally occurring microbes.

Why Genetic Engineering?

Why industry has turned to genetic engineering

The use of genetic engineering, especially for creation of plants and animals with desirable properties, has become increasingly controversial in recent years because of concerns about the hypothetical risks these new products pose for human health and the environment. The emphasis by antibiotechnology activists and the media on the negative side of the risk-benefit equation of biotechnology has tended to obscure the reasons that have put genetic engineering at the center of biotechnology in recent years. Genetic engineering has come to be recognized as the best solution to two problems confronting the biomedical and agricultural industries: the need for increased speed of development and increased safety of products. With bacteria and other microbes becoming increasingly resistant to existing antimicrobial compounds, it has become important for human scientists to move swiftly to counter the very rapid adaptive responses of microbial pathogens. Because genetic engineering and other molecular techniques speed up both the process of drug discovery and the production of large quantities of new compounds, such techniques naturally have attracted the attention of scientists responsible for developing new antimicrobial compounds. The same is true for vaccines.

Genetic engineering can speed up the production of new vaccines (see Chapter 6 for an example) and can make current vaccines safer. The old formulation for the vaccine against whooping cough, which consisted of killed bacteria, had many side effects. The newer formulation of the vaccine, which consists of purified proteins produced by genetic engineering, has fewer side effects and seems to be equally effective. Replacement of insulin from animal sources with genetically engineered human insulin produced by bacteria has provided major improvements in both the efficacy and safety of treatment for diabetes.

In agriculture, the challenge has been to improve crop yields and nutritional quality of important food plants and to produce animals that are resistant to pathogens currently controlled only by antibiotics. At present, if distribution problems could be overcome, the world produces enough to feed everyone. But increases in the world's population during the next 50 years will outpace today's food production. In addition to helping in the production of more and better food, genetic engineering may improve the distribution problem by working to create new breeds of plants and animals that allow farmers in developing countries to overcome local problems, such as drought and the expense of fertilizer and insecticides. With current methods of agriculture, the only solution to meeting the new needs for increased food production is to put more land under cultivation, an approach that would reduce further the wild areas that support much of Earth's biodiversity. In this chapter, we will explore some of the ways that genetic engineering is being applied to solve problems of the present and the future.

How does genetic engineering differ from traditional methods for development of new products?

The traditional means of developing new strains of bacteria for use in industry involve the introduction of random mutations and screening for mutants with desired traits. For example, antibiotic-producing bacteria or fungi isolated from nature tend to produce only very low levels of antibiotic. Scientists mutagenized the genomes of these organisms, sometimes extensively, in order to obtain mutants that produced high levels of antibiotics. Until late in the 20th century, when DNA sequencing came to be used widely, exactly what had been done to an individual strain could not be determined. Mutations could exist in a number of genes other than the ones responsible for the trait of interest. An example of the way in which such uncontrolled changes can go awry is provided by the tryptophan produced in the 1980s and 1990s by a Japanese firm, now defunct. The bacterial strain that overproduced the tryptophan was the result of classical mutagenesis procedures. Evidently, the bacteria produced toxic byproducts as well. This did not become apparent until the company decided to save money by shortcutting the process used to purify the tryptophan. Improperly purified tryptophan sold in Europe caused deaths and paralysis in a number of people before the product was removed from the market.

Similar problems can arise in plant breeding programs. When two breeds were crossed, the result was a throw of the genetic dice, in which the progeny acquired an unknown number of new genes. Because many crops were created in this manner, the process was not fraught with great danger if the progeny were screened carefully. Some troubling examples, however, reveal traditional breeding gone wrong. A classic case was the Lenape potato, which was developed in the early 1980s. This potato was bred for increased insect resistance. Unfortunately, one of the easiest ways for a plant to become resistant to insects is to heighten its production of toxic alkaloids, which are also toxic for humans. This proved to be the mechanism of resistance of the Lenape potato. The potato was marketed for a short time before its toxic properties were realized. There were no deaths or cases of serious illness, but this example served as a warning of what could happen in uncontrolled crosses. Similar experiences with celery and squash alerted plant breeders to look closely at insect-resistant breeds for toxicity before marketing them. Animal breeders faced similar problems.

Genetic engineering allowed plant and animal breeders to introduce specific genes with well-documented functions into plants and animals. This gave scientists much more control over what was happening in the bacterium, plant, or animal. Of course, specific changes made in an organism can affect other genes and in some cases can cause global changes in the organism's traits, but a detailed knowledge of what had been done to the organism made it much easier to evaluate these products for safety.

A second advantage of genetic engineering was that it allowed breeders to cross genus and species lines to introduce bacterial genes into plants or genes from one animal species into another. At first this type of gene crossing might seem unnatural. In fact, evidence is accumulating that the movement of bacterial genes into plants and animals has occurred more than once in evolution. It is too early in the genome sequencing of plants and animals to assess horizontal transfer events between plants and animals or between different species of plants and animals, but it will be surprising if such evidence is not found. Scientists know that such gene transfer events have occurred but not how. Genetic engineering provides an alternative to the natural form of cross-species, cross-genus transfers that have occurred by an as yet unknown mechanism.

Examples of the Products of Biotechnology

Production of human proteins by bacteria

Two of the first medical applications of genetic engineering were the production of human insulin and human growth hormone by bacteria. Before this, diabetics relied on porcine insulin. Growth hormone for the treatment of dwarfism was not available. To obtain proteins such as insulin or growth hormone from humans would have been prohibitively expensive, because the proteins are present in relatively small amounts in human blood. Although many features of human cell transcription and translation are similar to those of bacteria, enough differences existed to make it difficult to adapt human genes for accurate transcription and translation by bacteria. Such problems have now been overcome, although the process of adapting a human gene to a bacterial version can involve a number of steps, including construction of **expression vectors.** *Escherichia coli* is the bacterium most often used for production of foreign proteins.

One difference between human and bacterial genes is that human genes frequently have introns, whereas bacterial genes (except for some ribosomal RNA [rRNA] genes) do not. **Introns** are regions of noncoding DNA that are found interspersed in the coding region of a gene (Fig. 25.1).

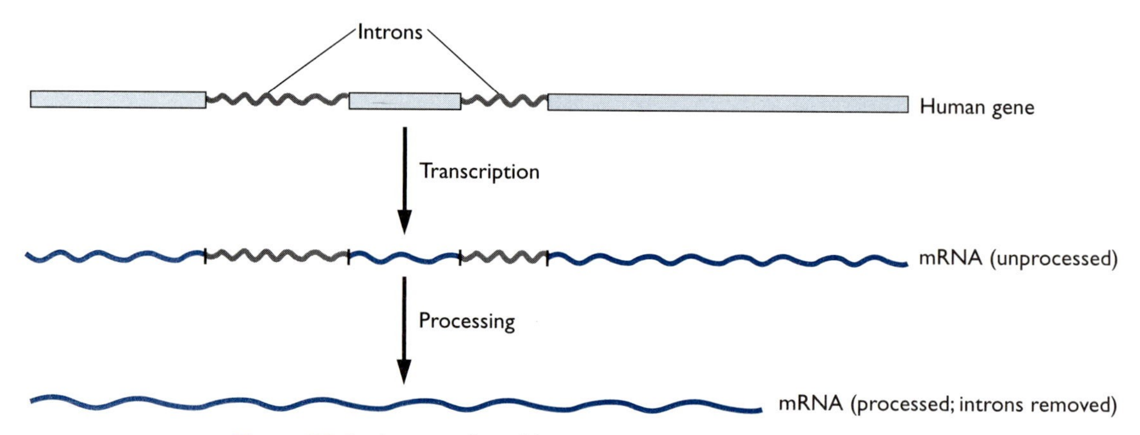

Figure 25.1 Introns found in many human genes are removed from the transcript of the gene to produce the form of the transcript that will be translated by ribosomes.

After the messenger RNA (mRNA) copy of the gene has been made, special enzymes splice out the intron, leaving an intronless copy of the gene. A strategy for producing a DNA copy of the human genes that lacks introns is shown in Figure 25.2. Reverse transcriptase (RT) is used to make a DNA copy of the mRNA (see Chapter 18). The DNA copy of the mRNA can then be cloned directly or amplified by polymerase chain reaction (PCR) for easier cloning. Both the RT and PCR steps can introduce errors into the sequence, although such unintentional mutations are uncommon. Thus, the scientist usually will have the gene sequenced at this point to assure that the cloned gene is an accurate copy of the original gene.

Human genes have promoters that are different from bacterial promoters. Thus, the next step is to give the gene a bacterial promoter. Note that the gene cloned by RT-PCR has no promoter, so it is now cloned downstream of a bacterial promoter (Fig. 25.2). The promoter will be a strong one, to ensure that many transcripts are made. The copy of the mRNA also must be given a bacterial ribosome-binding site so that it can be translated efficiently. The goal is for the bacterium to make as much of the desired protein as possible (overproduction of the protein).

Overproduction of a foreign protein can create two types of problems. First, overproduction of the protein is often toxic for *E. coli.* This problem is solved by making the bacterial promoter a regulated promoter. For example, the promoter of the bacterial gene *lacZ* is expressed only when the inducer, lactose, is present. Many such regulated promoters are now available. During production, the bacteria are first grown to high densities, then the inducer is added so that protein can be produced. At this point, toxicity for the bacteria is no longer a problem.

A second problem associated with overproduction of a protein is that the protein reaches such high concentrations inside the bacterial cell that it precipitates to form insoluble complexes called **inclusion bodies.** It is usually difficult to solubilize the proteins bound up in inclusion bodies. One solution to this problem is to give the protein a secretion signal at its amino terminal end that directs the bacteria to secrete the protein through the cytoplasmic membrane, clipping off the secretion signal peptide in the process. Even better, the protein can be engineered to be exported to the external medium, where it is not only dilute enough not to form inclusion bodies but is easier to harvest later.

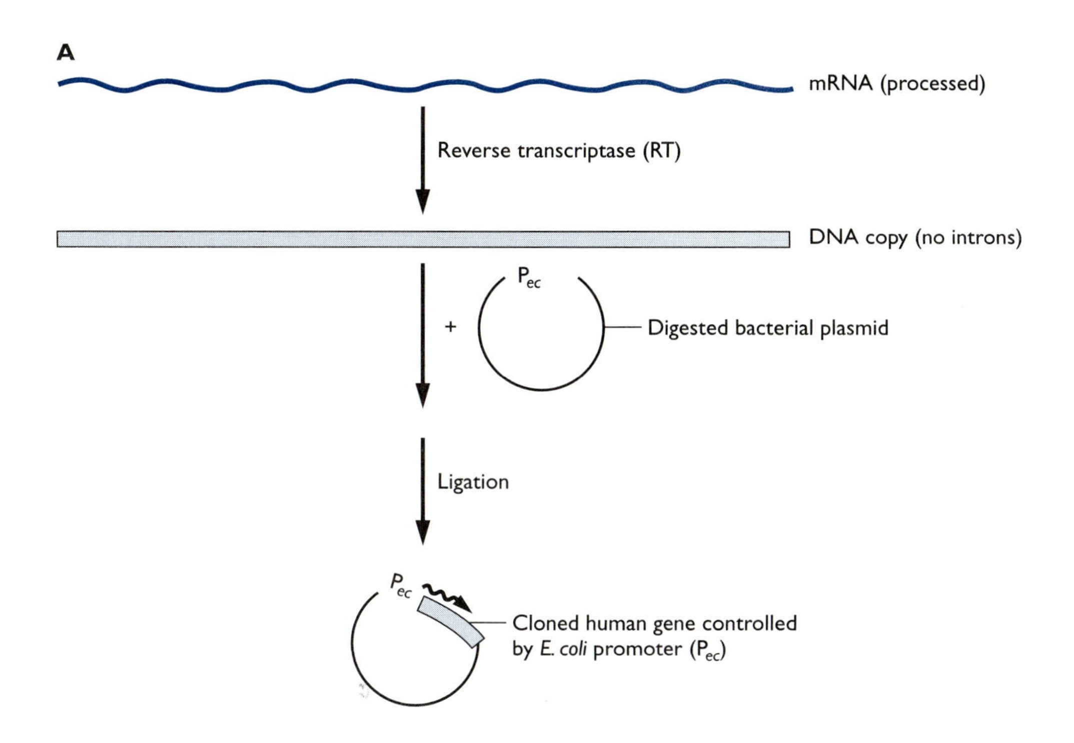

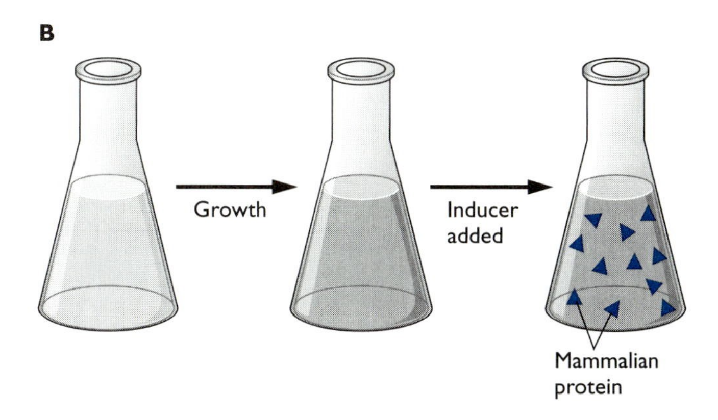

Figure 25.2 Preparation of an expression vector. (A) A DNA copy of the processed messenger RNA (mRNA) is made by reverse transcriptase (RT) and amplified for easier cloning. The DNA is cloned downstream of a strong regulated *Escherichia coli* promoter (P_{ec}). **(B)** The bacteria that contain this cloned gene will express only the gene and produce the protein when the inducer is provided.

Codon usage may be a problem, when the human cell uses a different codon preferentially for a particular amino acid than the bacterium. To check for this or other anomalies that may have occurred during the cloning process, the scientist will have the amino acid sequence of the protein determined. This is accomplished using an automated process comparable to DNA sequencing. In this process, amino acids are cleaved chemically one by one from the protein and identified by their migration on a chromatographic column. A second concern of scientists is that the bacteria might have altered one or more of the amino acids by adding a methyl group or some other chemical group. Such modifications may be missed in the amino acid sequencing. To check for such modifications, scientist use a technique called mass spectrometry, which measures the molecular weight of the protein so accurately that additions as small as a methyl group can be detected.

Vaccine production and gene therapy

Genetic engineering is becoming increasingly important for the production of vaccines, especially purified protein vaccines. Viral proteins, like human proteins, can be produced by bacteria. The same steps described for cloning a human protein apply to cloning a viral protein. An added advantage of using cloned proteins as vaccine is that it is much safer to produce a cloned viral protein than to grow large amounts of the virus, which are then processed to make the vaccine. Also, with this process it is impossible for live viruses to contaminate the final product if a mistake is made during production of the vaccine.

A very different biomedical application is the use of injected genes to cure human genetic defects, an approach called **gene therapy.** The goal is to place an effective copy of a defective gene in the appropriate type of human cell. The cloned effective copy of the human gene is either injected as naked DNA or cloned into a virus that integrates into the human cell genome. In this application, there is no need to change the introns or promoters of the human gene, because they are going into human cells. Gene therapy has been controversial, both because early attempts failed and because deaths were associated with some early clinical trials. The approach has such promising benefits for people suffering from genetic disorders that research in this area has continued despite early disappointments. As this book goes to press, a possible first successful gene therapy trial may have been made. Some children with a genetic disorder that causes them to have a malfunctioning immune system have been treated—apparently successfully—to make them immunocompetent.

Improving classical fermentation processes

In the past, antibiotics, amino acids, and vitamins were produced by bacteria or fungi that had been isolated from environmental settings. Scientists interested in producing antibiotics, for example, were likely to have shelves of soil samples from such exotic places as Bora Bora or Peoria, Illinois. They relied on the biodiversity of microbes found in different locations. Once a microbe was isolated that appeared to produce the desired compound, the strain had to be altered by mutation to a form that overproduced the desired product. Fresh isolates of soil bacteria or fungi produce only very low amounts of an antibiotic and must be induced by a combination of mutation and optimized growth conditions to produce higher amounts. In the case of amino acid-producing microbes, feedback inhibition must be overcome to increase the amount of amino acid produced.

Until the advent of genetic engineering, the process of adapting a new environmental isolate to high-level production of a compound had much in common with black magic. The ability to clone and study genes for antibiotic production or amino acid biosynthesis and to alter the promoters of the genes involved is beginning to change the approach of industry scientists to these classical fermentation reactions.

Genetic engineering of plants

Farmers continue to need crops with improved resistance to insects, drought, salt, and other environmental problems. They also need increased yields and plants with improved nutritional properties. Problems associated with overuse of herbicides and pesticides have spurred scientists to look for new plant breeds that reduce the need for chemical pesticides and herbicides. One of the first successful products of plant biotechnology was an insect-resistant, genetically engineered corn plant that contained an insecticidal bacterial protein, *Bacillus thuringiensis* toxin (BT toxin). **BT corn** provides a good example of the use of genetic engineering to introduce desirable traits into a plant. BT soybeans, potatoes, and other plants have also entered the market. BT toxin was an attractive choice, because it is selective for certain types of insect pests, such as the European corn borer (the problem BT corn was designed to solve), and leaves beneficial insects alone. One of the comments made by farmers who switched from using chemical insecticides to BT crops was that their fields, which once were cloaked in eerie silence, were now noisy with the sounds of insects. Moreover, BT toxin had a well-established safety record; organic farmers have sprayed toxin-producing *B. thuringiensis* on their crops for decades.

To create BT corn, the bacterial BT gene first had to be given a promoter and translation signals that would be recognized by a corn plant cell (Fig. 25.3). The final form of the plasmid contained the cloned BT toxin gene, the

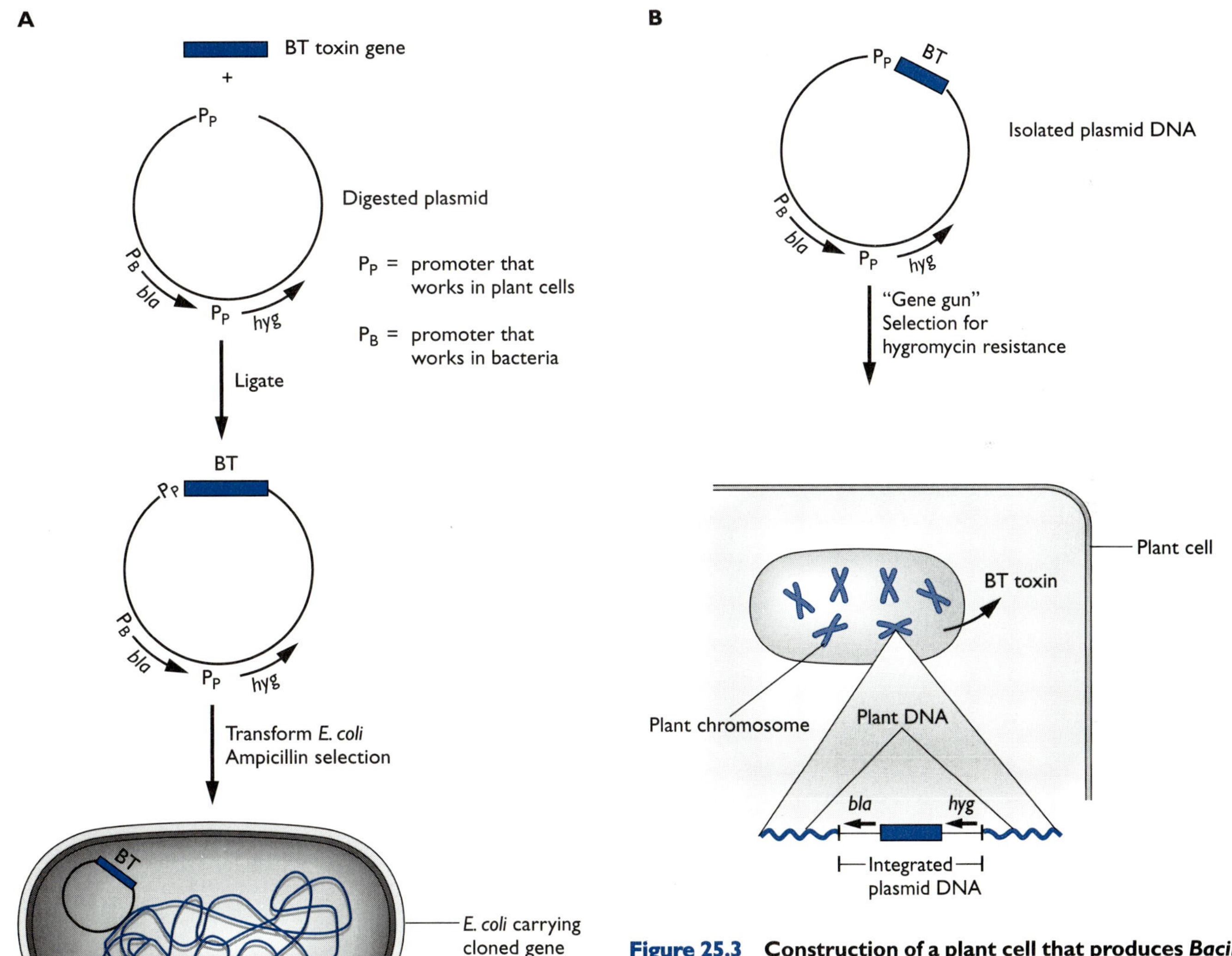

Figure 25.3 Construction of a plant cell that produces *Bacillus thuringiensis* (BT) toxin. The *bla* gene encodes a β-lactamase that confers resistance to ampicillin. This gene is used to select for bacteria that contain the plasmid during the initial cloning steps. The *hyg* gene confers resistance to hygromycin. This trait is used to select for plant cells that receive the plasmid. The promoters (P) that control expression of the BT gene and the *hyg* gene are ones that works in plant cells, not in bacteria.

antibiotic resistance gene used to select for bacteria that received the gene, and a gene that encoded a selectable marker, such as resistance to hygromycin, that made plant cells resistant to this otherwise toxic compound. There are two ways to introduce such a construct into plant cells. One way is to clone the important genes into the T-DNA segment of the *Agrobacterium tumefaciens* Ti plasmid (Chapter 24) and introduce the T-DNA into plant cells by conjugation. Unfortunately, T-DNA is not transferred to all plant cells. This limitation has caused scientists to resort to a second method for introducing DNA into plant cells, the gene gun. The gene gun is just what its name suggests, a high-pressure device that shoots gold or other particles coated with the DNA into plant cells in a section of plant tissue. Incredibly enough, a sufficient number of plant cells survive this onslaught to yield cells into which the plasmid DNA has been introduced. The plasmid DNA integrates into the plant cell genome by a mechanism that does not involve homologous recombination. Integration is random. Much was made earlier in this chapter of the fact that genetic engineering makes possible directed, controlled introduction of genes into organisms. One exception to this general principle is the fact that at present there is no way to control where the injected DNA enters the genome. The inserted DNA is most likely to enter a noncoding region, but it is possible that integration might activate a gene encoding a plant toxin or otherwise affect the physiology of the plant cell. For this reason, genetically engineered plants are tested exhaustively for safety and nutrient composition as well as for field performance. In spite of all of the extensive testing, some groups still oppose genetically modified plants (Box 25–1).

What Lies in the Future

A full description of all the biotechnology initiatives currently underway would fill a large book. Only a few examples will be given here to illustrate the variety of potential new applications.

Rapid response to new disease threats

DNA sequencing is a key part of biotechnology. It was not mentioned previously in this chapter because it was covered in Chapter 5, but it figures in an important new biotechnology application: rapid response to new emerging infectious diseases. In the past, if a new disease threat emerged, scientists first had to cultivate the responsible microbe, then determine its properties in order to develop an appropriate response. This could take precious months, if it could be accomplished at all. Today, a partial genome sequence of the microbe (~90%) can be obtained very rapidly. Such sequence information gives extensive information about the properties of the microbe and can be turned into a vaccine (see Chapter 6). Genome sequencing is easiest if the organism has been cultivated, but, especially in the case of viruses, PCR amplification using random primers can provide copies of segments of the genome in advance of cultivation.

More rapid production of vaccines

A serious problem mentioned in Chapter 14 is the long time it takes to produce many vaccines. The reason for the delay is that in whole-microbe vaccines, the microbe must be grown in quantity, then processed to create the final form of the vaccine. Using cloned genes encoding a vaccine protein

box 25–1 *Genetically Modified Plants Under Attack*

The 20th century was characterized by unprecedented changes in peoples' lives. The 21st century is likely to introduce an even greater rate of change. Not surprisingly, public understanding of scientific matters has not kept up with the pace of scientific innovation. As an indication of this gap, consider the results of a poll published in *Seed Trade News* (Ball Publishing, Batavia, IL) in 1999. The pollsters asked people in a number of different countries whether they thought that the main difference between traditionally bred and genetically engineered tomatoes was that genetically engineered tomatoes contained genes whereas traditionally bred plants did not. In Austria and Germany, more than 40% of people polled believed this statement to be true. Another 20%–22% were uncertain of the answer. U.S. citizens did better; only 10% of those polled agreed that the statement was true, although 45% were uncertain. Results of this poll show what a poor job scientists and politicians have done in educating the public and what a good job antibiotechnology activists have done in propagating misrepresentations about genetic engineering.

Some examples of concerns about the safety of genetically engineered plants and the current consensus about their validity are:

***Bacillus thuringiensis* (BT) toxin might be an allergen, causing some people to have severe allergic reactions.** This seemed unlikely, because organic farmers had been spraying BT on crops for years. Nonetheless, this is a concern for any foreign protein. Today, scientists are getting better at recognizing allergens from the amino acid sequence of a protein. Also, programs exist in which individuals who are allergic to various substances volunteer to be injected with minute amounts of the protein to test the new strains. One irony of the allegation about BT corn is that the method for testing traditionally grown crops for allergenicity is to sell them to the public and wait to see whether anyone has a reaction.

The antibiotic resistance marker gene in the corn might escape the corn in the digestive tracts of animals or humans and enter intestinal bacteria, making them resistant to antibiotics. After much discussion, a consensus has emerged among scientists that not only is this highly unlikely to happen but, even if it did happen, the event would have no medical significance. For a detailed analysis of this issue, see a paper posted on www.roar.antibiotic.org.

BT toxin is a threat to the monarch butterfly. This concern arose when a laboratory study found that feeding butterflies pollen from BT corn could kill them. Monarchs do not feed on corn, so the threat seemed remote in the real world. In fact, studies of the monarch population have shown that BT crops have had no effect on the population size.

Introducing bacterial genes into plants is unnatural. Chloroplasts, anyone? Chloroplasts are not the only evidence for a flow of bacterial genes into eukaryotic cells during evolution. Genome sequences of higher plants and animals are revealing the traces of many such events.

that enables bacteria to produce the final form of the vaccine should speed up this process considerably. Moreover, cloned vaccine proteins that might be needed only in special cases can be stored and retrieved more easily than whole microbes.

Better detergents

Okay, so in the context of all the grand possibilities already mentioned and those yet to come in this section, the lowly subject of detergents seems a bit out of place, but all of us—even college students—have to clean our clothes sometime. Such ordinary household products are not left out of the biotechnology revolution. Bioprospectors are sampling numerous microbial populations for new enzymes, especially proteases and lipases, that may be useful to the detergent industry. Such enzymes are found by random cloning of microbial genes (Fig. 25.4). The cloned genes, many thousands of them, are introduced into *E. coli* by artificial transformation, then the resulting bacterial strains are tested in a high throughput screening procedure for the ability to produce the enzyme of interest (for example, lipase that works best in cold water). In this application of cloning, scientists are really not interested in which microbe donated the cloned gene, only in the properties of the enzyme produced by the cloned gene in *E.coli.*

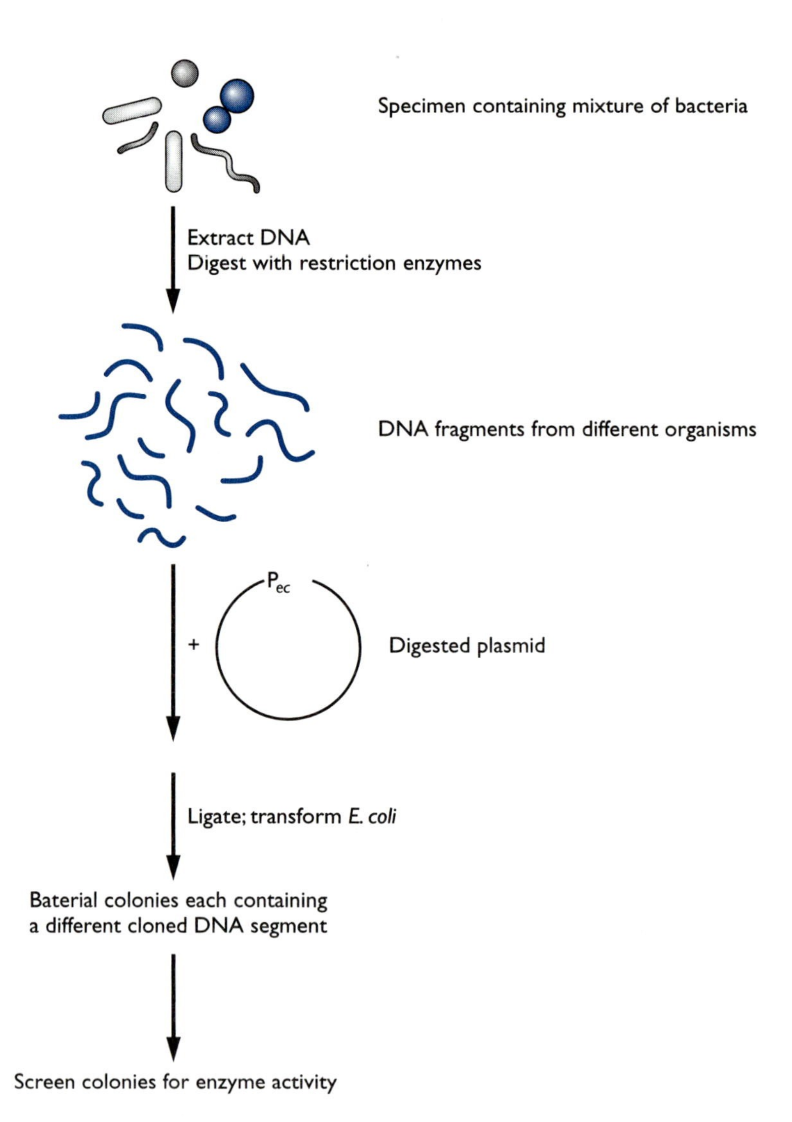

Figure 25.4 An approach to cloning genes of interest from a mixed microbial population. In this case, the scientist is looking for genes encoding a lipid-degrading enzyme (lipase). The assay for such enzymes uses a substrate that changes color (from colorless to blue) when lipase is present.

One problem with this approach is that if the gene comes from a bacterium that is only distantly related to *E. coli*, its genes may not have promoters that work in *E. coli*. Or the expression of the gene may be regulated and thus not occur under the general growth conditions used to produce cultures for screening. To circumvent this problem, scientists use cloning vectors that place the gene downstream of a promoter that works in *E. coli* (P_{ec}; Fig. 25.4). If the gene is cloned in the right direction, the promoter on the cloning vector should allow it to be expressed in *E. coli*. This does not always work, but it works often enough to be highly useful in applications of this type.

Plant-based vaccines

Genetic engineering of plants has focused primarily on improving the insect resistance or yield of a commercially important crop. Plants also might be engineered to produce edible vaccines. The genes encoding one or more vaccine proteins could be introduced into a widely consumed food, such as banana, that could be fed to infants. The fruit would contain the vaccine proteins that would immunize the infant via the mucosal immune system. The advantage of such a plant would be that it would be easy and cheap to distribute and, in theory, could reach far more people than conventional injected vaccines. Such plant-based vaccines are currently being constructed for testing.

Plants with improved nutritional quality: golden rice

In many parts of the developing world, a deficiency in vitamin A causes blindness in large numbers of children. Although the most obvious solution to this problem would be to distribute vitamin A capsules to the parents of children in danger, the economics of such distribution make it unlikely that such a program could be carried out. Certainly, despite much hand-wringing over this problem by the World Health Organization and other international agencies, no vitamin A distribution program has materialized. Another way to solve this problem is to genetically engineer rice to produce vitamin A. A strain of rice that does this is currently under construction in a scientific institute in the Philippines. This strain of rice has been dubbed "golden rice," because the vitamin A will give the rice a golden color. This development effort is unique, in that the development of the strain is occurring not in industry but through the cooperation of nonprofit institutions. This means that no licensing or patent issues will delay the final product or make its cost prohibitive. The only potential problem is that the rice is the wrong color, and people are often very sensitive to such nuances. Try selling green margarine! It is hoped that people in areas where vitamin A deficiencies are most acute will be willing to change their children's diet to stave off the threat of lifelong blindness.

Animal biotechnology

Animals, like plants, can be genetically engineered. This application of genetic engineering has encountered more public resistance than the genetic engineering of plants, but some applications are being discussed with such obvious potential benefits that public acceptance will be assured. One example of such a proposed application is genetically engineering animals to make them resistant to certain microbial pathogens for which no effective vaccines exist. Such animals could reduce the use of antibiotics to prevent microbial infections. If increasing antibiotic resistance renders antibiotics useless for preventing animal infections, this solution becomes even more critical. A variant on this application is engineering animals to resist colonization by such bacterial pathogens as *E. coli* O157:H7, *Salmonella* or *Listeria*. Such animals would reduce the chance of food-borne infections.

Feeding the world in 2050

Enough food is currently produced to feed the population of the world, if that food were properly distributed. Within a few decades, this production will no longer be sufficient unless significant increases in yields are achieved. Plant scientists believe that the "green revolution," begun in the 1950s and instrumental in helping a number of developing countries become self-sufficient in the production of some crops, has gone as far as it can to increase plant yields. Biotechnology is the only available method for improving yields further. Of course, more land could be brought under cultivation, but this would mean plowing under wild acreage we currently want to protect. An added advantage of biotechnology is that, if applied to producing crops specifically adapted to the problems of farmers in developing countries, it could go a long way to making these farmers more self-sufficient, thus solving the distribution problem.

chapter 25 at a glance

Producing foreign proteins in *E. coli*

Problems	Solutions
Introns in human genes but not bacterial genes	Splice introns out of messenger RNA (mRNA) Use RT to make DNA copy of mRNA Clone DNA copy
Different types of promoters in human and bacterial DNA	Clone DNA so that it has no mammalian promoter Insert promoterless DNA downstream from strong bacterial promoter
Different types of ribosome binding sites in human and bacterial mRNA transcripts	mRNA transcript has ribosome binding site recognized by bacterial ribosomes
Overproduction of foreign protein toxic to *E. coli*	Use regulated bacterial promoter that requires inducer Grow bacteria to high density, then induce protein production
Protein reaches such high concentrations in *E. coli* it precipitates (forms inclusion bodies)	Add secretion signal at N-terminus so that bacteria secrete protein out of cytoplasm or into external environment
Codon preferences different in bacterial and human systems	Check for different amino acids in protein by determining amino acid sequence
Amino acids may be altered by adding methyl or other groups	Check by determining molecular weight of protein by mass spectrometry

Construction of *Bacillus thuringiensis*-resistant corn (BT corn)

Initial problem	Solution	Resulting problem
Introducing into corn plant cells a plasmid with BT gene + promoter and translational signals recognized by corn cells	Clone important genes into T-DNA segment of *Agrobacterium tumefaciens* Ti plasmid Introduce into corn by conjugation	T-DNA not transferred to all types of plant cells
	Coat particles with DNA of interest Introduce into plant cells by gene gun	DNA enters plant genome randomly

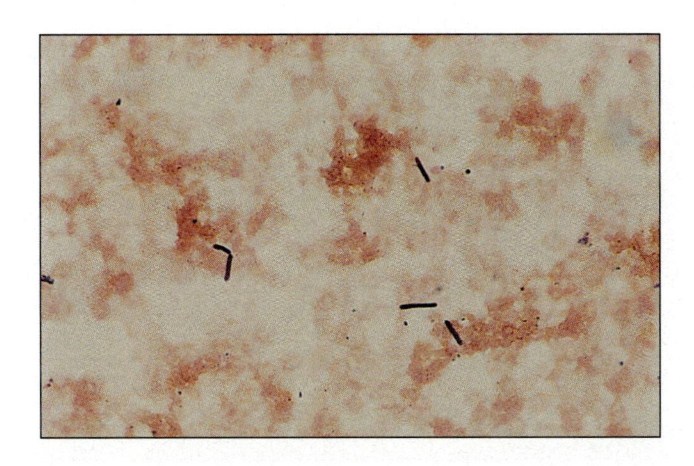

Color Plate 1 **Gram-positive rods in a patient's specimen.** These bacteria are purple in color. Courtesy of Peter Gilligan, University of North Carolina Hospitals and Medical School.

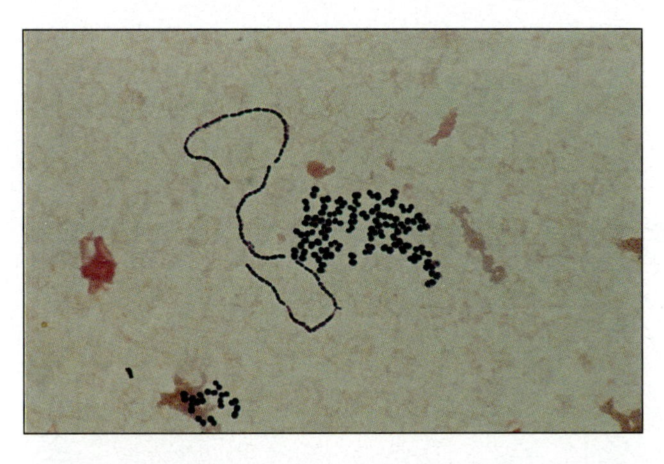

Color Plate 2 **Gram-positive cocci.** Depending on the species, the individual cells may occur in chains or in clusters. These cells stain purple. Courtesy of Peter Gilligan, University of North Carolina Hospitals and Medical School.

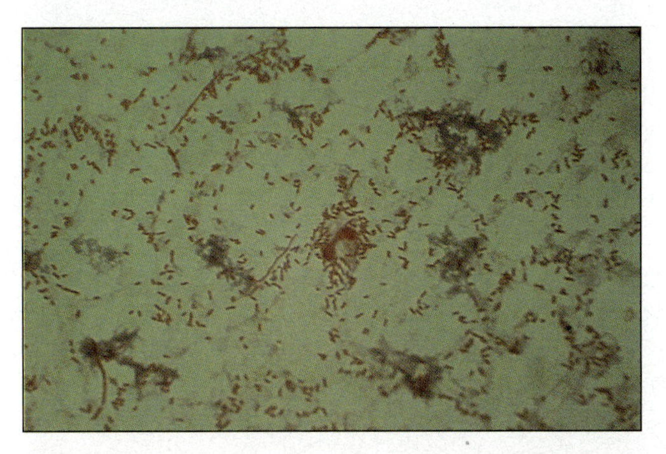

Color Plate 3 **Gram-negative rods.** These bacteria, which are stained pink, may occur as individual cells or may form short or long chains depending on the stage of growth. Courtesy of Peter Gilligan, University of North Carolina Hospitals and Medical School.

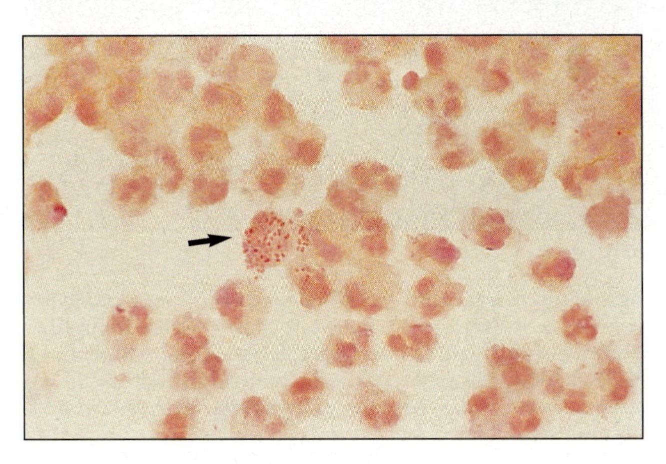

Color Plate 4 **Gram-negative cocci.** The pink-colored bacteria shown here (*Neisseria gonorrhoeae*, the cause of gonorrhea) occur in pairs (diplococci) and are found associated with neutrophils (arrow). Courtesy of Peter Gilligan, University of North Carolina Hospitals and Medical School.

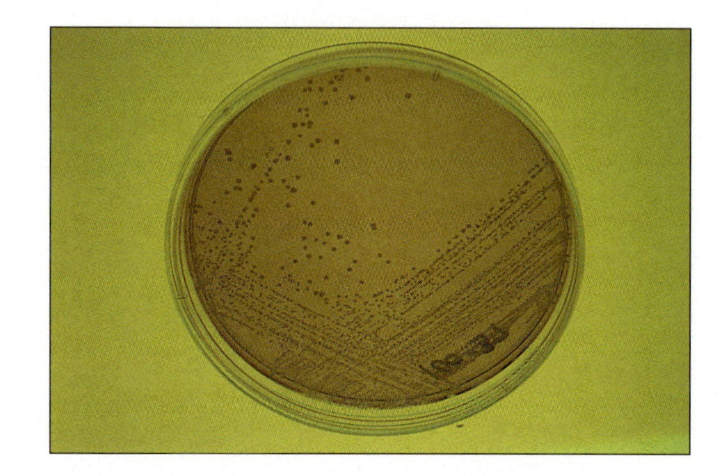

Color Plate 5 A plate of agar that has been streaked so that isolated colonies are formed.

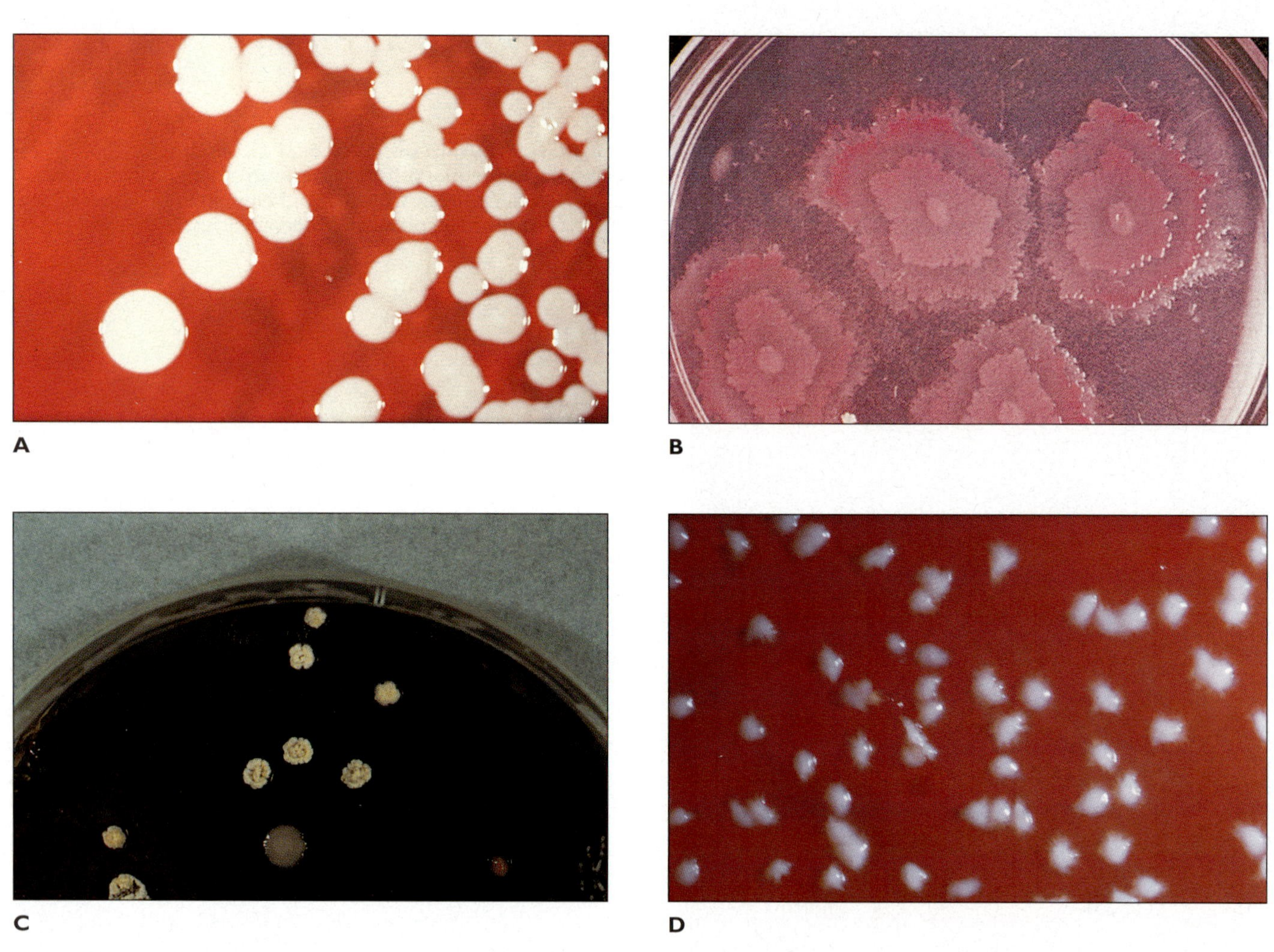

Color Plate 6 Examples of colonies of different types of microbes growing on agar plates. (A) *Staphylococcus epidermidis* forms white colonies that appear shiny because of the capsule that covers the cells. **(B)** Some bacteria such as *Proteus* are highly motile, and as they multiply they move rapidly outward from the center (swarming) and form a characteristic colony morphology. **(C)** More than one type of bacterium was present in the specimen streaked on this plate. *Nocardia*, which forms crumbly colonies, would be selected for further testing. **(D)** Some species of yeasts such as *Candida* form colonies similar to those of bacteria. Courtesy of Peter Gilligan, University of North Carolina Hospitals and Medical School.

Color Plate 7 ***Aspergillus* spp.** A common fungus, which may produce aflatoxin, forms woolly white colonies that eventually turn black. Courtesy of Peter Gilligan, University of North Carolina Hospitals and Medical School.

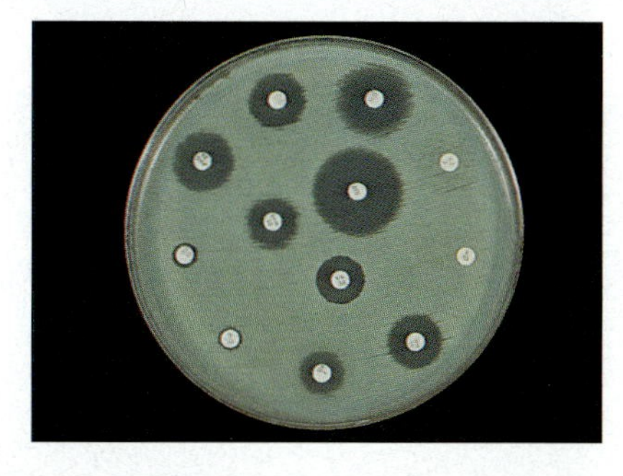

Color Plate 8 **Disk diffusion test for determining antimicrobial susceptibility.** Bacteria are spread on an agar plate to form a lawn. Disks containing a specific concentration of antibiotic are added and the plate is incubated. The diameter of the zones of inhibition that occur can be used to determine whether the microbe is susceptible or resistant to the antibiotics. Courtesy of Janet Hindler, University of California, Los Angeles.

Color Plate 9 **Microbial mats from the Great Sippewisset Salt Marsh (Cape Cod, MA).** Differently colored layers indicate different types of microbes. The top layer contains cyanobacteria and some diatoms. Lower colored layers contain nonoxygenic phototrophs. The bottom grey-black layer contains the sulfate-reducing bacteria. The sulfides these bacteria produce interact with iron to form iron-sulfides, which have a grey-black color. Courtesy of Rolf Schauder.

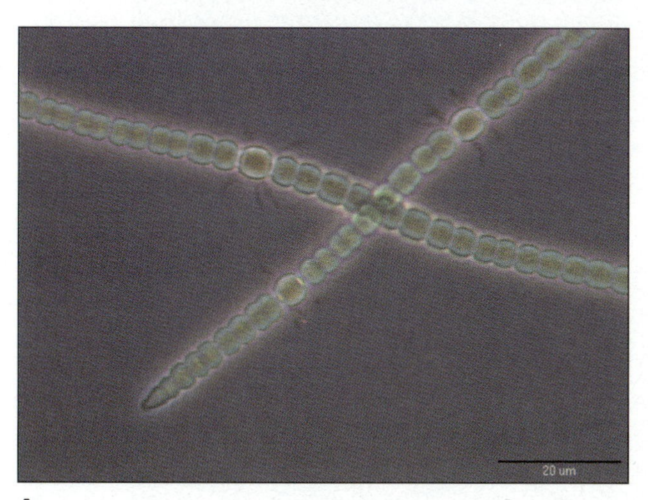

A

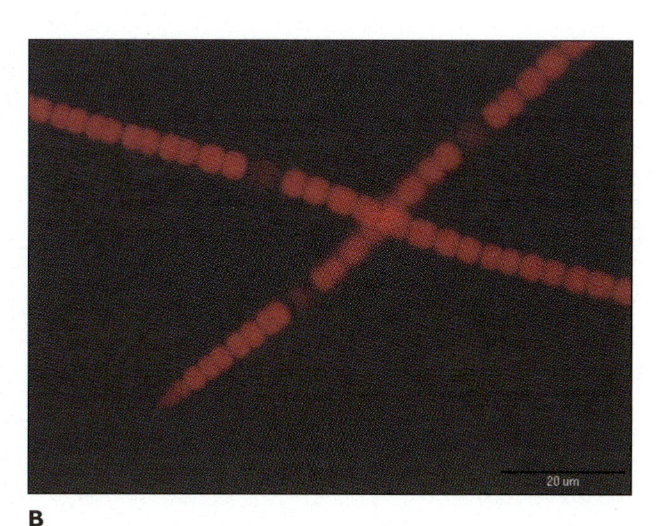

B

Color Plate 10 **(A)** Phase contrast view of the cyanobacterium *Anabaena*. If you look closely, you can see some "hairs" radiating from the different-looking cells, the heterocysts. The identity of these bacterial associates of the *Anabaena* heterocysts is not known, but they seem to be interested only in the nitrogen-fixing heterocysts and not in the photosynthetic cyanobacterial cells. **(B)** Fluorescence microscopic view of *Anabaena* showing the red fluorescence of the photosynthetic vegetative cells and the "dark" heterocysts, which do not perform photosynthesis.

Color Plate 11 The white film shown in this photograph is a layer of growth of a sulfide-oxidizing bacterium called *Thiovulum*. Layers ("veils") of this bacterium form only at the point in the gradients of oxygen (from top to bottom) and sulfide (bottom to top) that support the growth of this particular bacterium. This is an example of life in a gradient, where microbes preferentially migrate to the part of the complex gradients dealt them by nature in which they can grow optimally. Courtesy of Rolf Schauder.

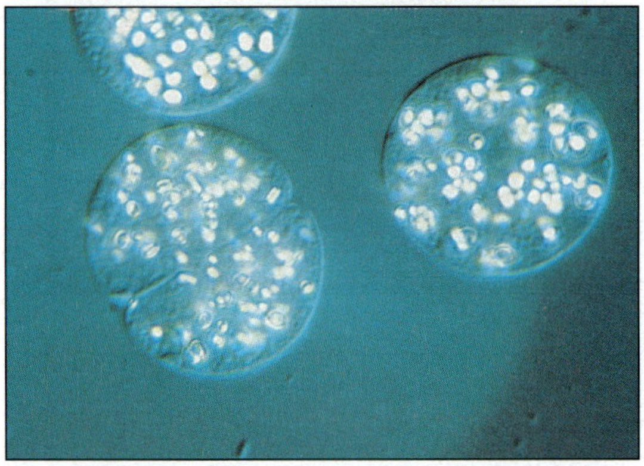

Color Plate 12 A close-up of *Thiovulum* species. The bright inclusion bodies are sulfur granules.

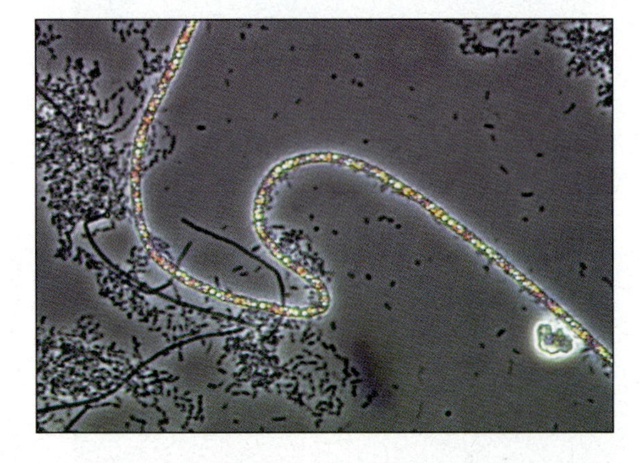

Color Plate 13 Another look at a bacterium that contains iridescent granules of sulfur and other compounds. One such bacterium is called *Beggiatoa*, and looks like a string of Christmas lights under a phase-contrast microscope. The identity of this particular bacterium is not known. Courtesy of Rolf Schauder.

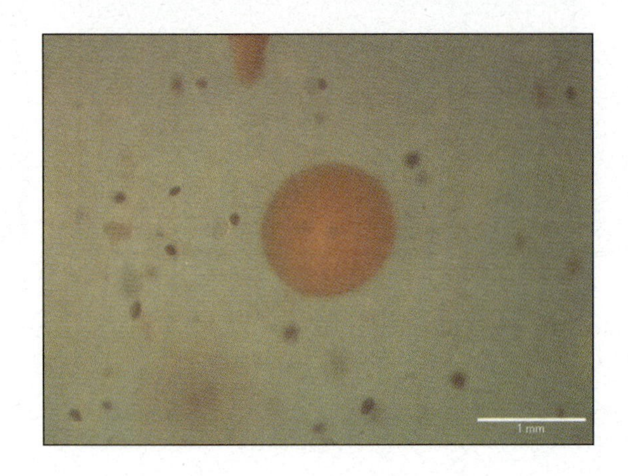

Color Plate 14 A closeup of some colonies of phototrophs growing in an agar shake culture. The large orange colony is surrounded by smaller red colonies suspended in the agar. The color comes from the photosynthetic pigments. Courtesy of Rolf Schauder.

key terms

biotechnology use of genetic engineering and related DNA-based techniques to develop new products and processes

BT corn genetically engineered corn that contains genes encoding insecticidal *Bacillus thuringiensis* toxin

expression vector cloning vector in which the human DNA is inserted such that it is under the control of a strong bacterial promoter and has a ribosome binding site in the messenger RNA (mRNA) transcript that is recognized by bacterial ribosomes

gene therapy introduction of a normal gene to cure human genetic defects

genetic engineering use of a variety of techniques to form a new DNA molecule, usually carried on a plasmid, that is introduced into a foreign host in which it can be transcribed and translated

introns regions of noncoding DNA found interspersed in coding regions of eukaryotic (but not bacterial) genes

Questions for Review and Reflection

1. What sort of change in the gene of interest would a scientist have to make to force a bacterium to secrete a protein out of the cytoplasm? What is an advantage other than preventing inclusion body formation of having a bacterium export a protein to the extracellular fluid?

2. How would you design the primers used for reverse transcriptase-polymerase chain reaction (RT-PCR) in Figure 25.2?

3. In Chapter 14, conjugated vaccines were described (carbohydrate residue linked to a protein). Could such a vaccine be produced by the same sort of cloning approach described for human insulin? Explain.

4. If you were trying to construct a strain of bacteria that produces a human protein, what are the steps in altering the gene that will affect the amount of protein produced? What step(s) will affect the ease with which the protein can be harvested from the bacterial culture?

5. A protein encoded by a cloned gene makes the bacteria producing the protein grow very slowly. Why would a scientist want the bacterium to produce this protein only when the bacterium enters the late exponential phase? What is a possible safety problem that could arise if the bacteria produce the protein throughout their growth?

6. What is an expression vector and under what conditions might you want to use one? Would it be useful to have an expression vector that contains a promoter that is expressed in mammalian cells? How and for what purpose would such a vector be used?

7. Most of the currently available plants containing *Bacillus thuringiensis* toxin also have an antibiotic resistance gene from the cloning vector. Will this gene make the plant cell resistant to the antibiotic? What would be the steps involved in transferring this gene to intestinal bacteria and why did scientists conclude that such a transfer event was highly improbable? (Hint: review Chapter 7 before answering this question.)

Glossary

A (aminoacyl) site site on ribosome where aminoacyl-tRNA initially binds

A-B type toxin protein toxin composed of two types of subunits: B portion of toxin is responsible for binding to target cell and A portion mediates enzymatic activity

abscess damaged, walled-off area containing pus

acellular pertussis vaccines (aPs) subunit vaccines consisting of inactivated pertussis toxin plus a surface adhesin of *Bordetella pertussis*

acellular vaccine vaccine made of proteins derived from microorganisms

acetogens bacteria in hindgut of termite that produce acetate for use by termite

achlorhydria decreased levels of HCl in the stomach result in decreased acidity

acid-fast staining a harsh method for staining; red dye fuchsin used

acid tolerance response after exposure to slightly acidic conditions, some bacteria become able to withstand much lower pH environments for prolonged periods; permits them to transit the stomach without being killed

acquired immunodeficiency syndrome (AIDS) caused by HIV

actin a major protein component of host cell cytoskeleton

activated macrophages macrophages with increased reactive oxygen intermediates and other toxic compounds; have increased killing capacity

activator protein that enhances binding of RNA polymerase to the promoter and initiation of transcription

active immunity specific immunity induced by infection or immunization

active transport process in which some bacteria take up sugars without phosphorylating them; sometimes protons are cotransported with the sugar; requires energy

acute illness abrupt, severe illness

acute infectious nonbacterial gastroenteritis (AING) a self-limited condition characterized by fever, vomiting, malaise, and diarrhea; caused by viruses

acute phase protein a group of serum proteins produced by the host in response to infection; a form of induced nonspecific immunity

acute respiratory distress syndrome (ARDS) accumulation of fluid and neutrophils in lung; lung damage and decreased gas exchange result

acyclovir nucleotide analog that can inhibit viral DNA synthesis; effective in controlling HSV

ADCC *see* antibody-dependent cell-mediated cytotoxicity

adenosine phosphosulfate (APS) energy-rich compound used to power reduction of sulfate to sulfite in dissimilatory sulfate reduction

adenosine triphosphate (ATP) high-energy phosphorylated compound used to store and provide energy to cellular processes

adenylate cyclase enzyme that catalyzes synthesis of cAMP; stimulated by LT of ETEC

adherence attachment to surfaces; can be mediated by pili, bacterial surface proteins, or capsular polysaccharides

adhesins components on microbial surfaces that bind to the host cell receptor

adhesive tip structure the tip of the pilus that binds to a receptor on host cells; often composed of peptides different from those in shaft of pilus

adjuvant a substance that enhances immune response to an antigen

ADP-ribosylation transfer of ADP-ribosyl from NAD to host cell protein; common action of A-B type toxins

aerobic oxygen is required for growth

aerosol a fine mist such as that produced by sneezing or coughing

aerotolerant anaerobes anaerobes that are not killed by oxygen; example is lactic acid bacteria

affinity strength of interaction of one antigen-binding site and one epitope

afimbrial adhesins surface proteins of bacteria important for adhesion but not organized in pilus-like structures

aflatoxin toxin produced by *Aspergillus* (a fungus); causes liver cancer

agglutination clumping of cells by specific antibody; a type of serological test

AIDS *see* acquired immunodeficiency syndrome; caused by HIV

AING *see* acute infectious nonbacterial gastroenteritis

algae photosynthetic eukaryotic microbes with rigid cell walls; recently evolved

alginate polysaccharide coating of *Pseudomonas aeruginosa*

α-hemolysis partial hemolysis of red blood cells in blood agar plates; results in greenish color in area surrounding bacterial colonies

α-hemolytic streptococci streptococci that are part of the microbiota of the mouth; partially hemolyze red blood cells in agar medium (greenish color surrounding bacterial colony)

alternative complement pathway one arm of the nonspecific immune system; complement components bind to components on the bacterial surface

alum an aluminum salt; only adjuvant licensed for use in the United States

alveolar macrophage a type of macrophage fixed in the alveoli of the lung; not activated

alveolus sac-like structure of the lung where gas exchange occurs

amantidine anti-influenza drug that inhibits viral uncoating

aminoglycosides type of antibiotics that prevent translation by binding to the 30S ribosomal subunit; includes kanamycin, gentamicin

aminoglycoside-modifying enzymes enzymes produced by bacteria that change aminoglycosides so they no longer bind to the ribosome

ammonia breath test test used to detect *Helicobacter pylori* in stomach; urease of *H. pylori* catalyzes formation of ammonia from urea in stomach

amoeba type of protozoa that moves by formation of pseudopods

amoeboflagellates an early evolving group of protists; have both amoeboid and flagellated forms; example is *Naegleria fowleri*

anaerobe microorganism that grows without oxygen; obligate anaerobes grow only in an environment totally free of oxygen

anaerobic complete lack of oxygen

anergy loss of reactivity to an antigen

anorexia loss of appetite

anoxygenic photosynthesis photosynthesis that transforms light energy into chemical energy using compounds such as sulfide as electron donors; used by purple and green bacteria

anthrax disease of farm animals, caused by spores of *Bacillus anthracis;* can cause fatal disease in humans; of interest to bioterrorists

antibiogram table showing antimicrobial susceptibility patterns of a group of microbes commonly isolated in a specific hospital or community

antibiotic chemical compound that can kill microorganisms or inhibit their growth

antibody protein complex (an immunoglobulin) that interacts with an antigen; produced by B cells; recognizes specific epitope on an antigen

antibody-dependent cell-mediated cytotoxicity (ADCC) phagocytes attracted to bacteria coated with IgG; release toxic compounds; kill bacteria

anticodon nucleotide triplet in tRNA that joins with complementary codon of mRNA during translation

antigen substance that interacts with a specific antibody (*see* immunogen); contains epitopes to which antibodies bind

antigen-binding site portion of the Fab region of an antibody that binds to an epitope

antigen-presenting cell (APC) a cell (such as a macrophage) that engulfs a microbe or its products, degrades the proteins and presents the resulting peptides on its surface

antigenic determinants site on an antigen that binds to the antigen binding site of an antibody

antigenic drift small changes in influenza virus genome that allow virus to partially avoid host defenses

antigenic shift reassortment of segments of influenza genome that make the virus completely unrecognizable by host immune system

antigenic variation ability of some bacteria to change the amino acid composition of their adhesins or other surface proteins

antimicrobial compound that inhibits the growth of or kills microorganisms

antimicrobial susceptibility testing means of determining whether a microbe is resistant or susceptible to a panel of antimicrobial agents (microbroth dilution MIC test, disk diffusion test, E test)

antiserum serum containing antibodies against a particular antigen

antitoxin antibody that will bind to a toxin

antiviral agents chemicals that interfere with viral replication cycle

aP *see* acellular pertussis vaccine

APC *see* antigen-presenting cell

apical portion of an epithelial cell that faces the lumen

apicomplexa group of protists with complex life cycle with both sexual and asexual forms

apoptosis programmed cell death in which cell shuts down biosynthetic functions; occurs naturally or may be triggered by cytotoxic cells; some viruses stimulate the process, others prevent it from occurring

APS *see* adenosine phosphosulfate

arabinogalactan-lipid complex component of mycobacterial cell wall; a glycolipid

archaea prokaryotic microbes with a cytoplasmic membrane that differs from that of bacteria and eukaryotes; includes microbes that can grow at extreme temperatures and those that make methane

archaezoa microbes that probably resemble earliest eukaryotes; some members (diplomonads and parabasalians) have lost mitochondria; may live under anaerobic conditions

arachidonic acid precursor of leukotrienes; component of inflammatory cascade

ARDS *see* acute respiratory distress syndrome

artificial transformation transformation induced by chemical or electrical shock in bacteria that do not normally take up free DNA

aseptic containing no microorganisms

aspiration introduction of fluids into, or removal of fluids from, body cavities

assimilatory sulfate reduction process carried out by many bacteria; used to produce cysteine; not used for energy production

asymptomatic carriage (or infection) disease-causing bacteria colonize but do not cause detectable symptoms

ATP *see* adenosine triphosphate

ATP synthase protein complex in the bacterial cell membrane; allows protons to reenter cytoplasm, driving phosphorylation of ADP to ATP

attachment first step in viral infection; virus binds to specific receptor on host cell

attack rate number of cases of clinically apparent disease divided by number of susceptible people in the population

attenuation (1) decrease in virulence of microorganisms used in a vaccine, or (2) genetic regulation involving stem-loop formation in RNA

AUG start codon on mRNA; encodes methionine

autoclaving sterilization by steam heat (121°C) under high pressure

autoimmune response immune system recognizes a host molecule as foreign

autoinducer signaling molecule in quorum sensing systems; in high concentrations binds repressor or activator and stimulates gene expression

autolysins bacterial enzymes that digest peptidoglycan; can cause lysis of bacteria

autotroph microbe that obtains carbon from carbon dioxide

auxotroph a microorganism that cannot synthesize some essential nutrient

avirulence genes bacterial genes encoding proteins recognized by plant R proteins that trigger the plant protective hypersensitive response

axial filament internal flagella of spirochetes located between cytoplasmic membrane and outer membrane

azido-deoxythymidine (AZT) an anti-HIV drug; acts by inhibiting viral reverse transcriptase

AZT *see* azido-deoxythymidine

B lymphocyte (B cell) lymphocyte that produces antibodies

bacillus rod-shaped bacterium

Bacillus of Calmette-Guerin attenuated *Mycobacterium bovis,* widely used outside the United States as a vaccine to prevent tuberculosis; stimulates activated macrophage response; efficacy variable

bacteremia bacteria present in the bloodstream

bacteria prokaryotic microbes with rigid cell walls

bacterial vaginosis vaginal inflammation caused by shift in members of vaginal microbiota; may cause premature births

bactericidal substance that kills bacteria

bacteriochlorophyll type of chlorophyll used by purple and green phototrophs

bacteriocin protein produced by one strain of bacteria that kills sensitive members of closely related strains

bacteroid form of *Rhizobium* and *Bradyrhizobium* species that are endosymbionts of plant root cells; differentiated to reduce nitrogen gas to ammonia in root cells

bacteriophage virus that infects bacteria

bacteriostatic substance that inhibits the growth of bacteria

bacteriuria bacteria present in the urine

bactoprenol complex lipid (containing 55 carbons) that transports peptidoglycan subunits from inner to outer face of cytoplasmic membrane; matrix on which second backbone sugar is added in peptidoglycan synthesis

bartonellosis common in immunocompromised patients; disease carried by cats; transmitted to humans by cat scratches contaminated with flea feces

basal lamina matrix of glycoproteins to which epithelial cells attach

basolateral surface the portions of epithelial cells in a confluent monolayer that are attached to each other and to the basal lamina; portions in contact with extracellular matrix

bc complex the electron transport system of purple bacteria

BCG *see* Bacillus of Calmette-Guerin

bead beater machine that uses glass beads to disrupt any type of microbe

β-galactosidase (LacZ) an enzyme commonly used as a reporter group

β-glucan polysaccharide component of the cell wall of many fungi

β-hemolytic complete hemolysis of red blood cells in blood agar; results in clear area around bacterial colonies

β-integrins proteins on surface of neutrophils that bind to endothelial cells

β-lactam antibiotics antibiotics that contain a β-lactam group and act by inhibiting peptidogylcan synthesis; include penicillins, cephalosporins, carbapenems, and monobactams

β-lactamase an enzyme that cleaves the β-lactam ring of β-lactam antibiotics and thus inactivates the antibiotics

bidirectional replication bacterial chromosomes and some plasmids are copied in two directions starting from the replication origin

bile salts detergents found in bile; disrupt bacterial membranes

binary fission division process in which each cell divides into two cells; population numbers increase exponentially

biofilm multilayer bacterial populations embedded in a polysaccharide matrix attached to some surface (such as a plastic implant, ship's hull)

bioprospector scientist who samples many diverse environments in a search for new microbes

biotechnology use of genetic engineering and related DNA-based techniques to develop new products and processes

bioterrorist person who uses microbes (or toxins) to disrupt normal societal functions

biotype species subtype with biochemical property that differs from other subtypes

bismuth metallic compound with some antibacterial activity that reduces gastric inflammation; used along with antibiotics in treatment of gastric ulcers

blackwater fever urine of malaria patients may turn dark brown because of excretion of high levels of hemoglobin from lysed red blood cells

blood-brain barrier membrane (meninges) covering of brain and spinal cord; prevents substances in blood from entering

boil abscess resulting from infection of a hair follicle

bone marrow stem cell precursor cells from which neutrophils and macrophages descend

botulinal toxin neurotoxin produced by *Clostridium botulinum;* causes botulism

botulism food-borne disease caused by toxin (not bacteria) present in food; caused by *Clostridium botulinum*

breakbone fever excessive muscle and joint pain that occurs in dengue fever

broad-host-range transfer transfer of DNA by bacteria across species and genus lines

broad-spectrum antimicrobial agent agent that is effective against a variety of species

bronchoscopy insertion of thin tubular instrument (bronchoscope) for examination, sampling, or treatment of trachea or bronchi

Brownian motion movement of microbes because of flow of liquid resulting from pressure on cover slip; may be mistaken as motility

brucellosis disease that targets reproductive system of domestic animals; caused by *Brucella* spp.; causes systemic disease in humans

bubonic plague disease caused by *Yersinia pestis;* transmitted from rodents (or other animals) to humans by fleas

budding (1) method of exit from host cell by many enveloped viruses, or (2) method of replication by yeasts

C3 a component of complement that is cleaved to C3a and C3b

C3 convertase complexes that convert C3 to C3a and C3b

C3a complement component that results from the proteolytic cleavage of C3; acts as a vasodilator

C3b complement component that results from the proteolytic cleavage of C3; opsonizes bacteria

C5 component of complement that is cleaved to C5a and C5b

C5 convertase complex consisting of C3 convertase + C3b; converts C5 to C5a and C5b

C5a complement component that results from the proteolytic cleavage of C5; acts as a chemoattractant for neutrophils; acts as a vasodilator

C5b complement component that results from cleavage of C5; recruits C6, C7, C8, and C9 to form MAC

Calvin cycle process used by microbes and plants to incorporate carbon dioxide into biomass; consumes large amounts of ATP and NADPH

capsid tightly packed proteins surrounding the viral nucleoprotein core; made up of capsomeres

capsomere polypeptides that form the capsid of viruses

capsule fibrous network (usually polysaccharide) that covers the cell wall of some bacteria

carbapenem a class of β-lactam antibiotics

carrageenan polysaccharide produced by algae; used as thickener in ice cream, salad dressing, and other products

carrier apparently healthy person who harbors pathogenic microorganisms

case fatality rate number of deaths caused by a disease divided by the number of clinically apparent cases of the disease

caseation tissue necrosis that resembles cheese; occurs in tuberculosis

caseous resembling cheese

catalase bacterial enzyme that helps convert superoxide to water; also activates isoniazid, an anti-tuberculosis drug

catheter tube for draining fluids from or introducing them into body cavities

CD4 receptor on helper T cells; facilitates response of T cell to epitopes displayed on MHC II

CD8 receptor on cytotoxic T cells; facilitates response of T cell to epitopes displayed on MHC I

CD14 receptor on surfaces of monocytes and macrophages that binds to compounds specific to bacteria such as LPS, peptidoglycan, and LTA

cell envelope peptidoglycan layer of gram-positive bacteria or peptidoglycan layer + outer membrane of gram-negative bacteria

cell line eukaryotic cells that have been "immortalized" so that they continue to divide

cell-mediated immunity (CMI) acquired immunity as a result of T cells and activated macrophages; does not involve antibodies

cell wall structure surrounding prokaryote cells; prevents cell from rupturing under osmotic pressure

cellulitis inflammation of connective tissues

cellulose plant polysaccharide that can be degraded by colonic bacteria

cephalosporins a class of β-lactam antibiotics

cGMP cyclic GMP; a signaling molecule; in excess amount causes cell to lose control of ion pumps

chancre primary lesion of syphilis

charging of tRNA covalent attachment of amino acid to tRNA; catalyzed by tRNA synthetase

chemoattractant a substance that stimulates motility in such a way that the bacteria move up a gradient of that substance

chemokines proteins produced by many human cell types (especially endothelial cells, macrophages, and cells of the specific defense system); organize the activities of cells of the specific and nonspecific defenses

chemoprophylaxis use of drugs to prevent disease

chemotaxis attraction of microbes or host cells (for example, neutrophils) to a specific substance

chemotherapy treatment of disease with drugs

chimera PCR products from different genes that have hybridized with each other

chloramphenicol antibiotic that inhibits translation by binding to the ribosome

chlorophyll green photosynthetic pigment used in the conversion of light energy to chemical energy by oxygenic phototrophs and green plants

chloroplast organelle of plant cells; site of photosynthesis and energy generation

cholera epidemic diarrheal disease; caused by *Vibrio cholerae*

cholera toxin A-B type exotoxin produced by *Vibrio cholerae*, which ADP-ribosylates a G protein that activates adenyl cyclase

chromogenic substrate substrate that changes color on plates or in liquid medium if a particular enzyme is present

chronic persisting over time

chytrid fungus type of fungus that damages the skin of frogs causing massive die-offs

cilia surface structures of eukaryotic cells that move mucus over surfaces in vertebrates; responsible for motility in some protozoa

ciliated cells mucosal cells with cilia on their apical surfaces; commonly found lining respiratory tract; action propels blobs of mucin containing trapped microbes out of the airway

circadian rhythm biological cycle with a periodicity of 24 hours; seen in some unicellular cyanobacteria as well as eukaryotes; allows the cyanobacteria to separate nitrogen fixation and photosynthesis temporally

cirrhosis chronic liver disease that interferes with normal liver function; ultimately fatal

citric acid cycle conversion of pyruvate to carbon dioxide and water; molecules of NAD and FAD reduced and one molecule of ATP produced

classical complement pathway complement pathway that is activated by antigen-antibody complexes

clavulanic acid an inhibitor of β-lactamase

clone (1) identical cells derived from a single cell, or (2) identical segments of DNA copied from a single piece of DNA

cloning of a gene making multiple copies of a piece of DNA containing a gene of interest

cloning vector circular plasmid designed to receive DNA segments for cloning; most have a high plasmid copy number

CMI cell-mediated immunity

coagulation pathway series of steps leading to the formation of clots; some components contribute to hypotension, shock

coccobacillus short oval bacterium

coccus sphere-shaped bacterium

codon nucleotide triplet in mRNA specifying an amino acid

cognate B cell B cell with MHC II-epitope complex identical to MHC II-epitope on a T helper cell

collagen protein fibers of connective tissue

collagenase enzyme released by lysosomes; degrades collagen

colitis inflammation of the colon

colonic fermentation fermentation of plant and human polysaccharides by colonic bacteria; end products are acetate, propionate, and butyrate

colonization ability of a bacterium to remain at a particular site and multiply there

colony discrete mass of cells derived from a single cell

columnar cells tall, thin epithelial cells

commensal association symbiotic association beneficial to one partner and neither beneficial nor harmful to the other

community-acquired infections infections acquired by otherwise healthy people who are not hospitalized

complement cascade series of steps in which various complement components are activated

complement system group of plasma proteins which, when activated, lead to proteolytic cleavage of complement components producing activated proteins that attract phagocytes, cause lysis of gram-negative bacteria, and opsonize bacteria; major components are C3b, C5a, and MAC

complementation test a test to determine whether two different mutants with the same phenotype are the result of mutations in the same gene

confluence cells cover an entire area with no space between them; used to describe tissue culture cells or bacterial cells growing on agar plates

conidiospores a type of spore found in sac-like structures of some fungi

conjugate vaccine polysaccharide covalently linked to protein; forces polysaccharide to be processed as a protein through T-dependent pathway

conjugation transfer of DNA from one bacterium to another via a conjugation bridge

conjugative plasmid plasmid that carries genes encoding proteins needed for the conjugation process

conjugative transposon transposon that can excise itself from the chromosome of the donor to form a circular intermediate that is transferred to the recipient, where it integrates into the chromosome; does not replicate itself

conjunctivitis inflammation of membranes of eyes and eyelids

consensus sequence sequence in promoter region on DNA to which RNA polymerase binds; different sequences found in different groups of microorganisms

consortia complex communities of microbes that cannot be isolated in pure culture; more than one microbe required to carry out a particular reaction

constitutive continuously produced; not regulated

convalescence time of recovery

cord factor mycolic acids from cell walls of *Mycobacterium* spp.

core enzyme the basic structure of RNA polymerase; consists of four subunits (α, α, β, β')

coreceptor secondary receptor proteins on host cell surfaces to which certain viruses, such as HIV, must bind in order to invade

crypts depressions in the intestinal mucosa in which stem cells divide and differentiate into mucosal cells

crypt stem cells progenitors of mucosal cells

cryptdins set of peptides toxic to bacteria; produced by Paneth cells found in intestinal crypts

culture microorganisms growing in liquid or on solid medium

cutaneous anthrax form of anthrax localized in skin; lesion may be the result of edema toxin produced by *Bacillus anthracis*

cutaneous infection infection that occurs in the upper layer of skin; not invasive

cyanobacteria first bacteria to practice oxygenic photosynthesis (water split, oxygen released into atmosphere); later became chloroplasts of plants

cyclic AMP (cAMP) regulatory molecule; controls ion pumps that keep water flowing in correct direction through cells

cyclic GMP (cGMP) regulatory molecule; increased levels cause ion pumps to cease functioning

cyst survival form of some protozoa; analogous to fungal spores and bacterial endospores

cystic fibrosis disease caused by defect in chloride secretion; characterized by production of thick mucin in lungs

cystitis infection of the urinary bladder

cytokines bioactive proteins produced by some mammalian cells, especially endothelial cells and macrophages, in response to stimuli; mediators of inflammation, septic shock

cytopathic effects characteristic changes in host tissues or tissue culture cells produced by viruses; may be helpful in identifying virus

cytoplasm interior of cell; contains genome, ribosomes, organelles, and enzymes that generate energy and perform biosynthetic reactions

cytoplasmic membrane phospholipid bilayer; in bacteria contains proteins involved in substrate transport, energy generation, peptidoglycan synthesis, secretion of proteins, and sensing of environmental signals

cytoskeleton complex array of proteins that give shape to eukaryotic cells

cytotoxic T cells T cells (with CD8 antigen on their surfaces) that kill host cells displaying specific foreign epitopes on their surfaces

cytotoxin a toxin that kills mammalian cells

database search search for sequence similarity of a given gene or protein to sequences on file in databases

dead zone area in ocean where oxygen levels have dropped so low that marine animals cannot survive

death phase stage of growth curve where bacteria begin to lyse and number of cells decreases

debilitation loss of health

debride remove dead tissue

defensins small lysosomal peptides that kill microbes by forming channels in the microbial membrane

deletion loss of genetic material

$\Delta G^{0\prime}$ free energy change; sum of energies of products minus energies of reactants; if it is negative, reaction will proceed; if it is positive, reaction will not proceed or will go in reverse

dendritic cells antigen-presenting cells of the dermis and lymphoid tissue

dengue fever viral disease transmitted from human to human by *Aedes aegypti* mosquitoes; dengue hemorrhagic fever is most serious form

denitrifiers bacteria that reduce nitrate and nitrite to nitrogen gas

denitrification reduction of nitrate to nitrous oxide and nitrogen gas by soil bacteria

deoxyribonucleic acid (DNA) double-stranded helix held together by hydrogen bonds between bases attached to a deoxyribose phosphate backbone; genetic material of all cellular organisms

dermis connective tissue layer below the epidermis of skin

desmosomes areas that serve as sites of adhesion of adjacent human cells, for example, epithelial cells

desquamation shedding of epithelial layers of skin

dexamethasone anti-inflammatory steroid; sometimes used in treatment of meningitis

diagnosis determination of the cause of an illness

diarrhea abnormal fluidity of stool

diatoms protists that have recently evolved; important component of phytoplankton; photosynthetic

DIC *see* disseminated intravascular coagulation

***Dictyostelium* (slime mold)** social amoebas that may associate into slug-like form that can eventually develop into stalk and fruiting body

dideoxynucleotide altered nucleotide that lacks a 3′-OH group; if incorporated into DNA, polymerase stops synthesizing DNA; used in DNA sequencing

differential medium medium that contains an indicator that changes color depending on the metabolic activity of microbes growing on it

dimorphic fungus a pathogen that grows in mycelial form in the environment (<30°C) but in yeast form in the body (37°C)

dinoflagellates protists with two flagella; many are photosynthetic; example is *Pfiesteria,* which produces a neurotoxin

diphtheria disease caused by toxin produced by *Corynebacterium diphtheriae;* may lead to heart failure

diphtheria toxin toxin producted by *Corynebacterium diphtheriae;* ADP-ribosylates mammalian elongation factor-2 (EF-2); stops protein synthesis

diplococci cocci arranged in pairs, for example, *Streptococcus pneumoniae*

diplomonads a type of protist that lacks both mitochondria and golgi bodies; reproduces asexually; characterized by two nuclei and two sets of four flagella; example is *Giardia*

disease specific symptoms arising from colonization of the body by a pathogenic microbe

disinfection treatment of inorganic surfaces to reduce number of microbes

disk diffusion test test for determining the relative susceptibility of a microbe to antibiotics; disks containing antibiotic are placed on a lawn of bacteria and incubated; clear zones indicate susceptibility

disseminated intravascular coagulation (DIC) formation of numerous small clots that obstruct peripheral blood vessels; symptom of septic shock

dissimilatory sulfate reduction process used to produce energy via electron transport system that uses sulfite as an electron acceptor; bacteria using this pathway are obligate anaerobes

DNA *see* deoxyribonucleic acid

DNA gyrase enzyme that mediates supercoiling of DNA

DNase deoxyribonuclease; an enzyme that degrades DNA

DNA hybridization test test in which a specimen is treated to release and denature DNA from cells; DNA is immobilized and single-stranded labeled DNA probe is added that will bind only to a DNA sequence found in a specific microbe

DNA ligase enzyme used by bacteria to connect segments of DNA produced during replication; seals gaps in DNA strands; in genetic engineering used to seal cloned DNA into cloning vector

DNA polymerase enzyme that catalyzes the synthesis of new DNA strands during DNA replication

DNA probe labeled single-stranded DNA from known gene; used in DNA hybridization assays

dot blot *see* spot blot

DPT vaccine trivalent vaccine against diphtheria, pertussis, and tetanus

dysentery a type of diarrhea in which stools contain blood and mucus

E'_0 oxidation-reduction potential of a redox couple; electrons flow from redox pair with more negative potential to one with more positive potential

E (exit) site site on ribosome from which empty tRNA leaves

E test quantitative variant of the disk diffusion test; gives MIC

ectopic pregnancy fetus starts developing in blocked fallopian tube or peritoneal cavity instead of uterus

edema excessive fluid in the tissues

edema toxin toxin produced by *Bacillus anthracis;* responsible for cutaneous anthrax

EF-2 *see* elongation factor-2

efflux mechanism cytoplasmic membrane protein mediates resistance to tetracycline, macrolides, and quinolones by pumping them out of the bacterial cytoplasm

EHEC *see* enterohemorrhagic *E. coli*

Ehrlichiosis acute febrile disease caused by *Ehrlichia* spp.; transmitted from a variety of animals (for example dogs) to humans by ticks

elastase enzyme that degrades elastin, a component of the extracellular matrix; may be important in causing lung damage in *Pseudomonas aeruginosa* infections

elastin protein that accounts for 30% of protein in lung tissue; also part of blood vessel walls

electron microscope microscope with greater resolving power than light microscope; can achieve 1000× magnification and higher

electron transport system complex of proteins and lipid quinones in bacterial cytoplasmic membrane that generates energy by separating protons from electrons and dumping electrons on terminal electron acceptor

electroporation introduction of DNA or proteins into a cell by using a sudden increase in electric field to increase permeability of cells

elementary body survival form of *Chlamydia* spp. that does not replicate; infective form

ELISA *see* enzyme-linked immunosorbent assay

elongation factor-2 protein that plays essential role in host cell protein synthesis; target of diphtheria toxin

emerging infectious diseases new diseases that appear; often due to increased human contact with microbe that has been around for years

emetic substance that induces vomiting

empiric therapy early treatment of a disease based on tentative diagnosis

endemic disease continually present at low levels in the community

endocarditis inflammation of heart valves

endocytosis engulfment of extracellular material into a vacuole by a eucaryotic cell

endometritis infection of the uterus

endoplasmic reticulum set of membranes in eukaryotic cells; studded with ribosomes; site of protein synthesis

endospore nonreplicating, metabolically inactive survival forms of some bacteria

endosymbiont microbe that lives inside the cells of its symbiotic partner

endothelial cell type of cell lining blood vessels and the heart; not tightly bound to each other; produces cytokines

endotoxemia endotoxin in the blood stream; symptoms include fever, chills, weight loss, shock

endotoxin *see* lipopolysaccharide

enrichment procedure use of selective liquid media to allow a subset of microbes with desired trait to proliferate

enteric relating to the gastrointestinal tract

enterohemorrhagic *E. coli* (EHEC) cause dysentery-like disease but do not invade host cells; may produce Shiga-like toxin which causes kidney damage

enteropathogenic *E. coli* (EPEC) bacteria produce ultrastructural changes in small intestinal mucosal cells; cause malabsorptive diarrhea; major killer of infants in developing countries

enterotoxin an exotoxin that acts specifically on the intestinal mucosa

enterotoxigenic *E. coli* (ETEC) produces two toxins, one a cholera-like protein toxin (LT) and the other a peptide hormone-like toxin (ST); causes secretory diarrhea

env polyprotein envelope proteins of HIV; cleaved to form gp41 and gp120

envelope layer of phospholipids and proteins surrounding some viruses

envelope proteins viral proteins embedded in a phospholipid bilayer

enveloped virus virus that has an envelope

enzyme-linked immunosorbent assay (EIA; ELISA) use of antibodies labeled with an enzyme to detect antigens or antibodies in the serum

eosinophil nonspecific cytotoxic cell that targets metazoal parasites

EPEC *see* enteropathogenic *E. coli*

epidemic disease that appears sporadically and affects many individuals in a community

epidemiology study of incidence, transmission, and prevention of disease

epidermis outermost layer of skin consisting of stratified squamous cells, mostly keratinocytes; a nonspecific defense

epididymis part of the sperm-collecting apparatus

epididymitis infection of the epididymis

epiglottitis disease caused by *Haemophilus influenzae* that can close airway and lead to suffocation

epithelium layers of cells that cover the surface of the body and body cavities

epitope portion of an antigen recognized by an antigen binding site of an antibody; usually 5–8 amino acids in the case of protein antigens

ergosterol major sterol found in the cell membrane of fungi

ergot substance produced by fungi; active ingredient in the hallucinogenic drug LSD

***erm* genes** encode a methylase that modifies bacterial ribosomes so that they are not inhibited by erythromycin

erythema red color resulting from dilatation of blood vessels

erythema migrans ring-shaped rash of Lyme disease that expands and moves outward from site of inoculation

eschar mass of necrotic tissue

ETEC *see* enterotoxigenic *E. coli*

ethambutol antibiotic effective against mycobacteria; probably inhibits mycolic acid synthesis

ethidium bromide stain that intercalates in DNA backbone; fluoresces under ultraviolet light

etiology study of factors that cause disease and their introduction into a host

euglenoids protists with a single nucleus and a single flagellum; photosynthetic members have an eyespot that mediates phototaxis

eukaryote organisms in which the DNA genome is enclosed in a nuclear membrane

exospore type of fungal spore that occurs outside the cell

exosymbiont microbe that lives outside the cells of its symbiotic partner

exotoxin protein toxin produced by bacteria; usually secreted into the extracellular fluid but can be intracellular

exotoxin A toxin produced by *Pseudomonas aeruginosa;* has same mechanism of action as diphtheria toxin

exponential phase phase of rapid division of bacteria that occurs after lag phase

extracellular matrix protein-polysaccharide material in which mammalian cells are embedded; contains collagen, hyaluronic acid, elastin, and other compounds

extravasation *see* transmigration

extremophiles organisms that can live at extreme temperatures (from near boiling to near freezing) or extreme pH values (0–10)

exudate fluid and cells that have escaped from blood vessels

eyespot organelle in some protists, such as the euglenoids, that mediates phototaxis

F protein a surface adhesin protein produced by *Streptococcus pyogenes;* binds fibronection

Fab portion of an antibody that contains the antigen-binding site

facultative bacteria those that can use either fermentation or respiration to obtain energy, depending on whether oxygen is present or not

facultative intracellular pathogen microbe that can grow extracellularly but during an infection grows intracellularly (protected from host defenses)

false negative negative test result when the looked-for microbe is present

false positive positive result when the looked-for microbe is not present

fastidious microorganism one that requires a rich mixture of sugars, vitamins, and amino acids

Fc portion of an antibody that binds complement and attaches the antibody-antigen complexes to phagocytes; mediates opsonization; region of antibody bound by some streptococcal and staphylococcal surface proteins

febrile having a fever

feedback inhibition form of post-translation control in which the end product of a series of enzymes in a pathway binds one of the enzymes and prevents it from acting; prevents further synthesis of end product

fermentation use of organic compound as an electron acceptor; used in ETS and some reduction reactions

ferredoxin electron carrier involved in photosynthesis

ferritin intracellular iron storage protein of mammalian cells

fertility incompatibility result of an infection of *Drosophila* by *Wolbachia;* makes infected males reproductively incompatible with uninfected females

fimH a protein on the tips of type 1 pili; recognize mannose receptors on epithelial cells

fimbriae fibrils on surface of bacteria; also called pili

FISH *see* fluorescent in situ hybridization

fixed macrophages stationary macrophages found in tissues that filter blood or lymph

flagellum long, flexible helically shaped protein fibrils projecting from surface of cell; rotation provides motility for cell

flavonoids compounds produced by plants that serve as signals to *Rhizobium* species to start nodulation process or *Agrobacterium tumefaciens* to initiate transfer of T-DNA from bacteria to plant

fluorescence microscopy used to view microbes that can be activated by short wavelength of light to emit longer wavelengths; used for photosynthetic bacteria and methanogens or microbes stained with fluorescent dyes

fluorescent in situ hybridization (FISH) microbes on a slide are treated to allow fluorescently labeled probes to enter cell and hybridize with rRNA sequences; used to identify specific types of microbes in a sample; gives quantitative information on a particular types of microbe in the sample

fluoroquinolone a member of the quinolone family of antibiotics; stops DNA replication by binding to DNA gyrase

fMet *see* formylmethionine

follicle M cells and lymphoid cells in the GALT; collections of follicles called Peyer's patches

forced transformation use of electrical or chemical shock to force uptake of DNA by bacteria that normally would not do so

formaldehyde toxic compound that is intermediate in pathway used by methane-oxidizing bacteria

formylmethionine first amino acid in a protein; forms the N-terminus

frameshift mutation insertion or deletion of base in DNA; changes reading frame of DNA resulting in different amino acid sequence and different termination point

Freund's adjuvant mixture of oil and other components, including mycobacterial cell walls; used to stimulate antibody response to protein antigens

fruiting body an aggregate of many individual cells of *Dictyostelium*

fuchsin red dye used in acid-fast staining procedure

fulminating sudden and intense development

fungi eukaryotic microbes with rigid cell walls; includes yeast form and mycelial form; may be dimorphic

fusogenic pathway means by which viruses can move directly from one host cell to another without being released into the external environment

$\Delta G^{0'}$ free energy change; sum of energies of products minus energies of reactants; if it is negative, reaction will proceed; if it is positive, reaction will not proceed or will go in reverse

gag polyprotein protein that contains the capsid proteins of HIV

GALT *see* gastrointestinal-associated lymphoid tissue

gametes sexual forms of some protozoa; occur as microgametes and macrogametes

ganglion collection of nerve cell bodies; some viruses such as HSV spend latent phase here

gangrene death of tissue usually associated with loss of blood supply, bacterial invasion, and putrefaction

GAS group A streptococci (*Streptococcus pyogenes*)

gas vacuole vacuoles inside microbes that reflect light in phase contrast microscopy; gives Christmas tree light appearance

gastric pits deep depressions in the stomach mucosa of mammals; cells at the base of the pits secrete acid; pits also contain mucin-producing cells

gastroenteritis inflammation of the lining of the stomach or small intestine, especially caused by bacterial infection

gastrointestinal-associated lymphoid tissue (GALT) mucosal immune system found in follicles and Peyer's patches; characterized by production of secretory IgA; a component of MALT

G-CSF *see* granulocyte colony stimulating factor

gene amplification single copy of a gene is replaced by multiple copies

gene cassette DNA circle that integrates into specific sites on integrons; contributes to formation of gene clusters that are controlled by a single promoter

gene disruption mutants mutants generated by using single crossover homologous recombination to interrupt a gene so it no longer functions

gene expression process of going from DNA (the gene) to mRNA (transcription) and from there to protein (translation); can be regulated at each level; target of many antibiotics

gene fusion hybrid gene with promoter of one gene fused to a promoterless structural gene encoding an assayable enzyme (often β-galactosidase or green fluorescent protein)

generalized transduction phage replicating lytically in a bacterial cell may inadvertently package bacterial DNA in a phage head; this DNA may be introduced into another bacterial cell that the phage infects

generation time amount of time it takes a bacterium to divide

genetic engineering complex variety of techniques ranging from DNA hybridization and DNA sequencing to PCR and cloning that use bacterial enzymes and processes to manipulate the composition of DNA molecules

genome complete set of genes of an organism

genotype genetic constitution of an organism

gentamicin an antibiotic in the aminoglycoside family

genus taxonomic group of related species

germination conversion of a bacterium from a spore to a vegetative state

giardiasis diarrheal disease caused by *Giardia intestinalis;* ranges from mild to severe

gingival crevice area where the tooth emerges from the gum

gliding motility motility along surfaces seen in the case of cytophagas and myxobacteria; not mediated by flagella

globobiose receptor on host cells recognized by P pili of uropathogenic *Escherichia coli*

glomerulus specialized capillaries in kidney that filter blood

glycolysis anaerobic degradation of glucose to pyruvate; results in a gain of 2 ATP molecules

glycopeptides a group of antibiotics that inhibit peptidoglycan synthesis; include vancomycin and teichoplanin

goblet cells cells imbedded in the epithelium that secrete mucus

Golgi system organelle of eukaryotic cells that processes proteins destined for excretion and determines what route they take to their ultimate destination

gonococcus *Neisseria gonorrhoeae;* the cause of gonorrhea

gonorrhea a sexually transmitted disease caused by *Neisseria gonorrhoeae;* causes urethral infections in men and cervical infections in women

gp41 protein embedded in envelope of HIV; thought to stabilize gp120; mediates fusion of viral and host cell membranes before viral entry and uncoating; results from cleaving of Env polyprotein

gp120 protein on surface of HIV that mediates attachment to CD4 on T helper cells and monocytes; results from cleavage of Env polyprotein

gram-negative bacteria with thin peptidoglycan layer and outer membrane (contains protein pores and LPS); do not retain crystal violet dye during the gram-staining process; pink color

gram-positive bacteria with thick peptidoglycan layer containing LTA and no outer membrane; sometimes have S-layer covering peptidoglycan; retain crystal violet dye during the gram-staining process; purple color

Gram stain series of dyes applied to bacterial samples to divide them into one of two types (gram-positive or gram-negative)

granulocytes white blood cells containing granules, for example, neutrophils and eosinophils

granulocyte colony stimulating factor (G-CSF) a cytokine appearing early in infection; triggers release of granulocytes from bone marrow into circulation

granuloma tubercle that has been walled off and calcified

green bacteria bacteria with colonies that have a green color and which use a type of photosynthesis that produces NADPH and PMF

green fluorescent protein a popular reporter group that fluoresces with correct wavelength of light; permits scientists to see expression of a gene in a single cell

greenhouse effect volcanic activity that kept Earth from freezing by spewing carbon dioxide into atmosphere

guanyl cyclase enzyme that mediates formation of cGMP

Guillain-Barre (G-B) syndrome rare neurological disease; commonly caused by *Campylobacter* spp.

H antigen flagellar antigen; used to distinguish different strains of a bacterial species

HAART *see* highly active antiretroviral therapy

HACCP *see* hazard analysis and critical control point program

Hadean period time when Earth's surface was very hot; time when life may have begun

hair follicle small invagination in the epidermis; site of development of hair root

harpins proteins that stimulate plant hypersensitve response; encoded in *hrp* genes

hazard analysis and critical control point (HACCP) program program to identify steps in food processing stream where contamination might occur

HBV hepatitis B virus; transmitted sexually or by contaminated blood products; causes chronic hepatitis and may lead to cirrhosis or liver cancer

heat-labile toxin A-B type toxin produced by ETEC; similar to cholera toxin in mechanism of action and sequence; leads to excess production of cAMP

heat-stable toxin peptide toxin produced by ETEC; contributes to secretory diarrhea by stimulating production of cGMP

heavy chain larger of two types of protein making up an antibody molecule

helicase enzyme that separates strands of DNA during DNA replication

helminths metazoal parasites (worms)

helper T cells class of T cells (CD4 antigen on surface) that stimulates proliferation of specific B cells, activation of macrophages and cytotoxic T cells

hemagglutination clumping of erythrocytes

hemagglutinin substance that agglutinates erythrocytes; found in envelope of influenza viruses and others

hematogenous derived from or spread by blood

hematuria blood in the urine

hemolysis lysis of erythrocytes

hepatitis infection of the liver; several very different viruses can cause the disease; may be chronic or acute; may predispose patient to liver cancer (HBV)

herd immunity nonimmune persons are protected from a disease because most other people in the population have been vaccinated or have had the disease

heterotroph microbes that obtain carbon from complex organic molecules

Hib conjugate vaccine of *Haemophilus influenzae* type b capsule linked to a protein; protects against childhood meningitis

high-energy phosphorylated compounds compounds used to store or provide energy; examples are ATP and PEP

highly active antiretroviral therapy triple-drug therapy used to treat HIV infections; targets reverse transcriptase and protease

histone-like proteins proteins that may function to organize bacterial DNA in supercoiled regions

HIV *see* human immunodeficiency virus

homeostatis maintenance of equilibrium (for example, temperature, enzyme levels) by the body

homologous recombination process by which identical segments of DNA pair with each other and cross over so that one strand of DNA from one double-stranded molecule now serves as a complementary strand to a single strand from a second DNA molecule

homology DNA sequence similarity between two genes or amino acid sequence similarity between two proteins

horizontal gene transfer transfer and incorporation of DNA from one organism into another organism; DNA may be from different species

hospital-acquired infections infections acquired in the hospital; microorganisms tend to be resistant to more antibiotics than those acquired in the community, *also called* nosomial

host defenses mechanisms used by animals and humans to keep microbes from harming them

HPV *see* human papilloma virus

***hrp* genes** bacterial genes encoding proteins that form type III secretion system that injects proteins into plant cell cytoplasm; also encode harpins that help elicit plant hypersensitive response

HSV herpes simplex virus; two types: HSV-1 which causes oral herpes and HSV-2 which causes genital herpes

human immunodeficiency virus (HIV) virus that causes AIDS

human papilloma virus (HPV) causes warts; some strains cause genital warts and cervical cancer

humic substance complex polymers that give soils and sediments their dark color; inhibit the PCR reaction

humoral immunity antibody-mediated immunity

HUS hemolytic uremic syndrome, a complication of dysentery (kidney failure) thought to be caused by Shiga toxin (from *Shigella* spp.) or Shiga-like toxin (from EHEC)

hyaluronic acid mucopolysaccharide component of extracellular matrix; forms ground substance of connective tissue

hydrogenase periplasmic enzyme that converts hydrogen to two protons and two electrons; found in microbes that use sulfate as electron acceptor

hydrogenosome organelle of parabasalians that makes acetate, carbon dioxide, and water from pyruvate, which yields an additional ATP during anaerobic growth

hydrophilic amino acid one that interacts well with aqueous environments; found on parts of proteins exposed to cytoplasm or external environment

hydrophobic amino acid one that does not interact well with aqueous environment; interacts with other hydrophobic amino acids or lipids; found folded inside proteins or embedded in membranes

hydroxyl radical reactive form of oxygen that can kill bacteria

hypersensitive response protective reaction by plants in which plant R proteins recognize bacterial avirulence proteins and harpins thus triggering local tissue damage that prevents bacteria from spreading

hypha cylindrical tube-like vegetative structure that grows from the tip of some fungi; may branch and form a mycelium

hypochlorous acid reactive form of chlorine toxic to bacteria

hypotension collapse of the circulatory system; drop in blood pressure

ICAM-1 *see* intercellular adhesion molecule-1

icosahedral capsid geometric shape of the capsid of many viruses

ICU *see* intensive care unit

ID_{50} *see* infectious dose

IFN-γ *see* interferon-γ

IgA the major class of antibody found in mucosal membranes; dimeric

IgE monomeric immunoglobulin; thought to play a role in metazoal parasite infections

IgG monomeric immunoglobulin; the major antibody type present in the serum; humans have four subtypes

IgM pentameric immunoglobulin found in serum; predominates in initial antibody response to an antigen

IL-1 cytokine that stimulates monocytes and granulocytes to leave the bloodstream and enter tissues

IL-2 cytokine produced by TH1 cells; stimulates cytotoxic T cell proliferation; in high levels, causes nausea, vomiting, malaise, fever

IL-4 cytokine that downregulates production of TNF-α; stimulates TH0 to differentiate to TH2

IL-6 cytokine that stimulates liver to release acute phase proteins

IL-8 cytokine that stimulates monocytes and granulocytes to migrate from bloodstream into tissues; stimulates PMNs to produce β-integrins

IL-10 cytokine that down-regulates production of TNF-α

IL-12 cytokine produced by PMNs; stimulates NK cells to produce IFN-α

IL-13 cytokine produced by TH2 cells; downregulates immune response

immortalized cell lines cells that can be maintained indefinitely in culture

immune system collection of phagocytic, cytotoxic, and antibody-producing cells that protect the body from infection

immunogen an antigen that induces an immune response

immunoglobulins antibodies

immunosuppression suppression of the normal immune response

inactivated polio vaccine (IPV) inactivated virus preparation; stimulates humoral immunity; must be given by injection

indigenous native to a particular site

indolyl-galactoside *see* X-gal

induced defenses host defenses produced only in response to exposure to specific antigens; includes antibodies, cytotoxic T cells, and activated macrophages

infant botulism disease resulting when botulinal toxin is produced by *Clostridium botulinum* colonizing colon of infants

infection colonization of the body by a microorganism capable of causing disease; may or may not cause symptoms

infection thread structure formed in root hair through which *Rhizobium meliloti* cells migrate to underlying root cells where they will become bacteroids

infectious capable of causing disease

infectious dose (ID_{50}) number of microorganisms required to cause infection in 50% of animals exposed to them; a measure of infectivity

infectivity ability of a microbe to infect the hosts it encounters

inflammation combination of complement activation, cytokine release, phagocyte activation, and production of a variety of inflammatory compounds; characterized by redness, swelling, pain, and heat

initiation complex ribosome-mRNA complex where translation begins

inoculum suspension of microorganisms introduced into tissue culture or culture medium

insertion sequence (IS element) DNA segment that integrates randomly into DNA; contains a gene encoding transposase, the enzyme that allows the IS element to move from one site to another; has promoters on each end; can activate gene expression or disrupt genes

integrase (1) enzyme that mediates site-specific recombination that integrates DNA segments to form gene clusters in integrons, or (2) enzyme that mediates integration of DNA copy of HIV genome into host cell genome

integron transposon that contains a gene for integrase, which facilitates integration of gene cassettes into a special site; responsible for evolution of plasmids carrying multiple antibiotic resistance genes under control of a single promoter

intensive care unit (ICU) units in hospitals where patients who are most ill are housed; common site for acquiring nosocomial infections

interferon class of proteins produced by animal cells in response to endotoxin or viruses

intercellular adhesion molecule-1 (ICAM-1) molecule of endothelial cells that binds to integrins on neutrophils causing them to flatten against blood vessel wall; results in margination

interferon-γ (IFN-γ) cytokine produced by TH1 cells in response to infection; activates macrophages and cytotoxic T cells and stimulates migration of monocytes and granulocytes from bloodstream to tissues

interleukins protein cytokines produced by monocytes and macrophages; in high levels, mediate inflammation, septic shock, fever

intracytoplasmic membrane extension of cytoplasmic membrane or membrane vesicles where photosynthetic systems of bacteria are located

intron segments of DNA encoding a portion of RNA that is eliminated from the final form of the mRNA transcript

invasive capable of penetrating the host's defenses; capable of entering host cells or passing through mucosal surfaces and spreading in body

in vitro outside the body (in a test tube)

in vivo occurring in a living organism

IPV *see* inactivated polio vaccine

IS element *see* insertion sequence

isoniazid antibiotic effective against mycobacteria; probably inhibits synthesis of mycolic acids

ischemia loss of blood supply to a body part

J chain peptide linking IgA monomers or IgM monomers

kanamycin an antibiotic in the aminoglycoside family

keratin protein produced by some skin cells; difficult for microbes to digest; important component of skin defenses

keratinocytes type of cell constituting up to 95% of epidermis; maintain acidic environment of skin; produce cytokines; synthesize keratin

kinetoplast unusual mitochondrion found in the kinetoplastid group of protists; consists of a few large chromosomes and many small DNA circles; target of antiprotozoal drugs

kinetoplastid protists with unusual mitochondrion (kinetoplast), a single nucleus, and a single flagellum attached to an undulating membrane; group contains serious pathogens such as *Trypanosoma* and *Leishmania*

Kirby-Bauer test disk diffusion antimicrobial susceptibility test

kleptochloroplasts chloroplasts that *Pfiesteria* acquires from true algae; allows *Pfiesteria* to be photosynthetic

Koch's postulates a set of postulates that must be met to prove a particular microbial pathogen causes a particular disease

Kupffer cells fixed macrophages of the liver

lactoferrin host protein than binds iron with high affinity

lactoperoxidase an enzyme found in mucus that produces superoxide radicals

lacZ gene that encodes β-galactosidase; used in gene fusions

lag phase adaptation period where bacteria do not divide after coming into a new environment

lagging strand strand of DNA on which DNA polymerase must synthesize very short segments of DNA; moves in opposite direction of replication fork; newly synthesized segments are joined by DNA ligase

lamina propria connective tissue below epithelial cells of mucous membranes

laminin component of host extracellular matrix

Langerhans cells antigen-presenting dendritic cells of the epidermis; part of SALT

latent infection one in which symptoms are not apparent but the microbe is present in the body

latency period of inactivity

LD_{50} *see* lethal dose

leader region upstream region where mRNA transcription starts

leading strand strand of DNA on which the DNA polymerase moves in the same direction as the replication fork as it synthesizes a complementary strand in the 5'→3' direction

leghemoglobin form of hemoglobin found in root nodules; composed of plant globin and bacterial hemin

lesion destruction of tissue resulting from host's inflammatory response

lethal dose (LD_{50}) the number of bacteria or the amount of toxin required to kill half of the animals experimentally inoculated

lethal toxin toxin produced by *Bacillus anthracis;* responsible for systemic anthrax, the most lethal form of the disease

leukocyte white blood cell

leukocyte esterase enzyme found in neutrophils; used as a quick test to determine whether neutrophils are present in urine indicating UTI

Lewis antigen glycoproteins found on human cells; may prevent bacterial attachment

lichens assemblages of fungi and photosynthetic microbes, either algae or cyanobacteria; fungi give macroscopic structure, photosynthetic microbes fix nitrogen used by the fungi

light chain smaller of two types of protein making up the basic antibody molecule

lignin polyphenolic compound that gives wood its tough rigid property; degraded by fungi

lincosamides a family of antibiotics that inhibits protein synthesis by binding to the 50S ribosomal subunit; examples are lincomycin, clindamycin

lipid A toxic portion of LPS

lipoarabinomannan mycobacterial cell wall glycolipid that suppresses T cell activation

lipooligosaccharide component of outer membrane of some gram-negative bacteria; similar to lipopolysaccharide but with a shorter O antigen

lipopolysaccharide a component of the gram-negative outer membrane; consists of lipid A (the toxic portion), a core made up of a series of sugars, and the O antigen, a long carbohydrate chain; activates complement and stimulates cytokine production; also called endotoxin

lipoteichoic acid (LTA) lipid-linked teichoic acid; lipid can be embedded in the cytoplasmic membrane or exposed on the bacterial surface

lithotrophs microbes that obtain energy by oxidizing or reducing inorganic compounds such as sulfide and sulfate

LOS *see* lipooligosaccharide

LPS *see* lipopolysaccharide or endotoxin

LT *see* heat-labile toxin

LTA *see* lipoteichoic acid

luciferase an enzyme that produces light when ATP is supplied

lumen cavity in an organ

luminescent bacteria bacteria that can produce light

Lyme disease systemic disease caused by *Borrelia burgdorferi;* transmitted from white-footed mice to humans by *Ixodes* ticks

lymph fluid moving through lymphatic vessels; contains lymphocytes

lymph node site where lymphatic vessels come together; contain APCs and lymphocytes; "cleanses" blood before returning it to bloodstream

lymphocytes T cells and B cells

lysis disruption of a cell

lysogeny bacteriophage genome integrated into the bacterial chromosome

lysosomal proteins degradative enzymes, defensins, and myeloperoxidase carried in lysosomal granules

lysosome (or lysosomal granule) mammalian cell organelle containing hydrolytic enzymes and other compounds toxic to bacteria

lysozyme enzyme that degrades peptidoglycan

lytic infection bacteriophage reproduces itself and ultimately kills the bacterial cell it has infected

M (microfold) cells cells in Peyer's patches that bring bacteria or fragments of bacteria into contact with antigen-presenting cells of the GALT

MAC *see* membrane attack complex

macrogametes one of the sexual forms of malaria parasites found in mosquitoes; fuse with microgametes and eventually form sporozoites which are the infective form injected by the mosquito bite

macrolide family of antibiotics that bind the 50S ribosomal subunit and prevent peptidyl transfer and/or translocation; an example is erythromycin

macrophage large tissue mononuclear cell having phagocytic activity; develops from monocyte

major histocompatibility complex (MHC) protein complex on the surface of mammalian cells; binds foreign (e.g., bacterial) peptides and displays them on the surface where they are recognized by T cells; two forms: MHC I found on most cell types in the body; MHC II found on antigen-presenting cells and B cells

malabsorptive diarrhea damage to mucosal cells impairs water uptake by the intestine

malaise not feeling well

malaria serious disease caused by *Plasmodium* spp.; transmitted from human to human by *Anopheles* mosquitoes

MALT *see* mucosa-associated lymphoid tissue

mannose receptor receptors on surface of many epithelial cell types; recognized by some bacterial pili

mast cells tissue cells that contain granules that contain histamine, heparin, and other substances that can attract phagocytes to the site of bacterial invasion; also can produce cytokines; important in allergic responses

mating bridge complex set of proteins that form the connection between mating cells in the conjugation process; mediates transfer of DNA from donor to recipient

matrix proteins proteins on the inner surface of a viral envelope that link the envelope to the capsid; stabilize the envelope

MBC minimal bactericidal concentration

membrane attack complex (MAC) complex consisting of complement components C5b-C9 which damages membranes of enveloped viruses and gram-negative bacteria

memory cells long-lived T or B cells produced during initial immune response; activated rapidly when body again encounters epitope they recognize

meninges membranes covering brain and spinal cord

meningitis inflammation of the meninges

merozoite replicative form of some protozoal parasites

messenger RNA (mRNA) RNA molecule transcribed from DNA; contains the coding sequence for at least one protein

metazoal parasites multicellular eukaryotes (for example, cestodes, trematodes, and nematodes)

methane-oxidizing bacteria bacteria that obtain energy and carbon by oxidizing methane; formaldehyde is intermediate in oxidation pathway

methanogenesis reduction of carbon dioxide to methane

methanogenic archaea species that use hydrogen to reduce carbon dioxide to methane; use carbon dioxide as carbon source

MHC *see* major histocompatibility complex

MHC I MHC type that when complexed with an epitope triggers activation and proliferation of cytotoxic T cells; found on most cells in body

MHC II MHC type that when complexed with an epitope leads to activation and proliferation of helper T cells; found on only a few cell types (e.g., B cells, antigen-presenting cells)

MIC *see* minimal inhibitory concentration

microarray chip containing thousands of squares, each with a portion of single-stranded DNA from a particular gene; used to detect expression of genes under a variety of conditions

microaerophile an organism that grows only at less than 1 atmosphere of oxygen

microbiota population of normally nonvirulent microorganisms found routinely in a specific site in the body of most normal adults

microbiota shift disease a shift in the population of microbes that normally resides in a particular site resulting in disease at that site or in an immediately adjacent site

microbroth dilution MIC test quantitative test using two-fold serial dilutions of an antibiotic to determine the MIC of that agent for a microbe

microflora *see* microbiota

microgametes one of the sexual forms of the malaria parasite found in mosquitoes; fuse with macrogametes to eventually produce sporozoites, the infective form injected by a mosquito bite

microglial cell APC of the brain

microsphere inert substance that slowly breaks down in the body releasing encapsulated vaccine antigens over a period of time

microvilli finger-like projections on the apical surface of mucosal absorptive cells

minimal bactericidal concentration the lowest concentration of an antibiotic that will kill a microbe

minimal inhibitory concentration lowest concentration of an antibiotic that will inhibit the growth of a microbe

missense mutation point mutation that results in an amino acid substitution in a protein; may or may not have an effect on protein function

mitochondrion energy-generating organelle in eukaryotes

mixed culture a culture that contains more than one type of microorganism

MMR vaccine vaccine against measles, mumps, and rubella

***mob* genes** genes on mobilizable plasmids that permit them to take advantage of transfer machinery of other, self-transmissible, plasmids in the same cell

mobilizable plasmid plasmid that is not self-transmissible but can be cotransferred with a self-transmissible plasmid

molecular yardstick use of sequences of highly conserved genes (for example, gene for 16S or 18S rRNA) to deduce relatedness of different organisms; used to estimate diversity in a particular group of organisms

monobactams a class of β-lactam antibiotics

monocyte mononuclear phagocyte circulating in blood; differentiates into macrophage

morbidity sickness

mortality fatality

motility ability to move through the environment; different types include flagella-driven, gliding, corkscrew movement, twitching, and intracellular actin polymerization

mRNA *see* messenger RNA

mucin same as mucus; complex, viscous, sticky mixture of proteins and carbohydrates covering mucosal membranes; produced by goblet cells

mucoid wet glistening appearance of colonies of bacteria that produce capsules

mucopolysaccharides polysaccharide "glue" that holds human cells together; degraded by colonic bacteria

mucosa-associated lymphoid tissue (MALT) specialized immune system that protects all mucosal surfaces; includes GALT

mucosal membranes layer(s) of cells that line internal parts of the body

multiple organ system failure cause of fatality in septic shock

mupirocin antibiotic that inhibits aminoacyl-tRNA synthetase

muramyl dipeptide cell wall component of *Mycobacterium;* stimulates immune system; triggers cytokine production

murein *see* peptidoglycan

mutualistic association symbiotic association beneficial to both partners

mycelium multicellular form of a fungus consisting of a fibrous network of hyphae

mycolic acid type of lipid found in cell walls of *Mycobacterium*

myeloperoxidase lysosomal enzyme that produces reactive forms of oxygen that are highly toxic to bacteria

NADPH oxidase enzyme system located in the membrane of phagosomes; produces reactive oxygen intermediates when membrane fuses to lysosome containing myeloperoxidase

naked DNA vaccine DNA encoding antigen is adsorbed to gold particles and injected into muscle; muscle cells transiently produce and display foreign antigen; evokes robust immune response

naked virus virus without an envelope

nalidixic acid member of the quinolone family of antibiotics; inhibits DNA gyrase

narrow host range transfer transfer of genes among very closely related bacteria such as members of the same species

narrow spectrum antimicrobial agent one that is effective against a limited number of species

natural killer cell type of phagocyte that targets microbes growing inside human cells; kills infected host cells; part of the nonspecific defenses

natural transformation uptake of free DNA by bacteria in nature

natural vaccination process of developing immunity by having a disease

necrosis death of tissue; characterized by loss of blood supply

nef regulatory protein of HIV

neonatal ICU ICU where high-risk infants are placed

neuraminidase enzyme that removes sialic acid residues from host cells; found in envelope of influenza viruses

neurotoxin toxin specific for nerve cells

neutralization antibodies bind to microbes or toxins thus preventing their attachment to host receptors

neutropenia decrease in the number of neutrophils in the blood

neutrophil leukocyte in which granules do not stain; polymorphonuclear leukocyte; engulf extracellular bacteria

NGU *see* nongonococcal urethritis

nitrate reductase enzyme commonly found in *Escherichia coli* and relatives; used in some tests as an indicator of UTI

nitrifiers bacteria that oxidize ammonia to nitrate and nitrite

nitrification oxidation of ammonia to nitrate and nitrite by soil bacteria

nitric oxide reactive nitrogen intermediate found in phagocytes; toxic to microbes

nitrogen cycle the means by which microbes control the balance of different nitrogenous compounds in soil

nitrogen fixation incorporation of atmospheric nitrogen into ammonia needed by microbes; catalyzed by nitrogenase

nitrogenase protein complex that catalyzes the reduction of atmospheric nitrogen to ammonia (nitrogen fixation); used by oxygenic phototrophs to acquire nitrogen

Nod factors *see* nodulation factors

nodulation infection of root hairs of some plants by bacteria capable of nitrogen fixation thus rendering the plant self-fertilizing

nodulation factors bacterial oligosaccharides that trigger nodulation in plants that attract nitrogen-fixing bacteria to enter their roots

nonenveloped virus virus that does not have an envelope; naked virus

nonfermenter organism that is not capable of fermentative metabolism, for example, *Pseudomonas aeruginosa*

nongonococcal urethritis disease with symptoms indistinguishable from gonorrhea; caused by *Chlamydia trachomatis* and other bacteria; may lead to PID

nonoxynol-9 spermicide that inhibits growth of lactobacilli; creates favorable environment for colonization of vagina by uropathogenic *Escherichia coli*

nonpilus adhesins adhesive cell surface components that do not form pilus-like structures

nonpolar amino acid *see* hydrophobic amino acids

nonsense mutation point mutation that results in a stop codon; early termination of translation

nonseptate hypha multicellular hypha in which individual cells are not completely separated

nonspecific defenses host defenses that are effective against most microorganisms and are always present; include physical barriers, complement, phagocytic cells, microbiota, and washing action of fluids

nosocomial infections acquired in the hospital

nuclear membrane membrane that surrounds the genome in eukaryotes; missing in prokaryotes

nucleoid prokaryotic genome that exists in a compact mass; not surrounded by a membrane

nucleoprotein core portion of virus that contains the genome and proteins needed to initiate the replication cycle

nucleotide analogs compounds that have a much higher affinity for viral replication enzymes than host replication enzymes; inhibit viral replication

nucleotide sequence sequence of bases in the genome of an organism; contains all information about what an organism is

O157:H7 the major group of EHEC; causes bloody diarrhea and HUS

O antigen polysaccharide side chains on LPS

obligate required

obligate anaerobe can only grow in absence of oxygen

obligate intracellular pathogens microbes that carry out most of their own biosynthetic reactions but require some essential nutrient present only inside a host cell

oncogenic cancer causing

oocyst survival form of some protozoa

open reading frame (*orf*) region between start and stop codons in a genome sequence; may correspond to the amino acid sequence of a protein

operator site region of DNA to which a repressor binds and prevents RNA synthesis

operon unit consisting of a series of structural genes that share the same promoter; mRNA transcript encodes multiple proteins

operon fusion *see* transcriptional fusion

ophthalmia neonatorum gonococcal infection of eyes of infant; can cause blindness

opines unusual carbon sources produced by crown gall tumor cells and used as a food source by *Agrobacterium tumefaciens*

opportunist an organism capable of infecting only when host defenses are compromised

opsonin antibody or complement component C3b that attach to the bacterial surface and enhance the ability of phagocytes to ingest the bacterium

opsonization process that helps neutrophils ingest microbes by attachment of antibody or complement component C3b to microbes

OPV *see* oral polio vaccine

oral microbiota microbes that normally live in the oral cavity; protect against lung infections

oral polio vaccine (OPV) live attenuated virus vaccine; stimulates mucosal immunity; administered orally; source of all poliovirus infections in the United States at this time

orf *see* open reading frame

ori gene involved in plasmid replication; site at which replication begins

oriT *see* transfer origin

Osp A-D outer surface proteins (lipoproteins) of *Borrelia burgdorferi,* the cause of Lyme disease

otitis media middle ear infection

outer membrane a membrane covering the surface of gram-negative bacteria; outer leaflet is made up of lipopolysaccharide

oxidase test enzyme produced by bacteria converts colorless chemical to purple color; used in process of identifying bacteria

oxidation loss of protons and electrons by a compound

oxidation-reduction reactions reactions in which one compound is oxidized and another is reduced

oxidative burst production of reactive oxygen intermediates by phagocytes

oxidative decarboxylation removal of carboxyl groups from pyruvate to produce a molecule of carbon dioxide coupled to an oxidation; involves acetyl-S-CoA

oxygenic photosynthesis process carried out by cyanobacteria where water is split and oxygen is released into the environment; provides oxygen atmosphere

ozone layer oxygen layer in atmosphere that protects Earth against ultraviolet light

P pili *see* "pyelonephritis-associated pili"

P (peptide) site site on ribosome where tRNA carrying growing peptide chain is located

pandemic epidemic involving many different countries

Panspermia theory theory that first life on Earth was brought from another planet (probably Mars)

pap *see* "pyelonephritis-associated pili"

PAPS *see* "phosphoadenosine phosphosulfate"

parabasalians protists that lack mitochondria but possess parabasal body (oddly shaped golgi body); undergo meiosis; have a single nucleus associated with 4-6 flagella; one example is *Trichomonas vaginalis,* which causes disease, another is *Trichonympha* which is an exosymbiont of termites

parenteral administered by injection

paroxysm severe attack of symptoms (associated with coughing in whooping cough)

particulate vaccine a complex vaccine (e.g., live viruses) needed to elicit an MHC I-mediated cytotoxic T cell response

passive immunization injecting preformed antibodies against a particular pathogen or toxin into an unimmunized patient

pathogen organism capable of causing disease

pathogenicity ability of a microbe to cause clinically apparent symptoms

pathogenicity factor characteristic of a microbe that allows it to cause disease; same as virulence factor

PCR *see* "polymerase chain reaction"

pectin plant polysaccharide that can be degraded by colonic bacteria

pelvic inflammatory disease (PID) infection of fallopian tubes; leads to permanent damage, ectopic pregnancy

penicillins class of β-lactam antibiotics

penicillin-binding proteins proteins (normally located on outer surface of cytoplasmic membrane) that bind penicillins; include enzymes for peptidoglycan synthesis and turnover

peptidoglycan polysaccharide backbone with peptide cross links that covers surface of cytoplasmic membrane and gives bacteria their shape

perforin protein found in granules of nonspecific cytotoxic cells; forms channels in target cell membrane

periodontal disease inflammation of gums due to shift in the composition of the microbiota of the gums

periplasm space between outer and cytoplasmic membranes in gram-negative bacteria; encloses peptidoglycan layer

peristalsis wave of contractions moving along intestinal tract; moves contents through intestine

peritrichous flagella flagella emerge from many locations on surface of bacteria

peroxidase enzyme that converts peroxide + NADH to water and NAD

peroxide oxidizing agent resulting from conversion of superoxide

pertussis disease caused by toxins produced by *Bordetella pertussis*

pertussis toxin a protein toxin of *Bordetella pertussis;* ADP-ribosylates a G factor controlling host cell adenyl cyclase activity; a component in the acellular pertussis vaccines

Petroff-Hauser counter device used to enumerate microbes by direct microscope count instead of cultivation

PERV *see* "porcine endogenous retroviruses"

Peyer's patch collection of follicles in the small intestine that contain cells of the GALT

PFGE *see* "pulsed field gel electrophoresis"

phagocyte host cell adapted specifically to engulf and destroy bacteria or other foreign particulate matter

phagocytosis ingestion of foreign particles by a cell

phagolysosome vacuole resulting from fusion of phagosome and lysosome in a host cell

phagosome endocytic membrane vesicle resulting from ingestion of particulate material by phagocytes: has low pH; contains toxic proteins

pharyngitis inflammation of pharynx; sore throat

phase contrast microscopy used to illuminate microbes with ordinary light; reveals shape, motility, granules, endospores; microbes are not stained

phase variation on-off control for some bacterial genes

phenotype structural and metabolic characteristics of an organism

phosphoadenosine phosphosulfate (PAPS) intermediate in the assimilatory sulfate reaction

phosphotransferase system bacterial process in which glucose is transported into the cell and phosphorylated at the same time

photosynthesis conversion of light energy into biomass

photosystem the component of the photosynthetic system that converts light energy to chemical energy in the form of an energized electron; photosystem I and photosystem II absorb light of different wavelengths

phototaxis movement of certain protists to a level of light where photosynthetic machinery works optimally; mediated by the eyespot

phototroph microbe that obtains energy from light; can be autotroph or heterotroph

phylogenetic group a cluster of organisms with related rRNA gene sequences

phytoplankton collection of photosynthetic microorganisms in the ocean; directly or indirectly feed all marine animals

PID *see* "pelvic inflammatory disease"

pilS silent version (i.e., not expressed) of *pilE* (the gene encoding the pilin protein) of *Neisseria gonorrhoeae*

pilin subunits proteins packed in helical array to form the shaft of a pilus

pilus rod-like protein structure projecting from the bacterial cell surface; often has special set of proteins at tip which mediate adhesion

plaque (1) clear areas formed in a monolayer of tissue culture cells when some of the cells are killed by infecting viruses; (2) term also applied to clear area formed by lytic bacteriophage on a confluent bacterial culture; (3) biofilms growing on surface of teeth

plasma noncellular portion of blood obtained by centrifuging whole blood; contains elements necessary for clot formation

plasmid autonomously replicating extrachromosomal DNA segment; most are circular but *Borrelia* has linear plasmids

plasmid addiction system mechanism by which plasmids maintain themselves in bacteria without providing any advantage to the bacteria

plasmid copy number number of copies of a particular plasmid in one cell; usually one to ten in nature but can be hundreds in laboratory strains

plastocyanin electron transfer protein that transfers electron to chlorophyll in photosystem I

PMF *see* "proton motive force"

PMN *see* "polymorphonuclear leukocyte"

pneumococcal surface protein A (PspA) protein exposed on the surface of the capsule of *Streptococcus pneumoniae;* possible vaccine candidate

pneumococcus *Streptococcus pneumoniae;* most common cause of pneumonia

pneumocyte lung cell

pneumonic plague a form of plague that can be transmitted from human to human by aerosols

pol polyprotein a large protein that contains enzymes involved in multiplication of HIV (protease, integrase, reverse transcriptase); must be cleaved for the enzymes to become active

polarized mammalian cells have different surface components on different faces of the cell; most epithelial cells are polar

polycistronic mRNA mRNA containing information from several genes; transcribed from a single promoter

polymerase chain reaction a method for amplifying DNA in vitro; involves oligonucleotide primers complementary to nucleotide sequences flanking a target sequence with subsequent replication of the target sequence

polymicrobic infection infection caused by more than one species of bacteria

polymorphonuclear leukocyte (neutrophil; PMN) short-lived professional phagocyte that circulates in the body

porcine endogenous retroviruses (PERV) viruses integrated into the genome of pigs; could possibly be transferred to the genome of human recipients of organs from pigs (xenotransplantation)

porin protein constituent of pores in the outer membrane of gram-negative bacteria that allows diffusion of nutrients

positive TB skin test raised, tough red area around site of PPD injection

post-transcriptional regulation regulation of genes at the level of translation or of protein activity by post-translational events

post-translational activation activation of protein by proteolytic nicking or covalent modification

PPD *see* "purified protein derivative"

primary stage of syphilis first stage of disease; characterized by lesion (chancre) at site of infection

probiotics live bacteria ingested with intention of repopulating resident microbiota disrupted by antibiotics or cancer chemotherapy

prokaryotes organisms that lack a nuclear membrane; usually have a single circular chromosome; genes often polycistronic; energy generated in cytoplasmic membrane

promoter site on DNA where RNA polymerase binds and initiates transcription

prophylaxis protection against disease

prostatitis infection of the prostate; common type of ascending UTI

protease the enzyme of HIV that cuts larger polyproteins into smaller active enzymes; target of some anti-HIV drugs

protective antigen the B subunit of lethal toxin of *Bacillus anthracis;* antibodies to protective antigen provide protection against anthrax

protein A surface protein of *Staphylococcus aureus* that binds the Fc portion of antibodies

protein G surface protein of *Streptococcus pyogenes* that binds the Fc portion of antibodies

protists eukaryotic microbes; great diversity; play roles in disease and in the environment

proton motive force (PMF) force that results from protons being pumped out of cell; used to make ATP

protozoa unicellular eukaryotes without a rigid cell wall; may have more than one developmental stage

pseudohypha short hyphal extensions from some yeasts; may contribute to invasiveness

pseudomembrane sheet-like layer of debris (fibrin, mucin, dead host cells) that covers a large area of the colon (pseudomembranous colitis) or the throat (diphtheria)

pseudomembranous colitis overgrowth of colon by *Clostridium difficile* after antibiotic treatment; toxins cause severe damage, form pseudomembrane; can result in death

pseudomurein constituent of archaeal cell wall; similar to peptidoglycan but contains an amino sugar that is different from muramic acid

PspA *see* "pneumococcal surface protein A"

puerperal related to the period after childbirth

pulsed field gel electrophoresis (PFGE) large pieces of DNA subjected to bursts of electrical charge in different directions migrate into gel; used for comparing bacterial strains in disease outbreaks

pure culture a culture that contains only one strain of bacteria

purified protein derivative (PPD) extract of *Mycobacterium tuberculosis* proteins used for skin test for tuberculosis

purine a type of base found in nucleic acids; adenine and guanine

purple bacteria bacteria with colonies that have a pink-purple color and which use a cyclic type of photosynthesis that does not concomitantly produce NADPH

purulent associated with formation of pus

pus accumulation of fibrin, neutrophils, and fragments of host cells; common sign of infection

pustule small lesion filled with pus

putrefaction decomposition of protein resulting in foul odors

pyelonephritis infection of the kidney

pyelonephritis-associated (P or Pap) pili major pilus type of uropathogenic *Escherichia coli*

pyogenic pus forming

pyrazinamide antibiotic effective against mycobacteria; mechanism of action unknown

pyrimidine a type of base found in nucleic acids; thymine, cytosine, and uracil

pyrimidine dimer pyrimidine bases in DNA crosslinked by ultraviolet radiation; can cause breaks in DNA

pyrogenic fever inducing

quinolones a family of antibiotics that inhibit DNA gyrase; an example is nalidixic acid

quorum-sensing system system that recognizes bacterial signal (autoinducer) thus sensing density of bacteria in area; controls either repressor or activator

R proteins (resistance proteins) plant proteins that recognize bacterial proteins; part of plant defense system

rabies fatal neurological viral disease transmitted to human primarily by bites of infected animals; vaccine available

radioimmunoassay (RIA) use of antibodies labeled with a radioactive compound to detect antigens or antibodies in a specimen

random walk movement of bacteria in different directions; allows microbes to sample environment for nutrients; mediated by flagella

reactive arthritis arthritis of peripheral joints that develops 2-6 weeks after a bacterial infection has cleared; thought to be caused by antibodies to bacterial antigens that cross-react with host antigens rather than a bacterial infection of the joints

reactivation tuberculosis *Mycobacterium tuberculosis* cells that have survived for long periods in walled-off lesions may break out of lesion due to immune suppression of host and cause active disease

red tides blooms of photosynthetic dinoflagellates that stain water red; organisms release toxins that kill fish and harm humans

reduction acquisition of protons and electrons by a compound

regulated expression genes encoding proteins needed under only some conditions are transcribed only under those conditions

rehydration therapy fluids administered either by mouth or intravenously in attempt to return water to tissues

replication fork unwound portion of DNA being copied; has fork-like appearance

replication origin site on chromosome where DNA synthesis begins

replicase RNA-dependent RNA polymerase encoded by viral genome; makes a +strand copy of -strand RNA

reporter gene structural gene encoding easily assayable enzyme; used in gene fusions

repressible enzyme one whose synthesis can be diminished by specific compounds

repressor protein protein that binds the operator; prevents transcription of adjacent genes

resident microbiota population of normally non-virulent bacteria found routinely in a particular site of the body of most normal adults

respiration use of an inorganic compound as an electron acceptor

restriction enzyme enzyme that cleaves DNA at a specific sequence

reticulate body form of *Chlamydia* spp. that replicates inside human cells

retrotransfer self-transmissible plasmid is transferred to recipient where it mobilizes a mobilizable plasmid to be transferred back to the donor

rev small protein that regulates gene expression of HIV

reverse transcriptase an enzyme that catalyzes the formation of a DNA copy of the RNA genome of certain viruses (retroviruses such as HIV); used extensively in biotechnology

rheumatic fever febrile illness that occurs several weeks after *Streptococcus pyogenes* sore throat; can be accompanied by damage to heart valves

rheumatic heart disease heart valve damage following strep throat

rhinitis inflammation of the nose

rhinovirus virus that causes at least half of all colds and many sore throats

rhizosphere space around roots of plants where concentrations of microbes are highest due to presence of organic compounds secreted by plant roots

rho protein that binds RNA polymerase and allows it to recognize termination sequences in DNA

RIA *see* "radioimmunoassay"

ribonucleic acid (RNA) polymers of ribonucleotides linked by phosphodiester bonds; genome of some viruses

ribosomal RNA (rRNA) RNA molecules found in ribosomes; three sizes; intermediate size (16S or 18S) used to determine relatedness of different organisms; molecular yardstick

ribosome RNA-protein complexes on which proteins are synthesized

ribosome binding site special sequence on 5' end of mRNA recognized by ribosome which attaches

ribosome protection bacterial protein binds to ribosome and prevents tetracycline from attaching

ribozyme catalytic RNA

rich medium one that contains many different components

rifampin an antibiotic that inhibits bacterial RNA polymerase by binding the β subunit; *also called* rifampicin

RNA *see* "ribonucleic acid"

RNA polymerase enzyme that moves along selected regions of DNA forming a complementary strand of mRNA; requires sigma and rho factors

RNA primer short segment of RNA that binds to single stranded DNA to creat double-stranded region where DNA polymerase can bind

RNase P enzyme of *Tetrahymena* that contains an RNA species that catalyzes the cleavage of RNA

rolling circle replication some plasmids are copied in one direction starting at replication; synthesis stops when DNA polymerase reaches origin

rubisco (ribulose bisphosphate carboxylase) enzyme used by Calvin cycle to combine carbon dioxide with ribulose bisphosphate to generate two molecules of 3-phosphoglyceraldehyde

SALT *see* "skin-associated lymphoid tissue"

sandwich-type serological tests antigens in patient's serum are trapped by specific antibodies bound to an inert substrate and then detected by adding specific labeled antibodies

sebaceous gland gland in the dermis that opens into hair follicle; secretes oily material

secondary infection a second infection that occurs after a first infection damages the host defenses (e.g., pneumococcal pneumonia often follows an influenza infection)

secondary stage of syphilis bacteria enter bloodstream; invade heart, musculoskeletal system, and CNS; symptoms include fever and rash

secretion system proteins in cytoplasmic membrane that secrete proteins made in cytoplasm into periplasm or extracellular fluid

secretory diarrhea excessive water and ion loss but no intestinal cell damage

secretory IgA (sIgA) dimerized IgA with two IgA molecules linked together; contains component called secretory piece; found in secretions, provides local protection to mucous membranes

secretory piece a portion of a receptor of mucosal cells that becomes attached to IgA as it passes through the mucosal cells; secretory piece + IgA becomes sIgA

segmented genome viral genome consisting of two or more distinct segments

selectable marker gene on a cloning vector that permits detection of cells that received the cloning vector; often an antibiotic resistance gene

selective medium one that contains dyes or antibiotics that inhibit the growth of normally occuring microbes; selects for pathogens

self-transmissible plasmid *see* "conjugative plasmid"

sensitivity (of a diagnostic test) likelihood that the test will be positive when the target microbe is present

sensor protein in the cytoplasmic membrane that detects a regulatory signal, changes conformation, and phosphorylates an activator protein; part of two-component regulatory system

sepsis condition resulting from microbes or microbial products in the blood

septate hyphae hyphal segments consisting of a chain of cells separated by cell walls (septa)

septic shock systemic reaction caused when bacterial cell wall components (LPS, LTA, peptidoglycan fragments) trigger release of cytokines that have a variety of effects on bodv temperature control and blood pressure; symptoms include fever, hypotension, DIC, acute respiratory disease, and multiple organ system failure

septicemia systemic disease in which microorganisms multiply in the blood or are continuously seeded into the bloodstream

sequela abnormal condition that develops following a particular disease

seroconversion induction of specific antibodies in the serum in response to an antigen

serological classification (serogroup; serotype) scheme based on reactivity of bacterial surface antigens with antibodies (e.g., O antigen of LPS, H antigen of flagella, C antigen of streptococci)

serological tests (1) tests that use antibodies to detect microbial antigens in a specimen, or (2) tests used to detect specific antibodies in a patient's serum

serum fluid portion of the blood without clotting factors

serum resistance lack of killing action of serum on gram-negative bacteria; results from alteration in O antigen of gram-negative bacteria so that MAC cannot form around LPS

serum sensitivity susceptibility of bacteria to killing by serum (i.e., lysis by MAC)

Shiga-like toxin toxin produced by EHEC that has an activity like that of Shiga toxin; cleaves rRNA and stops protein synthesis in host cells; responsible for HUS

sialic acid sugar found commonly on mammalian cell surfaces; some viruses such as influenza attach to sialic acid; some bacteria coat themselves with sialic acid to hide from host defenses

sialidase *see* "neuraminidase"

siderophores low molecular weight compounds produced by bacteria that chelate iron

sIgA *see* "secretory IgA"

sIgA protease enzyme that cleaves human sIgA at the hinge region

sigma factor protein that attaches to core enzyme of bacterial RNA polymerase thus enabling the RNA polymerase to recognize a particular class of promoter

sigma-70 sigma factor that recognizes most of the promoters in *Escherichia coli*

signs evidence of disease that can be observed by someone other than the patient

silent mutation change in DNA sequence but not amino acid sequence of proteins

simian immunodeficiency virus (SIV) immunodeficiency virus found in some African monkeys; may have given rise to HIV-1

simple epithelium epithelium consisting of a single layer of cells (e.g., intestinal epithelium)

skin-associated lymphoid tissue (SALT) specialized set of cells that confront bacterial invaders in the area immediately underlying the skin; Langerhans cells of epidermis are APCs of this system

S-layer cell wall of some archaea; composed of a dense layer of protein subunits

Southern blot a DNA hybridization assay in which restriction-digested DNA fragments are separated by electrophoresis and then transferred to a membrane; DNA fragments are made single-stranded, and labeled probe DNA is added

species taxonomic group containing most closely related strains (greater than 70% DNA sequence identity)

specific defenses defenses produced in response to invasion by specific bacteria or other infectious agents; includes antibodies, helper T cells, cytotoxic T cells, and activated macrophages

specificity (of a diagnostic test) ability of a test to discriminate between the target microbe and other microbes

sphincter circular muscle that closes the opening of the urethra

spirochete spiral-shaped organism with a cell wall consisting of two membranes with a peptidoglycan layer between them; internal flagella are also located between the two membranes

sporangiospore type of spore found in sac-like structures of some fungi

spore survival form of fungi comparable to bacterial endospores

sporozoite infective form of some protozoal parasites

spot blot single-stranded DNA being tested is applied as a spot on a membrane; single-stranded labeled probe DNA is added; if probe binds, original DNA specimen contained gene of interest

sputum material coughed up from infected lung

squamous cells flattened, scale-like cells on surfaces (e.g., stomach lining)

ST *see* "heat stable toxin"

start codon first codon used in translation; AUG; encodes methionine

stationary phase stage where bacteria stop dividing because of diminished nutrients and buildup of waste products

steatorrhea malabsorption of fats leading to an odorous, fatty diarrhea

stem-loop structure secondary structure in mRNA that forms when terminator sequence in DNA is transcribed; causes RNA polymerase to stop transcribing

sterilization complete killing of all microbes

sticky ends single-stranded ends left on DNA fragments after DNA is cleaved by restriction enzymes

stop codon last codon used in translation; UGA, UAA, UAG

stratified epithelium epithelium consisting of many layers of cells (e.g., skin)

streaking process in which a specimen is spread with a loop over the surface of an agar plate so that isolated colonies result

strep throat pharyngitis due to *Streptococcus pyogenes*

stromatolite large mounds formed by accumulations of cyanobacteria; become fossilized

strong promoter one to which RNA polymerase binds frequently

substrate level phosphorylation phosphate transferred directly from organic compound (e.g., PEP) to ADP to produce ATP

subunit vaccine vaccine consisting of one or a few purified proteins (e.g., diphtheria and tetanus vaccines)

sucking disk structure on ventral suraface of *Giardia intestinalis;* anchors parasite to mucosal surface

sulfate-reducing bacteria bacteria that use sulfate as a terminal electron acceptor in anoxic sediments; process is called dissimilatory sulfate reduction

sulfide-oxidizing bacteria bacteria that require oxygen to oxidize sulfide to sulfur or sulfate; may be photosynthetic

sulfonamides a family of antibiotics that inhibit an enzyme in the pathway that leads to synthesis of tetrahydrofolic acid

sulfur granules granules seen by phase contrast microscopy in bacteria that derive energy from sulfur compounds

superantigen toxins that stimulate large populations of T cells to proliferate and produce cytokines

supercoiling very tight winding of DNA; causes DNA to fold in on itself forming a compact mass

superoxide dismutase bacterial enzyme that helps convert superoxide to water

superoxide radical a toxic form of oxygen produced during phagocytosis

suppurative pus-forming

sweat gland coiled glands deep in skin; secrete sweat

symbiosis close association of two organisms; may be beneficial to one or both

symptoms aspects of an illness experienced by a patient

syndrome symptoms that characterize a specific disease

synovial fluid fluid found in joints

syntrophy interaction between two microbes that allows a reaction to proceed that is not energetically possible for one to carry out

systemic affecting the whole organism rather than a specific organ or tissue

systemic anthrax form of anthrax due to lethal toxin; causes rapidly developing shock

T cell thymus-dependent lymphocyte; T helper cells activate macrophages or stimulate antibody production by B cells; cytotoxic T cells kill host cells infected by specific intracellular pathogens

T cell receptor protein complex on surface of T cells; recognizes a specific epitope

T-dependent immune response specific interaction between APCs and T cells leading to antibody production or cell-mediated immunity; memory T cells formed

T-DNA segment of Ti plasmid that is transferred to the plant by conjugation; contains genes that encode plant hormones that direct the plant cells to multiply and produce opines

T helper cell aid in production of antibodies by B cells or activate macrophages; two major types (TH1 and TH2)

T-independent immune response B cells stimulated directly by polysaccharide or lipid antigens; antigen bypasses macrophages and T cells; only antibody-mediated immunity occurs; no memory cells formed

tachyzoite infectious form of *Toxoplasma gondii,* the cause of toxoplasmosis

Tamm-Horsfall glycoprotein most abundant protein in human urine; blocks binding of some types of pili

tat one of the regulatory proteins of HIV

TATAAT consensus sequence found at the –10 region of the promoter in *Escherichia coli*

teichoic acids polymers of sugar phosphate (or sugar alcohol phosphate) found interwoven in peptidoglycan of gram-positive bacteria

teichoplanin a glycopeptide antibiotic

temperate bacteriophages bacteriophages that can integrate in the bacterial chromosome (lysogeny) or enter the lytic cycle and kill the bacteria; temperate phages can encode toxin genes (e.g., diphtheria toxin)

–10 region region of promoter to which RNA polymerase binds in *Escherichia coli;* consensus sequence

tentative diagnosis one based on symptoms, signs, patient history, and epidemiological information

termination site point on bacterial chromosome where the two newly synthesized copies of the halves of the chromosome are joined, and the two chromosomes separate

terminator sequence DNA sequences that code a secondary structure in mRNA that stops transcription

tertiary syphilis final stage; symptoms include heart damage, neurological symptoms, fatigue, skin lesions

tetanus spastic paralysis caused by tetanus toxin

tetanus toxin an A-B type neurotoxin produced by *Clostridium tetani*

tetracyclines a family of antibiotics that bind to the 30S ribosomal

tetracycline efflux pump cytoplasmic membrane protein that pumps tetracycline out of the cell so that it never reaches an inhibitory level

TH0 T helper cell that can differentiate into TH1 or TH2

TH1 cell that develops from TH0 in response to IFN-γ production by NK cells; stimulates production of antibodies by B cells and activation of macrophagaes

TH2 cells that develop from TH0 in response to IL-4; lead to production of IgE and stimulation of eosinophils; important in allergic reactions

three-domain model model tree of life based on rRNA gene sequence analysis; divides all life into three domains: bacteria, archaea, and eukarya

–35 region region of promoter to which RNA polymerase binds in *Escherichia coli;* consensus sequence

Ti plasmid plasmid that encodes most of the genes needed for *Agrobacterium tumefaciens* to elicit formation of a crown gall tumor by a plant; also contains genes that allow *A. tumefaciences* to catabolize opines produced by plant tumor cells

tight junctions areas where epithelial cells join together; prevents fluids and microbes from moving between lumen and substratum

tissue culture cells animal cells that have been immortalized so that they continue to divide; used for cultivating viruses

TNF-α *see* "tumor necrosis factor-α"

topoisomerase enzyme that nicks supercoiled DNA so that helicases can unwind it

toxic shock-like syndrome (TSLS) disease caused by invasive strains of *Streptococcus pyogenes;* symptoms include fever, rash, exfoliation of palms and soles of feet, shock, death

toxic shock syndrome (TSS) disease caused by strains of *Staphylococcus aureus* that produce toxic shock syndrome toxin (TSST-1); symptoms include fever, rash, exfoliation of palms and soles of feet, shock, death

toxic shock syndrome toxin (TSST-1) exotoxin produced by some strains of *Staphylococcus aureus;* superantigen

toxin substance produced by microorganisms that can damage human tissue

toxin neutralization binding of antibodies to toxins thus preventing binding of toxin to host target cell

toxinoses diseases in which symptoms are due entirely to action of toxins

toxemia toxins in the bloodstream

toxoid protein toxin that has been treated to destroy its toxicity but retain its immunogenicity

toxoplasmosis invasive disease caused by *Toxoplasma gondii;* may cross placenta and infect the fetus

transcription formation of an mRNA strand complementary to a single-stranded DNA template

transcriptional activator protein that facilitates binding of RNA polymerase to promoter and initiation of transcription

transduction transfer of genetic information from one bacterial cell to another by a bacteriophage

transfer origin (*oriT*) point on plasmid where single-strand nick is made at the beginning of conjugal transfer

transfer RNA (tRNA) form of RNA that carries an amino acid to the ribosome; at the ribosome the anticodon region of the tRNA binds to a codon on the mRNA specific for the amino acid which the tRNA carries

transformation process in which bacteria take up free DNA from the environment; also used to describe process by which mammalian cells become tumorigenic

transglycosylation process by which N-acetylmuramic acid and N-acetylglucosamine residues of peptidoglycan are linked to form the polysaccharide backbone

translocation movement of A subunit of an A-B type toxin into the cytoplasm of a host cell; or movement of tRNA with peptide chain from A site to P site on ribosome during translation

transmigration movement of phagocytes and cytotoxic cells out of blood vessels into tissues

transpeptidation linking of peptide units on separate chains of peptidoglycan

transport proteins proteins in cytoplasmic membrane that admit molecules into cytoplasm; high degree of specificity

transposase enzyme that catalyzes transposition of transposons and insertion sequences

transposon segment of DNA containing one or more genes flanked by insertion sequences

transposon mutagenesis creation of mutations by inserting a transposon carrying a selectable marker into other genes

traveler's diarrhea adult version of secretory diarrhea caused by ETEC

trichomonad a type of parabasalian; example is *Trichomonas vaginalis,* a cause of vaginitis

trimethoprim an antibiotic that inhibits an enzyme in the tetrahydrofolic acid biosynthetic pathway

tRNA *see* "transfer RNA"

tRNA synthetase enzyme that attaches an amino acid to a tRNA with the appropriate anticodon

trophozoite replicating form of some protozoa

TTGACA consensus sequence found at the -35 region of the promoter in *Escherichia coli*

tubercle area of dead tissue containing *Mycobacterium tuberculosis* surrounded by layers of macrophages and other immune cells

tumor necrosis factor-α (TNF-α) a cytokine produced by monocytes and macrophages in response to LPS

twitching motility flagella bind bacteria to surface; bacteria move in jerky motion

two-component regulatory system one protein senses the signal, then phosphorylates second protein to produce the form that activates transcription

type I pili pili commonly found on many *Escherichia coli* strains, both resident flora and pathogens

typhoid fever systemic disease caused by *Salmonella typhi*

UAA, UAG, UGA stop codons where ribosome leaves mRNA after completion of translation

ubiquinone a compound that carries activated electrons from the light harvesting center to the bc complex during photosynthesis

ulcer circumscribed area of inflammation characterized by necrosis

uncoating release of the viral nucleoprotein core from the capsid inside the host cell cytoplasm

undulating membrane membrane attached to flagellum in some protists; flagellar movement causes membrane to have wave-like motion

UPEC uropathogenic *Escherichia coli*

urease an enzyme that hydrolyzes urea to ammonia and CO_2

ureters tubes through which urine passes on its way from the kidney to the bladder

urethra tube that leads from the bladder to the outside of the body

urethritis infection of the urethra

urinary tract infection (UTI) infection of any part of the tract, i.e., urethra, bladder, ureters, or kidney

urine fluid excreted by the kidneys; contains nitrogenous wastes and salt

uropathogenic strain strain that causes urinary tract infections

UTI *see* "urinary tract infection"

VacA *see* "vacuolating toxin A"

vaccination stimulation of a specific immune response by administering a vaccine

vaccine suspension of organisms (usually killed or attenuated) or nontoxic antigens or mixtures of antigen used to elicit a protective immune response against a microbial disease

vaccinia virus attenuated strain of cowpox virus used as smallpox vaccine

vacuolating toxin A (VacA) cytotoxin produced by *Helicobacter pylori* that causes vacuoles to form in cultured mammalian cells

vancomycin a glycopeptide antibiotic

vasodilator substance that causes dilatation of blood vessels

vascular containing a blood supply

vector transmitter of infectious microorganisms

vegetation loose clots of platelets and fibrin that form near damaged heart valves (e.g., heart valves damaged during rheumatic fever)

vegetative replicating form of a bacterium

ventilator-associated pneumonia pneumonia that occurs in seriously ill patients whose breathing must be assisted with ventilators; host defenses bypassed; antibiotic resistant organisms are common causes

vesicle blister

villi projections from the mucosa of the intestine; covered by a layer consisting of differentiated absorptive cells, goblet cells, and intraepithelial lymphocytes

virion an entire virus particle

virulence ability of an organism to cause severe disease

virulence cassette a region of the genome of a pathogenic bacterium that contains several virulence genes; cassette can be amplified during infection

virulence factor microbial product or strategy that contributes to the ability of the microbe to cause infection

virulence mechanism *see* "virulence factor"

virulence plasmid plasmid that contains genes for virulence factors; found in ETEC and *Salmonella*

virus microorganism that must replicate by using biosynthetic machinery of free-living organisms; genome may be DNA or RNA

walking pneumonia a debilitating but not usually lethal type of pneumonia; often caused by *Mycoplasma pneumoniae*

Wassermann test nonspecific test for syphilis

weak promoter one to which RNA polymerase binds inefficiently

Western blot procedure microbial proteins are separated by electrophoresis then transferred to a nylon membrane (blot); patient's serum is added to blot and any antibodies specific for the proteins will bind; bound antibodies are detected with labeled secondary antibodies

wet mount microbes suspended in drop of water covered by cover slip; viewed with phase-contrast microscope

whooping cough *see* "pertussis"

wound botulism caused by botulinal toxin produced by *Clostridium botulinum* growing in a wound

W strain strain of *Mycobacterium tuberculosis* resistant to isoniazid, rifampin, ethambutol, and streptomycin; especially rapid progress of disease; highly lethal

xenotransplantation transplantation of organs from animals to humans

X-gal a colorless substrate included in culture medium to detect β-galactosidase; turns blue if β-galactosidase cleaves it

xylan plant polysaccharide that can be degraded by colonic bacteria

yeast fungi that grow as single cells and reproduce by budding

yeast infection overgrowth of *Candida albicans* after antibiotic treatment reduces lactic acid bacteria in vagina

zoonosis animal disease that can be transmitted to humans

APPENDIX I
Answers to End-of-Chapter Questions

Chapter 1

1. The Earth was made habitable for higher life forms by microbial activities. Moreover, microbes are the one group of living things that could not be deleted from the world, as it currently exists, without bringing life to an end. The reason microbes are so essential is that they are the base of all the Earth's food chains and are responsible for the recycling of dead biomass that makes the growth and reproduction of new organisms possible. Finally, if prior occupation of a geographical location means anything about ownership, microbes are the Earth's natural owners because they were here first and were here for billions of years before other life forms materialized.

2. Since microbes were the only forms of life on Earth for billions of years and because we now know that microbial activities are an essential prerequisite for life as we know it, it makes sense to search for microbes. This is exactly the strategy that astrobiologists (scientists interested in life on other planets) have adopted. Water seems to be a prerequisite for life too, so the strategy is to find water and then look for microbes. There are other ways to look for traces of current or former microbial life, such as indications of various geochemical cycles. Such cycles will be covered in Chapters 21 and 22. Now that scientists have a better idea of what processes can be carried out by microorganisms, chemical data are easier to interpret.

3. These similarities will become more obvious as you move through later chapters, but some similarities are already evident. All microbes have a condensed DNA genome, a cytoplasm that contains enzymes necessary for life and a cytoplasmic membrane that contains the cytoplasm. Division by binary fission is common among microbes, although there are exceptions to this rule even in the prokaryotes. Many microbes are unicellular, although here again there are exceptions in both prokaryotes and eukaryotes. As scientists learn more about microbes, especially the most primitive eukaryotes (see Chapter 9), the differences between prokaryotes and the lower eukaryotes have begun to blur. There is even a report of a nuclear membrane in a bacterium, a *Planktomyces* species. That there would be many similarities is to be expected because as far as we know, eukaryotes were descended from prokaryotes and have recruited prokaryotes to serve as organelles (e.g., mitochondria, chloroplasts).

4. First, many plants depend on bacteria for nitrogen or on fungi for expanding their root network. The roots of plants are colonized by a population of microbes whose activities are not fully understood. Plants actually go to some trouble to recruit these bacteria, in that they exude phenolic compounds into the soil that attract bacteria and fungi. Our understanding of the services these microbes perform is still incomplete. The subject of plant–microbe interactions is covered in more detail in Chapter 24. Microbes contribute to generating soil from rocks and they recycle dead biomass, a prerequisite for growth of new plants. Finally, bacteria invented the type of photosynthesis used by plants (Chapter 22) and the photosynthetic organelles of plants (chloroplasts) were once bacteria. Although a plant biologist might not be convinced by these arguments, he or she would have to admit that microbes are mighty important to plant life.

5. The control microbes exert comes from their versatility and adaptability. Thus, when something new comes along, they will take advantage of it if they can, and they usually can take advantage with gusto. The higher plants and animals are very large compared to microbes. As such they offer many new surfaces for microbial colonization. Also, their metabolic activities (and their tissues, when they die or are unable to repell importunate microbes) provide new sources of nutrients. The purpose of this question is to make clear the type of "control" microbes exert. They dominate simply by being there, being an essential component of the biosphere, and being ready to move into any promising new niche (whether a new surface or a new nutrient opportunity). A dominant theme of this book will be that every change humans or other animals make in some environment is accompanied by some microbial response. Usually microbial responses are benign or neutral, but in cases like the "dead zone" or a new disease, their responses can be inimical to human interests.

6. The theme of the answer to question 5, that microbes take advantage of changes in human activities, is common to both these examples. Moreover, both of these examples show that microbial adaptation to changes we make can have an adverse effect on us. Also, in both cases we can see, in retrospect, that these developments could have, been predicted based on what we know about microbes. An important goal for future research is to do a better job of predicting such developments and averting them or at least being better prepared to respond to them.

7. As far as microbes are concerned, the hypoxic regions of the Gulf are very much "live zones," first because a bloom of microbes started the process, and second because the death of all this phytoplankton biomass created a big nutritional bonanza for microbes living in the sediment.

Chapter 2

1. The use of rRNA genes or any other trait to deduce evolutionary relationships between organisms is based on the assumption that these traits are transmitted vertically by inheritance. If a gene can be transferred horizontally from one bacterial species to another, and that gene is inadvertently used as a molecular yardstick, then two bacteria that are not closely related at all would appear to be very close relatives. One of the controversies surrounding the use of rRNA gene sequences as molecular yardsticks is based on the unanswered question of whether rRNA genes are transferred horizontally or not. There are two arguments against horizontal transfer of these genes. The first is that the groupings deduced from analysis of rRNA gene sequences usually (but not always) agree pretty well with analyses based on other gene sequences. Second, since rRNA interacts with proteins to form the ribosome, it seems likely that if an organism acquired a different rRNA than the one its ribosomal proteins were evolved to interact with, this acquisition might well be lethal. Fortunately, as more sequences of whole microbial genomes become available (allowing scientists to look at many gene sequences, not just those of rRNA genes) the picture should become clearer because all genes cannot be horizontally transferred to the same extent.

2. Your first step would be to compare the rRNA gene sequence of your isolate with sequences of rRNA genes from known bacteria. Computer programs and large databases allow you to learn, within minutes, which of the known microbes is the closest relative of your isolate (i.e., has the most similar rRNA gene sequence). If there is a sequence that is 97% identical (or more) to your sequence in the databases, your isolate is likely a member of that same species. If, however, your isolate's rRNA gene sequence is less than 97% identical to the sequence of rRNA genes from any known organism, your isolate probably represents a new genus and perhaps even a new phylogenetic group of microbes. There are large groups of bacteria and archaea that are represented only by rRNA gene sequences and have not been cultivated and characterized as living organisms. This is the basis for the statement that scientists have only begun to understand the true biodiversity of the microbial world. It is exciting to realize that whereas a scientist who studies birds is highly unlikely to find a new species of bird, microbiologists are stumbling daily over entirely new groups of organisms, the equivalent of a macrobiologist's discovering the whole category "bird" for the first time.

3. Microbes have been on Earth far longer than the larger forms of life. This, together with the short generation times of most microbes (hours or days), has given microbes the opportunity to diversify and adapt on a scale unthinkable for more recently adapted, longer-lived species. Macrobiologists have a hard time accepting the fact that there is far more diversity in the microbial world than in the world of plants and animals because microbes look so much simpler by comparison. But by taking an evolutionary view of biodiversity, it is evident that virtually all of the biodiversity on Earth resides in the microbial world.

4. It will be VERY difficult to prove the panspermia theory. If traces of microbes or even live microbes are found on Mars (assuming a Mars lander finally deigns to reply to NASA scientists), and if such a creature is clearly related to Earth microbes (but also clearly different, to account for the long evolutionary separation), this might be accepted as evidence for the panspermia theory. A major problem will be convincing people that the Mars microbes, if they are found, are not simply contaminants introduced by the instruments sent from Earth to Mars. Still, the panspermia theory is fun to think about and helps to broaden our perspective of the possible origins of life on Earth.

5. The most obvious contribution of cyanobacteria was the introduction of the first molecular oxygen into the Earth's atmosphere, with the concomitant creation of the ozone layer. Cyanobacteria made plants possible by providing chloroplasts, the photosynthetic organelle. Cyanobacteria may have been one of the first examples of multicellular life and of differentiated cells, although these features of organisms could have been invented by an earlier microbe. (Only some cyanobacteria have these features.)

6. It is odd that so many eukaryotic microbes, including some fungi, can grow in the absence of oxygen, whereas this trait seems to have been lost in higher animals and plants. More likely, we just haven't looked hard enough for higher life forms in anoxic environments.

7. PCR is an example of a new process made possible by microbial diversity. Of course, the idea behind PCR was more important than the enzyme itself, but the simple fact that a thermostable enzyme obtained from a previously obscure Yellowstone Park bacterium created a vast new multi-zillion dollar industry made bioprospectors wonder whether there were more PCR equivalents to be found by mining microbial diversity.

8. We leave this one up to you. It is a difficult question to answer. Diversa was well-qualified to exploit microbes isolated from the Park and it did agree to share profits with the park, but many scientists were very unhappy about having one company controlling all the potential richness (intellectual as well as commercial) that might be found in Yellowstone.

9. The DNA sequence of an organism contains the total blueprint for what the organism is and what it can do. As scientists poring over genome sequences are finding, however, extracting this sort of information from a nucleotide sequence is not as easy as some thought it might be. The problem lies in how to decode the sequence to extract information about how a living organism organizes its various activities. Scientists are beginning to realize that the limitation on our ability to decode sequence information is our comparatively small base of knowledge about various aspects of the physiology of different organisms and how this physiology is related to DNA sequence. It is going to be necessary to return to studies of microbial (and macrobial) physiology to obtain a more sophisticated understanding of how various activities of an organism interact with each other and are deployed in response to environmental stimuli before we can even begin to extract the information stored in a genome sequence.

Chapter 3

1. It is very tempting to bypass the streaking step because it takes at least a day for colonies to form on the plate. But if the culture is a mixed one, the results of biochemical tests can give misleading results. Suppose that the organism of interest does not grow on sucrose but a contaminant does. There would be growth in sucrose medium, indicating that the organism of interest was a sucrose utilizer, however it is not. Both the chemical tests used to identify microbes and the tests used to assess susceptibility to antibiotics are reliable only if the culture inoculated into them consists of a single microbe. A clinical microbiologist tempted to take a shortcut might realize that the culture was mixed if the cells in the culture had different sizes and shapes, especially if the appearance of the culture changed after it was grown on a particular substrate (indicating that one component of the mixture predominated under that particular condition).

2. Jet fuel is a complex substrate, which actually consists of a mixture of compounds. A single microbe might not be able to carry out all the reactions necessary to degrade such a complex substrate, whereas a mixture of microbes might be more successful. To assure that a jet fuel–degrading mixed culture was stable, the scientist would check the mixed culture at intervals under the microscope to determine whether the ratio of different components appeared to stay the same after repeated transfer of the mixed culture on jet fuel medium.

3. To use metabolic tests for species identification, it is necessary to have some idea of what species your isolate might be, so that the appropriate tests can be run. To identify species by rRNA gene sequence, however, all that is needed is to amplify the gene and obtain its sequence. This is a major strength of the rRNA gene sequence approach. PFGE is a different story. Its purpose is not to identify a species but to determine whether the same strain is responsible for all the cases under study. Even if you do not know from the PFGE profile what species you are dealing with, you can at least say whether all the isolates were the same (single source outbreak) or not.

4. The most dangerous people in a clinical laboratory are those who are scared of every step and those who are not worried about any of them. Prioritizing safety breaches is essential not only to allow scientists to give the most attention to the most serious breaches, but also to prevent the scientists from becoming terrified of all of them or careless about all of them.

5. The medium was incubated in air. Any microbe that was killed or inhibited by oxygen (as most colon bacteria proved to be) or any microbe that required a different atmosphere (e.g., elevated carbon dioxide) would not grow on the medium, even if it contains a rich mixture of nutrients. It is important to realize that all media are selective in some way. The important thing is to understand each medium's limitations and strengths.

6. The concentration in the original sample was 5×10^5 per mL (50×10^3/0.1 mL). The number of colonies on the plate inoculated with 0.1 mL of the 1:10,000 dilution should be about 5, since there has been another 10-fold dilution.

7. To count accurately with a Petroff-Hauser counter, it is necessary to have separated bacterial cells. Bacteria that clump are difficult to count accurately. Clumps can sometimes be broken up with mild detergents. Motile bacteria present a problem because they move around. Fixing them to the slide with heat or in some other way makes them immobile. Another source of inaccuracy occurs if there are surfaces in the original sample (e.g., particles of soil) to which bacteria adhere. Since only bacteria in the liquid transferred in the dilution series will be counted, any bacteria that did not come off the solid material will be missed. Getting adherent bacteria off surfaces is a real challenge and no single method works in all cases.

8. A virus must first attach to a receptor molecule on the eukaryotic cell surface. If a cell line does not have that receptor, the virus cannot enter and infect the cell. Some types of eukaryotic cells are more permissive for replication of a particular virus than others, so once the virus is inside a eukaryotic cell, it may still face barriers to replication.

Chapter 4

1. Pili are used by microbes to adhere to surfaces. Flagella can sometimes perform an adherence role, but they do not bring the microbe as close to the surface as pili. Also, flagella, being much longer than pili, tend to break more often and do not form as stable an attachment to a surface as pili.

2. The targets of most antibiotics are found in both gram-positive and gram-negative bacteria. Targets such as DNA gyrase, peptidoglycan-synthesizing enzymes, and ribosomes are highly conserved in bacteria, so an antibiotic that interferes with such a target will affect may different types of bacteria. One limitation is provided by the gram-negative outer membrane. An antibiotic that is too big to diffuse through outer membrane porins will not be able to get to its target.

This is the reason vancomycin does not inhibit most gram-negative bacteria: it is too large to diffuse through outer membrane porins of most gram-negative bacteria.

3. This question invites you to second-guess microbes, a risky undertaking in some cases. An educated guess is that bacteria living free in water can only move very short distances. If nutrients become limiting in that site, the bacteria are out of luck. Bacteria on a moving surface, such as a ship's hull, can experience many different areas. Also, they do not have to expend the energy to move themselves around. The problem with this explanation is that it assumes that the bacteria "know" the ship will be moving around. In fact, a completely stationary ship will attract a similar dense microbial biofilm. Thus, there must be some advantage to biofilm formation independent of transportation to new sites. Protection from phage or protozoa, or even nutrients in the ship's hull, may be the attraction of the biofilm way of life. Interest in biofilms is very recent and little is known about why prokaryotes form them.

4. Chemotaxis works only if bacteria can move. Movement is the way bacteria sample the gradient of nutrients or whatever chemical they are moving toward or away from.

5. The cytoplasmic membrane must keep the cytoplasm, which contains a high concentration of small molecules, inside the cell. If there were diffusion pores in the cytoplasmic membrane, small molecules in the cytoplasm would leak out and the bacterium would be unable to carry out essential metabolic processes.

6. The outer membrane prevents some toxic substances, such as some antibiotics (e.g., vancomycin) and lysozyme (an enzyme that attacks peptidoglycan), from reaching their targets. The outer membrane allows the gram-negative bacterium to have a thinner peptidoglycan layer, which may save the bacterium energy and carbon. The periplasm is a place to store proteins such as β-lactamase so that they can be outside the cytoplasmic membrane where peptidoglycan is located but are prevented from diffusing away from the cell.

7. The outer membrane is not a phospholipid bilayer but has an inner layer of phospholipid and an outer layer of LPS. The outer membrane has pores through which nutrients diffuse, whereas nutrients are transported through the cytoplasmic membrane by specialized transport proteins, most of which use energy to transport nutrients. The cytoplasmic membrane is the site of the electron transport system, if the bacterium has one, and it is the membrane across which the proton motive force forms. The cytoplasmic membrane also contains regulatory proteins and proteins that secrete proteins. Overall, the cytoplasmic membrane has a much more complex set of functions than the outer membrane.

8. Obviously, it would be impossible to make large sections of peptidoglycan in the cytoplasm, then try to transport them outside the cytoplasmic membrane. Making the peptidoglycan precursors outside the cytoplasmic membrane presents a different type of problem. The sugars and amino acids that are assembled into the peptide subunit would diffuse away from the cell if they were transported out of the cytoplasm. Note that the peptidoglycan subunit that is exported is tethered to the lipid bactoprenol, then handed off to the growing peptidoglycan chain, so that it is never free to diffuse away from the cell.

9. Both penicillin and vancomycin inhibit the cross-linking step in peptidoglycan synthesis but they do it in different ways. Penicillin inactivates the enzymes that make the cross-link, whereas vancomycin binds to the D-ala-D-ala portion of the peptidoglycan and inhibits the cross-linking reaction by steric hindrance. These different modes of action explain why mechanisms of resistance to the two antibiotics differ. One exception to this rule is that the porins of gram-negative bacteria can exclude both types of antibiotics, although in the case of the smaller penicillins, the porin usually needs to be mutated to reduce the size of molecule it admits. Vancomycin is large enough to be excluded by most gram-negative porins.

10. Since the targets of penicillin and vancomycin are different, the mechanism of action is also different. For this reason, bacteria that become resistant to one do not become resistant to the other (unless you count the porin defense)

Chapter 5

1. See the Chapter at a Glance section.

2. Fluoroquinolone antibiotics inhibit DNA gyrase, an enzyme that helps restore the correct degree of supercoiling after a DNA molecule is replicated. Since none of these processes uses DNA gyrase, fluoroquinolones would not affect them. This statement is not quite correct if you define cloning to include the step where the plasmid is introduced into a bacterium, which then replicates the plasmid. Fluoroquinolones would inhibit the growth of the bacterium containing the plasmid. So whether a fluoroquinolone would affect your cloning experiment or not would depend on when the antibiotic was snuck into your experiment.

3. This is a good exercise for reviewing the steps in DNA replication.

4. The DNA segments from the plant genome would most likely be linear segments because plants contain enzymes that degrade DNA. The most efficient way for a linear DNA segment to enter the bacterial chromosome would be by homologous recombination. If the resistance gene had no sequence identity with the genome of the bacterium, homologous recombination would not occur. If there was sequence homology, the bacterium already has the gene, so entry of the DNA into the chromosome would not make a significant difference in the traits of the bacterium.

5. If radiation or other DNA-damaging substances are present and are having an effect on an organism's DNA, there are probably multiple breaks, not just one. Since the microbe has no way of knowing where these breaks are, it makes sense to try to save both copies of the chromosome.

6. There are lots of possibilities. First, replacement of your primers with different ones would prevent you from getting the PCR product you want. Adding salt or altering the temperature of the annealing step could make hybridization of the primer with the target sequence less specific, producing more than one PCR product. Adding an inhibitor of DNA polymerase would stop the PCR amplification, and adding a DNA degrading enzyme would destroy copies of the region of interest as they are made. Altering the temperature of the step where DNA polymerase copies the DNA could also destroy your experiment. Unfortunately, even in the absence of a villain, contaminants in a sample, faulty equipment, and your own carelessness, can wreak havoc on the best-planned PCR experiment.

7. A major difference between PCR and DNA sequencing is that the goal of PCR is to get many copies of a specific segment of DNA, whereas in DNA sequencing, the polymerase is intentionally halted at intervals to create a mixture of fragments of different lengths, all with the same 5′ end but different 3′ ends. From the different sizes of the fragments, the sequence is read. In the case of PCR, all fragments are the same length. The primers for both PCR and DNA sequencing need to be at least 20 bp in length to give the required specificity of the primer for its target segment. If, for example, you used a 10 bp primer for PCR, you would get multiple bands instead of a single one because a 10 bp sequence will be found in more than one place in the bacterial genome.

8. The aim of sporulation is not to make two daughter cells but to make a single spore. The process is asymmetric because the spore is smaller than, and has a different composition from, a normal daughter cell. In normal cell division, replication has to be symmetric to ensure that the two daughter cells are the same size and have the same cell components.

9. The chromosome of a cell contains many essential genes but also many genes that are not essential under all conditions. Thus, just having nonessential genes does not make a segment of DNA a plasmid rather than a chromosome. Also, genes carried by a plasmid can be essential for life under some conditions (e.g., antibiotic resistance genes). At first the distinction between chromosome and plasmid was made on the basis of size because the first-discovered plasmids were so much smaller than a chromosome. More recently, however, plasmids nearly as large as a chromosome have been discovered. Although there is still debate about when to call a large "plasmid" a second chromosome, most people agree that if the "plasmid" carries genes like rRNA or tRNA, genes that are normally found on chromosomes, the "plasmid" can be considered a second chromosome. Part of the disagreement over what to call a chromosome and what to call a plasmid arose from the fact that bacteriologists had gotten used to the idea that bacteria had only one chromosome. So when scientists started finding examples of what might be considered second chromosomes, there was considerable resistance to this idea. The idea that bacteria can have more than one chromosome is now widely accepted.

10. The longer fragments with only one defined end arise from the original chromosome. Since there is only one copy of the original segment of DNA being amplified (2 strands), each time only two of the longer amplified fragments are produced. Thus, these fragments increase linearly in number. Once the first DNA segment with the two defined ends (the segment to be amplified) is formed, that segment, and each newly formed segment, produces 2 more of itself per round, and these segments in turn produce 2 more copies of themselves. Thus, the number of the short fragments increases exponentially. If it were not for this feature of PCR amplification, you would not end up with a single predominant fragment with a single size.

11. During cloning, it is possible (although fortunately not common) to clone a segment of DNA from a minor contaminant in your culture. To make sure this did not happen, obtain DNA from your culture, use it to make a Southern blot and probe the Southern blot with your clone. If your clone came from your culture, it should hybridize with the DNA from your culture. If it does not, your clone is probably DNA from a contaminant.

Chapter 6

1. The slightly different structure of RNA helps to insure that the ribosome does not confuse DNA and mRNA. Also, the enzymes that degrade mRNA after it has been used as a message are not likely to degrade the chromosome by mistake. There are many reasons for distinguishing RNA from DNA. DNA polymerase starts from a primer because its job is to replicate the entire chromosome. Thus, it can start at intervals around the DNA molecule. RNA polymerase has to start at a single specific point to make an mRNA copy of a gene. Thus, a different mechanism is needed to dictate the binding of RNA polymerase.

2. To save an extra step in the computer programs that predict the amino acid sequence of a protein from the DNA sequence of the gene, computer programs simply use the DNA triplet corresponding to each codon on the mRNA to identify the appropriate amino acid. A gene is a DNA segment that encodes a protein or RNA species. An open reading frame is technically only a DNA segment that might encode a protein. Thus, scientists often distinguish between open reading frames, which are identified by computer translation of DNA sequence and a DNA segment that has actually been shown to encode a protein or RNA species.

3. The different sequences of promoters and ribosome binding sites allow the bacterium to make different amounts of different proteins and to make them under different conditions. Sigma factors, the proteins that allow RNA polymerase to bind a particular type of protein, are regulatory proteins in the sense that they make it possible for RNA polymerase to transcribe some genes and not others under a certain set of conditions.

4. The promoters controlled by repressors are almost always strong promoters that would direct constitutive transcription of a gene if something (the repressor) did not interfere. Activators are needed to help RNA polymerase bind to what are usually very weak promoters. Activators usually bind upstream of RNA polymerase because that is the best position for a protein that helps RNA polymerase to bind without getting in the way of transcription, once it starts. Repressors usually bind downstream of the promoter because their job is to prevent RNA polymerase from binding to the DNA and starting transcription.

5. A quorum-sensing system detects and responds to small molecules produced by the bacteria themselves, which help the bacterium to sense the population density of other bacteria of its kind. The autoinducer binds an activator or repressor reversibly. A two-component regulatory system usually senses molecules from the environment. A reversible phosphorylation is involved in sensing, rather than binding, of a signal molecule to the activator or repressor. Two-component systems almost always involve activators rather than repressors. A quorum-sensing system would be best for controlling genes involved in biofilm formation because the bacteria have to reach a high concentration to construct the biofilm. A two-component system would be best for controlling genes involved in the synthesis of flagella because motility is a single cell phenomenon and the need for synthesis of flagella is dictated by environmental stimuli.

6. A newly acquired gene may not have a promoter that is recognized in its new host. Mutations in the promoter region can create a new promoter that now allows the gene to be expressed.

7. Some of the reasons were already given in the answer to question 1. Another reason for having an mRNA intermediate is that it gives the cell new ways to regulate the amount of protein made by regulating the number of transcripts and the number of proteins translated from the message.

8. Sometimes, a gene encoding a protein that is not supposed to be very abundant will contain codons that are not used preferentially, thus slowing translation and making it less likely that many proteins will be made. Using codons that are not recognized as well as the dominant one allows bacteria yet another level of control of translation.

9. The end of the tRNA that is attached to the amino acid has to transfer its amino acid to the growing peptide chain, whereas the anticodon region of tRNA recognizes the appropriate codon on the mRNA. Thus, these portions of the tRNA must be separated spatially. The enzymes that attach the amino acid to the tRNA (tRNA synthetases) have to recognize the anticodon part of tRNA, as well as the site to which the amino acid is to be added, so that they add the right amino acid to the tRNA with the right anticodon.

10. The different sites for tRNA on the ribosome are needed so that the aminoacyl-tRNA can bind to one and the peptidyl-tRNA can bind to one; then a translocation can occur when the new amino acid has been attached to the growing peptide chain. The E site receives the empty tRNA, which then exits. This process allows the ribosome to move along the mRNA, acquiring new charged tRNA molecules and discarding empty ones as translation proceeds.

11. The translocation process is more complex than the initiation step, so it is easier to interfere with it by using antibiotics that bind to many different sites on the ribosome. Possibly the initiation step is more complex than previously thought. If so, more antibiotics might be found that target it in different ways, but so far it has been easier to find antibiotics that inhibit translocation. Inhibitors of tRNA synthetases would block an important step in initiation and elongation, the production of an aminoacyl-tRNA. The example of the antibiotic mupirocin shows that inhibiting this step would kill cells. The problem now is to find a less toxic antibiotic that interferes with this step. The different tRNA synthetases that add different amino acids are not identical to each other but are similar enough in their structures that an antibiotic could interfere with many of them. It is only necessary to stop the charging of some tRNAs that carry commonly used amino acids such as alanine and valine to stop protein synthesis.

12. Varying the ribosome binding site sequence allows the bacterium to get different levels of protein production from an mRNA. This is another level of regulation. In the example given in the chapter, the bacterium is getting different levels of protein production from the same mRNA (which encodes two proteins).

13. Mutations in the promoter that eliminate expression of the protein, mutations that abolish the ribosome binding site, a mutation that eliminates the stop codon, and mutations that add or delete bases from the open reading frame would all stop production of the enzyme. Mutations that alter the stop codon or the terminator can also have this effect if they make the mRNA or protein unstable.

14. If a bacterium has to make a large protein complex or a series of enzymes in a pathway, this takes a lot of energy and carbon. One way to save carbon and energy is to inhibit this complex or pathway temporarily and reversibly so that it can be brought back into play later when conditions change. Regulation at earlier levels of transcription and translation may require transcripts or proteins to be degraded and remade every time conditions change.

15. The sequences that identify promoters were identified by comparing many different promoters to find conserved bases (a consensus promoter sequence). In some cases, the identification of the bases involved was checked by mutational analysis, but in many cases only a sequence comparison was used. Most of this work was done using *E. coli*. Thus, when one tries to use a consensus sequence to identify promoters, especially in an organism other than *E. coli,* the consensus may identify the wrong promoter region. Another consideration is that a very poor consensus sequence is characteristic of promoters that require activators for full expression of the gene. Similarly, predictions of protein structure from the deduced amino acid sequence (from the DNA sequence of a gene) are based on a relatively few examples of proteins for which the structure is known. In both cases, promoter consensus and structural features of proteins, there is still too little information available from well-studied examples to be confident about this type of prediction. This does not stop scientists from using sequence features to make predictions, however.

16. In microarray experiments, the scientist is looking for CHANGES in gene expression. Thus, expression under at least two conditions must be compared. If only one condition is used, you do not know if an expressed gene would have been expressed under all conditions or only under the conditions being tested. The other part of the question is about the limits of gene fusion studies, in which a possible promoter region is cloned upstream of a reporter gene, such as *lacZ.* If the region cloned upstream of the reporter gene is too small, no expression may be seen even if there is a promoter in that region. Sometimes scientists try to do this type of analysis by checking for expression of a promoter from some other bacterium in *E. coli.* If expression of the gene requires an activator that is not present in *E. coli,* or if *E. coli* does not recognize the consensus sequence of the other organism, or if *E. coli* does not have the appropriate sigma factor, there will be no expression of the reporter gene and there will appear to be no promoter even if one is actually present.

Chapter 7

1. Mutation exacts a toll on bacteria because so many mutations are either deleterious or neutral. Beneficial mutations are in the minority. Also, although it is possible for a single mutation to confer a desirable trait (e.g., resistance to fluoroquinolone antibiotics), usually multiple mutations are required. This means that many bacteria will die in the process of mutating to achieve some desirable trait. Acquiring a gene that has already been developed by some other bacterium is usually a much faster route to acquisition of a desirable trait. Generally, a gene acquired from another bacterium by horizontal gene transfer is either beneficial or neutral. In theory, such genes might occasionally be harmful. For example, if a bacterium acquired a new tRNA gene that was not recognized by its tRNA synthetases, the new gene could interfere with protein synthesis. A newly acquired gene encoding an enzyme involved in synthesis of peptidoglycan could also be disruptive. By sampling viable bacteria, we detect only the successful cases of gene transfer. No one knows how many deleterious gene transfer events occur.

2. Transfer of rRNA genes appears to be much less frequent than transfer of antibiotic resistance genes. This could be due to either of two factors. First, antibiotic resistance genes are more likely than rRNA genes to be located on transmissible elements such as plasmids or conjugative transposons. This means that the resistance genes are not only more likely to be transferred in the first place but are more likely to be maintained in their new host. Second, a newly acquired rRNA gene may make an rRNA that destabilizes the ribosome and thus renders the strain that receives it less competitive.

3. DNA is a charged molecule that does not move readily through a phospholipid membrane. Thus, specialized protein complexes are necessary for the transport of DNA molecules just as they are necessary for protein export. All of the known DNA transporting systems (transformation, conjugation, phage transduction) use complex protein channels to accomplish the movement of DNA across membranes. As more work on the proteins involved in such complexes is done, the complexes involved in the different types of DNA translocation appear more and more similar. In fact, the protein complexes involved in DNA translocation bear some resemblance to those that secrete or export proteins.

4. The toxin genes of *E. coli* O157:H7 appear to have come from *Shigella* species, a closely related genus. Phage transduction might have transferred these toxin genes if the phage could infect both of these bacteria. No such generalized transducing phage is known. Also, there is enough difference at the sequence level to make homologous recombination of the toxin genes into the *E. coli* genome unlikely. If the toxin genes came to *E. coli* on a lysogenic phage, there must be a *Shigella* phage that can infect *E. coli.* Since the genes were carried on a phage that is capable of integrating in the *E. coli* genome, homologous recombination was not necessary. In general, transfer by lysogeny is a more frequent and more likely process than generalized transduction. This may explain why so many transfers of toxin genes seem to have been mediated by lysogenic phage.

5. Transfer of plasmids or conjugative transposons starts with a single-stranded nick at the transfer origin (*oriT*). A single-stranded copy is transferred. If the transferred molecule is not a circle but a linear segment, only part of the DNA segment is transferred. Despite the wording of the question, DNA transferred by conjugation does NOT HAVE TO BE a circle, it's just that transfer of a circle is more likely to transfer the whole plasmid or conjugative transposon than transfer of a linear segment. In transformation and phage transduction, only a portion of a linear molecule may be transferred. The fact that in conjugation an entire element can be readily transferred may explain why bacteria preferentially transfer important DNA segments by conjugation.

6. One possible answer to this question is that the mating bridge formed in conjugation was originally a channel for export of proteins to the extracellular medium, and that DNA segments came to bind to an exported protein. Some protein export channels can export proteins from the bacterium and inject them into another cell. Such channels will be described in Chapter 15. This type of channel could have been the origin of conjugation. Another possibility is that retrotransfer was the original purpose of conjugation. Retrotransfer allows the donor cell to "shop around" for plasmids from other cells. In this case, the donor gets something out of the exchange, whereas in ordinary conjugation, the donor only donates DNA. A third possibility is that conjugal transfer systems developed from transformation systems. Transformation could be a means for a bacterium to obtain food in the form of DNA, as well as new genetic information. None of these explanations is very satisfactory, however. Clearly, there is a lot we do not understand about conjugal transfer of DNA.

7. Conjugative transposons are similar to plasmids in the way their circular forms are transferred from donor to recipient. They differ from plasmids in that they do not replicate but are normally integrated into the genome of the bacterium. Being integrated may extend their host range because they do not have to replicate to survive in a new host. Also, an integrated element may take less of an energy toll on its host than a plasmid does because replication of a plasmid, especially one that is present in multiple copies uses more energy than an integrated element that replicates once along with the rest of the genome.

8. Consider the differences between the different modes of gene transfer. If genes are being transferred by transformation, the DNA will be in the extracellular fluid at some point, due to lysis of the donor, before the DNA is taken up by the recipient. Thus, adding DNA-degrading enzymes to the mixture of donor and recipient should stop the transfer of genes. DNA-degrading enzymes will not stop phage transduction or conjugation because the DNA is protected at all times either by the proteins of the phage capsid or because it is inside the bacterial cell. If the donor is transferring DNA to the recipient by conjugation, they have to make tight contact with each other. Placing the donor on one side of a filter that has pores too small to let a bacterium pass and the recipient on the other side should stop conjugation. If the pores are large enough to admit phage, phage transduction should still occur.

9. If the X-carrying plasmid was the only plasmid in the strain to start with and then became larger and self-transmissible, something from the chromosome must have inserted into the plasmid. If a conjugative transposon previously integrated into the chromosome has left the chromosome and integrated into the plasmid, the conjugation machinery of the conjugative transposon will now be able to transfer the X-carrying plasmid as well as itself. Conjugative plasmids and other elements such as transposons or conjugative transposons occasionally merge with each other to form new chimeric forms. This is the source of new plasmids.

10. In convergent evolution, the amino acid sequence of the protein encoded by a gene remains virtually identical because selective pressures prevent changes in this sequence. Thus, two virtually identical proteins could arise independently without being transferred horizontally. Recall from Chapter 6 that many amino acids are specified by more than one codon. Thus, two proteins with identical amino acid sequences could be encoded by genes that vary by as much as 20–30% at the DNA sequence level. By insisting that there is >95% identity at the DNA sequence level between a gene in organism A and a gene in organism B before horizontal gene transfer is considered as an explanation, scientists rule out the possibility of convergent evolution.

11. Assume that the strain that carries the putative β-lactamase gene is resistant to penicillin (100 μg/mL). Clone an INTERNAL segment of the gene into a suicide vector that does not replicate in the strain of interest, introduce the clone in the vector, and select for the marker carried on the plasmid (which should NOT be penicillin resistance). Integration of the plasmid should disrupt the β-lactamase gene. Now test your bacterial strain. If it can no longer grow on medium containing penicillin, the gene you cloned must be responsible for the penicillin resistance of the wild type strain.

12. Introduce a transposon, carrying a selectable marker such as a tetracycline resistance gene into the strain and select for tetracycline resistant colonies. Each of these colonies should have a copy of the transposon somewhere in the bacterium's genome. You are looking for a colony with a transposon insertion that inactivated the penicillin resistance gene. Screen colonies for the ability to grow on agar containing penicillin. (You may have to screen thousands of colonies. Why?) If you find a colony that can no longer grow on penicillin, this is the colony you want. Use the transposon as a marker to clone the gene it has interrupted. For example, you could clone the DNA segment by selecting for a cloned DNA segment that made *E. coli* resistant to tetracycline. Of course, when you do the cloning, you will have to use a restriction enzyme that does not cut within the transposon. (Why?)

13. A transferred gene must be fixed in the recipient in order for the transfer event to be detected. The types of gene transfer most likely to be eliminated in a recombination-deficient strain are transformation and transduction because the incoming DNA is linear segments and must be integrated into the chromosome to survive. In conjugation, a plasmid or conjugative transposon is transferred and neither relies on homologous recombination for survival of the transferred DNA in the recipient. How could you use this information to answer question 8 in a different way than the answer given above?

14. Consider the steps you would be taking in doing the mutagenesis experiment. After you introduce the integrating element (transposon or IS) into the bacteria, only a fraction of the bacteria will actually have insertions in their genomes because the element does not integrate 100% of the time. It is useful to be able to select only those colonies that have insertions in their genomes. Using a transposon that carries a selectable marker (e.g., an antibiotic resistance gene) gives you that option. An IS element usually carries no such marker.

15. This subject was introduced in the answer to question 12. First pick a restriction enzyme that does not cut inside the transposon, because you want the DNA that flanks the transposon. Also, you do not want to cut into the selectable marker carried by the transposon. Digest chromosomal DNA with the restriction enzyme. Mix the DNA fragments with a plasmid digested with the same enzyme and ligate. Use artificial transformation to transform *E. coli* and select for the marker carried on the transposon. Any clone you obtain should contain the transposon plus a portion of the DNA into which the transposon inserted. The DNA flanking the transposon is what you want.

Chapter 8

1. Gram-negative bacteria have an outer membrane and a periplasm. Those that do not excrete extracellular enzymes can use their outer membranes to bind the polysaccharide and begin degrading it. The breakdown products are stored in the periplasm until they can be further degraded and transported into the cytoplasm. Gram-positive bacteria that use surface enzymes to break down polysaccharides must have some mechanism for sequestering the breakdown products of the polysaccharide until they are transported into the cytoplasm. Otherwise, they risk having them diffuse away from the cell surface. How they channel breakdown products to the cytoplasmic membrane transporters is not known.

2. NAD(P)H and NAD(P) participate in oxidation reduction reactions. These reactions may be in catabolic pathways like glycolysis or in biosynthetic pathways. NAD(P)H serves as the reductant in many electron transport systems. NAD(P)H donates the first protons to be pumped out of the cell and the electrons that will be dumped eventually on a terminal electron acceptor. Protons pumped out of the cell can be used to drive phosphorylation of an ADP to an ATP. ADP can also be phosphorylated in reactions such as some of the reactions in glycolysis (substrate level phosphorylation). ATP is consumed in biosynthetic reactions, just as NAD(P)H is, but it serves as a source of energy rather than as a source of electrons and protons.

3. It is true that other bacteria get more carbon and energy out of glucose than the lactic acid bacteria by oxidizing pyruvate to acetyl-S-CoA and using the citric acid cycle to oxidize acetyl-S-CoA to CO_2. Yet the lactic acid bacteria can grow faster than many of these bacteria, especially under anoxic conditions, for two reasons. First, bacteria that use the citric acid cycle obtain energy because this cycle is coupled to a proton-pumping electron transport system. Such electron transport systems have to have a terminal electron acceptor such as O_2 or NO_3^-. If both of these electron acceptors are absent, the energy advantage of the bacteria with a citric acid cycle disappears. Second, the lactic acid bacteria can perform glycolysis very efficiently and make up, by the volume of substrate pushed through the pathway, for any energy deficit inherent in their metabolism.

4. Many obligate anaerobes, unlike the lactic acid bacteria, do not produce lactate as a major end product. They get an extra ATP by oxidizing pyruvate to acetyl-S-CoA and then through acetyl-P to acetate. They still have to reoxidize those 2 NADH molecules, however. They do this by making succinate or propionate, gaining still more energy in the process. Thus, *Bacteroides* species are likely to be contributing to the acetate and propionate found in the colonic environment. You were asleep so you did not hear that *Bacteroides* do not produce butyrate, but now you know.

5. ATP is not the only energy storing and energy transferring molecule. PEP and other phosphorylated compounds also have a similar role. What these compounds have in common is that they have a high energy phosphoryl group. When this group is removed by hydrolysis to generate ADP, energy is released that can be used to drive another reaction. Having ATP as the "energy currency" of the cell allows the cell more flexibility in how it channels the energy in ATP and other energy-rich compounds. They can be used to drive biosynthetic reactions or to drive the first steps in glycolysis or to provide energy in any other reactions that consume energy. NAD(P)H has a similar role as a storage compound for protons and electrons. It is available for any reaction where a compound needs to be reduced.

6. Biosynthetic reactions require very specific substrates, so although general use compounds like ATP and NAD(P)H participate in such reactions, there are also reactants that are specific for a particular reaction. Glycolysis and the citric acid cycle provide some of the substrates that are involved in early steps in biosynthesis of a variety of compounds. In effect, biosynthetic reactions "bleed" these compounds off the catabolic, energy-producing reactions. The reason this does not bring the catabolic reactions to a stop is that only a small amount of any intermediate in glycolysis or the citric acid cycle is actually diverted to biosynthesis. A very high volume of carbon flows through the catabolic reactions. Thus, any compound used for biosynthetic reactions is quickly replaced.

7. They are using the term "breathe" facetiously to mean respiration. Respiration is the use of an inorganic compound as a terminal electron acceptor for an electron transport system. If oxygen is the terminal electron acceptor, the process is called aerobic respiration; if another compound is used as the electron acceptor, such as nitrate or arsenate or iron, the process is called anaerobic respiration.

8. $\Delta G'$ is the sum of $\Delta G^{0'}$ and a term that is 2.3RT times the logarithm of the ratio of products to substrates. In the case mentioned here, where $\Delta G^{0'} = +20$ mV, a log term that has a value of less than –0.33 will have a $\Delta G'$ that is negative. Thus, keeping the concentration of products very low compared to concentrations of substrates can have a profound effect on the $\Delta G'$ value and can take it into the range where the reaction becomes exergonic.

9. Bacteria take advantage of the fact that the $\Delta G^{0'}$ values of the reactions involved in moving protons and electrons from NAD(P)H/NAD(P) to a terminal electron acceptor cause the reactions to flow from more negative E_0 values to more positive ones, with NAD(P)H/NAD(P) being much more negative than the electron acceptor couple. By passing protons and electrons alternately to compounds that accept both and then to compounds that accept only electrons, the series of oxidation reduction reactions that compose the electron transport system move protons out of the cell across the cytoplasmic membrane and dispose of the leftover electrons ultimately by dumping them on the terminal electron acceptor. The protons can then do work to drive transport of nutrients or pass through the ATP synthase, converting an ADP to an ATP.

10. Conceivably, a bacterium that produces multiple end products (e.g., acetate, propionate, CO_2) might need to pair with more than one partner, each of which uses one of the products, in order to keep product concentrations as low as possible. There could even be linear chains of syntrophic partners, where the first one produced an end product like butyrate that could be used under anoxic conditions only if the butyrate user had a partner that consumed its products (e.g., H_2 and CO_2). No such syntrophic free-for-alls have been documented, but research in this area is still in a fairly primitive state and it is conceivable that more complex syntrophic reactions than just the two partner ones described in this chapter remain to be found.

Chapter 9

1. The archaezoa certainly seem to be very primitive in the sense that they are more like bacteria than more advanced protists such as fungi, but does this mean that they actually are accurate representatives of the first-evolving protists? rRNA analysis supports their ancient lineage but the fact that they are mostly obligate parasites of higher animals or arthropods today is somewhat troubling. This is a trait that would be expected of a more recently evolving organism. As you might have realized by reading between the lines in this chapter, microbiologists are limited by the lack of a geological record that can be used to check other methods of deducing evolution, such as rRNA analysis. Another limitation is that so little is known about the free-living protists. There may be many examples of free-living archaezoans that have not yet been found.

2. Basically, the protists covered in this chapter can be broken down into three metabolic types. One type can use nonliving biomass as a source of carbon and energy. Examples are the termite hindgut protists and fungi. These protists are recyclers of biomass. A second type of protist engulfs and digests other microbes. This type of protist requires the presence of prokaryotes or some other microbe; it functions mainly as a grazer on other microbes, keeping their numbers in check. A third type has a photosynthetic metabolism and could thus be, like the biomass recyclers, free-living without the need for other microbes. The biomass recyclers and photosynthetic protists are more likely to be at the base of food chains, whereas the grazers on other microbes are further up the food chain. Before the larger life forms appeared, these early protists established food chains, which may have included bacteria and archaea. These food chains were extended as the larger life forms later appeared.

3. Scientists are looking for ways to attack insect pests such as roaches and termites by eliminating important microbes from their intestinal tracts. The advantage of this approach is that antibacterial or antiprotist compounds are less likely to be toxic to humans and animals than insecticides that attack the insect's nervous system or other more advanced physiological systems that are similar to those of humans. One disadvantage is that antibacterial compounds would be introduced into the environment and thus possibly increase resistance to that compound in other bacteria, some of which might infect humans.

4. A photosynthetic organism requires sunlight and thus would not do well in most parts of the human body. For a killer diatom or alga to emerge that actually caused human infections, the organism would have to have an alternate energy metabolism. *Pfiesteria* and some euglenoids seem to be capable of an alternate, nonphotosynthetic metabolism, so it is conceivable—although unlikely—that a diatom or alga exists somewhere that also has this capacity. Given the metabolic flexibility of microbes, it is risky to rule anything out. The most likely scenario for a killer diatom or alga is the scenario we have seen in the past: production of toxins that kill fish and injure humans or blooms and die-offs that cause hypoxia in bodies of water.

5. Fungi have considerable metabolic flexibility and do not rely on photosynthesis. Since the fungi are recyclers of dead biomass, it is not much of a stretch for them to switch to using live biomass as a food source. A separate problem to be faced by any organism that tries to infect the human body is the defenses of the human body, which are described in detail in Chapters 11–13. An infecting organism also has to surmount these defenses. Scientists hope that by understanding more about the features of eukaryotic and prokaryotic microbes that cause infections it will become possible to predict emerging infectious diseases more accurately.

6. Very primitive protists such as *Giardia* and *Trichomonas* have ribosomes that are more similar to those of bacteria than ribosomes of more advanced protists such as *Plasmodium* species. Thus, it is not surprising that some of the protein synthesis inhibitors designed to inhibit the growth of bacteria also inhibit the growth of these primitive protists. Similarly, *Giardia* and *Trichomonas* share some other metabolic traits with bacteria, such as production of folic acid, that are targets for antibacterial agents.

7. Although the systems of the human body such as the nervous system and the circulatory system look very complex, it is important to remember that they are composed of individual cells, which are not much more complex than the single-celled protists. Also, the protists are the ancestors of human cells. Thus, it makes sense that simple protists might serve as good models for many features of human cells. So far, this has proven to be the case. The hope is that by figuring out how simple protists carry out aspects of metabolism and gene regulation, scientists will be able to recognize the genes encoding these functions in the human genome sequence. Once this base of fundamental cell biology is identified, attention can be turned to higher order functions that may be peculiar to the higher animals, such as organization of cells and intercellular communication.

Chapter 10

1. Since the proteins on the surface of the virus make very tight and specific contact with a molecule on the surface of the human cell the virus attacks, the attachment of the virus can be prevented by having some chemical bind just as tightly either to the viral surface proteins or to the human cell receptor. The presence of a bulky chemical group on either the viral protein or the human cell receptor would prevent the tight binding of viral proteins to the human cell and thus stop attachment. As you will see in Chapter 13, human blood proteins called antibodies bind very tightly and specifically to viral surface proteins and prevent viral attachment in just this way. This is the basis of one type of immunity to viral infections.

2. It might be possible to stop uncoating of viruses that are taken up in endocytic vesicles by preventing acidification of the vesicle. This could be risky for the host, however, since acidification is a normal process for human cells. Uncoating of enveloped vesicles involves the matrix proteins, the target of the anti-influenza virus drug amantidine. Since these proteins are different for different viruses, it seems unlikely that a single drug would target all matrix proteins. Similarly, uncoating of naked viruses seems to differ from that of enveloped viruses enough that a single compound is unlikely to stop uncoating of all viruses. There is still so much to be learned about uncoating and other steps in viral replication, however, that it is a bit premature to give up hope of broader spectrum antiviral compounds. Similarly, other steps in viral replication vary so much from one virus to another that so far, antiviral compounds that strike at replication have been fairly virus-specific. Once again, this may simply reflect our ignorance of common features that have been missed to date. The important point to be made here is that viruses, although they may seem very simple at first, exhibit a considerable amount of variety in all stages of their replication cycles and in their protein and nucleic acid components.

3. A virus whose RNA genome is composed of a negative strand has to carry a preformed replicase to make the positive (or mRNA) strand. A virus with a positive strand genome can have that genome translated directly to produce the enzymes and other proteins it needs for replication. In the case of retroviruses like HIV, the virus needs to provide the enzyme reverse transcriptase that is not present in human cells. See the text for other examples. This is a good way to review viral replication strategies. Note that some viruses such as the double-stranded RNA viruses that would seem not to have to carry preformed replicase may nonetheless do so. Having a preformed enzyme may give the virus a "head start" in taking over the human cell replication machinery.

4. A stripped down virus with a nucleoprotein core and a capsid can be built with only a few proteins to organize the nucleic acid genome. This is the reason it does not take much nucleic acid to encode the basic features of the simplest viruses. Enveloped viruses need more coding capacity to take care of the envelope and matrix proteins. Viruses that use preformed enzymes also need more genes in their genomes. Even this additional complexity does not account, however, for the extra nucleic acid in the genomes of viruses with very large genomes. Most of the extra coding capacity of these viruses goes to regulation of the steps of replication. More complex viruses produce their proteins in waves of early, medium, and late proteins. They need regulatory proteins to organize this process. Viruses such as herpes viruses that replicate actively in one cell type but have a latent stage in another need even more complex regulatory machinery.

5. Look at the Chapter at a Glance. At every stage in the viral life cycle, there are possibilities for variation. For example, a virus may be internalized with or without endocytosis. It may replicate in the cytoplasm or in the nucleus. It may bud through the nuclear membrane, the endoplasmic reticulum, or the cytoplasmic membrane, or it may lyse the cell. Given the number of possible variations at each point in what might otherwise seem a simple replication cycle, it is not surprising that there is such variation in viral replication cycles.

6. Latency is the state in which a virus is present in a cell but not causing symptoms. The virus is capable of returning to the actively replicating stage (usually in some other cell type) at a later time. Latency is poorly understood, but it is clear that for a virus to be in a latent state, its genome must be present and intact but it must not be actively producing copies of its genome or its proteins. As you will learn in later chapters, the immune system recognizes virally infected cells by the presence of viral proteins in the membrane of the cell, so a latent virus must limit such evidences of its presence if it wants to escape the immune system.

7. Latency is obviously a complex process and only some viruses are capable of it. As mentioned in the answer to question 4, viruses capable of a latent stage need to have the regulatory capacity to switch from actively replicating mode to latency and back. Only a limited number of viruses seem capable of this complex organization of replication steps. It is fortunate for us that most viruses do not have a latent stage because the ones that do are the most difficult ones to eliminate from the body.

8. This is a question that has been debated for years and there is still no agreement on the answer. In fact, the answer is probably different for different viruses. What can be said is that genetic differences in the population, differences in immune status (including prior encounter with the virus or a near relative), the presence of underlying conditions that affect immune status, age, the number of viruses encountered, and the route of infection all seem to make a contribution. Since a number of factors can affect the severity of an infection, it is not surprising that there is a very large number of combinations of these factors and thus a large number of possible outcomes. Most of the debate has been over which factors are the most important ones.

9. What symptoms a virus causes depends on what cell type the virus infects. This specificity is determined by what human cell receptor the virus recognizes and whether the virus can replicate successfully inside the cell it invades. The nature of the viral genome and the shape and size of the capsid may have no bearing on these aspects of infection. Although using symptoms of infection as a means of classification may seem appealing at first, this approach has proven not to be very useful for any type of microbe, viruses included. Scientists have found that it is more useful to use traits like capsid shape and size, presence of an envelope, and properties of the genome to classify viruses.

Chapter 11

1. During most of evolution, people have been a lot dirtier than they are today. The acidic environment of the skin may have been much more important in the past when the bacterial load experienced by most people was far greater than it is today in our overly clean society. Nonetheless, it is interesting to contemplate possible dermatological consequences of neutralizing skin pH. Acne and other skin conditions seem to be on the increase, and there could be a connection.

2. Gloves provide very good protection as long as they are not breached. In surgical procedures such as heart surgery, where cutting through the rib cage creates jagged bone fragments, or in other types of surgery where sharp edges of scalpels or metal sutures can cause breaches, gloves no longer provide an absolute barrier to prevent skin microbiota from contaminating implants. There have been enough cases of implant contamination to raise the question of whether the current surgical preventive practices are sufficient to protect implant patients from the consequences of a contaminated implant.

3. Some scientists have suggested that *P. acnes* is an opportunist in the sense that it causes the lesions of acne only after some breach of the skin defenses has occurred. This theory has never been proven conclusively but it is a viable explanation for the observation that some people seem to be more prone to acne than others. *S. epidermidis* is clearly an opportunist since it causes infections most commonly in people with indwelling venous catheters or with plastic implants. Whether a particular bacterium is an opportunist or not definitely changes with the available information about how that bacterium causes infections.

4. Surgery and cancer chemotherapy are two recent innovations that have made the microbiota of the human body more dangerous, because they provide openings in the normal defense systems that allow members of the microbiota (which are always on the scene) to take advantage of these new openings. Few would argue that surgery and cancer chemotherapy should be discontinued because of this. It is important, nonetheless, to take account of the dangers new and useful medical practices open up for bacteria. As has been reiterated in earlier chapters, knowledge is power, especially if knowledge can be used for prevention.

5. Technically, dental caries are an example of a microbiota-shift disease because build-up of a specific population of bacteria on the teeth is necessary for the disease. One could also view dental caries, however, as a metabolism-shift disease. Bacterial plaque has been accumulating on human teeth for thousands (millions?) of years. Fibrous and abrasive foods in the diet of earlier humans may have reduced plaque, but it has always been present. Plaque did not become a serious problem until sucrose was introduced into the diet. The same plaque that might have had very little effect now became dangerous, rotting teeth due to its capacity to produce prodigious amounts of lactic acid from sugar. Are there other examples of microbial communities that have coexisted benignly with the human body for millions of years only to be rendered dangerous by some change in human diet or changes in metabolism induced by pharmaceutical compounds? This is an intriguing but unanswered question.

6. This question raises another issue in connection with what should be considered "normal." Since most of the scientific studies of human health issues are currently conducted in developed countries, the tendency has been to assume that people in developed countries are the norm. In fact, in many ways, people in developed countries are profoundly abnormal (compared with what humans have experienced for the past million years) due to the changes in diet, cleanliness, and medical interventions that have occurred in these countries. In developing countries, people are colonized much earlier with *H. pylori,* and ulcers are much less common than in developing countries. This has led to the suggestion that ulcers might be the result of too-late colonization with *H. pylori* rather than to the ability of *H. pylori* to cause infection. This view of the *H. pylori*–human balance is quite controversial but is interesting to consider in view of the drastic changes in the balance between humans and microbes that has occurred in developing countries over the past century or so.

7. Infant botulism is, fortunately, a very rare disease. Thus, treating all infants with antibiotics to prevent the disease would be overkill—and dangerous too since antibiotics further disrupt the development of the microbiota of the human body. Prevention (by avoiding foods like honey that contain *C. botulinum* spores) and awareness of the early symptoms of the disease (leading to rapid and effective treatment) is the best strategy for eliminating this condition.

8. Oddly enough, few studies have been done to answer the question of whether each of us has his or her own designer microbiota. Pulsed field gel electrophoresis could definitely be used to answer this question as well as to answer the question of where our microbiota comes from. Do children "inherit" the microbiota of their parents or does it come from some other source? There are some studies that suggest that an infant's micobiota comes from its parents or other caregivers. Why is this an issue? If some people are colonized with bacteria that are resistant to many antibiotics, passing these bacteria on to their children could put their children at risk for later post-surgical infections that are incurable. This is an aspect of the antibiotic resistance picture that has received almost no attention.

9. Antibiotics are not, as the term "magic bullet" suggests, specific for disease-causing bacteria. Treatment of a person with antibiotics is actually a treatment of every member of that person's microbiota. Antibiotics can cause changes in the composition of the microbiota of various parts of the human body, a fact that explains why people treated with antibiotics often experience diarrhea or (in the case of women) vaginitis due to shifts in the normal microbiota. Antibiotics that reduce the numbers of the normally protective microbiota can leave a person transiently open to infection by pathogens that normally could not compete with the normal microbiota.

10. The original opportunistic infections were unquestionably wound infections. Until World War II, by far the greatest numbers of fatalities among soldiers were due to infections of battle wounds, rather than trauma resulting from the wound itself. High fatalities among women giving birth were also due to opportunistic infections by bacteria that gained access to the upper genital tract during births, especially those that were prolonged. Malnutrition, a constant threat to human survival, was also a recurrent theme in the infectious disease history of the human species. In a sense, we have conquered the opportunistic infections of the past, only to open up new opportunities for such infections in the future. This statement should not be interpreted as defeatist, but rather as an acknowledgment that progress occurs by increments. We control one type of opportunistic infection, but at the same time due to life-saving medical interventions we create other opportunistic infections. The positive aspect of this picture is that at least we now know WHY new medical interventions create new opportunities for bacterial infections. Knowledge is the best protection against future problems because it gives us preventive strategies that protect us from them.

Chapter 12

1. They are nonspecific in the sense that they offer a barrier of protection against many different microbes. This is the reason that they are such an important line of defense.

2. The nonspecific defenses are a very powerful protective barrier against disease. Few microbes can breach these defenses successfully. For those microbes that can breach the nonspecific defenses, whether on their own or as opportunists, the specific defenses provide the next line of defense. Since the specific defenses are the backup for the nonspecific defenses, it makes sense that the two lines of defense are closely coupled to each other. Also, keep in mind that the specific defenses are usually deployable (on first encounter with the pathogen) only after a week or so. This makes coordination of the nonspecific and specific defenses even more critical, so that the specific defenses are being initiated even as the nonspecific defenses go to work to repel the potential invader.

3. Mucus traps microbes and thus prevents them from gaining access to vulnerable mucosal membranes. Elevating mucus production is a way the body has of pouring more of the protective mucus into areas where it is most needed.

4. Complement and neutrophils are two essential partners in a defense response that is extremely effective against invading microbes. It is easy to get lost in the details of either the neutrophil or complement response to infection, but to really understand how they work, it is essential to realize that they work as a team. Complement activation is what labels the invading microbes as dangerous and targets them for destruction by the neutrophils. Complement itself can be toxic for some bacteria (especially gram-negative bacteria, because of the membrane attack complex) but the main effect of complement is to call in the neutrophils via C5a and identify the microbes that need ingesting and killing by the neutrophils by coating them with C3b. In terms of a police force, complement is the dispatcher component guiding the police to the site of infection and identifying the guilty parties.

5. Not being ingested by neutrophils is just as destructive as killing neutrophils, because failure of encapsulated bacteria to be phagocytosed by neutrophils causes neutrophils to release their destructive barrage on adjacent tissue cells. Failure of neutrophils to engulf and kill encapsulated bacteria does not stop the neutrophils from coming to the site; and the local tissue damage caused by frustrated neutrophils continues to grow, leading to greater tissue damage.

6. The fact that activated neutrophils in full killing mode can cause considerable tissue damage is evident from what happens when encapsulated bacteria frustrate the neutrophil defense: extensive tissue damage can occur as neutrophils release their toxic components into surrounding tissue. It is clearly in the best interest of the body to keep this killing capacity of neutrophils under strict control until it is needed. Keep in mind that thousands of neutrophils course through the average person's bloodstream daily. Activating these neutrophils to full killing capacity could be lethal. Thus, the complement cascade and the cytokine response are designed to activate neutrophils only as they approach a site where microbes are growing.

7. The systems that control the activities of neutrophils normally work very well, and the worst sign of neutrophil damage of tissue is localized inflammation that disappears with time. The neutrophil defense system is clearly very effective and it is understandable that this system has been selected over time. Such a system of balanced lethality, in which the neutrophils are only activated under certain conditions, has the potential for an imbalance that escalates into massive destruction of human tissue. This is the situation that develops when the neutrophil defense system is activated all over the body, a process known as septic shock. The danger of occasional episodes of septic shock is the price we pay for a defense system that is normally controlled and effective.

8. Natural killer cells recognize aberrant human cells, including those that are infected with viruses or bacteria and those that are tumorigenic. How they do this is still not well understood. But the killer cells, like the neutrophils, respond to cells affected in a number of ways, including cells that are infected with a variety of viruses or bacteria. Like the neutrophil defense, the natural killer cell defense appears to be tightly linked to its backup, the specific defense system that responds to specific pathogens.

9. The components of bacteria that trigger septic shock have been identified and characterized: lipoteichoic acid, peptidoglycan fragments, and lipopolysaccharide. These molecules serve as generalized signatures for "invading bacteria." Clearly, there are similar signals associated with fungi, but these have not so far been identified. Presumably these molecules, like those on bacteria, interact with the cytokine-producing cells of the body to initiate the septic shock cascade. Also, presumably, these unknown fungal molecules normally elicit an effective nonspecific defense against fungi unless the fungi become so widely disseminated that they trigger the septic shock response.

Chapter 13

1. Some functions of antibodies, such as neutralization of toxins or viruses only require that the antibody coats the toxin or virus, preventing it from binding to its target host cell. Other functions of antibodies require that the antibody not only bind to its target but to a host cell that will engulf and kill it. This is true of antibodies that opsonize bacteria and target them for destruction. The antibody binds the bacterial cell via its antigen binding sites, leaving free its Fc portion, which binds to receptors on phagocytic cells, facilitating the ingestion and killing of the bacterium. Similarly, binding by an antibody (via its antigen binding sites) to a bacterium facilitates the activation of complement (via the antibody's Fc portion) so that the bacterium is more efficiently killed. The Fc portion effectively communicates antibody binding via its antigen binding sites to the cells of the body that can take maximum advantage of this binding event.

2. Antibodies, like complement, opsonize bacteria. Antibodies, being more specific in their action, can opsonize encapsulated bacteria, something complement cannot do. Complement is more nonspecific and is thus useful for general targeting of bacteria for destruction. Antibodies can neutralize viruses and protein toxins, preventing them from binding to human cells. Complement does not play either of these roles. Finally, sIgA is secreted into mucus and helps prevent microbes from reaching mucosal cells. Complement is confined to blood and tissue. Both complement and antibodies have structures that allow them to bind to target microbes in a way that leaves another part of the molecule to interact with phagocytic cells.

3. Different classes of antibodies have different roles. sIgA is secreted into mucus and acts mainly by binding incoming microbes and preventing them from reaching mucosal cells. sIgA is also secreted in mother's milk and protects infants from disease during the first months of life. IgG and IgM activate complement as well as neutralize viruses and toxins and opsonize bacteria. These antibodies are confined to blood and tissue. IgM appears first after exposure to a pathogen, then IgG appears. IgG levels are higher than those of IgM. IgG crosses the placenta and gives an infant internal protection during the first months of life.

4. Since sIgA is secreted into mucus, where complement is not found, and since the main role of sIgA is to bind microbes to the mucin layer, there is no need for this antibody type to activate complement.

5. Since IgM appears transiently early in the response to an infection, a single measurement detecting this antibody type is a good indication that the patient has an active infection due to the microbe for which the IgM is specific. Since IgG rises to high levels later than IgM and persists in blood for long periods of time, two measurements of the IgG titer, showing a rise in IgG levels, are needed to establish that the patient has an active infection. If the IgG levels stay constant or decrease in the two sequential samples, the infection is probably a past event and the IgG is simply a remnant of the response to that infection. An antibody titer is established by testing sequential, two-fold dilutions of serum for the presence of the antibodies. The titer is the inverse of the highest dilution that tests positive for the antibody. Thus, a titer of 64 means that the concentration of antibody in the sample is higher than in a sample for which the titer is 8.

6. If a Western blot were stained with a general protein stain, the portion of the blot corresponding to proteins from each well would look like a smear, due to the numerous proteins contained in a sample. The labeled antibodies are used as a stain of the blot because they are specific for a particular protein and thus stain only one band in each column of the blot, assuming that the protein is in the sample. Sometimes, a mixture of labeled antibodies is used to detect more than one protein. In that case, multiple bands may be stained, but the pattern still consists of discrete bands.

7. The APCs that process antigens have to be phagocytic cells because they need to ingest antigens and digest them into peptide fragments that are displayed on the surface of the cell. Neutrophils are not capable of doing this and are very short-lived, so they would not make good APCs. The longer-lived macrophages, which are normally found fixed in tissues rather than moving through the bloodstream, are better-positioned to detect and process incoming antigens. (Technically, infected cells are antigen-presenting cells since they "present" the antigen of the virus or bacterium that is infecting them on their surfaces so that cytotoxic T cells can infect them. The term APC is reserved, however, for cells that are designed to present antigens.)

8. The recognition of T cells by APCs and of B cells by T helper cells works by a lock and key type of mechanism. Thus, the antigen presented on an MHC of an APC fits with the T cell receptor, which in turn (if the receptor is on a T helper cell) fits with the antibody being presented on a B cell. The MHC-antigen of an APC would not make this type of connection with an antibody presented on a B cell. Another reason for the existence of T cells is that some T cells are cytotoxic T cells, which kill infected cells; others are T helper cells that activate the antibody-producing B cells. Thus the T cell layer of the immune response provides the body with a decision point for determining the type of immune response that would be most effective in eliminating the infecting microbe.

9. sIgA that binds viral surface proteins would prevent the virus from reaching and infecting mucosal cells. As a backup, IgG that binds the virus would neutralize any virus that escaped the sIgA defense. A cytotoxic T cell response that killed virus-infected cells would serve as a further backup. A response that lacked the sIgA component is going to be less effective than one that had this component. A completely nonprotective response would be one in which the antibodies target an internal protein of the virus (e.g., capsid proteins of an enveloped virus or proteins of the nucleoprotein core). Since these proteins are not on the viral surface, antibodies directed against them would not neutralize the virus. Also, these proteins are much less likely than the envelope proteins (assuming the virus has an envelope) to be displayed on the surface of infected cells, so a cytotoxic T cell response against such proteins would be less effective than one directed against envelope proteins. Sometimes, if a whole virus is used to elicit an immune response, the internal proteins are more immunogenic than the surface proteins. If this happens, an inappropriate immune response may be elicited. Another type of nonprotective response arises in the case of viruses that infect phagocytic cells such as macrophages. In this case, antibodies that bind the surface of the virus may actually help the virus to be ingested by phagocytic cells. Uncoating of the virus removes the surface-bound antibodies and releases the nucleoprotein core of the virus to proceed to replicate.

10. The first decision point is which antigens the APC displays on its surface. Some antigens are more likely to be displayed (more immunogenic) than others. The next level of decision is between the T helper reponse and the cytotoxic T cell response. One factor that affects this decision is whether an infecting microbe stays in the phagocytic vacuole of an APC or invades its cytoplasm. There are probably other factors affecting the T helper/cytotoxic T cell decision, but if so these remain to be discovered. Another decision point occurs in the T helper cell pathway, where the decision is between the TH1 and TH2 response. In a nonprotective immune response, mistakes can be made at any of these decision points. A major challenge in vaccine formulation is to control the decision-making process so that an effective immune response is mounted. Cytokines play an important role in the decision-making process but this role is still not completely understood, as is evident from the failure of many attempts to use cytokines to control the type of immune response in experimental animal models. What types of cytokines are produced also constitutes a type of decision point for the body, but this remains to be elucidated.

Chapter 14

1. This is a complicated question that involves a variety of social and political factors. In addressing this for yourself, you might want to consider the issue from the viewpoint of a parent of small children, a person whose religion forbids vaccination, and a retired person with grown children. These different perspectives will help you to see the many ramifications of this issue, which are likely to surface in the future.

2. This question is intended to highlight an important feature of some viral diseases. Before the rubella vaccine became available, rubella was primarily a disease of children, and most children had it before the age of 10. These children were immune to reinfection. A subset of children did not contract the disease and were not immune. Among these were girls who later became pregnant and were exposed to infected children. In these young women, the disease had devastating effects on the children they were carrying, producing birth defects. The vaccine is, ideally, administered to everyone and not only makes everyone immune but eliminates the infected group of children who might transmit the disease to an unprotected pregnant woman. Other diseases like measles and mumps are far more serious in those who escape exposure in childhood and contract the disease as young adults. This is the reason the MMR vaccine is administered to older children, although it is ideally given to infants, to make sure older children who missed earlier vaccinations are protected from the more serious form of the disease.

3. This is, unfortunately, a no-brainer. The bacterium that causes whooping cough, *Bordetella pertussis,* was still around in the U.K. and once the number of protected children dropped below 50%, an outbreak of whooping cough occurred that caused a number of deaths. It also produced neurological damage in some of the survivors. This terrifying example convinced parents that vaccination was far preferable to the disease, and that rare side effects (if indeed they exist) are a small price to pay for avoidance of an outbreak of the real thing.

4. Although this chapter sings the praises of vaccines, it is important to realize that vaccines are not magic. There have been cases in which a vaccine actually made the disease worse. An experimental vaccine against respiratory syncytial virus (RSV), a common cause of lung disease in infants, had this effect, and clinical trials were rapidly suspended. Why such events can occur is still not clear. One possibility is that the vaccine inadvertently triggers a TH2 rather than a TH1 response and this misdirection of the immune response somehow weakens the normal response to infection. Another possible scenario is that the vaccine triggers an attack on human tissue (autoimmune response). Today, it is possible to avoid this latter outcome in most cases. There are probably other reasons for failed vaccines that will be discovered in the future, possibly by one of the readers of this book. Obviously, it is crucial to understand why a vaccine like the RSV vaccine, which looked very promising in animal trials and in trials involving human adults, malperformed so disastrously when tested in infants. Most of the infants were not harmed irreversibly, but even the prospect of worsening the symptoms in such a vulnerable population is frightening.

5. In many developing countries, vaccination clinics cannot afford to use a new needle on each person vaccinated. The same needle may be used to vaccinate many people. This enhances the chances of transmission of any blood-borne disease like HIV and Ebola virus. Such events are rare and most people agree that vaccination saves far more lives than are lost in such episodes. Still, this degree of risk is completely unacceptable. One alternative is for governments of developing countries to spend money on the health of their citizens rather than on ill-considered wars against their neighbors. Since this appears to be an unlikely development, the next best solution is to develop more vaccines that can be administered orally or nasally, thus bypassing the need for injections.

6. Antiviral vaccines usually elicit antibodies that neutralize the virus (prevent it from attaching to host cells) or cytotoxic T cells that attack infected cells. Many of the antibacterial vaccines enhance phagocytic killing of bacteria. Antibacterial vaccines can also prevent attachment of bacteria or their products to host cells. Thus, the vaccines against diphtheria and tetanus neutralize the bacterial toxins, preventing them from attaching to host cells just as antibodies neutralize viruses.

7. Poliovirus has been mutating much less rapidly than the influenza virus. Thus, vaccination with the polio vaccine makes a person immune for life. Another feature of poliovirus that has made its eradication a real possibility is that it seems to be a human-specific virus. As will be seen in a later chapter, influenza viruses of animals and birds can mix with human influenza virus to create hybrids that are not stopped by the immune response of people vaccinated with the older form of the virus.

8. Passive immunization is the injection of antibodies from another source into a person's bloodstream, thus mimicking temporarily the immune response of a person immune to the diseases. Since antibodies are injected, there are no memory APC, T and B cells to respond to the next encounter with the disease-causing microbe.

9. An example of this type of approach would be treatment of infections caused by strains of the bacterium *Staphylococcus aureus*. *S. aureus* is a common cause of wound infections, especially post-surgical infections (Chapter 19). There are strains of *S. aureus* that are resistant to so many antibiotics that they are almost untreatable. Some *S. aureus* strains produce protein toxins that contribute to disease. Antibodies that neutralize these toxins could help to suppress the disease. Antibodies that bind to the surface of *S. aureus* could help neutrophils to engulf and kill the bacteria. In other words, passive immunization could help replace failing antibiotics in the battle to control the infection. The reason this approach might be preferable to vaccination is, first, that there is no *S. aureus* vaccine, and even if there were, only a subset of people are at risk for the disease. There is a limit as to how many vaccines can be reasonably administered in a cost-effective way. Passive immunization makes it possible to have "designer" temporary immune responses to specific disease problems.

Chapter 15

1. We leave this one to you. It is important for you to practice your creativity since you are the generation that will confront many of the questions raised by this text.

2. It is easier mathematically to determine accurately the 50% value in a curve like those usually seen in experimental determinations of ID_{50} and LD_{50}. The 90% value can also be deduced and is used in some cases, but the 50% value is less ambiguous.

3. Diseases with an LD_{50} but not an ID_{50} are diseases caused by bacterial or fungal toxins, which kill but are not associated with an infection. An example would be botulism. Some toxin-associated diseases such as diphtheria and tetanus could be considered to have an ID_{50} (colonization of the throat or wound) as well as an LD_{50} (toxin necessary to cause death) in which colonization and toxin production are separate parts of the equation.

4. In the case of ID_{50} or LD_{50}, the saying is "less is worse," the lower the number the more infectious/lethal. The worst diseases have the lowest LD_{50}s. In practice, except for purely toxin-mediated diseases, the ID_{50} and the LD_{50} are closely linked (infection is required for lethality in low LD_{50} infections), so an infection with a high ID_{50} and a low LD_{50} could not exist. If we revert to the designation of attack rate and case fatality rate, a disease like anthrax could be considered to have a low attack rate but a high case fatality rate because the case fatality numbers are based on fatalities in those who develop symptoms of a disease.

5. The disease syphilis provides an example of how Koch's third postulate can be satisfied in a different way. Because the spirochete that causes syphilis (*Treponema pallidum*) has never been cultivated, and since the animal models do not adequately reproduce the disease, it is easy to say that Koch's postulates have not been satisfied for this bacterium. Yet few would argue that syphilis is caused by anything but *T. pallidum*. An alternative to Koch's postulate number 3 is to show that antibiotic therapy based on the assumption that *T. pallidum* is the cause of syphilis eliminates the bacterium and the symptoms of the disease. Similarly, a preventive measure (eliminating a disease-causing bacterium from the water supply or use of a vaccine directed against a specific microbe to prevent a disease) that worked to prevent the disease could also be used to satisfy Koch's third postulate indirectly. Koch would be happy to know that everyone today believes that *Vibrio cholerae* causes cholera despite difficulties in satisfying Koch's postulates directly, since this was the example that caused Koch to disavow his own third postulate. Sometimes, overwhelming public health considerations cause scientists to bypass Koch's third postulate if an improvement in public health is the likely result.

6. This is an example of the importance of intellectual flexibility. Let us use periodontal disease as an example. Koch's first postulate would be amended to state that a particular change

in the microbiota (from predominantly gram-positive to predominantly gram-negative) is associated with the lesions of the disease. Here, the second postulate, cultivating all members of the microbiota shift, may be a problem and scientists may want to go to an equivalent to the third postulate: when the normal balance of the microbiota is regained (through use of antibiotics or by some other means) the symptoms disappear. Koch's postulates are obeyed in spirit but not in literal fact because of the complexity of the microbiological situation.

7. Scientists are looking into ways of linking potent toxins such as diphtheria toxin to therapies for diseases ranging from AIDS to cancer. The basis of this approach is that because the B component of a two-component toxin determines what type of cell the A portion enters and kills, replacing the normal B component directs the toxin to cells that need killing. Thus, replacing the B component of diphtheria toxin with an antibody or other protein that binds to a tumor cell will direct the A component of diphtheria to kill tumor cells. This has worked *in vitro* but is still not ready for actual therapy. Such hybrid toxins, however, may have important therapeutic applications in the future.

8. As bacteria that cause post-surgical infections become more resistant to antibiotics, physicians worried about saving patients who contract such infections are using every advantage they have to do so. One is to gain early information about the MIC of different antibiotics for the organism causing the infection. Since the MIC gives quantitative information about what concentration of antibiotic might be effective, physicians can use this information to increase the dose they are giving, thus hopefully increasing the concentration of antibiotic experienced by the disease-causing microbe. Also, by introducing the antibiotic intravenously, a physician can maintain higher concentrations of an antibiotic in the tissues affected by the infection than could be achieved by intermittent oral doses. Of course, such increases in administration of antibiotics above the approved doses can increase toxicity of the antibiotic, a physician faced with a patient who may be dying has to make hard decisions about toxicity versus successful containment of the disease.

9. The Achilles' heel of the MIC test is whether the organism being tested grows in the MIC test medium. If there is no growth in any of the tubes of the MIC test, including the tube with no antibiotic, the organism of interest is clearly not growing under the medium condition used and the results of the test are meaningless.

Chapter 16

1. Macrophages are not only a first line of defense in the lung, but are also antigen-processing cells. The ability of bacteria to grow in macrophages gives them two advantages. First, they are not killed by the macrophage. Moreover, the macrophage provides a source of nutrients and eventually helps to create an area of dead tissue that protects the bacteria in the sense that it is much more difficult for cells of the immune system and antibiotics to enter this area, which now has no blood supply. Second, killing of macrophages by bacteria may interfere with the development of an effective immune response since it eliminates antigen-presenting cells. Bacteria that grow in and kill neutrophils have similar effects, except that neutrophils are not antigen-presenting cells. If the effects are similar, why does killing of alveolar macrophages produce a granulomatous response, whereas killing of neutrophils—although it can produce a zone of necrotic tissue—generally does not produce granulomas (hardened calcified lesions)? The fact that there is still no good answer to this question illustrates how much we need to learn about *M. tuberculosis* and other microes (such as lung fungal pathogens) that elicit the granulomatous response. There is something special about the way B and T cells are recruited to the site and about the resulting interactions among immune cells that is responsible for the granulomatous response. A granulomatous response is not specific for the lungs, since such a response can be elicited in other areas of the body.

2. The vaccine against influenza virus targets certain variants of the viral envelope proteins, hemagluttinin and neuraminidase, that are circulating at the current time. A big change in one or both of these antigens (antigenic shift) would render the vaccine ineffective. Since such a change could also render ineffective any other immunological memory resulting from past vaccinations or episodes of disease, even vaccinated people would be as unprotected as anyone who had never encountered the virus. The fact that antigenic shift has happened more than once in this century makes health officials worry that it could easily happen again. It is instructive to note that none of the antigenic shift episodes since the one that caused the 1918 pandemic that killed so many people has been as widely lethal. This shows that not all such events are equally dangerous. Yet the intervening antigenic shift events have been associated with a significant rise in mortality and thus are not to be taken lightly, especially in countries with aging populations. The available anti-influenza drugs, amantidine and rimatidine, are generally effective against all serotypes of influenza A virus, but they merely lessen the severity of the disease, not cure it completely. To date, these drugs have been reserved for treatment of cases in the elderly and other groups with a high risk for contracting severe, lethal infections. How they would perform if used to stop an epidemic in the general population is not clear.

3. Most strains of *Streptococcus pneumoniae* are still treatable with several antibiotics. Even *Haemophilus influenzae* and *Klebsiella pneumoniae,* though more resistant to antibiotics, are still treatable with some antibiotics. Yet, the trend toward increasing resistance in all of these species has made it clear that something must be done to prepare for those cases of untreatable infections that have already begun to appear. A better vaccine that elicits antibodies that bind the bacterial surface is an obvious solution for people with healthy immune systems. As mentioned in this chapter and Chapter 14, however, the variety of capsular antigens makes development of a polysaccharide vaccine, even a conjugated one, problematic. Attempts are being made to find surface proteins, such as

PspA, which are more highly conserved and are better antigens than polysaccharides. For immunocompromised people, vaccination may be ineffective. An alternative is to use passive immunization, introduction of protective antibodies obtained from another source. These foreign antibodies would aid temporarily the phagocytic cells, which are still active in most immunocompromised patients, to ingest and kill the bacteria. Of course, this strategy requires that scientists find antigens to target that are more conserved than the polysaccharide capsular antigens of *S. pneumoniae.* Alternatively, a rapid test to determine which antigenic type is causing the infection could be used to choose the appropriate antibody. Maintaining lots of antibodies against even several capsular serotypes, however, would be a very expensive proposition for hospitals. This strategy is unlikely to be feasible in an epidemic that involved many people. Antibiotics remain the best and most efficient treatment. The best strategy would be to stop the increased resistance of bacterial lung pathogens. Two major forces for selection of resistant strains is use of antibiotics by physicians to treat viral infections such as cold, flu, or sore throat. The antibiotic does nothing against the virus, but it provides a selective pressure on such bacteria as *S. pneumoniae* that are commonly found in the mouth and throat that encourages the development of resistance to antibiotics. Another common practice that has come under attack is the use of antibiotics to treat earache in children. Most of the strains of *S. pneumoniae* are already resistant to penicillin and other older antibiotics, thus encouraging physicians to turn to antibiotics that should be saved for life-threatening infections. Since most earaches resolve spontaneously within a day or two, the recommended strategy is to wait for 24–48 hr before considering antibiotic treatment.

4. This question brings up a very important point: how can the same bacterium be benign in some parts of the body yet cause life-threatening infections in others? The answer goes back to the basic principle that symptoms of an infectious disease are often caused by the body's response to the bacteria rather than direct action by the bacteria themselves. The reason *S. pneumoniae, K. pneumoniae,* and *H. influenzae* cause very similar diseases is that they evoke the same type of inflammatory response. Similarly, some fungal lung pathogens (not covered here) cause a disease very similar to tuberculosis (despite the fact that they are obviously very different from *M. tuberculosis* in a number of ways) because they evoke the same type of granulomatous response. In the mouth, *S. pneumoniae* colonizes oral tissues without eliciting an inflammatory response and thus does not cause disease in that site. Only if *S. pneumoniae* can reach the lungs or bloodstream does it cause disease.

5. The main problem with the skin test is that it does not distinguish between those with active disease, those with prior disease that has been resolved, those in danger of reactivation, and those who have been vaccinated with BCG. Moreover, it takes several weeks after exposure for a person to become positive, and some immunocompromised people become negative on the test. Thus, there are both false positive and false negative responses. Despite this, the test has two major advantages. It is simple and cheap to perform, and results are available within a day or two. Assuming that the person interpreting the test is aware of its limitations, the test can be very useful in identifying people at risk for the disease. Given that the alternative tests, acid-fast staining (which requires an expert to perform and interpret), and cultivation of the bacterium (which can take weeks), the simple skin test becomes very attractive indeed. Scientists are currently working to develop a skin test that is more discriminatory, but so far no such test is available.

6. This is a very difficult question to answer. Certainly, anyone who tests positive on the TB skin test should be given further tests to look for signs of active disease. The question arises as to what to do with those who have recently converted to skin-test positive but show no signs of active disease. Most infectious disease specialists are inclined to take the wait-and-see approach in such cases, reserving treatment only in cases where active disease appears. The problem with treating people without active disease is that the anti-tuberculosis drugs can have unpleasant side effects, and people are understandably reluctant to be submitted to a months-long therapy with side effects if there is no evidence of active disease. Emotional and political factors can enter the picture, however. The authors have seen several cases on their own college campus, where a student is found to be skin test positive. If this student has been in large classes or a dorm with many other students, there can be tremendous pressure brought by parents and politicians to take draconian measures, including forced treatment of the student and any close contacts. As usual, public education is the best way to handle such emergencies, but university officials confronted with lurid news reports and demands for immediate action are not necessarily in the best position to undertake such an educational effort.

7. *P. carinii* appears to exert its effect by a direct contact-mediated toxicity for lung cells, rather than producing a capsule or killing macrophages. This may explain the different pathology of *P. carinii* infections (honeycomb lung tissue rather than areas of necrosis or granulomas). Another difference is that whereas the other lung pathogens covered in this chapter can cause infections in otherwise healthy people, *P. carinii* infections are seen exclusively in immunocompromised people, especially AIDS patients. The fact that most children have developed an immune response against *P. carinii* by age 4 years, with no sign of infection, shows the immune response of healthy people is sufficient to take care of *P. carinii* very handily.

8. The answer to this question was introduced in the answer to question 2, but it is worth reiterating. For a long time, physicians have known that antibiotics were ineffective against viruses and were thus of no value in treating colds, flu, and sore throats except in those rare cases where a secondary bacterial infection was likely to occur (e.g., in elderly patients with influenza). They felt, however, that since there were no adverse effects of the antibiotic and the antibiotics made patients happy, there was no problem with prescribing a treatment known to be ineffective. In recent years, physicians and scientists have become aware that there is a significant danger in this approach to prescribing antibiotics: it creates a strong selection pressure for development of resistance in bacteria that can later cause serious infections. Any single person administered antibiotics for a cold will probably not have resistant bacteria arise, but on a population level, especially in view of the massive amounts of antibiotics prescribed in this way, the selection pressure has clearly created new generations of resistant bacteria. This is emerging as a serious public health problem. It can also operate on an individual basis. A person who is in the habit of seeking antibiotics for treatment of colds and sore throat may well develop a resistant microbiota, which includes such significant potential pathogens as *S. pneumoniae.* Such people could be considered walking time bombs in the sense that if they develop bacterial pneumonia or a blood stream infection caused by their own microbiota (such as a post-surgical infection), they will be at a significantly higher risk of suffering an infection that is difficult to cure, or incurable.

9. An effective antiviral vaccine target is the surface proteins that mediate adhesion to host cells, the first step in the infection process. Unfortunately, the number of antigenic types of these surface proteins in viruses that cause colds and sore throats is so numerous that a vaccine that targets surface proteins could only be effective against a tiny fraction of disease-causing viruses. Scientists are now considering using the receptor for many of these viruses, ICAM-1, as a target instead. The most likely strategy will be to administer small molecules that block binding of viruses to this receptor. Eliciting antibodies to such a widespread human protein could be dangerous if it elicited an autoimmune response. Even the small molecule approach could have adverse side effects unless it could be limited to areas of the body that are most affected (the upper respiratory tract). Numerous attempts have been made to design antiviral compounds that would target this and other steps in viral replication, which are common to most cold-causing viruses. Although some promising candidates have been found, they all have the same drawback: unless they are administered very early in the infection process, they do little to reduce the duration of symptoms. Thus, people who wait too long to start taking them experience no relief.

10. These tiny bacteria have very small genomes and also lack a peptidoglycan cell wall. *M. pneumoniae* has cholesterol in its cytoplasmic membrane. It associates very tightly with host cells, but does not enter them. Like *P. carinii, M. pneumoniae* seems to exert its toxic effect by an unknown contact-dependent mechanism rather than by eliciting an inflammatory response. *C. pneumoniae* is a newly discovered respiratory pathogen about which little is known. It is an obligate intracellular pathogen that invades cells and replicates inside them, releasing survival forms called elementary bodies that are metabolically inactive but can infect new host cells to spread the infection. *C. pneumoniae* appears to lack a peptidoglycan cell wall, although it has a gram-negative type cytoplasmic and outer membrane. *C. pneumoniae* elicits an inflammatory response and this may be the way it causes symptoms.

Chapter 17

1. Initially, scientists tested different antibiotics for their killing effect on *H. pylori* by using the MIC test described in Chapter 15. They found that low concentrations of many antibiotics would kill *H. pylori* in this test. In addition to the MIC test, scientists also tested each of the concentrations of antibiotics where no growth occurred to determine whether the antibiotic was bacteriostatic or bactericidal. They wanted antibiotics that were bactericidal because bacteriostatic antibiotics, which only inhibit the growth of the bacteria, not kill them, need the help of phagocytic cells to clear the infection. The fact that the inflammatory response to *H. pylori* in the stomach was not clearing the bacteria was a clear warning that bacteriostatic antibiotics were likely to fail in this case. A limitation of the MIC test is that it only tests the effects of an antibiotic on bacteria growing in laboratory media. These media are a far cry from the conditions experienced by *H. pylori* in the stomach. Physiological differences induced in bacteria by different growth conditions can have a profound effect on the activity of antibiotics on that bacterium. This may be the reason why the MIC results proved to be such poor predictors of how antibiotics work on *H. pylori* in the stomach. The MIC results were not completely useless but a bit of trial and error and the use of antibiotic combinations proved to be the key to successful treatment of *H. pylori* infections in the end. One reason for learning more about the conditions bacteria experience at various points in the infection process is that this information might aid in the development of MIC-type tests with more predictive value.

2. In some ways this is a question without a good answer. Yet it is worth considering because its answer has profound consequences for the food industry. It is virtually impossible to render meat, especially ground meat, totally free of bacteria. Surface sterilization of foods by irradiation may come close to that goal for food that is not ground, but the prospect of food totally free of bacteria is probably unattainable unless consumers are willing to pay prices that are at least 10-fold higher than those at present. This means that agencies whose job it is to insure a safe food supply must set allowable contamination levels for various bacteria. Animal trials used to be the gold standard, and these indicated, at least in the case of *Salmonella* and *E.coli,* that the ID_{50} was on the order of 1000 bacteria or more. This number was called into question by very crude estimates based on analyses of human outbreaks, where the estimated ID_{50} was closer to 10 or fewer. In the case of *Listeria* and *E. coli* O157:H7, the debate takes on another feature because although the attack rate of these infections in the human population is clearly low, indicating a high ID_{50}, when such cases occur, they are more likely to be lethal than cases of other food-borne infections. Should there be a lower allowance for contamination for this reason alone? This is currently being debated.

One thing that seems clear from analyses of human outbreaks of disease, *Salmonella* and *Campylobacter* have a high attack rate but a very low case fatality rate. By contrast, the case fatality rate for *Listeria* infections is very high, even though the attack rate is quite low.

3. Most of the available information on food-borne infections is based on the analysis of large outbreaks. In the case of *Salmonella, Campylobacter,* and *Listeria,* the question has been raised as to how often isolated infections occur, and if they are common, what is their effect on public health. There is very little information on this subject. If an outbreak of listeriosis has been reported in the newspapers, a physician is likely to identify stillbirths occurring during that time as possibly caused by *Listeria,* however, if the same physician saw a stillbirth in the absence of such news coverage, how likely would he or she be to make the *Listeria* association, especially since there are many causes of stillbirth? Because *Listeria* is a preventable cause of stillbirth, physicians are counseling pregnant women on how to avoid infection. The CDC has inaugurated a new program that will attempt to assess isolated cases or small clusters of cases of salmonellosis and campylobacteriosis.

4. As mentioned in the previous answer, the CDC has decided to try to monitor cases of salmonellosis (and to a lesser extent campylobacteriosis) that are not associated with large epidemics. This is going to be difficult to do because people experiencing uncomplicated gastroenteritis may not seek medical attention. And if they do, it is unlikely that cultures of the causative organisms will be obtained, identified, and kept for further analysis. Thus, the CDC has identified areas of the U.S with well-equipped facilities and aggressive infectious disease prevention programs. In these areas, efforts are being made to increase the number of isolations of causative strains and prompt submission of these isolates to the state laboratories for further analysis. Among other tests, pulsed field gel electrophoresis (PFGE; Chapter 3) is being used to characterize the isolates to determine if there are certain isolates that are more successful at causing disease nationwide.

5. Normally, HACCP programs are based on measurements of contamination taken at the plant in question and on a detailed analysis of the steps in processing the meat. The purpose of this question is to get you to think about what steps might be the ones to monitor. Monitoring chunks of meat that come into the plant for surface contamination with food-borne pathogens might be an appropriate step, because in the processing of meat to make hamburger, any surface contaminants would become internalized by the grinding process and might even increase in numbers depending on how the meat was stored. Contaminated meat will contaminate grinding machines so that uncontaminated lots of meat will be contaminated when ground. Monitoring pieces of machinery that are in contact with the meat after daily cleansing could help to detect faulty cleaning practices by workers. It would also be a good idea to check lots of ground meat to make sure that contamination did not arise during processing. Rapid tests for food-borne pathogens are now being developed and tested. Such tests should make monitoring much easier and more effective.

6. There is, in fact, an epidemic of hepatitis. This question is worded from the perspective of a member of the general public who would not know that hepatitis, especially due to hepatitis B and hepatitis C is a growing problem. Although there is no hepatitis C vaccine yet, the hepatitis B vaccine can help to prevent hepatitis cases. In some areas of the country, hepatitis A infections are becoming a problem. Certainly for travelers to developing countries where hepatitis A infections are rife, receiving the hepatitis A vaccine is a wise precaution. The question also refers to the fact that exposure to hepatitis viruses is common, as indicated by the number of people with antibodies to hepatitis A virus. Since many of the people exposed to hepatitis viruses do not develop liver damage, it is clear that such people have mounted an effective defense that protects them from hepatitis. In the people who do not mount such a defense, however, hepatitis can be a serious and life-threatening disease. Keep in mind that the treatment for severe hepatitis is a liver transplant, an expensive and difficult operation. The need for vaccination with hepatitis A and B vaccines is to prevent hepatitis cases that might occur and cause serious health problems for infected people, not to mention increasing health care costs.

7. The purpose of this question is to encourage you to think about what factors are most important for determining which pathogens should receive the most attention and resources. Factors to consider include potential seriousness of the disease, economic impact of the disease, number of cases, and age groups affected. If you were in a developing country, there is not much question that targeting rotavirus infections in infants and young children would be a major priority since this is a common cause of childhood death in these countries. In a developed country, *Salmonella* and *Campylobacter* might move to the top of the list because they are common causes of infections that cause people to lose days at work and can become systemic. For the same reason, *Cryptosporidium* might become a priority. Also, these food and water-borne pathogens are indicators of the safety of the food and water supply, so there is some reason to focus on them for this reason. *Listeria* is always a quandary because, although the number of cases is small compared to the other food-borne infections, the case fatality rate is far higher than for the other infections. In terms of setting up a program, public education usually rates highest in priority because prevention is the best defense. To mount an effective public education program, however, it is necessary to have a basis of accurate scientific information. If, for example, most cases of food-borne disease are coming from centralized processing plants, an education program aimed at proper preparation of home grown produce would be ineffective. A perhaps surprising culprit is organic produce, probably due to the use of manure as fertilizer. A disproportionate number of cases of *E. coli* O157:H7 have been associated with organic produce, although the largest outbreaks have so far been associated with mass-produced meat products such as hamburger. Developing an effective program to prevent food-borne infections is not as simple as it might at first appear.

8. *Giardia* uses a sucking disk to adhere to the lining of the small intestine. Apparently, the defenses of the human body are not able to counter this adherence strategy, at least in some people. By contrast, the defense systems of the human body seem to be able to handle most cases of diarrheal disease caused by bacteria.

9. A variety of factors affect the severity of the disease. First, the number of organisms encountered is important. People who do not develop symptomatic disease may not have ingested enough organisms to develop symptoms. The infected person's defenses is another major factor. Some people repel intestinal pathogens better than others. By contrast, in people with immune defects, the microbes can become persistent or even invasive enough to cause more serious disease. Age is also a factor. Infants dehydrate more readily than adults and thus often have more severe symptoms than adults.

10. Any water-borne infection can become a food-borne infection if fruits or vegetables are sprayed with or are preserved in ice made from contaminated water. The examples of the police divers and the triathlon participants show that you don't have to actively drink water to acquire a water-borne infection. Involuntary ingestion and contact with cuts may be sufficient. There have even been cases of transmission associated with water rides in amusement parks. Oral-anal transmission of "food–borne" pathogens has also been documented in homosexuals.

Chapter 18

1. HIV might be considered an equal opportunity infectious microbe except that the danger of mother-to-infant transmission during or after birth by intimate contact is a female-specific concern. Hepatitis B virus is another virus that affects men and women equally because it attacks the liver. In fact, since hepatitis B is spread by sexual transmission primarily among homosexuals, you might argue that this is a virus that is more likely to infect men than women. Finally, herpes simplex virus has similar effects on men and women, especially in view of the fact that mother to infant transmission is rare. The other pathogens described in this chapter, however, cause more problems for women than for men. (Some might argue that *Neisseria gonorrhoeae* and *Chlamydia* are more likely to cause symptomatic disease in men than women, but it is difficult to know how to balance this aspect of the disease against the more serious conditions caused by these pathogens in women.) The point of this question is that, because of the different anatomical features of the male and female urogenital tract and because women give birth, urogenital tract infections tend to have more serious consequences in women.

2. Knowing that these two viruses cause cancer makes it possible to prevent the cancer they cause. Early detection and removal by cauterization of papilloma virus–caused warts from the cervix is a very effective way of preventing cervical cancer. Similarly, the hepatitis B vaccine will not only prevent cases of liver disease but will also prevent some liver cancers.

3. We leave the answer to this question up to you. The questions raised here about azithromycin have also been raised about the anti-HIV drugs. Recently, pharmaceutical companies volunteered to cut the cost of the anti-HIV drug AZT drastically for developing countries, in order to help prevent the mother-to-infant spread of HIV infection. Unfortunately, at the same time this announcement was made, the president of a large African country let it be known that he was not convinced that HIV caused AIDS. For many years, countries in the developing world have denied that AIDS exists within their borders. So there are political problems as well as economic ones to be overcome by those who want to bring sexually transmitted diseases under control.

4. In theory, antibodies that bind the surface of the elementary body and prevent it from infecting human cells should be protective. Since little is known about the interaction between the elementary body and the cells it infects, designing a vaccine that would elicit such a response is more difficult than it might seem at first. If chlamydial proteins are exposed on the surface of cells in which *Chlamydia* are growing, a cytotoxic T cell response might be protective, but so far no chlamydial antigens have been found to be exposed on the surfaces of infected cells.

5. There are several problems with designing an effective anti-HIV vaccine. Antibodies that prevent HIV from binding to its CD4 receptor on human cells should be protective, although as statisticians have pointed out, this type of defense has proven at best to be only about 95% effective. In a person exposed repeatedly to HIV, neutralizing antibodies might not be a sufficient barrier to infection. A vaccine that elicits a cytotoxic T cell response as well as an antibody response should be the most effective type of vaccine, although for people who were not protected by such a vaccine and developed AIDS, the cytotoxic T cell response, triggered late in the infection, could actually help the virus destroy the cells of the immune system. The discovery that a coreceptor for HIV is essential for infection, and that people who lack this coreceptor are either immune to HIV infection or develop AIDS much more slowly than people with the coreceptor is encouraging. This discovery tells us not only that blocking the coreceptor would help a lot in preventing HIV infection but that this coreceptor is not an essential protein. A vaccine that elicits an immune response against the coreceptor is likely to have bad side effects since it might well trigger an autoimmune response that affects many cells of the body. Information about how the coreceptor interacts with the HIV envelope proteins might make it possible to target a vaccine to produce an antibody response that more effectively neutralizes the virus. For reasons described in this chapter and in Chapter 14, developing a vaccine against HIV is one of the most difficult challenges scientists have faced to date.

6. In developed countries, the generally accepted ethical design of studies such as the one involving AZT is that the control group is not given a placebo but rather is given the current therapy of choice for controlling disease transmission. Thus, ethicists argued, women in the control groups should have been given the long-term AZT therapy (starting in mid-pregnancy and continuing for months to years after the infant is born) that is accepted medical practice in the United States and Europe rather than receiving pills that had no therapeutic value. Scientists in the countries involved defended this design—are you ready for this?—by arguing that the standard of medical practice is different in developing countries and, further, that if one group of women got the full course of therapy and did better than those who got the short course, then in the future women who got the short course would feel cheated and would complain. Fortunately, the short course of AZT performed as well as reported for the longer-term therapy routinely used in developed countries, cutting the transmission from mother to infant by about 50%. Unfortunately for those women in the control group, many of their babies are now infected with HIV.

7. We leave the answer of this question to you. Ethicists have argued that the criteria for design of all clinical trials, regardless of the severity of the disease involved, should be the same. That is, the control group should get the currently approved medical treatment prescribed in developed countries. (The companies supporting such studies are all located in developed countries, so this is the justification for using developed countries as the standard.) Their fear is that if some exceptions were to be made for diseases like herpes simplex infections or yeast infections, which are mostly nuisance-type infections rather than life-threatening ones, such exceptions would lead to an erosion of the central principle of protecting the people who end up in the control group.

8. As pointed out in this chapter, there are a lot of ways to get hepatitis B infections without participating in what some would characterize as immoral behavior. Liver damage and liver cancer are very serious conditions. Those who see a possible operation or use of blood or blood products in their own or their children's future should be running not walking to the nearest source of the hepatitis B vaccine. There is no reason to risk liver disease when one source of it can be avoided. Making this choice for a child is even less defensible.

Chapter 19

1. The point of this question was to get you to think about all the routine procedures in a hospital that could potentially serve as vehicles for introduction of infectious agents. Some examples of such incursions that have actually been documented are bacteria on plants brought to cheer sick patients, infections carried by visitors, bacteria spread by air-conditioning systems, bacteria that contaminate food or intravenous solutions, and viruses and bacteria carried by physicians (especially surgeons). Perhaps if you have visited someone in a hospital recently, you can think of still other possibilities.

2. PFGE is a very useful technique for determining whether an outbreak in a hospital or in some other setting was from a single source or from multiple sources. Interpreting PFGE results can be difficult, however, if the DNA fragment patterns are not identical for the different isolates or are not completely different. Difference in migration distance for a single band can be caused by technical problems (e.g., the restriction enzyme being used did not cut the DNA completely) or because a different strain, which is very similar to other strains in the study, is present. Usually, such problems are resolved by digesting the DNA with other restriction enzymes and comparing the resulting profiles. Such additional experiments are routinely done when it is important to be sure whether the same strain of bacteria is responsible for an outbreak.

3. The first step would be to isolate the bacterium from infected patients as a pure culture. Usually, this means plating the organism on agar medium and obtaining a single colony. (If the single colony contains two different microbes, as can happen in some cases, the following analysis will be compromised. This is another consideration in the question raised in question 2.) The next step would be to obtain DNA from each isolate and digest that DNA with an enzyme that cuts infrequently enough in the organisms' genome to generate a series of fragments to give a clear PFGE pattern. Choosing the restriction enzyme that gives the most easily interpretable pattern is not always easy and takes some fine-tuning. As indicated in the answer to question 2, scientists usually use more than one restriction enzyme in their analyses. In the case of the anesthetic-associated infections described in this chapter, the PFGE patterns of most isolates would have been completely different. In fact, PFGE analysis would probably not have been necessary if scientists analyzing the isolates from infected patients noticed that the isolates had different colony morphologies and different gram-staining characteristics. Simple metabolic tests might then have been done to establish that the isolates were members of different genera, in which case there is no need to do PFGE analysis, which is designed to differentiate strains of bacteria that are members of the same species.

4. This is a very difficult question to answer and many hospitals are grappling with this decision. The first question is whether a hospital has the right to test for and disclose the bacteriological status of its employees. Clearly, such testing is in the interest of patients and HMOs, but may violate privacy rights of employees. One answer to this question is to move a person known to be shedding dangerous multi-drug resistant strains of bacteria to wards where the person is less likely to be a threat to patients. That is, to move the person from an intensive care unit to some other part of the hospital. This problem is compounded by the fact that a person colonized with multi-drug resistant bacteria is a threat to himself or herself as well as to family members. Do hospitals that place workers in settings that might cause them to be colonized by multi-drug resistant bacteria have responsibility for this risk? This whole area of risk and risk management is so new that no one has formulated answers that are acceptable to all parties concerned.

5. We leave the answer to this question up to you. Suffice it to say that these examples have been much debated from various legal and health-related points of view and an answer that is acceptable to all parties is still not in view.

6. Organ transplant patients receive immunosuppressive drugs to prevent them from rejecting the grafted organ. These same immunosuppressive drugs place them at higher than average risk for contracting microbial infections. An interesting twist to this story, which arises in connection with transplanting organs from animals to humans, is described in Chapter 20.

7. Hospital food was mentioned in the answer to question 1. This is potentially a significant source of infection or colonization, not only by microbes that cause the usual gastrointestinal problems associated with food-borne pathogens but also by potential pathogens that are resistant to antibiotics. For a hospital patient recovering from a surgical procedure, exposure to multi-drug resistant *Staphylococcus aureus* or *Enterococcus* species from any source is a serious threat. Contamination of foods brought into a hospital patient's room is something that is not often considered by infection control specialists, but it is a possible source of wound contamination.

8. Frankly, we think that terms like "nosocomial" and "iatrogenic" are completely unnecessary and actually impede acceptance of the fact that physicians' offices and hospitals have environmental risks that need to be confronted openly. Nonetheless, the fact that such terms are so widely used indicates that some people think that they are useful.

Chapter 20

1. Most of the diseases described in this chapter were transmitted by arthopods from animals to humans. In fact, in many cases, humans were an accidental host in a cycle that did not much affect the normal host. Malaria and dengue virus infection were exceptions to this "rule" in that they are largely diseases that are transmitted from human to human with possibly other mammalian hosts. The point of this question is to emphasize the fact that there is no "typical" arthropod-transmitted disease. Many different kinds of microbes can be involved and the transmission patterns can vary considerably.

2. The microbes covered in this chapter varied considerably with respect to their characteristics. Some were bacteria, some viruses, some protozoa. Yet all had in common that they were able to replicate in arthropods as well as in non-arthropod hosts.

3. The most serious form of malaria is *P. falciparum* malaria. It is the most serious form because this species of protozoa can infect more different types of red blood cells than other plasmodia that cause malaria. It is the easiest to cure because it targets red blood cells primarily and only cycles once in liver cells. In effect, *P. falciparum* has no "latent phase" in liver cells, whereas other types of *Plasmodium* species cycle many times in liver cells. As with other microbes that can parasitize more than one cell type, plasmodia that can continue to cycle in more than one cell type are more difficult to eliminate from the body. The ability of *P. falciparum* to cause more serious disease than other species is due to its ability to infect many stages of red blood cells, a feature that has little to do with its susceptibility to antimalarial drugs.

4. Once it was known that the virus was spread from animals to humans by aerosols from mouse droppings, the control strategy was evident: control the access of rodents to human dwellings. Understanding the pinon nut connection gave public health authorities yet another level of control. In years when the pinon nut harvest is highest, remind people of the need to control rodent access to their houses.

5. The Lyme disease example serves as a good model for using information about the cause and ecology of a disease to inform the public about effective prevention strategies. Although Lyme disease has not been eliminated and cases continue to be reported, prevention strategies have been effective in limiting the spread of this disease.

6. Rabies is unusual in that it takes a long time for the disease to develop. Presumably, the rabies virus has to spread from the site of the bite to neurons, then to the brain. In the early stages of the disease, before the virus is established in neurons (which are protected from the immune system), the virus is susceptible to an immune response. This slow development of the disease explains why vaccination immediately after exposure to the virus would act in time to prevent the disease. In many people, HIV infections develop very slowly. This feature of the disease might make it possible to use a strategy similar to that used for rabies virus, immunizing people after they have been exposed. The problem in the case of HIV is to establish what is an effective antiviral response and what is an effective form of the vaccine.

7. A vaccine that clears cats of carriage of *Bartonella* is an obvious solution. So far, no such vaccine has been developed. The attachment of terminally ill patients to companion animals is an important factor to consider. Possibly, treatment of an infected cat with antibiotics might reduce the likelihood that the disease will be spread from the cat to its owner. Probably the easiest solution at present is to keep the cat free of fleas and keep it indoors so that it does not acquire fleas, the likely vector.

8. This has proven a difficult issue to resolve. Clearly, *Brucella* in buffalo poses a threat to cattle in the vicinity of the buffalo herd. How great a threat is not clear, but farmers' concerns about possible transmission of infection cannot be ignored. Culling buffalo herds and other measures that have worked to free cattle and swine from this type of infection would be advisable for those who want to raise buffalo. This would protect both the buffalo themselves and any livestock in the vicinity of the buffalo herd.

Chapter 21

1. The cloning step is necessary because PCR amplification of rRNA genes from a mixture of microbes will yield a mixture of sequences. The cloning step separates the different amplified segments so that each transformed *E. coli* cell contains a single cloned PCR product with a single sequence.

2. There are primers that are labeled "universal," which are supposed to amplify most prokaryotic sequences. In practice, these primers miss some microbes. This is the reason why scientists interested in amplifying sequences from all microbes in a particular site often resort to more specific primers that target specific groups (archaea, bacteria, or subsets of bacteria or archaea).

3. Amplifying genes such as the genes encoding enzymes in the sulfide oxidation pathway or genes encoding proteins involved in photosynthesis can give additional information about the types of microbes in a particular site. It is sometimes difficult to reconcile the rRNA gene sequences with DNA sequences from genes that encode known metabolic functions, but codon usage, %G+C information, and matches with genes from known prokaryotic species can help to sort out the information.

4. Phase microscopy gives information about the shape of microbes in a particular site, whether they are producing spores or contain sulfur granules, and how they are moving if they are motile. Fluorescence microscopy uses limited wavelengths of light to excite microbes to emit specific wavelengths of light that give information about their metabolism. For example, photosynthetic microbes exposed to green light emit a characteristic bright red fluorescence. In this type of fluorescence microscopy, the natural fluorescence of microbes in the specimen is detected. In FISH and in DAPI-stained preparations, the microbes are first stained with some fluorescent compound. In the case of FISH, the fluorescent compound is a hybridization probe that interacts with rRNA from certain microbes. In the case of DAPI staining, the fluorescently labeled dye intercalates with double-stranded DNA molecules.

5. FISH shows whether the sequences obtained from PCR/cloning are representative of the population as a whole. FISH also gives information about the spatial propinquity of different microbes. FISH has some of the same limitations as PCR/sequencing. For example, failure to permeabilize a microbe will lead to failure to visualize that microbe even if the labeled probe for that microbe is used as a stain. Specificity is also a problem. The power of FISH identification is only as good as the specificity of the probe. Conditions of hybridization also have to be optimized for maximum specificity. These include temperature and salt concentration of the hybridization mixture. A special limitation of FISH is background fluorescence. Using a label that fluoresces red will not work on photosynthetic microbes because of their natural red fluorescence.

6. A special limitation of FISH was mentioned in the answer to question 5 above, background fluorescence. If material in a specimen fluoresces at one wavelength, the solution is to use a fluorescent label that fluoresces at another wavelength.

7. Methane production is carried out by archaea, whereas sulfate reduction is carried out by bacteria. In both processes, a low molecular weight compound is reduced to provide energy. In the case of methane reduction, the pathway can also provide a carbon source. In both cases, an electron transport system is coupled to the reduction pathway to generate energy. Another shared feature of these two pathways is that microbes that use these processes can do so only in the absence of oxygen. Sulfate reduction is unique in that the sulfate must first be activated by phosphorylation before it can serve as an electron acceptor.

8. Both of these pathways require molecular oxygen. The methane oxidation pathway serves not only as an energy-generating pathway but as a source of carbon compounds for biosynthesis. This is not the case for sulfide oxidation, which is exclusively a pathway for energy generation. Both methane and sulfide are oxidized spontaneously if the oxygen concentration is high enough. This is the reason why microbes that rely on oxidation of these compounds are found at or near the anoxic/oxic interface where oxygen levels are high enough for the microbes to oxidize methane or sulfide but low enough to eliminate competition from spontaneous oxidation.

9. Chemical analysis of material in a sample being studied can give clues as to what processes are being used by microbes in that location. Finding evidence for methane production alerts a scientist studying a site that archaea should be present. A limitation of the chemical identification method is that if a compound is being consumed as fast as it is being produced, there will be no record of these coupled activities. Usually, however, there is a net accumulation of enough of a metabolic end-product to be detected.

10. Obtaining the rRNA gene sequence from the microbes you isolated will allow you to design a probe that detects that microbe specifically. Using FISH, you can determine how abundant the microbe you isolated is in the actual specimen you are studying.

11. Microbes form communities in the sense that they interact with each other metabolically. Those that consume oxygen make an anoxic environment that allows the sulfate reducers to grow. Similarly, end products of some, such as the sulfides produced by the sulfate reducers, serve as substrates for others, such as the sulfide oxidizers. Whether microbes interact in other ways remains to be determined.

Chapter 22

1. Photosynthesis is best understood as a photosystem (or two, in the case of oxygenic phototrophs), which converts light energy into a high energy electron that is coupled to an electron transport system that uses the energized electron to pump protons and create a PMF. Chorophyll (or bacteriochlorophyll) molecules allow photons of light to excite electrons to E_0' values that allow them to transfer electrons to components of an electron transport system that pumps protons. The main difference between the photosynthetic systems described in this chapter is whether the flow of electrons is cyclic or not and whether the electrons associated with chlorophyll molecules reach low enough E_0 values to produce NAD(P)H. See the Chapter at a Glance section for a comparison of the different types of photosystems.

2. In both cases, electrons are passed from one compound to another in a set of oxidation-reduction reactions, the end result of which is to pump protons and create a PMF that can be used to do work. The main differences are that the source of electrons is not NAD(P)H and the flow of electrons in photosynthesis may be cyclic rather than linear. The NAD(P)H to terminal electron acceptors were linear pathways as far as the electrons were concerned. Of course, each reaction was cyclic in the sense that a component was oxidized and then reduced to recycle it. The "circular" aspect of some photosynthetic systems is that the same electron is recycled to chlorophyll (bacteriochlorophyll) rather than being dumped on a terminal electron acceptor.

3. An electron associated with two chlorophyll (bacteriochlorophyll) molecules absorbs the energy in a photon of light. The result is that the E_0' value of the electron is made much more negative, which means that the electron can now reduce compounds in a set of redox reactions.

4. The most self-sufficient of the phototrophs are the oxygenic phototrophs, which can use nitrogen gas as a source of nitrogen and carbon dioxide as a source of carbon. Nitrogenase and the Calvin cycle, which allow the oxygenic phototrophs to be so self-sufficient, require a lot of ATP. Thus, it makes sense that these fixation systems are found in an organism that has an almost unlimited energy source for creating PMF and ATP— sunlight. Some of the green and purple phototrophs can fix carbon dioxide, but most of them prefer organic carbon sources. They also need sources of nitrogen other than nitrogen gas. The need for more complex substrates and the inability (in the case of the purple phototrophs) to make NAD(P)H at the same time as PMF in the photosynthetic pathway, is probably the reason why the anoxygenic phototrophs are less abundant than the oxygenic phototrophs. The anoxygenic phototrophs do pretty well, however, and are widely distributed in nature.

5. Basically, it underscores the fact that most of the features of the higher organisms were invented during evolution by prokaryotes. At one time, much was made of the differences between prokaryotes and eukaryotes, but as more is learned about both, the more it is becoming clear that even such supposedly eukaryotic features as cytoskeleton and sexual reproduction have their more primitive equivalents, and probably their origins, in the prokaryotic world.

6. The major difficulty has been duplicating the conversion of light energy to chemical energy by chlorophyll (bacteriochlorophyll). Also, the complexity of the electron transport reactions makes duplicating photosynthesis in the laboratory a challenge. Another barrier to duplicating photosynthesis in the laboratory is the necessity of providing so many proteins and other molecules that carry out the oxidation-reduction reactions. No wonder plants decided to opt for the simplest way to become photosynthetic—enslave a photosynthetic prokaryote. The animals of the coral reefs and lichens used a similar strategy but in their case, they annexed photosynthetic eukaryotes (algae).

7. As already mentioned, the systems for fixing nitrogen and carbon dioxide require copious amounts of ATP and NAD(P)H. The photosynthetic system of the oxygenic phototrophs produces large amounts of both of these compounds. The ability of oxygenic phototrophs to grow on air and water has made them invaluable as the biomass that forms the base of marine and terrestrial food chains.

8. To reinforce the answer to question 7, consider what happens when cyanobacteria or algae no longer have to fix atmospheric nitrogen and can instead use forms of nitrogen that do not require such a big energy expenditure. Many of the oxygenic phototrophs can use such forms of nitrogen if they are available. In the laboratory, if you are trying to cultivate cyanobacteria and you want instant gratification, throw a little nitrate into the medium when no one is looking and you will have colonies of cyanobacterium in a week, whereas the purists will have to wait at least an extra week for theirs to form. They will claim that they have cultivated more exotic beasts, not the weeds you got—and there is some truth to this—but your approach will have demonstrated why blooms occur when more easily used forms of nitrogen are made available to the oxygenic phototrophs.

9. Does your roommate like whales? Dolphins? Well, your roommate could kiss those loveable marine mammals goodbye if it were not for the phytoplankton that serves as food for the fish that feed the whales and dolphins. Of course, a toxin-producing cyanobacterium might be responsible for the demise of that cute little puppy your roommate has sneaked into your dorm room. Better not to mention that until after the final exam. Seriously, photosynthesis can be made to be a very dull subject, but in fact, the conversion of light energy to chemical energy is what made the Earth habitable for us and so many other creatures. The photosynthetic prokaryotes and eukaryotes are not called "primary producers" of biomass for nothing. You might also try a small guilt trip. Since the acquisition of a photosynthetic alga is what makes coral animals (and thus coral reefs) healthy, and since the loss of this photosynthetic endosymbiont is what causes bleaching of corals, you could point out that your roommate's short-sighted view of photosynthesis is responsible in part for the loss of coral reefs. Even more compelling, point out to him/her that a knowledge of photosynthesis could well translate into a job that involves diving and snorkeling in some tropical paradise.

Chapter 23

1. Oxygen will oxidize sulfide spontaneously if the oxygen concentration is high enough. The hemoglobin keeps the oxygen and sulfide separate until the endosymbiont can use it to generate energy.

2. The worm *Heterhabditis* and its intestinal exosymbiont, *Photorhabdus* have a somewhat similar association to that of the termite and its hindgut exosymbionts, in the sense that the bacterium produces nutrients (in the case of *Photorhabdus,* a cysteine and lysine rich protein) that are essential to its invertebrate host. Also, once in the caterpillar larva, the bacteria help to break down the larval biomass to provide nutrients for the worm to use during its reproductive stage. In the case of the *Vibrio fischeri*–squid symbiosis, new information showing that this association has no beneficial effect on the squid after all would change the description of this symbiosis from mutualistic to commensal. Conversely, if the effect of *Wolbachia* on some insect turned out to be beneficial, the nature of this interaction would have to be reevaluated.

3. What is unusual about the *Euprymna–V. fischeri* interaction is that this interaction between bacterium and invertebrate host is not nutritional. Most of the other interactions (except the *Wolbachia*–arthropod interaction) involve production of nutrients by a prokaryotic endo- or exosymbiont for the benefit of the invertebrate host. Although the squid provides nutrients for *V. fischeri,* the bacteria provide light and thus protection from predators for their squid host.

4. Again, the *Photorhabdus–Heterorhaditis* interaction comes to mind. The bacteria are exosymbionts that provide nutritional support for their animal host, as was the case for the bacterial populations of the human body. It is important to keep in mind, however, that the bacterial populations of the human body also play a protective role. Thus, in a sense, their interaction with their human host is not too dissimilar to that of *V. fischeri* and its squid host, even though the nature of the protection is quite different.

5. The animals of the coral reef have photosynthetic algal endosymbionts that apparently provide nitrogen and carbon for their animal hosts. Bleaching of coral reefs occurs when the algal endosymbiont (which gives the coral reef its color) is lost. Not much is known about this aspect of the bleaching of corals, but it is a lethal event for the coral.

6. Any microbe that colonizes the sex organs of an invertebrate can exert effects ranging from infertility to reproductive incompatibility to hormonal aberrations. *Wolbachia* seems to exert each of these effects in different arthropod hosts. A recently discovered effect of *Wolbachia* is that it is essential for the survival of the helminth *Onchocercus,* which causes river blindness. An important discovery was that antibiotics that eliminate the bacteria also eliminate the worm. The use of antibiotic therapy is now being explored as a possible therapy for river blindness, a serious disease in Africa.

7. In the case of *V. fischeri,* the luminescence of the bacteria seems to have a clearly protective function for its invertebrate host. In the case of *Photorhabdus luminescens,* the role of luminescence of the bacteria in the caterpillar larva is not so clearly beneficial. If the speculations of scientists about the possible role of luminescence in attracting other caterpillar larvae are correct, luminescence could well have a nutritional role in the sense of attracting new prey for the nematode. It is also possible, however, that the luminescence is simply a manifestation of the fact that for bacteria growing inside the caterpillar larva, there is ATP to burn.

Chapter 24

1. Both require an electron transport system that oxidizes or reduces the nitrogen compound, then hands off electrons to a final oxygen acceptor. In the case of nitrate reduction, nitrate is the terminal electron acceptor. In the case of ammonia oxidation, oxygen is the final electron acceptor.

2. It is striking that over one-tenth of the photosynthetically generated carbon and energy of a plant is exuded from its roots. This apparent throwing away of carbon so hard-won from the energy-expensive Calvin cycle is hard to explain unless the plants gain some very big advantage from it. An obvious benefit is attracting microbes that could aid the plant, either by feeding it some essential nutrient or protecting it from pathogens. Little is known about the roles of rhizosphere bacteria or fungi in benefiting plants, but plants obviously have a vested interest in attracting and maintaining the rhizosphere microbes. Understanding this relationship better could lead to improved growth of plants and improved fertility of soil.

3. Bacteroids differ in many ways from their free-living *Rhizobium* (or *Bradyrhizobium*) progenitors. They are nonmotile. They are larger and no longer rod-shaped. They fix nitrogen, and they appear to be unable to divide in standard laboratory media that support the growth of the free-living form. They probably differ in other, as-yet-undiscovered, ways as well. In the soil, the ability to move is important for finding new sources of carbon and energy. Nitrogen fixation is less important to free-living bacteria because there are plenty of nitrogen compounds available to them. Once the bacteria enter a plant, however, their metabolism becomes wedded with that of the plant and they begin to fix nitrogen in return for the organic compounds the plant provides. They even cooperate with the plant cells to create the leghemoglobin that provides bound oxygen to the bacteroid to give it an essential electron acceptor (oxygen) without imperiling the nitrogenase complex that is reducing nitrogen gas to ammonia.

4. Most scientists prefer to see the interaction between the plant and the *Rhizobium* species as mutualistic, although it could be seen as parasitic on the part of the plant. The bacteria seem to have evolved to participate in this interaction and produce some of the factors that induce the plant roots to form nodules. The nodules provide the bacteroids with nutrients and with a protected environment.

5. Scientists had hoped to be able to use genetic engineering to make normally non-nodulating plants such as corn able to harbor *Rhizobium* species so that they would become self-fertilizing. Such an accomplishment would be a major step forward and would reduce significantly the need for fertilizer. However, the effort to engineer nodulation has foundered on the complexity of the process. The number of genes involved in the process is daunting. Early genetically engineered plants had only one foreign gene introduced. Recently, the successful construction of "golden rice," a strain of rice that produces vitamin A—a process that required introduction of multiple genes—has given new hope that more complex constructions would be possible. Another issue beside the number of genes involved in the nodulation process is the question of how important it is that there be high specificity between plant and nodulating bacterium. Clearly, it would not be a good idea to open up a plant's roots to colonization by many different types of bacteria, but the question remains of whether there are reasons that have not yet been discovered for the high specificity seen in the nodulation interactions so far studied. It might be that only certain species would be suitable for corn. Will any old Nod factor induce nodulation in the plant?

6. Some of the determinants of specificity are known. For example, the bacterium responds to certain plant phenolic compounds by producing its own specific nodulation (Nod) factors, which induce nodulation in the plant. Binding of the bacterium to the plant root hair is also a specific process that involves an interaction between specific carbohydrates on the bacterial surface and specific proteins on the surface of the root hair. What is not yet clear is whether the "infection" process that follows, in which bacteria follow the infection thread into the cells that will become the nodule is also a specific bacterium-plant interaction. The determinants of specificity are still under study and it is likely that more will be discovered in the future.

7. The interaction between *A. tumefaciens* and plants is now considered to be a parasitic or a commensal one. The bacteria clearly benefit from this reaction, but what about the plant? The designation "parasitic" comes from an esthetic perception of the crown gall as reducing the plant's attractiveness, but clearly there is not enough damage to the plant to impair its survival. Plants seem unaffected by the crown gall they produce and carry. Scientists have suspected that there might be some benefit to the plant and have looked for such benefits but have not found any so far. Finding that induction of a crown gall by *A. tumefaciens* is beneficial would change the designation of the interaction to "mutualistic." The point of this question is to underscore the fact that designations such as "mutualistic" can change as more information about the interaction becomes available. We realize that this fact is not much comfort to the poor student preparing for an examination in which the teacher might ask about whether a particular interaction is parasitic or mutualistic.

8. At first, the hypersensitive reaction seems like a damaging one because it produces the areas of dead tissue that are identified by farmers and scientists as the symptoms of the disease. Yet, killing of plant cells and production of an area of dry dead tissue can keep the invading bacteria from getting into the plant's circulatory system and killing it. Recall from Chapter 12 that the defenses of the human body are often responsible for the symptoms of human disease because they cause collateral damage in the effort to contain an infection and eliminate the invaders. Although in this respect the plant hypersensitive response resembles the nonspecific defenses of the human body, which are the cause of inflammation, other aspects of the hypersensitive response resemble the specific defenses. In partricular the specificity of the plant's R proteins for particular proteins produced by the bacteria is reminiscent of the specificity of an antibody or cytotoxic T cell receptor for its protein target. Plant pathogens also resemble human pathogens in that they have type III secretion systems that allow them to inject proteins and other compounds directly into the plant cell cytoplasm. This strategy for infection has now been found in a number of human pathogens, and the secretion proteins used by human pathogens resemble those used by plant pathogens at the amino acid sequence level.

9. Avirulence genes are genes encoding proteins that are injected by microbes into the plant cell cytoplasm and are recognized by plant-produced R factors. A point of confusion is that proteins called avirulence proteins can actually be toxins or other factors that would normally be called virulence factors. The "avirulence" label refers to the fact that the plant R proteins recognize these proteins and use their presence as a signal for induction of the hypersensitive response, which is usually successful in stopping the infection from becoming systemic. Just as human pathogens can mutate their genes to get around the defenses of the human body, some plant pathogens have mutated avirulence genes so that they are no longer recognized by the plant R proteins. However, if such mutations make the plant pathogen less able to cause disease, they may be deleterious to the plant pathogen and thus become less likely to be maintained in the population of that plant pathogenic species.

10. Nodulation is an "infection" in the sense that bacteria invade plant tissues and evoke a response by the plant that contains them in a particular site. The quotation marks around the term infection are there because the bacterial invasion is clearly beneficial for the plant, something that cannot be said for a true infection. In the case of *A. tumefaciens,* there is no infection. The bacteria remain at the surface of the crown gall tumor. What invades the plant cells and causes the formation of a crown gall tumor is the T-DNA from the Ti plasmid, which is injected into plant cells by conjugation. Thus, although both of these interactions produce a "tumor," they do so in very different ways.

Chapter 25

1. A signal sequence would have to be added to the amino terminal end of the gene. A signal sequence is a short stretch of hydrophobic amino acids that directs the secretion machinery to secrete the protein across the cytoplasmic membrane (Chapter 6). A DNA sequence specifying such a signal sequence would be added to the 5′ end of the gene. Since the signal sequence is cleaved off after secretion, the secreted protein would not have extra amino acids attached to it. Even better would be excretion of the protein out of the cell completely. The signals that mediate export through the outer as well as the inner membrane of *E. coli* are still only poorly understood, so this is not an easy thing to direct a protein to do. The advantage of excretion to the outside of the outer membrane is that the protein is much easier to harvest. The bacteria do not have to be disrupted to release the protein. Another advantage of not disrupting the bacteria is that the final protein preparation is much less likely to be contaminated with lipopolysaccharide, a toxic molecule that has to be removed before a protein preparation can be used internally (e.g., injected, as is done with insulin). An alternative approach is to use a gram-positive bacterium as the producer of the cloned protein. In this case, secretion of the protein through the cytoplasmic membrane effectively places the protein in the extracellular fluid.

2. You want to clone the entire open reading frame, so your primers would have to anneal to the very ends of the message. One primer primes synthesis of the first strand of DNA by RT, and the other primes synthesis of the copy of that copy. Once there are two strands, PCR takes over and makes many copies of the fragment. A higher concentration of DNA molecules makes the fragment easy to clone. To make cloning easier, some scientists would add a segment to the outer end of each primer that contains a restriction site. Cutting the PCR amplified DNA fragments with the restriction enzyme generates sticky ends that make annealing with the plasmid sticky ends more efficient.

3. Carbohydrate residues are added by specialized enzymes in the cell after a protein is translated or are added by an *in vitro* chemical reaction (as is done with the conjugated vaccines). Carbohydrate residues are not specified in the DNA sequence. Many human proteins have carbohydrate residues on them. Recreating these residues using bacterial enzymes has not worked. Yeast or some other eukaryotic cell can be used to produce such proteins if the pattern of carbohydrate substitutions is the same as that for the human protein.

4. The strength of the promoter is an important consideration. Recall that scientists usually want to use a regulated promoter to delay expression of the gene until the cell density is high enough to maximize protein production. The induced level of expression from the promoter should be high. The ribosome binding site should also be a strong one to maximize translation of the message. There is a limitation, however. Too high a level of protein production may result in inclusion body formation. Since most methods of resolubliz-ing an inclusion body reduce or eliminate the activity of the protein you are producing, inclusion body formation is undesirable (unless you have come up with a gentler way to solubilize the protein or if the protein of interest is not inactivated by the solubilization method). Thus, a balance between maximizing protein production and minimizing inclusion body formation must be reached. As mentioned in the answer to question 1, one way to prevent inclusion body formation and at the same time increase the ease with which the protein of interest is harvested and purified, is to have the protein secreted out of the cytoplasm. Here too, a balance needs to be struck because too high a level of production of the protein in the cytoplasm will result in jamming of the secretion machinery and trapping of much of the protein inside the cell.

5. Any time a cloned gene that has been introduced into a bacterium makes the bacterium grow more slowly, there is the possibility that the protein being produced is toxic for the cell or that the level of production is high enough to take a severe energy toll. In either case, there is the likelihood that the bacteria will mutate the cloned gene to make it less toxic. Changes that make the protein less toxic to the bacteria could make the protein inactive or more toxic in humans. To prevent such mutations from occurring, scientists use a tightly regulated promoter that is induced only after the bacteria have reached a high population density. The shorter the phase during which the protein is being produced, the less chance the bacteria have to introduce mutations into the gene. And, of course, each batch of bacteria starts with a fresh inoculum so that the same culture is not diluted and used over and over again. The integrity of the cloned DNA and the resulting protein are also monitored very carefully for changes that may occur.

6. An expression vector is a plasmid that has a promoter next to the site where the restriction enzymes cut the plasmid. Thus, an open reading frame in a segment of DNA that is cloned into that site has a 50–50 chance of having the promoter upstream of its 5′ end. Scientists use expression vectors if they want to improve the chances that genes from an organism that might not have promoters recognized by *E. coli* will be expressed in *E. coli.* The example given in this chapter was production of enzymes of interest to the detergent industry. The DNA being cloned and screened for protein production came from a mixture of microbes that were very diverse. Using a mammalian promoter would not help if screening is being done using bacteria, but a mammalian promoter is essential if the scientist wants to produce a protein in a tissue culture cell line that is mammalian in origin.

7. The resistance genes used in cloning have bacterial promoters and thus would not be expressed in the plant. Even if these genes were expressed, they would not make the plant resistant to antibiotics. Plants are naturally resistant to antibiotics because the antibiotics do not interfere with plant cell growth. Antibiotics target bacterial functions. One of the safety issues raised in connection with BT plants was the concern that the antibiotic resistance genes might get out of the plant cells during digestion in the human or animal intestinal tract and might then be taken up by intestinal bacteria. Scientists concluded that this was highly unlikely because it would require a series of low probability steps. DNA-degrading enzymes from the disrupted plant cell or intestinal contents would cut most of the released DNA into fragments too small to carry an intact resistance gene, and the resistance gene is only a tiny fraction of the genome. Few human or animal intestinal bacteria are naturally transformable, so uptake of the DNA by intestinal bacteria is unlikely. Even if the DNA were taken up, it would have to be incorporated in the bacterial chromosome by homologous recombination. Then it would have to be expressed. Finally, there would have to be a selective pressure applied that would cause that one cell to increase in number. The genes used in cloning vectors are already widespread in natural isolates of bacteria. This was the source of the original genes. Thus, even in the unlikely event the series of steps outlined above resulted in a bacterium carrying one of the resistance genes, it is unlikely that this event would even be noticeable against the background of naturally occurring resistant strains.

APPENDIX II

Resources in Microbiology: References & Web Sites

References

Textbooks

We have listed four titles that we find especially helpful. Each of these texts has a different emphasis, so you might wish to look at all of them.

Madigan, M. T., J. M. Martinko, and J. Parker. 1999. *Brock's Biology of Microorganisms,* 9th ed. Prentice Hall, Inc.

Prescott, L. M., J. P. Harley, and D. A. Klein. 1999. *Microbiology,* 4th ed. WCB/McGraw-Hill, Inc.

Perry, J. J. and J. Staley. 1999. *Microbiology: Dynamics and Diversity,* 2nd ed. Harcourt College Publishers

Schaechter, M., C. N. Engleberg, B. Eisenstein, and G. Medoff. 1998. *Mechanisms of Microbial Disease.* 3rd ed., Lippincott Williams&Wilkins

Books about microbiology and microbiologists

If you are interested in history, or in finding out how microbiology became the exciting field it is today, you might want to check out some of the books below.

Atlas, R. M., ed. 2000. *Many Faces—Many Microbes: Personal Reflections in Microbiology.* ASM Press, Washington, D. C.

Brock, T. D., ed. 1999. *Milestones in Microbiology: 1546–1940.* ASM Press, Washington, D.C.

DeKruif, P. 1941. *Microbe Hunters.* Harcourt, Brace and Co., New York.

Dubos, R. 1998. *Pasteur and Modern Science.* ASM Press, Washington, D. C.

McNeill, W. H. 1998. *Plagues and People.* Anchor Books/Doubleday.

Needham, C., M. Hoagland, K. McPherson, and B. Dodson. 2000. *Intimate Strangers: Unseen Life on Earth.* ASM Press, Washington, D. C.

Zinsser, H. 1996. *Rats, Lice, and History.* Black Dog and Leventhal Pub.

Journals

The journals listed below provide review articles that summarize the latest findings in microbiology and related scientific fields.

Immunology Today

Parasitology Today

Science News

Scientific American

Trends in Biotechnology

Trends in Cell Biology

Trends in Microbiology

A different type of resource

The American Society for Microbiology (ASM) is a great source of information on all aspects of microbiology. ASM recognizes that students are the future of microbiology and so provides many benefits to students. The best of these is a special student membership rate. Any student joining ASM will be eligible for all of the products and services of the society at reduced rates. Membership includes *ASM News,* a monthly newsmagazine that has short review articles on exciting topics in microbiology as well as updates on events that are of interest to microbiologists. Another advantage, free to members, is the Employment Clearinghouse—something to keep in mind when it is time to look for a job! If you are interested in joining ASM, visit www.asmusa.org for an application form and to check other benefits of membership.

Web Sites

Chapter 1
Earth: Planet of the Microbes

WWW Virtual Library: Microbiology and Virology

http://microbiol.org/vl_micro/index.htm

A site ". . . devoted to the science, practice, and application of microbiology and virology. Subjects covered include culture collections, databases, educational sites, journals, government and regulatory sites, images, medical microbiology, university departments, virology, protista, and references to newsgroups, E-mail lists, and other resources for the microbiologist." An extensive list of Web links concerning all topics microbiological is included.

Significant Events in Microbiology of the Last 125 Years

http://www.asmusa.org/mbrsrc/archive/SIGNIFICANT.htm

From Pasteur's studies on yeast to the complete genome of hantavirus, this site highlights the major historical developments in microbiology. Maintained by the American Society for Microbiology.

National Center for Biotechnology Information Coffee Break

http://www.ncbi.nlm.nih.gov/Coffeebreak/

Short articles of general interest to scientists posted daily and designed to be read over the morning cup of coffee. Many of these articles concern microbiological topics.

PubMed

http://www.ncbi.nlm.nih.gov/entrez/query.fcgi

A search engine from The National Library of Medicine providing instant access to more than 10 million citations in the biomedical literature. Simply type in a term and get back a host of abstracts to relevant research articles. If you are looking for a review article on a topic, simply include the word "review" in the search.

The Mississippi River Basin: The Issue of Hypoxia

http://www.epa.gov/msbasin/issue.html

Maintained by the Environmental Protection Agency, this site describes the problems of hypoxia in the Gulf of Mexico, Chesapeake Bay, Long Island Sound, Sarasota Bay, and the Baltic Sea, and describes new federal legislation on hypoxia in the Gulf of Mexico.

Chapter 2
Diversity and History of Microorganisms

The Three Domains of Life

http://www.ucmp.berkeley.edu/alllife/threedomains.html

Links to the fossil record, life history and ecology, systematics, and morphology of the three major groups of living organisms: the archaea, the bacteria, and the eukarya. Produced by students at the Museum of Paleontology at the University of California, Berkeley.

The NCBI Taxonomy Homepage

http://www3.ncbi.nlm.nih.gov/Taxonomy/taxonomyhome.html

This page, maintained by the National Center for Biotechnology Information, contains the names of all organisms represented in the genetic databases with at least one nucleotide or protein sequence. Use the "Taxonomy Browser" to explore the taxonomic structure of microorganisms or to retrieve sequence data for a particular group of organisms.

Astrobiology Web

http://www.reston.com/astro/index.html

A list of links to dozens of sites and news reports on the topic of exobiology, the search for life elsewhere in the universe.

Chapter 3
Cultivation and Identification of Microorganisms

Meningitis Research Foundation

http://www.meningitis.org/

A site designed for anyone with a personal or professional interest in bacterial meningitis.

ATCC

http://www.atcc.org/

The American Type Culture Collection, a global nonprofit resource center that provides pure cultures of bacteria, viruses, protists, and fungi to industrial, governmental, and academic institutions.

List of Bacterial Names with Standing in Nomenclature

http://www-sv.cict.fr/bacterio/

A French site useful in definitively deciding issues of bacterial nomenclature. "Includes, alphabetically and chronologically, the nomenclature of bacteria and nomenclatural changes as cited in the *Approved Lists of Bacterial Names,* or validly published in the *International Journal of Systematic Bacteriology* or in the *International Journal of Systematic and Evolutionary Microbiology.* It is extensively annotated to clarify the rules which govern the scientific nomenclature."

Chapter 4
Structural Features of Prokaryotes

Characteristics of the Prokaryotes and Eukaryotes

http://esg-www.mit.edu:8001/esgbio/cb/prok_euk.html

Part of the MIT Biology Hypertextbook produced by the Experimental Study Group at the Massachusetts Institute of Technology, this site lists the essential differences between these two groups, with lots of hyperlinks to the rest of the Hypertextbook.

Prokaryotic and Eukaryotic Cell Structure

http://nauonline.nau.edu/welcome/tdrive/bio220/lesson.html

Part of a microbiology course at Northern Arizona University, this outline includes numerous links to other sites on bacterial structure.

Biofilms and Biodiversity

http://www.mdsg.umd.edu/Education/biofilm/

Prepared by the Maryland Sea Grant, this site examines many aspects of bacterial biofilm formation, with a focus on the use of biofilm cultures grown on Plexiglass discs suspended in the Baltimore Inner Harbor. These discs were then used to monitor water quality and biodiversity. Provides real data for students to calculate microbial diversity and a good list of links.

How Penicillin Kills Bacteria

http://www.cellsalive.com/pen.htm

Watch a QuickTime movie of penicillin killing a bacterium. Explore the "Cells Alive" site for a variety of beautiful images of microbes.

The Challenge of Antibiotic Resistance

http://www.sciam.com/1998/0398issue/0398levy.html

An article in *Scientific American* by Stuart Levy concerning the mechanisms of bacterial resistance and pharmaceutical strategies to keep us one step ahead of the bacteria.

Reservoirs of Antibiotic Resistance (ROAR) Network

http://www.healthsci.tufts.edu/apua/roarhome.htm

Jointly sponsored by the University of Illinois and the Alliance for the Prudent Use of Antibiotics (APUA), this site seeks "to create the first network and interactive database organized around antibiotic resistance in commensal bacteria from humans, farm animals, plants, fish, food, soil, and water [and] to act as the definitive source of information on bacteria that may not be pathogens themselves but can function as reservoirs for resistance genes that can potentially be transferred to human pathogens."

Chapter 5
Bacterial Genetics I: DNA Replication and Genetic Engineering

DNA Structure: An Interactive Animated Nonlinear Tutorial

http://www.umass.edu/microbio/chime/dna/index.htm

After downloading a free copy of the Chime 2.0 program (http://www.mdli.com/cgi/dynamic/downloadsect.html?uid=$uid&key=$key&id=1) viewers can rotate three-dimensional DNA models and follow this basic tutorial. Written by Eric Martz of the University of Massachusetts Department of Microbiology.

DNA from the Beginning

http://vector.cshl.org/dnaftb/

"An animated primer on the basics of DNA, genes, and heredity . . . organized around key concepts. The science behind each concept is explained by animations, an image gallery, video interviews, problems, biographies, and links." Maintained by the DNA Learning Center at the Cold Spring Harbor Laboratory.

Access Excellence at the National Health Museum

http://www.accessexcellence.org/

Originally developed by Genentech, Inc., now ceded to the nonprofit National Health Museum, this site includes a variety of activities and teaching modules about all aspects of biotechnology. Designed for advanced placement high school students, the material is nonetheless appropriate for a college audience.

Recombinant DNA

http://esg-www.mit.edu:8001/esgbio/rdna/rdnadir.html

Part of the MIT Biology Hypertextbook produced by the Experimental Study Group at the Massachusetts Institute of

Technology, this site describes basic techniques involved in gene cloning into plasmids, the polymerase chain reaction, and DNA fingerprinting.

Chapter 6
Bacterial Genetics II: Transcription, Translation, and Regulation

Cell and Molecular Biology On-line

http://cellbio.com

A detailed and frequently-updated list of categorized links to all aspects of cell and molecular biology.

A Guide to Molecular Sequence Analysis

http://www.sequenceanalysis.com/

This easily accessible site guides the neophyte through the use of online molecular biology databases to analyze DNA sequences *in silico.*

Microbial Genome Program, United States Department of Energy

http://www.ornl.gov/microbialgenomes/

An offshoot of the Human Genome Project, the Microbial Genome Program is charged with sequencing the genomes of a number of nonpathogeneic microbes with potential benefit in a variety of applications. This site asks us to imagine . . . "A future in which we can use 'super bugs' to detect chemical contamination in soil, air, and water and clean up oil spills and chemicals in landfills; cook and heat with natural gas collected from a backyard septic tank or bottled at a local waste-treatment facility; obtain affordable alcohol-based fuels and solvents from cornstalks, wood chips, and other plant by-products; and produce new classes of antibiotics and process food and chemicals more efficiently."

Chapter 7
Bacterial Genetics III: Horizontal Gene Transfer Among Bacteria

Bacteriophage Homepage

http://www.evergreen.edu/user/T4/home.html

A categorized list of links to issues in bacteriophage biology, including current research, information about the T4 coliphage, and news from the growing field of phage therapy. Maintained by Evergreen State College, Olympia, Washington

Bacterial Conjugation

http://www.mun.ca/biochem/courses/4103/topics/conjugation.html

A good review of the historical development of bacteria as sexual creatures.

Transformation of *E. coli* with pGAL™

http://www.edvotek.com/experiments/microbiology-transformation/221.html

The simplicity of bacterial transformation is made clear by the availability of commercial kits for laboratory experiments. This is one of many such kits.

BioProtocol

http://www.bioprotocol.com/beta/index.jsp

After registering, the reader has access to a growing number of molecular biology protocols. The site allows the reader to maintain his or her own personal set of protocols.

Chapter 8
Bacterial Energetics

Interactive Glycolysis

http://biotech.icmb.utexas.edu/glycolysis/glycohome.html

Readers can click through the reactions of glycolysis and be tested on their mastery of the subject.

Design-It-Yourself Glycolysis

http://www.gvsu.edu/acad/chm/diygly/home.htm

Starting with glucose, readers work through a set of multiple choice questions until lactate is produced. Excellent review on the chemical mechanisms involved in glycolysis.

Cellular Energy References (Sugar Sugar)

http://www.people.virginia.edu/~rjh9u/enrgysum.html

A humorous look at the processes of glycolysis and the citric acid cycle, using chemical animations. Also contains a list of valuable links.

Chemical Reactions of the Krebs Cycle

http://www.stark.kent.edu/~cearley/pchem/Krebs/Krebs.htm

An interactive Web page illustrating the reactions of the citric acid cycle.

Animation of Chemiosmotic ATP Synthesis in Bacteria

http://virtual.class.uconn.edu/~terry/229sp98/ATPsynthbact.html

Chemiosmosis, although simple in concept, is often difficult to understand in words. This animation helps the reader visualize the dynamic nature of the proton-motive force.

Direct Observation of the Rotation of F1-ATPase

http://www.res.titech.ac.jp/seibutu/nature/f1rotate.html

By immobilizing the F1-ATPase to a cover slip and attaching fluorescently-labeled actin filaments, the rotation of the gamma subunit can be observed directly with an epifluorescent microscope. An utterly amazing set of downloadable videos.

Microbiology Webbed Out

http://www.bact.wisc.edu/microtextbook/TOC.html

A "microtextbook" written by Timothy Paustian of the University of Wisconsin-Madison, the pages on bacterial metabolism are particularly strong.

Chapter 9
Introduction to Eukaryotic Microorganisms

The Internet Resource Guide for Zoology

http://www.york.biosis.org/zrdocs/zoolinfo/grp_prot.htm

Part of a larger project concerning all zoological taxa and produced by BIOSIS, publishers of the *Zoological Record,* this page lists numerous links to anything and everything concerning the protista.

The Harmful Algae Page

http://www.redtide.whoi.edu/hab/

A site on algal blooms (red tides) maintained by the Woods Hole Oceanographic Institution.

Natural Perspective: The Fungus Kingdom

http://www.perspective.com/nature/fungi/index.html

Loaded with lots of pictures, this site explores the world of fungi through the genuine enthusiasm generated by the site's naturalist authors.

Medical Mycology

http://www.medsch.wisc.edu/medmicro/myco/mycology.html

An alphabetical list of fungal species that cause human disease. Excellent photographs and a diagnostic key are included.

Chapter 10
Viruses of Mammalian Cells

All the Virology on the WWW

http://www.virology.net/garryfavweb.html

"All the Virology on the WWW seeks to be the best single site for Virology information on the Internet. We have collected all the virology-related Web sites that might be of interest to our fellow virologists, and others interested in learning more about viruses."

An Electronic Introduction to Molecular Virology

http://www.uct.ac.za/microbiology/tutorial/virtut1.html

A South African site that is a virtual textbook on many aspects of molecular virology. Contains excellent integrated links to other virology sites.

Institute for Molecular Virology

http://www.bocklabs.wisc.edu/Welcome.html

A huge award-winning site with excellent links, computer-enhanced graphics, and recent research in plant and animal virology.

Chapter 11
Microbial Populations of the Human Body

Skin Savvy

http://www.aad.org/ss99/index.html

Articles on skin care produced by the American Academy of Dermatology. Readers can search the archives by subject.

Feces: Biochemical and Microbiological Aspects

http://www.newcastle.edu.au/department/bi/birjt/cpruis/Faeces.html

The Bioanalytical Research Group (BRG) is a group of faculty from multiple disciplines at the University of Newcastle in the Department of Biological Sciences. This group supports research and development of novel biotechnologies for early detection of disease processes. This site includes detailed information on the microbiota of the human colon.

Chapter 12
Nonspecific Defenses of the Human Body

Complement

http://ntri.tamuk.edu/immunology/complement.html

Maintained by the Natural Toxins Research Center at Texas A&M University, Kingsville, this site has a wealth of full-color drawings on the details of complement activation, either by antibodies or bacterial surfaces.

Acute Inflammation

http://medweb.bham.ac.uk/http/mod/3/1/a/acute.html

Part of MedWeb, sponsored by the Department of Pathology in the School of Medicine of Birmingham University, U.K., this site provides detailed discussion and pictures of all pathological aspects of acute inflammation.

Bacteremia and Septic Shock

http://www.merck.com/pubs/mmanual/section13/chapter156/156c.htm

Chapter 156 of the *Merck Manual of Diagnosis and Therapy.* This famous desk reference is designed for practicing physicians but is quite accessible to the nonphysician. Use the search function to explore the medical aspects of any microbial infection.

Chapter 13
Specific Defenses of the Human Body

Biology Links: Immunology

http://mcb.harvard.edu/BioLinks/Immunology.html

A list of links to a variety of immunology topics, immunology journals, and research institutions with strong immunology programs. Maintained and updated frequently by the Department of Molecular and Cellular Biology at Harvard University.

Understanding the Immune System

http://rex.nci.nih.gov/PATIENTS/INFO_TEACHER/bookshelf/NIH_immune/index.html

A site written by the staff of the National Cancer Institute provides a general overview written for the public on all aspects of the immune system.

Mike's Immunoglobulin Structure/Function Home Page

http://www.path.cam.ac.uk/~mrc7/mikeimages.html

Web pages on immunoglobulin structure and function prepared by Mike Clark, PhD, of Cambridge University. Links to computerized images of antibody molecules are based on Professor Clark's own research and teaching interests and are very useful for understanding the molecular details of antibody structure and function and the therapeutic uses of antibodies.

Theoretical Immunology

http://post.queensu.ca/~forsdyke/theorimm.htm

Professor Donald Forsdyke of Queens University, Canada, has provided full texts of and commentary on many of the seminal papers in the history of immunology. To quote Professor Forsdyke from his introduction, "To really understand immunology we must first understand the historical development of ideas on immunology. But to really understand its history, we must first understand immunology."

Chapter 14
Vaccination

The Vaccine Page

http://vaccines.org/

Provides access to up-to-the-minute news about vaccines and an annotated database of vaccine resources on the Internet. This site is maintained by Daily University Science News UniSci, the first science daily news site on the Web.

DNAvaccine.com

http://www.dnavaccine.com/

A site for professionals working in the field of DNA vaccine research. Two links, Research Tools & Literature and Resources, are especially valuable to the student.

National Network for Immunization Information

http://www.infoinc.com/imnews2/

Maintained by the Infectious Diseases Society of America, a private group of academic physicians "dedicated to providing education and communications about immunization issues."

The International Vaccine Institute

http://www.ivi.org/

Located in Seoul, Republic of Korea, this organization is ". . . dedicated to improving health through vaccine science. . ." The "vaccines" link allows readers to search for detailed information on different vaccines.

Chapter 15
Introduction to Infectious Diseases

Centers for Disease Control and Prevention

http://www.cdc.gov/

With offices in Altanta, Georgia, the CDC maintains a huge site full of detailed information on specific diseases, governmental reports, traveler's advisories, prevention guidelines, and epidemiological data. Of particular interest is the Morbidity and Mortality Weekly Report (MMWR), with weekly updates on infectious disease activity. CDC Wonder http://wonder.cdc.gov/ allows the reader to search the CDC databases for precise data on particular diseases.

The World Health Organization

http://www.who.int/

Established on April 7, 1948, World Health Day, WHO's mission statement is "the attainment by all peoples of the highest possible levels of health." This huge site contains information on specific diseases, position papers by the WHO, and detailed statistical information on world health. Of particular interest is the *Weekly Epidemiology Record,* a worldwide analog of the CDC's *Morbidity and Mortality Weekly Report.*

The National Institutes of Allergy and Infectious Diseases

http://www.niaid.nih.gov/

The branch of the National Institutes of Health (NIH) responsible for supporting most of the research activities on infectious disease in the United States. The Division of Microbiology and Infectious Diseases and the Division of Acquired Immunodeficiency Disease are linked to the homepage.

Johns Hopkins Infectious Diseases

http://www.hopkins-id.edu/

A huge site from the world famous hospital in Baltimore includes a variety of interesting sites about clinical aspects of infectious disease (check out the Top 10 page) and numerous and frequently updated news reports. This site also contains links to other huge sites at Hopkins.

The Wonderful World of Diseases

http://www.diseaseworld.com/disease.htm

This whimsically named site contains links to a number of common human diseases, many of them infectious. The site is also linked to chat rooms and bulletin boards that readers might find useful. Readers who wish to add their own links are encouraged to e-mail the author. There's even a bookstore!

Outbreak

http://www.outbreak.org/cgi-unreg/dynaserve.exe/index.html

Maintained by an interested member of the public, this site brings together a wealth of information about newly emerging infectious diseases. Readers are asked to register and to participate in the development of the site. The pessimistic tone of this site can be seen in the chosen epigraph on the visitor's page: "In the fight between you and the world, back the world" —Frank Zappa.

Outbreak!

http://www.nbif.org/outbreak/

Not to be confused with the site described above, Outbreak!© is an interactive teaching tool for use by students and science educators, produced by the National Biotechnology Information Facility at New Mexico State University. Players must use standard microbial identification techniques to identify the causative agent of an illness outbreak.

Chapter 16
Respiratory Tract Infections

The American Lung Association

http://www.lungusa.org

The oldest voluntary health organization in the United States was founded in 1904 to fight tuberculosis. The ALA today fights lung disease in all its forms, with special emphasis on asthma, tobacco control, and environmental health. This site provides general information on all varieties of lung diseases and provides data on their prevalence and treatment.

National Jewish Medical and Research Center

http://www.njc.org/

This world famous hospital in Denver specializes in research and treatment of lung, allergic, and immune diseases. Their Web site focuses on the services offered to their patients as well as the research conducted by their staff.

FluNet: The WHO Global Influenza Surveillance

http://oms2.b3e.jussieu.fr/flunet/

Developed for the World Health Organization in collaboration with the Institute for Medical Research and Health, Paris, France, this site keeps track of influenza outbreaks by linking collaborating influenza centers worldwide.

The Common Cold

http://www.nlm.nih.gov/medlineplus/commoncold.html

MedlinePlus is a site maintained by the National Library of Medicine that provides links to a variety of sites on medical topics.

Chapter 17
GI Tract Infections

***Helicobacter pylori* and Peptic Ulcer Disease**

http://www.cdc.gov/ncidod/dbmd/hpylori.htm

From the Centers for Disease Control and Prevention, a site dedicated to many aspects of the bacterium responsible for peptic ulcers.

Fight Bac!

http://www.fightbac.org/

The Partnership for Food Safety Education is a public-private partnership created to reduce the incidence of food-borne illness by educating Americans about safe food handling practices.

Pathogens in Water

http://inweh1.uwaterloo.ca/447/lectures/pathog1.htm

Lecture notes from the Environmental Microbiology course at the University of Waterloo, Canada, with lots of excellent links interspersed among the notes. Back up to the homepage for the course for lots of other links to environmental microbiology topics.

Food-borne Pathogenic Microorganisms and Natural Toxins Handbook: The "Bad Bug Book"

http://vm.cfsan.fda.gov/~mow/intro.html

A handbook of basic facts regarding food-borne pathogenic microorganisms (viral, bacterial, and eukaryotic) and natural toxins. Maintained by the United States Food and Drug Administration.

Chapter 18
Urogenital Tract Infections

Urinary Tract Infections in Adults

http://www.niddk.nih.gov/health/urolog/pubs/utiadult/utiadult.htm

Part of the Web page for the The National Institute of Diabetes and Digestive and Kidney Diseases (NIDDK).

The Johns Hopkins University STD Research Group

http://ww2.med.jhu.edu/jhustd/frame3.htm

A research group at The Johns Hopkins University School of Medicine in the Department of Medicine Division of Infectious Diseases that focuses on the epidemiology, prevention, and behavioral aspects of sexually transmitted diseases (STDs). Within the table of contents, organized in part around specific infections, are a huge number of links to STD topics.

Chapter 19
Nosocomial and Iatrogenic Infections

The National Nosocomial Infections Surveillance

http://www.cdc.gov/ncidod/hip/Surveill/nnis.htm

A branch of the Centers for Disease Control and Prevention (CDC), the NNIS collects nosocomial infection surveillance data into a national database. "The database is used to describe the epidemiology of nosocomial infections in hospitals in the United States and to produce nosocomial infection rates that can be used for comparison purposes by hospitals following NNIS methodology."

The American Iatrogenic Association

http://www.houstoncorealestate.com/iatrogenic/index.html

An advocacy group with the self-described goal of "promoting accountability for medical professionals and institutions." Abstracts of news reports submitted by readers demonstrate the full range of iatrogenic diseases.

Chapter 20
Arthropod-Borne Diseases and More Zoonoses

Arthropod-Borne Virus Information Exchange

http://lablink.utmb.edu/arbovirus/

An on-line journal of submissions from around the world. *"The Arbovirus Information Exchange"* contains preliminary reports, summaries, observations, and comments submitted voluntarily by qualified agencies and individual investigators."

The American Lyme Disease Foundation, Inc.

http://www.aldf.com/

A national nonprofit organization dedicated to "advancing the prevention, diagnosis, treatment, and control of Lyme disease and other tick-borne infections." The ALDF was formed in 1990 by a group of citizens concerned about the rising incidence of Lyme disease in the tri-state area of New York, New Jersey, and Connecticut.

The *Borrelia burgdorferi* sensu lato Molecular Genetics Server

http://www.pasteur.fr/recherche/borrelia/Welcome.html

Contains extensive lists of the addresses and Web pages of spirochete researchers. Maintained by the Pasteur Institute in Paris, France.

The Malaria Foundation International

http://www.malaria.org/

An extensive and well-organized group of links to a variety of educational and research topics concerning malaria.

Daniel Shapiro's Zoonosis Homepage

http://medicine.bu.edu/dshapiro/zoo1.htm

Maintained by a professor of pathology at Boston University School of Medicine, this sites lists zoonoses by animal. Learn for example, about contracting conjunctivitis from seals.

Chapter 21
Analyzing Microbial Communities

Digital Learning Center for Microbial Ecology

http://commtechlab.msu.edu/sites/dlc-me/

A science education project developed at Michigan State University, this site contains a variety of fascinating links, including the "microbe zoo." For the public.

The Center for Microbial Ecology

http://www.cme.msu.edu/CME/index.html

The Center for Microbial Ecology was founded in 1989 at Michigan State University by the National Science Foundation (NSF) as one of the first 11 Science and Technology Centers in the nation. " The intellectual focus of the Center for Microbial Ecology is to understand factors that influence the competitiveness, diversity, and function of microorganisms in their natural and managed habitats. This knowledge is important because microorganisms have major roles in determining global warming, ground water quality, plant and animal health, and organic matter cycling." This site includes a link to the ribosome database project at MSU, very useful to the researchers active in the analysis of microbial communities through rRNA gene sequence data.

Molecular Expressions Microscopy Primer: Virtual Microscopy

http://micro.magnet.fsu.edu/primer/virtual/virtual.html

A novel site that provides the reader with the opportunity to operate virtual microscopes useful in the identification of microbes, including electron scanning and transmission, phase contrast, polarized light, Hoffman modulation contrast, Rheinberg illumination, differential interference, and fluorescence microscopy.

The Environmental Protection Agency

http://www.epa.gov/

Founded in 1970 for the "establishment and enforcement of environmental protection standards consistent with national environmental goals . . ." the EPA is responsible for monitoring the state of our environment and for promulgating standards, enforcing legislation, and sponsoring research. Use the "search" function for topics such as bioremediation and global warming.

Chapter 22
Phototrophic Microbes: Food Sources for Planet Earth

ASU Photosynthesis Center

http://photoscience.la.asu.edu/photosyn/

The Arizona State University Center for the Study of Early Events in Photosynthesis consists of students, postdoctoral associates and research scientists, and faculty members in the Department of Chemistry and Biochemistry and the Department of Plant Biology. These research groups share the common goal of "understanding the process of photosynthesis, which is responsible for producing all of our food and filling the vast majority of our energy and fiber needs. The impetus for development of the Center was the premise that photosynthesis is a complex problem that will only yield to an investigation using a wide variety of approaches and techniques." Includes an educational resources section, with general information on plants and photosynthesis and a variety of articles and links geared to students from early childhood to college level.

Photosynthesis

http://esg-www.mit.edu:8001/esgbio/ps/psdir.html

Part of the MIT Biology Hypertextbook produced by the Experimental Study Group at the Massachusetts Institute of Technology, this site describes the physics of light and the photosystem, the light and "dark" reactions, and alternate modes of photosynthesis.

Tutorial on the Photosynthetic Reaction Center of *Rhodopseudomonas viridis*

http://indycc1.agri.huji.ac.il/~marder/rc_view/

After downloading a free copy of Chime 2.0, viewers can follow this tutorial on the single photosystem of the purple bacterium. Written by Jonathan B. Marder at The Hebrew University in Jerusalem.

Cyanosite: A Webserver for Cyanobacterial Research

http://www-cyanosite.bio.purdue.edu/

Includes a large bibliographic database, a video gallery, research protocols, taxonomic relationships, and links to other sites.

Chapter 23
Invertebrate–Microbe Interactions

Ocean Planet

http://seawifs.gsfc.nasa.gov/OCEAN_PLANET/HTML/oceanography_recently_revealed1.html

Maintained by the Smithsonian Institution, this site contains a large number of links on hydrothermal vent communities.

The Bioluminescence Webpage

http://lifesci.ucsb.edu/~biolum/

Fundamentals and current research on marine bioluminescence. An award-winning site maintained at the University of California, Santa Barbara.

Chapter 24
Plant–Microbe Interactions

Microbial Ecology of the Nitrogen Cycle

http://burgundy.uwaterloo.ca/biol446/chapter8.htm

Chapter 8 of a virtual textbook from the University of Waterloo, this site contains a wealth of good illustrations and tabular data on nitrogen cycles. One page only.

***Rhizobium* Research Laboratory**

http://www.rhizobium.umn.edu/

Part of The Department of Soil, Water, and Climate at the University of Minnesota. While focused on the problems of nodulation and nitrogen fixation in grain, prairie, and pasture legumes, this site includes a set of FAQs, a glossary, research bibliography, and a list of strains maintained by the center with photographs.

Biology and Control of Crown Gall (*Agrobacterium tumefaciens*)

http://helios.bto.ed.ac.uk/bto/microbes/crown.htm

Part of the Microbial World Web site at the University of Edinburgh, this site contains links, diagrams, and photographs on the infective process in the formation of crown gall tumors.

Chapter 25
Biotechnology

The Biotechnology Center

http://www.biotech.wisc.edu/

The unique aspect of this large site is a series of video lectures offered at the center on a variety of biotechnology subjects by researchers from around the world.

Genetic Engineering News

http://www.genwire.com

This publication is focused on the commercial and financial aspects of the biotechnology revolution. At the same time, frequent articles on the biology of biotech developments are included. Qualified persons can receive a free hardcopy subscription.

Genetic Engineering and Its Dangers

http://online.sfsu.edu/~rone/gedanger.htm

No study of biotechnology would be complete without considering the arguments posed by those who see more threat than promise in this brave new world. This site is comprised of numerous articles, compiled by Ron Epstein of the Philosophy Department at San Francisco State University, that argue that genetic engineering is dangerous.

Index

J

K

W